Introductory Algebra

7975.

Introductory Algebra
Seventh Edition

Margaret L. Lial
American River College

John Hornsby
University of New Orleans

Terry McGinnis

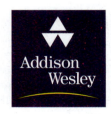

Addison
Wesley

Boston   San Francisco   New York
London   Toronto   Sydney   Tokyo   Singapore   Madrid
Mexico City   Munich   Paris   Cape Town   Hong Kong   Montreal

Publisher	Jason A. Jordan
Executive Editor	Maureen O'Connor
Editorial Project Management	Ruth Berry and Suzanne Alley
Editorial Assistant	Melissa Wright
Managing Editor/Production Supervisor	Ron Hampton
Text and Cover Design	Dennis Schaefer
Supplements Production	Sheila C. Spinney
Production Services	Elm Street Publishing Services, Inc.
Media Producer	Lorie Reilly
Associate Producer	Rebecca Martin
Marketing Manager	Dona Kenly
Marketing Coordinator	Heather Rosefsky
Prepress Services Buyer	Caroline Fell
Technical Art Supervisor	Joseph K. Vetere
First Print Buyer	Hugh Crawford
Art Creation & Composition Services	Pre-Press Company, Inc.
Cover Photo Credits	Jan Halaska/Index Stock Imagery;
	Walter Bibikow/Index Stock Imagery;
	Tina Buckman/Index Stock Imagery

Photo Credits All photos from PhotoDisk except the following:
Bill Bachman/PhotoResearchers, Inc., p.155 left; Bettmann/CORBIS, p. 383; Greg Crisp/SportsChrome USA, p. 462 left; Bob Daemmrich/The Image Works, p. 504 bottom; Tony Freeman/PhotoEdit, p. 185; Jeff Greenberg/PhotoEdit, p. 154; Arles Gupton/Stock Boston, p. 616; James R. Holland/Stock Boston, pp. 545, 594, 598; Robert Holmes/CORBIS, p. 536; Courtesy, John Hornsby, pp. 105, 147, 150, 181, 556; James Marshall/CORBIS, p. 465; Michael Newman/PhotoEdit, pp. 516, 648; NASA, pp. 317 left, 317 right, 578, 628, 636; Chuck Pefley/Stock Boston, p. 611; Reuters News/Media Inc./CORBIS, pp. 128, 155 right, 399, 454; John Sarkissian/Parkes Observatory, Australia, pp. 613, 643; David Shopper/Stock Boston, p. 610; Brian Spurlock/SportsChrome USA, p. 462 right; Rob Tringali, Jr./SportsChrome USA, pp. 25, 52; Nik Wheeler/CORBIS, pp. 584 top, 584 bottom; Ben Woods/CORBIS, p. 602 top.

Library of Congress Cataloging-in-Publication Data

Lial, Margaret L.

    Introductory algebra.—7 ed./Margaret L. Lial, John Hornsby, Terry McGinnis.

       p.   cm.

    Includes index.

    ISBN 0-321-06458-5 (Student Edition)

    ISBN 0-321-08869-7 (Annotated Instructor's Edition)

    ISBN 0-321-09727-0 (Hardback)

1. Algebra.   I. Hornsby, John.   II. McGinnis, Terry.   III. Title.

QA152.2 .L54 2002

512.9—dc21   00-054324

2 3 4 5 6 7 8 9 10   WC   04 03 02 01

Contents

List of Applications

List of Focus on Real-Data Applications

Preface

The seventh edition of *Introductory Algebra* continues our ongoing commitment to provide the best possible text and supplements package to help instructors teach and students succeed. To that end, we have tried to address the diverse needs of today's students through an attractive design, updated figures and graphs, helpful features, careful explanations of topics, and an expanded package of supplements and study aids. We have also taken special care to respond to the suggestions of users and reviewers and have added many new examples and exercises based on their feedback. Students who have never studied algebra—as well as those who require further review of basic algebraic concepts before taking additional courses in mathematics, business, science, nursing, or other fields—will benefit from the text's student-oriented approach.

This text is part of a series that also includes the following books:

- *Essential Mathematics*, by Lial and Salzman
- *Basic College Mathematics*, Sixth Edition, by Lial, Salzman, and Hestwood
- *Prealgebra*, Second Edition, by Lial and Hestwood
- *Intermediate Algebra with Early Functions and Graphing*, Seventh Edition, by Lial, Hornsby, and McGinnis
- *Introductory and Intermediate Algebra*, Second Edition, by Lial, Hornsby, and McGinnis.

WHAT'S NEW IN THIS EDITION?

We believe students and instructors will welcome the following new features.

▶ *New, Real-Life Applications* We are always on the lookout for interesting data to use in real-life applications. As a result, we have included many new or updated example and exercises throughout the text that focus on real-life applications of mathematics These applied problems provide a modern flavor that will appeal to and motivate students. (See pp. 184, 207, and 462.) A comprehensive List of Applications appears at the beginning of the text.

▶ *New Figures and Photos* Today's students are more visually oriented than ever. The we have made a concerted effort to add mathematical figures, diagrams, tables, a graphs whenever possible. (See pp. 195, 249, and 518.) Many of the graphs use a sty similar to that seen by students in today's print and electronic media. Photos have been i corporated to enhance applications in examples and exercises. (See pp. 229, 378, and 392

▶ *Increased Emphasis on Problem Solving* Introduced in Chapter 2, our six-ste problem-solving method has been refined and integrated throughout the text. The si steps, *Read*, *Assign a Variable*, *Write an Equation*, *Solve*, *State the Answer*, and *Check* are emphasized in boldface type and repeated in examples and exercises to reinforce the problem-solving process for students. (See pp. 132, 456, and 516.)

Study Skills

⊙ Study Skills Component A desk-light icon at key points in the text directs students to a separate *Study Skills Workbook* containing activities correlated directly to the text. (See pp. 33 and 481.) This unique workbook explains *how* the brain actually learns, so students understand *why* the study tips presented will help them succeed in the course. Students are introduced to the workbook in an updated To the Student section at the beginning of the text.

⊙ Focus on Real-Data Applications These one-page activities present a relevant and in-depth look at how mathematics is used in the real world. Designed to help instructors answer the often-asked question, "When will I ever use this stuff?," these activities ask students to read and interpret data from newspaper articles, the Internet, and other familiar, real sources. (See pp. 90, 460, and 520.) The activities are well-suited to collaborative work and can also be completed by individuals or used for open-ended class discussions. Instructor teaching notes and extensions for the activities are provided in the *Printed Test Bank and Instructor's Resource Guide.*

⊙ Diagnostic Pretest A diagnostic pretest is now included on p. xxxi and covers all the material in the book, much like a sample final exam. This pretest can be used to facilitate student placement in the correct chapter according to skill level.

⊙ Chapter Openers New chapter openers feature real-world applications of mathematics that are relevant to students and tied to specific material within the chapters. Examples of topics include credit card debt, the Olympics, and U.S. trade. (See pp. 105, 399, and 485—Chapters 3, 6, and 8.)

⊙ Calculator Tips These optional tips, marked with calculator icons, offer basic information and instruction for students using calculators in the course. (See pp. 28, 315, and 551.) In addition, a new Introduction to Calculators has been included at the beginning of the text.

⊙ Test Your Word Power To help students understand and master mathematical vocabulary, this new feature has been incorporated in each chapter summary. Key terms from the chapter are presented along with four possible definitions in a multiple-choice format. Answers and examples illustrating each term are provided. (See pp. 94, 320, and 531.)

WHAT FAMILIAR FEATURES HAVE BEEN RETAINED?

We have retained the popular features of previous editions of the text, some of which follow.

⊙ Learning Objectives Each section begins with clearly stated, numbered objectives, and the included material is directly keyed to these objectives so that students know exactly what is covered in each section. (See pp. 106, 332, and 546.)

⊙ Cautions and Notes One of the most popular features of previous editions, Caution and Note boxes warn students about common errors and emphasize important ideas throughout the exposition. (See pp. 33, 72, and 228.) There are more of these in the seventh edition than in the sixth, and the new text design makes them easier to spot; Cautions are highlighted in bright yellow and Notes are highlighted in green.

⊙ Margin Problems Margin problems, with answers immediately available at the bottom of the page, are found in every section of the text. (See pp. 43, 188, and 263.) This key feature allows students to immediately practice the material covered in the examples in preparation for the exercise sets. Based on reviewer feedback, we have added more margin exercises to the seventh edition.

⊙ Ample and Varied Exercise Sets The text contains a wealth of exercises to provide students with opportunities to practice, apply, connect, and extend the algebraic skills they are learning. Numerous illustrations, tables, graphs, and photos have been added to the exercise sets to help students visualize the problems they are solving. Problem types include writing, estimation, and calculator exercises as well as applications and multiple-choice, matching, true/false, and fill-in-the-blank problems. In the *Annotated Instructor's Edition* of the text, writing exercises are marked with ✍ icons so that instructors may assign these problems at their discretion. Exercises suitable for calculator work are marked in both the student and instructor editions with calculator icons ▦ . (See pp. 407, 415, and 553.)

▶ *Relating Concepts Exercises* Formerly titled Mathematical Connections, these sets of exercises help students tie together topics and develop problem-solving skills as they compare and contrast ideas, identify and describe patterns, and extend concepts to new situations. (See pp. 254, 268, and 588.) These exercises make great collaborative activities for pairs or small groups of students.

▶ *Summary Exercises* Four sets of in-chapter summary exercises on problem solving, factoring, rational expressions, and quadratic equations provide students with the all-important *mixed* practice problems they need to master these typically difficult topics. (See pp. 157, 363, 451, and 637.)

▶ *Ample Opportunity for Review* Each chapter concludes with a Chapter Summary that features Key Terms with definitions and helpful graphics, New Symbols, Test Your Word Power, and a Quick Review of each section's content with additional examples. A comprehensive set of Chapter Review Exercises, keyed to individual sections, is included, as are Mixed Review Exercises and a Chapter Test. Beginning with Chapter 2, each chapter concludes with a set of Cumulative Review Exercises that cover material going back to Chapters R and 1. (See pp. 245, 385, and 599.)

WHAT CONTENT CHANGES HAVE BEEN MADE?

We have worked hard to fine-tune and polish presentations of topics throughout the text based on user and reviewer feedback. Some of the content changes include the following:

- Multiplication and division of real numbers are consolidated in Section 1.6.

- A new set of summary exercises on solving applied problems is included in Chapter 2.

- Topics on graphing linear equations and inequalities in two variables, which were formerly in Chapter 6, are now presented in Chapter 3.

- Chapter 4, Exponents and Polynomials, has been reorganized for increased continuity.

- Factoring trinomials by grouping and factoring trinomials using FOIL are treated in separate sections in Chapter 5.

- Variation is covered in a new Section 6.8.

- In Chapter 7, Systems of Equations and Inequalities, the substitution method (Section 7.2) is covered before the elimination method (Section 7.3).

- Functions are introduced in a new Section 9.5.

WHAT SUPPLEMENTS ARE AVAILABLE?

Our extensive supplements package includes an *Annotated Instructor's Edition*, testing materials, solutions manuals, tutorial software, videotapes, and a state-of-the-art Web site. For more information about any of the following supplements, please contact your Addison-Wesley sales consultant.

FOR THE STUDENT

▶ *Student's Solutions Manual* (ISBN 0-321-09105-1) The *Student's Solutions Manual* provides detailed solutions to the odd-numbered section exercises and to all margin, Relating Concepts, Summary, Chapter Review, Chapter Test, and Cumulative Review exercises

⦿ *Study Skills Workbook* **(ISBN 0-321-09242-2)** A desk-light icon at key points in the text directs students to correlated activities in this unique workbook by Diana Hestwood and Linda Russell. The activities in the workbook teach students how to use the textbook effectively, plan their homework, take notes, make mind maps and study cards, manage study time, and prepare for and take tests. Students find out *how* their brains actually learn and what research tells us about how to study effectively. An updated To the Student section at the beginning of the text introduces students to the *Study Skills Workbook*.

⦿ *Addison-Wesley Math Tutor Center* The Addison-Wesley Math Tutor Center is staffed by qualified college mathematics instructors who tutor students on examples and exercises from the textbook. Tutoring is provided via toll-free telephone, toll-free fax, e-mail, and the Internet. White Board technology allows tutors and students to actually see problems being worked while they "talk" in real time over the Internet during tutoring sessions. The Math Tutor Center is accessed through a registration number that may be bundled free with a new textbook or purchased separately.

⦿ *InterAct Math® Tutorial Software* **(ISBN 0-321-09109-4)** Available on a dual-platform, Windows/Macintosh CD-ROM, this interactive tutorial software provides algorithmically generated practice exercises that are correlated at the objective level to the content of the text. Every exercise in the program is accompanied by an example and a guided solution designed to involve students in the solution process. For Windows users, selected problems also include a video clip to help students visualize concepts. The software tracks student activity and scores and can generate printed summaries of students' progress. Instructors can use the InterAct Math® Plus course-management software to create, administer, and track tests and monitor student performance during practice sessions. (See For the Instructor.)

⦿ *InterAct MathXL:* **www.mathxl.com** InterAct MathXL is a Web-based tutorial system that enables students to take practice tests and receive personalized study plans based on their results. Practice tests are correlated directly to the section objectives in the text, and once a student has taken a practice test, the software scores the test and generates a study plan that identifies strengths, pinpoints topics where more review is needed, and links directly to InterAct Math® tutorial software for additional practice and review. A course-management feature allows instructors to create and administer tests and view students' test results, study plans, and practice work. Students gain access to the InterAct MathXL Web site through a password-protected subscription, which can either be bundled free with a new copy of the text or purchased separately.

⦿ *Real-to-Reel Videotape Series* **(ISBN 0-321-09202-3)** This series of videotapes, created specifically for *Introductory Algebra*, Seventh Edition, features an engaging team of lecturers who provide comprehensive lessons on every objective in the text. The videos include a stop-the-tape feature that encourages students to pause the video, work through the example presented on their own, and then resume play to watch the video instructor go over the solution.

⦿ *Digital Video Tutor* **(ISBN 0-321-09203-1)** This supplement provides the entire set of Real-to-Reel videotapes for the text in digital format on CD-ROM, making it easy and convenient for students to watch video segments from a computer, either at home or on campus. Available for purchase with the text at minimal cost, the Digital Video Tutor is ideal for distance learning and supplemental instruction.

⦿ *MathPass* MathPass helps students succeed in their developmental mathematics courses by creating customized study plans based on diagnostic test results from ACT, Inc.'s Computer-Adaptive Placement Assessment and Support System (COMPASS®). MathPass pinpoints topics where the student needs in-depth study or targeted review and correlates these topics with the student's textbook and related supplements. The study plan can be saved as an HTML file that, when viewed on the Internet, links directly to text-specific, on-line resources. Instructors can add their own custom Web links to the HTML study plan. The MathPass learning system provides diagnostic assessment, focused instruction, and exit placement all in one package.

Web Site: www.MyMathLab.com Ideal for lecture-based, lab-based, and on-line courses, MyMathLab.com provides students with a centralized point of access to the wide variety of on-line resources available with this text. The pages of the actual book are loaded into MyMathLab.com, and as students work through a section of the on-line text, they can link directly from the pages to supplementary resources (such as tutorial software, interactive animations, and audio and video clips) that provide instruction, exploration, and practice beyond what is offered in the printed book. MyMathLab.com generates personalized study plans for students and allows instructors to track all student work on tutorials, quizzes, and tests.

FOR THE INSTRUCTOR

Annotated Instructor's Edition (ISBN 0-321-08869-7) For immediate access, the *Annotated Instructor's Edition* provides answers to all text exercises in color next to the corresponding problems. To assist instructors in assigning homework problems, icons identify writing and calculator exercises.

Instructor's Solutions Manual (ISBN 0-321-09106-X) The *Instructor's Solutions Manual* provides complete solutions to all even-numbered section exercises.

Answer Book (ISBN 0-321-09240-6) The *Answer Book* provides answers to all the exercises in the text.

Printed Test Bank and Instructor's Resource Guide (ISBN 0-321-09107-8) The *Printed Test Bank* portion of this manual contains two diagnostic pretests, six free-response and two multiple-choice test forms per chapter, and two final exams. The *Instructor's Resource Guide* portion of the manual contains teaching suggestions for each chapter, additional practice exercises for every objective of every section, a correlation guide from the sixth to the seventh edition, phonetic spellings for all key terms in the text, and teaching notes and extensions for the Focus on Real-Data Applications in the text.

TestGen-EQ with QuizMaster EQ (ISBN 0-321-09108-6) This fully networkable software enables instructors to create, edit, and administer tests using a computerized test bank of questions organized according to the chapter content of the text. Six question formats are available, and a built-in question editor allows the user to create graphs, import graphics, and insert mathematical symbols and templates, variables, or text. An "Export to HTML" feature allows practice tests to be posted to the Internet, and instructors can use QuizMaster-EQ to post quizzes to a local computer network so that students can take them on-line and have their results tracked automatically.

InterAct Math® Plus (ISBN 0-201-72140-6) This networkable software provides course-management capabilities and on-line test administration for Addison-Wesley's Inter-Act Math® tutorial software. (See For the Student.) InterAct Math® Plus enables instructors to create and administer tests, summarize students' results, and monitor students' progress in the tutorial software, providing an invaluable teaching and tracking resource.

Web Site: www.MyMathLab.com In addition to providing a wealth of resources for lecture-based courses, MyMathLab.com gives instructors a quick and easy way to create a complete on-line course based on *Introductory Algebra*, Seventh Edition. MyMathLab.com is hosted nationally at no cost to instructors, students, or schools, and it provides access to an interactive learning environment where all content is keyed directly to the text. Using a customized version of Blackboard™ as the course-management platform, MyMathLab.com lets instructors administer preexisting tests and quizzes or create their own. It provides detailed tracking of all student work as well as a wide array of communication tools for course participants. Within MyMathLab.com, students link directly from on-line pages of their text to supplementary resources such as tutorial software, interactive animations, and audio and video clips.

ACKNOWLEDGMENTS

The comments, criticisms, and suggestions of users, nonusers, instructors, and students have positively shaped this textbook over the years, and we are most grateful for the many responses we have received. The feedback gathered for this revision of the text was particularly helpful, and we especially wish to thank the following individuals who provided invaluable suggestions:

Randall Allbritton, *Daytona Beach Community College*
Jannette Avery, *Monroe Community College*
Linda Beattie, *Western New Mexico University*
Jean Bolyard, *Fairmont State College*
Tim C. Caldwell, *Meridian Community College*
Bill Dunn, *Las Positas College*
J. Lloyd Harris, *Gulf Coast Community College*
Edith Hays, *Texas Woman's University*
Karen Heavin, *Morehead State University*
Christine Heinecke Lehmann, *Purdue University—North Central*
Elizabeth Heston, *Monroe Community College*
Harriet Kiser, *Floyd College*
Valerie Lazzara, *Palm Beach Community College*
Valerie H. Maley, *Cape Fear Community College*
Susan McClory, *San Jose State University*
Jeffrey Mills, *Ohio State University*
Linda J. Murphy, *Northern Essex Community College*
Elizabeth Ogilvie, *Horry-Georgetown Technical College*
Larry Pontaski, *Pueblo Community College*
Diann Robinson, *Ivy Tech State College—Lafayette*
Rachael Schettenhelm, *Southern Connecticut State University*
Lee Ann Spahr, *Durham Technical Community College*
Cora S. West, *Florida Community College at Jacksonville*
Gabriel Yimesghen, *Community College of Philadelphia*

We especially wish to thank Valerie Maley of Cape Fear Community College, who provided a set of new exercises on order of operations for Section 1.1.

Our sincere thanks go to these dedicated individuals at Addison-Wesley who worked long and hard to make this revision a success: Maureen O'Connor, Ruth Berry, Ron Hampton, Dennis Schaefer, Dona Kenly, Suzanne Alley, and Melissa Wright.

While Kitty Pellissier did her usual outstanding job checking the answers to all the exercises, she also reviewed the entire manuscript and provided invaluable content suggestions during both the writing and production processes. Steven Pusztai of Elm Street Publishing Services provided his customary excellent production work. We are most grateful to Peg Crider for researching and writing the Focus on Real-Data Applications feature; Paul Van Erden for his accurate and useful index; Becky Troutman for preparing the comprehensive List of Applications; Abby Tanenbaum for writing the new Diagnostic Pretest; and Randall Allbritton, Linda Buchanan, Scott Higinbotham, and Laurie Semarne for accuracy checking the manuscript.

As an author team, we are committed to the goal stated earlier in this Preface—to provide the best possible text and supplements package to help instructors teach and students succeed. We are most grateful to all those over the years who have aspired to this goal with us. As we continue to work toward it, we would welcome any comments or suggestions you might have. Please feel free to send your comments via e-mail to math@awl.com.

Margaret L. Lial
John Hornsby
Terry McGinnis

Feature Walk-Through

New! Chapter Openers New chapter openers feature real-world applications of mathematics that are relevant to students and tied to specific material within the chapters.

Graphs of Linear Equations and Inequalities in Two Variables — 3

- **3.1** Reading Graphs; Linear Equations in Two Variables
- **3.2** Graphing Linear Equations in Two Variables
- **3.3** Slope
- **3.4** Equations of Lines
- **3.5** Graphing Linear Inequalities in Two Variables

U.S. debt from credit cards continues to increase. In recent years, college campuses have become fertile territory as credit card companies pitch their plastic to students at bookstores, student unions, and sporting events. As a result, three out of four undergrads now have at least one credit card and carry an average balance of $2748. (*Source:* Nellie Mae.) In Example 6 of Section 3.2, we use the concepts of this chapter to investigate credit card debt.

MyMathLab.com
You're Connected

183

---

37. Use the results of Exercises 35(b) and 36(b) to determine the target heart rate zone for age 30.

38. Should the graphs of the target heart rate zone in the Section 3.1 exercises be used to estimate the target heart rate zone for ages below 20 or above 80? Why or why not?

39. Per capita consumption of carbonated soft drinks increased for the years 1992 through 1997 as shown in the graph. If $x = 0$ represents 1992, $x = 1$ represents 1993, and so on, per capita consumption can be modeled by the linear equation

$$y = .8x + 49,$$

where y is in gallons.

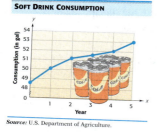

SOFT DRINK CONSUMPTION

Source: U.S. Department of Agriculture.

40. The income generated by the Walgreen Company from 1994 through 1998 is shown in the graph. If $x = 0$ corresponds to 1994, $x = 1$ corresponds to 1995, and so on, the income can be modeled by the linear equation

$$y = 57.3x + 270,$$

where y is in billions of dollars.

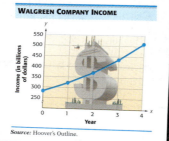

WALGREEN COMPANY INCOME

Source: Hoover's Outline.

Figures and Photos Today's students are more visually oriented than ever. Thus, a concerted effort has been made to add mathematical figures, diagrams, tables, and graphs whenever possible. Many of the graphs use a style similar to that seen by students in today's print and electronic media. Photos have been incorporated to enhance applications in examples and exercises.

Relating Concepts   Formerly titled *Mathe-matical Connections,* these sets of exercises help students tie together topics and develop problem-solving skills as they compare and contrast ideas, identify and describe patterns, and extend concepts to new situations. These exercises make great collaborative activities for pairs or small groups of students.

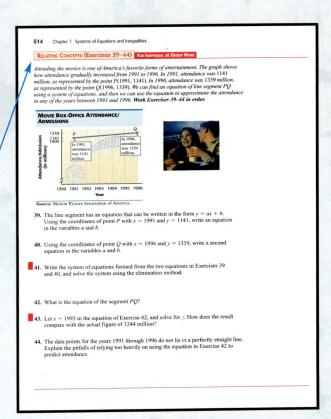

514   Chapter 7 Systems of Equations and Inequalities

RELATING CONCEPTS (Exercises 39–44) FOR INDIVIDUAL OR GROUP WORK

Attending the movies is one of America's favorite forms of entertainment. The graph shows how attendance gradually increased from 1991 to 1996. In 1991, attendance was 1141 million, as represented by the point $P(1991, 1141)$. In 1996, attendance was 1339 million, as represented by the point $Q(1996, 1339)$. We can find an equation of line segment PQ using a system of equations, and then we can use the equation to approximate the attendance in any of the years between 1991 and 1996. **Work Exercises 39–44 in order.**

MOVIE BOX-OFFICE ATTENDANCE/ ADMISSIONS

In 1991, attendance was 1141 million.

In 1996, attendance was 1339 million.

Source: Motion Picture Association of America.

39. The line segment has an equation that can be written in the form $y = ax + b$. Using the coordinates of point P with $x = 1991$ and $y = 1141$, write an equation in the variables a and b.

40. Using the coordinates of point Q with $x = 1996$ and $y = 1339$, write a second equation in the variables a and b.

41. Write the system of equations formed from the two equations in Exercises 39 and 40, and solve the system using the elimination method.

42. What is the equation of the segment PQ?

43. Let $x = 1993$ in the equation of Exercise 42, and solve for y. How does the result compare with the actual figure of 1244 million?

44. The data points for the years 1991 through 1996 do not lie in a perfectly straight line. Explain the pitfalls of relying too heavily on using the equation in Exercise 42 to predict attendance.

Real-Data Applications

Focus on

Linear or Nonlinear? That Is the Question about Windchill

The **windchill factor** measures the cooling effect that the wind has on one's skin. The table gives the windchill factor for various wind speeds and temperatures.

WINDCHILL FACTOR

| Wind Speed (mph) | Air Temperature (°Fahrenheit) | | | | | | | | | | | | | | |
	35	30	25	20	15	10	5	0	−5	−10	−15	−20	−25	−30	−35
4	35	30	25	20	15	10	5	0	−5	−10	−15	−20	−25	−30	−35
5	32	27	22	16	11	6	0	−5	−10	−15	−21	−26	−31	−36	−42
10	22	16	10	3	−3	−9	−15	−22	−27	−34	−40	−46	−52	−58	−64
15	16	9	2	−5	−11	−18	−25	−31	−38	−45	−51	−58	−65	−72	−78
20	12	4	−3	−10	−17	−24	−31	−39	−46	−53	−60	−67	−74	−81	−88
25	8	1	−7	−15	−22	−29	−36	−44	−51	−59	−66	−74	−81	−88	−96
30	6	−2	−10	−18	−25	−33	−41	−49	−56	−64	−71	−79	−86	−93	−101
35	4	−4	−12	−20	−27	−35	−43	−52	−58	−67	−74	−82	−89	−97	−105
40	3	−5	−13	−21	−29	−37	−45	−53	−60	−69	−76	−84	−92	−100	−107
45	2	−6	−14	−22	−30	−38	−46	−54	−62	−70	−78	−85	−93	−102	−109

Source: USA Today.

The data in the table represents the relationships between two different sets of variable quantities: Windchill versus Air Temperature and Windchill versus Wind Speed. The question is whether either of these relationships is linear, that is, whether the data points, when graphed, could be connected to form a straight line.

Example 1 Windchill versus Air Temperature
Choose one measure of wind speed to keep constant, such as 15 mph. Complete the table with Air Temperature (AT) as the *input* and Windchill (WC) as the *output*. Both variables are measured in degrees Fahrenheit.

AT	35	30	25	20	15	10	5	0	−5	−10	−15				
WC	16	9	2	−5	−11	−18	−25	−31	−38						

Example 2 Windchill versus Wind Speed
Choose one measure of air temperature to keep constant, such as 10°F. Complete the table with Wind Speed (WS) as the *input* and Windchill (WC) as the *output*. Wind speed is measured in mph.

WS	4	5	10	15	20	25	30				
WC	10	6	−9	−18	−24						

For Group Discussion
1. Refer to the Windchill versus Air Temperature data (Example 1).
   (a) Write the data as ordered pairs.
   (b) On a sheet of graph paper, draw and label a rectangular coordinate system. Use a scale of 5 on the *x*-axis and the *y*-axis. Make a scatter diagram of the data. Does the graph represent a linear relationship?
2. Repeat Problem 1 using the Windchill versus Wind Speed data (Example 2).

194

New! Focus on Real-Data Applications   These one-page activities found throughout the text present even more relevant and in-depth looks at how mathematics is used in the real world. Designed to help instructors answer the often-asked question, "When will I ever use this stuff?," these activities ask students to read and interpret data from newspaper articles, the Internet, and other familiar, real sources. The activities are well suited to collaborative work and can also be completed by individuals or used for open-ended class discussions.

In $\sqrt[n]{a}$, the number n is the **index** or **order** of the radical. It would be possible to write $\sqrt[2]{a}$ instead of $\sqrt{a}$, but the simpler symbol $\sqrt{a}$ is customary since the square root is the most commonly used root.

Calculator Tip   A calculator that has a key marked $\sqrt{x}$, x^y, or y^x (again perhaps in conjunction with the INV or 2nd key) can be used to find higher roots.

When working with cube roots or fourth roots, it is helpful to memorize the first few *perfect cubes* ($2^3 = 8$, $3^3 = 27$, and so on) and the first few perfect fourth powers ($2^4 = 16$, $3^4 = 81$, and so on).

New! Calculator Tips   These optional tips, marked with calculator icons, offer basic information and instruction for students using calculators in the course.

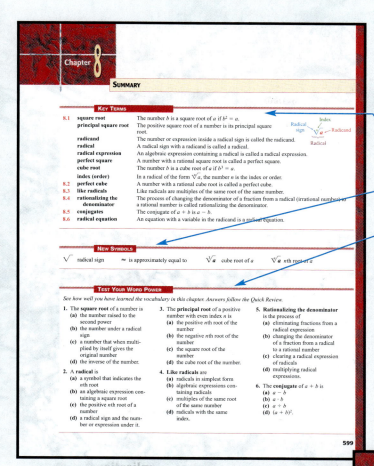

End-of-Chapter Material One of the most admired features of the Lial textbooks is the extensive and well-thought-out end-of-chapter material. At the end of each chapter, students will find:

Key Terms are listed, defined, and referenced back to the appropriate section number.

New Symbols are listed for easy reference and study.

New! Test Your Word Power To help students understand and master mathematical vocabulary, Test Your Word Power has been incorporated in each Chapter Summary. Students are quizzed on Key Terms from the chapter in a multiple-choice format. Answers and examples illustrating each term are provided.

A Chapter Test helps students practice for the real thing.

New! Study Skills Component A desk-light icon at key points in the text directs students to a separate *Study Skills Workbook* containing activities correlated directly to the text. This unique workbook explains how the brain actually learns, so students understand *why* the study tips presented will help them succeed in the course.

Quick Review sections give students not only the main concepts from the chapter (referenced back to the appropriate section), but also an adjacent example of each concept.

Review Exercises are keyed to the appropriate sections so that students can refer to examples of that type of problem if they need help.

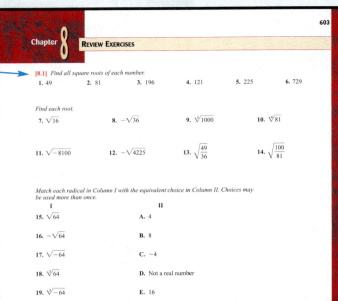

Chapter 8   REVIEW EXERCISES

[8.1] *Find all square roots of each number.*

1. 49   2. 81   3. 196   4. 121   5. 225   6. 729

Find each root.

7. $\sqrt{16}$ 8. $-\sqrt{36}$ 9. $\sqrt[3]{1000}$ 10. $\sqrt[4]{81}$

11. $\sqrt{-8100}$ 12. $-\sqrt{4225}$ 13. $\sqrt{\dfrac{49}{36}}$ 14. $\sqrt{\dfrac{100}{81}}$

Match each radical in Column I with the equivalent choice in Column II. Choices may be used more than once.

I	II
15. $\sqrt{64}$	A. 4
16. $-\sqrt{64}$	B. 8
17. $\sqrt{-64}$	C. -4
18. $\sqrt[3]{64}$	D. Not a real number
19. $\sqrt[3]{-64}$	E. 16

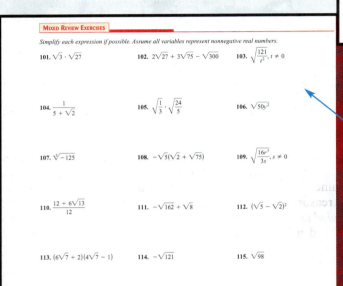

MIXED REVIEW EXERCISES

Simplify each expression if possible. Assume all variables represent nonnegative real numbers.

101. $\sqrt{3} \cdot \sqrt{27}$ 102. $2\sqrt{27} + 3\sqrt{75} - \sqrt{300}$ 103. $\sqrt{\dfrac{121}{t^2}}, t \neq 0$

104. $\dfrac{1}{5 + \sqrt{2}}$ 105. $\sqrt{\dfrac{1}{3}} \cdot \sqrt{\dfrac{24}{5}}$ 106. $\sqrt{50y^2}$

107. $\sqrt[3]{-125}$ 108. $-\sqrt{5}(\sqrt{2} + \sqrt{75})$ 109. $\sqrt{\dfrac{16r^3}{3s}}, s \neq 0$

110. $\dfrac{12 + 6\sqrt{13}}{12}$ 111. $-\sqrt{162} + \sqrt{8}$ 112. $(\sqrt{5} - \sqrt{2})^2$

113. $(6\sqrt{7} + 2)(4\sqrt{7} - 1)$ 114. $-\sqrt{121}$ 115. $\sqrt{98}$

Mixed Review Exercises require students to solve problems without the help of section references.

Cumulative Review Exercises gather various types of exercises from preceding chapters to help students remember and retain what they are learning throughout the course.

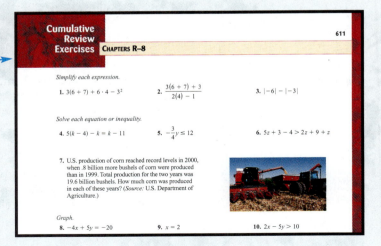

Cumulative Review Exercises   CHAPTERS R–8

Simplify each expression.

1. $3(6 + 7) + 6 \cdot 4 - 3^2$ 2. $\dfrac{3(6 + 7) + 3}{2(4) - 1}$ 3. $|-6| - |-3|$

Solve each equation or inequality.

4. $5(k - 4) - k = k - 11$ 5. $-\dfrac{3}{4}y \leq 12$ 6. $5z + 3 - 4 > 2z + 9 + z$

7. U.S. production of corn reached record levels in 2000, when .8 billion more bushels of corn were produced than in 1999. Total production for the two years was 19.6 billion bushels. How much corn was produced in each of these years? (*Source:* U.S. Department of Agriculture.)

Graph.

8. $-4x + 5y = -20$ 9. $x = 2$ 10. $2x - 5y > 10$

An Introduction to Calculators

There is little doubt that the appearance of handheld calculators three decades ago and the later development of scientific and graphing calculators have changed the methods of learning and studying mathematics forever. For example, computations with tables of logarithms and slide rules made up an important part of mathematics courses prior to 1970. Today, with the widespread availability of calculators, these topics are studied only for their historical significance.

Most consumer models of calculators are inexpensive. At first, however, they were costly. One of the first consumer models available was the Texas Instruments SR-10, which sold for about $150 in 1973. It could perform the four operations of arithmetic and take square roots, but could do very little more.

Today, calculators come in a large array of different types, sizes, and prices. *For the course for which this textbook is intended, the most appropriate type is the scientific calculator*, which costs $10–$20.

In this introduction, we explain some of the features of scientific and graphing calculators. However, remember that calculators vary among manufacturers and models, and that while the methods explained here apply to many of them, they may not apply to your specific calculator. For this reason, it is important to remember that *this introduction is only a guide and is not intended to take the place of your owner's manual*. Always refer to the manual in the event you need an explanation of how to perform a particular operation.

SCIENTIFIC CALCULATORS

Scientific calculators are capable of much more than the typical four-function calculator that you might use for balancing your checkbook. Most scientific calculators use *algebraic logic*. (Models sold by Texas Instruments, Sharp, Casio, and Radio Shack, for example, use algebraic logic.) A notable exception is Hewlett-Packard, a company whose calculators use *Reverse Polish Notation* (RPN). In this introduction, we explain the use of calculators with algebraic logic.

Arithmetic Operations To perform an operation of arithmetic, simply enter the first number, press the operation key (+, −, ×, or ÷), enter the second number, and then press the = key. For example, to add 4 and 3, use the following keystrokes.

Change Sign Key The key marked +/− allows you to change the sign of a display. This is particularly useful when you wish to enter a negative number. For example, to enter −3, use the following keystrokes.

Memory Key Scientific calculators can hold a number in memory for later use. The label of the memory key varies among models; two of these are Ⓜ and (STO). The (M+) and (M−) keys allow you to add to or subtract from the value currently in memory. The memory recall key, labeled (MR), (RM), or (RCL), allows you to retrieve the value stored in memory.

   Suppose that you wish to store the number 5 in memory. Enter 5, then press the key for memory. You can then perform other calculations. When you need to retrieve the 5, press the key for memory recall.

   If a calculator has a constant memory feature, the value in memory will be retained even after the power is turned off. Some advanced calculators have more than one memory. It is best to read the owner's manual for your model to see exactly how memory is activated.

Clearing/Clear Entry Keys These keys allow you to clear the display or clear the last entry entered into the display. They are usually marked Ⓒ and (CE). In some models, pressing the Ⓒ key once will clear the last entry, while pressing it twice will clear the entire operation in progress.

Second Function Key This key is used in conjunction with another key to activate a function that is printed *above* an operation key (and not on the key itself). It is usually marked (2nd). For example, suppose you wish to find the square of a number, and the squaring function (explained in more detail later) is printed above another key. You would need to press (2nd) before the desired squaring function can be activated.

Square Root Key Pressing the square root key, (√x), will give the square root (or an approximation of the square root) of the number in the display. For example, to find the square root of 36, use the following keystrokes.

The square root of 2 is an example of an irrational number (Chapter 8). The calculator will give an approximation of its value, since the decimal for $\sqrt{2}$ never terminates and never repeats. The number of digits shown will vary among models. To find an approximation of $\sqrt{2}$, use the following keystrokes.

An approximation for $\sqrt{2}$

Squaring Key The (x²) key allows you to square the entry in the display. For example, to square 35.7, use the following keystrokes.

The squaring key and the square root key are often found on the same key, with one of them being a second function (that is, activated by the second function key previously described).

Reciprocal Key The key marked (1/x) is the reciprocal key. (When two numbers have a product of 1, they are called *reciprocals*. See Chapter R.) Suppose that you wish to find the reciprocal of 5. Use the following keystrokes.

Inverse Key Some calculators have an inverse key, marked (INV). Inverse operations are operations that "undo" each other. For example, the operations of squaring and taking the square root are inverse operations. The use of the (INV) key varies among different models of calculators, so read your owner's manual carefully.

Exponential Key The key marked  allows you to raise a number to a power. For example, if you wish to raise 4 to the fifth power (that is, find 4^5, as explained in Chapter 1), use the following keystrokes.

Root Key Some calculators have this key specifically marked $\sqrt[x]{\ }$ or $\sqrt[y]{\ }$; with others, the operation of taking roots is accomplished by using the inverse key in conjunction with the exponential key. Suppose, for example, your calculator is of the latter type and you wish to find the fifth root of 1024. Use the following keystrokes.

Notice how this "undoes" the operation explained in the exponential key discussion.

Pi Key The number π is an important number in mathematics. It occurs, for example, in the area and circumference formulas for a circle. By pressing the π key, you can display the first few digits of π. (Because π is irrational, the display shows only an approximation.) One popular model gives the following display when the π key is pressed.

    An approximation for π

Methods of Display When decimal approximations are shown on scientific calculators, they are either *truncated* or *rounded*. To see how a particular model is programmed, evaluate 1/18 as an example. If the display shows .0555555 (last digit 5), it truncates the display. If it shows .0555556 (last digit 6), it rounds the display.

When very large or very small numbers are obtained as answers, scientific calculators often express these numbers in scientific notation (Chapter 4). For example, if you multiply 6,265,804 by 8,980,591, the display might look like this:

$$5.6270623 \ 13$$

The 13 at the far right means that the number on the left is multiplied by 10^{13}. This means that the decimal point must be moved 13 places to the right if the answer is to be expressed in its usual form. Even then, the value obtained will only be an approximation: 56,270,623,000,000.

GRAPHING CALCULATORS

Graphing calculators are becoming increasingly popular in mathematics classrooms. While you are not expected to have a graphing calculator to study from this book, we include the following as background information and reference should your course or future courses require the use of graphing calculators.

BASIC FEATURES

Graphing calculators provide many features beyond those found on scientific calculators. In addition to the typical keys found on scientific calculators, they have keys that can be used to create graphs, make tables, analyze data, and change settings. One of the major differences between graphing and scientific calculators is that a graphing calculator has a larger viewing screen with graphing capabilities. The screens below illustrate the graphs of $y = x$ and $y = x^2$.

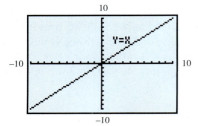
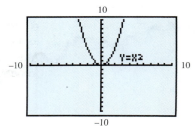

If you look closely at the screens, you will see that the graphs appear to be jagged rather than smooth, as they should be. The reason for this is that graphing calculators have much lower resolution than computer screens. Because of this, graphs generated by graphing calculators must be interpreted carefully.

EDITING INPUT

The screen of a graphing calculator can display several lines of text at a time. This feature allows you to view both previous and current expressions. If an incorrect expression is entered, an error message is displayed. The erroneous expression can be viewed and corrected by using various editing keys, much like a word-processing program. You do not need to enter the entire expression again. Many graphing calculators can also recall past expressions for editing or updating. The screen on the left below shows how two expressions are evaluated. The final line is entered incorrectly, and the resulting error message is shown in the screen on the right.

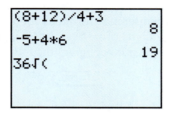
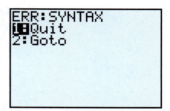

ORDER OF OPERATIONS

Arithmetic operations on graphing calculators are usually entered as they are written in mathematical expressions. For example, to evaluate $\sqrt{36}$ on a typical scientific calculator, you would first enter 36 and then press the square root key. As seen above, this is not the correct syntax for a graphing calculator. To find this root, you would first press the square root key, and then enter 36. See the screen on the left below. The order of operations on a graphing calculator is also important, and current models assist the user by inserting parentheses when typical errors might occur. The open parenthesis that follows the square root symbol is automatically entered by the calculator so that an expression such as $\sqrt{2 \times 8}$ will not be calculated incorrectly as $\sqrt{2} \times 8$. Compare the two entries and their results in the screen on the right.

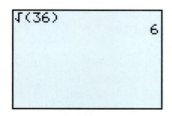
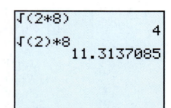

VIEWING WINDOWS

The viewing window for a graphing calculator is similar to the viewfinder in a camera. A camera usually cannot take a photograph of an entire view of a scene. The camera must be centered on some object and can capture only a portion of the available scenery. A camera with a zoom lens can photograph different views of the same scene by zooming in and out. Graphing calculators have similar capabilities. The xy-coordinate plane is infinite. The calculator screen can only show a finite, rectangular region in the plane, and it must be specified before the graph can be drawn. This is done by setting both minimum and maximum values for the x- and y-axes. The scale (distance between tick marks) is usually specified as well. Determining an appropriate viewing window for a graph is often a challenge, and many times it will take a few attempts before a satisfactory window is found.

The screen on the left shows a standard viewing window, and the graph of $y = 2x + 1$ is shown on the right. Using a different window would give a different view of the line.

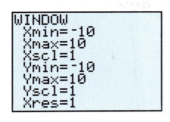

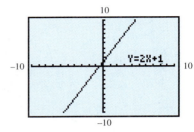

LOCATING POINTS ON A GRAPH: TRACING AND TABLES

Graphing calculators allow you to trace along the graph of an equation and display the co-ordinates of points on the graph. See the screen on the left below, which indicates that the point $(2, 5)$ lies on the graph of $y = 2x + 1$. Tables for equations can also be displayed. The screen on the right shows a partial table for this same equation. Note the middle of the screen, which indicates that when $x = 2$, $y = 5$.

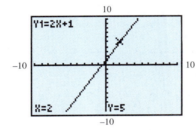

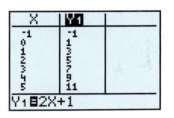

ADDITIONAL FEATURES

There are many features of graphing calculators that go far beyond the scope of this book. These calculators can be programmed, much like computers. Many of them can solve equations at the stroke of a key, analyze statistical data, and perform symbolic algebraic manipulations. Mathematicians from the past would have been amazed by today's calculators. Many important equations in mathematics cannot be solved by hand. However, their solutions can often be approximated using a calculator. Calculators also provide the opportunity to ask "What if . . . ?" more easily. Values in algebraic expressions can be altered and conjectures tested quickly.

FINAL COMMENTS

Despite the power of today's calculators, they cannot replace human thought. ***In the entire problem-solving process, your brain is the most important component.*** Calculators are only tools and, like any tool, they must be used appropriately in order to enhance our ability to understand mathematics. Mathematical insight may often be the quickest and easiest way to solve a problem; a calculator may neither be needed nor appropriate. By applying mathematical concepts, you can make the decision whether or not to use a calculator.

To the Student: Success in Algebra

There are two main reasons students have difficulty with mathematics:

- Students start in a course for which they do not have the necessary background knowledge.

- Students don't know how to study mathematics effectively.

Your instructor can help you decide whether this is the right course for you. We can give you some study tips.

Studying mathematics *is* different from studying subjects like English and history. The key to success is regular practice. This should not be surprising. After all, can you learn to play the piano or ski well without a lot of regular practice? The same is true for learning mathematics. Working problems nearly every day is the key to becoming successful. Here is a list of things that will help you succeed in studying algebra.

1. *Attend class regularly.* Pay attention to what your instructor says and does in class, and take careful notes. In particular, note the problems the instructor works on the board and copy the complete solutions. Keep these notes separate from your homework to avoid confusion when you review them later.

2. Don't hesitate to *ask questions in class.* It is not a sign of weakness but of strength. There are always other students with the same question who are too shy to ask.

3. *Read your text carefully.* Many students read only enough to get by, usually only the examples. Reading the complete section will help you solve the homework problems. Most exercises are keyed to specific examples or objectives that will explain the procedures for working them.

4. Before you start on your homework assignment, *rework the problems the teacher worked in class.* This will reinforce what you have learned. Many students say, "I understand it perfectly when you do it, but I get stuck when I try to work the problem myself."

5. Do your homework assignment only *after reading the text* and reviewing your notes from class. Check your work against the answers in the back of the book. If you get a problem wrong and are unable to understand why, mark that problem and ask your instructor about it. Then practice working additional problems of the same type to reinforce what you have learned.

6. *Work as neatly as you can.* Write your symbols clearly, and make sure the problems are clearly separated from each other. Working neatly will help you to think clearly and also make it easier to review the homework before a test.

7. After you complete a homework assignment, *look over the text again.* Try to identify the main ideas that are in the lesson. Often they are clearly highlighted or boxed in the text.

8. *Use the chapter test at the end of each chapter as a practice test.* Work through the problems under test conditions, without referring to the text or the answers until you are finished. You may want to time yourself to see how long it takes you. When you finish, check your answers against those in the back of the book, and study the problems you missed.

9. *Keep all quizzes and tests that are returned to you,* and use them when you study for future tests and the final exam. These quizzes and tests indicate what concepts your instructor considers to be most important. Be sure to correct any problems on these tests that you missed, so you will have the corrected work to study.

10. *Don't worry if you do not understand a new topic right away.* As you read more about it and work through the problems, you will gain understanding. Each time you review a topic you will understand it a little better. Few people understand each topic completely right from the start.

Reading a list of study tips is a good start, but you may need some help actually *applying* the tips to your work in this math course.

Watch for this icon as you work in this textbook, particularly in the first **Study Skills** few chapters. It will direct you to one of 12 activities in the *Study Skills Workbook* that comes with this text. Each activity helps you to actually *use* a study skills technique. These techniques will greatly improve your chances for success in this course.

- Find out *how your brain learns new material.* Then use that information to set up effective ways to learn math.

- Find out *why short-term memory is so short* and what you can do to help your brain remember new material weeks and months later.

- Find out *what happens when you "blank out" on a test* and simple ways to prevent it from happening.

All the activities in the *Study Skills Workbook* are practical ways to enjoy and succeed at math. Whether you need help with note taking, managing homework, taking tests, or preparing for a final exam, you'll find specific, clearly explained ideas that really work because they're based on research about how the brain learns and remembers.

Diagnostic Pretest

 Study Skills Workbook
Study Skills **Activity 1**

[Chapter R]

1. Find the quotient and write it in lowest terms.

$$\frac{42}{5} \div \frac{7}{15}$$

2. Find the sum and write it in lowest terms.

$$6\frac{7}{8} + 3\frac{2}{3}$$

3. Subtract $38 - 9.678$.

4. **(a)** Convert .99% to a decimal.

   (b) Convert 4.72 to a percent.

[Chapter 1]

5. Select the smaller number from this pair: $|-35|$, $-|35|$.

Perform the indicated operations.

6. $(-3)(-8) - 4(-2)^3$

7. $\dfrac{-6 + |-11 + 5|}{4^2 - (-9)}$

8. Evaluate $\dfrac{6r - 2s^2}{-3t}$ if $r = -5$, $s = -3$, and $t = 4$.

[Chapter 2]

9. Solve and check $-8(x - 3) + 12x = 15 - (2x + 3)$.

10. The two largest cities in the United States are New York and Los Angeles. On July 1, 1998, the population of New York was about 3,822,610 greater than the population of Los Angeles, and there were a total of about 11,017,722 people living in these two cities. Find the population of each city, using these population estimates. Do not round your answer. (*Source: The World Almanac and Book of Facts, 2000.*)

11. Find the measure of each marked angle.

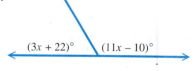

$(3x + 22)°$ $(11x - 10)°$

1. _____

2. _____

3. _____

4. (a) _____

   (b) _____

5. _____

6. _____

7. _____

8. _____

9. _____

10. _____

11. _____

12.

12. Solve $-4x + 8 \geq -12$, and graph the solutions.

[Chapter 3]

13. x-intercept: _____

13. Graph $2x - 5y = 10$. Give the x- and y-intercepts.

y-intercept: _____

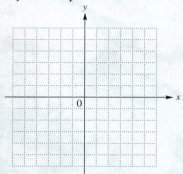

14. _____

14. Find the slope of the line through $(-2, 5)$ and $(1, -7)$.

15. _____

15. Write an equation in slope-intercept form for the line through $(-5, 6)$ and $(1, 0)$.

16.

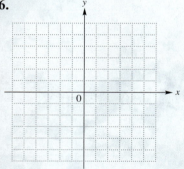

16. Graph $2x + y > 4$.

[Chapter 4]

17. _____

17. Subtract $(4m^3 - 5m^2 + m - 8) - (6m^3 - 5m^2 + 10m - 3)$.

18. _____

18. Multiply $(7z + 3w)^2$.

19. _____

19. Evaluate the expression $5^{-1} + 2^{-1} - 3^0$.

20. (a) _____

20. (a) Write 445,000,000 in scientific notation.

(b) Write 2.34×10^{-4} without exponents.

(b) _____

[Chapter 5]

21. _____

21. Factor $3x^2 + 2x - 8$.

22. _____

22. Factor $16n^2 - 49$.

23. _____

23. Solve $t^2 - 2t = 15$.

24. _____

24. The length of the cover of a road atlas is 4 in. more than the width. The area is 165 in.^{2}. Find the dimensions of the cover.

[Chapter 6]

25. Write $\dfrac{x^2 + x - 20}{x^2 - 16}$ in lowest terms.

25. _____

26. Divide. Write your answer in lowest terms.

$$\frac{2r + 1}{r - 4} \div \frac{6r^2 + 3r}{4 - r}$$

26. _____

27. Subtract. Write your answer in lowest terms.

$$\frac{z^2}{z - 3} - \frac{z}{z + 3}$$

27. _____

28. Simplify.

$$\frac{\dfrac{1}{y} + \dfrac{1}{y + 2}}{\dfrac{1}{y} - \dfrac{1}{y + 2}}$$

28. _____

[Chapter 7]

Solve each system of equations.

29. $3x + y = 9$
$x - y = -1$

29. _____

30. $-4x + 7y = 3$
$12x - 21y = 9$

30. _____

31. Write a system of equations and use it to solve the problem.

Marla and Rick left from the same place at the same time and traveled in opposite directions. Marla drove 8 mph faster than Rick. After 2 hr, they were 228 mi apart. Find Marla and Rick's speeds.

31. _____

32. Graph the solution of the system of inequalities.

$$3x - 5y < 15$$
$$y \geq -2x$$

32.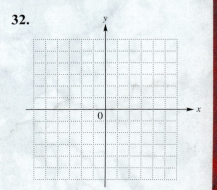

[Chapter 8]

Simplify when possible.

33. _____

33. $4\sqrt{300} - 8\sqrt{75}$

34. _____

34. $(\sqrt{5} - \sqrt{7})^2$

35. _____

35. Rationalize the denominator.

$$\frac{4\sqrt{5}}{\sqrt{2}}$$

36. _____

36. Solve $\sqrt{3r} + 6 = 15$.

[Chapter 9]

37. _____

37. Solve $(x - 3)^2 = 20$.

38. _____

38. Solve $2r^2 + 3r - 1 = 0$.

39. vertex: _____

39. Sketch the graph of $y = -x^2 + 5$. Identify the vertex.

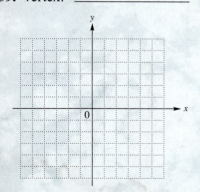

40. _____

domain: _____

range: _____

40. Decide whether the relation

$$\{(-2, 4), (-1, 1), (0, 0), (1, 1), (2, 4)\}$$

represents a function. Give the domain and range.

Prealgebra Review

R.1 FRACTIONS

Studying algebra requires good arithmetic skills. Most people do not get much practice using fractions, so we review the rules for fractions in this section.

The numbers used most often in everyday life are the **whole numbers,**

$$0, 1, 2, 3, 4, 5, \ldots$$

and **fractions,** such as

$$\frac{1}{3}, \frac{5}{4}, \quad \text{and} \quad \frac{11}{12}.$$

The parts of a fraction are named as follows.

$$\text{Fraction bar} \longrightarrow \frac{4 \; \leftarrow \textbf{Numerator}}{7 \; \leftarrow \textbf{Denominator}}$$

If the numerator of a fraction is smaller than the denominator, we call it a **proper fraction.** A proper fraction has a value less than 1. If the numerator is greater than the denominator, the fraction is an **improper fraction.** An improper fraction, which has a value greater than 1, is often written as a **mixed number.** For example, $\frac{12}{5}$ may be written as $2\frac{2}{5}$. In algebra, we prefer to use the improper form because it is easier to work with. In applications, we usually convert answers to mixed number form, which is more meaningful.

 **Identify prime numbers.** In work with fractions, we will need to write the numerators and denominators as products. A **product** is the answer to a multiplication problem. When 12 is written as the product $2 \cdot 6$, for example, 2 and 6 are called **factors** of 12. Other factors of 12 are 1, 3, 4, and 12. A whole number is **prime** if it has exactly two different factors (itself and 1). The first dozen primes are listed here.

$$2, 3, 5, 7, 11, 13, 17, 19, 23, 29, 31, 37$$

A whole number greater than 1 that is not prime is called a **composite number.** For example, 4, 6, 8, 9, and 12 are composite numbers. The number 1 is neither prime nor composite.

OBJECTIVES

1 Identify prime numbers.
2 Write numbers in prime factored form.
3 Write fractions in lowest terms.
4 Multiply and divide fractions.
5 Add and subtract fractions.

 Study Skills Workbook
Activity 2

❶ Tell whether each number is prime or composite.

(a) 12

(b) 13

(c) 27

(d) 59

(e) 1806

❷ Write each number in prime factored form.

(a) 70

(b) 72

(c) 693

(d) 97

Example 1 **Distinguishing between Prime and Composite Numbers**

Decide whether each number is prime or composite.

(a) 33
33 has factors of 3 and 11 as well as 1 and 33, so it is composite.

(b) 43
Since there are no numbers other than 1 and 43 itself that divide *evenly* into 43, the number 43 is prime.

(c) 9832
9832 can be divided by 2, giving 2 · 4916, so it is composite.

Work Problem ❶ at the Side.

2 ▭ **Write numbers in prime factored form.** As mentioned earlier, to factor a number means to write it as the product of two or more numbers. Factoring is just the reverse of multiplying two numbers to get the product.

Multiplication	Factoring
6 · 3 = 18	18 = 6 · 3
↑ ↑ ↑	↑ ↑ ↑
Factors Product	Product Factors

In algebra, a dot is used instead of the × symbol to indicate multiplication because × may be confused with the letter *x*. Each composite number can be written as the product of prime numbers in only one way (disregarding the order of the factors). A number written using factors that are all prime numbers is in **prime factored form.** We will write the factored form with the prime factors in order by size, although any ordering is correct.

Example 2 **Writing Numbers in Prime Factored Form**

Write each number in prime factored form.

(a) 35
Factor 35 as the product of the prime factors 5 and 7, or as
$$35 = 5 \cdot 7.$$

(b) 24
A handy way to keep track of the factors is to use a tree, as shown below. The prime factors are circled.

Divide by the smallest prime, 2, to get   24 = 2 · 12.   ②· 12

Now divide 12 by 2 to find factors of 12.   24 = 2 · 2 · 6   ②· 6

Since 6 can be factored as 2 · 3, 24 = 2 · 2 · 2 · 3,   24 = 2 · 2 · 2 · 3   ②·③
where all factors are prime.

24

Work Problem ❷ at the Side.

3 ▭ **Write fractions in lowest terms.** We use prime factors to write fractions in *lowest terms*. A fraction is in **lowest terms** when the numerator and denominator have no factors in common (other than 1). We write a fraction in this form by using the following facts.

ANSWERS
1. (a) composite **(b)** prime **(c)** composite
    (d) prime **(e)** composite
2. (a) 2 · 5 · 7 **(b)** 2 · 2 · 2 · 3 · 3
    (c) 3 · 3 · 7 · 11 **(d)** 97 is prime.

Properties of 1

Any nonzero number divided by itself is equal to 1.

Any number multiplied by 1 remains the same.

For example,

$$\frac{3}{3} = 1, \quad \frac{8}{8} = 1, \quad \text{and} \quad 17 \cdot 1 = 17.$$

Writing a Fraction in Lowest Terms

Step 1   Write the numerator and denominator in prime factored form.

Step 2   Replace each pair of factors common to the numerator and denominator with 1.

Step 3   Multiply the remaining factors in the numerator and in the denominator.

(This procedure is sometimes called "simplifying the fraction.")

Example 3   Writing Fractions in Lowest Terms

Write each fraction in lowest terms.

(a) $\dfrac{10}{15} = \dfrac{2 \cdot 5}{3 \cdot 5} = \dfrac{2}{3} \cdot \dfrac{5}{5} = \dfrac{2}{3} \cdot 1 = \dfrac{2}{3}$

Since 5 is a common factor of 10 and 15, we use the first property of 1 to replace $\frac{5}{5}$ with 1.

(b) $\dfrac{15}{45} = \dfrac{3 \cdot 5}{3 \cdot 3 \cdot 5} = \dfrac{1 \cdot 3 \cdot 5}{3 \cdot 3 \cdot 5} = \dfrac{1}{3} \cdot \dfrac{3}{3} \cdot \dfrac{5}{5} = \dfrac{1}{3} \cdot 1 \cdot 1 = \dfrac{1}{3}$

Multiplying by 1 in the numerator does not change the value of the numerator and makes it possible to rewrite the expression as the product of three fractions in the next step.

(c) $\dfrac{150}{200}$

It is not always necessary to factor into *prime* factors in Step 1. Here, if you see that 50 is a common factor of the numerator and the denominator, factor as follows:

$$\frac{150}{200} = \frac{3 \cdot 50}{4 \cdot 50} = \frac{3}{4} \cdot 1 = \frac{3}{4}.$$

NOTE

When you are writing a fraction in lowest terms, look for the largest common factor in the numerator and the denominator. If none is obvious, factor the numerator and the denominator into prime factors. *Any* common factor can be used, and the fraction can be simplified in stages. For example,

$$\frac{150}{200} = \frac{15 \cdot 10}{20 \cdot 10} = \frac{3 \cdot 5 \cdot 10}{4 \cdot 5 \cdot 10} = \frac{3}{4}.$$

Work Problem **❸** at the Side.

❸ Write each fraction in lowest terms.

(a) $\dfrac{8}{14}$

(b) $\dfrac{35}{42}$

(c) $\dfrac{120}{72}$

❹ Find each product, and write it in lowest terms.

(a) $\dfrac{5}{8} \cdot \dfrac{2}{10}$

(b) $\dfrac{1}{10} \cdot \dfrac{12}{5}$

(c) $\dfrac{7}{9} \cdot \dfrac{12}{14}$

(d) $3\dfrac{1}{3} \cdot 1\dfrac{3}{4}$

4 ▭ Multiply and divide fractions.

Multiplying Fractions

To multiply two fractions, multiply the numerators to get the numerator of the product, and multiply the denominators to get the denominator of the product. The product must be written in lowest terms.

In practice, we will show the products of the numerator and the denominator in factored form to make it easier to write the product in lowest terms. We often simplify before performing the multiplication, as shown in the next example.

Example 4 Multiplying Fractions

Find each product, and write it in lowest terms.

(a) $\dfrac{3}{8} \cdot \dfrac{4}{9}$

$$\dfrac{3}{8} \cdot \dfrac{4}{9} = \dfrac{3 \cdot 4}{8 \cdot 9} \qquad \text{Multiply numerators.}$$
$$\text{Multiply denominators.}$$
$$= \dfrac{3 \cdot 4}{2 \cdot 4 \cdot 3 \cdot 3} \qquad \text{Factor.}$$
$$= \dfrac{1}{2 \cdot 3} = \dfrac{1}{6} \qquad \text{Write in lowest terms.}$$

(b) $2\dfrac{1}{3} \cdot 5\dfrac{1}{2}$

$$2\dfrac{1}{3} \cdot 5\dfrac{1}{2} = \dfrac{7}{3} \cdot \dfrac{11}{2} \qquad \text{Write as improper fractions.}$$
$$= \dfrac{77}{6} \text{ or } 12\dfrac{5}{6} \qquad \text{Multiply numerators and denominators.}$$

Work Problem ❹ at the Side.

Two fractions are **reciprocals** of each other if their product is 1. For example, $\frac{3}{4}$ and $\frac{4}{3}$ are reciprocals because

$$\dfrac{3}{4} \cdot \dfrac{4}{3} = 1.$$

The numbers $\frac{7}{11}$ and $\frac{11}{7}$ are reciprocals also. Other examples are $\frac{1}{5}$ and 5, $\frac{4}{9}$ and $\frac{9}{4}$, and 16 and $\frac{1}{16}$.

Because division is the opposite or inverse of multiplication, we use reciprocals to divide fractions.

Dividing Fractions

To divide two fractions, multiply the first fraction by the reciprocal of the second. The result is called the **quotient.**

The reason this method works will be explained in Section 1.6. However, as an example, we know that $20 \div 10 = 2$, and $20 \cdot \frac{1}{10} = 2$.

Example 5   Dividing Fractions

Find each quotient, and write it in lowest terms.

(a) $\dfrac{3}{4} \div \dfrac{8}{5} = \dfrac{3}{4} \cdot \dfrac{5}{8} = \dfrac{3 \cdot 5}{4 \cdot 8} = \dfrac{15}{32}$

— Multiply by the reciprocal of the second fraction.

(b) $\dfrac{3}{4} \div \dfrac{5}{8} = \dfrac{3}{4} \cdot \dfrac{8}{5} = \dfrac{3 \cdot 8}{4 \cdot 5} = \dfrac{3 \cdot 4 \cdot 2}{4 \cdot 5} = \dfrac{6}{5}$

(c) $\dfrac{5}{8} \div 10 = \dfrac{5}{8} \div \dfrac{10}{1} = \dfrac{5}{8} \cdot \dfrac{1}{10} = \dfrac{1}{16}$

Write 10 as $\frac{10}{1}$.

(d) $1\dfrac{2}{3} \div 4\dfrac{1}{2}$

$1\dfrac{2}{3} \div 4\dfrac{1}{2} = \dfrac{5}{3} \div \dfrac{9}{2}$ Write as improper fractions.

$\phantom{1\dfrac{2}{3} \div 4\dfrac{1}{2}} = \dfrac{5}{3} \cdot \dfrac{2}{9}$ Multiply by the reciprocal of the second fraction.

$\phantom{1\dfrac{2}{3} \div 4\dfrac{1}{2}} = \dfrac{10}{27}$

CAUTION

Notice that *only* the second fraction (the divisor) is replaced by its reciprocal in the multiplication.

Work Problem ⑤ at the Side.

5   **Add and subtract fractions.**   The result of adding two numbers is called the **sum** of the numbers. For example, since $2 + 3 = 5$, the sum of 2 and 3 is 5.

Adding Fractions

To find the sum of two fractions with the *same* denominator, add their numerators and keep the *same* denominator.

Example 6   Adding Fractions with the Same Denominator

Add. Write sums in lowest terms.

(a) $\dfrac{3}{7} + \dfrac{2}{7} = \dfrac{3 + 2}{7} = \dfrac{5}{7}$ Denominator does not change.

(b) $\dfrac{2}{10} + \dfrac{3}{10} = \dfrac{2 + 3}{10} = \dfrac{5}{10} = \dfrac{1}{2}$ Write in lowest terms.

Work Problem ⑥ at the Side.

⑤ Find each quotient, and write it in lowest terms.

(a) $\dfrac{3}{10} \div \dfrac{2}{7}$

(b) $\dfrac{3}{4} \div \dfrac{7}{16}$

(c) $\dfrac{4}{3} \div 6$

(d) $3\dfrac{1}{4} \div 1\dfrac{2}{5}$

⑥ Add. Write sums in lowest terms.

(a) $\dfrac{3}{5} + \dfrac{4}{5}$

(b) $\dfrac{5}{14} + \dfrac{3}{14}$

If the fractions to be added do not have the same denominator, the procedure above can still be used, but only *after* the fractions are rewritten with a common denominator. For example, to rewrite $\frac{3}{4}$ as a fraction with a denominator of 32,

$$\frac{3}{4} = \frac{?}{32},$$

we must find the number that can be multiplied by 4 to give 32. Since $4 \cdot 8 = 32$, we use the number 8. By the second property of 1, we can multiply the numerator and the denominator by 8.

$$\frac{3}{4} = \frac{3}{4} \cdot 1 = \frac{3}{4} \cdot \frac{8}{8} = \frac{3 \cdot 8}{4 \cdot 8} = \frac{24}{32}$$

Finding the Least Common Denominator (LCD)

Step 1 Factor all denominators to prime factored form.

Step 2 The LCD is the product of every (different) factor that appears in any of the factored denominators. If a factor is repeated, use the largest number of repeats as factors of the LCD.

Step 3 Write each fraction with the LCD as the denominator, using the second property of 1.

Example 7 Adding Fractions with Different Denominators

Add. Write sums in lowest terms.

(a) $\frac{4}{15} + \frac{5}{9}$

Step 1 To find the LCD, we first factor both denominators to prime factored form.

$$15 = 5 \cdot 3 \quad \text{and} \quad 9 = 3 \cdot 3$$

3 is a factor of both denominators.

$$\overset{\displaystyle 15 \quad 9}{\wedge \wedge}$$

Step 2 $\qquad\qquad\qquad \text{LCD} = 5 \cdot 3 \cdot 3 = 45$

In this example, the LCD needs one factor of 5 and two factors of 3 because the second denominator has two factors of 3.

Step 3 Now we can use the second property of 1 to write each fraction with 45 as the denominator.

$$\frac{4}{15} = \frac{4}{15} \cdot \frac{3}{3} = \frac{12}{45} \quad \text{and} \quad \frac{5}{9} = \frac{5}{9} \cdot \frac{5}{5} = \frac{25}{45}$$

Now add the two equivalent fractions to get the required sum.

$$\frac{4}{15} + \frac{5}{9} = \frac{12}{45} + \frac{25}{45} = \frac{37}{45}$$

Continued on Next Page

(b) $3\dfrac{1}{2} + 2\dfrac{3}{4}$

$$3\dfrac{1}{2} + 2\dfrac{3}{4} = \dfrac{7}{2} + \dfrac{11}{4} \qquad \text{Change to improper fractions.}$$

$$= \dfrac{14}{4} + \dfrac{11}{4} \qquad \text{Get a common denominator.}$$

$$= \dfrac{25}{4} \text{ or } 6\dfrac{1}{4} \qquad \text{Add.}$$

(c) $45\dfrac{2}{3} + 73\dfrac{1}{2}$

We could use an alternative vertical method here, adding the whole numbers and the fractions separately.

$$45\dfrac{2}{3} = 45\dfrac{4}{6}$$

$$+\ 73\dfrac{1}{2} = 73\dfrac{3}{6}$$

$$\overline{\phantom{+\ 73\dfrac{1}{2} = 73\dfrac{3}{6}}}$$

$$118\dfrac{7}{6} = 118 + \left(1 + \dfrac{1}{6}\right) = 119\dfrac{1}{6}$$

══════════ **Work Problem ➐ at the Side.**

The **difference** between two numbers is found by subtracting the numbers. For example, $9 - 5 = 4$, so the difference between 9 and 5 is 4. We find the difference between two fractions as follows.

Subtracting Fractions

To find the difference between two fractions with the *same* denominator, subtract their numerators and keep the *same* denominator.
   If the fractions have *different* denominators, write them with a common denominator first.

Example 8   **Subtracting Fractions**

Subtract. Write differences in lowest terms.

(a) $\dfrac{15}{8} - \dfrac{3}{8} = \dfrac{15 - 3}{8} = \dfrac{12}{8} = \dfrac{3}{2} \qquad$ Lowest terms

(b) $\dfrac{15}{16} - \dfrac{4}{9}$

Since $16 = 2 \cdot 2 \cdot 2 \cdot 2$ and $9 = 3 \cdot 3$ have no common factors, the LCD is $16 \cdot 9 = 144$.

$$\dfrac{15}{16} - \dfrac{4}{9} = \dfrac{15 \cdot 9}{16 \cdot 9} - \dfrac{4 \cdot 16}{9 \cdot 16} \qquad \text{Get a common denominator.}$$

$$= \dfrac{135}{144} - \dfrac{64}{144}$$

$$= \dfrac{71}{144} \qquad \begin{array}{l}\text{Subtract numerators; keep}\\\text{the same denominator.}\end{array}$$

═════ **Continued on Next Page**

➐ Add. Write sums in lowest terms.

(a) $\dfrac{7}{30} + \dfrac{2}{45}$

(b) $\dfrac{17}{10} + \dfrac{8}{27}$

(c) $2\dfrac{1}{8} + 1\dfrac{2}{3}$

(d) $132\dfrac{4}{5} + 28\dfrac{3}{4}$

ANSWERS

7. (a) $\dfrac{5}{18}$ **(b)** $\dfrac{539}{270}$ **(c)** $\dfrac{91}{24}$ or $3\dfrac{19}{24}$

(d) $161\dfrac{11}{20}$

8 Subtract.

(a) $\dfrac{9}{11} - \dfrac{3}{11}$

(b) $\dfrac{13}{15} - \dfrac{5}{6}$

(c) $2\dfrac{3}{8} - 1\dfrac{1}{2}$

(d) $50\dfrac{1}{4} - 32\dfrac{2}{3}$

9 To make a three-piece outfit from the same fabric, Wei Jen needs $1\dfrac{1}{4}$ yd for the blouse, $1\dfrac{2}{3}$ yd for the skirt, and $2\dfrac{1}{2}$ yd for the jacket. How much fabric does she need?

(c) $2\dfrac{1}{2} - 1\dfrac{3}{4}$

First, change the mixed numbers $2\frac{1}{2}$ and $1\frac{3}{4}$ to improper fractions.

$$2\dfrac{1}{2} - 1\dfrac{3}{4} = \dfrac{5}{2} - \dfrac{7}{4} \qquad \text{Write as improper fractions.}$$

$$= \dfrac{10}{4} - \dfrac{7}{4} \qquad \text{Get a common denominator.}$$

$$= \dfrac{3}{4} \qquad \text{Subtract.}$$

Work Problem 8 at the Side.

We often see mixed numbers used in applications of mathematics.

Example 9   Solving an Applied Problem Requiring Addition of Fractions

The diagram below appears in the book *Woodworker's 39 Sure-Fire Projects.* It is a view of a corner bookcase/desk. Add the fractions shown in the diagram to find the height of the bookcase/desk to the top of the writing surface.

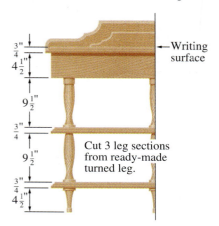

We must find the following sum (" means inches).

$$\dfrac{3}{4} + 4\dfrac{1}{2} + 9\dfrac{1}{2} + \dfrac{3}{4} + 9\dfrac{1}{2} + \dfrac{3}{4} + 4\dfrac{1}{2}$$

Change the mixed numbers to improper fractions.

$$\dfrac{3}{4} + \dfrac{9}{2} + \dfrac{19}{2} + \dfrac{3}{4} + \dfrac{19}{2} + \dfrac{3}{4} + \dfrac{9}{2}$$

The LCD is 4. Change all fractions to fourths.

$$\dfrac{3}{4} + \dfrac{18}{4} + \dfrac{38}{4} + \dfrac{3}{4} + \dfrac{38}{4} + \dfrac{3}{4} + \dfrac{18}{4}$$

Now we can add and simplify the answer.

$$\dfrac{3}{4} + \dfrac{18}{4} + \dfrac{38}{4} + \dfrac{3}{4} + \dfrac{38}{4} + \dfrac{3}{4} + \dfrac{18}{4} = \dfrac{121}{4} \text{ or } 30\dfrac{1}{4}$$

The height is $30\frac{1}{4}$ in.

Work Problem 9 at the Side.

Example 10 **Solving an Applied Problem Requiring Division of Fractions**

An upholsterer needs $2\frac{1}{4}$ yd of fabric to recover a chair. How many chairs can be covered with $23\frac{2}{3}$ yd of fabric?

It helps to understand the problem if we replace the fractions with whole numbers. Suppose each chair requires 2 yd, and we have 24 yd of fabric. Dividing 24 by 2 gives the number of chairs (12) that can be recovered. To solve the original problem, we must divide $23\frac{2}{3}$ by $2\frac{1}{4}$.

$$23\frac{2}{3} \div 2\frac{1}{4} = \frac{71}{3} \div \frac{9}{4}$$

$$= \frac{71}{3} \cdot \frac{4}{9}$$

$$= \frac{284}{27} \text{ or } 10\frac{14}{27}$$

Thus, 10 chairs can be recovered with some fabric left over.

══════════════ **Work Problem 10 at the Side.**

10 A gallon of paint covers 500 ft^2. (ft^2 means square feet.) To paint his house, Tram needs enough paint to cover 4200 ft^2. How many gallons of paint should he buy?

Real-Data Applications

Quantity of Fabric Needed to Sew a Quilt

Quilting is a popular craft. Quilts are often based on an underlying square grid, called a **block.** Blocks are typically sewn from square pieces forming patterns of 4, 9, 16, 25, and so on. The square pieces (or triangular pieces that form squares) are cut from various colored fabrics and are arranged to create a pleasing design.

The quilt design *Infinity*, shown on the left, is based solely on squares (no triangles). The enlarged underlying block design is shown on the right. Each block is constructed of 4-in. finished squares. Pieces of fabric are cut into $4\frac{1}{2}$ in. squares, allowing $\frac{1}{4}$ in. for seams. The quilt top is made of 9 blocks. The inner border is made of four strips finished to size 4 in. by 84 in., and the backing is folded to form the 4-in. outer border. You should be able to verify that the finished quilt measures 100 in. by 100 in.

Infinity Quilt

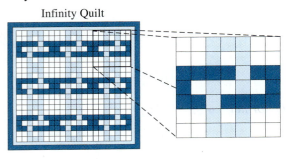

Block Design

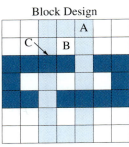

For Group Discussion

Cotton fabrics used in quilting are typically sold by the yard in widths of 45 in. The selvage edges are discarded, so allow 44 in. maximum width as you work these problems.

1. For each block, how many squares make up

   Color A? _____   Color B? _____   Color C? _____

2. Determine the amount of fabric needed to sew the inner border (color A). Sketch and label a diagram to illustrate how the fabric would be cut from the yardage purchased. Also show the unused portion of fabric.

3. Determine the amount of fabric needed to sew the backing and outer border (Color C). Assume that the backing is cut large enough to fold over to make the outer border. Sketch and label a diagram to illustrate how the fabric would be cut from the yardage purchased, including the unused portion of fabric.

4. Sketch a diagram to show the number of $4\frac{1}{2}$ in. squares that can be cut from a 9 in. by 18 in. piece of fabric.

5. Determine the amount of fabric of each color needed to construct the *Infinity Quilt* top (minus the border). (*Hint:* Use the sketches from Problems 2, 3, and 4.)

6. What is your recommendation for the total amount of fabric of each color needed for this quilt?

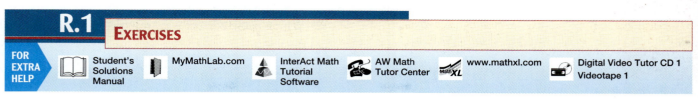

R.1 **EXERCISES**

FOR EXTRA HELP

Student's Solutions Manual   MyMathLab.com   InterAct Math Tutorial Software   AW Math Tutor Center   www.mathxl.com   Digital Video Tutor CD 1 Videotape 1

Decide whether each statement is true *or* false. *If it is* false, *say why.*

Study Skills Workbook **Activity 3**

1. In the fraction $\frac{3}{7}$, 3 is the numerator and 7 is the denominator.

2. The mixed number equivalent of $\frac{41}{5}$ is $8\frac{1}{5}$.

3. The fraction $\frac{17}{51}$ is in lowest terms.

4. The reciprocal of $\frac{8}{2}$ is $\frac{4}{1}$.

5. The product of 8 and 2 is 10.

6. The difference between 12 and 2 is 6.

Identify each number as prime, composite, or neither. See Example 1.

7. 19

8. 29

9. 52

10. 99

11. 2468

12. 3125

13. 1

14. 14

Write each number in prime factored form. See Example 2.

15. 30

16. 40

17. 252

18. 168

19. 124

20. 165

21. 29

22. 31

Write each fraction in lowest terms. See Example 3.

23. $\frac{8}{16}$

24. $\frac{4}{12}$

25. $\frac{15}{18}$

26. $\frac{16}{20}$

27. $\frac{15}{75}$

28. $\frac{24}{64}$

29. $\frac{144}{120}$

30. $\frac{132}{77}$

31. For the fractions $\dfrac{p}{q}$ and $\dfrac{r}{s}$, which can serve as a common denominator?

    A. $q \cdot s$

    B. $q + s$

    C. $p \cdot r$

    D. $p + r$

32. Which is the correct way to write $\dfrac{16}{24}$ in lowest terms?

    A. $\dfrac{16}{24} = \dfrac{8 + 8}{8 + 16} = \dfrac{8}{16} = \dfrac{1}{2}$

    B. $\dfrac{16}{24} = \dfrac{4 \cdot 4}{4 \cdot 6} = \dfrac{4}{6}$

    C. $\dfrac{16}{24} = \dfrac{8 \cdot 2}{8 \cdot 3} = \dfrac{2}{3}$

    D. $\dfrac{16}{24} = \dfrac{14 + 2}{21 + 3} = \dfrac{2}{3} + \dfrac{2}{3} = \dfrac{4}{3}$

Find each product or quotient, and write it in lowest terms. See Examples 4 and 5.

33. $\dfrac{4}{5} \cdot \dfrac{6}{7}$

34. $\dfrac{5}{9} \cdot \dfrac{10}{7}$

35. $\dfrac{1}{10} \cdot \dfrac{12}{5}$

36. $\dfrac{6}{11} \cdot \dfrac{2}{3}$

37. $\dfrac{15}{4} \cdot \dfrac{8}{25}$

38. $\dfrac{4}{7} \cdot \dfrac{21}{8}$

39. $2\dfrac{2}{3} \cdot 5\dfrac{4}{5}$

40. $3\dfrac{3}{5} \cdot 7\dfrac{1}{6}$

41. $\dfrac{5}{4} \div \dfrac{3}{8}$

42. $\dfrac{7}{6} \div \dfrac{9}{10}$

43. $\dfrac{32}{5} \div \dfrac{8}{15}$

44. $\dfrac{24}{7} \div \dfrac{6}{21}$

45. $\dfrac{3}{4} \div 12$

46. $\dfrac{2}{5} \div 30$

47. $2\dfrac{5}{8} \div 1\dfrac{15}{32}$

48. $2\dfrac{3}{10} \div 7\dfrac{4}{5}$

49. In your own words, explain how to divide two fractions.

50. In your own words, explain how to add two fractions that have different denominators.

Find each sum or difference, and write it in lowest terms. See Examples 6–8.

51. $\dfrac{7}{12} + \dfrac{1}{12}$

52. $\dfrac{3}{16} + \dfrac{5}{16}$

53. $\dfrac{5}{9} + \dfrac{1}{3}$

54. $\dfrac{4}{15} + \dfrac{1}{5}$

55. $3\dfrac{1}{8} + \dfrac{1}{4}$

56. $5\dfrac{3}{4} + \dfrac{2}{3}$

57. $\dfrac{7}{12} - \dfrac{1}{9}$

58. $\dfrac{11}{16} - \dfrac{1}{12}$

59. $6\dfrac{1}{4} - 5\dfrac{1}{3}$

60. $8\dfrac{4}{5} - 7\dfrac{4}{9}$

61. $\dfrac{5}{3} + \dfrac{1}{6} - \dfrac{1}{2}$

62. $\dfrac{7}{15} + \dfrac{1}{6} - \dfrac{1}{10}$

Use the chart, which appears on a package of Quaker Quick Grits, to answer the questions in Exercises 63 and 64.

63. How many cups of water would be needed for eight microwave servings?

64. How many tsp of salt would be needed for five stove top servings? (*Hint:* 5 is halfway between 4 and 6.)

	Microwave		Stove Top		
Servings	**1**		**1**	**4**	**6**
Water	$\dfrac{3}{4}$ cup		1 cup	3 cups	4 cups
Grits	3 Tbsp		3 Tbsp	$\dfrac{3}{4}$ cup	1 cup
Salt (optional)	Dash		Dash	$\dfrac{1}{4}$ tsp	$\dfrac{1}{2}$ tsp

Solve each applied problem. See Examples 9 and 10.

65. A motel owner has decided to expand his business by buying a piece of property next to the motel. The property has an irregular shape, with five sides as shown in the figure. Find the total distance around the piece of property. This is called the *perimeter* of the figure.

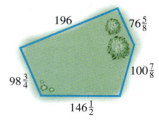

Measurements in feet

66. A triangle has sides of lengths $5\dfrac{1}{4}$ ft, $7\dfrac{1}{2}$ ft, and $10\dfrac{1}{8}$ ft. Find the perimeter of the triangle. See Exercise 65.

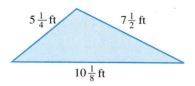

67. A hardware store sells a 40-piece socket wrench set. The measure of the largest socket is $\frac{3}{4}$ in., while the measure of the smallest socket is $\frac{3}{16}$ in. What is the difference between these measures?

68. Two sockets in a socket wrench set have measures of $\frac{9}{16}$ in. and $\frac{3}{8}$ in. What is the difference between these two measures?

69. Under existing standards, most of the holes in Swiss cheese must have diameters between $\frac{11}{16}$ and $\frac{13}{16}$ in. To accommodate new high-speed slicing machines, the USDA wants to reduce the minimum size to $\frac{3}{8}$ in. How much smaller is $\frac{3}{8}$ in. than $\frac{11}{16}$ in.? (*Source:* U.S. Department of Agriculture.)

70. Tex's favorite recipe for barbecue sauce calls for $2\frac{1}{3}$ cups of tomato sauce. The recipe makes enough barbecue sauce to serve 7 people. How much tomato sauce is needed for 1 serving?

More than 8 million immigrants were admitted to the United States between 1990 and 1997. The pie chart gives the fractional number from each region of birth for these immigrants. Use the chart to answer the following questions.

71. What fractional part of the immigrants were from other regions?

72. What fractional part of the immigrants were from Latin America or Asia?

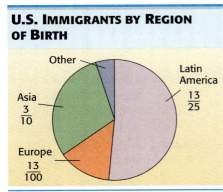

U.S. Immigrants by Region of Birth

Other

Latin America $\frac{13}{25}$

Asia $\frac{3}{10}$

Europe $\frac{13}{100}$

Source: U.S. Bureau of the Census.

73. How many (in millions) were from Europe?

R.2 DECIMALS AND PERCENTS

Fractions are one way to represent parts of a whole. Another way is with a **decimal fraction** or **decimal,** a number written with a decimal point, such as 9.4. Each digit in a decimal number has a place value, as shown below.

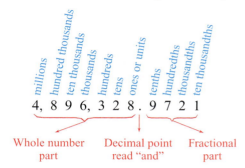

Each successive place value is ten times larger than the place value to its right and is one-tenth as large as the place value to its left.

Prices are often written as decimals. The price $14.75 means 14 dollars and 75 cents, or 14 dollars and $\frac{75}{100}$ of a dollar.

1 **Write decimals as fractions.**   Place value is used to write a decimal number as a fraction. For example, since the last digit (that is, the digit farthest to the right) of .67 is in the *hundredths* place,

$$.67 = \frac{67}{\mathbf{100}}.$$

Similarly, $.9 = \frac{9}{10}$ and $.25 = \frac{25}{100}$. Digits to the left of the decimal point indicate whole numbers, so 12.342 is the sum of 12 and .342, and

$$12.342 = 12 + .342 = 12 + \frac{342}{1000} = \frac{12,000}{1000} + \frac{342}{1000} = \frac{12,342}{1000}.$$

These examples suggest the following rule.

Converting a Decimal to a Fraction

Read the name using the correct place value. Write it in fraction form just as you read it. The denominator will be a **power of 10,** a number like 10, 100, 1000, and so on.

For example, we read .16 as "sixteen hundredths" and write it in fraction form as $\frac{16}{100}$. The same thing is accomplished by counting the number of digits to the right of the decimal point, then writing the given number without a decimal point over a denominator of 1 followed by that number of zeros.

Example 1 Writing Decimals as Fractions

Write each decimal as a fraction. Do not write in lowest terms.

(a) .95

We read .95 as 95 hundredths, so the fraction form is $\frac{95}{100}$. Using the shortcut method, since there are two places to the right of the decimal point, there will be two zeros in the denominator.

Continued on Next Page

1 Write each decimal as a fraction. Do not write in lowest terms.

(a) .8

(b) .431

(c) 20.58

2 Add or subtract as indicated.

(a) 68.9
     42.72
 +   8.973

(b)   32.5
     − 21.72

(c) 42.83 + 71 + 3.074

(d) 351.8 − 2.706

$$.95 = \frac{95}{100}$$

2 places 2 zeros

(b) $.056 = \frac{56}{1000}$

3 places 3 zeros

(c) $4.2095 = 4 + .2095 = 4 + \frac{2095}{10,000} = \frac{42,095}{10,000}$

4 places   4 zeros

Work Problem 1 at the Side.

2 ▭ **Add and subtract decimals.** In the next example, we explain addition and subtraction of decimals.

Example 2 Adding and Subtracting Decimals

Add or subtract as indicated.

(a) 6.92 + 14.8 + 3.217

Place the digits of the numbers in columns, with decimal points lined up so that tenths are in one column, hundredths in another column, and so on.

      6.92   Decimal points lined up
     14.8
 +   3.217
     24.937

A good way to avoid errors is to attach zeros to make all the numbers the same length. For example,

      6.92   6.920   Attach zeros.
     14.8   becomes   14.800
 +   3.217   1   3.217
                            24.937.

(b) 47.6 − 32.509

Write the numbers in columns, attaching zeros to 47.6.

     47.6   47.600
 − 32.509   becomes   −   32.509
                             15.091

(c) 3 − .253

A whole number is assumed to have the decimal point at the right of the number. Write 3 as 3.000; then subtract.

      3.000
 −   .253
      2.747

Work Problem 2 at the Side.

3 ▭ **Multiply and divide decimals.** We multiply decimals by slightly modifying multiplication of whole numbers. We will sometimes use the times symbol, ×, instead of a dot to avoid confusion with the decimal point.

ANSWERS

1. (a) $\frac{8}{10}$ (b) $\frac{431}{1000}$ (c) $\frac{2058}{100}$
2. (a) 120.593 (b) 10.78
   (c) 116.904 (d) 349.094

Multiplying Decimals

Ignore the decimal points and multiply as if the numbers were whole numbers. Then add together the number of **decimal places** (digits to the *right* of the decimal point) in each number being multiplied. Place the decimal point in the answer that many digits from the right.

Example 3 **Multiplying Decimals**

Multiply.

(a) 29.3×4.52

Multiply as if the numbers were whole numbers.

```
      29.3      1 decimal place in first number
×     4.52      2 decimal places in second number
      586       1 + 2 = 3
     1465
    1172
  132.436       3 decimal places in answer
```

(b) 7.003×55.8

```
     7.003      3 decimal places
×     55.8      1 decimal place
     56024      3 + 1 = 4
    35015
   35015
  390.7674      4 decimal places
```

(c) 31.42×65

```
     31.42      2 decimal places
×       65      0 decimal places
     15710      2 + 0 = 2
    18852
   2042.30      2 decimal places
```

The final 0 here can be dropped and the result can be expressed as 2042.3.

Work Problem **3** at the Side.

To divide decimals, convert the divisor to a whole number.

Dividing Decimals

Change the **divisor** (the number you are dividing *by*) into a whole number by moving the decimal point as many places as necessary to the right. Move the decimal point in the **dividend** (the number you are dividing *into*) to the right by the same number of places. Move the decimal point straight up and then divide as with whole numbers.

$$\text{Divisor} \longrightarrow 25\overline{)125} \quad \underset{\longleftarrow \text{ Quotient}}{5}$$

Dividend

3 Multiply.

(a) $2.13 \times .05$

(b) 9.32×1.4

(c) $300.2 \times .052$

(d) $42,001 \times .012$

4 Divide.

(a) $14.9\overline{)451.47}$

(b) $.37\overline{)5.476}$

(c) $375.1 \div 3.001$

Example 4   **Dividing Decimals**

Divide.

(a) $233.45 \div 11.5$

Write the problem as follows.

$$11.5\overline{)233.45}$$

To change 11.5 into a whole number, move the decimal point one place to the right. Move the decimal point in 233.45 the same number of places to the right, to get 2334.5.

$$11.5.\overline{)233.4.5} \qquad \text{Move one decimal place to the right.}$$

To see why this works, write the division in fraction form and multiply by $\frac{10}{10}$ or 1.

$$\frac{233.45}{11.5} \cdot \frac{10}{10} = \frac{2334.5}{115}$$

The result is the same as when we moved the decimal point one place to the right in the divisor and the dividend.

Move the decimal point straight up and divide as with whole numbers.

$$
\begin{array}{r}
20.3 \\
115\overline{)2334.5} \qquad \text{Move decimal point straight up.} \\
\underline{230} \\
345 \\
\underline{345} \\
0
\end{array}
$$

In the second step of the division, 115 does not divide into 34, so we used zero as a placeholder in the quotient.

(b) $73.85\overline{)1852.882}$ (Round the answer to two decimal places.)

Move the decimal point two places to the right in 73.85, to get 7385. Do the same thing with 1852.882, to get 185288.2.

$$73.85.\overline{)1852.88.2}$$

Move the decimal point straight up and divide as with whole numbers.

$$
\begin{array}{r}
25.089 \\
7385\overline{)185288.200} \\
\underline{14770} \\
37588 \\
\underline{36925} \\
66320 \\
\underline{59080} \\
72400 \\
\underline{66465} \\
5935
\end{array}
$$

We carried out the division to three decimal places so that we could round to two decimal places, getting the quotient 25.09.

Work Problem 4 at the Side.

ANSWERS
4. (a) 30.3   (b) 14.8
   (c) 124.99 (rounded)

A shortcut can be used when multiplying or dividing by powers of 10.

Multiplying or Dividing by Powers of 10

To *multiply* by a power of 10, move the decimal point to the *right* as many places as the number of zeros.

To *divide* by a power of 10, move the decimal point to the *left* as many places as the number of zeros.

In both cases, insert 0s as placeholders if necessary.

Example 5 Multiplying and Dividing by Powers of 10

Multiply or divide as indicated.

(a) $48.731 \times 100 = 48.73\underset{\llcorner\uparrow}{.}1 = 4873.1$

We moved the decimal point two places to the right because 100 has two zeros.

(b) $48.7 \div 1000 = \underset{\uparrow\llcorner\!\llcorner}{.}048.7 = .0487$

We moved the decimal point three places to the left because 1000 has three zeros. We needed to insert a zero in front of the 4 to do this.

=============== **Work Problem ⑤ at the Side.**

To avoid misplacing the decimal point, check your work by estimating the answer. To get a quick estimate, round the numbers so that only the first digit is not zero, using the rule for rounding. For more accurate estimates, the numbers could be rounded to the first two or even three nonzero digits.

Rule for Rounding

If the digit to become 0 or be dropped is 5 or more, round up by adding 1 to the final digit to be kept.

If the digit to become 0 or be dropped is 4 or less, do not round up.

For example, to estimate the answer to Example 2(a), round

$$6.92 \text{ to } 7, \quad 14.8 \text{ to } 10, \quad \text{and} \quad 3.217 \text{ to } 3.$$
$$\hphantom{6.92 \text{ to } 7,}\;\uparrow\qquad\quad\;\uparrow\qquad\qquad\qquad\;\uparrow$$
5 or more   4 or less   4 or less

Since $7 + 10 + 3 = 20$, the answer of 24.937 is reasonable. In Example 4(a), round 233.45 to 200 and 11.5 to 10. Since $200 \div 10 = 20$, the answer of 20.3 is reasonable.

4  Write fractions as decimals.

Writing a Fraction as a Decimal

Because a fraction bar indicates division, write a fraction as a decimal by dividing the denominator into the numerator.

⑤ Multiply or divide as indicated.

(a) 294.72×10

(b) 19.5×1000

(c) $4.793 \div 100$

(d) $960.1 \div 10$

⑥ Convert to decimals. For repeating decimals, write the answer two ways: using the bar notation and rounding to the nearest thousandth.

(a) $\dfrac{2}{9}$

(b) $\dfrac{17}{20}$

(c) $\dfrac{1}{11}$

Example 6 Writing Fractions as Decimals

Write each fraction as a decimal.

(a) $\dfrac{19}{8}$

$$8)\overline{19.000}$$
$$\begin{array}{r} 2.375 \\ \underline{16} \\ 30 \\ \underline{24} \\ 60 \\ \underline{56} \\ 40 \\ \underline{40} \\ 0 \end{array}$$

$\dfrac{19}{8} = 2.375$

(b) $\dfrac{2}{3}$

$$3)\overline{2.0000\ldots}$$
$$\begin{array}{r} .6666\ldots \\ \underline{18} \\ 20 \\ \underline{18} \\ 20 \\ \underline{18} \\ 20 \\ \underline{18} \\ 20 \end{array}$$

The remainder in the division in part (b) is never 0. Because 2 is always left after the subtraction, this quotient is a **repeating decimal.** A convenient notation for a repeating decimal is a bar over the digit (or digits) that repeats. For instance, we can write .6666 . . . as $.\overline{6}$. In applications, we often round repeating decimals to as many places as needed. For example, rounding to the *nearest thousandth,*

$$\frac{2}{3} = .66\mathbf{7}. \qquad \text{An approximation}$$

CAUTION

When rounding, be careful to distinguish between *thousandths* and *thousands* or between *hundredths* and *hundreds,* and so on.

Work Problem ⑥ at the Side.

5 Convert percents to decimals and decimals to percents. An important application of decimals is in work with percents. The word **percent** means "per one hundred." Percent is written with the sign %. One percent means "one per one hundred" or "one one-hundredth."

$$1\% = .01 \quad \text{or} \quad 1\% = \frac{1}{100}$$

Example 7 Converting Percents and Decimals

(a) Write 73% as a decimal.
Since $1\% = .01$,

$$73\% = 73 \cdot 1\% = 73 \times .01 = .73.$$

Also, 73% can be written as a decimal using the fraction form $1\% = \frac{1}{100}$.

$$73\% = 73 \cdot 1\% = 73 \cdot \left(\frac{1}{100}\right) = \frac{73}{100} = .73$$

(b) Write 125% as a decimal.

$$125\% = 125 \cdot 1\% = 125 \times .01 = 1.25$$

Continued on Next Page

(c) Write $3\frac{1}{2}\%$ as a decimal.

First write the fractional part as a decimal.

$$3\frac{1}{2}\% = (3 + .5)\% = 3.5\%$$

Now change the percent to decimal form.

$$3.5\% = 3.5 \times .01 = .035$$

(d) Write .32 as a percent.

Since .32 means 32 hundredths, write .32 as $32 \times .01$. Finally, replace .01 with 1%.

$$.32 = 32 \times \mathbf{.01} = 32 \times \mathbf{1\%} = 32\%$$

(e) Write 2.63 as a percent.

$$2.63 = 263 \times .01 = 263 \times 1\% = 263\%$$

NOTE

A quick way to change from a percent to a decimal is to move the decimal point two places to the left. To change from a decimal to a percent, move the decimal point two places to the right.

Divide by 100;
Move 2 places left.

Decimal ⟷ Percent

Multiply by 100;
Move 2 places right.

Example 8   **Converting Percents and Decimals by Moving the Decimal Point**

Convert each percent to a decimal and each decimal to a percent.

(a) $45\% = .45$ **(b)** $250\% = 2.50$

(c) $.57 = 57\%$ **(d)** $1.5 = 1.50 = 150\%$

(e) $.327 = 32.7\%$

=========== **Work Problem 7 at the Side.**

🖩 **Calculator Tip**   In this book, we do not use 0 in the ones place for decimal fractions between 0 and 1. Many calculators (and other books) will show 0.45 instead of just .45 to emphasize that there is a 0 in the ones place. Graphing calculators do *not* show 0 in the ones place. Either way is correct.

---

7 Convert as indicated.

(a) 23% to a decimal

(b) 310% to a decimal

(c) .71 to a percent

(d) 1.32 to a percent

(e) .685 to a percent

Real-Data Applications

Decimalization of Stock Prices

When the New York Stock Exchange (NYSE) was founded in 1792, Thomas Jefferson suggested that stock prices be based on a decimal system. Instead, stock prices were based on the Spanish milled dollar. Even before the United States began minting its own coinage, the Spanish *eight-reales* coin, or the Spanish milled dollar, had been a commonly used currency and continued to be legal currency until 1857. The bits of the Spanish coin—the four reales, two reales, and one real—were the legendary *pieces of eight* famous in pirate and treasure lore.

The NYSE decision to base stock prices on an archaic coin resulted in the continuing practice of representing a **price per share** as a mixed number with fractional parts of $\frac{1}{2}$, $\frac{1}{4}$, $\frac{1}{8}$, or $\frac{1}{16}$. After two centuries, the Securities Exchange Commission (SEC) and the NYSE proposed a change in stock pricing to a decimal system consistent with that used by foreign stock exchanges. If they had only listened to Thomas Jefferson!

On August 28, 2000, the U.S. stock markets began the 8-month process of decimalization, to be completed by April 9, 2001. The pilot program included seven stocks on the New York Stock Exchange and six stocks on the American Stock Exchange. *Bond markets will continue to use a fraction-based pricing scheme.*

One of the primary issues had been whether to price stocks in one-cent increments or five-cent increments. Under the fraction-based pricing scheme, there are only 16 price changes per $1.00. At five-cent increments, there would be 20 price changes per $1.00 compared to 100 price changes per dollar for one-cent increments. Trading in smaller increments will increase competition because it will lower the spread between "bid" and "ask" prices—the difference between what a buyer is willing to pay for a security and what the seller is offering for a security. The volumes of trades could increase dramatically with one-cent price increments.

For the purpose of discussion only, assume that stocks are priced in five-cent increments. Assume that on April 9, 2001, you own 50 shares of Allied Technology, priced at $65\frac{7}{16}$ or \$65.4375 per share, resulting in an equity value of

$$\$65.4375 \times 50 = \$3271.875.$$

With decimalization, the price per share must be adjusted to the nearest nickel less than the original price, or \$65.40. Since the equity is unchanged, the new number of shares is

$$\$3271.875 \div \$65.40 = 50.02866972,$$

which is reported to the nearest thousandth as 50.029 shares. For a one-cent incremental scheme, the adjusted price would have been \$65.43 and the adjusted number of shares 50.006.

For Group Discussion

For each stock given, calculate the equity, the adjusted price per share, and the number of shares (to the nearest thousandth).

| | Before Decimalization | | | After Decimalization | |
Stock	Number of Shares	Price per Share	Equity	Price per Share	Number of Shares
1. Aeroflex (five-cent pricing)	50	$61\frac{7}{8}$	_____	_____	_____
2. Philadelphia Suburban (one-cent pricing)	100	$23\frac{1}{16}$	_____	_____	_____

R.2 **EXERCISES**

 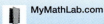

1. In the decimal 367.9412, name the digit that is in each place value.

   (a) tens   **(b)** tenths   **(c)** thousandths

   (d) ones or units   **(e)** hundredths

2. Write a numeral that has 5 in the thousands place, 0 in the tenths place, and 4 in the ten thousandths place.

3. For the decimal number 46.249, round to the place value indicated.

   (a) hundredths   **(b)** tenths

   (c) ones or units   **(d)** tens

4. Round each decimal to the nearest thousandth.

   (a) $.\overline{8}$ **(b)** $.\overline{5}$

   (c) .9762   **(d)** .8642

5. For the sum 35.89 + 24.1, which is the best estimate?
   A. 40   **B.** 50   **C.** 60   **D.** 70

6. For the difference 119.83 − 52.4, which is the best estimate?
   A. 40   **B.** 50   **C.** 60   **D.** 70

7. For the product 84.9 × 98.3, which is the best estimate?
   A. 7000   **B.** 8000   **C.** 80,000   **D.** 70,000

8. For the quotient 9845.3 ÷ 97.2, which is the best estimate?
   A. 10   **B.** 1000   **C.** 100   **D.** 10,000

Write each decimal as a fraction. Do not write in lowest terms. See Example 1.

9. .4   **10.** .6   **11.** .64   **12.** .82

13. .138   **14.** .104   **15.** 3.805   **16.** 5.166

Add or subtract as indicated. Make sure that your answer is reasonable by estimating first. Show your estimate, then the exact answer. See Example 2.

17. 25.32 + 109.2 + 8.574

18. 90.527 + 32.43 + 589.83 + 399.327

19. 28.73 − 3.12   **20.** 46.88 − 13.45

21. 43.5 − 28.17   **22.** 345.1 − 56.31

23. 32.56 + 47.356 + 1.8

24. 75.22 + 123.96 + 3.897

25. 18 − 2.789

26. 29 − 8.582

Multiply or divide as indicated. Make sure that your answer is reasonable by estimating first. Show your estimate, then the exact answer. See Examples 3–5.

27. .2 × .03   **28.** .07 × .004   **29.** 12.8 × 9.1   **30.** 34.04 × .56

31. 57.2 ÷ 8   **32.** 73.36 ÷ 14   **33.** 19.967 ÷ 9.74   **34.** 44.4788 ÷ 5.27

35. 57.116×100 **36.** $.094 \times 1000$ **37.** $1.62 \div 10$ **38.** $24.03 \div 100$

39. Explain in your own words how to add or subtract decimals.

40. Explain in your own words how to **(a)** multiply decimals and **(b)** divide decimals.

Write each fraction as a decimal. For repeating decimals, write the answer two ways: using the bar notation and rounding to the nearest thousandth. See Example 6.

41. $\dfrac{1}{8}$ **42.** $\dfrac{7}{8}$ **43.** $\dfrac{1}{4}$ **44.** $\dfrac{3}{4}$

45. $\dfrac{5}{9}$ **46.** $\dfrac{8}{9}$ **47.** $\dfrac{1}{6}$ **48.** $\dfrac{5}{6}$

49. In your own words, explain how to convert a decimal to a percent.

50. In your own words, explain how to convert a percent to a decimal.

Convert each percent to a decimal. See Examples 7(a)–(c), 8(a), and 8(b).

51. 54%   **52.** 39%   **53.** 117%   **54.** 189%   **55.** 2.4%

56. 3.1%   **57.** $6\dfrac{1}{4}\%$ **58.** $5\dfrac{1}{2}\%$ **59.** .8%   **60.** .9%

Convert each decimal to a percent. See Examples 7(d), 7(e), and 8(c)–(e).

61. .75   **62.** .83   **63.** .004   **64.** .005

65. 1.28   **66.** 2.35   **67.** .3   **68.** .6

One method of converting a fraction to a percent is to first convert the fraction to a decimal, as shown in Example 6, and then convert the decimal to a percent, as shown in Examples 7 and 8. Convert each fraction to a percent in this way.

69. $\dfrac{3}{4}$ **70.** $\dfrac{1}{4}$ **71.** $\dfrac{5}{6}$ **72.** $\dfrac{11}{16}$

73. Brand new tires have a tread of about $\dfrac{10}{32}$ in. By law, if 80% of the tread is worn off, the tire needs to be replaced.

 (a) What fraction of an inch of tread (in $\dfrac{1}{32}$ of an inch) indicates that a new tire is needed?

 (b) How much tread wear (in $\dfrac{1}{32}$ of an inch) remains (according to the legal limit) if the tread depth is $\dfrac{4}{32}$?

The Real Number System

1

Positive and negative numbers indicate position with respect to a starting point. A common example is the thermometer—temperatures are either "above zero" (positive) or "below zero" (negative). Another example is the number of yards gained (positive) or lost (negative) from the line of scrimmage in football. Exercises 63 and 64 of Section 1.4 are familiar examples of our need for positive and negative numbers to keep track of finances.

You're Connected

1.1 EXPONENTS, ORDER OF OPERATIONS, AND INEQUALITY

OBJECTIVES

1 — Use exponents.
2 — Use the order of operations guidelines.
3 — Use more than one grouping symbol.
4 — Know the meanings of $\neq$, $<$, $>$, $\leq$, and $\geq$.
5 — Translate word statements to symbols.
6 — Reverse the direction of inequality statements.

Study Skills Workbook
Activity 2

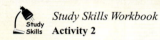

1 Find the value of each exponential expression.

(a) 6^2

(b) 3^5

(c) $\left(\dfrac{3}{4}\right)^2$

(d) $\left(\dfrac{1}{2}\right)^4$

(e) $(.4)^3$

In preparation for the study of algebra, we begin by introducing some basic ideas and vocabulary that we will be using throughout the course.

1 ▢ **Use exponents.** In algebra, we use a raised dot for multiplication as shown in Chapter R, where we factored a number as the product of its prime factors. For example, 81 is written in prime factored form as

$$81 = 3 \cdot 3 \cdot 3 \cdot 3,$$

where the factor 3 appears four times. Repeated factors are written in an abbreviated form by using an *exponent*. The prime factored form of 81 is written with an exponent as

$$\underbrace{3 \cdot 3 \cdot 3 \cdot 3}_{\text{4 factors of 3}} = 3^{\overset{\text{Exponent}}{4}}.$$

Base

The number 4 is the **exponent** and 3 is the **base** in the **exponential expression** 3^4. Exponents are also called **powers**. We read 3^4 as "3 to the fourth power" or simply "3 to the fourth."

Example 1 Finding Values of Exponential Expressions

Find the value of each exponential expression.

(a) 5^2

$$\underbrace{5 \cdot 5}_{} = 25$$

5 is used as a factor 2 times.

Read 5^2 as "5 to the second power" or, more commonly, "5 squared."

(b) 6^3

$$\underbrace{6 \cdot 6 \cdot 6}_{} = 216$$

6 is used as a factor 3 times.

Read 6^3 as "6 to the third power" or, more commonly, "6 cubed."

(c) 2^5

$$2 \cdot 2 \cdot 2 \cdot 2 \cdot 2 = 32 \qquad \text{2 is used as a factor 5 times.}$$

Read 2^5 as "2 to the fifth power."

(d) 7^4

$$7 \cdot 7 \cdot 7 \cdot 7 = 2401 \qquad \text{7 is used as a factor 4 times.}$$

Read 7^4 as "7 to the fourth power."

(e) $\left(\dfrac{2}{3}\right)^3$

$$\frac{2}{3} \cdot \frac{2}{3} \cdot \frac{2}{3} = \frac{8}{27} \qquad \frac{2}{3} \text{ is used as a factor 3 times.}$$

Work Problem 1 at the Side.

2   **Use the order of operations guidelines.**   Many problems involve more than one operation. To indicate the order in which the operations should be performed, we often use *grouping symbols.* If no grouping symbols are used, we apply the order of operations discussed below.

Consider the expression $5 + 2 \cdot 3$. To show that the multiplication should be performed before the addition, parentheses can be used to write

$$5 + (2 \cdot 3) = 5 + 6 = 11.$$

If addition is to be performed first, the parentheses should group $5 + 2$ as follows.

$$(5 + 2) \cdot 3 = 7 \cdot 3 = 21$$

Other grouping symbols used in more complicated expressions are brackets [   ], braces {   }, and fraction bars. (For example, in $\frac{8-2}{3}$, the expression $8 - 2$ is considered to be grouped in the numerator.)

To work problems with more than one operation, use the following **order of operations.** This order is used by most calculators and computers.

Order of Operations

If grouping symbols are present, simplify within them, innermost first (and above and below fraction bars separately), in the following order.

Step 1   Apply all exponents.

Step 2   Do any multiplications or divisions in the order in which they occur, working from left to right.

Step 3   Do any additions or subtractions in the order in which they occur, working from left to right.

If no grouping symbols are present, start with Step 1.

A dot has been used to show multiplication; another way to show multiplication is with parentheses. For example, 3(7) means $3 \cdot 7$ or 21. Also, $3(4 + 5)$ means 3 times the sum of 4 and 5. By the order of operations, the sum in parentheses must be found first, then the product.

Example 2   **Using the Order of Operations**

Find the value of each expression.

(a) $4 \cdot 5 - 6$

Using the order of operations given in the box, first multiply 4 and 5, then subtract 6 from the product.

$$\begin{aligned} \mathbf{4 \cdot 5} - 6 &= \mathbf{20} - 6 && \text{Multiply.} \\ &= 14 && \text{Subtract.} \end{aligned}$$

(b) $9(6 + 11)$

Work first inside the parentheses.

$$\begin{aligned} 9(\mathbf{6 + 11}) &= 9(\mathbf{17}) && \text{Add inside parentheses.} \\ &= 153 && \text{Multiply.} \end{aligned}$$

Continued on Next Page

❷ Find the value of each expression.

(a) $7 + 3 \cdot 8$

(b) $2 \cdot 9 + 7 \cdot 3$

(c) $7 \cdot 6 - 3(8 + 1)$

(d) $2 + 3^2 - 5$

❸ Find the value of each expression.

(a) $9[(4 + 8) - 3]$

(b) $\dfrac{2(7 + 8) + 2}{3 \cdot 5 + 1}$

(c) $6 \cdot 8 + 5 \cdot 2$

Perform any multiplications from left to right, then add.

$$6 \cdot 8 + 5 \cdot 2 = 48 + 10 \quad \text{Multiply.}$$
$$= 58 \quad \text{Add.}$$

(d) $2(5 + 6) + 7 \cdot 3 = 2(11) + 7 \cdot 3$ Add inside parentheses.
$$= 22 + 21 \quad \text{Multiply.}$$
$$= 43 \quad \text{Add.}$$

(e) $9 + 2^3 - 5$

Following the order of operations, calculate 2^3 first.

$$9 + 2^3 - 5 = 9 + 8 - 5 \quad \text{Use the exponent.}$$
$$= 12 \quad \text{Add, then subtract.}$$

Work Problem ❷ at the Side.

3   **Use more than one grouping symbol.** An expression with double parentheses, such as the expression $2(8 + 3(6 + 5))$, can be confusing. We avoid confusion by using square brackets, [ ], in place of one pair of parentheses.

Example 3   **Using Brackets and Fraction Bars**

Find the value of each expression.

(a) $2[8 + 3(6 + 5)]$

Begin inside the parentheses. Then follow the order of operations.

$$2[8 + 3(6 + 5)] = 2[8 + 3(11)] \quad \text{Add.}$$
$$= 2[8 + 33] \quad \text{Multiply.}$$
$$= 2[41] \quad \text{Add.}$$
$$= 82 \quad \text{Multiply.}$$

(b) $\dfrac{4(5 + 3) + 3}{2(3) - 1}$

Simplify the numerator and denominator separately.

$$\frac{4(5 + 3) + 3}{2(3) - 1} = \frac{4(8) + 3}{2(3) - 1} \quad \text{Add inside parentheses.}$$
$$= \frac{32 + 3}{6 - 1} \quad \text{Multiply.}$$
$$= \frac{35}{5} \quad \text{Add and subtract.}$$
$$= 7 \quad \text{Divide.}$$

Work Problem ❸ at the Side.

Calculator Tip Calculators follow the order of operations given in this section. You may want to try some of the examples to see that your calculator gives the same answers. Be sure to use the parentheses keys to insert parentheses where they are needed. To work Example 3(b) with a calculator, you must put parentheses around the numerator and the denominator.

ANSWERS
2. (a) 31   **(b)** 39   **(c)** 15   **(d)** 6
3. (a) 81   **(b)** 2

4 ▭ **Know the meanings of ≠, <, >, ≤, and ≥.** So far we have used only the symbols of arithmetic, such as +, −, ·, and ÷ and the equality symbol =. The equality symbol with a slash through it means "is *not* equal to." For example,

$$7 \neq 8$$

indicates that 7 is not equal to 8.

If two numbers are not equal, then one of the numbers must be less than the other. The symbol < represents "is less than," so "7 is less than 8" is written as

$$7 < 8.$$

Also, we write "6 is less than 9" as $6 < 9$.

The symbol > means "is greater than." We write "8 is greater than 2" as

$$8 > 2.$$

The statement "17 is greater than 11" becomes $17 > 11$.

Keep the meanings of the symbols < and > clear by remembering that the symbol always points to the *smaller* number.

Smaller number → **8** < 15

$15 >$ **8** ← Smaller number

Work Problem ④ at the Side.

Two other symbols, ≤ and ≥, also represent the idea of inequality. The symbol ≤ means "is less than or equal to," so

$$5 \leq 9$$

means "5 is less than or equal to 9." If either the < part or the = part is true, then the inequality ≤ is true. The statement $5 \leq 9$ is true because $5 < 9$ is true. Also, $8 \leq 8$ is true because $8 = 8$ is true. But $13 \leq 9$ is not true because neither $13 < 9$ nor $13 = 9$ is true.

The symbol ≥ means "is greater than or equal to";

$$9 \geq 5$$

is true because $9 > 5$ is true.

▭ **Example 4** **Using the Symbols ≤ and ≥**

Tell whether each statement is *true* or *false*.

(a) $15 \leq 20$ The statement $15 \leq 20$ is true because $15 < 20$.

(b) $12 \geq 12$ Since $12 = 12$, this statement is true.

(c) $\dfrac{6}{15} \geq \dfrac{2}{3}$

To compare fractions, write them with a common denominator. Here, 15 is a common denominator and $\frac{2}{3} = \frac{10}{15}$. Now decide whether $\frac{6}{15} \geq \frac{10}{15}$ is true or false. Both statements $\frac{6}{15} > \frac{10}{15}$ and $\frac{6}{15} = \frac{10}{15}$ are false; therefore, $\frac{6}{15} \geq \frac{2}{3}$ is false.

Work Problem ⑤ at the Side.

5 ▭ **Translate word statements to symbols.** Word phrases or statements often must be converted to symbols in algebra. The next example illustrates this.

④ Write each statement in words, then decide whether it is *true* or *false*.

(a) $7 < 5$

(b) $12 > 6$

(c) $4 \neq 10$

(d) $28 \neq 4 \cdot 7$

⑤ Tell whether each statement is *true* or *false*.

(a) $30 \leq 40$

(b) $25 \geq 10$

(c) $40 \leq 10$

(d) $21 \leq 21$

(e) $3 \geq 3$

6 Write in symbols.

(a) Nine equals eleven minus two.

(b) Seventeen is less than thirty.

(c) Eight is not equal to ten.

(d) Fourteen is greater than twelve.

(e) Thirty is less than or equal to fifty.

(f) Two is greater than or equal to two.

7 Write each statement with the inequality symbol reversed.

(a) $8 < 10$

(b) $3 > 1$

(c) $9 \leq 15$

(d) $6 \geq 2$

Example 5 Converting Words to Symbols

Write each word statement in symbols.

(a) Twelve **equals** ten **plus** two. $12 = 10 + 2$

(b) Nine **is less than** ten. $9 < 10$
Compare this with 9 less than 10, which is written $10 - 9$.

(c) Fifteen **is not equal to** eighteen. $15 \neq 18$

(d) Seven **is greater than** four. $7 > 4$

(e) Thirteen **is less than or equal to** forty. $13 \leq 40$

(f) Six **is greater than or equal to** six. $6 \geq 6$

Work Problem 6 at the Side.

6 **Reverse the direction of inequality statements.**   Any statement with $<$ can be converted to one with $>$, and any statement with $>$ can be converted to one with $<$. We do this by reversing both the order of the numbers and the direction of the symbol. For example, the statement $6 < 10$ can be written as $10 > 6$.

$6 < 10$ becomes $10 > 6$

Example 6 Converting between $<$ and $>$

Parts (a)–(d) show the same statements written in two equally correct ways.

(a) $9 < 16$, $16 > 9$ **(b)** $5 > 2$, $2 < 5$

(c) $3 \leq 8$, $8 \geq 3$ **(d)** $12 \geq 5$, $5 \leq 12$

Note that in each inequality, the point of the symbol is directed toward the smaller number.

Work Problem 7 at the Side.

Here is a summary of the symbols of equality and inequality.

SYMBOLS OF EQUALITY AND INEQUALITY

Symbol	Meaning	Example
$=$	Is equal to	$.5 = \frac{1}{2}$ means .5 is equal to $\frac{1}{2}$.
$\neq$	Is not equal to	$3 \neq 7$ means 3 is not equal to 7.
$<$	Is less than	$6 < 10$ means 6 is less than 10.
$>$	Is greater than	$15 > 14$ means 15 is greater than 14.
$\leq$	Is less than or equal to	$4 \leq 8$ means 4 is less than or equal to 8.
$\geq$	Is greater than or equal to	$1 \geq 0$ means 1 is greater than or equal to 0.

CAUTION

The symbols of equality and inequality are used to write mathematical *sentences*. They differ from the symbols for operations ($+$, $-$, $\cdot$, and $\div$), discussed earlier, which are used to write mathematical *expressions* that represent a number. For example, compare the sentence $4 < 10$, which gives the relationship between 4 and 10, with the expression $4 + 10$, which tells how to operate on 4 and 10 to get the number 14. This distinction between sentences and expressions will be important throughout your study of algebra.

1.1 EXERCISES

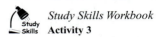

Study Skills Workbook
Activity 3

Decide whether each statement is true *or* false. *If it is* false, *explain why.*

1. Exponents are also called powers.

2. Some grouping symbols are $+$, $-$, $\cdot$, and $\div$.

3. When evaluated, $4 + 3(8 - 2)$ is equal to 42.

4. $3^3 = 9$

5. The statement "4 is 12 less than 16" is interpreted $4 = 12 - 16$.

6. The statement "6 is 4 less than 10" is interpreted $6 < 10 - 4$.

Find the value of each exponential expression. See Example 1.

7. 7^2

8. 4^2

9. 12^2

10. 14^2

11. 4^3

12. 5^3

13. 10^3

14. 11^3

15. 3^4

16. 6^4

17. 4^5

18. 3^5

19. $\left(\dfrac{2}{3}\right)^4$

20. $\left(\dfrac{3}{4}\right)^3$

21. $(.04)^3$

22. $(.05)^4$

23. When evaluating $(4^2 + 3^3)^4$, what is the *last* exponent that would be applied? Explain your answer.

24. Which are not grouping symbols—parentheses, brackets, fraction bars, exponents?

Find the value of each expression. See Examples 2 and 3.

25. $13 + 9 \cdot 5$

26. $11 + 7 \cdot 6$

27. $20 - 4 \cdot 3 + 5$

28. $18 - 7 \cdot 2 + 6$

29. $9 \cdot 5 - 13$

30. $7 \cdot 6 - 11$

31. $18 - 2 + 3$

32. $22 - 8 + 9$

33. $\dfrac{1}{4} \cdot \dfrac{2}{3} + \dfrac{2}{5} \cdot \dfrac{11}{3}$

34. $\dfrac{9}{4} \cdot \dfrac{2}{3} + \dfrac{4}{5} \cdot \dfrac{5}{3}$

35. $9 \cdot 4 - 8 \cdot 3$

36. $11 \cdot 4 + 10 \cdot 3$

37. $2.5(1.9) + 4.3(7.3)$

38. $4.3(1.2) + 2.1(8.5)$

39. $10 + 40 \div 5 \cdot 2$

40. $12 + 8^2 \div 8 - 4$

41. $18 - 2(3 + 4)$

42. $30 - 3(4 + 2)$

43. $5[3 + 4(2^2)]$

44. $6\left[\dfrac{3}{4} + 8\left(\dfrac{1}{2}\right)^3\right]$

45. $\left(\dfrac{3}{2}\right)^2\left[\left(11 + \dfrac{1}{3}\right) - 6\right]$

46. $4^2[(13 + 4) - 8]$

47. $\dfrac{8 + 6(3^2 - 1)}{3 \cdot 2 - 2}$

48. $\dfrac{8 + 2(8^2 - 4)}{4 \cdot 3 - 10}$

49. $\dfrac{4(7+2)+8(8-3)}{6(4-2)-2^2}$

50. $\dfrac{6(5+1)-9(1+1)}{5(8-4)-2^3}$

Tell whether each statement is true *or* false. *In Exercises 53–62, first simplify each expression involving an operation. See Example 4.*

51. $8 \geq 17$

52. $10 \geq 41$

53. $17 \leq 18 - 1$

54. $12 \geq 10 + 2$

55. $6 \cdot 8 + 6 \cdot 6 \geq 0$

56. $4 \cdot 20 - 16 \cdot 5 \geq 0$

57. $6[5 + 3(4 + 2)] \leq 70$

58. $6[2 + 3(2 + 5)] \leq 135$

59. $\dfrac{9(7-1)-8 \cdot 2}{4(6-1)} > 3$

60. $\dfrac{2(5+3)+2 \cdot 2}{2(4-1)} > 1$

61. $8 \leq 4^2 - 2^2$

62. $10^2 - 8^2 > 6^2$

Write each word statement in symbols. See Example 5.

63. Fifteen is equal to five plus ten.

64. Twelve is equal to twenty minus eight.

65. Nine is greater than five minus four.

66. Ten is greater than six plus one.

67. Sixteen is not equal to nineteen.

68. Three is not equal to four.

69. Two is less than or equal to three.

70. Five is less than or equal to nine.

Write each statement in words and decide whether it is true *or* false. *(Hint: To compare fractions, write them with the same denominator.)*

71. $7 < 19$

72. $9 < 10$

73. $\dfrac{1}{3} \neq \dfrac{3}{10}$

74. $\dfrac{10}{7} \neq \dfrac{3}{2}$

75. $8 \geq 11$

76. $4 \leq 2$

Write each statement with the inequality symbol reversed. See Example 6.

77. $5 < 30$

78. $8 > 4$

79. $12 \geq 3$

80. $25 \leq 41$

The table shows results of a science literacy survey by Jon Miller of the International Center for the Advancement of Science Literacy in Chicago. Use this table to answer Exercises 81–83.

Country	Science Literacy Index
United States	56
Netherlands	52
France	50
Canada	45
Greece	38
Japan	36

81. Which countries scored more than 50?

82. Which countries scored at most 40?

83. For which countries were scores not less than 50?

1.2   VARIABLES, EXPRESSIONS, AND EQUATIONS

To make general statements about numbers in algebra, letters called **variables** are used to represent numbers. Different numbers can replace the variables to form specific statements. For example, in Section 1.7 we will see the statement

$$a + b = b + a.$$

This statement is true for any replacements of the variables a and b, such as 2 for a and 5 for b, which gives the true statement

$$2 + 5 = 5 + 2.$$

An **algebraic expression** is a collection of numbers, variables, operation symbols, and grouping symbols, such as parentheses, square brackets, or fraction bars. For example,

$$x + 5, \quad 2m - 9, \quad \text{and} \quad 8p^2 + 6(p - 2)$$

are all algebraic expressions. In $2m - 9$, the expression $2m$ means $2 \cdot m$, the product of 2 and m, and $8p^2$ represents the product of 8 and p^2. Also, $6(p - 2)$ means the product of 6 and $p - 2$.

1   Evaluate algebraic expressions, given values for the variables.   An algebraic expression has different numerical values for different values of the variables.

Example 1   Evaluating Expressions Given Values of the Variable

Find the value of each algebraic expression if $m = 5$ and then if $m = 9$.

(a) $8m$

$$8m = 8 \cdot \mathbf{5} \qquad \text{Let } m = 5. \qquad\qquad 8m = 8 \cdot \mathbf{9} \qquad \text{Let } m = 9.$$
$$ = 40 \qquad \text{Multiply.} \qquad\qquad = 72 \qquad \text{Multiply.}$$

(b) $3m^2$

$$3m^2 = 3 \cdot \mathbf{5}^2 \qquad \text{Let } m = 5. \qquad\qquad 3m^2 = 3 \cdot \mathbf{9}^2 \qquad \text{Let } m = 9.$$
$$ = 3 \cdot 25 \qquad \text{Square.} \qquad\qquad = 3 \cdot 81 \qquad \text{Square.}$$
$$ = 75 \qquad \text{Multiply.} \qquad\qquad = 243 \qquad \text{Multiply.}$$

CAUTION

In Example 1(b), notice that $3m^2$ means $3 \cdot m^2$; it *does not* mean $3m \cdot 3m$. Unless parentheses are used, the exponent refers only to the variable or number just before it. To write $3m \cdot 3m$ with exponents, use parentheses: $3m \cdot 3m = (3m)^2$.

Work Problem ➊ at the Side.

Example 2   Evaluating Expressions with More Than One Variable

Find the value of each expression if $x = 5$ and $y = 3$.

(a) $2x + 5y$

$$2x + 5y = 2 \cdot \mathbf{5} + 5 \cdot \mathbf{3} \qquad \text{Replace } x \text{ with 5 and } y \text{ with 3.}$$
$$ = 10 + 15 \qquad \text{Multiply.}$$
$$ = 25 \qquad \text{Add.}$$

Continued on Next Page

OBJECTIVES

1   Evaluate algebraic expressions, given values for the variables.

2   Convert phrases from words to algebraic expressions.

3   Identify solutions of equations.

4   Translate word statements to equations.

5   Distinguish between expressions and equations.

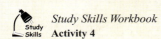

Study Skills Workbook
Study Skills   Activity 4

➊   Find the value of each expression if $p = 3$.

(a) $6p$

(b) $p + 12$

(c) $5p^2$

ANSWERS
1. **(a)** 18   **(b)** 15   **(c)** 45

② Find the value of each expression if $x = 6$ and $y = 9$.

(a) $4x + 7y$

(b) $\dfrac{9x - 8y}{2x - y}$

$$\dfrac{9x - 8y}{2x - y} = \dfrac{9 \cdot 5 - 8 \cdot 3}{2 \cdot 5 - 3} \qquad \text{Replace } x \text{ with 5 and } y \text{ with 3.}$$

$$= \dfrac{45 - 24}{10 - 3} \qquad \text{Multiply.}$$

$$= \dfrac{21}{7} \qquad \text{Subtract.}$$

$$= 3 \qquad \text{Divide.}$$

(c) $x^2 - 2y^2$

$$x^2 - 2y^2 = 5^2 - 2 \cdot 3^2 \qquad \text{Replace } x \text{ with 5 and } y \text{ with 3.}$$

$$= 25 - 2 \cdot 9 \qquad \text{Use the exponents.}$$

$$= 25 - 18 \qquad \text{Multiply.}$$

$$= 7 \qquad \text{Subtract.}$$

Work Problem ② at the Side.

(b) $\dfrac{4x - 2y}{x + 1}$

🖩 **Calculator Tip** An Introduction to Calculators in the front of this book explains how to perform arithmetic operations and evaluate exponentials with a calculator.

2▭ Convert phrases from words to algebraic expressions.

Problem Solving

Sometimes variables must be used to change word phrases into algebraic expressions. This process will be important later for solving applied problems.

(c) $2x^2 + y^2$

Example 3 Using Variables to Change Word Phrases into Algebraic Expressions

Change each word phrase to an algebraic expression. Use x as the variable to represent the number.

(a) The **sum** of a number and 9
 "Sum" is the answer to an addition problem. This phrase translates as

$$x + 9 \quad \text{or} \quad 9 + x.$$

(b) 7 **minus** a number
 "Minus" indicates subtraction, so the translation is

$$7 - x.$$

Note that $x - 7$ would *not* be correct because we cannot subtract in either order and get the same results.

(c) A number **subtracted from 12**
 Since a number is subtracted *from* 12, write this as

$$12 - x.$$

Compare this result with "12 is subtracted from a number," which is $x - 12$.

Continued on Next Page

(d) The **product** of 11 and a number

$$11 \cdot x \quad \text{or} \quad 11x$$

(e) 5 **divided by** a number

$$\frac{5}{x}$$

(f) The **product of** 2 and the **difference** between a number and 8

$$2(x - 8)$$

CAUTION

Notice that in translating the words "the difference between a number and 8" the order is kept the same: $x - 8$. "The difference between 8 and a number" would be written $8 - x$.

Work Problem **❸** at the Side.

3 ▭ **Identify solutions of equations.** An **equation** is a statement that two expressions are equal. Examples of equations are

$$x + 4 = 11, \quad 2y = 16, \quad \text{and} \quad 4p + 1 = 25 - p.$$

To **solve** an equation, we must find all values of the variable that make the equation true. Such values of the variable are called the **solutions** of the equation.

Example 4 **Deciding Whether a Number Is a Solution of an Equation**

Decide whether the given number is a solution of the equation.

(a) Is 7 a solution of $5p + 1 = 36$?

$$5p + 1 = 36$$
$$5 \cdot 7 + 1 = 36 \qquad \text{Replace } p \text{ with 7.}$$
$$35 + 1 = 36 \qquad \text{Multiply.}$$
$$36 = 36 \qquad \text{True}$$

The number 7 is a solution of the equation.

(b) Is $\frac{14}{3}$ a solution of $9m - 6 = 32$?

$$9m - 6 = 32$$
$$9 \cdot \frac{14}{3} - 6 = 32 \qquad \text{Replace } m \text{ with } \frac{14}{3}.$$
$$42 - 6 = 32 \qquad \text{Multiply.}$$
$$36 = 32 \qquad \text{False}$$

The number $\frac{14}{3}$ is not a solution of the equation.

Work Problem **❹** at the Side.

4 ▭ **Translate word statements to equations.** We have seen how to translate phrases from words to expressions. Sentences given in words are translated as equations.

---

❸ Write as an algebraic expression. Use x as the variable.

(a) The sum of 5 and a number

(b) A number minus 4

(c) A number subtracted from 48

(d) The product of 6 and a number

(e) 9 multiplied by the sum of a number and 5

❹ Decide whether the given number is a solution of the equation.

(a) $p - 1 = 3; 2$

(b) $2k + 3 = 15; 7$

(c) $8p - 11 = 5; 2$

5 Change each sentence to an equation. Let x represent the number.

(a) Three times the sum of a number and 13 is 19.

(b) Five times a number is subtracted from 21, giving 15.

6 Decide whether each is an equation or an expression.

(a) $2x + 5y - 7$

(b) $\dfrac{3x - 1}{5}$

(c) $2x + 5 = 7$

(d) $\dfrac{x}{y - 3} = 4x$

Example 5 Translating Word Sentences to Equations

Change each word sentence to an equation. Let x represent the number.

(a) Twice the sum of a number and four is six.

"Twice" means two times. The word *is* suggests equals. With x representing the number, translate as follows.

$$
\begin{array}{ccccc}
\text{Twice} & \text{the sum of} & & & \\
 & \text{a number and four} & \text{is} & \text{six.} \\
\downarrow & \downarrow & \downarrow & \downarrow \\
2 \cdot & (x + 4) & = & 6
\end{array}
$$

$$2(x + 4) = 6$$

(b) Nine more than five times a number is 49.

"Nine more than" means "nine is added to." Use x to represent the unknown number.

$$
\begin{array}{ccccc}
\text{Nine} & \text{more than} & \text{five times a number} & \text{is} & \text{49.} \\
\downarrow & \downarrow & \downarrow & \downarrow & \downarrow \\
9 & + & 5x & = & 49
\end{array}
$$

$$9 + 5x = 49$$

(c) Seven less than three times a number is eleven.

Here, 7 is *subtracted* from three times a number to get 11.

$$
\begin{array}{ccccc}
\text{Three times} & & & & \\
\text{a number} & \text{less} & \text{seven} & \text{is} & \text{eleven.} \\
\downarrow & \downarrow & \downarrow & \downarrow & \downarrow \\
3x & - & 7 & = & 11
\end{array}
$$

$$3x - 7 = 11$$

Work Problem 5 at the Side.

5 **Distinguish between expressions and equations.** Students often have trouble distinguishing between equations and expressions. Remember that an equation is a sentence (with an $=$ symbol); an expression is a phrase that represents a number.

$$
\begin{array}{cc}
4x + 5 = 9 & 4x + 5 \\
\uparrow & \uparrow \\
\text{Equation} & \text{Expression}
\end{array}
$$

Example 6 Distinguishing between Equations and Expressions

Decide whether each is an equation or an expression.

(a) $2x - 5y$

There is no equals sign, so this is an expression.

(b) $2x = 5y$

Because of the equals sign, this is an equation.

Work Problem 6 at the Side.

1.2 EXERCISES

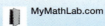

Fill in each blank with the correct response.

1. If $x = 3$, then the value of $x + 7$ is _____.

2. If $x = 1$ and $y = 2$, then the value of $4xy$ is _____.

3. "The sum of 12 and x" is represented by the expression _____. If $x = 9$, the value of that expression is _____.

4. Will the equation $x = x + 4$ ever have a solution? _____

5. $2x + 3$ is an _____, while $2x + 3 = 8$ is an _____.
   (equation/expression)   (equation/expression)

Exercises 6–10 cover some of the concepts introduced in this section. Give a short explanation for each.

6. Why is $2x^3$ not the same as $2x \cdot 2x \cdot 2x$? Explain, using an exponent to write $2x \cdot 2x \cdot 2x$.

7. If the words *more than* in Example 5(b) were changed to *less than,* how would the equation be changed?

8. Explain in your own words why, when evaluating the expression $4x^2$ for $x = 3$, 3 must be squared *before* multiplying by 4.

9. There are many pairs of values of x and y for which $2x + y$ will equal 6. Name two such pairs and describe how you determined them.

10. Suppose that for the equation $3x - y = 9$, the value of x is given as 4 . What would be the corresponding value of y? How do you know this?

*Find the numerical value of each expression if **(a)** x = 4 and **(b)** x = 6. See Example 1.*

11. $4x^2$

   (a)   **(b)**

12. $5x^2$

   (a)   **(b)**

13. $\dfrac{3x - 5}{2x}$

   (a)   **(b)**

14. $\dfrac{4x - 1}{3x}$

   (a)   **(b)**

15. $\dfrac{6.459x}{2.7}$ (to the nearest thousandth)

   (a)   **(b)**

16. $\dfrac{.74x^2}{.85}$ (to the nearest thousandth)

   (a)   **(b)**

17. $3x^2 + x$

   (a)   **(b)**

18. $2x + x^2$

   (a)   **(b)**

*Find the numerical value of each expression if **(a)** x = 2 and y = 1 and **(b)** x = 1 and y = 5. See Example 2.*

19. $3(x + 2y)$

   (a)   **(b)**

20. $2(2x + y)$

   (a)   **(b)**

21. $x + \dfrac{4}{y}$

   (a)   **(b)**

22. $y + \dfrac{8}{x}$

   (a)   **(b)**

23. $\dfrac{x}{2} + \dfrac{y}{3}$

   (a)   **(b)**

24. $\dfrac{x}{5} + \dfrac{y}{4}$

   (a)   **(b)**

25. $\dfrac{2x + 4y - 6}{5y + 2}$

   (a)   **(b)**

26. $\dfrac{4x + 3y - 1}{2x + y}$

   (a)   **(b)**

27. $2y^2 + 5x$

   (a)   **(b)**

28. $6x^2 + 4y$

   (a)   **(b)**

29. $\dfrac{3x + y^2}{2x + 3y}$

   (a)   **(b)**

30. $\dfrac{x^2 + 1}{4x + 5y}$

   (a)   **(b)**

31. $.841x^2 + .32y^2$

    (a)   **(b)**

32. $.941x^2 + .2y^2$

    (a)   **(b)**

*Change each word phrase to an algebraic expression. Use x to represent the number.
See Example 3.*

33. Twelve times a number

34. Thirteen added to a number

35. Two subtracted from a number

36. Eight subtracted from a number

37. Four times a number, subtracted from seven

38. Three times a number, subtracted from fourteen

39. The difference between twice a number and 6

40. The difference between 6 and half a number

41. 12 divided by the sum of a number and 3

42. The difference between a number and 5, divided by 12

43. The product of 6 and four less than a number

44. The product of 9 and five more than a number

45. In the phrase "four more than the product of a number and 6," does the word *and* signify the operation of addition? Explain.

46. Suppose that the directions on a test read "Solve the following expressions." How would you politely correct the person who wrote these directions?

Decide whether the given number is a solution of the equation. See Example 4.

47. Is 7 a solution of $p - 5 = 12$?

48. Is 10 a solution of $x + 6 = 15$?

49. Is 1 a solution of $5m + 2 = 7$?

50. Is 1 a solution of $3x + 5 = 8$?

51. Is $\dfrac{1}{5}$ a solution of $6p + 4p + 9 = 11$?

52. Is $\dfrac{12}{5}$ a solution of $2x + 3x + 8 = 20$?

53. Is 3 a solution of
$2y + 3(y - 2) = 14$?

54. Is 2 a solution of
$6a + 2(a + 3) = 14$?

55. Is $\dfrac{1}{3}$ a solution of
$\dfrac{z + 4}{2 - z} = \dfrac{13}{5}$?

56. Is $\dfrac{13}{4}$ a solution of
$\dfrac{x + 6}{x - 2} = \dfrac{37}{5}$?

57. Is 4.3 a solution of
$3r^2 - 2 = 53.47$?

58. Is 3.7 a solution of
$2x^2 + 1 = 28.38$?

Change each sentence to an equation. Use x to represent the number. See Example 5.

59. The sum of a number and 8 is 18.

60. A number minus three equals 1.

61. Five more than twice a number is 5.

62. The product of 2 and the sum of a number and 5 is 14.

63. Sixteen minus three-fourths of a number is 13.

64. The sum of six-fifths of a number and 2 is 14.

65. Three times a number is equal to 8 more than twice the number.

66. Twelve divided by a number equals $\dfrac{1}{3}$ times that number.

Identify each as an expression *or an* equation. *See Example 6.*

67. $3x + 2(x - 4)$

68. $5y - (3y + 6)$

69. $7t + 2(t + 1) = 4$

70. $9r + 3(r - 4) = 2$

RELATING CONCEPTS (Exercises 71–74) | **FOR INDIVIDUAL OR GROUP WORK**

A mathematical model is an equation that describes the relationship between two quantities. For example, based on data from the U.S. Bureau of Labor Statistics, average hourly earnings of production workers in manufacturing industries in the United States from 1990 through 1997 are approximated by the equation $y = .319x - 624.31$, where x represents the year and y represents the hourly earnings in dollars. Use this model to approximate the hourly earnings during each year. Compare with the actual earnings given in parentheses.

71. 1990 ($10.83)

72. 1994 ($12.07)

73. 1996 ($12.37)

74. 1997 ($13.17)

1.3   REAL NUMBERS AND THE NUMBER LINE

In Chapter R, we introduced the set of whole numbers. We use set braces, { }, to enclose the elements of a set.

Whole Numbers

$$\{0, 1, 2, 3, 4, 5, \ldots\}$$

The numbers used for counting are called the **natural numbers.**

Natural Numbers

$$\{1, 2, 3, 4, 5, \ldots\}$$

These numbers, along with many others, can be represented on **number lines** like the one in Figure 1. We draw a number line by choosing any point on the line and labeling it 0. Choose any point to the right of 0 and label it 1. The distance between 0 and 1 gives a unit of measure used to locate other points, as shown in Figure 1. The points labeled in Figure 1 correspond to the first few whole numbers.

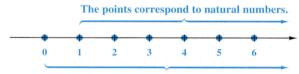

Figure 1

1   **Use integers to express numbers in applications.**   The natural numbers are located to the right of 0 on the number line. But numbers may also be placed to the left of 0. For each natural number we can place a corresponding number to the left of 0. These numbers, written $-1, -2, -3, -4$, and so on, are shown in Figure 2. Each is the **opposite** or **negative** of a natural number. The natural numbers, their opposites, and 0 form a new set of numbers called the **integers.**

Integers

$$\{\ldots, -3, -2, -1, 0, 1, 2, 3, \ldots\}$$

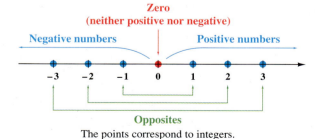

Figure 2

OBJECTIVES

1   Use integers to express numbers in applications.

2   Graph rational numbers on the number line.

3   Tell which of two real numbers is smaller.

4   Find the opposite of a real number.

5   Find the absolute value of a real number.

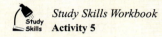

Study Skills Workbook
Activity 5

There are many practical applications of negative numbers. For example, a Fahrenheit temperature on a cold January day might be $-10°$, and a business that spends more than it takes in has a negative "profit."

Example 1 **Using Negative Numbers in Applications**

Use an integer to express the number in each application.

(a) The lowest Fahrenheit temperature ever recorded in meteorological records was 129° below zero at Vostok, Antartica, on July 21, 1983. (*Source: The World Almanac and Book of Facts, 2000.*)
Use $-129°$ because "below zero" indicates a negative number.

(b) The shore surrounding the Dead Sea is 1340 ft below sea level. (*Source: Microsoft Encarta Encyclopedia 2000.*)
Again, "below sea level" indicates a negative number, -1340.

Work Problem ❶ at the Side.

❷ **Graph rational numbers on the number line.** Not all numbers are integers. For example, $\frac{1}{2}$ is not; it is a number halfway between the integers 0 and 1. Also, $3\frac{1}{4}$ is not an integer. These numbers and others that are quotients of integers are **rational numbers.** (The name comes from the word *ratio*, which indicates a quotient.)

Rational Numbers

{numbers that can be written as quotients of integers, with denominators not 0}

Since any integer can be written as the quotient of itself and 1, all integers are also rational numbers. For example, $-5 = \frac{-5}{1}$. A decimal number that comes to an end (terminates), such as .23, is a rational number: $.23 = \frac{23}{100}$. Decimal numbers that repeat in a fixed block of digits, such as $.3333 \ldots = .\overline{3}$ and $.454545 \ldots = .\overline{45}$, are also rational numbers. For example, $.\overline{3} = \frac{1}{3}$.

As shown in Figures 1 and 2, to **graph** a number, we place a dot on the number line at the point that corresponds to the number. The number is called the **coordinate** of the point. Think of the graph of a set of numbers as a picture of the set.

Example 2 **Graphing Rational Numbers**

Graph each number on the number line.

$$-\frac{3}{2}, \ -\frac{2}{3}, \frac{1}{2}, \ 1\frac{1}{3}, \frac{23}{8}, 3\frac{1}{4}$$

To locate the improper fractions on the number line, write them as mixed numbers or decimals. The graph is shown in Figure 3.

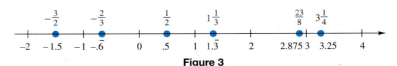

Figure 3

Work Problem ❷ at the Side.

---

❶ Use an integer to express the number(s) in each application.

(a) Erin discovers that she has spent $53 more than she has in her checking account.

(b) The record high Fahrenheit temperature in the United States was 134° in Death Valley, California on July 10, 1913. (*Source: The World Almanac and Book of Facts, 2000.*)

(c) A football team gained 5 yd, then lost 10 yd on the next play.

❷ Graph each number on the number line.

$$-3, \frac{17}{8}, -2.75, 1\frac{1}{2}, -\frac{3}{4}$$

Although many numbers are rational, not all are. For example, a square that measures one unit on a side has a diagonal whose length is the square root of 2, written $\sqrt{2}$. See Figure 4. It can be shown that $\sqrt{2}$ cannot be written as a quotient of integers. Because of this, $\sqrt{2}$ is not rational; it is **irrational**. Other examples of irrational numbers are $\sqrt{3}$, $\sqrt{7}$, $-\sqrt{10}$, and π (the ratio of the circumference of a circle to its diameter).

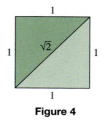

Figure 4

Irrational Numbers

{nonrational numbers represented by points on the number line}

The decimal form of an irrational number neither terminates nor repeats. Irrational numbers are discussed in Chapter 8.

Both rational and irrational numbers can be represented by points on the number line and are called **real numbers.**

Real Numbers

{all numbers that are either rational or irrational}

All the numbers mentioned so far are real numbers. The relationships between the various types of numbers are shown in Figure 5. Notice that any real number is either a rational number or an irrational number.

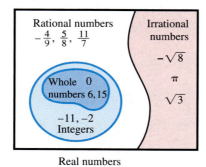

Real numbers

Figure 5

3 **Tell which of two real numbers is smaller.** Given any two whole numbers, we can tell which number is smaller. But what about two negative numbers, as in the set of integers? Moving from 0 to the right along a number line, the positive numbers corresponding to the points on the number line *increase*. For example, $8 < 12$, and 8 is to the left of 12 on a number line. We extend this ordering to all real numbers.

❸ Tell whether each statement is *true* or *false*.

(a) $-2 < 4$

(b) $6 > -3$

(c) $-9 < -12$

(d) $-4 \geq -1$

(e) $-6 \leq 0$

Ordering of the Real Numbers

For any two real numbers a and b, **a is less than b** if a is to the left of b on a number line.

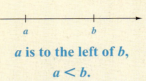

a is to the left of b,
$a < b$.

This means that any negative number is smaller than 0, and any negative number is smaller than any positive number. Also, 0 is smaller than any positive number.

Example 3 Determining the Order of Real Numbers

Is it true that $-3 < -1$?

To find out, locate -3 and -1 on a number line, as shown in Figure 6. Because -3 is to the left of -1 on the number line, -3 is smaller than -1. The statement $-3 < -1$ is true.

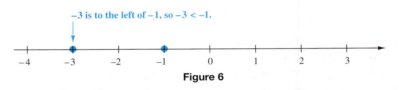

-3 is to the left of -1, so $-3 < -1$.

Figure 6

Work Problem ❸ at the Side.

4 Find the opposite of a real number. Earlier, we saw that every positive integer has a negative integer that is its opposite or negative. This is true for every real number except 0, which is its own opposite.* A characteristic of pairs of opposites is that they are the same distance from 0 on the number line but in opposite directions. See Figure 7.

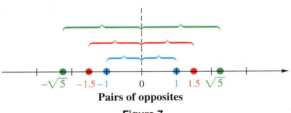

$-\sqrt{5}$ -1.5 -1 0 1 1.5 $\sqrt{5}$

Pairs of opposites

Figure 7

We indicate the opposite of a number by writing the symbol $-$ in front of the number. For example, the opposite of 7 is -7 (read "negative 7"). We could write the opposite of -4 as $-(-4)$, but we know that 4 is the opposite of -4. Since a number can have only one opposite, $-(-4)$ and 4 must represent the same number, so

$$-(-4) = 4.$$

This idea can be generalized.

---

*The opposite (or negative) of a number is also called the *additive inverse* of the number, as we shall see in Section 1.7.

Double Negative Rule

For any real number a,
$$-(-a) = a.$$

The following chart shows several numbers and their opposites.

Number	Opposite
-4	$-(-4)$, or 4
-3	3
0	0
5	-5
19	-19

The chart suggests the following rule.

Except for 0, the opposite of a number is found by changing the sign of the number.

Work Problem ④ at the Side.

5   **Find the absolute value of a real number.**   As previously mentioned, opposites are numbers the same distance from 0 on the number line but on opposite sides of 0. Another way to say this is to say that opposites have the same *absolute value*. The **absolute value** of a number is the undirected distance between 0 and the number on the number line. The symbol for the absolute value of the number a is $|a|$, read "the absolute value of a." For example, the distance between 2 and 0 on the number line is 2 units, so
$$|2| = 2.$$
Also, the distance between -2 and 0 on the number line is 2, so
$$|-2| = 2.$$

Since distance is a physical measurement, which is never negative, we can make the following statement.

The absolute value of a number can never be negative.

For example,
$$|12| = 12 \quad \text{and} \quad |-12| = 12$$
because both 12 and -12 lie at a distance of 12 units from 0 on the number line. Since the distance of 0 from 0 is 0 units, we have
$$|0| = 0.$$

Example 4   **Evaluating Absolute Value**

Simplify.

(a) $|5| = 5$

(b) $|-5| = 5$

(c) $-|-5| = -(5) = -5$ Replace $|-5|$ with 5.

Continued on Next Page

④ Find the opposite of each number.

(a) 6

(b) 15

(c) -9

(d) -12

(e) 0

❺ Simplify.

(a) $|-6|$

(b) $|9|$

(c) $-|15|$

(d) $-|-9|$

(e) $|9 - 4|$

(f) $-|32 - 2|$

(d) $-|-13| = -(13) = -13$

(e) $|8 - 5|$

Simplify within the absolute value bars first.

$$|8 - 5| = |3| = 3$$

(f) $-|8 - 5| = -|3| = -3$

(g) $-|12 - 3| = -|9| = -9$

Parts (e)–(g) in Example 5 show that absolute value bars also act as grouping symbols. You must perform any operations within absolute value bars before finding the absolute value.

Work Problem ❺ at the Side.

1.3 EXERCISES

In Exercises 1–6, give an example of a number that satisfies each given condition.

1. An integer between 3.5 and 4.5

2. A rational number between 3.8 and 3.9

3. A whole number that is not positive and is less than 1

4. A whole number greater than 4.5

5. An irrational number that is between $\sqrt{11}$ and $\sqrt{13}$

6. A real number that is neither negative nor positive

List all numbers from each set that are (a) natural numbers, (b) whole numbers, (c) integers, (d) rational numbers, (e) irrational numbers, (f) real numbers.

7. $\left\{ -9, -\sqrt{7}, -1\frac{1}{4}, -\frac{3}{5}, 0, \sqrt{5}, 3, 5.9, 7 \right\}$

8. $\left\{ -5.3, -5, -\sqrt{3}, -1, -\frac{1}{9}, 0, 1.2, 4, \sqrt{12} \right\}$

Use an integer to express each number representing a change in the following applications. See Example 1.

9. In February 1998, the number of housing starts in the United States increased from the previous month by 93,000 units. (*Source: Wall Street Journal.*)

10. The Wolfsburg, Germany, Volkswagen plant turns out 1550 fewer cars per day than it did in 1991. (*Source:* Klebnikov, P., "Bringing Back the Beetle," *Forbes,* April 7, 1997.)

11. Between 1980 and 1990, the population of the District of Columbia decreased by 31,532. (*Source:* U.S. Bureau of the Census.)

12. In 1994, Taiwan produced 159,376 more passenger cars than commercial vehicles. (*Source:* American Automobile Manufacturers Association.)

Graph each group of numbers on a number line. See Example 2.

13. $0, 3, -5, -6$

14. $2, 6, -2, -1$

15. $-2, -6, -4, 3, 4$

16. $-5, -3, -2, 0, 4$

17. $\frac{1}{4}, 2\frac{1}{2}, -3\frac{4}{5}, -4, -\frac{13}{8}$

18. $5\frac{1}{4}, \frac{41}{9}, -2\frac{1}{3}, 0, -3\frac{2}{5}$

Select the smaller number in each pair. See Example 3.

19. $-11, -4$

20. $-9, -16$

21. $-21, 1$

22. $-57, 3$

23. $0, -100$

24. $-215, 0$

25. $-\dfrac{2}{3}, -\dfrac{1}{4}$

26. $-\dfrac{3}{8}, -\dfrac{9}{16}$

Decide whether each statement is true *or* false. *See Example 3.*

27. $8 < -16$

28. $12 < -24$

29. $-3 < -2$

30. $-10 < -9$

For each number, **(a)** *find its opposite and* **(b)** *find its absolute value.*

31. -2

32. -8

33. 6

34. 11

35. $-\dfrac{3}{4}$

36. $-\dfrac{1}{3}$

Simplify. See Example 4.

37. $|-7|$

38. $|-3|$

39. $-|12|$

40. $-|23|$

41. $-|-14|$

42. $-|-19|$

43. $|13 - 4|$

44. $|8 - 7|$

Decide whether each statement is true *or* false.

45. $|-8| < 7$

46. $|-6| \geq -|6|$

47. $4 \leq |4|$

48. $-|-3| > 2$

49. Students often say "The absolute value of a number is always positive." Is this true? If not, explain.

50. If the absolute value of a number is equal to the number itself, what must be true about the number?

To answer the questions in Exercises 51–54, refer to the table, which gives the changes in producer price indexes for two recent years.

Commodity	Change from 1995 to 1996	Change from 1996 to 1997
Food	4.9	4.0
Transportation	3.9	1.3
Apparel	−.5	.9
Video/Audio equipment	−2.6	−2.2
Shelter	5.3	5.3

Source: U.S. Bureau of Labor Statistics.

51. What commodity for which years represents the greatest decrease?

52. What commodity for which years represents the least change?

53. Which has smaller absolute value, the change for video/audio equipment from 1995 to 1996 or from 1996 to 1997?

54. Which has greater absolute value, the change for apparel from 1995 to 1996 or from 1996 to 1997?

1.4 ADDITION OF REAL NUMBERS

1 **Add two numbers with the same sign.** We can use the number line to explain addition of real numbers. Later, we will give the rules for addition. Recall that the answer to an addition problem is called the **sum.**

OBJECTIVES

1 Add two numbers with the same sign.

2 Add numbers with different signs.

3 Add mentally.

4 Use the order of operations with real numbers.

5 Translate words and phrases that indicate addition.

Example 1 **Adding with the Number Line**

Use the number line to find the sum $2 + 3$.

Add the positive numbers 2 and 3 by starting at 0 and drawing an arrow two units to the *right,* as shown in Figure 8. This arrow represents the number 2 in the sum $2 + 3$. Next, from the right end of this arrow draw another arrow three units to the right. The number below the end of this second arrow is 5, so $2 + 3 = 5$.

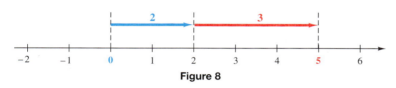

Figure 8

Example 2 **Adding with the Number Line**

Use the number line to find the sum $-2 + (-4)$. (Parentheses are placed around the -4 to avoid the confusing use of $+$ and $-$ next to each other.)

To add the negative numbers -2 and -4 on the number line, we start at 0 and draw an arrow two units to the *left,* as shown in Figure 9. From the left end of this first arrow, we draw a second arrow four units to the left. We draw the arrow to the left to represent the addition of the *negative* number, -4. The number below the end of this second arrow is -6, so $-2 + (-4) = -6$.

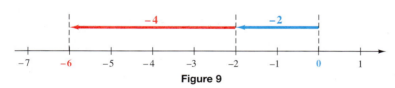

Figure 9

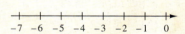
Work Problem 1 at the Side.

In Example 2, we found that the sum of the two negative numbers -2 and -4 is a negative number whose distance from 0 is the sum of the distance of -2 from 0 and the distance of -4 from 0. That is, *the sum of two negative numbers is the negative of the sum of their absolute values.*

$$-2 + (-4) = -(|-2| + |-4|) = -(2 + 4) = -6$$

To add two numbers having the same sign, add the absolute values of the numbers. Give the result the same sign as the numbers being added.

Example: $-4 + (-3) = -7$.

1 Use a number line to find each sum.

(a) $1 + 4$

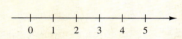

(b) $-2 + (-5)$

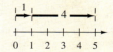

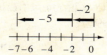

❷ Find each sum.

(a) $-7 + (-3)$

(b) $-12 + (-18)$

(c) $-15 + (-4)$

---

❸ Use a number line to find each sum.

(a) $6 + (-3)$

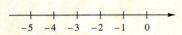

(b) $-5 + 1$

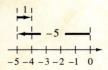

Example 3　Adding Two Negative Numbers

Find each sum.

(a) $-2 + (-9) = -11$　The sum of two negative numbers is negative.

(b) $-8 + (-12) = -20$　　　　**(c)** $-15 + (-3) = -18$

Work Problem ❷ at the Side.

2　Add numbers with different signs.　We use the number line again to illustrate the sum of a positive number and a negative number.

Example 4　Adding Numbers with Different Signs

Use the number line to find the sum $-2 + 5$.

　We find the sum $-2 + 5$ on the number line by starting at 0 and drawing an arrow two units to the left. From the left end of this arrow, we draw a second arrow five units to the right, as shown in Figure 10. The number below the end of this second arrow is 3, so $-2 + 5 = 3$.

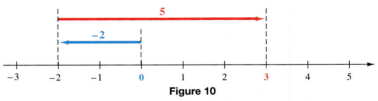

Figure 10

Work Problem ❸ at the Side.

　Addition of numbers with different signs also can be defined using absolute value.

> To add numbers with different signs, first find the difference between the absolute values of the numbers. Give the answer the same sign as the number with the larger absolute value.
>
> *Example:* $-12 + 6 = -6$.

　For example, to add -12 and 5, we find their absolute values: $|-12| = 12$ and $|5| = 5$; then we find the difference between these absolute values: $12 - 5 = 7$. Since $|-12| > |5|$, the sum will be negative, so $-12 + 5 = -7$.

Calculator Tip　The ⊖ or ⊕∕⊖ key is used to input a negative number in some scientific calculators. Try using your calculator to add negative numbers.

3　Add mentally.　While a number line is useful in showing the rules for addition, it is important to be able to find sums mentally.

Example 5　Adding a Positive Number and a Negative Number

Check each answer, trying to work the addition mentally. If you have trouble, use a number line.

(a) $7 + (-4) = 3$

(b) $-8 + 12 = 4$

Continued on Next Page

(c) $-\dfrac{1}{2} + \dfrac{1}{8} = -\dfrac{4}{8} + \dfrac{1}{8} = -\dfrac{3}{8}$ Remember to find a common denominator first.

(d) $\dfrac{5}{6} + \left(-1\dfrac{1}{3}\right) = \dfrac{5}{6} + \left(-\dfrac{4}{3}\right) = \dfrac{5}{6} + \left(-\dfrac{8}{6}\right) = -\dfrac{3}{6} = -\dfrac{1}{2}$

(e) $-4.6 + 8.1 = 3.5$

=== **Work Problem ❹ at the Side.**

The rules for adding signed numbers are summarized below.

Adding Signed Numbers

Same sign   Add the absolute values of the numbers. Give the sum the same sign as the numbers being added.

Different signs   Find the difference between the larger absolute value and the smaller. Give the answer the sign of the number having the larger absolute value.

4 **Use the order of operations with real numbers.**   Sometimes a problem involves square brackets, [ ]. As we mentioned earlier, brackets are treated just like parentheses. We do the calculations inside the brackets until a single number is obtained. Remember to use the order of operations given in Section 1.1 for adding more than two numbers.

Example 6   Adding with Brackets

Find each sum.

(a) $-3 + [4 + (-8)]$
First work inside the brackets. Follow the order of operations given in Section 1.1.

$$-3 + \mathbf{\color{blue}[4 + (-8)]} = -3 + \mathbf{\color{blue}(-4)} = -7$$

(b) $8 + [\mathbf{\color{blue}(-2 + 6)} + (-3)] = 8 + [\mathbf{\color{blue}4} + (-3)] = 8 + 1 = 9$

=== **Work Problem ❺ at the Side.**

5 **Translate words and phrases that indicate addition.**   Let's now look at the interpretation of words and phrases that involve addition. Problem solving often requires translating words and phrases into symbols. We began this process with translating simple phrases in Section 1.1.

The word *sum* indicates addition. There are other key words and phrases that also indicate addition. Some of these are given in the chart below.

Word or Phrase	Example	Numerical Expression and Simplification
Sum of	The *sum of* -3 and 4	$-3 + 4 = 1$
Added to	5 *added to* -8	$-8 + 5 = -3$
More than	12 *more than* -5	$(-5) + 12 = 7$
Increased by	-6 *increased by* 13	$-6 + 13 = 7$
Plus	3 *plus* 14	$3 + 14 = 17$

❹ Check each answer, trying to work the addition mentally. If you have trouble, use a number line.

(a) $-8 + 2 = -6$

(b) $-15 + 4 = -11$

(c) $17 + (-10) = 7$

(d) $\dfrac{3}{4} + \left(-1\dfrac{3}{8}\right) = -\dfrac{5}{8}$

(e) $-9.5 + 3.8 = -5.7$

❺ Find each sum.

(a) $2 + [7 + (-3)]$

(b) $6 + [(-2 + 5) + 7]$

(c) $-9 + [-4 + (-8 + 6)]$

6 Write a numerical expression for each phrase, and simplify the expression.

(a) 4 more than -12

(b) The sum of 6 and -7

(c) -12 added to -31

(d) 7 increased by the sum of 8 and -3

7 A football team lost 8 yd on first down, lost 5 yd on second down, and then gained 7 yd on third down. How many yards did the team gain or lose altogether?

Example 7   **Translating Words and Phrases**

Write a numerical expression for each phrase, and simplify the expression.

(a) The **sum of** -8 and 4 and 6

$$-8 + 4 + 6 = (-8 + 4) + 6 = -4 + 6 = 2$$

Notice that parentheses were placed around $-8 + 4$, and this addition was done first, using the order of operations given earlier.

(b) 3 **more than** -5, **increased by** 12

$$-5 + 3 + 12 = (-5 + 3) + 12 = -2 + 12 = 10$$

Work Problem 6 at the Side.

Gains (or increases) and losses (or decreases) sometimes appear in applied problems. When they do, the gains may be interpreted as positive numbers and the losses as negative numbers.

Example 8   **Interpreting Gains and Losses**

The Tennessee Titans football team gained 3 yd on first down, lost 12 yd on second down, and then gained 13 yd on third down. How many yards did the team gain or lose altogether?

The gains are represented by positive numbers and the loss by a negative number.

$$3 + (-12) + 13$$

Add from left to right.

$$3 + (-12) + 13 = [3 + (-12)] + 13 = (-9) + 13 = 4$$

The team gained 4 yd altogether.

Work Problem 7 at the Side.

1.4 EXERCISES

FOR EXTRA HELP

 Student's Solutions Manual   MyMathLab.com   InterAct Math Tutorial Software   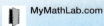 AW Math Tutor Center    www.mathxl.com   MathXL   Digital Video Tutor CD 1 Videotape 2

By the order of operations, what is the first step you would use to simplify each expression?

1. $4[3(-2 + 5) - 1]$ **2.** $[-4 + 7(-6 + 2)]$ **3.** $9 + ([-1 + (-3)] + 5)$ **4.** $[(-8 + 4) + (-6)] + 5$

Find each sum. See Examples 1–6.

5. $6 + (-4)$

6. $8 + (-5)$

7. $12 + (-15)$

8. $4 + (-8)$

9. $-7 + (-3)$

10. $-11 + (-4)$

11. $-10 + (-3)$

12. $-16 + (-7)$

13. $-12.4 + (-3.5)$

14. $-21.3 + (-2.5)$

15. $10 + [-3 + (-2)]$

16. $13 + [-4 + (-5)]$

17. $5 + [14 + (-6)]$

18. $7 + [3 + (-14)]$

19. $-3 + [5 + (-2)]$

20. $-7 + [10 + (-3)]$

21. $-8 + [3 + (-1) + (-2)]$

22. $-7 + [5 + (-8) + 3]$

23. $\dfrac{9}{10} + \left(-\dfrac{3}{5}\right)$

24. $\dfrac{5}{8} + \left(-\dfrac{17}{12}\right)$

25. $-\dfrac{1}{6} + \dfrac{2}{3}$

26. $-\dfrac{6}{25} + \dfrac{19}{20}$

27. $2\dfrac{1}{2} + \left(-3\dfrac{1}{4}\right)$

28. $-4\dfrac{3}{8} + 6\dfrac{1}{2}$

29. $7.8 + (-9.4)$

30. $14.7 + (-10.1)$

31. $-7.1 + [3.3 + (-4.9)]$

32. $-9.5 + [-6.8 + (-1.3)]$

33. $[-8 + (-3)] + [-7 + (-7)]$

34. $[-5 + (-4)] + [9 + (-2)]$

Work each problem. (Source: Population Reference Bureau.) See Example 8.

35. Based on census population projections for 2020, New York will lose 5 seats in the U.S. House of Representatives, Pennsylvania will lose 4 seats, and Ohio will lose 3. Write a signed number that represents the total number of seats these three states are projected to lose.

36. Michigan is projected to lose 3 seats in the U.S. House of Representatives and Illinois 2 in 2020. The states projected to gain the most seats are California with 9, Texas with 5, Florida with 3, Georgia with 2, and Arizona with 2. Write a signed number that represents the algebraic sum of these changes.

Perform each operation, and then determine whether the statement is true *or* false. *Try to do all work mentally. See Examples 5 and 6.*

37. $-11 + 13 = 13 + (-11)$

38. $16 + (-9) = -9 + 16$

39. $-10 + 6 + 7 = -3$

40. $-12 + 8 + 5 = -1$

41. $18 + (-6) + (-12) = 0$

42. $-5 + 21 + (-16) = 0$

43. $|-8 + 10| = -8 + (-10)$

44. $|-4 + 6| = -4 + (-6)$

45. $2\dfrac{1}{5} + \left(-\dfrac{6}{11}\right) = -\dfrac{6}{11} + 2\dfrac{1}{5}$

46. $-1\dfrac{1}{2} + \dfrac{5}{8} = \dfrac{5}{8} + \left(-1\dfrac{1}{2}\right)$

47. $-7 + [-5 + (-3)] = [(-7) + (-5)] + 3$

48. $6 + [-2 + (-5)] = [(-4) + (-2)] + 5$

RELATING CONCEPTS (Exercises 49–52)   **FOR INDIVIDUAL OR GROUP WORK**

Recall the rules for adding signed numbers introduced in this section, and **work** *Exercises 49–52 in order.*

49. Suppose that the sum of two numbers is negative, and you know that one of the numbers is positive. What can you conclude about the other number?

50. If you are asked to solve the equation $x + 5 = -7$ from a set of numbers, why could you immediately eliminate any positive numbers as possible solutions? (Remember how you answered Exercise 49.)

51. Suppose that the sum of two numbers is positive, and you know that one of the numbers is negative. What can you conclude about the other number?

52. If you are asked to solve the equation $x + (-8) = 2$ from a set of numbers, why could you immediately eliminate any negative numbers as possible solutions? (Remember how you answered Exercise 51.)

---

53. In your own words, explain how to add two negative numbers.

54. In your own words, explain how to add a positive number and a negative number. Give two cases.

Write a numerical expression for each phrase, and simplify the expression. See Example 7.

55. The sum of -5 and 12 and 6

56. The sum of -3 and 5 and -12

57. 14 added to the sum of −19 and −4

58. −2 added to the sum of −18 and 11

59. The sum of −4 and −10, increased by 12

60. The sum of −7 and −13, increased by 14

61. 4 more than the sum of 8 and −18

62. 10 more than the sum of −4 and −6

Solve each problem. See Example 8.

63. Kramer owed Jerry $10 for snacks raided from the refrigerator. Kramer later borrowed $70 from George to finance his latest get-rich scheme. What positive or negative number represents Kramer's financial status?

64. Shalita's checking account balance is $54.00. She then takes a gamble by writing a check for $89.00. What is her new balance? (Write the balance as a signed number.)

65. The surface, or rim, of a canyon is at altitude 0. On a hike down into the canyon, a party of hikers stops for a rest at 130 m below the surface. They then descend another 54 m. What is their new altitude? (Write the altitude as a signed number.)

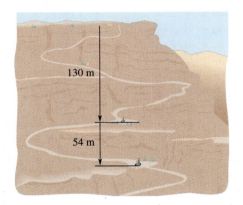

66. A pilot announces to the passengers that the current altitude of their plane is 34,000 ft. Because of some unexpected turbulence, the pilot is forced to descend 2100 ft. What is the new altitude of the plane? (Write the altitude as a signed number.)

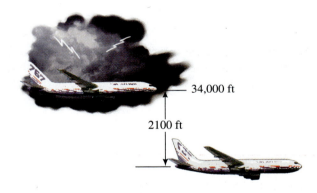

67. On three consecutive passes, Troy Aikman of the Dallas Cowboys passed for a gain of 6 yd, was sacked for a loss of 12 yd, and passed for a gain of 43 yd. What positive or negative number represents the total net yardage for the plays?

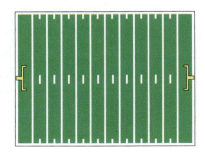

68. On a series of three consecutive running plays, Peyton Manning of the Indianapolis Colts gained 4 yd, lost 3 yd, and lost 2 yd. What positive or negative number represents his total net yardage for the series of plays?

69. The lowest temperature ever recorded in Arkansas was −29°F. The highest temperature ever recorded there was 149°F more than the lowest. What was this highest temperature? (*Source: World Almanac and Book of Facts,* 2000.)

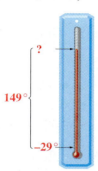

70. On January 23, 1943, the temperature rose 49°F in two minutes in Spearfish, South Dakota. If the starting temperature was −4°F, what was the temperature two minutes later?

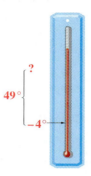

71. Jennifer owes $153 to a credit card company. She makes a $14 purchase with the card, and then pays $60 on the account. What is her current balance as a signed number?

72. A female polar bear weighed 660 lb when she entered her winter den. She lost 45 lb during each of the first two months of hibernation, and another 205 lb before leaving the den with her two cubs in March. How much did she weigh when she left the den?

73. Jim Yee owes $870.00 on his MasterCard account. He returns two items costing $35.90 and $150.00 and receives credits for these on the account. Next, he makes a purchase of $82.50, and then two more purchases of $10.00 each. He finally makes a payment of $500.00. What is his new account balance?

74. A welder working with stainless steel must use precise measurements. Suppose a welder attaches two pieces of steel that are each 3.60 in. long, and then attaches an additional three pieces that are each 9.10 in. long. She finally cuts off a piece that is 7.60 in. long. Find the length of the welded piece of steel.

1.5 SUBTRACTION OF REAL NUMBERS

OBJECTIVES

1 Find a difference.

2 Use the definition of subtraction.

3 Work subtraction problems that involve brackets.

4 Translate words and phrases that indicate subtraction.

1 **Find a difference.** As we mentioned earlier, the answer to a subtraction problem is called a **difference.** Differences between signed numbers can be found by using a number line. Addition and subtraction are opposite operations. Thus, because *addition* of a positive number on the number line is shown by drawing an arrow to the *right, subtraction* of a positive number is shown by drawing an arrow to the *left.*

Example 1 Subtracting with the Number Line

Use the number line to find the difference $7 - 4$.

To find the difference $7 - 4$ on the number line, begin at 0 and draw an arrow 7 units to the *right*. From the right end of this arrow, draw an arrow 4 units to the *left,* as shown in Figure 11. The number at the end of the second arrow shows that $7 - 4 = 3$.

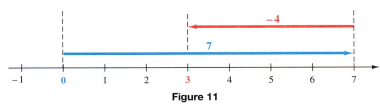

Figure 11

Work Problem **1** at the Side.

2 **Use the definition of subtraction.** The procedure used in Example 1 to find $7 - 4$ is exactly the same procedure that would be used to find $7 + (-4)$, so

$$7 - 4 = 7 + (-4).$$

This shows that *subtracting* a positive number from a larger positive number is the same as *adding* the opposite of the smaller number to the larger. We use this idea to define subtraction for all real numbers.

Subtraction

For any real numbers a and b,

$$a - b = a + (-b).$$

Example: $4 - 9 = 4 + (-9) = -5$.

That is, to *subtract* b from a, *add the opposite* (or *negative*) of b to a. This definition leads to the following procedure for subtracting signed numbers.

Subtracting Signed Numbers

Step 1 Change the subtraction symbol to addition, and change the sign of the number being subtracted.

Step 2 Add, as in the previous section.

1 Use the number line to find each difference.

(a) $5 - 1$

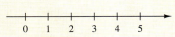

(b) $6 - 2$

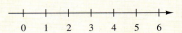

ANSWERS
1. (a) $5 - 1 = 4$

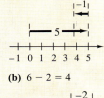

(b) $6 - 2 = 4$

2 Subtract.

(a) $6 - 10$

(b) $-2 - 4$

(c) $3 - (-5)$

(d) $-8 - (-12)$

(e) $\dfrac{5}{4} - \left(-\dfrac{3}{7}\right)$

Example 2 Using the Definition of Subtraction

Subtract.

(a) $12 - 3 = 12 + (-3) = 9$
- No change
- Change − to +.
- Opposite of 3

(b) $5 - 7 = 5 + (-7) = -2$

(c) $8 - 15 = 8 + (-15) = -7$

(d) $-3 - (-5) = -3 + (5) = 2$
- No change
- Change − to +.
- Opposite of −5

(e) $-6 - (-9) = -6 + (9) = 3$

(f) $\dfrac{3}{8} - \left(-\dfrac{4}{5}\right) = \dfrac{15}{40} - \left(-\dfrac{32}{40}\right) = \dfrac{15}{40} + \dfrac{32}{40} = \dfrac{47}{40}$

Work Problem 2 at the Side.

Subtraction can be used to reverse the result of an addition problem. For example, if 4 is added to a number and then subtracted from the sum, the original number is the result.

$$12 + 4 = 16 \quad \text{and} \quad 16 - 4 = 12$$

The symbol − has now been used for three purposes:

1. to represent subtraction, as in $9 - 5 = 4$;

2. to represent negative numbers, such as -10, -2, and -3;

3. to represent the opposite (or negative) of a number, as in "the opposite (or negative) of 8 is -8."

We may see more than one use in the same problem, such as $-6 - (-9)$, where -9 is subtracted from -6. The meaning of the symbol depends on its position in the algebraic expression.

3 Work subtraction problems that involve brackets. As before, with problems that have both parentheses and brackets, first do any operations inside the parentheses and brackets. Work from the inside out. Because subtraction is defined in terms of addition, the order of operations from Section 1.1 can still be used.

Example 3 Subtracting with Grouping Symbols

Perform each operation.

(a) $-6 - [2 - (8 + 3)] = -6 - [2 - 11]$
$$= -6 - [2 + (-11)] \quad \text{Change − to +.}$$
$$= -6 - (-9)$$
$$= -6 + (9) = 3$$

Continued on Next Page

(b) $5 - \left[\left(-\dfrac{1}{3} - \dfrac{1}{2}\right) - (4 - 1)\right] = 5 - \left[\left(-\dfrac{1}{3} + \left(-\dfrac{1}{2}\right)\right) - 3\right]$

$$= 5 - \left[\left(-\dfrac{5}{6}\right) - 3\right]$$

$$= 5 - \left[\left(-\dfrac{5}{6}\right) + (-3)\right]$$

$$= 5 - \left(-\dfrac{23}{6}\right)$$

$$= 5 + \dfrac{23}{6} = \dfrac{53}{6}$$

=== **Work Problem ❸ at the Side.**

4▭ **Translate words and phrases that indicate subtraction.**   Now we translate words and phrases that involve subtraction of real numbers. *Difference* is one of them. Some others are given in the chart below.

Word or Phrase	Example	Numerical Expression and Simplification
Difference between	The *difference between* −3 and −8	$-3 - (-8) = -3 + 8 = 5$
Subtracted from	12 *subtracted from* 18	$18 - 12 = 6$
Less than	6 *less than* 5	$5 - 6 = 5 + (-6) = -1$
Decreased by	9 *decreased by* −4	$9 - (-4) = 9 + 4 = 13$
Minus	−8 *minus* 5	$-8 - 5 = -8 + (-5) = -13$

CAUTION

When you are subtracting two numbers, it is important that you write them in the correct order, because, in general, $a - b \neq b - a$. For example, $5 - 3 \neq 3 - 5$. For this reason, it is important to *think carefully before interpreting an expression involving subtraction!* (This problem does not arise for addition.)

Example 4   **Translating Words and Phrases**

Write a numerical expression for each phrase, and simplify the expression.

(a) The **difference between** −8 and 5
When "difference between" is used, write the numbers in the order they are given.

$$-8 - 5 = -8 + (-5) = -13$$

(b) 4 **subtracted from** the sum of 8 and −3
Here the operation of addition is also used, as indicated by the word *sum*. First, add 8 and −3. Next, subtract 4 from this sum.

$$[8 + (-3)] - 4 = 5 - 4 = 1$$

=== **Continued on Next Page**

---

❸ Perform each operation.

(a) $2 - [(-3) - (4 + 6)]$

(b) $[(5 - 7) + 3] - 8$

(c) $6 - [(-1 - 4) - 2]$

4 Write a numerical expression for each phrase, and simplify the expression.

(a) The difference between −5 and −12

(b) −2 subtracted from the sum of 4 and −4

(c) 7 less than −2

(d) 9, decreased by 10 less than 7

5 The highest elevation in Argentina is Mt. Aconcagua, which is 6960 m above sea level. The lowest point in Argentina is the Valdes Peninsula, 40 m below sea level. Find the difference between the highest and lowest elevations.

(c) 4 **less than** −6

Be careful with order here. 4 must be taken *from* −6, so write −6 first.

$$-6 - 4 = -6 + (-4) = -10$$

Notice that "4 less than −6" differs from "4 *is less than* − 6." The statement "4 *is less than* −6" is symbolized as $4 < -6$ (which is a false statement).

(d) 8, **decreased by** 5 **less than** 12

First, write "5 less than 12" as $12 - 5$. Next, subtract $12 - 5$ from 8.

$$8 - (12 - 5) = 8 - 7 = 1$$

Work Problem 4 at the Side.

We have seen a few applications of signed numbers in earlier sections. The next example involves subtraction of signed numbers.

Example 5 Solving a Problem Involving Subtraction

The record high temperature of 134°F in the United States was recorded at Death Valley, California, in 1913. The record low was −80°F, at Prospect Creek, Alaska, in 1971. See Figure 12. What is the difference between these highest and lowest temperatures? (*Source: World Almanac and Book of Facts,* 2000.)

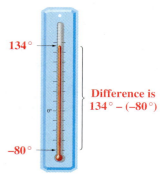

Figure 12

We must subtract the lowest temperature from the highest temperature.

$$134 - (-80) = 134 + 80 \quad \text{Use the definition of subtraction.}$$
$$= 214 \quad \text{Add.}$$

The difference between the two temperatures is 214°F.

Work Problem 5 at the Side.

1.5 EXERCISES

FOR EXTRA HELP

 Student's Solutions Manual

 MyMathLab.com

 InterAct Math Tutorial Software

 AW Math Tutor Center

 www.mathxl.com

 Digital Video Tutor CD 1 Videotape 3

Fill in each blank with the correct response.

1. By the definition of subtraction, in order to perform the subtraction $-6 - (-8)$, we must add the opposite of _____ to _____ .

2. By the order of operations, to simplify $8 - [3 - (-4 - 5)]$, the first step is to subtract _____ from _____ .

3. "The difference between 7 and 12" translates as _____ , while "the difference between 12 and 7" translates as _____ .

4. $-9 - (-3) = -9 +$ _____

5. $-8 - 4 = -8 +$ _____

6. $-19 - 22 = -19 +$ _____

Find each difference. See Examples 1–3.

7. $-7 - 3$

8. $-12 - 5$

9. $-10 - 6$

10. $-13 - 16$

11. $7 - (-4)$

12. $9 - (-6)$

13. $6 - (-13)$

14. $13 - (-3)$

15. $-7 - (-3)$

16. $-8 - (-6)$

17. $3 - (4 - 6)$

18. $6 - (7 - 14)$

19. $-3 - (6 - 9)$

20. $-4 - (5 - 12)$

21. $\dfrac{1}{2} - \left(-\dfrac{1}{4}\right)$

22. $\dfrac{1}{3} - \left(-\dfrac{4}{3}\right)$

23. $-\dfrac{3}{4} - \dfrac{5}{8}$

24. $-\dfrac{5}{6} - \dfrac{1}{2}$

25. $\dfrac{5}{8} - \left(-\dfrac{1}{2} - \dfrac{3}{4}\right)$

26. $\dfrac{9}{10} - \left(\dfrac{1}{8} - \dfrac{3}{10}\right)$

27. $4.4 - (-9.2)$

28. $6.7 - (-12.6)$ **29.** $-7.4 - 4.5$ **30.** $-5.4 - 9.6$

31. $-5.2 - (8.4 - 10.8)$ **32.** $-9.6 - (3.5 - 12.6)$

33. $[(-3.1) - 4.5] - (.8 - 2.1)$ **34.** $[(-7.8) - 9.3] - (.6 - 3.5)$

35. $-12 - [(9 - 2) - (-6 - 3)]$ **36.** $-4 + [(-6 - 9) - (-7 + 4)]$ **37.** $-8 + [(-3 - 10) - (-4 + 1)]$

38. $\left(-\dfrac{3}{4} - \dfrac{5}{2}\right) - \left(-\dfrac{1}{8} - 1\right)$ **39.** $\left(-\dfrac{3}{8} - \dfrac{2}{3}\right) - \left(-\dfrac{9}{8} - 3\right)$

40. $[-34.99 + (6.59 - 12.25)] - 8.33$ **41.** $[-12.25 - (8.34 + 3.57)] - 17.88$

42. Explain in your own words how to subtract signed numbers.

43. We know that, in general, $a - b \neq b - a$. Find two pairs of values for a and b so that $a - b = b - a$.

Simplify each expression. Use the order of operations.

44. $-3 - (-4) - 5$ **45.** $8 - (-3) - 9 + 6$ **46.** $-5 - 2 + 4 - 8 - (-6)$

47. Make up a subtraction problem so that the difference between two negative numbers is a negative number.

48. Make up a subtraction problem so that the difference between two negative numbers is a positive number.

Write a numerical expression for each phrase and simplify. See Example 4.

49. The difference between 4 and −8

50. The difference between 7 and −14

51. 8 less than −2

52. 9 less than −13

53. The sum of 9 and −4, decreased by 7

54. The sum of 12 and −7, decreased by 14

55. 12 less than the difference between 8 and −5

56. 19 less than the difference between 9 and −2

Solve each problem. See Example 5.

57. The coldest temperature recorded in Chicago, Illinois, was −35°F in 1996. The record low in South Dakota was set in 1936 and was 23°F lower than −35°F. What was the record low in South Dakota? (*Source: World Almanac and Book of Facts,* 2000.)

58. No one knows just why humpback whales love to heave their 45-ton bodies out of the water, but leap they do. Mark and Debbie, two researchers based on the island of Maui, noticed that one of their favorite whales, "Pineapple," leaped 15 ft above the surface of the ocean while her mate cruised 12 ft below the surface. What is the difference between these two heights?

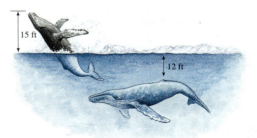

59. The top of Mount Whitney, visible from Death Valley, has an altitude of 14,494 ft above sea level. The bottom of Death Valley is 282 ft below sea level. Using 0 as sea level, find the difference between these two elevations. (*Source: World Almanac and Book of Facts,* 2000.)

60. A chemist is running an experiment under precise conditions. At first, she runs it at $-174.6°F$. She then lowers the temperature by $2.3°F$. What is the new temperature for the experiment?

61. Chris owed his brother $10. He later borrowed $70. What positive or negative number represents his present financial status?

62. Francesca has $15 in her purse, and Emilio has a debt of $12. Find the difference between these amounts.

63. For the year 1999, one health club showed a profit of $76,000, while another showed a loss of $29,000. Find the difference between these amounts.

64. At 1:00 A.M., a plant worker found that a dial reading was 7.904. At 2:00 A.M., she found the reading to be -3.291. Find the difference between these two readings.

The average sales prices of new single-family homes in the United States for the years 1990 through 1995 are shown in the table. Complete the table, determining the change from one year to the next by subtraction.

	Year	Average Sales Price	Change from Previous Year
	1990	$149,800	
	1991	$147,200	−$2600
65.	1992	$144,100	_____
66.	1993	$147,700	_____
67.	1994	$154,500	_____
68.	1995	$158,700	_____

Source: U.S. Bureau of the Census.

In Exercises 69–72, suppose that x represents a positive number and y represents a negative number. Determine whether the given expression must represent a positive number or a negative number.

69. $x - y$ **70.** $y - x$ **71.** $x + |y|$ **72.** $y - |x|$

1.6 MULTIPLICATION AND DIVISION OF REAL NUMBERS

In this section we learn how to multiply positive and negative numbers. The result of multiplication is called the **product.** We already know how to multiply positive numbers and that the product of two positive numbers is positive. We also know that the product of 0 and any positive number is 0, and we extend that property to all real numbers.

Multiplication Property of 0

For any real number a,

$$a \cdot 0 = 0 \cdot a = 0.$$

1   **Find the product of numbers with different signs.**   To define the product of numbers with different signs so that the result is consistent with multiplication of positive numbers, look at the following pattern.

$$3 \cdot 5 = 15$$
$$3 \cdot 4 = 12$$
$$3 \cdot 3 = 9$$
$$3 \cdot 2 = 6$$
$$3 \cdot 1 = 3$$
$$3 \cdot 0 = 0$$
$$3 \cdot (-1) = ?$$

The products decrease by 3.

What should $3(-1)$ equal? Since multiplication can also be considered repeated addition, the product $3(-1)$ represents the sum

$$-1 + (-1) + (-1) = -3,$$

so the product should be -3, which fits the pattern. Also,

$$3(-2) = -2 + (-2) + (-2) = -6.$$

Work Problem ❶ at the Side.

The results from Problem 1 maintain the pattern in the list above, which suggests the following rule.

The product of a positive number and a negative number is negative.

Example: $6(-3) = -18$.

Example 1   **Multiplying a Positive Number and a Negative Number**

Find each product using the multiplication rule.

(a) $8(-5) = -(8 \cdot 5) = -40$ **(b)** $-7(2) = -(7 \cdot 2) = -14$

(c) $-9\left(\dfrac{1}{3}\right) = -3$ **(d)** $-6.2(4.1) = -25.42$

Work Problem ❷ at the Side.

OBJECTIVES

1   Find the product of numbers with different signs.

2   Find the product of two negative numbers.

3   Use the reciprocal of a number to apply the definition of division.

4   Use the order of operations when multiplying and dividing signed numbers.

5   Evaluate expressions involving variables.

6   Translate words and phrases involving multiplication and division.

7   Translate simple sentences into equations.

---

❶ Find each product by finding the sum of three numbers.

(a) $3(-3)$

(b) $3(-4)$

(c) $3(-5)$

❷ Find each product.

(a) $2(-6)$

(b) $7(-8)$

(c) $-9(2)$

(d) $-16\left(\dfrac{5}{32}\right)$

(e) $4.56(-10)$

ANSWERS
1. **(a)** -9 **(b)** -12 **(c)** -15
2. **(a)** -12 **(b)** -56 **(c)** -18
   (d) $-\dfrac{5}{2}$ **(e)** -45.6

❸ Find each product.

(a) $-5(-6)$

(b) $-7(-3)$

(c) $-8(-5)$

(d) $-11(-2)$

(e) $-17(3)(-7)$

(f) $-41(2)(-13)$

2 **Find the product of two negative numbers.** The product of two positive numbers is positive, and the product of a positive number and a negative number is negative. What about the product of two negative numbers? Look at another pattern.

$$-5(\mathbf{4}) = -20$$
$$-5(\mathbf{3}) = -15$$
$$-5(\mathbf{2}) = -10 \qquad \text{The products increase by 5.}$$
$$-5(\mathbf{1}) = -5$$
$$-5(\mathbf{0}) = 0$$
$$-5(\mathbf{-1}) = ?$$

The numbers on the left of the equals signs (in color) decrease by 1 for each step down the list. The products on the right increase by 5 for each step down the list. To maintain this pattern, $-5(-1)$ should be 5 more than $-5(0)$, or 5 more than 0, so

$$-5(\mathbf{-1}) = 5.$$

The pattern continues with

$$-5(\mathbf{-2}) = 10$$
$$-5(\mathbf{-3}) = 15$$
$$-5(\mathbf{-4}) = 20$$
$$-5(\mathbf{-5}) = 25,$$

and so on. This pattern suggests the next rule.

> The product of two negative numbers is positive.
>
> *Example:* $-5(-4) = 20.$

Example 2 **Multiplying Two Negative Numbers**

Find each product using the multiplication rule.

(a) $-9(-2) = 18$ (b) $-6(-12) = 72$

(c) $-2(4)(-1) = -8(-1) = 8$ (d) $3(-5)(-2) = -15(-2) = 30$

Work Problem ❸ at the Side.

Here is a summary of the results for multiplying signed numbers.

Multiplying Signed Numbers

The product of two numbers having the *same* sign is *positive,* and the product of two numbers having *different* signs is *negative.*

3 **Use the reciprocal of a number to apply the definition of division.** Recall that the result of division is called the **quotient.** In the previous section we saw that the difference between two numbers is found by adding the opposite of the second number to the first. Similarly, the *quotient* of two numbers involves multiplying by the *reciprocal* of the second number.

Reciprocals

Pairs of numbers whose product is 1 are called **reciprocals** of each other.

Since $8 \cdot \dfrac{1}{8} = \dfrac{8}{8} = 1$ and $\dfrac{5}{4} \cdot \dfrac{4}{5} = \dfrac{20}{20} = 1,$

the reciprocal of 8 is $\frac{1}{8}$, and that of $\frac{5}{4}$ is $\frac{4}{5}$. The following table shows several numbers and their reciprocals.

Number	Reciprocal
4	$\frac{1}{4}$
-5	$\frac{1}{-5}$ or $-\frac{1}{5}$
$\frac{3}{4}$	$\frac{4}{3}$
$-\frac{5}{8}$	$-\frac{8}{5}$
0	None

By definition, the product of a number and its reciprocal is 1. But the multiplication property of 0 says that the product of 0 and any number is 0. Thus,

0 has no reciprocal.

Work Problem ❹ at the Side.

By definition, the quotient of a and b is the product of a and the reciprocal of b.

Division

The quotient $\frac{a}{b}$ of real numbers a and b, with $b \neq 0$, is

$$\frac{a}{b} = a \cdot \frac{1}{b}.$$

Example: $\dfrac{8}{-4} = 8\left(-\dfrac{1}{4}\right) = -2.$

This definition indicates that b, the number to divide by, cannot be 0. Since 0 has no reciprocal,

$\frac{1}{0}$ **is not a number and** *division by 0 is undefined.* **If a division problem requires division by 0, write "undefined."**

NOTE

While division *by* 0 is undefined, we may divide 0 by any nonzero number.

If $a \neq 0,$ then $\dfrac{0}{a} = 0.$

Because division is defined in terms of multiplication, all the rules for multiplying signed numbers also apply to dividing them.

❹ Complete the table.

Number	Reciprocal
(a) 6	
(b) -2	
(c) $\dfrac{2}{3}$	
(d) $-\dfrac{1}{4}$	
(e) 0	

⑤ Find each quotient.

(a) $\dfrac{42}{7}$

(b) $\dfrac{-36}{(-2)(-3)}$

(c) $\dfrac{-12.56}{-.4}$

(d) $\dfrac{10}{7} \div \left(-\dfrac{24}{5}\right)$

(e) $\dfrac{-3}{0}$

(f) $\dfrac{0}{-53}$

⑥ Find each quotient.

(a) $\dfrac{-8}{-2}$

(b) $\dfrac{-16.4}{2.05}$

(c) $\dfrac{1}{4} \div \left(-\dfrac{2}{3}\right)$

Example 3 **Using the Definition of Division**

Find each quotient.

(a) $\dfrac{12}{3} = 12 \cdot \dfrac{1}{3} = 4$

(b) $\dfrac{5(-2)}{2} = -10 \cdot \dfrac{1}{2} = -5$

(c) $\dfrac{-1.47}{-7} = -1.47 \cdot \left(-\dfrac{1}{7}\right) = .21$

(d) $-\dfrac{2}{3} \div \left(-\dfrac{5}{4}\right) = -\dfrac{2}{3} \cdot \left(-\dfrac{4}{5}\right) = \dfrac{8}{15}$

(e) $\dfrac{-10}{0}$ Undefined

(f) $\dfrac{0}{13} = 0$ $\dfrac{0}{a} = 0$ $(a \neq 0)$

Work Problem ⑤ at the Side.

When dividing fractions, multiplying by the reciprocal works well. However, using the definition of division directly with integers is awkward. It is easier to divide in the usual way, then determine the sign of the answer. The following rule for division can be used instead of multiplying by the reciprocal.

Dividing Signed Numbers

The quotient of two numbers having the *same* sign is *positive;* the quotient of two numbers having *different* signs is *negative.*

Examples: $\dfrac{-15}{-5} = 3$ and $\dfrac{-15}{5} = -3.$

Example 4 **Dividing Signed Numbers**

Find each quotient.

(a) $\dfrac{8}{-2} = -4$

(b) $\dfrac{-4.5}{-.09} = 50$

(c) $-\dfrac{1}{8} \div \left(-\dfrac{3}{4}\right) = -\dfrac{1}{8} \cdot \left(-\dfrac{4}{3}\right) = \dfrac{1}{6}$

Work Problem ⑥ at the Side.

From the definitions of multiplication and division of real numbers,

$$\dfrac{-40}{8} = -40 \cdot \dfrac{1}{8} = -5 \quad \text{and} \quad \dfrac{40}{-8} = 40\left(\dfrac{1}{-8}\right) = -5, \text{ so}$$

$$\dfrac{-40}{8} = \dfrac{40}{-8}.$$

Based on this example, the quotient of a positive number and a negative number can be written in any of the following three forms.

For any positive real numbers a and b,

$$\dfrac{-a}{b} = \dfrac{a}{-b} = -\dfrac{a}{b}.$$

The form $\dfrac{a}{-b}$ is seldom used.

Similarly, the quotient of two negative numbers can be expressed as the quotient of two positive numbers.

For any positive real numbers a and b,

$$\frac{-a}{-b} = \frac{a}{b}.$$

4 Use the order of operations when multiplying and dividing signed numbers.

Example 5 Using the Order of Operations

Simplify.

(a) $-9(2) - (-3)(2)$
First find all products, working from left to right.

$$-9(2) - (-3)(2) = -18 - (-6)$$
$$= -18 + 6$$
$$= -12$$

(b) $-6(-2) - 3(-4) = 12 - (-12)$
$$= 12 + 12$$
$$= 24$$

(c) $\dfrac{5(-2) - 3(4)}{2(1 - 6)}$

Follow the order of operations. Simplify the numerator and denominator separately. Then divide or write in lowest terms.

$$\frac{5(-2) - 3(4)}{2(1 - 6)} = \frac{-10 - 12}{2(-5)} \qquad \text{Multiply in numerator.}$$
$$\text{Subtract in denominator.}$$
$$= \frac{-22}{-10} \qquad \text{Subtract in numerator.}$$
$$\text{Multiply in denominator.}$$
$$= \frac{11}{5} \qquad \text{Write in lowest terms.}$$

Work Problem ❼ at the Side.

The rules for operations with signed numbers are summarized here.

Operations with Signed Numbers

Addition
Same sign Add the absolute values of the numbers. The sum has the same sign as the numbers.

$$-4 + (-6) = -10$$

Different signs Subtract the number with the smaller absolute value from the one with the larger. Give the sum the sign of the number having the larger absolute value.

$$4 + (-6) = -(6 - 4) = -2$$

(continued)

❼ Perform the indicated operations.

(a) $-3(4) - 2(6)$

(b) $-8[-1 - (-4)(-5)]$

(c) $\dfrac{6(-4) - 2(5)}{3(2 - 7)}$

(d) $\dfrac{-6(-8) + 3(9)}{-2[4 - (-3)]}$

8 Evaluate each expression.

(a) $2x - 7(y + 1)$
if $x = -4$ and $y = 3$

Subtraction
Add the opposite of the second number to the first number.
$$8 - (-3) = 8 + 3 = 11$$

Multiplication and Division
Same sign   The product or quotient of two numbers with the same sign is positive.

$$-5(-6) = 30 \quad \text{and} \quad \frac{-36}{-12} = 3$$

Different signs   The product or quotient of two numbers with different signs is negative.

$$-5(6) = -30 \quad \text{and} \quad \frac{18}{-6} = -3$$

Division by 0 is undefined.

(b) $2x^2 - 4y^2$
if $x = -2$ and $y = -3$

5 ▭ **Evaluate expressions involving variables.**   The next examples show numbers substituted for variables where the rules for operating with signed numbers must be used.

Example 6   **Evaluating Expressions for Numerical Values**

Evaluate each expression, given that $x = -1$, $y = -2$, and $m = -3$.

(a) $(3x + 4y)(-2m)$
First substitute the given values for the variables. Then use the order of operations to find the value of the expression.

$$(3x + 4y)(-2m) = [3(-1) + 4(-2)][-2(-3)] \quad \text{Put parentheses around the number for each variable.}$$

$$= [-3 + (-8)][6] \quad \text{Find the products.}$$
$$= (-11)(6) \quad \text{Add inside the brackets.}$$
$$= -66 \quad \text{Multiply.}$$

(b) $2x^2 - 3y^2$
Use parentheses as shown.

$$2(-1)^2 - 3(-2)^2 = 2(1) - 3(4) \quad \text{Substitute, then apply the exponents.}$$
$$= 2 - 12 \quad \text{Multiply.}$$
$$= -10 \quad \text{Subtract.}$$

(c) $\dfrac{4x - 2y}{-3x}$
if $x = 2$ and $y = -1$

(c) $\dfrac{4y^2 + x}{m}$

$$\frac{4(-2)^2 + (-1)}{-3} = \frac{4(4) + (-1)}{-3} \quad \text{Substitute, then apply the exponent.}$$

$$= \frac{16 + (-1)}{-3} \quad \text{Multiply.}$$

$$= \frac{15}{-3} \quad \text{Add.}$$

$$= -5 \quad \text{Divide.}$$

Notice how the fraction bar was used as a grouping symbol.

Work Problem 8 at the Side.

6 Translate words and phrases involving multiplication and division. Just as there are words and phrases that indicate addition or subtraction, certain words and phrases indicate multiplication or division. The chart gives some phrases indicating multiplication.

Word or Phrase	Example	Numerical Expression and Simplification
Product of	The *product of* −5 and −2	$-5(-2) = 10$
Times	13 *times* −4	$13(-4) = -52$
Twice (meaning "2 times")	*Twice* 6	$2(6) = 12$
Of (used with fractions)	$\frac{1}{2}$ *of* 10	$\frac{1}{2}(10) = 5$
Percent of	12**%** *of* −16	$.12(-16) = -1.92$

Example 7 **Translating Words and Phrases**

Write a numerical expression for each phrase and simplify. Use the order of operations.

(a) The **product of** 12 and the sum of 3 and −6
Here 12 is multiplied by "the sum of 3 and −6."

$$12[3 + (-6)] = 12(-3) = -36$$

(b) **Three times** the difference between 4 and −11

$$3[4 - (-11)] = 3(4 + 11) = 3(15) = 45$$

(c) Two-thirds **of** the sum of −5 and −3

$$\frac{2}{3}[-5 + (-3)] = \frac{2}{3}(-8) = -\frac{16}{3}$$

(d) 15**%** **of** the difference between 14 and −2
Remember that 15% = .15.

$$.15[14 - (-2)] = .15(14 + 2) = .15(16) = 2.4$$

━━━━━━━━━ **Work Problem ❾ at the Side.**

The word *quotient* refers to the answer in a division problem. In algebra, a quotient is usually represented with a fraction bar; the symbol ÷ is seldom used. When translating an applied problem involving division, use a fraction bar. The chart gives some key phrases associated with division.

Word or Phrase	Example	Numerical Expression and Simplification
Quotient of	The *quotient of* −24 and 3	$\frac{-24}{3} = -8$
Divided by	−16 *divided by* −4	$\frac{-16}{-4} = 4$
Ratio of	The *ratio of* 2 to 3	$\frac{2}{3}$

When translating a phrase involving division, we write the first number named as the numerator and the second as the denominator.

❾ Write a numerical expression for each phrase and simplify.

(a) The product of 6 and the sum of −5 and −4

(b) Twice the difference between 8 and −4

(c) Three-fifths of the sum of 2 and −7

(d) 20% of the sum of 9 and −4

10 Write a numerical expression for each phrase, and simplify the expression.

(a) The quotient of 20 and the sum of 8 and −3

(b) The product of −9 and 2, divided by the difference between 5 and −1

11 Write each sentence in symbols, using x to represent the number.

(a) Twice a number is −6.

(b) The difference between −8 and a number is −11.

(c) The sum of 5 and a number is 8.

(d) The quotient of a number and −2 is 6.

Example 8 **Translating Words and Phrases**

Write a numerical expression for each phrase, and simplify the expression.

(a) The **quotient** of 14 and the sum of −9 and 2

"Quotient" indicates division. The number 14 is the numerator and "the sum of −9 and 2" is the denominator.

$$\frac{14}{-9 + 2} = \frac{14}{-7} = -2$$

(b) The product of 5 and −6, **divided by** the difference between −7 and 8

The numerator of the fraction representing the division is obtained by multiplying 5 and −6. The denominator is found by subtracting −7 and 8.

$$\frac{5(-6)}{-7 - 8} = \frac{-30}{-15} = 2$$

Work Problem 10 at the Side.

7 **Translate simple sentences into equations.** In this section and the previous two sections, important words and phrases involving the four operations of arithmetic have been introduced. We can use these words and phrases to translate sentences into equations. This skill will be useful later when solving applied problems in Section 2.4.

Example 9 **Translating Sentences into Equations**

Write each sentence with symbols, using x to represent the number.

(a) Three **times** a number **is** −18.

The word *times* indicates multiplication, and the word *is* translates as the equals sign (=).

$$3x = -18$$

(b) The **sum** of a number and 9 **is** 12.

$$x + 9 = 12$$

(c) The **difference between** a number and 5 **is** 0.

$$x - 5 = 0$$

(d) The **quotient of** 24 and a number **is** −2.

$$\frac{24}{x} = -2$$

Work Problem 11 at the Side.

CAUTION

It is important to recognize the distinction between the types of problems found in Example 8 and Example 9. In Example 8, the phrases translate as *expressions,* while in Example 9, the sentences translate as *equations.* Remember that an equation is a sentence, while an expression is a phrase.

$$\frac{5(-6)}{-7 - 8} \qquad\qquad 3x = -18$$

Expression   Equation

ANSWERS

10. **(a)** $\dfrac{20}{8 + (-3)}$; 4

(b) $\dfrac{-9(2)}{5 - (-1)}$; −3

11. **(a)** $2x = -6$
(b) $-8 - x = -11$
(c) $5 + x = 8$
(d) $\dfrac{x}{-2} = 6$

1.6 **EXERCISES**

FOR EXTRA HELP

 Student's Solutions Manual   MyMathLab.com   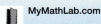 InterAct Math Tutorial Software   AW Math Tutor Center   www.mathxl.com   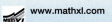 Digital Video Tutor CD 1 Videotape 3

Fill in each blank with one of the following: greater than 0, less than 0, equal to 0.

1. The product or the quotient of two numbers with the same sign is _____.

2. The product or the quotient of two numbers with different signs is _____.

3. If three negative numbers are multiplied together, the product is _____.

4. If two negative numbers are multiplied and then their product is divided by a negative number, the result is _____.

5. If a negative number is squared and the result is added to a positive number, the final answer is _____.

6. The reciprocal of a negative number is _____.

Find each product. See Examples 1 and 2.

7. $-7(4)$

8. $-8(5)$

9. $5(-6)$

10. $-4(-20)$

11. $-8(0)$

12. $0(-12)$

13. $-\dfrac{3}{8}\left(-\dfrac{20}{9}\right)$

14. $-\dfrac{5}{4}\left(-\dfrac{6}{25}\right)$

15. $-6.8(.35)$

16. $-4.6(.24)$

17. $-6\left(-\dfrac{1}{4}\right)$

18. $-8\left(-\dfrac{1}{2}\right)$

Find each quotient. See Examples 3 and 4.

19. $\dfrac{-15}{5}$

20. $\dfrac{-18}{6}$

21. $\dfrac{20}{-10}$

22. $\dfrac{28}{-4}$

23. $\dfrac{-160}{-10}$

24. $\dfrac{-260}{-20}$

25. $\dfrac{0}{-3}$

26. $\dfrac{-6}{0}$

27. $\dfrac{-10.252}{-.4}$

28. $\dfrac{-29.584}{-.8}$

29. $\left(-\dfrac{3}{4}\right) \div \left(-\dfrac{1}{2}\right)$

30. $\left(-\dfrac{3}{16}\right) \div \left(-\dfrac{5}{8}\right)$

31. Which expression is undefined?

A. $\dfrac{5-5}{5+5}$ B. $\dfrac{5+5}{5+5}$ C. $\dfrac{5-5}{5-5}$ D. $\dfrac{5-5}{5}$

32. What is the reciprocal of .4?

Perform each indicated operation. See Example 5.

33. $\dfrac{-5(-6)}{9-(-1)}$

34. $\dfrac{-12(-5)}{7-(-5)}$

35. $\dfrac{-21(3)}{-3-6}$

36. $\dfrac{-40(3)}{-2-3}$

37. $\dfrac{-10(2)+6(2)}{-3-(-1)}$

38. $\dfrac{8(-1)+6(-2)}{-6-(-1)}$

39. $\dfrac{-27(-2)-(-12)(-2)}{-2(3)-2(2)}$

40. $\dfrac{-13(-4)-(-8)(-2)}{(-10)(2)-4(-2)}$

41. $\dfrac{3^2-4^2}{7(-8+9)}$

42. Explain the method you would use to evaluate $3x+2y$ if $x=-3$ and $y=4$.

43. If x and y are both replaced by negative numbers, is the value of $4x+8y$ positive or negative? What about $4x-8y$?

Evaluate each expression if $x=6$, $y=-4$, and $a=3$. See Example 6.

44. $5x-2y+3a$

45. $6x-5y+4a$

46. $(2x+y)(3a)$

47. $(5x-2y)(-2a)$

48. $\left(\dfrac{1}{3}x-\dfrac{4}{5}y\right)\left(-\dfrac{1}{5}a\right)$

49. $\left(\dfrac{5}{6}x+\dfrac{3}{2}y\right)\left(-\dfrac{1}{3}a\right)$

50. $(-5+x)(-3+y)(3-a)$

51. $(6-x)(5+y)(3+a)$

52. $-2y^2+3a$

53. $5x - 4a^2$

54. $\dfrac{2y^2 - x}{a - 3}$

55. $\dfrac{xy + 8a}{x - y}$

Write a numerical expression for each phrase and simplify. See Examples 7 and 8.

56. The product of -9 and 2, added to 9

57. The product of 4 and -7, added to -12

58. Twice the product of -1 and 6, subtracted from -4

59. Twice the product of -8 and 2, subtracted from -1

60. The product of 12 and the difference between 9 and -8

61. The product of -3 and the difference between 3 and -7

62. Four-fifths of the sum of -8 and -2

63. Three-tenths of the sum of -2 and -28

64. The quotient of -12 and the sum of -5 and -1

65. The quotient of -20 and the sum of -8 and -2

66. The sum of 15 and -3, divided by the product of 4 and -3

67. The sum of -18 and -6, divided by the product of 2 and -4

68. The product of $-\dfrac{1}{2}$ and $\dfrac{3}{4}$, divided by $-\dfrac{2}{3}$

69. The product of $-\dfrac{2}{3}$ and $-\dfrac{1}{5}$, divided by $\dfrac{1}{7}$

Write each sentence with symbols, using x to represent the number. See Example 9.

70. Six times a number is -42.

71. Four times a number is -36.

72. The quotient of a number and 3 is -3.

73. The quotient of a number and 4 is -1.

74. 6 less than a number is 2.

75. 7 less than a number is 5.

76. When 15 is divided by a number, the result is -5.

77. When 6 is divided by a number, the result is -3.

RELATING CONCEPTS (Exercises 78–83)   **FOR INDIVIDUAL OR GROUP WORK**

To find the average of a group of numbers, we add the numbers and then divide the sum by the number of terms added. **Work Exercises 78–81 in order,** *to find the average of* 23, 18, 13, -4, *and* -8. *Then find the averages in Exercises 82 and 83.*

78. Find the sum of the given group of numbers.

79. How many numbers are in the group?

80. Divide your answer for Exercise 78 by your answer for Exercise 79. Give the quotient as a mixed number.

81. What is the average of the given group of numbers?

82. What is the average of all integers between -10 and 14, including both -10 and 14?

83. What is the average of the integers between -15 and -10, including -15 and -10?

1.7 PROPERTIES OF REAL NUMBERS

If you are asked to find the sum

$$3 + 89 + 97,$$

you might mentally add $3 + 97$ to get 100, and then add $100 + 89$ to get 189. While the order of operations guidelines say to add (or multiply) from left to right, the fact is we may change the order of the terms (or factors) and group them in any way we choose without affecting the sum (or product). This is an example of a shortcut we use in everyday mathematics that is justified by the properties of real numbers introduced in this section. In the following statements, *a*, *b*, and *c* represent real numbers.

1 Use the commutative properties. The word *commute* means to go back and forth. Many people commute to work or to school. If you travel from home to work and follow the same route from work to home, you travel the same distance each time. The **commutative properties** say that if two numbers are added or multiplied in any order, they give the same result.

$$a + b = b + a \quad \text{Addition}$$
$$ab = ba \quad \text{Multiplication}$$

Example 1 Using the Commutative Properties

Use a commutative property to complete each statement.

(a) $-8 + 5 = 5 + \underline{}$
  By the commutative property for addition, the missing number is -8 because $-8 + 5 = 5 + (-8)$.

(b) $-2(7) = \underline{} (-2)$
  By the commutative property for multiplication, the missing number is 7, since $-2(7) = 7(-2)$.

━━━ Work Problem **1** at the Side.

2 Use the associative properties. When we *associate* one object with another, we tend to think of those objects as being grouped together. The **associative properties** say that when we add or multiply three numbers, we can group them in any manner and get the same answer.

$$(a + b) + c = a + (b + c) \quad \text{Addition}$$
$$(ab)c = a(bc) \quad \text{Multiplication}$$

Example 2 Using the Associative Properties

Use an associative property to complete each statement.

(a) $8 + (-1 + 4) = (8 + \underline{}) + 4$
  The missing number is -1.

(b) $[2(-7)]6 = 2\underline{}$
  The missing expression on the right should be $[(-7)6]$.

━━━ Work Problem **2** at the Side.

OBJECTIVES

1 Use the commutative properties.
2 Use the associative properties.
3 Use the identity properties.
4 Use the inverse properties.
5 Use the distributive property.

1 Complete each statement. Use a commutative property.

(a) $x + 9 = 9 + \underline{}$

(b) $-12(4) = \underline{} (-12)$

(c) $5x = x \cdot \underline{}$

2 Complete each statement. Use an associative property.

(a) $(9 + 10) + (-3)$
  $= 9 + [\underline{} + (-3)]$

(b) $-5 + (2 + 8)$
  $= (\underline{}) + 8$

(c) $10[-8(-3)] = \underline{}$

❸ Decide whether each statement is an example of a commutative property, an associative property, or both.

(a) $2(4 \cdot 6) = (2 \cdot 4)6$

(b) $(2 \cdot 4)6 = (4 \cdot 2)6$

(c) $(2 + 4) + 6 = 4 + (2 + 6)$

By the associative property of addition, the sum of three numbers will be the same no matter how the numbers are "associated" in groups. For this reason, parentheses can be left out in many addition problems. For example, both

$$(-1 + 2) + 3 \quad \text{and} \quad -1 + (2 + 3)$$

can be written as

$$-1 + 2 + 3.$$

In the same way, parentheses also can be left out of many multiplication problems.

Example 3 Distinguishing between the Associative and Commutative Properties

(a) Is $(2 + 4) + 5 = 2 + (4 + 5)$ an example of the associative or the commutative property?

The order of the three numbers is the same on both sides of the equals sign. The only change is in the grouping, or association, of the numbers. Therefore, this is an example of the associative property.

(b) Is $6(3 \cdot 10) = 6(10 \cdot 3)$ an example of the associative or the commutative property?

The same numbers, 3 and 10, are grouped on each side. On the left, however, 3 appears first in $(3 \cdot 10)$. On the right, 10 appears first. Since the only change involves the order of the numbers, this statement is an example of the commutative property.

(c) Is $(8 + 1) + 7 = 8 + (7 + 1)$ an example of the associative or the commutative property?

In the statement, both the order and the grouping are changed. On the left, the order of the three numbers is 8, 1, and 7. On the right it is 8, 7, and 1. On the left, 8 and 1 are grouped, and on the right, 7 and 1 are grouped. Therefore, both the associative and the commutative properties are used.

Work Problem ❸ at the Side.

We can use the commutative and associative properties to simplify expressions.

❹ Find the sum:

$$5 + 18 + 29 + 31 + 12.$$

Example 4 Using the Commutative and Associative Properties

The commutative and associative properties make it possible to choose pairs of numbers that are easy to add or multiply.

(a) $23 + 41 + 2 + 9 + 25 = (41 + 9) + (23 + 2) + 25$
$$= 50 + 25 + 25$$
$$= 100$$

(b) $25(69)(4) = 25(4)(69)$
$$= 100(69)$$
$$= 6900$$

Work Problem ❹ at the Side.

3   **Use the identity properties.**   The identity or value of a real number is left unchanged when identity properties are applied. The **identity properties** say that the sum of 0 and any number equals that number, and the product of 1 and any number equals that number.

$$a + 0 = a \qquad \text{and} \qquad 0 + a = a$$
$$a \cdot 1 = a \qquad \text{and} \qquad 1 \cdot a = a$$

The number 0 leaves the identity, or value, of any real number unchanged by addition. For this reason, 0 is called the **identity element for addition** or the **additive identity.** Since multiplication by 1 leaves any real number unchanged, 1 is the **identity element for multiplication** or the **multiplicative identity.**

Example 5   **Using the Identity Properties**

These statements are examples of the identity properties.

(a) $-3 + 0 = -3$ **(b)** $1 \cdot 25 = 25$

=========== **Work Problem 5 at the Side.** ===========

We use the identity property for multiplication to write fractions in lowest terms and to get common denominators.

Example 6   **Using the Identity Element for Multiplication to Simplify Expressions**

Simplify each expression.

(a) $\dfrac{49}{35}$

$$\frac{49}{35} = \frac{7 \cdot 7}{5 \cdot 7} \qquad \text{Factor.}$$

$$= \frac{7}{5} \cdot \frac{7}{7} \qquad \text{Write as a product.}$$

$$= \frac{7}{5} \cdot 1 \qquad \text{Property of 1}$$

$$= \frac{7}{5} \qquad \text{Identity property}$$

(b) $\dfrac{3}{4} + \dfrac{5}{24}$

$$\frac{3}{4} + \frac{5}{24} = \frac{3}{4} \cdot 1 + \frac{5}{24} \qquad \text{Identity property}$$

$$= \frac{3}{4} \cdot \frac{6}{6} + \frac{5}{24} \qquad \text{Use } 1 = \tfrac{6}{6} \text{ to get a common denominator.}$$

$$= \frac{18}{24} + \frac{5}{24} \qquad \text{Multiply.}$$

$$= \frac{23}{24} \qquad \text{Add.}$$

=========== **Work Problem 6 at the Side.** ===========

---

5 Use an identity property to complete each statement.

(a) $9 + 0 =$ _____

(b) _____ $+ (-7) = -7$

(c) _____ $\cdot 1 = 5$

---

6 Use an identity property to simplify each expression.

(a) $\dfrac{85}{105}$

(b) $\dfrac{9}{10} - \dfrac{53}{50}$

ANSWERS
5. (a) 9   **(b)** 0   **(c)** 5
6. (a) $\dfrac{17}{21}$ **(b)** $-\dfrac{4}{25}$

7 Complete each statement so that it is an example of either an identity property or an inverse property. Tell which property is used.

(a) $-6 +$ _____ $= 0$

(b) $\dfrac{4}{3} \cdot$ _____ $= 1$

(c) $-\dfrac{1}{9} \cdot$ _____ $= 1$

(d) $275 +$ _____ $= 275$

---

4 ▭ **Use the inverse properties.** Each day before you go to work or school, you probably put on your shoes before you leave. Before you go to sleep at night, you probably take them off, and this leads to the same situation that existed before you put them on. These operations from everyday life are examples of inverse operations. The **inverse properties** of addition and multiplication lead to the additive and multiplicative identities, respectively. The opposite of a, $-a$, is the **additive inverse** of a and the reciprocal of a, $\frac{1}{a}$, is the **multiplicative inverse** of the nonzero number a. The sum of the numbers a and $-a$ is 0, and the product of the nonzero numbers a and $\frac{1}{a}$ is 1.

$$a + (-a) = 0 \quad \text{and} \quad -a + a = 0$$

$$a \cdot \frac{1}{a} = 1 \quad \text{and} \quad \frac{1}{a} \cdot a = 1 \quad (a \neq 0)$$

Example 7   Using the Inverse Properties

The following statements are examples of the inverse properties.

(a) $\dfrac{2}{3} \cdot \dfrac{3}{2} = 1$

(b) $(-5)\left(-\dfrac{1}{5}\right) = 1$

(c) $-\dfrac{1}{2} + \dfrac{1}{2} = 0$

(d) $4 + (-4) = 0$

Work Problem 7 at the Side.

5 ▭ **Use the distributive property.** The everyday meaning of the word *distribute* is "to give out from one to several." An important property of real number operations involves this idea.

Look at the value of the following expressions.

$$2(5 + 8) = 2(13) = 26$$
$$2(5) + 2(8) = 10 + 16 = 26$$

Since both expressions equal 26,

$$2(5 + 8) = 2(5) + 2(8).$$

This result is an example of the *distributive property*, the only property involving *both* addition and multiplication. With this property, a product can be changed to a sum or difference. This idea is illustrated by the divided rectangle in Figure 13.

The area of the left part is $2(5) = 10$.
The area of the right part is $2(8) = 16$.
The total area is $2(5 + 8) = 26$ or the total area is
$2(5) + 2(8) = 10 + 16 = 26$.
Thus, $2(5 + 8) = 2(5) + 2(8)$.

Figure 13

The **distributive property** says that multiplying a number a by a sum of numbers $b + c$ gives the same result as multiplying a by b and a by c and then adding the two products.

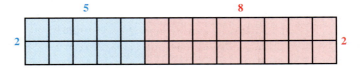

$$a(b + c) = ab + ac \quad \text{and} \quad (b + c)a = ba + ca$$

As the arrows show, the a outside the parentheses is "distributed" over the b and c inside. The distributive property is also valid for subtraction.

$$a(b - c) = ab - ac \quad \text{and} \quad (b - c)a = ba - ca$$

The distributive property also can be extended to the sum (or difference) of more than two numbers.

$$a(b + c + d) = ab + ac + ad$$

Example 8 **Using the Distributive Property**

Use the distributive property to rewrite each expression.

(a) $5(9 + 6) = 5 \cdot 9 + 5 \cdot 6$ ⠀⠀Distributive property

$\qquad\qquad = 45 + 30$ ⠀⠀⠀Multiply.

$\qquad\qquad = 75$ ⠀⠀⠀⠀⠀Add.

(b) $4(x + 5 + y) = 4x + 4 \cdot 5 + 4y$ ⠀⠀Distributive property

$\qquad\qquad\quad = 4x + 20 + 4y$ ⠀⠀Multiply.

(c) $-2(x + 3) = -2x + (-2)(3)$ ⠀⠀Distributive property

$\qquad\qquad = -2x + (-6)$ ⠀⠀Multiply.

$\qquad\qquad = -2x - 6$

(d) $3(k - 9) = 3k - 3 \cdot 9$ ⠀⠀Distributive property

$\qquad\qquad = 3k - 27$ ⠀⠀Multiply.

(e) $6 \cdot 8 + 6 \cdot 2$

The distributive property says that $a(b + c) = ab + ac$. This can be reversed to read $ab + ac = a(b + c)$. We use this form of the distributive property to write a sum like $6 \cdot 8 + 6 \cdot 2$ with a common factor (of 6) as a product.

$$ab + ac = a(b + c) \quad \text{Distributive property}$$

$$6 \cdot 8 + 6 \cdot 2 = 6(8 + 2) \quad \text{Let } a = 6, b = 8, \text{ and } c = 2.$$

$$\qquad\qquad = 6(10)$$

$$\qquad\qquad = 60$$

(f) $8(3r + 11t + 5z) = 8(3r) + 8(11t) + 8(5z)$ ⠀⠀Distributive property

$\qquad\qquad = (8 \cdot 3)r + (8 \cdot 11)t + (8 \cdot 5)z$ ⠀⠀Associative property

$\qquad\qquad = 24r + 88t + 40z$

Work Problem **8** at the Side.

The symbol $-a$ may be interpreted as $-1 \cdot a$. Similarly, when a negative sign precedes an expression within parentheses, it may also be interpreted as a factor of -1. Thus, we can use the distributive property to remove the parentheses from expressions such as $-(2y + 3)$. We do this by first writing $-(2y + 3)$ as $-1 \cdot (2y + 3)$.

$$-(2y + 3) = -1 \cdot (2y + 3) \quad -a = -1 \cdot a$$

$$= -1 \cdot (2y) + (-1) \cdot (3) \quad \text{Distributive property}$$

$$= -2y - 3 \quad \text{Multiply.}$$

8 Use the distributive property to rewrite each expression.

(a) $2(p + 5)$

(b) $-4(y + 7)$

(c) $5(m - 4)$

(d) $9 \cdot k + 9 \cdot 5$

(e) $3a - 3b$

(f) $7(2y + 7k - 9m)$

9 Write without parentheses.

(a) $-(3k - 5)$

(b) $-(2 - r)$

(c) $-(-5y + 8)$

(d) $-(-z + 4)$

Example 9 **Using the Distributive Property to Remove Parentheses**

Write without parentheses.

(a) $-(7r - 8) = \textbf{-1}(7r) + (\textbf{-1})(-8)$ Distributive property
$$= -7r + 8 \qquad \text{Multiply.}$$

(b) $-(-9w + 2) = -1(-9w + 2)$
$$= 9w - 2$$

Work Problem ❾ at the Side.

The properties discussed here are the basic properties of real numbers that justify how we add and multiply in algebra. You should know them by name because we will be referring to them frequently. Here is a summary of these properties.

Properties of Addition and Multiplication

For any real numbers a, b, and c, the following properties hold.

Commutative properties $a + b = b + a \qquad ab = ba$

Associative properties $(a + b) + c = a + (b + c)$
$$(ab)c = a(bc)$$

Identity properties   There is a real number 0 such that
$$a + 0 = a \quad \text{and} \quad 0 + a = a.$$
There is a real number 1 such that
$$a \cdot 1 = a \quad \text{and} \quad 1 \cdot a = a.$$

Inverse properties   For each real number a, there is a single real number $-a$ such that
$$a + (-a) = 0 \quad \text{and} \quad (-a) + a = 0.$$
For each nonzero real number a, there is a single real number $\frac{1}{a}$ such that
$$a \cdot \frac{1}{a} = 1 \quad \text{and} \quad \frac{1}{a} \cdot a = 1.$$

Distributive property $a(b + c) = ab + ac$
$$(b + c)a = ba + ca$$

ANSWERS
9. (a) $-3k + 5$ (b) $-2 + r$
(c) $5y - 8$ (d) $z - 4$

1.7   **EXERCISES**

FOR EXTRA HELP

Student's Solutions Manual

MyMathLab.com

InterAct Math Tutorial Software

AW Math Tutor Center

www.mathxl.com

Digital Video Tutor CD 1 Videotape 3

Match each item in Column I with the correct choice from Column II. Choices may be used once, more than once, or not at all.

I

1. Identity element for addition

2. Identity element for multiplication

3. Additive inverse of a

4. Multiplicative inverse, or reciprocal, of the nonzero number a

5. The only number that has no multiplicative inverse

6. An example of the associative property

7. An example of the commutative property

8. An example of the distributive property

II

A. $(5 \cdot 4) \cdot 3 = 5 \cdot (4 \cdot 3)$

B. 0

C. $-a$

D. -1

E. $5 \cdot 4 \cdot 3 = 60$

F. 1

G. $(5 \cdot 4) \cdot 3 = 3 \cdot (5 \cdot 4)$

H. $5(4 + 3) = 5 \cdot 4 + 5 \cdot 3$

I. $\dfrac{1}{a}$

Decide whether each statement is an example of the commutative, associative, identity, inverse, or distributive property. See Examples 1, 2, 3, and 5–8.

9. $\dfrac{2}{3}(-4) = -4\left(\dfrac{2}{3}\right)$

10. $6\left(-\dfrac{5}{6}\right) = \left(-\dfrac{5}{6}\right)6$

11. $-6 + (12 + 7) = (-6 + 12) + 7$

12. $(-8 + 13) + 2 = -8 + (13 + 2)$

13. $-6 + 6 = 0$

14. $12 + (-12) = 0$

15. $\left(\dfrac{2}{3}\right)\left(\dfrac{3}{2}\right) = 1$

16. $\left(\dfrac{5}{8}\right)\left(\dfrac{8}{5}\right) = 1$

17. $2.34 \cdot 1 = 2.34$

18. $-8.456 \cdot 1 = -8.456$

19. $(4 + 17) + 3 = 3 + (4 + 17)$

20. $(-8 + 4) + (-12) = -12 + (-8 + 4)$

21. $6(x + y) = 6x + 6y$

22. $14(t + s) = 14t + 14s$

23. $-\dfrac{5}{9} = -\dfrac{5}{9} \cdot \dfrac{3}{3} = -\dfrac{15}{27}$

24. $\dfrac{13}{12} = \dfrac{13}{12} \cdot \dfrac{7}{7} = \dfrac{91}{84}$

25. $5(2x) + 5(3y) = 5(2x + 3y)$

26. $3(5t) - 3(7r) = 3(5t - 7r)$

27. What number(s) satisfy each condition? **(a)** a number that is its own additive inverse **(b)** two numbers that are their own multiplicative inverses

28. The distributive property holds for multiplication with respect to addition. Is there a distributive property for addition with respect to multiplication? That is, does $a + b \cdot c = (a + b)(a + c)$? If not, give an example to show why.

29. Evaluate $25 - (6 - 2)$ and $(25 - 6) - 2$. Use the results to explain why subtraction is or is not associative.

30. Suppose that a student shows you the following work.
$$-2(5 - 6) = -2(5) - 2(6) = -10 - 12 = -22$$
The student has made a very common error. Explain the error and then work the problem correctly.

Write a new expression that is equal to the given expression, using the given property. Then simplify the new expression if possible. See Examples 1, 2, 5, 7, and 8.

31. $r + 7$; commutative

32. $t + 9$; commutative

33. $s + 0$; identity

34. $w + 0$; identity

35. $-6(x + 7)$; distributive

36. $-5(y + 2)$; distributive

37. $(w + 5) + (-3)$; associative

38. $(b + 8) + (-10)$; associative

39. Explain how the procedure of changing $\frac{3}{4}$ to $\frac{9}{12}$ requires the use of the multiplicative identity element, 1.

Use the properties of this section to simplify each expression. See Example 4.

40. $26 + 8 - 26 + 12$

41. $-\frac{3}{8} + \frac{2}{5} + \frac{8}{5} + \frac{3}{8}$

42. $\frac{9}{7}(-.38)\left(\frac{7}{9}\right)$

Use the distributive property to rewrite each expression. Simplify if possible. See Example 8.

43. $5 \cdot 3 + 5 \cdot 17$

44. $15 \cdot 6 + 5 \cdot 6$

45. $4(t + 3)$

46. $5(w + 4)$

47. $-8(r + 3)$

48. $-11(x + 4)$

49. $-5(y - 4)$

50. $-9(g - 4)$

51. $-\frac{4}{3}(12y + 15z)$

52. $-\frac{2}{5}(10b + 20a)$

53. $8 \cdot z + 8 \cdot w$

54. $4 \cdot s + 4 \cdot r$

55. $7(2v) + 7(5r)$

56. $13(5w) + 13(4p)$

57. $8(3r + 4s - 5y)$

58. $2(5u - 3v + 7w)$

59. $-3(8x + 3y + 4z)$

60. $-5(2x - 5y + 6z)$

Use the distributive property to write each expression without parentheses. See Example 9.

61. $-(4t + 5m)$

62. $-(9x + 12y)$

63. $-(-5c - 4d)$

64. $-(-13x - 15y)$

65. $-(-3q + 5r - 8s)$

66. $-(-4z + 5w - 9y)$

67. "Getting out of bed" and "taking a shower" are not commutative. Give an example of another pair of everyday activities that are not commutative.

68. Are "going upstairs" and "going downstairs" commutative?

69. True or false: "preparing a meal" and "eating a meal" are commutative.

70. The phrase "dog biting man" has two different meanings, depending on how the words are associated.

$$(\text{dog biting}) \text{ man} \qquad \text{or} \qquad \text{dog (biting man)}$$

Give another example of a three-word phrase that has different meanings depending on how the words are associated.

71. Use parentheses to show how the associative property can be used to give two different meanings to "foreign sales clerk."

72. Use parentheses to show two different meanings for "new cook book."

RELATING CONCEPTS (Exercises 73–76)   **FOR INDIVIDUAL OR GROUP WORK**

In Section 1.6 we used a pattern to see that the product of two negative numbers is a positive number. In the exercises that follow, we show another justification for determining the sign of the product of two negative numbers. **Work Exercises 73–76 in order.**

73. Evaluate the expression $-3[5 + (-5)]$ by using the order of operations.

74. Write the expression in Exercise 73 using the distributive property. Do not simplify the products.

75. The product $-3(5)$ should be one of the terms you wrote when answering Exercise 74. Based on the results in Section 1.6, what is this product?

76. In Exercise 73, you should have obtained 0 as the answer. Now, consider the following, using the results of Exercises 73 and 75.

$$-3[5 + (-5)] = -3(5) + (-3)(-5)$$
$$0 = -15 + ?$$

The question mark represents the product $-3(-5)$. When added to -15, it must give a sum of 0. Therefore, $-3(-5)$ must equal what?

1.8 SIMPLIFYING EXPRESSIONS

1 **Simplify expressions.** In this section, we show how to simplify expressions using the properties of addition and multiplication introduced in the previous section.

OBJECTIVES

1 Simplify expressions.

2 Identify terms and numerical coefficients.

3 Identify like terms.

4 Combine like terms.

5 Simplify expressions from word phrases.

Example 1 **Simplifying Expressions**

Simplify each expression.

(a) $4x + 8 + 9$
Since $8 + 9 = 17$, $4x + \mathbf{8 + 9} = 4x + \mathbf{17}$.

(b) $4(3m - 2n)$
Use the distributive property.

$$4(3m - 2n) = \mathbf{4}(3m) - \mathbf{4}(2n)$$
$$= 12m - 8n$$

(c) $6 + 3(4k + 5) = 6 + 3(4k) + 3(5)$ Distributive property
$$= 6 + 12k + 15 \qquad \text{Multiply.}$$
$$= 21 + 12k \qquad \text{Add.}$$

(d) $5 - (2y - 8) = 5 - \mathbf{1}(2y - 8)$ $-a = -1 \cdot a$
$$= 5 - 2y + 8 \qquad \text{Distributive property}$$
$$= 13 - 2y \qquad \text{Add.}$$

NOTE

Although the steps were not shown, in Examples 1(c) and 1(d) we mentally used the commutative and associative properties to add in the last step. In practice, these steps are usually left out, but we should realize that they are used whenever the ordering in a sum is rearranged.

Work Problem ❶ at the Side.

2 **Identify terms and numerical coefficients.** A **term** is a number, a variable, or a product or quotient of a number and one or more variables raised to powers. Examples of terms include

$$-9x^2, \quad 15y, \quad -3, \quad 8m^2n, \quad \frac{2}{p}, \quad \text{and} \quad k.$$

The **numerical coefficient,** or simply coefficient, of the term $9m$ is 9; the numerical coefficient of $-15x^3y^2$ is -15; the numerical coefficient of x is 1; and the numerical coefficient of 8 is 8. In the expression $\frac{x}{3}$, the numerical coefficient of x is $\frac{1}{3}$ since $\frac{x}{3} = \frac{1x}{3} = \frac{1}{3}x$.

CAUTION

It is important to be able to distinguish between *terms* and *factors*. For example, in the expression $8x^3 + 12x^2$, there are two terms, $8x^3$ and $12x^2$. Terms are separated by a $+$ or $-$ sign. On the other hand, in the one-term expression $(8x^3)(12x^2)$, $8x^3$ and $12x^2$ are *factors*. Factors are multiplied.

❶ Simplify each expression.

(a) $9k + 12 - 5$

(b) $7(3p + 2q)$

(c) $2 + 5(3z - 1)$

(d) $-3 - (2 + 5y)$

ANSWERS
1. (a) $9k + 7$ **(b)** $21p + 14q$
   (c) $15z - 3$ **(d)** $-5 - 5y$

② Give the numerical coefficient of each term.

(a) $15q$

(b) $-2m^3$

(c) $-18m^7q^4$

(d) $-r$

(e) $\dfrac{5x}{4}$

③ Identify each pair of terms as *like* or *unlike*.

(a) $9x, 4x$

(b) $-8y^3, 12y^2$

(c) $5x^2y^4, 5x^4y^2$

(d) $7x^2y^4, -7x^2y^4$

(e) $13kt, 4tk$

Here are some examples of terms and their numerical coefficients.

Term	Numerical Coefficient
$-7y$	-7
$34r^3$	34
$-26x^5yz^4$	-26
$-k$	-1
r	1
$\dfrac{3x}{8} = \dfrac{3}{8}x$	$\dfrac{3}{8}$

Work Problem ② at the Side.

3 ▭ **Identify like terms.** Terms with exactly the same variables (including the same exponents) are called **like terms.** For example, $9m$ and $4m$ have the same variables and are like terms. Also, $6x^3$ and $-5x^3$ are like terms. The terms $-4y^3$ and $4y^2$ have different exponents and are **unlike terms.** Here are some additional examples.

     Like terms $5x$ and $-12x$ $3x^2y$ and $5x^2y$
   Unlike terms $4xy^2$ and $5xy$ $8x^2y^3$ and $7x^3y^2$

Work Problem ③ at the Side.

4 ▭ **Combine like terms.** Recall the distributive property:

$$x(y + z) = xy + xz.$$

As seen in the previous section, this statement can also be written "backward" as

$$xy + xz = x(y + z).$$

This form of the distributive property may be used to find the sum or difference of like terms. For example,

$$3x + 5x = (3 + 5)x = 8x.$$

This process is called **combining like terms.**

CAUTION

Remember that *only like terms* may be combined.

Example 2   **Combining Like Terms**

Combine like terms in each expression.

(a) $9m + 5m$
   Use the distributive property as given above.

$$9m + 5m = (9 + 5)m = 14m$$

(b) $6r + 3r + 2r = (6 + 3 + 2)r = 11r$ Distributive property

(c) $\dfrac{3}{4}x + x = \dfrac{3}{4}x + 1x = \left(\dfrac{3}{4} + 1\right)x = \dfrac{7}{4}x$ (Note: $x = 1x$.)

Continued on Next Page

(d) $16y^2 - 9y^2 = (16 - 9)y^2 = 7y^2$

(e) $32y + 10y^2$ cannot be combined because $32y$ and $10y^2$ are unlike terms. The distributive property cannot be used here to combine coefficients.

━━━━━━━━━ Work Problem ❹ at the Side.

When an expression involves parentheses, the distributive property is used both "forward" and "backward" to combine like terms, as shown in the following example.

Example 3 Simplifying Expressions Involving Like Terms

Simplify each expression.

(a) $14y + \mathbf{2(6 + 3y)} = 14y + \mathbf{2(6)} + \mathbf{2(3y)}$ Distributive property

$\qquad\qquad\qquad = 14y + \mathbf{12} + \mathbf{6y}$ Multiply.

$\qquad\qquad\qquad = \mathbf{20y} + 12$ Combine like terms.

(b) $9k - 6 \mathbf{- 3(2 - 5k)} = 9k - 6 \mathbf{- 3(2)} \mathbf{- 3(-5k)}$ Distributive property

$\qquad\qquad\qquad = 9k - 6 - 6 + 15k$ Multiply.

$\qquad\qquad\qquad = 24k - 12$ Combine like terms.

(c) $\mathbf{-(2 - r)} + 10r = \mathbf{-1(2 - r)} + 10r$ $-(2 - r) = -1(2 - r)$

$\qquad\qquad\qquad = -1(2) - 1(-r) + 10r$ Distributive property

$\qquad\qquad\qquad = -2 + r + 10r$ Multiply.

$\qquad\qquad\qquad = -2 + 11r$ Combine like terms.

(d) $5(2a^2 - 6a) - 3(4a^2 - 9) = 10a^2 - 30a - 12a^2 + 27$ Distributive property

$\qquad\qquad\qquad\qquad\qquad = -2a^2 - 30a + 27$ Combine like terms.

━━━━━━━━━ Work Problem ❺ at the Side.

5   **Simplify expressions from word phrases.** Earlier we saw how to translate words, phrases, and statements into expressions and equations. Now we can simplify translated expressions by combining like terms.

Example 4 Translating Words into a Mathematical Expression

Write the following phrase as a mathematical expression and simplify: four times a number, subtracted from the sum of twice the number and 4.

Let x represent the number.

The sum of twice the number and 4   Four times the number

$(2x + 4) - 4x$ Write with symbols.

which simplifies to

$-2x + 4$ Combine like terms.

━━━━━━━━━ Work Problem ❻ at the Side.

CAUTION

In Example 4, we are dealing with an expression to be simplified, *not* an equation to be solved.

---

❹ Combine like terms.

(a) $4k + 7k$

(b) $4r - r$

(c) $5z + 9z - 4z$

(d) $8p + 8p^2$

(e) $5x - 3y + 2x - 5y - 3$

❺ Simplify.

(a) $10p + 3(5 + 2p)$

(b) $7z - 2 - (1 + z)$

(c) $-(3k^2 + 5k) + 7(k^2 - 4k)$

❻ Write each phrase as a mathematical expression, and simplify by combining like terms.

(a) Three times a number, subtracted from the sum of the number and 8

(b) Twice a number added to the sum of 6 and the number

Answers
4. (a) $11k$ **(b)** $3r$ **(c)** $10z$
    (d) cannot be combined **(e)** $7x - 8y - 3$
5. (a) $16p + 15$ **(b)** $6z - 3$ **(c)** $4k^2 - 33k$
6. (a) $(x + 8) - 3x; -2x + 8$
    (b) $2x + (6 + x); 3x + 6$

Real-Data Applications

Algebraic Expressions and Tuition Costs

Algebraic expressions are useful in real-life scenarios in which the same set of instructions are repeated for different choices of numbers. Below is the description of how tuition and fees are calculated for "Resident of District" students at North Harris Montgomery Community College District (NHMCCD) in Texas for 2000–2001. The information is given in the college's schedule and can be found at the Web site www.nhmccd.edu.[*]

> ***Fees Required at NHMCCD***
>
> [*Residents of the district pay*] tuition at the rate of $26 per credit hour, a $4 per credit hour technology fee, and a registration fee of $12.

For Group Discussion

1. Calculate the tuition and fees for a student who is a resident of the district and who enrolls at NHMCCD for the specified number of credit hours. Let x represent the number of credit hours. Pay attention to the *process* used in your calculations so that you can write the algebraic expression for x credit hours.

   (a) 3 credit hours: _____   **(b)** 9 credit hours: _____

   (c) 12 credit hours: _____   **(d)** x credit hours: _____ dollars

Write the algebraic expression that represents the tuition and fees for each institution for one semester. Let x represent the number of credit hours. If you have difficulty, first calculate the costs for 3 or 9 credit hours and focus on the process that you used to get the answer.

2. American River College, California (nonresident student) www.arc.losrios.cc.ca.us[*]
   Enrollment: $11 per unit; parking: $30 per semester; an additional nonresident enrollment: $134 per unit; other fees: $8

3. Austin Community College, Texas (out-of-district student) www.austin.cc.tx.us[*]
   Tuition: $31 per credit hour; parking: $10 per year; an additional out-of-district tuition: $75 per credit hour; student service fee: $3

4. Valdosta State University, Georgia (in-state student) www.valdosta.edu[*]
   Matriculation: $78 per credit hour; health fee: $66; student services fee: $78; athletics fees: $97; technology fee: $38; parking fee: $25; special fees also apply.

5. Your college tuition and fees

---

[*]**Note** that URLs sometimes change, although that is unlikely for academic institutions. If the Web address given does not work, use a search engine, such as www.yahoo.com, to find the new URL.

1.8 EXERCISES

Decide whether each statement is true *or* false.

1. $6t + 5t^2 = 11t^3$

2. $9xy^2 - 3x^2y = 6xy$

3. $8r^2 + 3r - 12r^2 + 4r = -4r^2 + 7r$

4. $4 + 3t^3 = 7t^3$

In Exercises 5–8, choose the letter of the correct response.

5. Which is true for all real numbers x?
  A. $6 + 2x = 8x$ **B.** $6 - 2x = 4x$
  C. $6x - 2x = 4x$ **D.** $3 + 8(4x - 6) = 11(4x - 6)$

6. Which is an example of a pair of like terms?
  A. $6t, 6w$ **B.** $-8x^2y, 9xy^2$
  C. $5ry, 6yr$ **D.** $-5x^2, 2x^3$

7. Which is an example of a term with numerical coefficient 5?
  A. $5x^3y^7$ **B.** x^5 **C.** $\dfrac{x}{5}$ **D.** 5^2xy^3

8. Which is a correct translation for "six times a number, subtracted from the product of eleven and the number" (if x represents the number)?
  A. $6x - 11x$ **B.** $11x - 6x$
  C. $(11 + x) - 6x$ **D.** $6x - (11 + x)$

Simplify each expression. See Example 1.

9. $4r + 19 - 8$

10. $7t + 18 - 4$

11. $5 + 2(x - 3y)$

12. $8 + 3(s - 6t)$

13. $-2 - (5 - 3p)$

14. $-10 - (7 - 14r)$

Give the numerical coefficient of each term.

15. $-12k$ **16.** $-23y$ **17.** $5m^2$ **18.** $-3n^6$ **19.** xw

20. pq **21.** $-x$ **22.** $-t$ **23.** 74 **24.** 98

25. Give an example of a pair of like terms with the variable x, such that one of them has a negative numerical coefficient, one has a positive numerical coefficient, and their sum has a positive numerical coefficient.

26. Give an example of a pair of unlike terms such that each term has x as the only variable.

Identify each group of terms as like *or* unlike.

27. $8r, -13r$ **28.** $-7a, 12a$ **29.** $5z^4, 9z^3$ **30.** $8x^5, -10x^3$

31. $4, 9, -24$ **32.** $7, 17, -83$ **33.** x, y **34.** t, s

35. There is an old saying "You can't add apples and oranges." Explain how this saying can be applied to Objective 3 in this section.

36. Explain how the distributive property is used in combining $6t + 5t$ to get $11t$.

Simplify each expression. See Examples 2 and 3.

37. $5 - 2(x - 3)$

38. $-8 - 3(2x + 4)$

39. $-\dfrac{4}{3} + 2t + \dfrac{1}{3}t - 8 - \dfrac{8}{3}t$

40. $-\dfrac{5}{6} + 8x + \dfrac{1}{6}x - 7 - \dfrac{7}{6}$

41. $-5.3r + 4.9 - (2r + .7) + 3.2r$

42. $2.7b + 5.8 - (3b + .5) - 4.4b$

43. $2y^2 - 7y^3 - 4y^2 + 10y^3$

44. $9x^4 - 7x^6 + 12x^4 + 14x^6$

45. $13p + 4(4 - 8p)$

46. $5x + 3(7 - 2x)$

47. $-\dfrac{4}{3}(y - 12) - \dfrac{1}{6}y$

48. $-\dfrac{7}{5}(t - 15) - \dfrac{3}{2}$

49. $-5(5y - 9) + 3(3y + 6)$

50. $-3(2t + 4) + 8(2t - 4)$

Write each phrase as a mathematical expression. Use x to represent the number.
Combine like terms when possible. See Example 4.

51. Five times a number, added to the sum of the number and three

52. Six times a number, added to the sum of the number and six

53. A number multiplied by -7, subtracted from the sum of 13 and six times the number

54. A number multiplied by 5, subtracted from the sum of 14 and eight times the number

55. Six times a number added to -4, subtracted from twice the sum of three times the number and 4

56. Nine times a number added to 6, subtracted from triple the sum of 12 and 8 times the number

57. Write the expression $9x - (x + 2)$ using words, as in Exercises 51–56.

58. Write the expression $2(3x + 5) - 2(x + 4)$ using words, as in Exercises 51–56.

RELATING CONCEPTS (Exercises 59–62)   **FOR INDIVIDUAL OR GROUP WORK**

A manufacturer has fixed costs of $1000 to produce widgets. Each widget costs $5 to make. The fixed cost to produce gadgets is $750, and each gadget costs $3 to make.
Work Exercises 59–62 in order.

59. Write an expression for the cost to make x widgets. (*Hint:* The cost will be the sum of the fixed cost and the cost per item times the number of items.)

60. Write an expression for the cost to make y gadgets.

61. Write an expression for the total cost to make x widgets and y gadgets.

62. Simplify the expression you wrote in Exercise 61.

SUMMARY

Study Skills Workbook
Activity 6

KEY TERMS

1.1	**exponent**	An exponent, or **power**, is a number that indicates how many times a factor is repeated.

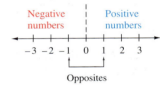

	base	The base is the number that is a repeated factor when written with an exponent.
	exponential expression	A number written with an exponent is an exponential expression.
1.2	**variable**	A variable is a symbol, usually a letter, used to represent an unknown number.
	algebraic expression	An algebraic expression is a collection of numbers, variables, operation symbols, and grouping symbols.
	equation	An equation is a statement that says two expressions are equal.
	solution	A solution of an equation is any value of the variable that makes the equation true.
1.3	**whole numbers**	The set of whole numbers is $\{0, 1, 2, 3, 4, 5, \ldots\}$.
	natural numbers	The set of natural numbers is $\{1, 2, 3, 4, \ldots\}$.
	number line	The number line shows the ordering of the real numbers on an infinite line.
	opposite	The opposite of a number a is the number that is the same distance from 0 on the number line as a, but on the opposite side of 0. This number is also called the **negative** of a or the **additive inverse** of a.
	integers	The set of integers is $\{\ldots, -3, -2, -1, 0, 1, 2, 3, \ldots\}$.
	negative number	A negative number is located to the *left* of 0 on the number line.
	positive number	A positive number is located to the *right* of 0 on the number line.
	rational numbers	A rational number is a number that can be written as the quotient of two integers, with denominator not 0.
	coordinate	The number that corresponds to a point on the number line is the coordinate of that point.
	irrational numbers	An irrational number is a real number that is not a rational number.
	real numbers	Real numbers are numbers that can be represented by points on the number line, or all rational and irrational numbers.
	absolute value	The absolute value of a number is the distance between 0 and the number on the number line.
1.4	**sum**	The answer to an addition problem is called the sum.
1.5	**difference**	The answer to a subtraction problem is called the difference.
1.6	**product**	The answer to a multiplication problem is called the product.
	reciprocal	Pairs of numbers whose product is 1 are called reciprocals or **multiplicative inverses** of each other.
	quotient	The answer to a division problem is called the quotient.
1.7	**identity element for addition**	When the identity element for addition, which is 0, is added to a number, the number is unchanged.
	identity element for multiplication	When a number is multiplied by the identity element for multiplication, which is 1, the number is unchanged.
1.8	**term**	A term is a number, a variable, or a product or quotient of a number and one or more variables raised to powers.
	numerical coefficient	The numerical factor in a term is its numerical coefficient.
	like terms	Terms with exactly the same variables (including the same exponents) are called like terms.

NEW SYMBOLS

a^n	n factors of a	$a(b), (a)b, (a)(b), a \cdot b,$ or ab	a times b
$=$	is equal to	$\dfrac{a}{b}, a/b,$ or $a \div b$	a divided by b
$\neq$	is not equal to	$\{\ \}$	set braces
$<$	is less than	$|x|$	absolute value of x
$\leq$	is less than or equal to	$\dfrac{1}{x}$	the multiplicative inverse or reciprocal of x ($x \neq 0$)
$>$	is greater than		
$\geq$	is greater than or equal to		

TEST YOUR WORD POWER

See how well you have learned the vocabulary in this chapter. Answers follow the Quick Review.

1. The **product** is
   (a) the answer in an addition problem
   (b) the answer in a multiplication problem
   (c) one of two or more numbers that are added to get another number
   (d) one of two or more numbers that are multiplied to get another number.

2. A number is **prime** if
   (a) it cannot be factored
   (b) it has just one factor
   (c) it has only itself and 1 as factors
   (d) it has at least two different factors.

3. An **exponent** is
   (a) a symbol that tells how many numbers are being multiplied
   (b) a number raised to a power
   (c) a number that tells how many times a factor is repeated
   (d) one of two or more numbers that are multiplied.

4. A **variable** is
   (a) a symbol used to represent an unknown number
   (b) a value that makes an equation true
   (c) a solution of an equation
   (d) the answer in a division problem.

5. An **integer** is
   (a) a positive or negative number
   (b) a natural number, its opposite, or zero
   (c) any number that can be graphed on a number line
   (d) the quotient of two numbers.

6. A **coordinate** is
   (a) the number that corresponds to a point on a number line
   (b) the graph of a number
   (c) any point on a number line
   (d) the distance from 0 on a number line.

7. The **absolute value** of a number is
   (a) the graph of the number
   (b) the reciprocal of the number
   (c) the opposite of the number
   (d) the distance between 0 and the number on a number line.

8. A **term** is
   (a) a numerical factor
   (b) a number, a variable, or a product or quotient of numbers and variables raised to powers
   (c) one of several variables with the same exponents
   (d) a sum of numbers and variables raised to powers.

9. A **numerical coefficient** is
   (a) the numerical factor in a term
   (b) the number of terms in an expression
   (c) a variable raised to a power
   (d) the variable factor in a term.

Concepts	*Examples*

1.1 *Exponents, Order of Operations, and Inequality*

Order of Operations

If necessary, simplify within parentheses or brackets and above and below fraction bars, using the following steps.

Step 1   Apply all exponents.

Step 2   Do any multiplications or divisions from left to right.

Step 3   Do any additions or subtractions from left to right.

$$\frac{9(2+6)}{2} - 2(2^3 + 3) = 36 - 2(8 + 3)$$
$$= 36 - 2(11)$$
$$= 36 - 22$$
$$= 14$$

1.2 *Variables, Expressions, and Equations*

Evaluate an expression with a variable by substituting a given number for the variable.

Evaluate $2x + y^2$ if $x = 3$ and $y = -4$.
$$2x + y^2 = 2(3) + (-4)^2$$
$$= 6 + 16$$
$$= 22$$

Values of a variable that make an equation true are solutions of the equation.

Is 2 a solution of $5x + 3 = 18$?
$$5(2) + 3 = 18$$
$$13 = 18 \quad \text{False}$$

2 is not a solution.

1.3 *Real Numbers and the Number Line*

Ordering Real Numbers

a is less than b if a is to the left of b on the number line.

$$-2 < 3 \qquad 3 > 0 \qquad 0 < 3$$

The opposite or additive inverse of a is $-a$.

$$-(5) = -5 \qquad -(-7) = 7 \qquad -0 = 0$$

The absolute value of a, $|a|$, is the distance between a and 0 on the number line.

$$|13| = 13 \qquad |0| = 0 \qquad |-5| = 5$$

1.4 *Addition of Real Numbers*

To add two numbers with the same sign, add their absolute values. The sum has that same sign.

$$9 + 4 = 13$$
$$-8 + (-5) = -13$$

To add two numbers with different signs, subtract their absolute values. The sum has the sign of the number with larger absolute value.

$$7 + (-12) = -5$$
$$-5 + 13 = 8$$

1.5 *Subtraction of Real Numbers*

To subtract signed numbers:

Change the subtraction symbol to addition, and change the sign of the number being subtracted. Add as in the previous section.

$$5 - (-2) = 5 + 2 = 7$$
$$-3 - 4 = -3 + (-4) = -7$$
$$-2 - (-6) = -2 + 6 = 4$$

Concepts	Examples

1.6 Multiplication and Division of Real Numbers

The product (or quotient) of two numbers having the *same sign* is *positive*; the product (or quotient) of two numbers having *different signs* is *negative*.

$6 \cdot 5 = 30$ $(-7)(-8) = 56$

$\dfrac{10}{2} = 5$ $\dfrac{-24}{-6} = 4$

$-6(5) = -30$ $6(-5) = -30$

$-18 \div 9 = \dfrac{-18}{9} = -2$ $49 \div (-7) = \dfrac{49}{-7} = -7$

To divide a by b, multiply a by the reciprocal of b.

$\dfrac{\frac{10}{2}}{3} = 10 \div \dfrac{2}{3} = 10 \cdot \dfrac{3}{2} = 15$

Division *by* 0 is undefined.

$\dfrac{0}{5} = 0$ $\dfrac{5}{0}$ is undefined.

1.7 Properties of Real Numbers

Commutative Properties

$a + b = b + a$

$ab = ba$

$7 + (-1) = -1 + 7$

$5(-3) = (-3)5$

Associative Properties

$(a + b) + c = a + (b + c)$

$(ab)c = a(bc)$

$(3 + 4) + 8 = 3 + (4 + 8)$

$[-2(6)](4) = -2[6(4)]$

Identity Properties

$a + 0 = a$ $0 + a = a$

$a \cdot 1 = a$ $1 \cdot a = a$

$-7 + 0 = -7$ $0 + (-7) = -7$

$9 \cdot 1 = 9$ $1 \cdot 9 = 9$

Inverse Properties

$a + (-a) = 0$ $-a + a = 0$

$a \cdot \dfrac{1}{a} = 1$ $\dfrac{1}{a} \cdot a = 1 \ (a \neq 0)$

$7 + (-7) = 0$ $-7 + 7 = 0$

$-2\left(-\dfrac{1}{2}\right) = 1$ $-\dfrac{1}{2}(-2) = 1$

Distributive Properties

$a(b + c) = ab + ac$

$(b + c)a = ba + ca$

$5(4 + 2) = 5(4) + 5(2)$

$(4 + 2)5 = 4(5) + 2(5)$

1.8 Simplifying Expressions

Only like terms may be combined.

$-3y^2 + 6y^2 + 14y^2 = 17y^2$

$-8a^5b^3 + 2a^3b^5 - 6a^5b^3 + 5a^3b^5 = -14a^5b^3 + 7a^3b^5$

$4(3 + 2x) - 6(5 - x) = 12 + 8x - 30 + 6x$

$\qquad\qquad\qquad\qquad\quad = 14x - 18$

ANSWERS TO TEST YOUR WORD POWER

1. (b) *Example:* The product of 2 and 5, or 2 times 5, is 10. **2. (c)** *Examples:* 2, 3, 11, 41, 53 **3. (c)** *Example:* In 2^3, the number 3 is the exponent (or power), so 2 is a factor three times; $2^3 = 2 \cdot 2 \cdot 2 = 8$. **4. (a)** *Examples:* a, b, c **5. (b)** *Examples:* $-9, 0, 6$ **6. (a)** *Example:* The point graphed three units to the right of 0 on a number line has coordinate 3. **7. (d)** *Examples:* $|2| = 2$ and $|-2| = 2$ **8. (b)** *Examples:* $6, \frac{x}{2}, -4ab^2$ **9. (a)** *Examples:* The term 3 has numerical coefficient 3, $8z$ has numerical coefficient 8, and $-10x^4y$ has numerical coefficient -10.

Chapter 1

REVIEW EXERCISES

If you need help with any of these Review Exercises, look in the section indicated in brackets.

[1.1] *Find the value of each exponential expression.*

1. 5^4

2. $(.03)^4$

3. $.21^3$

4. $\left(\dfrac{5}{2}\right)^3$

Find the value of each expression.

5. $-13 + 8 \cdot 5$

6. $5[4^2 + 3(2^3)]$

7. $\dfrac{7(3^2 - 5)}{2 \cdot 6 - 16}$

8. $\dfrac{3(9 - 4) + 5(8 - 3)}{2^3 - (5 - 3)}$

Write each word statement in symbols.

9. Thirteen is less than seventeen.

10. Five plus two is not equal to ten.

11. Write $6 < 15$ in words.

12. Construct a false statement that involves addition on the left side, the symbol $\geq$, and division on the right side.

[1.2] *Evaluate each expression if $x = 6$ and $y = 3$.*

13. $2x + 6y$

14. $4(3x - y)$

15. $\dfrac{x}{3} + 4y$

16. $\dfrac{x^2 + 3}{3y - x}$

Change each word phrase to an algebraic expression. Use x to represent the number.

17. Six added to a number

18. A number subtracted from eight

19. Nine subtracted from six times a number

20. Three-fifths of a number added to 12

Decide whether the given number is a solution of the equation.

21. $5x + 3(x + 2) = 22$; 2

22. $\dfrac{t + 5}{3t} = 1$; 6

Change each word sentence to an equation. Use x to represent the number.

23. Six less than twice a number is 10.

24. The product of a number and 4 is 8.

Identify each of the following as either an equation or an expression.

25. $5r - 8(r + 7) = 2$

26. $2y + (5y - 9) + 2$

[1.3] *Graph each group of numbers on a number line.*

27. $-4, -\dfrac{1}{2}, 0, 2.5, 5$

28. $-2, -3, |-3|, |-1|$

29. $-3\dfrac{1}{4}, \dfrac{14}{5}, -1\dfrac{1}{8}, \dfrac{5}{6}$

30. $|-4|, -|-3|, -|-5|, -6$

Select the smaller number in each pair.

31. $-10, 5$

32. $-8, -9$

33. $-\dfrac{2}{3}, -\dfrac{3}{4}$

34. $0, -|23|$

Decide whether each statement is true *or* false.

35. $12 > -13$

36. $0 > -5$

37. $-9 < -7$

38. $-13 > -13$

Simplify by finding the absolute value.

39. $-|3|$

40. $-|-19|$

41. $-|9 - 2|$

42. $|15 - 6|$

[1.4] *Find each sum.*

43. $-10 + 4$

44. $14 + (-18)$

45. $-8 + (-9)$

46. $\dfrac{4}{9} + \left(-\dfrac{5}{4}\right)$

47. $[-6 + (-8) + 8] + [9 + (-13)]$

48. $(-4 + 7) + (-11 + 3) + (-15 + 1)$

Write a numerical expression for each phrase, and simplify the expression.

49. 19 added to the sum of -31 and 12

50. 13 more than the sum of -4 and -8

Solve each problem.

51. Tri Nguyen has $18 in his checking account. He then writes a check for $26. What negative number represents his balance?

52. The temperature at noon on an August day in Houston was 93°F. After a thunderstorm, it dropped 6°. What was the new temperature?

[1.5] *Find each difference.*

53. $-7 - 4$

54. $-12 - (-11)$

55. $5 - (-2)$

56. $-\dfrac{3}{7} - \dfrac{4}{5}$

57. $2.56 - (-7.75)$

58. $(-10 - 4) - (-2)$

59. $(-3 + 4) - (-1)$

60. $|5 - 9| - |-3 + 6|$

Write a numerical expression for each phrase, and simplify the expression.

61. The difference between -4 and -6

62. Five less than the sum of 4 and -8

63. In the 1969–1970 school year, the percent of high school graduates among 17-year-olds reached a maximum of 76.9%. This percent decreased by 5.5% in 1979–1980 and then increased by 2.8% in 1989–1990. What percent of 17-year-olds graduated in 1990? (*Source:* National Center for Education Statistics, U.S. Department of Education.)

64. The 1988 Women's Olympic Downhill Skiing champion, Marina Kiehl, from West Germany, finished the course in 1 min, 25.86 sec. The winning time of the 1994 champion, Katja Seizinger of Germany, increased by 10.07 sec. Seizinger won again in 1998. Her time decreased by 7.04 sec. What was Seizinger's winning time in 1998? (*Source: World Almanac and Book of Facts,* 1999.)

65. Explain in your own words how the subtraction problem $-8 - (-6)$ is performed.

66. Can the difference of two negative numbers be positive? Explain with an example.

The bar graph shows the federal budget outlays for national defense for the years 1990–1998. Use a signed number to represent the change in outlay for each time period. For example, the change from 1995 to 1996 was $253.3 - $259.6 = -$6.3 billion.

67. 1991–1992

68. 1993–1994

69. 1996–1997

70. 1997–1998

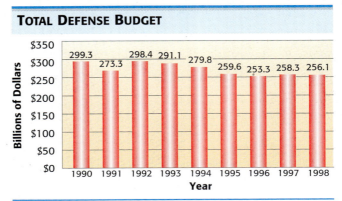

TOTAL DEFENSE BUDGET

Billions of Dollars

299.3 273.3 298.4 291.1 279.8 259.6 253.3 258.3 256.1

1990 1991 1992 1993 1994 1995 1996 1997 1998

Year

Source: U.S. Office of Management and Budget.

[1.6] *Perform the indicated operations.*

71. $(-12)(-3)$

72. $15(-7)$

73. $\left(-\dfrac{4}{3}\right)\left(-\dfrac{3}{8}\right)$

74. $(-4.8)(-2.1)$

75. $5(8-12)$

76. $(5-7)(8-3)$

77. $2(-6)-(-4)(-3)$

78. $3(-10)-5$

79. $\dfrac{-36}{-9}$

80. $\dfrac{220}{-11}$

81. $-\dfrac{1}{2} \div \dfrac{2}{3}$

82. $-33.9 \div (-3)$

83. $\dfrac{-5(3)-1}{8-4(-2)}$

84. $\dfrac{5(-2)-3(4)}{-2[3-(-2)]+10}$

85. $\dfrac{10^2-5^2}{8^2+3^2-(-2)}$

86. $\dfrac{(.6)^2+(.8)^2}{(-1.2)^2-(-.56)}$

Evaluate each expression if $x = -5$, $y = 4$, and $z = -3$.

87. $6x - 4z$

88. $5x + y - z$

89. $5x^2$

90. $z^2(3x - 8y)$

Write a numerical expression for each phrase, and simplify the expression.

91. Nine less than the product of -4 and 5

92. Five-sixths of the sum of 12 and -6

93. The quotient of 12 and the sum of 8 and -4

94. The product of -20 and 12, divided by the difference between 15 and -15

Translate each sentence to an equation, using x to represent the number.

95. The quotient of a number and the sum of the number and 5 is -2.

96. 3 less than 8 times a number is -7.

[1.7] *Decide whether each statement is an example of the commutative, associative, identity, inverse, or distributive property.*

97. $6 + 0 = 6$

98. $5 \cdot 1 = 5$

99. $-\dfrac{2}{3}\left(-\dfrac{3}{2}\right) = 1$

100. $17 + (-17) = 0$

101. $5 + (-9 + 2) = [5 + (-9)] + 2$

102. $w(xy) = (wx)y$

103. $3x + 3y = 3(x + y)$

104. $(1 + 2) + 3 = 3 + (1 + 2)$

Use the distributive property to rewrite each expression. Simplify if possible.

105. $7y + y$

106. $-12(4 - t)$

107. $3(2s) + 3(4y)$

108. $-(-4r + 5s)$

[1.8] *Use the distributive property as necessary and combine like terms.*

109. $16p^2 - 8p^2 + 9p^2$

110. $4r^2 - 3r + 10r + 12r^2$

111. $-8(5k - 6) + 3(7k + 2)$

112. $2s - (-3s + 6)$

113. $-7(2t - 4) - 4(3t + 8) - 19(t + 1)$

114. $3.6t^2 + 9t - 8.1(6t^2 + 4t)$

Translate each phrase into a mathematical expression. Use x to represent the number, and combine like terms when possible.

115. Seven times a number, subtracted from the product of -2 and three times the number

116. The quotient of 9 more than a number and 6 less than the number

117. In Exercise 115, does the word *and* signify addition? Explain.

118. Write the expression $3(4x - 6)$ using words, as in Exercises 115 and 116.

Perform the indicated operations.

119. $[(-2) + 7 - (-5)] + [-4 - (-10)]$

120. $\left(-\dfrac{5}{6}\right)^2$

121. $-|(-7)(-4)| - (-2)$

122. $\dfrac{6(-4) + 2(-12)}{5(-3) + (-3)}$

123. $\dfrac{3}{8} - \dfrac{5}{12}$

124. $\dfrac{12^2 + 2^2 - 8}{10^2 - (-4)(-15)}$

125. $\dfrac{8^2 + 6^2}{7^2 + 1^2}$

126. $-16(-3.5) - 7.2(-3)$

127. $2\dfrac{5}{6} - 4\dfrac{1}{3}$

128. $-8 + [(-4 + 17) - (-3 - 3)]$

129. $-\dfrac{12}{5} \div \dfrac{9}{7}$

130. $(-8 - 3) - 5(2 - 9)$

131. $[-7 + (-2) - (-3)] + [8 + (-13)]$

132. $\dfrac{15}{2} \cdot \left(-\dfrac{4}{5}\right)$

Write a numerical expression or an equation for each problem, and simplify it if possible. Use x as the variable, and specify what it represents.

133. In 2000, a company spent $1400 less on advertising than in the previous year. The total spent for this purpose over these two years was $25,800. How much was spent in 2000?

134. The quotient of a number and 14 less than three times the number

---

Chapter TEST

Study Skills Workbook
Activity 7

Decide whether the statement is true *or* false.

1. $4[-20 + 7(-2)] \leq -135$

        1. _____

2. $(-3)^2 + 2^2 = 5^2$

        2. _____

3. Graph numbers -1, -3, $|-4|$, and $|-1|$ on the number line.

        3.
```
  ┼──┼──┼──┼──┼──┼──┼──┼──►
 -3 -2 -1  0  1  2  3  4
```

Select the smaller number from each pair.

4. 6, $-|-8|$

        4. _____

5. $-.742$, -1.277

        5. _____

6. Write in symbols: The quotient of -6 and the sum of 2 and -8. Simplify the expression.

        6. _____

7. If a and b are both negative, is $\dfrac{a + b}{a \cdot b}$ positive or negative?

        7. _____

Perform the indicated operations whenever possible.

8. $-2 - (5 - 17) + (-6)$

        8. _____

9. $-5\dfrac{1}{2} + 2\dfrac{2}{3}$

        9. _____

10. $-6 - [-7 + (2 - 3)]$

        10. _____

11. $4^2 + (-8) - (2^3 - 6)$

        11. _____

12. $(-5)(-12) + 4(-4) + (-8)^2$

        12. _____

13. $\dfrac{-7 - |-6 + 2|}{-5 - (-4)}$

        13. _____

14. _____

$$14. \frac{30(-1-2)}{-9[3-(-2)]-12(-2)}$$

In Exercises 15 and 16, evaluate each expression if x = −2 and y = 4.

15. _____

15. $3x - 4y^2$

16. _____

$$16. \frac{5x + 7y}{3(x + y)}$$

17. _____

17. The highest Fahrenheit temperature ever recorded in Idaho was 118°, while the lowest was −60°. What is the difference between these highest and lowest temperatures? (*Source: World Almanac and Book of Facts*, 2000.)

Match each example in Column I with a property in Column II.

I	II

18. _____

18. $3x + 0 = 3x$ **A.** Commutative

19. _____

19. $(5 + 2) + 8 = 8 + (5 + 2)$ **B.** Associative

20. _____

20. $-3(x + y) = -3x + (-3y)$ **C.** Inverse

21. _____

21. $-5 + (3 + 2) = (-5 + 3) + 2$ **D.** Identity

22. _____

22. $-\frac{5}{3}\left(-\frac{3}{5}\right) = 1$ **E.** Distributive

23. _____

23. Simplify $-2(3x^2 + 4) - 3(x^2 + 2x)$ by using the distributive property and combining like terms.

24. _____

24. What properties are used to show that $-(3x + 1) = -3x - 1$?

25. (a) _____
    (b) _____

25. Consider the expression $-6[5 + (-2)]$.
    (a) Evaluate it by first working within the brackets.
    (b) Evaluate it by using the distributive property.
    (c) Why must the answers in items (a) and (b) be the same?

    (c) _____

Equations, Inequalities, and Applications

2

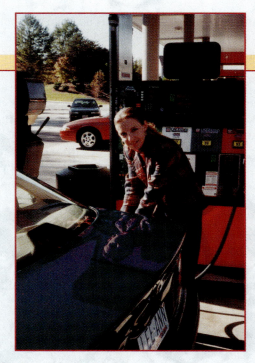

After Lee Ann Spahr pumped 5.0 gal of gasoline, the display showing the price read $7.90. When she finished pumping the gasoline, the price display read $21.33. How many gallons did she pump?

During the course of a day, it is likely that you use the concepts of ratio, proportion, and percent to solve simple problems. In Example 5 of Section 2.6, we use a proportion to answer the question posed here.

2.1 THE ADDITION PROPERTY OF EQUALITY

OBJECTIVES

1. Identify linear equations.
2. Use the addition property of equality.
3. Simplify equations, and then use the addition property of equality.
4. Solve equations that have no solution or infinitely many solutions.

To solve applied problems, we must be able to solve equations. The simplest type of equation is a *linear equation*. Methods for solving linear equations will be introduced in this section. We will be using the definitions and properties of real numbers that we learned in Chapter 1.

1 **Identify linear equations.** We begin with a definition.

Linear Equation in One Variable

A **linear equation in one variable** can be written in the form

$$Ax + B = C$$

for real numbers A, B, and C, with $A \neq 0$.

For example,

$$4x + 9 = 0, \quad 2x - 3 = 5, \quad \text{and} \quad x = 7$$

are linear equations in one variable (x). The final two can be written in the specified form using properties developed in this chapter. However,

$$x^2 + 2x = 5, \quad \frac{1}{x} = 6, \quad \text{and} \quad |2x + 6| = 0$$

are *not* linear equations.

As we saw in Chapter 1, a solution of an equation is a number that makes the equation true when it replaces the variable. Equations that have exactly the same solutions are **equivalent equations**. Linear equations are solved by using a series of steps to produce a simpler equivalent equation of the form

$$x = \text{a number.}$$

2 **Use the addition property of equality.** In the equation $x - 5 = 2$, both $x - 5$ and 2 represent the same number because this is the meaning of the equals sign. To solve the equation, we change the left side from $x - 5$ to just x. This is done by adding 5 to $x - 5$. To keep the two sides equal, we must also add 5 to the right side.

$$
\begin{aligned}
x - 5 &= 2 && \text{Given equation} \\
x - 5 + 5 &= 2 + 5 && \text{Add 5 to each side.} \\
x + 0 &= 7 && \text{Additive inverse property} \\
x &= 7 && \text{Additive identity property}
\end{aligned}
$$

The solution of the given equation is 7. Check by replacing x with 7 in the original equation.

Check:
$$
\begin{aligned}
x - 5 &= 2 && \text{Original equation} \\
7 - 5 &= 2 \quad ? && \text{Let } x = 7. \\
2 &= 2 && \text{True}
\end{aligned}
$$

Since the final equation is true, 7 checks as the solution.

To solve the equation, we added the same number to each side. The **addition property of equality** justifies this step.

Addition Property of Equality

If A, B, and C are real numbers, then the equations

$$A = B \quad \text{and} \quad A + C = B + C$$

are equivalent equations.

In words, we can add the same number to each side of an equation without changing the solution.

In the addition property, C represents a real number. This means that any quantity that represents a real number can be added to each side of an equation to change it to an equivalent equation.

Example 1 Using the Addition Property of Equality

Solve $x - 16.2 = 7.5$.

If the left side of this equation were just x, the solution would be known. Get x alone by using the addition property of equality, adding 16.2 to each side.

$$x - 16.2 = 7.5$$
$$x - 16.2 + 16.2 = 7.5 + 16.2 \qquad \text{Add 16.2 to each side.}$$
$$x = 23.7 \qquad \text{Combine terms.}$$

Here we combined the steps that change $x - 16.2 + 16.2$ to $x + 0$ and $x + 0$ to x. We will combine these steps from now on. Check by substituting 23.7 for x in the original equation.

Check:
$$x - 16.2 = 7.5 \qquad \text{Original equation}$$
$$23.7 - 16.2 = 7.5 \quad ? \qquad \text{Let } x = 23.7.$$
$$7.5 = 7.5 \qquad \text{True}$$

Since the check results in a true statement, 23.7 is the solution.

Work Problem ❶ at the Side.

The addition property of equality says that the same number may be *added* to each side of an equation. In Chapter 1, subtraction was defined as addition of the opposite. Thus, we can also use the following rule when solving an equation.

The same number may be subtracted from each side of an equation without changing the solution.

For example, to solve $x + 5 = 10$, subtract 5 from each side to get $x = 5$.

Work Problem ❷ at the Side.

Example 2 Subtracting a Variable Expression

Solve $\frac{3}{5}k + 17 = \frac{8}{5}k$.

Get all terms with variables on the same side of the equation. One way to do this is to subtract $\frac{3}{5}k$ from each side.

Continued on Next Page

---

❶ Solve.

(a) $m - 2.9 = -6.4$

(b) $y - 4.1 = 6.3$

❷ Solve.

(a) $a + 2 = -3$

(b) $r + 16 = 22$

❸ (a) Solve $\dfrac{7}{2}m + 1 = \dfrac{9}{2}m$.

$$\dfrac{3}{5}k + 17 = \dfrac{8}{5}k$$

$$\dfrac{3}{5}k + 17 - \dfrac{3}{5}k = \dfrac{8}{5}k - \dfrac{3}{5}k \qquad \text{Subtract } \tfrac{3}{5}k \text{ from each side.}$$

$$17 = 1k \qquad \text{Combine terms; } \tfrac{5}{5}k = 1k.$$

$$17 = k \qquad \text{Identity property}$$

The solution is 17. From now on we will skip the step that changes $1k$ to k. Check the solution by replacing k with 17 in the original equation.

(b) What is the solution of $-x = 6$?

Another way to solve the equation in Example 2 is to first subtract $\tfrac{8}{5}k$ from each side.

$$\dfrac{3}{5}k + 17 = \dfrac{8}{5}k$$

$$\dfrac{3}{5}k + 17 - \dfrac{8}{5}k = \dfrac{8}{5}k - \dfrac{8}{5}k \qquad \text{Subtract } \tfrac{8}{5}k \text{ from each side.}$$

$$17 - k = 0 \qquad \text{Combine terms.}$$

$$17 - k - 17 = 0 - 17 \qquad \text{Subtract 17 from each side.}$$

$$-k = -17 \qquad \text{Combine terms; additive inverse}$$

(c) What is the solution of $-x = -12$?

This result gives the value of $-k$, but not of k itself. However, it does say that the additive inverse of k is -17, which means that k must be 17, the same result we obtained in Example 2.

$$-k = -17$$

$$k = 17$$

(This result can also be justified using the multiplication property of equality, covered in Section 2.2.) We can make the following generalization.

❹ Solve.

(a) $-(5 - 3r) + 4(-r + 1)$
$= 1$

If a is a number and $-x = a$, then $x = -a$.

Work Problem ❸ at the Side.

3 Simplify equations, and then use the addition property of equality. Sometimes an equation must be simplified as a first step in its solution.

Example 3 Using the Distributive Property to Simplify an Equation

Solve $3(2 + 5x) - (1 + 14x) = 6$.

$$3(2 + 5x) - (1 + 14x) = 6$$

$$3(2 + 5x) - 1(1 + 14x) = 6 \qquad -(1 + 14x) = -1(1 + 14x)$$

$$6 + 15x - 1 - 14x = 6 \qquad \text{Distributive property}$$

$$x + 5 = 6 \qquad \text{Combine terms.}$$

$$x + 5 - 5 = 6 - 5 \qquad \text{Subtract 5 from each side.}$$

$$x = 1 \qquad \text{Combine terms.}$$

(b) $-3(m - 4) + 2(5 + 2m)$
$= 29$

The solution is 1. Check by substituting 1 for x in the original equation.

Work Problem ❹ at the Side.

4 Solve equations that have no solution or infinitely many solutions.   Every equation solved so far has had exactly one solution. Sometimes this is not the case, as shown in the next examples.

Example 4 Solving an Equation That Has Infinitely Many Solutions

Solve $5x - 15 = 5(x - 3)$.

$$5x - 15 = 5(x - 3)$$
$$5x - 15 = 5x - 15 \qquad \text{Distributive property}$$
$$5x - 15 \mathbf{+ 15} = 5x - 15 \mathbf{+ 15} \qquad \text{Add 15 to each side.}$$
$$5x = 5x \qquad \text{Combine terms.}$$
$$5x \mathbf{- 5x} = 5x \mathbf{- 5x} \qquad \text{Subtract } 5x \text{ from each side.}$$
$$0 = 0$$

The final step leads to an equation that contains no variables ($0 = 0$ in this case). Whenever such a statement is true, as it is in this example, *any* real number is a solution. (Try several replacements for x in the given equation to see that they all satisfy the equation.) An equation with both sides exactly the same, like $0 = 0$, is called an **identity.** An identity is true for all replacements of the variables. We indicate this by writing *all real numbers*.

CAUTION

When you are solving an equation like the one in Example 4, do not write "0" as the solution. While 0 is a solution, there are infinitely many other solutions.

Example 5 Solving an Equation That Has No Solution

Solve $2x + 3(x + 1) = 5x + 4$.

$$2x + 3(x + 1) = 5x + 4$$
$$2x + 3x + 3 = 5x + 4 \qquad \text{Distributive property}$$
$$5x + 3 = 5x + 4 \qquad \text{Combine terms.}$$
$$5x + 3 \mathbf{- 5x} = 5x + 4 \mathbf{- 5x} \qquad \text{Subtract } 5x \text{ from each side.}$$
$$3 = 4 \qquad \text{Combine terms.}$$

Again, the variable has disappeared, but this time a false statement ($3 = 4$) results. Whenever this happens in solving an equation, it is a signal that the equation has no solution and we write *no solution.*

Work Problem 5 at the Side.

---

5 Solve each equation.

(a) $2(x - 6) = 2x - 12$

(b) $3x + 6(x + 1) = 9x - 4$

Real-Data Applications

The Magic Number in Sports

National League			
East Division			
	W	**L**	**Pct.**	**GB**
Atlanta	86	60	.589	—
New York	84	62	.575	2
Florida	69	76	.476	16.5
Montreal	61	84	.421	24.5
Philadelphia	60	85	.414	25.5
Central Division			
	W	**L**	**Pct.**	**GB**
St. Louis	86	61	.585	—
Cincinnati	75	72	.510	11
Milwaukee	64	82	.438	21.5
Houston	63	83	.432	22.5
Pittsburgh	61	84	.421	24
Chicago	60	86	.411	25.5
West Division			
	W	**L**	**Pct.**	**GB**
San Francisco	87	58	.600	—
Arizona	78	66	.542	8.5
Colorado	75	70	.517	12
Los Angeles	75	71	.514	12.5
San Diego	71	76	.483	17

American League			
East Division			
	W	**L**	**Pct.**	**GB**
New York	84	59	.587	—
Boston	76	68	.528	8.5
Toronto	75	70	.517	10
Baltimore	66	80	.452	19.5
Tampa Bay	61	85	.418	24.5
Central Division			
	W	**L**	**Pct.**	**GB**
Chicago	87	58	.600	—
Cleveland	77	65	.542	8.5
Detroit	71	74	.490	16
Kansas City	68	78	.466	19.5
Minnesota	63	82	.434	24
West Division			
	W	**L**	**Pct.**	**GB**
Seattle	80	66	.548	—
Oakland	77	67	.535	2
Anaheim	74	72	.507	6
Texas	66	80	.452	14

The climax of any sports season is the playoffs. Baseball fans eagerly debate predictions of which team will win the pennant for their division. The *magic number* for each first place team is often used to predict the division winner. The **magic number** is the required number of additional wins by the first-place team that would exceed by one the maximum number of wins of the second-place team.

The baseball league standings on September 15, 2000 are shown at the left. There were 162 regulation games in the 2000 baseball season.

For Group Discussion

To calculate the magic number, consider the following conditions.

The number of wins for the first-place team (W_1) plus the magic number (M) is one more than the sum of the number of wins to date (W_2) and the number of games remaining in the season (N_2) for the second-place team.

1. First, use the variable definitions to write an equation involving the magic number. Second, solve the equation for the magic number. Write the formula for the magic number.

2. Find the magic number for each team. The number of games remaining in the season for the second-place team is calculated as

$$N_2 = 162 - (W_2 + L_2).$$

(a) NL East: Atlanta vs New York
   Magic No. _____

(b) NL Central: St. Louis vs Cincinnati
   Magic No. _____

(c) NL West: San Francisco vs Arizona
   Magic No. _____

(d) AL East: New York vs Boston
   Magic No. _____

(e) AL Central: Chicago vs Cleveland
   Magic No. _____

(f) AL West: Seattle vs Oakland
   Magic No. _____

2.1 EXERCISES

FOR EXTRA HELP

 Student's Solutions Manual

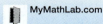 MyMathLab.com

 InterAct Math Tutorial Software

 AW Math Tutor Center

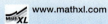 www.mathxl.com
MathXL

Digital Video Tutor CD 2 Videotape 4

1. Which of the pairs of equations are equivalent equations?

   A. $x + 2 = 6$ and $x = 4$ **B.** $10 - x = 5$ and $x = -5$

   C. $x + 3 = 9$ and $x = 6$ **D.** $4 + x = 8$ and $x = -4$

2. Decide whether each is an expression or an equation. If it is an expression, simplify it. If it is an equation, solve it.

   (a) $5x + 8 - 4x + 7$ **(b)** $-6y + 12 + 7y - 5$

   (c) $5x + 8 - 4x = 7$ **(d)** $-6y + 12 + 7y = -5$

3. Which of the following are not linear equations in one variable?

   A. $x^2 - 5x + 6 = 0$ **B.** $x^3 = x$

   C. $3x - 4 = 0$ **D.** $7x - 6x = 3 + 9x$

4. Refer to the definition of linear equation in one variable given in this section. Why is the restriction $A \neq 0$ necessary?

Solve each linear equation by inspection (that is, do not write out any steps).

   5. $x + 1 = 5$ **6.** $x - 2 = 4$ **7.** $x - 10 = 0$ **8.** $x + 7 = 0$

Solve each equation by using the addition property of equality. Check each solution.
See Examples 1, 2, 4, and 5.

   9. $x - 4 = 8$ **10.** $x - 8 = 9$ **11.** $7 + r = -3$

   12. $8 + k = -4$ **13.** $\dfrac{9}{7}r - 3 = \dfrac{2}{7}r$ **14.** $\dfrac{8}{5}w - 6 = \dfrac{3}{5}w$

   15. $5.6x + 2 = 4.6x$ **16.** $9.1x - 5 = 8.1x$ **17.** $3p + 6 = 10 + 2p$

   18. $8x - 4 = -6 + 7x$ **19.** $1.2x - 4 = .2x - 4$ **20.** $7.7r + 6 = 6.7r + 6$

   21. $3x + 9 = 3x + 8$ **22.** $-2x + 5 = -2x$

   23. $8x + 1 = 1 + 8x$ **24.** $4w - 5 = -5 + 4w$

Solve each equation. First simplify each side of the equation as much as possible. Check each solution. See Examples 3, 4, and 5.

25. $10x + 5x + 7 - 8 = 12x + 3 + 2x$

26. $7p + 4p + 13 - 7 = 7p + 9 + 3p$

27. $6x + 5 - 7x + 3 = 5x - 6x - 4$

28. $4x - 3 - 8x + 1 = 5x - 9x + 7$

29. $5.2q - 4.6 - 7.1q = -2.1 - 1.9q - 2.5$

30. $-4.0x + 2.7 - 1.6x = 1.3 - 5.6x + 1.4$

31. $\dfrac{5}{7}x + \dfrac{1}{3} = \dfrac{2}{5} - \dfrac{2}{7}x + \dfrac{2}{5}$

32. $\dfrac{6}{7}s - \dfrac{3}{4} = \dfrac{4}{5} - \dfrac{1}{7}s + \dfrac{1}{6}$

33. $(5x + 6) - (3 + 4x) = 10$

34. $(8r - 3) - (7r + 1) = -6$

35. $2(p + 5) - (9 + p) = -3$

36. $4(k - 6) - (3k + 2) = -5$

37. $-6(2x + 1) + (13x - 7) = 0$

38. $-5(3w - 3) + (1 + 16w) = 0$

39. $10(-2x + 1) = -14(x + 2) + 38 - 6x$

40. $2(2 - 3r) = 5(1 - r) - r - 1$

41. $-2(8p + 2) - 3(2 - 7p) = 2(4 + 2p)$

42. $-5(1 - 2z) + 4(3 - z) = 7(3 + z)$

43. $4(7x - 1) + 3(2 - 5x) = 4(3x + 5) - 6$

44. $9(2m - 3) - 4(5 + 3m) = 5(4 + m) - 3$

45. In your own words, state how you would find the solution of a linear equation if your next-to-last step reads "$-x = 5$."

46. If the final step in solving a linear equation leads to the statement $0 = 0$, why is it incorrect to say that 0 is the solution of the equation? What are the solutions of the equation?

47. Write an equation where 6 must be added to each side to solve the equation, and the solution is a negative number.

48. Write an equation where $\dfrac{1}{2}$ must be subtracted from each side, and the solution is a positive number.

2.2   THE MULTIPLICATION PROPERTY OF EQUALITY

The addition property of equality alone is not enough to solve some equations, such as $3x + 2 = 17$.

$$3x + 2 = 17$$
$$3x + 2 - \mathbf{2} = 17 - \mathbf{2} \qquad \text{Subtract 2 from each side.}$$
$$3x = 15 \qquad \text{Combine terms.}$$

Notice that the coefficient of x on the left side is 3, not 1 as desired. We must develop a method that leads to an equation of the form

$$x = \text{a number.}$$

1   **Use the multiplication property of equality.** If $3x = 15$, then $3x$ and 15 both represent the same number. Multiplying both $3x$ and 15 by the same number will also result in an equality. The **multiplication property of equality** states that we can multiply each side of an equation by the same nonzero number without changing the solution.

Multiplication Property of Equality

If A, B, and C ($C \neq 0$) represent real numbers, then the equations

$$A = B \quad \text{and} \quad AC = BC$$

have exactly the same solution.
    In words, we can multiply each side of an equation by the same nonzero number without changing the solution.

    This property can be used to solve $3x = 15$. The $3x$ on the left must be changed to $1x$, or x, instead of $3x$. To isolate x, multiply each side of the equation by $\frac{1}{3}$. We use $\frac{1}{3}$ because $\frac{1}{3}$ is the reciprocal of 3, and $\frac{1}{3} \cdot 3 = \frac{3}{3} = 1$.

$$3x = 15$$
$$\frac{1}{3}(3x) = \frac{1}{3} \cdot 15 \qquad \text{Multiply each side by } \tfrac{1}{3}.$$
$$\left(\frac{1}{3} \cdot 3\right)x = \frac{1}{3} \cdot 15 \qquad \text{Associative property}$$
$$1x = 5 \qquad \text{Multiplicative inverse property}$$
$$x = 5 \qquad \text{Multiplicative identity property}$$

The solution of the equation is 5. We can check this result in the original equation. We will sometimes combine the last two steps shown in the preceding example.

Work Problem ❶ at the Side.

    Just as the addition property of equality permits *subtracting* the same number from each side of an equation, the multiplication property of equality permits *dividing* each side of an equation by the same nonzero number. For example, the equation $3x = 15$, which we just solved by multiplication, could also be solved by dividing each side by 3, as follows.

$$3x = 15$$
$$\frac{3x}{3} = \frac{15}{3} \qquad \text{Divide each side by 3.}$$
$$x = 5$$

OBJECTIVES

1   Use the multiplication property of equality.

2   Use the multiplication property of equality to solve equations with decimals.

3   Simplify equations, and then use the multiplication property of equality.

4   Use the multiplication property of equality to solve equations such as $-r = 4$.

❶   Check that 5 is the solution of $3x = 15$.

ANSWERS
1. Since $3(5) = 15$, the solution of $3x = 15$ is 5.

❷ Solve.

(a) $-6p = -14$

(b) $3r = -12$

(c) $-2m = 16$

We can divide each side of an equation by the same nonzero number without changing the solution. Do not, however, divide each side by a variable, as that may result in losing a valid solution.

NOTE

In practice, it is usually easier to multiply on each side if the coefficient of the variable is a fraction, and divide on each side if the coefficient is an integer. For example, to solve

$$-\frac{3}{4}x = 12,$$

it is easier to multiply by $-\frac{4}{3}$, the reciprocal of $-\frac{3}{4}$, than to divide by $-\frac{3}{4}$. On the other hand, to solve

$$-5x = -20,$$

it is easier to divide by -5 than to multiply by $-\frac{1}{5}$.

Example 1 Dividing Each Side of an Equation by a Nonzero Number

Solve $25p = 30$.

Transform the equation so that p (instead of $25p$) is on the left by using the multiplication property of equality. Divide each side of the equation by 25, the coefficient of p.

$$25p = 30$$

$$\frac{25p}{25} = \frac{30}{25} \qquad \text{Divide by 25.}$$

$$p = \frac{30}{25} = \frac{6}{5} \qquad \text{Write in lowest terms.}$$

To check, substitute $\frac{6}{5}$ for p in the original equation.

Check: $\qquad 25p = 30$

$$\frac{25}{1}\left(\frac{6}{5}\right) = 30 \quad ? \qquad \text{Let } p = \frac{6}{5}.$$

$$30 = 30 \qquad \text{True}$$

The solution is $\frac{6}{5}$.

Work Problem ❷ at the Side.

In the next two examples, multiplication produces the solution more quickly than division would.

Example 2 Using the Multiplication Property of Equality

Solve $\frac{a}{4} = 3$.

Replace $\frac{a}{4}$ by $\frac{1}{4}a$, since division by 4 is the same as multiplication by $\frac{1}{4}$. To get a alone on the left, multiply each side by 4, the reciprocal of the coefficient of a.

Continued on Next Page

$$\frac{a}{4} = 3$$

$$\frac{1}{4}a = 3 \qquad \text{Change } \tfrac{a}{4} \text{ to } \tfrac{1}{4}a.$$

$$4 \cdot \frac{1}{4}a = 4 \cdot 3 \qquad \text{Multiply by 4.}$$

$$a = 12 \qquad \begin{array}{l}\text{Multiplicative inverse property;}\\ \text{multiplicative identity property}\end{array}$$

Check that 12 is the solution.

Check:

$$\frac{a}{4} = 3 \qquad \text{Original equation}$$

$$\frac{12}{4} = 3 \quad ? \quad \text{Let } a = 12.$$

$$3 = 3 \qquad \text{True}$$

The correct solution is 12.

============ Work Problem **3** at the Side.

Example 3   **Using the Multiplication Property of Equality**

Solve $\frac{3}{4}h = 6$.

Transform the equation so that h is alone on the left by multiplying each side of the equation by $\frac{4}{3}$. Use $\frac{4}{3}$ because $\frac{4}{3} \cdot \frac{3}{4}h = 1 \cdot h = h$.

$$\frac{3}{4}h = 6$$

$$\frac{4}{3}\left(\frac{3}{4}h\right) = \frac{4}{3} \cdot 6 \qquad \text{Multiply by } \tfrac{4}{3}.$$

$$1 \cdot h = \frac{4}{3} \cdot \frac{6}{1} \qquad \text{Multiplicative inverse property}$$

$$h = 8 \qquad \begin{array}{l}\text{Multiplicative identity property;}\\ \text{multiply fractions.}\end{array}$$

The solution is 8. Check the solution by substitution in the original equation.

============ Work Problem **4** at the Side.

2 ▭   Use the multiplication property of equality to solve equations with decimals.

Example 4   **Solving an Equation with Decimals**

Solve $2.1x = 6.09$.

Divide each side by 2.1.

$$2.1x = 6.09$$

$$\frac{2.1x}{2.1} = \frac{6.09}{2.1}$$

You may use a calculator to simplify the work at this point.

$$x = 2.9 \qquad \text{Divide.}$$

Check that the solution is 2.9.

============ Work Problem **5** at the Side.

---

3 Solve.

(a) $\dfrac{y}{5} = 5$

(b) $\dfrac{p}{4} = -6$

4 Solve.

(a) $-\dfrac{5}{6}t = -15$

(b) $\dfrac{3}{4}k = -21$

5 Solve.

(a) $-.7m = -5.04$

(b) $12.5k = -63.75$

6 Solve.

(a) $4r - 9r = 20$

(b) $7m - 5m = -12$

3 ◼ **Simplify equations, and then use the multiplication property of equality.** In the next example, it is necessary to simplify the equation before using the multiplication property of equality.

Example 5 Simplifying an Equation

Solve $5m + 6m = 33$.

$$5m + 6m = 33$$
$$11m = 33 \qquad \text{Combine terms.}$$
$$\frac{11m}{\mathbf{11}} = \frac{33}{\mathbf{11}} \qquad \text{Divide by 11.}$$
$$1m = 3 \qquad \text{Divide.}$$
$$m = 3 \qquad \text{Multiplicative identity property}$$

The solution is 3. Check this solution.

Work Problem 6 at the Side.

4 ◼ **Use the multiplication property of equality to solve equations such as** $-r = 4$. The following example shows how to solve equations where the coefficient of the variable is understood to be -1.

7 Solve.

(a) $-m = 2$

Example 6 Using the Multiplication Property of Equality When the Coefficient of the Variable Is -1

Solve $-r = 4$.

On the left side, change $-r$ to r by first writing $-r$ as $-1 \cdot r$.

$$-r = 4$$
$$-\mathbf{1} \cdot r = 4 \qquad -r = -1 \cdot r$$
$$-\mathbf{1}(-1 \cdot r) = -\mathbf{1} \cdot 4 \qquad \text{Multiply by } -1, \text{ since } -1(-1) = 1.$$
$$[-1(-1)] \cdot r = -4 \qquad \text{Associative property}$$
$$1 \cdot r = -4 \qquad \text{Multiplicative inverse property}$$
$$r = -4 \qquad \text{Multiplicative identity property}$$

Check this solution.

Check:
$$-r = 4 \qquad \text{Original equation}$$
$$-(-\mathbf{4}) = 4 \qquad ? \qquad \text{Let } r = -4.$$
$$4 = 4 \qquad \text{True}$$

The solution, -4, checks.

Work Problem 7 at the Side.

(b) $-p = -7$

ANSWERS
6. (a) -4 (b) -6
7. (a) -2 (b) 7

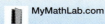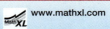

Solve each equation by inspection.

1. $3x = 12$

2. $4x = 36$

3. $\dfrac{1}{2}x = -4$

4. $\dfrac{1}{3}x = -2$

By what number is it necessary to multiply each side of each equation in order to obtain just x on the left side? Do not actually solve these equations.

5. $\dfrac{2}{3}x = 8$

6. $\dfrac{4}{5}x = 6$

7. $.1x = 3$

8. $.01x = 8$

9. $-\dfrac{9}{2}x = -4$

10. $-\dfrac{8}{3}x = -11$

11. $-x = .36$

12. $-x = .29$

By what number is it necessary to divide each side of each equation in order to obtain just x on the left side? Do not actually solve these equations.

13. $6x = 5$

14. $7x = 10$

15. $-4x = 13$

16. $-13x = 6$

17. $.12x = 48$

18. $.21x = 63$

19. $-x = 23$

20. $-x = 49$

21. In the statement of the multiplication property of equality in this section, there is a restriction that $C \neq 0$. What would happen if you multiplied each side of an equation by 0?

22. Which equation does not require the use of the multiplication property of equality?

A. $3x - 5x = 6$ **B.** $-\dfrac{1}{4}x = 12$ **C.** $5x - 4x = 7$ **D.** $\dfrac{x}{3} = -2$

Solve each equation, and check your solution. See Examples 1–6.

23. $2m = 15$

24. $3m = 10$

25. $3a = -15$

26. $5k = -70$

27. $10t = -36$

28. $4s = -34$

29. $-6x = -72$

30. $-8x = -64$

31. $2r = 0$ **32.** $5x = 0$ **33.** $-y = 12$ **34.** $-t = 14$

35. $.2t = 8$ **36.** $.9x = 18$ **37.** $\frac{1}{4}y = -12$ **38.** $\frac{1}{5}p = -3$

39. $\frac{x}{7} = -5$ **40.** $\frac{k}{8} = -3$ **41.** $-\frac{7}{9}c = \frac{3}{5}$ **42.** $-\frac{5}{6}d = \frac{4}{9}$

43. $4x + 3x = 21$ **44.** $9x + 2x = 121$ **45.** $3r - 5r = 10$ **46.** $9p - 13p = 24$

47. $5m + 6m - 2m = 63$ **48.** $11r - 5r + 6r = 168$

49. $5x + 2 = 8x + 8$ **50.** $2y - 4 = 7y + 1$

51. $9w - 5w + 3 = -w$ **52.** $7 + 6k - 2k = 8$

53. Write an equation that requires the use of the multiplication property of equality, where each side must be multiplied by $\frac{2}{3}$, and the solution is a negative number.

54. Write an equation that requires the use of the multiplication property of equality, where each side must be divided by 100, and the solution is not an integer.

Write an equation using the information given in the problem. Use x to represent the number. Then solve the equation, and give the required number.

55. Three times a number is 18 more than five times the number. Find the number.

56. If four times a number is added to three times the number, the result is the sum of five times the number and 10. Find the number.

2.3   MORE ON SOLVING LINEAR EQUATIONS

OBJECTIVES

1 Learn the four steps for solving a linear equation, and apply them.

2 Solve equations by clearing fractions and decimals.

1 Learn the four steps for solving a linear equation, and apply them.   In this section, we use the addition and multiplication properties together to solve more complicated equations. We will use the following four-step method.

Solving Linear Equations

Step 1   **Simplify each side separately.** Clear parentheses using the distributive property, if needed, and combine terms.

Step 2   **Isolate the variable term on one side.** Use the addition property if necessary so that the variable term is on one side of the equation and a number is on the other.

Step 3   **Isolate the variable.** Use the multiplication property if necessary to get the equation in the form $x =$ a number.

Step 4   **Check.** Check the proposed solution by substituting into the *original* equation.

The check is used only to catch errors in carrying out the steps.

Example 1   Using the Four Steps to Solve an Equation

Solve the equation $3r + 4 - 2r - 7 = 4r + 3$.
   We use the four steps described above.

Step 1 $3r + 4 - 2r - 7 = 4r + 3$

$r - 3 = 4r + 3$ Combine terms.

Step 2 $r - 3 + 3 = 4r + 3 + 3$ Add 3.

$r = 4r + 6$

$r - 4r = 4r + 6 - 4r$ Subtract $4r$.

$-3r = 6$ Combine terms.

Step 3 $\dfrac{-3r}{-3} = \dfrac{6}{-3}$ Divide by -3.

$r = -2$ $\frac{-3}{-3} = 1; 1r = r$

Step 4   Substitute -2 for r in the original equation to check.

$3r + 4 - 2r - 7 = 4r + 3$

$3(-2) + 4 - 2(-2) - 7 = 4(-2) + 3$?   Let $r = -2$.

$-6 + 4 + 4 - 7 = -8 + 3$?   Multiply.

$-5 = -5$ True

The solution of the equation is -2.

NOTE

In Step 2 of Example 1, we added and subtracted the terms in such a way that the variable term ended up on the left side of the equation. Choosing differently would have put the variable term on the right side of the equation.

1 Solve.

(a) $5y - 7y + 6y - 9 = 3 + 2y$

(b) $-3k - 5k - 6 + 11 = 2k - 5$

ANSWERS

1. **(a)** 6   **(b)** 1

Work Problem **1** at the Side.

❷ Solve.

(a) $7(p - 2) + p = 2p + 4$

(b) $3(m + 5) - 1 + 2m$
$= 5(m + 2)$

Example 2   Using the Four Steps to Solve an Equation

Solve the equation $4(k - 3) - k = k - 6$.

Step 1 $4(k - 3) - k = k - 6$

$\qquad 4k - 12 - k = k - 6$ Distributive property

$\qquad\quad 3k - 12 = k - 6$ Combine terms.

Step 2 $3k - 12 + 12 = k - 6 + 12$ Add 12.

$\qquad\qquad\quad 3k = k + 6$ Combine terms.

$\qquad\quad 3k - k = k + 6 - k$ Subtract k.

$\qquad\qquad\quad 2k = 6$ Combine terms.

Step 3 $\dfrac{2k}{2} = \dfrac{6}{2}$ Divide by 2.

$\qquad\qquad\quad k = 3$

Step 4   Check this answer by substituting 3 for k in the original equation. Remember to do all the work inside the parentheses first.

$\qquad 4(k - 3) - k = k - 6$

$\qquad 4(3 - 3) - 3 = 3 - 6$?   Let $k = 3$.

$\qquad\quad 4(0) - 3 = 3 - 6$? $3 - 3 = 0$

$\qquad\qquad 0 - 3 = 3 - 6$? $4(0) = 0$

$\qquad\qquad\quad -3 = -3$ True

The solution of the equation is 3.

Work Problem ❷ at the Side.

Example 3   Using the Four Steps to Solve an Equation

Solve the equation $8a - (3 + 2a) = 3a + 1$.

Step 1   Simplify.

$\qquad 8a - (3 + 2a) = 3a + 1$

$\qquad 8a - 1(3 + 2a) = 3a + 1$ Multiplicative identity property

$\qquad\quad 8a - 3 - 2a = 3a + 1$ Distributive property

$\qquad\qquad 6a - 3 = 3a + 1$ Combine terms.

Step 2   First, add 3 to each side; then subtract $3a$.

$\qquad 6a - 3 + 3 = 3a + 1 + 3$ Add 3.

$\qquad\qquad\quad 6a = 3a + 4$ Combine terms.

$\qquad 6a - 3a = 3a + 4 - 3a$ Subtract $3a$.

$\qquad\qquad\quad 3a = 4$ Combine terms.

Step 3 $\dfrac{3a}{3} = \dfrac{4}{3}$ Divide by 3.

$\qquad\qquad\quad a = \dfrac{4}{3}$

Step 4   Check that the solution is $\frac{4}{3}$.

CAUTION

Be very careful with signs when solving an equation like the one in Example 3. When clearing parentheses in the expression

$$8a - (3 + 2a),$$

remember that the $-$ sign acts like a factor of -1, changing the sign of *every* term in the parentheses. Thus,

$$8 - (3 + 2a) = 8 - 3 - 2a.$$

Change to $-$ in both terms.

Work Problem ❸ at the Side.

❸ Solve.

(a) $7m - (2m - 9) = 39$

(b) $4x + 2(3 - 2x) = 6$

Example 4   **Using the Four Steps to Solve an Equation**

Solve the equation $4(8 - 3t) = 32 - 8(t + 2)$.

Step 1 $4(8 - 3t) = 32 - 8(t + 2)$

$32 - 12t = 32 - 8t - 16$ Distributive property

$32 - 12t = 16 - 8t$ Combine terms.

Step 2 $32 - 12t \mathbf{- 32} = 16 - 8t \mathbf{- 32}$ Subtract 32.

$-12t = -16 - 8t$ Combine terms.

$-12t \mathbf{+ 8t} = -16 - 8t \mathbf{+ 8t}$ Add $8t$.

$-4t = -16$ Combine terms.

Step 3 $\dfrac{-4t}{-4} = \dfrac{-16}{-4}$ Divide by -4.

$t = 4$

Step 4   Check this solution in the original equation.

$4(8 - 3t) = 32 - 8(t + 2)$

$4(8 - 3 \cdot 4) = 32 - 8(4 + 2)$?   Let $t = 4$.

$4(8 - 12) = 32 - 8(6)$?

$4(-4) = 32 - 48$?

$-16 = -16$ True

The solution, 4, checks.

Work Problem ❹ at the Side.

❹ Solve.

(a) $2(4 + 3r)$
$= 3(r + 1) + 11$

(b) $2 - 3(2 + 6z)$
$= 4(z + 1) + 18$

2▭   **Solve equations by clearing fractions and decimals.**   We clear an equation of fractions by multiplying each side by the LCD of all the fractions in the equation. It is a good idea to do this before starting the four-step method to avoid working with fractions.

Example 5   **Solving an Equation with Fractions as Coefficients**

Solve $\frac{2}{3}x - \frac{1}{2}x = -\frac{1}{6}x - 2$.

The least common denominator of all the fractions in the equation is 6. Start by multiplying each side of the equation by 6.

Continued on Next Page

ANSWERS
3. (a) 6   **(b)** all real numbers
4. (a) 2   **(b)** $-\dfrac{13}{11}$

❺ Solve $\frac{1}{4}x - 4 = \frac{3}{2}x + \frac{3}{4}x$.

$$\frac{2}{3}x - \frac{1}{2}x = -\frac{1}{6}x - 2$$

$$6\left(\frac{2}{3}x - \frac{1}{2}x\right) = 6\left(-\frac{1}{6}x - 2\right) \qquad \text{Multiply by 6.}$$

$$6\left(\frac{2}{3}x\right) + 6\left(-\frac{1}{2}x\right) = 6\left(-\frac{1}{6}x\right) + 6(-2) \qquad \text{Distributive property}$$

$$4x - 3x = -x - 12$$

Now use the four steps to solve this equivalent equation.

Step 1	$x = -x - 12$	Combine terms.
Step 2	$x + x = -x - 12 + x$	Add x.
	$2x = -12$	Combine terms.
Step 3	$\dfrac{2x}{2} = \dfrac{-12}{2}$	Divide by 2.
	$x = -6$	

Step 4   Check by substituting -6 for x in the original equation.

$$\frac{2}{3}(-6) - \frac{1}{2}(-6) = -\frac{1}{6}(-6) - 2 \qquad ? \qquad \text{Let } x = -6.$$

$$-4 + 3 = 1 - 2 \qquad\qquad ?$$

$$-1 = -1 \qquad\qquad \text{True}$$

The solution of the equation is -6.

CAUTION

When clearing an equation of fractions, be sure to multiply *every* term on each side of the equation by the LCD.

Work Problem ❺ at the Side.

The multiplication property is also used to clear an equation of decimals.

❻ Solve $.06(100 - x) + .04x = .05(92)$.

Example 6   Solving an Equation with Decimals as Coefficients

Solve $.1t + .05(20 - t) = .09(20)$.

    The decimals are expressed as tenths and hundredths. Choose the smallest exponent on 10 needed to eliminate the decimals; in this case, use $10^2 = 100$. A number can be multiplied by 100 by moving the decimal point two places to the right.

$$.10t + .05(20 - t) = .09(20) \qquad .1 = .10$$

$$10t + 5(20 - t) = 9(20) \qquad \text{Multiply by 100.}$$

Now use the four steps.

Step 1	$10t + 5(20) + 5(-t) = 180$	Distributive property
	$10t + 100 - 5t = 180$	
	$5t + 100 = 180$	Combine terms.
Step 2	$5t + 100 - 100 = 180 - 100$	Subtract 100.
	$5t = 80$	Combine terms.
Step 3	$\dfrac{5t}{5} = \dfrac{80}{5}$	Divide by 5.
	$t = 16$	

Step 4   Check to see that 16 is the solution.

Work Problem ❻ at the Side.

2.3 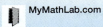 **EXERCISES**

Solve each equation, and check your solution. See Examples 1–4.

1. $5m + 8 = 7 + 4m$

2. $4r + 2 = 3r - 6$

3. $10p + 6 = 12p - 4$

4. $-5x + 8 = -3x + 10$

5. $7r - 5r + 2 = 5r - r$

6. $9p - 4p + 6 = 7p - 3p$

7. $x + 3 = -(2x + 2)$

8. $2x + 1 = -(x + 3)$

9. $4(2x - 1) = -6(x + 3)$

10. $6(3w + 5) = 2(10w + 10)$

11. $6(4x - 1) = 12(2x + 3)$

12. $6(2x + 8) = 4(3x - 6)$

13. $3(2x - 4) = 6(x - 2)$

14. $3(6 - 4x) = 2(-6x + 9)$

15. After correctly working through several steps of the solution of a linear equation, a student obtains the equation $7x = 3x$. Then the student divides each side by x to get $7 = 3$ and gives "no solution" as the answer. Is this correct? If not, explain why.

16. Which linear equation does *not* have all real numbers as solutions?

   A. $5x = 4x + x$ **B.** $2(x + 6) = 2x + 12$ **C.** $\dfrac{1}{2}x = .5x$ **D.** $3x = 2x$

17. Explain in your own words the major steps used in solving a linear equation that does not contain fractions or decimals as coefficients.

18. Explain in your own words the major steps used in solving a linear equation that contains fractions or decimals as coefficients.

Solve each equation by first clearing it of fractions or decimals. See Examples 5 and 6.

19. $-\dfrac{2}{7}r + 2r = \dfrac{1}{2}r + \dfrac{17}{2}$

20. $\dfrac{3}{5}t - \dfrac{1}{10}t = t - \dfrac{5}{2}$

21. $\dfrac{1}{9}(y + 18) + \dfrac{1}{3}(2y + 3) = y + 3$

22. $-\dfrac{1}{4}(x - 12) + \dfrac{1}{2}(x + 2) = x + 4$

23. $-\dfrac{5}{6}q - \left(q - \dfrac{1}{2}\right) = \dfrac{1}{4}(q + 1)$

24. $\dfrac{2}{3}k - \left(k + \dfrac{1}{4}\right) = \dfrac{1}{12}(k + 4)$

25. $.30(30) + .15x = .20(30 + x)$

26. $.20(60) + .05x = .10(60 + x)$

27. $.92x + .98(12 - x) = .96(12)$

28. $1.00x + .05(12 - x) = .10(63)$

29. $.02(5000) + .03x = .025(5000 + x)$

30. $.06(10,000) + .08x = .072(10,000 + x)$

RELATING CONCEPTS (Exercises 31–36)   **FOR INDIVIDUAL OR GROUP WORK**

Work Exercises 31–36 in order.

31. Evaluate the term $100ab$ for $a = 2$ and $b = 4$.

32. Will you get the same answer as in Exercise 31 if you evaluate $(100a)b$ for $a = 2$ and $b = 4$? Why or why not?

33. Is the term $(100a)(100b)$ equivalent to $100ab$? Why or why not?

34. If your answer to Exercise 33 is *no,* explain why the distributive property is not involved.

35. The simplest way to solve the equation $.05(x + 2) + .10x = 2.00$ is to begin by multiplying each side by 100. If we do this, the first term on the left becomes $100(.05)(x + 2)$. Is this expression equivalent to $[100(.05)](x + 2)$? Explain. (*Hint:* Compare to Exercises 31 and 32 with $a = .05$ and $b = x + 2$.)

36. Students often want to "distribute" the 100 to both .05 and $(x + 2)$ in the expression $100(.05)(x + 2)$. Is this correct? (*Hint:* See Exercises 34 and 35.)

Solve each equation, and check your solution. See Examples 1–6.

37. $-3(5z + 24) + 2 = 2(3 - 2z) - 4$

38. $-2(2s - 4) - 8 = -3(4s + 4) - 1$

39. $-(6k - 5) - (-5k + 8) = -3$

40. $-(4y + 2) - (-3y - 5) = 3$

41. $\frac{1}{3}(x + 3) + \frac{1}{6}(x - 6) = x + 3$

42. $\frac{1}{2}(x + 2) + \frac{3}{4}(x + 4) = x + 5$

43. $.30(x + 15) + .40(x + 25) = 25$

44. $.10(x + 80) + .20x = 14$

45. $4(x + 3) = 2(2x + 8) - 4$

46. $4(x + 8) = 2(2x + 6) + 20$

47. $8(t - 3) + 4t = 6(2t + 1) - 10$

48. $9(v + 1) - 3v = 2(3v + 1) - 8$

Write the answer to each problem as an algebraic expression.

49. Two numbers have a sum of 12. One number is q. Find the other number.

50. The product of two numbers is 13. One number is k. What is the other number?

51. A bank teller has t dollars in ten-dollar bills. How many ten-dollar bills does the teller have?

52. A plane ticket costs b dollars for an adult and d dollars for a child. Find the total cost of 5 adult and 3 child tickets.

2.4 AN INTRODUCTION TO APPLICATIONS OF LINEAR EQUATIONS

1 **Learn the six steps for solving applied problems.** We now look at how algebra is used to solve applied problems. It must be emphasized that many *meaningful* applications of mathematics require concepts that are beyond the level of this book. Some of the problems you will encounter will seem "contrived," and to some extent they are. But the skills you will develop in solving simple problems will help you in solving more realistic problems in chemistry, physics, biology, business, and other fields.

In earlier sections we learned how to translate words, phrases, and sentences into mathematical expressions and equations. Now we will use these translations to solve applied problems using algebra. While there is no specific method that enables you to solve all kinds of applied problems, the following six-step method is suggested.

Solving an Applied Problem

Step 1 **Read** the problem carefully until you understand what is given and what is to be found.

Step 2 **Assign a variable** to represent the unknown value, using diagrams or tables as needed. Write down what the variable represents. If necessary, express any other unknown values in terms of the variable.

Step 3 **Write an equation** using the variable expression(s).

Step 4 **Solve** the equation.

Step 5 **State the answer.** Does it seem reasonable?

Step 6 **Check** the answer in the words of the *original* problem.

The third step in solving an applied problem is often the hardest. Begin to translate the problem into an equation by writing the given phrases as mathematical expressions. In transforming an applied problem into an algebraic equation, replace any words that mean *equal* or *same* with an = sign. Other forms of the verb "to be," such as *is, are, was,* and *were*, also translate this way. The = sign leads to an equation to be solved.

2 **Solve problems involving unknown numbers.** Some of the simplest applied problems involve unknown numbers.

Example 1 Finding the Value of an Unknown Number

The product of 4, and a number decreased by 7, is 100. What is the number?

Step 1 **Read** the problem carefully. Decide what you are being asked to find.

Step 2 **Assign a variable** to represent the unknown quantity. In this problem, we are asked to find a number, so we write

$$\text{Let } x = \text{the number.}$$

There are no other unknown quantities to find.

Continued on Next Page

❶ Use the six steps to solve the problem. Give the equation, using x as the variable, and give the answer.

　If 5 is added to the product of 9 and a number, the result is 19 less than the number. Find the number.

Step 3 **Write an equation.**

The product of 4,	and	a number	decreased by	7,	is	100.
↓		↓	↓	↓	↓	↓
$4 \cdot$		$(x$	$-$	$7)$	$=$	100

　Because of the commas in the given problem, writing the equation as $4x - 7 = 100$ is incorrect. The equation $4x - 7 = 100$ corresponds to the statement "The product of 4 and a number, decreased by 7, is 100."

Step 4 **Solve** the equation.

$$4(x - 7) = 100$$
$$4x - 28 = 100 \qquad \text{Distributive property}$$
$$4x - 28 + 28 = 100 + 28 \qquad \text{Add 28.}$$
$$4x = 128 \qquad \text{Combine terms.}$$
$$x = 32 \qquad \text{Divide by 4.}$$

Step 5 **State the answer.** The number is 32.

Step 6 **Check.** When 32 is decreased by 7, we get $32 - 7 = 25$. If 4 is multiplied by 25, we get 100, as required. The answer, 32, is correct.

Work Problem ❶ at the Side.

3 Solve problems involving sums of quantities. A common type of problem in elementary algebra involves finding two quantities when the sum of the quantities is known.

　In general, to solve such problems, choose a variable to represent one of the unknowns and then represent the other quantity in terms of the same variable, using information contained in the problem. Then write an equation based on the words of the problem. The next example illustrates these ideas.

Example 2 Finding the Numbers of Olympic Medals Won by the United States

In the 2000 Olympics, U.S. contestants won 14 more gold than silver medals. They won a total of 64 gold and silver medals. How many of each type medal were won? (*Source:* United States Olympic Committee.)

Step 1 **Read** the problem. We are given information about the total number of gold and silver medals, and we are asked to find the number of each kind.

Step 2 **Assign a variable.**

Let　x = the number of silver medals.

　　Then　$x + 14$ = the number of gold medals.

Step 3 **Write an equation.**

The total	is	the number of silver	plus	the number of gold.
↓	↓	↓	↓	↓
64	$=$	x	$+$	$(x + 14)$

Continued on Next Page

Step 4 **Solve** the equation.

$$64 = 2x + 14 \qquad \text{Combine terms.}$$
$$64 - 14 = 2x + 14 - 14 \qquad \text{Subtract 14.}$$
$$50 = 2x \qquad \text{Combine terms.}$$
$$25 = x \qquad \text{Divide by 2.}$$

Step 5 **State the answer.** Because x represents the number of silver medals, the U.S. athletes won 25 silver medals. Because $x + 14$ represents the number of gold medals, they won $25 + 14 = 39$ gold medals.

Step 6 **Check.** Since there were 39 gold and 25 silver medals, the total number of medals was $39 + 25 = 64$. Because $39 - 25 = 14$, there were 14 more gold medals than silver medals. This information agrees with what is given in the problem, so the answers check.

=========================== **Work Problem ❷ at the Side.**

NOTE

The problem in Example 2 could also have been solved by letting x represent the number of gold medals. Then $x - 14$ would represent the number of silver medals. The equation would then be

$$64 = x + (x - 14).$$

The solution of this equation is 39, which is the number of gold medals. The number of silver medals would then be $39 - 14 = 25$. The answers are the same, whichever approach is used.

Sometimes it is necessary to find three unknown quantities in an applied problem. Frequently the three unknowns are compared in *pairs*. When this happens, it is usually easiest to let the variable represent the unknown found in both pairs. The next example illustrates this.

Example 3 **Dividing a Board into Pieces**

The instructions for a woodworking project call for three pieces of wood. The longest piece must be twice the length of the middle-sized piece, and the shortest piece must be 10 in. shorter than the middle-sized piece. Maria Gonzales has a board 70 in. long that she wishes to use. How long can each piece be?

Step 1 **Read** the problem. There will be three answers.

Step 2 **Assign a variable.** Since the middle-sized piece appears in both pairs of comparisons, let x represent the length, in inches, of the middle-sized piece. We have

$$x = \text{the length of the middle-sized piece,}$$
$$2x = \text{the length of the longest piece, and}$$
$$x - 10 = \text{the length of the shortest piece.}$$

A sketch is helpful here. See Figure 1.

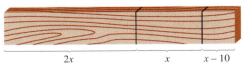

2x x x − 10

Figure 1

================ **Continued on Next Page**

❷ Solve the problem.
   On one day of their vacation, Annie drove three times as far as Jim. Altogether they drove 84 mi that day. Find the number of miles driven by each.

❸ Solve the problem.

A piece of pipe is 50 in. long. It is cut into three pieces. The longest piece is 10 in. more than the middle-sized piece, and the shortest piece measures 5 in. less than the middle-sized piece. Find the lengths of the three pieces.

Step 3 **Write an equation.**

Longest		Middle-sized		Shortest	is	Total length

$$2x + x + (x - 10) = 70$$

Step 4 **Solve.**

$$4x - 10 = 70 \qquad \text{Combine terms.}$$
$$4x - 10 + 10 = 70 + 10 \qquad \text{Add 10.}$$
$$4x = 80 \qquad \text{Combine terms.}$$
$$x = 20 \qquad \text{Divide by 4.}$$

Step 5 **State the answer.** The middle-sized piece is 20 in. long, the longest piece is $2(20) = 40$ in. long, and the shortest piece is $20 - 10 = 10$ in. long.

Step 6 **Check.** The sum of the lengths is 70 in. All conditions of the problem are satisfied.

Work Problem ❸ at the Side.

Example 4 Analyzing a Gasoline/Oil Mixture

A lawn trimmer uses a mixture of gasoline and oil. The mixture contains 16 oz of gasoline for each ounce of oil. If the tank holds 68 oz of the mixture, how many ounces of oil and how many ounces of gasoline does it require when it is full?

Step 1 **Read** the problem. We must find how many ounces of oil and gasoline are needed to fill the tank.

Step 2 **Assign a variable.** Let $x =$ the number of ounces of oil required. Then $16x =$ the number of ounces of gasoline required.

❹ Solve the problem.

At a meeting of the local stamp club, each member brought two nonmembers. If a total of 27 people attended, how many were members and how many were nonmembers?

Step 3 **Write an equation.**

Amount of gasoline		Amount of oil	is	Total amount in tank

$$16x + x = 68$$

Step 4 **Solve.**

$$17x = 68 \qquad \text{Combine terms.}$$
$$x = 4 \qquad \text{Divide by 17.}$$

Step 5 **State the answer.** The lawn trimmer requires 4 oz of oil and $16(4) = 64$ oz of gasoline when full.

Step 6 **Check.** Since $4 + 64 = 68$, and 64 is 16 times 4, the answers check.

Work Problem ❹ at the Side.

4 Solve problems involving supplementary and complementary angles. The next example deals with concepts from geometry. An angle can be measured by a unit called the degree (°). Two angles whose sum is 90° are said to be **complementary,** or complements of each other. An angle that measures 90° is a **right angle.** Two angles whose sum is 180° are said to be **supplementary,** or supplements of each other. One angle *supplements* the other to form a **straight angle** of 180°. See Figure 2. If x represents the degree measure of an angle, then

$90 - x$ represents the degree measure of its complement, and

$180 - x$ represents the degree measure of its supplement.

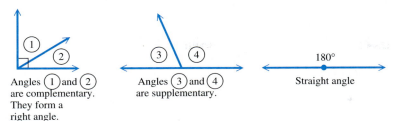

Angles ① and ②
are complementary.
They form a
right angle.

Angles ③ and ④
are supplementary.

180°

Straight angle

Figure 2

⑤ Find each angle measure.

  (a) The supplement of an angle that measures 92°

Example 5   **Finding the Measure of an Angle**

Find the measure of an angle whose supplement is 10° more than twice its complement.

Step 1 **Read** the problem. We are to find the measure of an angle, given information about its complement and its supplement.

Step 2 **Assign a variable.**

Let x = the degree measure of the angle.

Then $90 - x$ = the degree measure of its complement;

                $180 - x$ = the degree measure of its supplement.

Step 3 **Write an equation.**

Supplement   is   10   more than   twice   its complement.

$$180 - x \quad = \quad 10 \quad + \quad 2 \quad \cdot \quad (90 - x)$$

Step 4 **Solve.**

$$180 - x = 10 + 180 - 2x \qquad \text{Distributive property}$$
$$180 - x = 190 - 2x \qquad \text{Combine terms.}$$
$$180 - x + 2x = 190 - 2x + 2x \qquad \text{Add } 2x.$$
$$180 + x = 190 \qquad \text{Combine terms.}$$
$$180 + x - 180 = 190 - 180 \qquad \text{Subtract 180.}$$
$$x = 10$$

Step 5 **State the answer.** The measure of the angle is 10°.

Step 6 **Check.** The complement of 10° is 80° and the supplement of 10° is 170°. 170° is equal to 10° more than twice 80° ($170 = 10 + 2(80)$ is true); therefore, the answer is correct.

  (b) An angle whose complement has twice its measure

  (c) An angle such that twice its complement is 30° less than its supplement

Work Problem ⑤ at the Side.

5   **Solve problems involving consecutive integers.**   Two integers that differ by 1 are called **consecutive integers.** For example, 3 and 4, 6 and 7, and -2 and -1 are pairs of consecutive integers. In general, if x represents an integer, $x + 1$ represents the next larger consecutive integer.

Consecutive *even* integers, such as 8 and 10, differ by 2. Similarly, consecutive *odd* integers, such as 9 and 11, also differ by two. In general, if x represents an even integer, $x + 2$ represents the next larger consecutive even integer. The same holds true for odd integers; that is, if x is an odd integer, $x + 2$ is the next larger odd integer.

❻ Solve the problem.
Find two consecutive integers whose sum is −45.

Example 6 Finding Consecutive Integers

Two pages that face each other in this book have 569 as the sum of their page numbers. What are the page numbers?

Step 1 **Read** the problem. Because the two pages face each other, they must have page numbers that are consecutive integers.

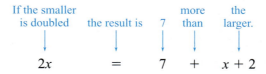

Step 2 **Assign a variable.**

Let x = the smaller page number.

Then $x + 1$ = the larger page number.

Step 3 **Write an equation.** Because the sum of the page numbers is 569, the equation is

$$x + (x + 1) = 569.$$

Step 4 **Solve.**
$$
\begin{aligned}
x + (x + 1) &= 569 \\
2x + 1 &= 569 \quad &\text{Combine terms.} \\
2x &= 568 \quad &\text{Subtract 1.} \\
x &= 284 \quad &\text{Divide by 2.}
\end{aligned}
$$

Step 5 **State the answer.** The smaller page number is 284, and the larger page number is $284 + 1 = 285$.

Step 6 **Check.** The sum of 284 and 285 is 569. The answers are correct.

Work Problem ❻ at the Side.

In the final example, we do not number the steps. See if you can identify them.

❼ Solve the problem.
Find two consecutive even integers such that six times the smaller added to the larger gives a sum of 86.

Example 7 Finding Consecutive Odd Integers

If the smaller of two consecutive odd integers is doubled, the result is 7 more than the larger of the two integers. Find the two integers.

Let x be the smaller integer. Since the two numbers are consecutive *odd* integers, then $x + 2$ is the larger. Now write an equation.

If the smaller is doubled	the result is	7	more than	the larger.
$2x$	$=$	7	$+$	$x + 2$

Solve the equation.
$$
\begin{aligned}
2x &= 7 + x + 2 \\
2x &= 9 + x \quad &\text{Combine terms.} \\
2x - x &= 9 + x - x \quad &\text{Subtract } x. \\
x &= 9 \quad &\text{Combine terms.}
\end{aligned}
$$

The first integer is 9 and the second is $9 + 2 = 11$. To check the answers, we see that when 9 is doubled, we get 18, which is 7 more than the larger odd integer, 11. The answers are correct.

Work Problem ❼ at the Side.

ANSWERS

6. −23, −22

7. 12, 14

2.4 **EXERCISES**

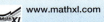
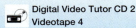
1. In your own words, write the general procedure for solving applications as outlined in this section.

2. List some of the words that translate as "=" when writing an equation to solve an applied problem.

3. Suppose that a problem requires you to find the number of cars on a dealer's lot. Which one of the following would not be a reasonable answer? Justify your answer.

   A. 0   **B.** 45   **C.** 1   **D.** $6\frac{1}{2}$

4. Suppose that a problem requires you to find the number of hours a light bulb is on during a day. Which one of the following would not be a reasonable answer? Justify your answer.

   A. 0   **B.** 4.5   **C.** 13   **D.** 25

Solve each problem. See Example 1.

5. If 2 is subtracted from a number and this difference is tripled, the result is 6 more than the number. Find the number.

6. If 3 is added to a number and this sum is doubled, the result is 2 more than the number. Find the number.

7. The sum of three times a number and 7 more than the number is the same as the difference between −11 and twice the number. What is the number?

8. If 4 is added to twice a number and this sum is multiplied by 2, the result is the same as if the number is multiplied by 3 and 4 is added to the product. What is the number?

Solve each problem. See Examples 2–4.

9. The U.S. Senate has 100 members. During the 106th session (1999–2001), there were 10 more Republicans than Democrats. How many Democrats and Republicans were there in the Senate? (*Source: The World Almanac and Book of Facts,* 2000.)

10. The total number of Democrats and Republicans in the U.S. House of Representatives during the 106th session was 434. There were 12 fewer Democrats than Republicans. How many members of each party were there? (*Source: The World Almanac and Book of Facts,* 2000.)

11. There were 2783 more men than women competing in the 1996 Olympic Games in Atlanta. The total number of competitors was 10,341. How many men and how many women competed? (*Source: The Universal Almanac*, 1997.)

12. In the first NFL Championship, played in 1967, Green Bay and Kansas City scored a total of 45 points. Green Bay won by 25 points. What was the score? (*Source:* www.nfl.com.)

13. The largest recorded dog is an English mastiff named Zorba, who weighs 63 lb more than an average lioness. The sum of the two animals' weights is 623 lb. How much do the dog and the lioness each weigh? (*Source: The Guinness Book of Records*, 1996.)

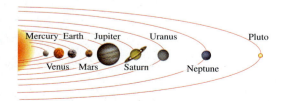

14. In the 1998 U.S. Senior Open, Hale Irwin finished with a score of 1 more than the winner, Vicente Fernandez. The sum of their scores was 571. Find their scores. (*Source:* Television coverage.)

15. In one day, Akilah Cadet received 13 packages. Federal Express delivered three times as many as Airborne Express, while United Parcel Service delivered 2 fewer than Airborne Express. How many packages did each service deliver to Akilah?

16. In his job at the post office, Eddie Thibodeaux works a 6.5-hr day. He sorts mail, sells stamps, and does supervisory work. One day he sold stamps twice as long as he sorted mail, and he supervised .5 hr longer than he sorted mail. How many hours did he spend at each task?

17. Venus is 31.2 million mi farther from the sun than Mercury, while Earth is 57 million mi farther from the sun than Mercury. If the total of the distances from these three planets to the sun is 196.2 million mi, how far away from the sun is Mercury? (All distances given here are *mean* (*average*) distances.) (*Source: The Universal Almanac*, 1997.)

18. Together, Saturn, Jupiter, and Mars have a total of 36 known satellites (moons). Jupiter has 2 fewer satellites than Saturn, and Mars has 16 fewer satellites than Saturn. How many known satellites does Mars have? (*Source: The World Almanac and Book of Facts*, 2000.)

19. The sum of the measures of the angles of any triangle is 180°. In triangle *ABC*, angles *A* and *B* have the same measure, while the measure of angle *C* is 60° larger than each of *A* and *B*. What are the measures of the three angles?

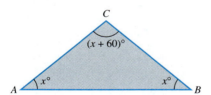

20. Nagaraj Nanjappa has a party-length submarine sandwich 59 in. long. He wants to cut it into three pieces so that the middle piece is 5 in. longer than the shortest piece and the shortest piece is 9 in. shorter than the longest piece. How long should the three pieces be?

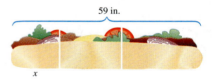

Solve each problem. See Example 4.

21. The September 11, 2000, issue of *Coin World* listed the value of a "Mint State-65" (uncirculated) 1950 Jefferson nickel minted at Denver as $\frac{7}{6}$ the value of a similar condition 1945 nickel minted at Philadelphia. Together the total value of the two coins is $26.00. What is the value of each coin?

22. The largest sheep ranch in the world is located in Australia. The number of sheep on the ranch is $\frac{8}{3}$ the number of uninvited kangaroos grazing on the pastureland. Together, herds of these two animals number 88,000. How many sheep and how many kangaroos roam the ranch? (*Source: The Guinness Book of Records,* 1996.)

23. In 1988, a dairy in Alberta, Canada, created a sundae with approximately 1 lb of topping for every 83.2 lb of ice cream. The total of the two ingredients weighed approximately 45,225 lb. To the nearest tenth of a pound, how many pounds of ice cream and how many pounds of topping were there? (*Source: The Guinness Book of Records,* 1996.)

24. A husky running the Iditarod (a thousand-mile race between Anchorage and Nome, Alaska) burns $5\frac{3}{8}$ calories in exertion for every 1 calorie burned in thermoregulation in extreme cold. According to one scientific study, a husky in top condition burns an amazing total of 11,200 calories per day. How many calories are burned for exertion, and how many are burned for regulation of body temperature? Round answers to the nearest whole number.

Solve each problem. See Example 5.

25. Find the measure of an angle whose complement is four times its measure.

26. Find the measure of an angle whose supplement is three times its measure.

27. Find the measure of an angle whose supplement measures 39° more than twice its complement.

28. Find the measure of an angle whose supplement measures 38° less than three times its complement.

29. Find the measure of an angle such that the difference between the measures of its supplement and three times its complement is 10°.

30. Find the measure of an angle such that the sum of the measures of its complement and its supplement is 160°.

Solve each problem. See Examples 6 and 7.

31. The numbers on two consecutively numbered gym lockers have a sum of 137. What are the locker numbers?

32. The sum of two consecutive checkbook check numbers is 357. Find the numbers.

33. Find two consecutive even integers such that the smaller added to three times the larger gives a sum of 46.

34. Find two consecutive odd integers such that twice the larger is 17 more than the smaller.

35. Two pages that are back-to-back in this book have 203 as the sum of their page numbers. What are the page numbers?

36. Two houses on the same side of the street have house numbers that are consecutive even integers. The sum of the integers is 58. What are the two house numbers?

37. When the smaller of two consecutive integers is added to three times the larger, the result is 43. Find the integers.

38. If five times the smaller of two consecutive integers is added to three times the larger, the result is 59. Find the integers.

Apply the ideas of this section to solve Exercises 39 and 40, based on the graphs.

39. In a recent year, the funding for Head Start programs increased by .55 billion dollars from the funding in the previous year. The following year, the increase was .20 billion dollars more. For those three years, the total funding was 9.64 billion dollars. How much was funded in each of these years? (*Source:* U.S. Department of Health and Human Services.)

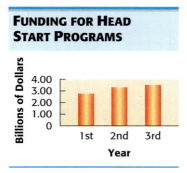

40. According to data provided by the National Safety Council for a recent year, the number of serious injuries per 100,000 participants in football, bicycling, and golf is illustrated in the graph. There were 800 more in bicycling than in golf, and there were 1267 more in football than in bicycling. Altogether there were 3179 serious injuries per 100,000 participants. How many such serious injuries were there in each sport?

2.5 FORMULAS AND APPLICATIONS FROM GEOMETRY

Many applied problems can be solved with formulas. For example, formulas exist for geometric figures such as squares and circles, for distance, for money earned on bank savings, and for converting English measurements to metric measurements. The formulas used in this book are given on the inside covers.

1 **Solve a formula for one variable, given the values of the other variables.** Given the values of all but one of the variables in a formula, we can find the value of the remaining variable by using the methods introduced in this chapter.

In Example 1, we use the idea of *area*. The **area** of a plane (two-dimensional) geometric figure is a measure of the surface covered by the figure.

OBJECTIVES

1 Solve a formula for one variable, given the values of the other variables.

2 Use a formula to solve an applied problem.

3 Solve problems involving vertical angles and straight angles.

4 Solve a formula for a specified variable.

Example 1 **Using a Formula to Evaluate a Variable**

Find the value of the remaining variable in each formula.

(a) $A = LW; A = 64, L = 10$

As shown in Figure 3, this formula gives the area of a rectangle with length L and width W.

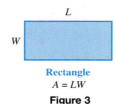

Rectangle
$A = LW$

Figure 3

Substitute the given values into the formula and then solve for W.

$$A = LW$$
$$64 = 10W \qquad \text{Let } A = 64 \text{ and } L = 10.$$
$$6.4 = W \qquad \text{Divide by 10.}$$

Check that the width of the rectangle is 6.4.

(b) $A = \frac{1}{2}h(b + B); A = 210, B = 27, h = 10$

This formula gives the area of a trapezoid with parallel sides of lengths b and B and distance h between the parallel sides. See Figure 4.

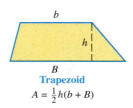

Trapezoid
$A = \frac{1}{2}h(b + B)$

Figure 4

Again, begin by substituting the given values into the formula.

$$A = \frac{1}{2}h(b + B)$$

$$210 = \frac{1}{2}(10)(b + 27) \qquad A = 210, h = 10, B = 27$$

Continued on Next Page

① Find the value of the remaining variable in each formula.

(a) $I = prt$; $I = \$246$, $r = .06$, $t = 2$

(b) $P = 2L + 2W$; $P = 126$, $W = 25$

② Solve the problem.

A farmer has 800 m of fencing material to enclose a rectangular field. The width of the field is 175 m. Find the length of the field.

Now solve for b.

$$210 = 5(b + 27) \qquad \text{Multiply.}$$
$$210 = 5b + 135 \qquad \text{Distributive property}$$
$$210 - \mathbf{135} = 5b + 135 - \mathbf{135} \qquad \text{Subtract 135.}$$
$$75 = 5b \qquad \text{Combine terms.}$$
$$\frac{75}{\mathbf{5}} = \frac{5b}{\mathbf{5}} \qquad \text{Divide by 5.}$$
$$15 = b$$

Check that the length of the shorter parallel side, b, is 15.

Work Problem ① at the Side.

2 ▭ **Use a formula to solve an applied problem.** As the next examples show, formulas are often used to solve applied problems. *It is a good idea to draw a sketch when a geometric figure is involved.* Example 2 uses the idea of *perimeter*. The **perimeter** of a plane (two-dimensional) geometric figure is the distance around the figure, that is, the sum of the lengths of its sides. We use the six steps introduced in the previous section.

Example 2 Finding the Width of a Rectangular Lot

A rectangular lot has perimeter 80 m and length 25 m. Find the width of the lot.

Step 1 **Read.** We are told to find the width of the lot.

Step 2 **Assign a variable.** Let W = the width of the lot in meters. See Figure 5.

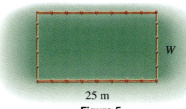

25 m

Figure 5

Step 3 **Write an equation.** The formula for the perimeter of a rectangle is

$$P = 2L + 2W.$$

Find the width by substituting 80 for P and 25 for L in the formula.

$$\mathbf{80} = 2(\mathbf{25}) + 2W \qquad P = 80, L = 25$$

Step 4 **Solve** the equation.

$$80 = 50 + 2W \qquad \text{Multiply.}$$
$$80 - \mathbf{50} = 50 + 2W - \mathbf{50} \qquad \text{Subtract 50.}$$
$$30 = 2W \qquad \text{Combine terms.}$$
$$15 = W \qquad \text{Divide by 2.}$$

Step 5 **State the answer.** The width is 15 m.

Step 6 **Check.** If the width is 15 m and the length is 25 m, the distance around the rectangular lot (perimeter) is $2(25) + 2(15) = 50 + 30 = 80$ m, as required.

Work Problem ② at the Side.

Example 3   **Finding the Height of a Triangular Sail**

The area of a triangular sail of a sailboat is 126 ft^2. (Recall that ft^2 means "square feet.") The base of the sail is 12 ft. Find the height of the sail.

Step 1   **Read.** We must find the height of the triangular sail.

Step 2   **Assign a variable.** Let h = the height of the sail in feet. See Figure 6.

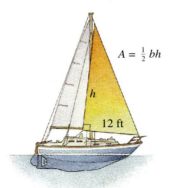

$A = \frac{1}{2} bh$

h

12 ft

Figure 6

Step 3   **Write an equation.** The formula for the area of a triangle is $A = \frac{1}{2} bh$, where A is the area, b is the base, and h is the height. Using the information given in the problem, substitute 126 for A and 12 for b in the formula.

$$A = \frac{1}{2} bh$$

$$\mathbf{126} = \frac{1}{2}(\mathbf{12})h \quad A = 126, b = 12$$

Step 4   **Solve** the equation.

$$126 = 6h \quad \text{Multiply.}$$
$$21 = h \quad \text{Divide by 6.}$$

Step 5   **State the answer.** The height of the sail is 21 ft.

Step 6   **Check** to see that the values $A = 126$, $b = 12$, and $h = 21$ satisfy the formula for the area of a triangle.

Work Problem ❸ at the Side.

❸ Solve the problem.
   The area of a triangle is 120 m^2. The height is 24 m. Find the length of the base of the triangle.

3▭   **Solve problems involving vertical angles and straight angles.**   Figure 7 shows two intersecting lines forming angles that are numbered ①, ②, ③, and ④. Angles ① and ③ lie "opposite" each other. They are called **vertical angles.** Another pair of vertical angles is ② and ④. In geometry, it is shown that vertical angles have equal measures.
   Now look at angles ① and ②. When their measures are added, we get the measure of a **straight angle,** which is 180°. There are three other such pairs of angles: ② and ③, ③ and ④, and ① and ④.
   The next example uses these ideas.

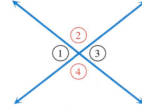

②
①   ③
④

Figure 7

❹ Find the measure of each marked angle.

(a)

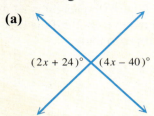

$(2x + 24)°$ $(4x - 40)°$

(b)

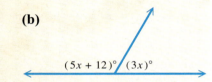

$(5x + 12)°$ $(3x)°$

Example 4 Finding Angle Measures

Refer to the appropriate figure in each part.

(a) Find the measure of each marked angle in Figure 8.

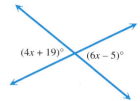

$(4x + 19)°$ $(6x - 5)°$

Figure 8

Since the marked angles are vertical angles, they have equal measures. Set $4x + 19$ equal to $6x - 5$ and solve.

$$4x + 19 = 6x - 5$$
$$-4x + 4x + 19 = -4x + 6x - 5 \quad \text{Add } -4x.$$
$$19 = 2x - 5$$
$$19 + 5 = 2x - 5 + 5 \quad \text{Add } 5.$$
$$24 = 2x$$
$$12 = x \quad \text{Divide by 2.}$$

Since $x = 12$, one angle has measure $4(12) + 19 = 67$ degrees. The other has the same measure, since $6(12) - 5 = 67$ as well. Each angle measures $67°$.

(b) Find the measure of each marked angle in Figure 9.

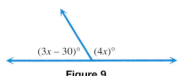

$(3x - 30)°$ $(4x)°$

Figure 9

The measures of the marked angles must add to $180°$ because together they form a straight angle. The equation to solve is

$$(3x - 30) + 4x = 180.$$
$$7x - 30 = 180 \quad \text{Combine terms.}$$
$$7x - 30 + 30 = 180 + 30 \quad \text{Add 30.}$$
$$7x = 210$$
$$x = 30 \quad \text{Divide by 7.}$$

To find the measures of the angles, replace x with 30 in the two expressions.

$$3x - 30 = 3(30) - 30 = 90 - 30 = 60$$
$$4x = 4(30) = 120$$

The two angle measures are $60°$ and $120°$.

Work Problem ❹ at the Side.

4▬▬ **Solve a formula for a specified variable.** Sometimes it is necessary to solve a large number of problems that use the same formula. For example, a surveying class might need to solve several problems that involve the formula for the area of a rectangle, $A = LW$. Suppose that in each problem the area (A) and the length (L) of a rectangle are given, and the width (W) must be found. Rather than solving for W each time the formula is used, it would be simpler to rewrite the *formula* so that it is solved for W. This process is called **solving for a specified variable.**

In solving a formula for a specified variable, we treat the specified variable as if it were the *only* variable in the equation, and treat the other variables as if they were numbers. We use the same steps to solve the equation for the specified variable that we have used to solve equations with just one variable.

Example 5 **Solving for a Specified Variable**

Solve $A = LW$ for W.

Think of undoing what has been done to W. Since W is multiplied by L, undo the multiplication by dividing each side of $A = LW$ by L.

$$A = LW$$

$$\frac{A}{L} = \frac{LW}{L} \qquad \text{Divide by } L.$$

$$\frac{A}{L} = W \qquad \frac{L}{L} = 1; 1W = W$$

The formula is now solved for W.

Work Problem ❺ at the Side.

Example 6 **Solving for a Specified Variable**

Solve $P = 2L + 2W$ for L.

We want to get L alone on one side of the equation. We begin by subtracting $2W$ from each side.

$$P = 2L + 2W$$

$$P - 2W = 2L + 2W - 2W \qquad \text{Subtract } 2W.$$

$$P - 2W = 2L \qquad \text{Combine terms.}$$

$$\frac{P - 2W}{2} = \frac{2L}{2} \qquad \text{Divide by 2.}$$

$$\frac{P - 2W}{2} = L \qquad \frac{2}{2} = 1; 1L = L$$

The last step gives the formula solved for L, as required.

---

❺ (a) Solve $I = prt$ for t.

(b) Solve $P = a + b + c$ for a.

---

6 (a) Solve $A = p + prt$ for t.

Example 7 Solving for a Specified Variable

Solve $F = \frac{9}{5}C + 32$ for C. (This is the formula for converting from Celsius to Fahrenheit.)

We need to isolate C on one side of the equation. First undo the addition of 32 to $\frac{9}{5}C$ by subtracting 32 from each side.

$$F = \frac{9}{5}C + 32$$

$$F - 32 = \frac{9}{5}C + 32 - 32 \qquad \text{Subtract 32.}$$

$$F - 32 = \frac{9}{5}C$$

Now multiply each side by $\frac{5}{9}$. Use parentheses on the left.

$$\frac{5}{9}(F - 32) = \frac{5}{9} \cdot \frac{9}{5}C \qquad \text{Multiply by } \frac{5}{9}.$$

$$\frac{5}{9}(F - 32) = C$$

This last result is the formula for converting temperatures from Fahrenheit to Celsius.

Work Problem 6 at the Side.

(b) Solve $Ax + By = C$ for y.

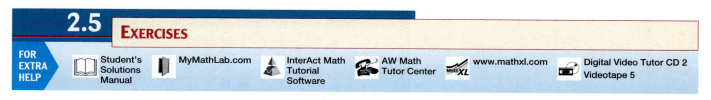

2.5 **EXERCISES**

FOR EXTRA HELP   Student's Solutions Manual   MyMathLab.com   InterAct Math Tutorial Software   AW Math Tutor Center   www.mathxl.com   Digital Video Tutor CD 2 Videotape 5

1. In your own words, explain what is meant by each term.
   (a) Perimeter of a plane geometric figure

   (b) Area of a plane geometric figure

2. Perimeter is to a polygon as _____ is to a circle.

3. If a formula has exactly five variables, how many values would you need to be given in order to find the value of any one variable?

4. Look at the drawings of a rectangle and a trapezoid at the beginning of this section. Discuss their similarities and their differences.

Decide whether perimeter or area would be used to solve a problem concerning the measure of the quantity.

5. Sod for a lawn

6. Carpeting for a bedroom

7. Baseboards for a living room

8. Fencing for a yard

9. Fertilizer for a garden

10. Tile for a bathroom

11. Determining the cost of planting rye grass in a lawn for the winter

12. Determining the cost of replacing a linoleum floor with a wood floor

In the following exercises a formula is given, along with the values of all but one of the variables in the formula. Find the value of the variable that is not given. (When necessary, use 3.14 as an approximation for π.) See Example 1.

13. $P = 2L + 2W$ (perimeter of a rectangle); $L = 8$, $W = 5$

14. $P = 2L + 2W$; $L = 6$, $W = 4$

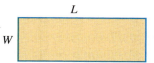

15. $A = \dfrac{1}{2}bh$ (area of a triangle); $b = 8$, $h = 16$

16. $A = \dfrac{1}{2}bh$; $b = 10$, $h = 14$

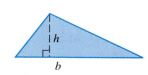

17. $P = a + b + c$ (perimeter of a triangle); $P = 12$, $a = 3$, $c = 5$

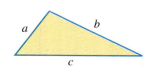

18. $P = a + b + c$; $P = 15$, $a = 3$, $b = 7$

19. $d = rt$ (distance formula); $d = 252$, $r = 45$

20. $d = rt$; $d = 100$, $t = 2.5$

21. $I = prt$ (simple interest); $p = 7500$, $r = .035$, $t = 6$

22. $I = prt$; $p = 5000$, $r = .025$, $t = 7$

23. $C = 2\pi r$ (circumference of a circle); $C = 16.328$

24. $C = 2\pi r$; $C = 8.164$

25. $A = \pi r^2$ (area of a circle); $r = 4$

26. $A = \pi r^2$; $r = 12$

The **volume** of a three-dimensional object is a measure of the space occupied by the object. For example, we would need to know the volume of a gasoline tank in order to know how many gallons of gasoline it would take to completely fill the tank. In the following exercises, a formula for the volume (V) of a three-dimensional object is given, along with values for the other variables. Evaluate V. (Use 3.14 as an approximation for π.) See Example 1.

27. $V = LWH$ (volume of a rectangular box); $L = 10$, $W = 5$, $H = 3$

28. $V = LWH$; $L = 12$, $W = 8$, $H = 4$

29. $V = \dfrac{1}{3}Bh$ (volume of a pyramid); $B = 12$, $h = 13$

30. $V = \dfrac{1}{3}Bh$; $B = 36$, $h = 4$

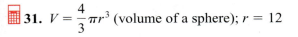

31. $V = \dfrac{4}{3}\pi r^3$ (volume of a sphere); $r = 12$

32. $V = \dfrac{4}{3}\pi r^3$; $r = 6$

Use a formula to write an equation for each application, and then use the problem-solving method of Section 2.4 to solve. (Use 3.14 as an approximation for π.) Formulas are found on the inside covers of this book. See Examples 2 and 3.

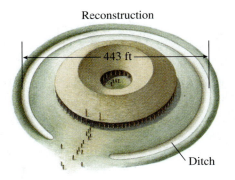

Reconstruction

443 ft

Ditch

33. Recently, a prehistoric ceremonial site dating to about 3000 B.C. was discovered at Stanton Drew in southwestern England. The site, which is larger than Stonehenge, is a nearly perfect circle, consisting of nine concentric rings that probably held upright wooden posts. Around this timber temple is a wide, encircling ditch enclosing an area with a diameter of 443 ft. Find this enclosed area to the nearest thousand square feet. (*Source: Archaeology,* vol. 51, no. 1, Jan./Feb. 1998.)

34. The Skydome in Toronto, Canada, is the first stadium with a hard-shell, retractable roof. The steel dome is 630 ft in diameter. To the nearest foot, what is the circumference of this dome? (*Source:* www.4ballparks.com.)

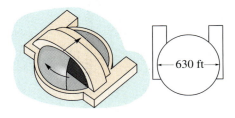

630 ft

35. The *Daily Banner,* published in Roseberg, Oregon, in the 19th century, had page size 3 in. by 3.5 in. What was the perimeter? What was the area? (*Source: The Guinness Book of Records,* 1994.)

36. The newspaper *The Constellation,* printed in 1859 in New York City as part of the Fourth of July celebration, had length 51 in. and width 35 in. What was the perimeter? What was the area? (*Source: The Guinness Book of Records,* 1994.)

37. The largest drum ever constructed was played at the Royal Festival Hall in London in 1987. It had a diameter of 13 ft. What was the area of the circular face of the drum? (*Hint:* Use $A = \pi r^2$.) (*Source: The Guinness Book of Records,* 1994.)

38. What was the circumference of the drum described in Exercise 37? (*Hint:* Use $C = 2\pi r$.)

39. The survey plat depicted here shows two lots that form a trapezoid. The measures of the parallel sides are 115.80 ft and 171.00 ft. The height of the trapezoid is 165.97 ft. Find the combined area of the two lots. Round your answer to the nearest hundredth of a square foot.

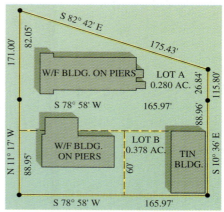

Source: Property survey in New Roads, Louisiana.

40. Lot A in the figure is in the shape of a trapezoid. The parallel sides measure 26.84 ft and 82.05 ft. The height of the trapezoid is 165.97 ft. Find the area of Lot A. Round your answer to the nearest hundredth of a square foot.

41. The U.S. Postal Service requires that any box sent through the mail have length plus girth (distance around) totaling no more than 108 in. The maximum volume that meets this condition is contained by a box with a square end 18 in. on each side. What is the length of the box? What is the maximum volume?

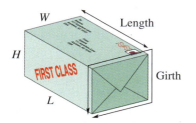

42. The largest box of popcorn was filled by students in Jacksonville, Florida. The box was approximately 40 ft long, $20\frac{2}{3}$ ft wide, and 8 ft high. To the nearest cubic foot, what was the volume of the box? (*Source: The Guinness Book of Records,* 1998.)

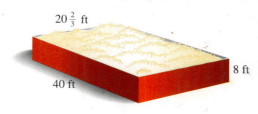

Find the measure of each marked angle. See Example 4.

43.

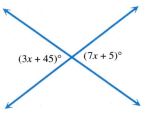

$(x + 1)°$ $(4x - 56)°$

44.

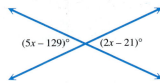

$(10x + 7)°$ $(7x + 3)°$

45.

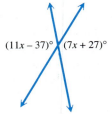

$(5x - 129)°$ $(2x - 21)°$

46.

$(3x + 45)°$ $(7x + 5)°$

47.

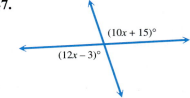

$(10x + 15)°$

$(12x - 3)°$

48.

$(11x - 37)°$ $(7x + 27)°$

Solve each formula for the specified variable. See Examples 5–7.

49. $d = rt$ for t

50. $d = rt$ for r

51. $V = LWH$ for H

52. $A = LW$ for L

53. $P = a + b + c$ for b

54. $P = a + b + c$ for a

55. $I = prt$ for r

56. $I = prt$ for p

57. $A = \dfrac{1}{2}bh$ for h

58. $A = \dfrac{1}{2}bh$ for b

59. $P = 2L + 2W$ for W

60. $A = p + prt$ for r

61. $V = \dfrac{1}{3}\pi r^2 h$ for h

62. $V = \pi r^2 h$ for h

63. $C = \dfrac{5}{9}(F - 32)$ for F

2.6   RATIO, PROPORTION, AND PERCENT

1   **Write ratios.**   A **ratio** is a comparison of two quantities using a quotient.

OBJECTIVES

1   Write ratios.

2   Solve proportions.

3   Solve applied problems using proportions.

4   Find percentages and percents.

Ratio

The ratio of the number a to the number b is written

$$a \text{ to } b, \quad a:b, \quad \text{or} \quad \frac{a}{b}.$$

This last way of writing a ratio is most common in algebra.

Percents are ratios where the second number is always 100. For example, 50% represents the ratio of 50 to 100, 27% represents the ratio of 27 to 100, and so on.

1 Write each ratio.

(a) 9 women to 5 women

Example 1   Writing a Word Phrase as a Ratio

Write a ratio for each word phrase.

(a) The ratio of 5 hr to 3 hr is

$$\frac{5 \text{ hr}}{3 \text{ hr}} = \frac{5}{3}.$$

(b) To find the ratio of 6 hr to 3 days, first convert 3 days to hours.

$$3 \text{ days} = 3 \cdot 24$$
$$= 72 \text{ hr}$$

The ratio of 6 hr to 3 days is thus

$$\frac{6 \text{ hr}}{3 \text{ days}} = \frac{6 \text{ hr}}{72 \text{ hr}} = \frac{6}{72} = \frac{1}{12}.$$

Work Problem 1 at the Side.

An example of the use of a ratio is in unit pricing, to see which size of an item offered in different sizes produces the best price per unit. To do this, set up the ratio of the price of the item to the number of units on the label. Then divide to obtain the price per unit.

(b) 4 in. to 1 ft

Example 2   Finding the Price per Unit

The Winn-Dixie supermarket in Mandeville, Louisiana, charges the following prices for a box of trash bags.

Size	Price
10-count	$1.28
20-count	$2.68
30-count	$3.88

Which size is the best buy? That is, which size has the lowest unit price?

Continued on Next Page

❷ Solve the problem.

The local supermarket charges the following prices for a popular brand of pancake syrup.

Size	Price
36-oz	$3.89
24-oz	$2.79
12-oz	$1.89

Which size is the best buy? What is the unit cost for that size?

To find the best buy, write ratios comparing the price for each box size to the number of units (bags) per box. The results in the following table are rounded to the nearest thousandth.

Size	Unit Cost (dollars per bag)	
10-count	$\dfrac{\$1.28}{10} = \$.128$	← The best buy
20-count	$\dfrac{\$2.68}{20} = \$.134$	
30-count	$\dfrac{\$3.88}{30} = \$.129$	

Because the 10-count size produces the lowest unit cost, it is the best buy. This example shows that buying the largest size does not always provide the best buy, although this is often true.

Work Problem ❷ at the Side.

2 ▬▬▬ **Solve proportions.** A ratio is used to compare two numbers or amounts. A **proportion** says that two ratios are equal. For example,

$$\frac{3}{4} = \frac{15}{20}$$

is a proportion that says that the ratios $\frac{3}{4}$ and $\frac{15}{20}$ are equal. In the proportion

$$\frac{a}{b} = \frac{c}{d},$$

$a, b, c,$ and d are the **terms** of the proportion. Beginning with the proportion

$$\frac{a}{b} = \frac{c}{d}$$

and multiplying each side by the common denominator, bd, gives

$$bd \cdot \frac{a}{b} = bd \cdot \frac{c}{d}$$

$$\frac{b}{b}(d \cdot a) = \frac{d}{d}(b \cdot c) \qquad \text{Associative and commutative properties}$$

$$ad = bc. \qquad \text{Commutative and identity properties}$$

The products ad and bc are found by multiplying diagonally, as shown below.

$$\frac{a}{b} \underset{ad}{\overset{bc}{=}} \frac{c}{d}$$

For this reason, ad and bc are called **cross products.**

In the discussion that follows, we assume that no denominators are 0.

If $\dfrac{a}{b} = \dfrac{c}{d}$, then the cross products ad and bc are equal.

Also, if $ad = bc$, then $\dfrac{a}{b} = \dfrac{c}{d}$.

Answers
2. 36-oz; $.108 per oz

From this rule, if $\frac{a}{b} = \frac{c}{d}$ then $ad = bc$. However, if $\frac{a}{c} = \frac{b}{d}$, then $ad = cb$, or $ad = bc$. This means that the two proportions are equivalent, and

$$\text{the proportion } \frac{a}{b} = \frac{c}{d} \text{ can always be written as } \frac{a}{c} = \frac{b}{d}.$$

Sometimes one form is more convenient to work with than the other.

Four numbers are used in a proportion. If any three of these numbers are known, the fourth can be found.

Example 3 Finding an Unknown in a Proportion

Solve the proportion

$$\frac{5}{9} = \frac{x}{63}.$$

The cross products must be equal.

$$\begin{aligned} 5 \cdot 63 &= 9 \cdot x && \text{Cross products} \\ 315 &= 9x && \text{Multiply.} \\ 35 &= x && \text{Divide by 9.} \end{aligned}$$

Work Problem 3 at the Side.

CAUTION

The cross product method cannot be used directly if there is more than one term on either side.

Example 4 Solving an Equation Using Cross Products

Solve the equation

$$\frac{m-2}{5} = \frac{m+1}{3}.$$

Find the cross products.

$$\begin{aligned} 3(m-2) &= 5(m+1) && \text{Be sure to use parentheses.} \\ 3m - 6 &= 5m + 5 && \text{Distributive property} \\ 3m &= 5m + 11 && \text{Add 6.} \\ -2m &= 11 && \text{Subtract } 5m. \\ m &= -\frac{11}{2} && \text{Divide by } -2. \end{aligned}$$

Work Problem 4 at the Side.

NOTE

When you set cross products equal to each other, you are really multiplying each ratio in the proportion by a common denominator.

3 Solve applied problems using proportions. Proportions are useful in many practical applications. We continue to use the six-step method, although the steps are not numbered here.

3 Solve each proportion.

(a) $\dfrac{y}{6} = \dfrac{35}{42}$

(b) $\dfrac{a}{24} = \dfrac{15}{16}$

4 Solve each equation.

(a) $\dfrac{z}{2} = \dfrac{z+1}{3}$

(b) $\dfrac{p+3}{3} = \dfrac{p-5}{4}$

5 Solve the problem.

Twelve gal of diesel fuel costs $20.88. How much would 16.5 gal of the same fuel cost?

Example 5 Applying Proportions

After Lee Ann Spahr pumped 5.0 gal of gasoline, the display showing the price read $7.90. When she finished pumping the gasoline, the price display read $21.33. How many gallons did she pump?

We will solve this problem by setting up a proportion, with prices in the numerators and gallons in the denominators. Make sure that the corresponding numbers appear together.

Let x = the number of gallons she pumped. Then

$$\text{Price} \longrightarrow \frac{\$7.90}{5.0} = \frac{\$21.33}{x} \longleftarrow \text{Price}$$
$$\text{Gallons} \longrightarrow \qquad\qquad \longleftarrow \text{Gallons}$$

$$7.90x = 5.0(21.33) \qquad \text{Cross products}$$
$$7.90x = 106.65 \qquad \text{Multiply.}$$
$$x = 13.5. \qquad \text{Divide by 7.90.}$$

She pumped a total of 13.5 gal. Check this answer. Notice that the way the proportion was set up uses the fact that the unit price is the same, no matter how many gallons are purchased.

Calculator Tip Using a calculator to perform the arithmetic in Example 5 reduces the possibility of errors.

Work Problem 5 at the Side.

4 **Find percentages and percents.** We can use the techniques for solving proportions to solve percent problems. Recall, the decimal point is moved two places to the left to change a percent to a decimal number.

Calculator Tip Many calculators have a percent key that does this automatically.

We can solve a percent problem by writing it as the proportion

$$\frac{amount}{base} = \frac{percent}{100} \quad \text{or} \quad \frac{a}{b} = \frac{p}{100}.$$

The amount, or **percentage**, is compared to the **base** (the whole amount). Since *percent* means *per 100,* we compare the numerical value of the percent to 100. Thus, we write 50% as

$$\frac{p}{100} = \frac{50}{100}.$$

Example 6   **Finding Percentages**

Solve each problem.

(a) Find 15% of 600.

Here, the base is 600, the percent is 15, and we must find the percentage.

$$\frac{a}{b} = \frac{p}{100}$$

$$\frac{a}{600} = \frac{15}{100}$$

$$100a = 600(15) \qquad \text{Cross products}$$

$$a = \frac{600(15)}{100} \qquad \text{Divide by 100.}$$

$$a = 90$$

Thus, 15% of 600 is 90.

(b) A DVD with a regular price of $18 is on sale this week at 22% off. Find the amount of the discount and the sale price of the disc.

The discount is 22% of $18. We want to find a, given b is 18 and p is 22.

$$\frac{a}{b} = \frac{p}{100}$$

$$\frac{a}{18} = \frac{22}{100}$$

$$100a = 18(22) \qquad \text{Cross products}$$

$$100a = 396$$

$$a = 3.96 \qquad \text{Divide by 100.}$$

The amount of the discount on the DVD is $3.96, and the sale price is $18.00 − $3.96 = $14.04.

──────── **Work Problem ⑥ at the Side.**

Example 7   **Solving an Applied Percent Problem**

A newspaper ad offered a set of tires at a sale price of $258. The regular price was $300. What percent of the regular price were the savings?

The savings amounted to $300 − $258 = $42. We can now restate the problem: What percent of 300 is 42? Substitute into the percent proportion. We have $a = 42$, $b = 300$, and p is to be found.

$$\frac{a}{b} = \frac{p}{100}$$

$$\frac{42}{300} = \frac{p}{100}$$

$$300p = 4200 \qquad \text{Cross products}$$

$$p = 14 \qquad \text{Divide by 300.}$$

The sale price represented a 14% savings.

──────── **Work Problem ⑦ at the Side.**

⑥ Solve each problem.

(a) Find 20% of 70.

(b) Find the discount on a television set with a regular price of $270 if the set is on sale at 25% off. Find the sale price of the set.

⑦ Solve each problem.

(a) 90 is what percent of 360?

(b) The interest in 1 yr on deposits of $11,000 was $682. What percent interest was paid?

ANSWERS
6. (a) 14   **(b)** $67.50; $202.50
7. (a) 25%   **(b)** 6.2%

Real-Data Applications

Currency Exchange

When you travel between countries, you need to exchange your U.S. dollars for the local currency. The exchange rate between currencies changes daily, and you can easily find the updated rates using the Internet. The table shown here was taken from the Bloomberg Currency Calculator Web page.

WESTERN EUROPE CURRENCY RATES

Currency	Symbol	Currency per 1 unit of USD		
		Value	Net Chg	Pct Chg
British Pound	GBP	.6614	+.003	+.4557
Euro	EUR	1.0453	−.0052	−.4950
Danish Krone	DKK	7.7996	−.0367	−.4683
German Mark (based on Euro vs. dollar)	DEM	2.0445	−.0101	−.4915

Source: Bloomberg L.P.

On June 30, 2000, the currency exchange rate from U.S. dollars to British pounds was given as:

$1.00 U.S. was equivalent to £.6614 (British pounds).

You can set up a proportion to convert dollars to British pounds. For example, suppose you want to determine how many British pounds is equivalent to $50.00.

$$\frac{\$1}{£.6614} = \frac{\$50}{£x} \quad \text{or} \quad \frac{1}{.6614} = \frac{50}{x}$$

$$1(x) = .6614(50)$$

$$x = 33.07$$

So 33.07 British pounds is equivalent to 50 U.S. dollars.

For Group Discussion

1. Based on the currency exchange rates in the table above, find the amount of the local currency equivalent to $50 U.S. and find the number of U.S. dollars equivalent to 200 units of the local currency.

   (a) $50 = _____ Danish Krone and 200 Krone = _____ dollars

   (b) $50 = _____ German Marks and 200 Marks = _____ dollars

   (c) $50 = _____ Euros and 200 Euros = _____ dollars

2. Set up a proportion to find the number of U.S. dollars equivalent to £1 (British pound).

   £1 (British) was equivalent to $_____ (U.S.).

3. From problem 2, you should recognize the conversion rate based on £1 as the expression $\frac{1}{.6614}$. What is the mathematical term that describes the relationship between the conversion rates .6614 and $\frac{1}{.6614}$?

2.6 EXERCISES

FOR EXTRA HELP | Student's Solutions Manual | MyMathLab.com | InterAct Math Tutorial Software | AW Math Tutor Center | MathXL www.mathxl.com | Digital Video Tutor CD 2 Videotape 5

1. Match each ratio in Column I with the ratio equivalent to it in Column II.

I	II
(a) 75 to 100	**A.** 80 to 100
(b) 5 to 4	**B.** 50 to 100
(c) $\dfrac{1}{2}$	**C.** 3 to 4
(d) 4 to 5	**D.** 15 to 12

2. Give three different, equivalent forms of the ratio $\dfrac{4}{3}$.

Write a ratio for each word phrase. In Exercises 7–10, first write the amounts with the same units. Write fractions in lowest terms. See Example 1.

3. 60 ft to 70 ft

4. 40 mi to 30 mi

5. 72 dollars to 220 dollars

6. 120 people to 90 people

7. 30 in. to 8 ft

8. 20 yd to 8 ft

9. 16 min to 1 hr

10. 24 min to 2 hr

A supermarket was surveyed to find the prices charged for items in various sizes. Find the best buy (based on price per unit) for each item. See Example 2.

11. Seasoning mix

8-oz size: $1.75

17-oz size: $2.88

12. Red beans

1-lb package: $.89

2-lb package: $1.79

13. Prune juice

32-oz can: $1.95

48-oz can: $2.89

64-oz can: $3.29

14. Corn oil

24-oz bottle: $2.08

64-oz bottle: $3.94

128-oz bottle: $7.65

15. Artificial sweetener packets

50-count: $1.19

100-count: $1.85

250-count: $3.79

500-count: $6.38

16. Chili (no beans)

7.5-oz can: $1.19

10.5-oz can: $1.29

15-oz can: $1.78

25-oz can: $2.59

17. Extra crunchy peanut butter

12-oz size: $1.49

28-oz size: $1.99

40-oz size: $3.99

18. Tomato ketchup

14-oz size: $.93

32-oz size: $1.19

44-oz size: $2.19

19. Explain how percent and ratio are related.

20. Explain the distinction between *ratio* and *proportion*.

Solve each equation. See Examples 3 and 4.

21. $\dfrac{k}{4} = \dfrac{175}{20}$

22. $\dfrac{x}{6} = \dfrac{18}{4}$

23. $\dfrac{49}{56} = \dfrac{z}{8}$

24. $\dfrac{z}{80} = \dfrac{20}{100}$

25. $\dfrac{3y - 2}{5} = \dfrac{6y - 5}{11}$

26. $\dfrac{2r + 8}{4} = \dfrac{3r - 9}{3}$

27. $\dfrac{5k + 1}{6} = \dfrac{3k - 2}{3}$

28. $\dfrac{2p + 7}{3} = \dfrac{p - 1}{4}$

29. $\dfrac{3m - 2}{5} = \dfrac{4 - m}{3}$

Solve each problem involving proportion. See Example 5.

30. A chain saw requires a mixture of 2-cycle engine oil and gasoline. According to the directions on a bottle of Oregon 2-cycle Engine Oil, for a 50 to 1 ratio requirement, approximately 2.5 fluid oz of oil are required for 1 gal of gasoline. For 2.75 gal, how many fluid ounces of oil are required?

31. The directions on the bottle mentioned in Exercise 30 indicate that if the ratio requirement is 24 to 1, approximately 5.5 oz of oil are required for 1 gal of gasoline. If gasoline is to be mixed with 22 oz of oil, how much gasoline is to be used?

32. In a recent year, the average exchange rate between British pounds and U.S. dollars was 1 pound to $1.6762. Margaret went to London and exchanged her U.S. currency for British pounds, and received 400 pounds. How much in U.S. dollars did Margaret exchange?

33. If 3 U.S. dollars can be exchanged for 4.5204 Swiss francs, how many Swiss francs can be obtained for $49.20? (Round to the nearest hundredth.)

34. If 6 gal of premium unleaded gasoline cost $11.34, how much would it cost to completely fill a 15-gal tank?

35. If sales tax on a $16.00 compact disc is $1.32, how much would the sales tax be on a $120.00 compact disc player?

36. The distance between Kansas City, Missouri, and Denver is 600 mi. On a certain wall map, this is represented by a length of 2.4 ft. On the map, how many feet would there be between Memphis and Philadelphia, two cities that are actually 1000 mi apart?

37. The distance between Singapore and Tokyo is 3300 mi. On a certain wall map, this distance is represented by 11 in. The actual distance between Mexico City and Cairo is 7700 mi. How far apart are they on the same map?

38. Biologists tagged 500 fish in Willow Lake on October 5. At a later date they found 7 tagged fish in a sample of 700. Estimate the total number of fish in Willow Lake to the nearest hundred.

39. On May 13 researchers at Argyle Lake tagged 840 fish. When they returned a few weeks later, their sample of 1000 fish contained 18 that were tagged. Give an approximation of the fish population in Argyle Lake to the nearest hundred.

Answer each question about percent. See Example 6.

40. What is 48.6% of 19?

41. What is 26% of 480?

42. What percent of 48 is 96?

43. What percent of 30 is 36?

44. 12% of what number is 3600?

45. 25% of what number is 150?

46. 78.84 is what percent of 292?

47. .392 is what percent of 28?

Use mental techniques to answer the questions in Exercises 48–50. Try to avoid using paper and pencil or a calculator.

48. Jane Gunton bought a boat five years ago for $5000 and sold it this year for $2000. What percent of her original purchase did she lose on the sale?

  A. 40% **B.** 50% **C.** 20% **D.** 60%

49. The 1990 U.S. Census showed that the population of Alabama was 4,040,587, with 25.3% represented by African-Americans. What is the best estimate of the African-American population in Alabama? (*Source:* U.S. Bureau of the Census.)

  A. 500,000   **B.** 750,000

  C. 1,000,000   **D.** 1,500,000

50. The 1990 U.S. Census showed that the population of New Mexico was 1,515,069, with 38.2% being Hispanic. What is the best estimate of the Hispanic population of New Mexico? (*Source:* U.S. Bureau of the Census.)

  A. 600,000   **B.** 60,000

  C. 750,000   **D.** 38,000

Work each problem. Round all money amounts to the nearest dollar and percents to the nearest tenth. See Examples 6 and 7.

51. In 1998, the U.S. civilian labor force consisted of 137,673,000 persons. Of this total, 6,210,000 were unemployed. What was the percent of unemployment? (*Source:* U.S. Bureau of Labor Statistics.)

52. In 1998, the U.S. labor force (excluding agricultural employees, self-employed persons, and the unemployed) consisted of 116,730,000 persons. Of this total, 16,211,000 were union members. What percent of this labor force belonged to unions? (*Source:* U.S. Bureau of Labor Statistics.)

53. During former President George Bush's tenure, he vetoed a total of 44 bills. Fifteen of these were pocket vetoes. What percent of his vetoes were *not* pocket vetoes? (*Source:* Senate Library.)

54. During 1996 and 1997, the total public and private school enrollment in the United States was 51,375,000. Of this total, 12.7% of the students were enrolled in private schools. How many students were enrolled in public schools? (*Source:* National Center for Education Statistics, U.S. Department of Education.)

55. A family of four with a monthly income of $3800 plans to spend 8% of this amount on entertainment. How much will be spent on entertainment?

56. Quinhon Dac Ho earns $3200 per month. He wants to save 12% of this amount. How much will he save?

57. The 1916 dime minted in Denver is quite rare. The 1979 edition of *A Guide Book of United States Coins* listed its value in Extremely Fine condition as $625. The 1997 value had increased to $2400. What was the percent increase in the value of this coin?

58. Here is a common business problem. If the sales tax rate is 6.5% and I have collected $3400 in sales tax, how much were my sales?

The Consumer Price Index, issued by the U.S. Bureau of Labor Statistics, provides a means of determining the purchasing power of the U.S. dollar from one year to the next. Using the period from 1982 to 1984 as a measure of 100.0, the Consumer Price Index in each year from 1990 to 1998 is shown here. To use the Consumer Price Index to predict a price in a particular year, we can set up a proportion and compare it with a known price in another year, as follows:

$$\frac{\text{price in year } A}{\text{index in year } A} = \frac{\text{price in year } B}{\text{index in year } B}.$$

Year	Consumer Price Index
1990	130.7
1991	136.2
1992	140.3
1993	144.5
1994	148.2
1995	152.4
1996	156.9
1997	160.5
1998	163.0

Source: U.S. Bureau of Labor Statistics.

Use the Consumer Price Index figures in the table to find the amount that would be charged for the use of the same amount of electricity that cost $225 in 1990. Give your answer to the nearest dollar.

59. in 1995   **60.** in 1996   **61.** in 1997   **62.** in 1998

RELATING CONCEPTS (Exercises 63–66)   **FOR INDIVIDUAL OR GROUP WORK**

In Section 2.3 we solved equations with fractions by first multiplying each side of the equation by the common denominator. A proportion with a variable is this kind of equation. **Work Exercises 63–66 in order.** *The steps justify the method of solving a proportion by cross products.*

63. What is the LCD of the fractions in the equation $\dfrac{x}{6} = \dfrac{2}{5}$?

64. Solve the equation in Exercise 63 as follows.

  (a) Multiply each side by the LCD. What equation do you get?

  (b) Solve the equation from part (a) by dividing each side by the coefficient of *x*.

65. Solve the equation in Exercise 63 using cross products.

66. Compare your solutions from Exercises 64 and 65. What do you notice?

Summary Exercises on SOLVING APPLIED PROBLEMS

The following problems are of the various types discussed earlier in this chapter. Solve each problem. The problem-solving steps from Section 2.4 are repeated here.

Step 1   **Read** the problem carefully until you understand what is given and what is to be found.

Step 2   **Assign a variable** to represent the unknown value, using diagrams or tables as needed. Write down what the variable represents. If necessary, express any other unknown values in terms of the variable.

Step 3   **Write an equation** using the variable expression(s).

Step 4   **Solve** the equation.

Step 5   **State the answer.** Does it seem reasonable?

Step 6   **Check** the answer in the words of the *original* problem.

1. Nevaraz and Smith were opposing candidates in the school board election. Nevaraz received 30 more votes than Smith, with 516 votes cast. How many votes did Smith receive?

2. On an algebra test, the highest grade was 42 points more than the lowest grade. The sum of the two grades was 138. Find the lowest grade.

3. A certain lawn mower uses 3 tanks of gas to cut 10 acres of lawn. How many tanks of gas would be needed for 30 acres?

4. If 2 lb of fertilizer will cover 50 ft^2 of garden, how many pounds would be needed for 225 ft^2?

5. The perimeter of a certain square is seven times the length of a side, decreased by 12. Find the length of a side.

6. The perimeter of a certain rectangle is 16 times the width. The length is 12 cm more than the width. Find the width of the rectangle.

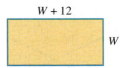

7. Find the measure of an angle whose measure is 70° more than its complement.

8. Find the measure of an angle whose measure is 20° more than its supplement.

9. If 2 is added to five times a number, the result is equal to 5 more than four times the number. Find the number.

10. If four times a number is added to 8, the result is three times the number added to 5. Find the number.

11. The smallest of three consecutive integers is added to twice the largest, producing a result 15 less than four times the middle integer. Find the smallest integer.

12. If the middle of three consecutive even integers is added to 100, the result is 42 more than the sum of the largest integer and twice the smallest. Find the smallest integer.

13. A store has 39 qt of milk, some in pint cartons and some in quart cartons. There are six times as many quart cartons as pint cartons. How many quart cartons are there? (*Hint:* 1 qt = 2 pt.)

14. A rectangular table is three times as long as it is wide. If it were 3 ft shorter and 3 ft wider, it would be square (with all sides equal). How long and how wide is it?

15. Find the measures of the marked angles.

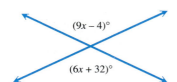

16. Find the measures of the marked angles.

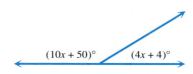

17. If the radius of a certain circle is tripled, and then 8.2 cm are added, the result is the circumference of the circle. Find the radius of the circle. (Use 3.14 as the approximation for π.) Round your answer to the nearest tenth.

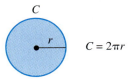

$C = 2\pi r$

18. A fully inflated professional basketball has a circumference of 78 cm. What is the radius of a circular cross section through the center of the ball? (Use 3.14 as the approximation for π.) Round your answer to the nearest hundredth.

78 cm

19. Mr. Silvester is 5 years older than his wife. Five years ago his age was $\frac{4}{3}$ her age. What are their ages now?

20. Chris is 10 years older than Josh. Next year, Chris will be twice as old as Josh. What are their ages now?

21. A DVD player that normally sells for $180 is on sale for $150. What is the percent discount on the player?

22. A grocer marks up cereal boxes by 20%. If a box of Raisin Bran costs the grocer $1.80, what is the grocer's selling price?

23. If nine pairs of jeans cost $121.50, find the cost of five pairs. (Assume all are equally priced.)

24. The distance between two cities on a road map is 11 in. The cities are actually 308 mi apart. The distance between two other cities on the map is 15 in. What is the actual distance between those cities?

25. In a recent year, General Motors announced that it would raise prices on its next year's vehicles by an average of 1.6%. If a certain vehicle had an original price of $10,526 and this price was raised 1.6%, what would the new price be? Round your answer to the nearest dollar.

26. Forty percent of Florence Griffith Joyner's Olympic medals are not gold medals. She has two medals that are not gold. How many gold medals does she have?

27. Two slices of bacon contain 85 calories. How many calories are there in twelve slices of bacon?

28. Three ounces of liver contain 22 g of protein. How many ounces of liver provide 121 g of protein?

In Exercises 29 and 30, find the best buy based on unit pricing.

29. Spaghetti sauce

$15\frac{1}{2}$-oz size: $1.19

32-oz size: $1.69

48-oz size: $2.69

30. Black pepper

4-oz size: $1.57

8-oz size: $2.27

31. According to *The Guinness Book of World Records,* the longest recorded voyage in a paddle-boat is 2226 mi in 103 days; the boat was propelled down the Mississippi River by the foot power of two boaters. Assuming a constant rate, how far would they have gone in 120 days? Round your answer to the nearest mile.

32. In the 2000 Olympics held in Sydney, Australia, Russian athletes earned 88 medals. Four of every 11 medals were gold. How many gold medals did Russia earn? (*Source: Times Picayune,* October 2, 2000.)

2.7 SOLVING LINEAR INEQUALITIES

The addition and multiplication properties can be extended to inequalities. **Inequalities** are statements with algebraic expressions related in the following ways:

$$< \quad \text{"is less than"}$$
$$\leq \quad \text{"is less than or equal to"}$$
$$> \quad \text{"is greater than"}$$
$$\geq \quad \text{"is greater than or equal to."}$$

We solve an inequality by finding all real number solutions for it. For example, the solutions of $x \leq 2$ include all *real numbers* that are less than or equal to 2, not just the *integers* less than or equal to 2.

1 **Graph the solutions of inequalities on a number line.** Graphing is a good way to show the solutions of an inequality. To graph all real numbers satisfying $x \leq 2$, we place a closed circle at 2 on a number line and draw an arrow extending from the closed circle to the left (to represent the fact that all numbers less than 2 are also part of the graph). The graph is shown in Figure 10.

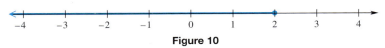

Figure 10

Example 1 **Graphing the Solutions of an Inequality**

Graph $x > -5$.

The statement $x > -5$ says that x can take any value greater than -5, but x cannot equal -5 itself. We show this on a graph by placing an open circle at -5 and drawing an arrow to the right, as in Figure 11. The open circle at -5 shows that -5 is not part of the graph.

Figure 11

Example 2 **Graphing the Solutions of an Inequality**

Graph $3 > x$.

The statement $3 > x$ means the same as $x < 3$. The graph of $x < 3$ is shown in Figure 12.

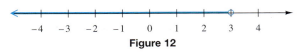

Figure 12

NOTE

It is usually easier to graph an inequality like the one in Example 2 by first rewriting it with the variable on the left. Fewer errors occur this way.

Work Problem ❶ at the Side.

OBJECTIVES

1 Graph the solutions of inequalities on a number line.

2 Use the addition property of inequality.

3 Use the multiplication property of inequality.

4 Solve inequalities using both properties of inequality.

5 Use inequalities to solve applied problems.

---

❶ Graph each inequality.

(a) $x \leq 3$

_____

(b) $x > -4$

_____

(c) $-4 \geq x$

_____

(d) $0 < x$

_____

ANSWERS

1. (a)

-4 -2   0   2   4

(b)

-8 -6 -4 -2

(c)

-8 -6 -4 -2   0

(d)

-4 -2   0   2   4

❷ Graph each inequality.

(a) $-7 < x < -2$

—————————————→

(b) $-6 < x \le -4$

—————————————→

 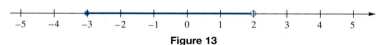
Example 3 **Graphing the Solutions of an Inequality**

Graph $-3 \le x < 2$.

The statement $-3 \le x < 2$ is read "-3 is less than or equal to x and x is less than 2." We graph the solutions of this inequality by placing a closed circle at -3 (because -3 is part of the graph) and an open circle at 2 (because 2 is not part of the graph), then drawing a line segment between the two circles. Notice that the graph includes all points *between* -3 and 2 and includes -3 as well. See Figure 13.

$-5 \quad -4 \quad -3 \quad -2 \quad -1 \quad 0 \quad 1 \quad 2 \quad 3 \quad 4 \quad 5$

Figure 13

Work Problem ❷ at the Side.

2   **Use the addition property of inequality.** Inequalities such as $x + 4 \le 9$ can be solved in much the same way as equations. Consider the inequality $2 < 5$. If 4 is added to each side of this inequality, the result is

$$2 + 4 < 5 + 4$$
$$6 < 9,$$

a true sentence. Now subtract 8 from each side:

$$2 - 8 < 5 - 8$$
$$-6 < -3.$$

The result is again a true sentence. These examples suggest the **addition property of inequality,** which states that the same real number can be added to each side of an inequality without changing the solutions.

Addition Property of Inequality

For any real numbers A, B, and C, the inequalities

$$A < B \quad \text{and} \quad A + C < B + C$$

have exactly the same solutions.

In words, the same number may be added to each side of an inequality without changing the solutions.

We can replace $<$ in the addition property of inequality with $>$, $\le$, or $\ge$. Also, as with the addition property of equality, the same number may be *subtracted* from each side of an inequality.

The following examples show how the addition property is used to solve inequalities.

Example 4 **Using the Addition Property of Inequality**

Solve $7 + 3k > 2k - 5$.

$$7 + 3k > 2k - 5$$
$$7 + 3k - 2k > 2k - 5 - 2k \qquad \text{Subtract } 2k.$$
$$7 + k > -5 \qquad \text{Combine terms.}$$
$$7 + k - 7 > -5 - 7 \qquad \text{Subtract 7.}$$
$$k > -12 \qquad \text{Combine terms.}$$

Continued on Next Page

A graph of the solutions, $k > -12$, is shown in Figure 14.

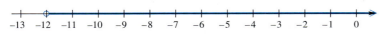

Figure 14

====== Work Problem ❸ at the Side.

3▭ **Use the multiplication property of inequality.** The addition property of inequality alone cannot be used to solve inequalities such as $4y \geq 28$. These inequalities require the *multiplication property of inequality.* To see how this property works, we look at some examples.

First, write the inequality $3 < 7$ and then multiply each side by the positive number 2.

$$3 < 7$$
$$\mathbf{2}(3) < \mathbf{2}(7) \quad \text{Multiply by 2.}$$
$$6 < 14 \quad \text{True}$$

Now multiply each side of $3 < 7$ by the negative number -5.

$$3 < 7$$
$$\mathbf{-5}(3) < \mathbf{-5}(7) \quad \text{Multiply by } -5.$$
$$-15 < -35 \quad \text{False}$$

To get a true statement when multiplying each side by -5 requires reversing the direction of the inequality symbol.

$$3 < 7$$
$$\mathbf{-5}(3) > \mathbf{-5}(7) \quad \text{Multiply by } -5;\ \text{reverse the symbol.}$$
$$-15 > -35 \quad \text{True}$$

Take the inequality $-6 < 2$ as another example. Multiply each side by the positive number 4.

$$-6 < 2$$
$$\mathbf{4}(-6) < \mathbf{4}(2) \quad \text{Multiply by 4.}$$
$$-24 < 8 \quad \text{True}$$

Multiplying each side of $-6 < 2$ by -5 *and at the same time reversing the direction of the inequality symbol* gives

$$-6 < 2$$
$$\mathbf{-5}(-6) > \mathbf{-5}(2) \quad \text{Multiply by } -5;\ \text{reverse the symbol.}$$
$$30 > -10. \quad \text{True}$$

====== Work Problem ❹ at the Side.

---

❸ Solve each inequality, and graph the solutions.

(a) $-1 + 8r < 7r + 2$

_____→

(b) $5m - \dfrac{4}{3} \leq 4m$

_____→

❹ **(a)** Multiply each side of $-2 < 8$ by 6 and then by -5. Reverse the direction of the inequality symbol if necessary.

(b) Multiply each side of $-4 > -9$ by 2 and then by -8. Reverse the direction of the inequality symbol if necessary.

In summary, the multiplication property of inequality has two parts.

> **Multiplication Property of Inequality**
>
> For any real numbers A, B, and C $(C \neq 0)$,
>
> **1.** if C is *positive,* then the inequalities
>
> $$A < B \quad \text{and} \quad AC < BC$$
>
> have exactly the same solutions;
>
> **2.** if C is *negative,* then the inequalities
>
> $$A < B \quad \text{and} \quad AC > BC$$
>
> have exactly the same solutions.
>
> In words, each side of an inequality may be multiplied by the same positive number without changing the solutions. If the multiplier is negative, we must reverse the direction of the inequality symbol.

We can replace $<$ in the multiplication property of inequality with $>$, $\leq$, or $\geq$. As with the multiplication property of equality, the same nonzero number may be divided into each side.

It is important to remember the differences in the multiplication property for positive and negative numbers.

1. When each side of an inequality is multiplied or divided by a positive number, the direction of the inequality symbol *does not change.* (Adding or subtracting terms on each side also does not change the symbol.)

2. When each side of an inequality is multiplied or divided by a negative number, the direction of the symbol *does change.* ***Reverse the direction of the inequality symbol only when multiplying or dividing each side by a negative number.***

Example 5 **Using the Multiplication Property of Inequality**

Solve $3r < -18$.

Using the multiplication property of inequality, we divide each side by 3. Since 3 is a positive number, the direction of the inequality symbol *does not* change. *It does not matter that the number on the right side of the inequality is negative.*

$$3r < -18$$

$$\frac{3r}{3} < \frac{-18}{3} \qquad \text{Divide by 3.}$$

$$r < -6$$

The graph of the solutions is shown in Figure 15.

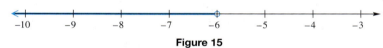

Figure 15

Example 6   Using the Multiplication Property of Inequality

Solve $-4t \geq 8$.

Here each side of the inequality must be divided by -4, a negative number, which *does* change the direction of the inequality symbol.

$$-4t \geq 8$$

$$\frac{-4t}{-4} \leq \frac{8}{-4} \qquad \text{Divide by } -4; \text{ reverse the symbol.}$$

$$t \leq -2$$

The solutions are graphed in Figure 16.

Figure 16

Work Problem ⑤ at the Side.

④ **Solve inequalities using both properties of inequality.**   The steps in solving an inequality are summarized below. (Remember that $<$ can be replaced with $>$, $\leq$, or $\geq$ in this summary.)

Solving Inequalities

Step 1   **Simplify each side separately.** Clear parentheses and combine terms on each side.

Step 2   **Isolate the variable term on one side.** Use the addition property to write the inequality in the form $ax < b$.

Step 3   **Isolate the variable.** Use the multiplication property to write the inequality in the form $x < c$ or $x > c$.

Notice how these steps are used in the next example.

Example 7   Solving an Inequality

Solve $5(k - 3) - 7k \geq 4(k - 3) + 9$.

Step 1   Use the distributive property to clear parentheses; then combine like terms.

$$5(k - 3) - 7k \geq 4(k - 3) + 9$$

$$5k - 15 - 7k \geq 4k - 12 + 9 \qquad \text{Distributive property}$$

$$-2k - 15 \geq 4k - 3 \qquad \text{Combine terms.}$$

Step 2   Use the addition property.

$$-2k - 15 \mathbf{- 4k} \geq 4k - 3 \mathbf{- 4k} \qquad \text{Subtract } 4k.$$

$$-6k - 15 \geq -3 \qquad \text{Combine terms.}$$

$$-6k - 15 \mathbf{+ 15} \geq -3 \mathbf{+ 15} \qquad \text{Add 15.}$$

$$-6k \geq 12 \qquad \text{Combine terms.}$$

Step 3   Divide each side by -6, a negative number. Change the direction of the inequality symbol.

$$\frac{-6k}{-6} \leq \frac{12}{-6} \qquad \text{Divide by } -6; \text{ reverse the symbol.}$$

$$k \leq -2$$

Continued on Next Page

---

⑤ Solve each inequality. Graph the solutions.

(a) $9y < -18$

(b) $-2r > -12$

(c) $-5p \leq 0$

6 Solve each inequality. Graph the solutions.

(a) $5r - r + 2 < 7r - 5$

A graph of the solutions is shown in Figure 17.

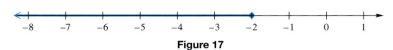

Figure 17

Work Problem **6** at the Side.

5 **Use inequalities to solve applied problems.** The chart below gives some of the more common phrases that suggest inequality, along with examples and translations.

Phrase	Example	Inequality
Is greater than	A number *is greater than* 4	$x > 4$
Is less than	A number *is less than* -12	$x < -12$
Is at least	A number *is at least* 6	$x \geq 6$
Is at most	A number *is at most* 8	$x \leq 8$

(b) $4(y - 1) - 3y > -15 - (2y + 1)$

CAUTION

Do not confuse phrases like "5 less than a number" and statements like "5 *is* less than a number." The first of these is expressed as $x - 5$ while the second is expressed as $5 < x$.

The next example shows an application that is important to anyone who has ever asked, "What score can I make on my next test and have a (particular grade) in this course?" It uses the idea of finding the average of a number of grades. In general, to find the average of n numbers, add the numbers and divide by n.

7 Solve the problem.

Maggie has scores of 98, 86, and 88 on her first three tests in algebra. If she wants an average of at least 90 after her fourth test, what score must she make on her fourth test?

Example 8 Finding an Average Test Score

Brent has test grades of 86, 88, and 78 on his first three tests in geometry. If he wants an average of at least 80 after his fourth test, what score must he make on his fourth test?

Let x represent Brent's score on his fourth test. To find the average of the four scores, add them and divide by 4.

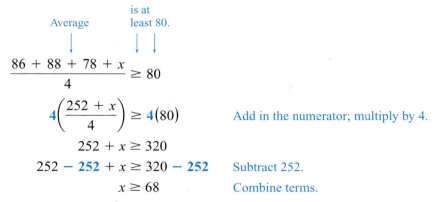

$$\frac{86 + 88 + 78 + x}{4} \geq 80$$

$$4\left(\frac{252 + x}{4}\right) \geq 4(80) \qquad \text{Add in the numerator; multiply by 4.}$$

$$252 + x \geq 320$$

$$252 - 252 + x \geq 320 - 252 \qquad \text{Subtract 252.}$$

$$x \geq 68 \qquad \text{Combine terms.}$$

He must score 68 or more on the fourth test to have an average of *at least* 80.

Work Problem **7** at the Side.

ANSWERS

6. (a) $r > \dfrac{7}{3}$

$\dfrac{7}{3}$

−1 0 1 2 3 4 5

(b) $y > -4$

−6 −5 −4 −3 −2 −1 0 1

7. 88 or more

2.7   EXERCISES

Write an inequality using the variable x that corresponds to each graph of solutions on a number line.

1.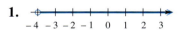

2.

3.

4.

5.

6.

7.

8.

9. How can you determine whether to use an open circle or a closed circle at an endpoint when graphing an inequality on a number line?

10. How does the graph of $t \geq -7$ differ from the graph of $t > -7$?

Graph each inequality on the given number line. See Examples 1–3.

11. $k \leq 4$

12. $r \leq -11$

13. $x < -3$

14. $y < 3$

15. $8 \leq x \leq 10$

16. $3 \leq x \leq 5$

17. $0 < y \leq 10$

18. $-3 \leq x < 5$

19. Why is it *wrong* to write $3 < x < -2$ to indicate that x is between -2 and 3?

20. Your friend tells you that when solving the inequality $6x < -42$, he reversed the direction of the inequality because of the presence of -42. How would you respond?

Solve each inequality, and graph the solutions. See Example 4.

21. $z - 8 \geq -7$

22. $p - 3 \geq -11$

23. $2k + 3 \geq k + 8$

24. $3x + 7 \geq 2x + 11$

25. $3n + 5 < 2n - 6$

26. $5x - 2 < 4x - 5$

27. Under what conditions must the inequality symbol be reversed when using the multiplication property of inequality?

28. Explain the steps you would use to solve the inequality $-5x > 20$.

Solve each inequality, and graph the solutions. See Examples 5 and 6.

29. $3x < 18$

30. $5x < 35$

31. $2y \geq -20$

32. $6m \geq -24$

33. $-8t > 24$

34. $-7x > 49$

35. $-x \geq 0$

36. $-k < 0$

37. $-\dfrac{3}{4}r < -15$

38. $-\dfrac{7}{8}t < -14$

39. $-.02x \leq .06$

40. $-.03v \geq -.12$

Solve each inequality, and graph the solutions. See Example 7.

41. $5r + 1 \geq 3r - 9$

42. $6t + 3 < 3t + 12$

43. $6x + 3 + x < 2 + 4x + 4$

44. $-4w + 12 + 9w \geq w + 9 + w$

45. $-x + 4 + 7x \le -2 + 3x + 6$

46. $14y - 6 + 7y > 4 + 10y - 10$

47. $5(x + 3) - 6x \le 3(2x + 1) - 4x$

48. $2(x - 5) + 3x < 4(x - 6) + 1$

49. $\frac{2}{3}(p + 3) > \frac{5}{6}(p - 4)$

50. $\frac{7}{9}(y - 4) \le \frac{4}{3}(y + 5)$

51. $4x - (6x + 1) \le 8x + 2(x - 3)$

52. $2y - (4y + 3) < 6y + 3(y + 4)$

53. $5(2k + 3) - 2(k - 8) > 3(2k + 4) + k - 2$

54. $2(3z - 5) + 4(z + 6) \ge 2(3z + 2) + 3z - 15$

Solve each application of inequalities. See Example 8.

55. John Douglas has grades of 84 and 98 on his first two history tests. What must he score on his third test so that his average is at least 90?

56. Elizabeth Gainey has scores of 74 and 82 on her first two algebra tests. What must she score on her third test so that her average is at least 80?

57. When 2 is added to the difference between six times a number and 5, the result is greater than 13 added to 5 times the number. Find all such numbers.

58. When 8 is subtracted from the sum of three times a number and 6, the result is less than 4 more than the number. Find all such numbers.

59. The formula for converting Celsius temperature to Fahrenheit is

$$F = \frac{9}{5}C + 32.$$

The Fahrenheit temperature of Providence, Rhode Island, has never exceeded 104°. How would you describe this using Celsius temperature?

60. The formula for converting Fahrenheit temperature to Celsius is

$$C = \frac{5}{9}(F - 32).$$

If the Celsius temperature on a certain day in San Diego, California, is never more than 25°, how would you describe the corresponding Fahrenheit temperature?

61. For what values of x would the rectangle have perimeter of at least 400?

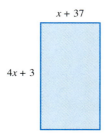

62. For what values of x would the triangle have perimeter of at least 72?

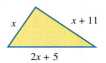

63. A long-distance phone call costs $2.00 for the first three minutes plus $.30 per minute for each minute or fractional part of a minute after the first three minutes. If x represents the number of minutes of the length of the call after the first three minutes, then $2 + .30x$ represents the cost of the call. If Jorge has $5.60 to spend on a call, what is the maximum total time he can use the phone?

64. At the Speedy Gas'n Go, a car wash costs $3.00, and gasoline is selling for $1.50 per gal. Lee Ann Spahr has $17.25 to spend, and her car is so dirty that she must have it washed. What is the maximum number of gallons of gasoline that she can purchase?

RELATING CONCEPTS (Exercises 65–68)   **FOR INDIVIDUAL OR GROUP WORK**

Work Exercises 65–68 in order, to see the connection between the solution of an equation and the solutions of the corresponding inequalities. *Graph the solutions in Exercises 65–67.*

65. $3x + 2 = 14$

66. $3x + 2 < 14$

67. $3x + 2 > 14$

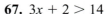

68. Now graph all the solutions together on the following number line.

How would you describe the graph?

SUMMARY

2.1 **linear equation** A linear equation in one variable is an equation that can be written in the form $Ax + B = C$, for real numbers A, B, and C, with $A \neq 0$.

equivalent equations Equations that have the same solutions are equivalent equations.

identity An identity is an equation that is true for all replacements of the variable.

2.4 **complementary angles** Two angles whose measures have a sum of 90° are complementary angles.

right angle A right angle measures 90°.

supplementary angles Two angles whose measures have a sum of 180° are supplementary angles.

straight angle A straight angle measures 180°.

consecutive integers Two integers that differ by 1 are consecutive integers.

2.5 **area** The area of a plane geometric figure is a measure of the surface covered by the figure.

perimeter The perimeter of a plane geometric figure is the distance around the figure, that is, the sum of the length of its sides.

vertical angles Vertical angles are angles formed by intersecting lines. They have the same measure.

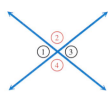

2.6 **ratio** A ratio is a comparison of two quantities using a quotient.

proportion A proportion is a statement that two ratios are equal.

cross products The method of cross products provides a way of determining whether a proportion is true.

$$\frac{a}{b} = \frac{c}{d}$$

terms In the proportion $\frac{a}{b} = \frac{c}{d}$, a, b, c, and d are the terms.

2.7 **inequality** Inequalities are statements with algebraic expressions related by $<$, $\leq$, $>$, or $\geq$.

a **to** b, $a : b$, **or** $\dfrac{a}{b}$ the ratio of a to b

---

TEST YOUR WORD POWER

See how well you have learned the vocabulary in this chapter. Answers follow the Quick Review.

1. A **solution** of an equation is a number that
   (a) makes an expression undefined
   (b) makes the equation false
   (c) makes the equation true
   (d) makes an expression equal to 0.

2. **Complementary angles** are angles
   (a) formed by two parallel lines
   (b) whose sum is 90°
   (c) whose sum is 180°
   (d) formed by perpendicular lines.

3. **Supplementary angles** are angles
   (a) formed by two parallel lines
   (b) whose sum is 90°
   (c) whose sum is 180°
   (d) formed by perpendicular lines.

4. A **ratio**
   (a) compares two quantities using a quotient
   (b) says that two quotients are equal
   (c) is a product of two quantities
   (d) is a difference between two quantities.

5. A **proportion**
   (a) compares two quantities using a quotient
   (b) says that two ratios are equal
   (c) is a product of two quantities
   (d) is a difference between two quantities.

6. An **inequality** is
   (a) a statement that two algebraic expressions are equal
   (b) a point on a number line
   (c) an equation with no solutions
   (d) a statement with algebraic expressions related by $<$, $\leq$, $>$, or $\geq$.

QUICK REVIEW

Concepts	Examples

2.1 The Addition Property of Equality
The same number may be added to (or subtracted from) each side of an equation without changing the solution.

Solve.
$$x - 6 = 12$$
$$x - 6 + 6 = 12 + 6 \qquad \text{Add 6.}$$
$$x = 18 \qquad \text{Combine terms.}$$

2.2 The Multiplication Property of Equality
Each side of an equation may be multiplied (or divided) by the same nonzero number without changing the solution.

Solve.
$$\frac{3}{4}x = -9$$
$$\frac{4}{3} \cdot \frac{3}{4}x = \frac{4}{3}(-9) \qquad \text{Multiply by } \tfrac{4}{3}.$$
$$x = -12$$

2.3 More on Solving Linear Equations
Step 1   Simplify each side separately.

Step 2   Isolate the variable term on one side.

Step 3   Isolate the variable.

Step 4   Check.

Solve.
$$2x + 2(x + 1) = 14 + x$$
$$2x + 2x + 2 = 14 + x \qquad \text{Distributive property}$$
$$4x + 2 = 14 + x \qquad \text{Combine terms.}$$
$$4x + 2 - x - 2 = 14 + x - x - 2 \qquad \text{Subtract } x; \text{ subtract 2.}$$
$$3x = 12 \qquad \text{Combine terms.}$$
$$\frac{3x}{3} = \frac{12}{3} \qquad \text{Divide by 3.}$$
$$x = 4$$

Check:
$$2(4) + 2(4 + 1) = 14 + 4 \qquad ? \qquad \text{Let } x = 4.$$
$$18 = 18 \qquad \text{True}$$

The solution is 4.

Concepts	Examples
2.4 *An Introduction to Applications of Linear Equations*	One number is 5 more than another. Their sum is 21. What are the numbers?
Step 1 Read.	We are looking for two numbers.
Step 2 Assign a variable.	Let x represent the smaller number. Then $x + 5$ represents the larger number.
Step 3 Write an equation.	$$x + (x + 5) = 21$$
Step 4 Solve the equation.	$2x + 5 = 21$ Combine terms.
	$2x + 5 - 5 = 21 - 5$ Subtract 5.
	$2x = 16$ Combine terms.
	$\dfrac{2x}{2} = \dfrac{16}{2}$ Divide by 2.
	$x = 8$
Step 5 State the answer.	The numbers are 8 and 13.
Step 6 Check.	13 is 5 more than 8, and $8 + 13 = 21$. It checks.
2.5 *Formulas and Applications from Geometry*	Find L if $A = LW$, given that $A = 24$ and $W = 3$.
To find the value of one of the variables in a formula, given values for the others, substitute the known values into the formula.	$24 = L \cdot 3$ $A = 24, W = 3$
	$\dfrac{24}{3} = \dfrac{L \cdot 3}{3}$ Divide by 3.
	$8 = L$
To solve a formula for one of the variables, isolate that variable by treating the other variables as numbers and using the steps for solving equations.	Solve $P = 2L + 2W$ for W.
	$P - 2L = 2L + 2W - 2L$ Subtract $2L$.
	$P - 2L = 2W$ Combine terms.
	$\dfrac{P - 2L}{2} = \dfrac{2W}{2}$ Divide by 2.
	$\dfrac{P - 2L}{2} = W$ or $W = \dfrac{P - 2L}{2}$
2.6 *Ratio, Proportion, and Percent*	
To write a ratio, express quantities in the same units.	4 ft to 8 in. $=$ 48 in. to 8 in. $= \dfrac{48}{8} = \dfrac{6}{1}$
To solve a proportion, use the method of cross products.	Solve $\dfrac{x}{12} = \dfrac{35}{60}.$
	$60x = 12 \cdot 35$ Cross products
	$60x = 420$ Multiply.
To solve a percent problem, use the proportion	$\dfrac{60x}{60} = \dfrac{420}{60}$ Divide by 60.
$$\dfrac{\text{amount}}{\text{base}} = \dfrac{\text{percent}}{100}.$$	$x = 7$

Concepts	Examples
2.7 Solving Linear Inequalities	Solve and graph the solutions.

2.7 Solving Linear Inequalities

Step 1 Simplify each side separately.

Step 2 Isolate the variable term on one side.

Step 3 Isolate the variable.

(Be sure to reverse the direction of the inequality symbol when multiplying or dividing by a negative number.)

Solve and graph the solutions.

$$3(1 - x) + 5 - 2x > 9 - 6$$

$$3 - 3x + 5 - 2x > 9 - 6 \qquad \text{Distributive property}$$

$$8 - 5x > 3 \qquad \text{Combine terms.}$$

$$8 - 5x \mathbf{- 8} > 3 \mathbf{- 8} \qquad \text{Subtract 8.}$$

$$-5x > -5 \qquad \text{Combine terms.}$$

$$\frac{-5x}{\mathbf{-5}} < \frac{-5}{\mathbf{-5}} \qquad \text{Divide by } -5; \text{ change} > \text{to} <.$$

$$x < 1$$

ANSWERS TO TEST YOUR WORD POWER

1. (c) *Example:* 8 is the solution of $2x + 5 = 21$.   **2. (b)** *Example:* Angles with measures 35° and 55° are complementary angles.   **3. (c)** *Example:* Angles with measures 112° and 68° are supplementary angles.

4. (a) *Example:* $\dfrac{7 \text{ in.}}{12 \text{ in.}} = \dfrac{7}{12}$ **5. (b)** *Example:* $\dfrac{2}{3} = \dfrac{8}{12}$ **6. (d)** *Examples:* $x < 5, 7 + 2y \geq 11$

Chapter 2   **REVIEW EXERCISES**

[2.1–2.3] *Solve each equation. Check the solution.*

1. $x - 7 = 2$

2. $4r - 6 = 10$

3. $5x + 8 = 4x + 2$

4. $8t = 7t + \dfrac{3}{2}$

5. $(4r - 8) - (3r + 12) = 0$

6. $7(2x + 1) = 6(2x - 9)$

7. $-\dfrac{6}{5}y = -18$

8. $\dfrac{1}{2}r - \dfrac{1}{6}r + 3 = 2 + \dfrac{1}{6}r + 1$

9. $3x - (-2x + 6) = 4(x - 4) + x$

10. $.10(x + 80) + .20x = 8 + .30x$

[2.4] *Solve each problem.*

11. If 7 is added to five times a number, the result is equal to three times the number. Find the number.

12. If 4 is subtracted from twice a number, the result is 36. Find the number.

13. The land area of Hawaii is 5213 mi² greater than that of Rhode Island. Together, the areas total 7637 mi². What is the area of each state?

14. The height of Seven Falls in Colorado is $\dfrac{5}{2}$ the height (in feet) of Twin Falls in Idaho. The sum of the heights is 420 ft. Find the height of each.

15. The supplement of an angle measures 10 times the measure of its complement. What is the measure of the angle (in degrees)?

16. Find two consecutive odd integers such that when the smaller is added to twice the larger, the result is 24 more than the larger integer.

[2.5] *A formula is given in each exercise, along with the values for all but one of the variables. Find the value of the variable that is not given. (For Exercises 19 and 20, use* 3.14 *as an approximation for* π.)

17. $A = \dfrac{1}{2}bh$; $A = 44, b = 8$

18. $A = \dfrac{1}{2}h(b + B)$; $b = 3, B = 4, h = 8$

19. $C = 2\pi r$; $C = 29.83$

20. $V = \dfrac{4}{3}\pi r^3$; $r = 6$

Solve each formula for the specified variable.

21. $A = LW$ for L

22. $A = \dfrac{1}{2}h(b + B)$ for h

Find the measure of each marked angle.

23.

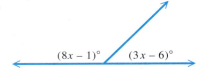

$(8x - 1)°$ $(3x - 6)°$

24.

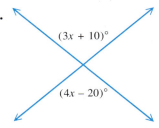

$(3x + 10)°$

$(4x - 20)°$

Solve each application of geometry.

25. A cinema screen in Indonesia has length 92.75 ft and width 70.5 ft. What is the perimeter? What is the area? (*Source: The Guinness Book of Records*, 1994.)

26. The Ziegfield Room in Reno, Nevada, has a circular turntable on which its showgirls dance. The circumference of the turntable is 62.5 ft. What is the diameter of the turntable? What is the radius? What is its area? (Use 3.14 as an approximation for π.) (*Source: The Guinness Book of Records*, 1994.)

[2.6]   *Write a ratio for each word phrase. Write fractions in lowest terms.*

27. 60 cm to 40 cm

28. 5 days to 2 weeks

29. 90 in. to 10 ft

30. 3 mo to 3 yr

Solve each proportion.

31. $\dfrac{p}{21} = \dfrac{5}{30}$

32. $\dfrac{5 + x}{3} = \dfrac{2 - x}{6}$

33. $\dfrac{y}{5} = \dfrac{6y - 5}{11}$

34. Explain how 40% can be expressed as a ratio of two whole numbers.

Solve each problem involving proportion.

35. If 2 lb of fertilizer will cover 150 ft^2 of lawn, how many pounds would be needed to cover 500 ft^2?

36. If 8 oz of medicine must be mixed with 20 oz of water, how many ounces of medicine must be mixed with 90 oz of water?

37. An enlarged version of the chair used by George Washington at the Constitutional Convention casts a shadow 18 ft long at the same time a vertical pole 12 ft high casts a shadow 4 ft long. How tall is the chair? (*Source: The Guinness Book of Records*, 1994.)

38. The distance between two cities on a road map is 32 cm. The two cities are actually 150 km apart. The distance on the map between two other cities is 80 cm. How far apart are these cities?

39. What is 23% of 76?

40. What percent of 12 is 21?

41. 6 is what percent of 18?

42. 36% of what number is 900?

43. Gwen and John paid $25,407.00 for their 1999 Chevrolet conversion van. The sales tax rate was 8.75%, and the tax was added to that amount. What was the final price for the van? (*Source:* Author Hornsby's sales receipt.)

44. Ruth, from the mathematics editorial division of Addison-Wesley, took a community college faculty out to dinner. The bill was $304.75. Ruth added a 15% tip, and paid for the meal with her corporate credit card. What was the total price she paid?

[2.7] *Graph each inequality on the number line provided.*

45. $p \geq -4$

46. $x < 7$

47. $-5 \leq y < 6$

48. $r \geq \dfrac{1}{2}$

Solve each inequality. Graph the solutions.

49. $y + 6 \geq 3$

50. $5t < 4t + 2$

51. $-6x \leq -18$

52. $8(k - 5) - (2 + 7k) \geq 4$

53. $4x - 3x > 10 - 4x + 7x$

54. $3(2w + 5) + 4(8 + 3w) < 5(3w + 2) + 2w$

55. Carlotta Valdes has grades of 94 and 88 on her first two calculus tests. What possible scores on a third test will give her an average of at least 90?

56. If nine times a number is added to 6, the result is at most 3. Find all such numbers.

MIXED REVIEW EXERCISES

Solve each problem.

57. $\dfrac{y}{7} = \dfrac{y - 5}{2}$

58. $I = prt$ for r

59. $-2x > -4$

60. $2k - 5 = 4k + 13$

61. $.05x + .02x = 4.9$

62. $2 - 3(y - 5) = 4 + y$

63. $9x - (7x + 2) = 3x + (2 - x)$

64. $\dfrac{1}{3}s + \dfrac{1}{2}s + 7 = \dfrac{5}{6}s + 5 + 2$

65. One of the tallest candles ever constructed was exhibited at the 1897 Stockholm Exhibition. If it cast a shadow 5 ft long at the same time a vertical pole 32 ft high cast a shadow 2 ft long, how tall was the candle? (*Source: The Guinness Book of Records,* 1994.)

66. Two-thirds of a number added to the number is 10. What is the number?

67. Rita and Bobby commute to work. Rita travels three times as far as Bobby each day, and together they travel 112 mi. How far does each travel?

68. Mike defeated William in an election. Mike had twice as many votes as William, and together they had 1800 votes. How many votes did each of the candidates receive?

69. In the 2000 Olympic Games in Sydney, Australia, the United States and Russia earned a total of 185 medals. The United States earned 9 more medals than Russia. How many medals did each country earn? (*Source: Times Picayune,* October 2, 2000.)

70. Of the 58 medals earned by the host country, Australia, in the 2000 Olympics, there were 9 more silver than gold medals, and 8 fewer bronze than silver medals. How many of each medal did Australia earn? (*Source: Times Picayune,* October 2, 2000.)

71. The perimeter of a triangle is 96 m. One side is twice as long as another, and the third side is 30 m long. What is the length of the longest side?

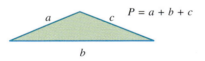

72. The perimeter of a rectangle is 288 ft. The length is 4 ft longer than the width. Find the width.

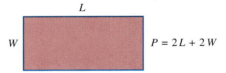

73. The perimeter of a rectangle is 75 in. The width is 17 in. What is the length?

74. The area of a triangle is 182 in.^{2}. The height is 14 in. Find the length of the base.

75. Latarsha has grades of 82 and 96 on her first two English tests. What must she make on her third test so that her average will be at least 90?

76. Find the measure of each marked angle.

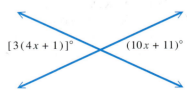

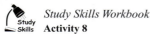

Chapter 2 TEST

Study Skills Workbook
Activity 8

Solve each equation, and check the solution.

1. $3x - 7 = 11$

2. $5x + 9 = 7x + 21$

3. $2 - 3(x - 5) = 3 + (x + 1)$

4. $2.3x + 13.7 = 1.3x + 2.9$

5. $7 - (m - 4) = -3m + 2(m + 1)$

6. $-\dfrac{4}{7}x = -12$

7. $.06(x + 20) + .08(x - 10) = 4.6$

8. $-8(2x + 4) = -4(4x + 8)$

Solve each problem.

9. This problem refers to the 1997 All-Star basketball game held February 9 in Cleveland. (*Source: The ESPN 1998 Information Please Sports Almanac.*)
   (a) The East won the game by 12 points. The total of the East and West team scores was 252. What was the final score of the game?
   (b) The high scorer of the game was Glenn Rice, playing for the Eastern Conference. He scored 7 more points than Latrell Sprewell, the high scorer on the Western Conference team. The total number of points scored by both players was 45. How many points did Rice score?

10. The three largest islands in the Hawaiian island chain are Hawaii (the Big Island), Maui, and Kauai. Together, their areas total 5300 mi². The island of Hawaii is 3293 mi² larger than the island of Maui, and Maui is 177 mi² larger than Kauai. What is the area of each island?

Kauai
Oahu   Molokai
Lanai   Maui
The Big Island
HAWAII

11. Find the measure of an angle if its supplement measures 10° more than three times its complement.

1. _____

2. _____

3. _____

4. _____

5. _____

6. _____

7. _____

8. _____

9. (a) _____

(b) _____

10. _____

11. _____

12. (a) _____

   (b) _____

13. _____

14. _____

15. _____

16. _____

17. _____

18. _____

19. _____

20. (a) _____

   (b) _____

21.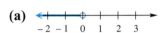

22. _____

23. _____

24. _____

25. _____

12. The formula for the perimeter of a rectangle is $P = 2L + 2W$.

  (a) Solve for W.

  (b) If $P = 116$ and $L = 40$, find the value of W.

Find the measure of each marked angle.

13.

$(3x + 55)°$ $(7x - 25)°$

14.

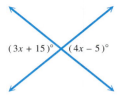

$(3x + 15)°$ $(4x - 5)°$

Solve each proportion.

15. $\dfrac{z}{8} = \dfrac{12}{16}$

16. $\dfrac{y + 5}{3} = \dfrac{y - 3}{4}$

17. Which is the better buy for processed cheese slices: 8 slices for \$2.19 or 12 slices for \$3.30?

18. The distance between Milwaukee and Boston is 1050 mi. On a certain map, this distance is represented by 42 in. On the same map, Seattle and Cincinnati are 92 in. apart. What is the actual distance between Seattle and Cincinnati?

19. In 1997, the Tampa Bay Lightning hockey team had debt of about \$177 million, with a franchise value of only \$75 million. What percent of the franchise value was the debt? (*Source:* Ozanian, M. K., "Fields of Debt," *Forbes,* December 15, 1997, vol. 160, no. 13.)

20. Write an inequality involving x that describes the numbers graphed.

  (a)    **(b)**

Solve each inequality, and graph the solutions.

21. $-3x > -33$

22. $-4x + 2(x - 3) \geq 4x - (3 + 5x) - 7$

23. $-.04x \leq .12$

24. Twylene Johnson has grades of 76 and 81 on her first two algebra tests. If she wants an average of at least 80 after her third test, what score must she make on her third test?

25. Write a short explanation of the additional (extra) rule that must be remembered when solving an inequality (as opposed to solving an equation).

Beginning with this chapter, each chapter in the text will conclude with a set of cumulative review exercises designed to cover the major topics from the beginning of the course. This feature will allow you to constantly review topics that have been introduced up to that point.

Write each fraction in lowest terms.

1. $\dfrac{15}{40}$

2. $\dfrac{108}{144}$

Perform the indicated operations.

3. $\dfrac{5}{6} + \dfrac{1}{4} + \dfrac{7}{15}$

4. $16\dfrac{7}{8} - 3\dfrac{1}{10}$

5. $\dfrac{9}{8} \cdot \dfrac{16}{3}$

6. $\dfrac{3}{4} \div \dfrac{5}{8}$

7. $4.8 + 12.5 + 16.73$

8. $56.3 - 28.99$

9. $67.8(.45)$

10. $236.46 \div 4.2$

11. In making dresses, Earth Works uses $\dfrac{5}{8}$ yd of trim per dress. How many yards of trim would be used to make 56 dresses?

12. A cook wants to increase a recipe for Quaker Quick Grits that serves 4 to make enough for 10 people. The recipe calls for 3 cups of water. How much water will be needed to serve 10?

13. Pythagoras weighs $71\dfrac{1}{4}$ lb and Fred weighs $28\dfrac{3}{8}$ lb. How much do the two dogs weigh together?

14. A purchasing agent bought 3 Executive Single-Pedestal Desks at $1099.99 each and 3 chairs for $159.99, $189.99, and $199.99. What was the final bill (without tax)? (*Source:* Office Depot catalog "The Big Book," 2000.)

Tell whether each inequality is true or false.

15. $\dfrac{8(7) - 5(6 + 2)}{3 \cdot 5 + 1} \geq 1$

16. $\dfrac{4(9 + 3) - 8(4)}{2 + 3 - 3} \geq 2$

Perform the indicated operations.

17. $-11 + 20 + (-2)$

18. $13 + (-19) + 7$

19. $9 - (-4)$

20. $-2(-5)(-4)$

21. $\dfrac{4 \cdot 9}{-3}$

22. $\dfrac{8}{7 - 7}$

23. $(-5 + 8) + (-2 - 7)$

24. $(-7 - 1)(-4) + (-4)$

25. $\dfrac{-3 - (-5)}{1 - (-1)}$

26. $\dfrac{6(-4) - (-2)(12)}{3^2 + 7^2}$

27. $\dfrac{(-3)^2 - (-4)(2^4)}{5 \cdot 2 - (-2)^3}$

28. $\dfrac{-2(5^3) - 6}{4^2 + 2(-5) + (-2)}$

Find the value of each expression when $x = -2$, $y = -4$, and $z = 3$.

29. $xz^3 - 5y^2$

30. $\dfrac{xz - y^3}{-4z}$

Name the property illustrated by each equation.

31. $7(k + m) = 7k + 7m$

32. $3 + (5 + 2) = 3 + (2 + 5)$

33. $7 + (-7) = 0$

34. $3.5(1) = 3.5$

Simplify each expression by combining terms.

35. $4p - 6 + 3p - 8$

36. $-4(k + 2) + 3(2k - 1)$

Solve each equation, and check the solution.

37. $2r - 6 = 8$

38. $2(p - 1) = 3p + 2$

39. $4 - 5(a + 2) = 3(a + 1) - 1$

40. $2 - 6(z + 1) = 4(z - 2) + 10$

41. $-(m - 1) = 3 - 2m$

42. $\dfrac{y - 2}{3} = \dfrac{2y + 1}{5}$

43. $\dfrac{2x + 3}{5} = \dfrac{x - 4}{2}$

44. $\dfrac{2}{3}y + \dfrac{3}{4}y = -17$

Solve each formula for the indicated variable.

45. $P = a + b + c$ for c

46. $P = 4s$ for s

Solve each inequality. Graph the solutions.

47. $-5z \geq 4z - 18$

48. $6(r - 1) + 2(3r - 5) < -4$

Solve each problem.

49. The purchasing agent in Exercise 14 paid a sales tax of $6\frac{1}{4}\%$ on his purchase. What was the final bill, including tax?

50. A car has a price of $5000. For trading in her old car, Shannon D'hemecourt will get 25% off. Find the price of the car with the trade-in.

51. Jennifer Johnston bought textbooks at the college bookstore for $244.33, including 6% sales tax. What did the books cost?

52. Carter Fenton received a bill from his credit card company for $104.93. The bill included interest at $1\frac{1}{2}\%$ per month for one month and a $5.00 late charge. How much did his purchases amount to?

53. The perimeter of a rectangle is 98 cm. The width is 19 cm. Find the length.

?

19 cm

54. The area of a triangle is 104 in.². The base is 13 in. Find the height.

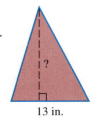

?

13 in.

Graphs of Linear Equations and Inequalities in Two Variables

3

U.S. debt from credit cards continues to increase. In recent years, college campuses have become fertile territory as credit card companies pitch their plastic to students at bookstores, student unions, and sporting events. As a result, three out of four undergrads now have at least one credit card and carry an average balance of $2748. (*Source:* Nellie Mae.) In Example 6 of Section 3.2, we use the concepts of this chapter to investigate credit card debt.

ADDISON - WESLEY
MyMathLab.com
You're Connected

183

3.1 READING GRAPHS; LINEAR EQUATIONS IN TWO VARIABLES

OBJECTIVES

1. Interpret graphs.
2. Write a solution as an ordered pair.
3. Decide whether a given ordered pair is a solution of a given equation.
4. Complete ordered pairs for a given equation.
5. Complete a table of values.
6. Plot ordered pairs.

We live in an age of information. Graphs provide a quick way to organize and communicate much of this information. They can also be used to analyze data, make predictions, or simply entertain us. To prepare for the material in this chapter, we begin by looking at some graphs typically seen in newspapers, magazines, and other print and electronic media.

1 **Interpret graphs.** There are many ways to represent the relationship between two quantities. *Circle graphs, bar graphs,* and *line graphs* are often used for this purpose.

In a **circle graph** or **pie chart,** a circle is used to indicate the total of all the categories represented. The circle is divided into *sectors,* or wedges (like pieces of a pie), whose sizes show the relative magnitudes of the categories. The sum of all the fractional parts of the graph must be 1 (for 1 whole circle).

Example 1 Interpreting a Circle Graph

The 1999 market share for satellite-TV home subscribers is shown in the circle graph in Figure 1.

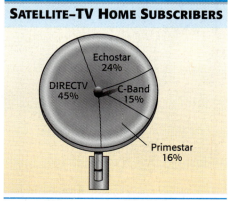

SATELLITE–TV HOME SUBSCRIBERS

Echostar 24%
DIRECTV 45%
C-Band 15%
Primestar 16%

Source: Skyreport.com; *USA Today.*

Figure 1

The number of subscribers reached 12 million in August 1999.

(a) Which provider had the largest share of the home subscriber market in August 1999? What was that share?

In the circle graph, the sector for DIRECTV is the largest, so DIRECTV had the largest market share, 45%.

(b) Estimate the number of home subscribers to DIRECTV in August 1999.

A market share of 45% can be rounded to 50%, or .5. We multiply .5 (or $\frac{1}{2}$) by the total number of subscribers, 12 million. A good estimate for the number of DIRECTV subscribers would be

$$.5(12) = 6 \text{ million.}$$

Continued on Next Page

(c) How many actual home subscribers to DIRECTV were there?

To find the answer, we multiply the actual percent from the graph for DIRECTV, 45% or .45, by the number of subscribers, 12 million:

$$.45(12) = 5.4.$$

Thus, 5.4 million homes subscribed to DIRECTV. This is reasonable given our estimate in part (b).

═══════════ **Work Problem ❶ at the Side.**

A **bar graph** is used to show comparisons. It consists of a series of bars arranged either vertically or horizontally. In a bar graph, values from two categories are paired with each other.

Example 2 Interpreting a Bar Graph

The bar graph in Figure 2 compares average monthly savings, including re-tirement plans, for five countries.

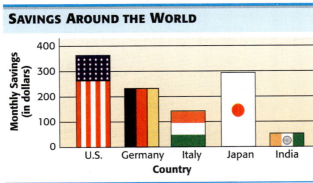

SAVINGS AROUND THE WORLD

Source: Taylor Nelson–Sofres for American Express.

Figure 2

(a) Which country has the largest average monthly savings? the smallest?

The tallest bar corresponds to the United States, so the United States has the largest average monthly savings. India (with the shortest bar) has the smallest savings.

(b) Which countries have average monthly savings greater than $200?

Locate 200 on the vertical scale and follow the line across to the right. Three countries, the United States, Germany, and Japan, have bars that extend above the line for 200, so they have average monthly savings greater than $200.

(c) Estimate the average monthly savings for Japan and Italy.

Locate the top of the bar for Japan and move horizontally across to the vertical scale to see that average monthly savings in Japan is about $300.

Follow the top of the bar for Italy across to the vertical scale to see that this bar is almost halfway between $100 and $200. Average monthly savings in Italy is about $150.

(d) Find the difference between average monthly savings in Japan and Italy. Interpret this result.

Use the results from part (c) to see that the difference is about $300 − 150 = $150. Average monthly savings in Japan is about twice as much as that in Italy.

═══════════ **Work Problem ❷ at the Side.**

❶ Refer to the circle graph in Figure 1.

(a) Which provider had the smallest market share in August 1999?

(b) Estimate the number of home subscribers to Echostar.

(c) How many actual home subscribers to Echostar were there?

❷ Refer to the bar graph in Figure 2.

(a) Which countries have average monthly savings less than $150?

(b) Estimate the average monthly savings for Germany and India.

Answers
1. **(a)** C-Band
    (b) .25 or $\frac{1}{4}$ of 12 million = 3 million
    (c) .24(12) = 2.88 million
2. **(a)** Italy and India
    (b) Germany: about $230; India: about $50

❸ Refer to the line graph in Figure 3.

(a) Which year had the highest average PC price?

(b) Estimate the average price of a PC in 1994.

(c) About how much did average PC prices decline from 1994 to 1999?

A **line graph** is used to show changes or trends in data over time. To form a line graph, we connect a series of points representing data with line segments.

Example 3 Interpreting a Line Graph

The line graph in Figure 3 shows average prices for personal computers (PCs) for the years 1993 through 1999.

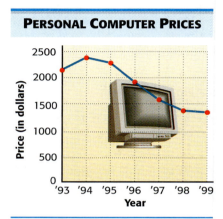

PERSONAL COMPUTER PRICES

Source: CNW Marketing/Research; *USA Today.*

Figure 3

(a) Between which years did the average price of a PC increase?

The line between 1993 and 1994 rises, so PC prices increased from 1993 to 1994.

(b) What has been the general trend in average PC prices since 1994?

The line graph falls from 1994 to 1999, so PC prices have been decreasing over these years.

(c) Estimate average PC prices in 1996 and 1999. About how much did PC prices decline between 1996 and 1999?

Move up from 1996 on the horizontal scale to the point plotted for 1996. Then move across to the vertical scale. The average price of a PC in 1996 was about $2000.

The point for 1999 is a little more than halfway between the lines for $1000 and $1500, so estimate the average price in 1999 at about $1300.

Between 1996 and 1999, PC prices declined about

$$\$2000 - 1300 = \$700.$$

Work Problem ❸ at the Side.

Many everyday situations, such as those illustrated in Examples 2 and 3, involve two quantities that are related. The equations and applications we discussed in Chapter 2 had only one variable. In this chapter, we extend those ideas to *linear equations in two variables.*

Linear Equation in Two Variables

A **linear equation in two variables** is an equation that can be written in the form

$$Ax + By = C,$$

where A, B, and C are real numbers and A and B are not both 0.

Some examples of linear equations in two variables in this form, called *standard form,* are

$$3x + 4y = 9, \quad x - y = 0, \quad \text{and} \quad x + 2y = -8.$$

NOTE

> Other linear equations in two variables, such as
>
> $$y = 4x + 5 \quad \text{and} \quad 3x = 7 - 2y,$$
>
> are not written in standard form but could be. We will discuss the forms of linear equations in more detail in Section 3.4.

2 ▄▄▄ **Write a solution as an ordered pair.**   Recall that a *solution* of an equation is a number that makes the equation true when it replaces the variable. For example, the linear equation in one variable $x - 2 = 5$ has solution 7, since replacing x with 7 gives a true statement.

A solution of a linear equation in *two* variables requires *two* numbers, one for each variable. For example, a true statement results when we replace x with 2 and y with 13 in the equation $y = 4x + 5$ since

$$13 = 4(2) + 5. \quad \text{Let } x = 2, y = 13.$$

The pair of numbers $x = 2$ and $y = 13$ gives one solution of the equation $y = 4x + 5$. The phrase "$x = 2$ and $y = 13$" is abbreviated

x-value ──┐ ┌── y-value

$$(2, 13)$$

Ordered pair

with the x-value, 2, and the y-value, 13, given as a pair of numbers written inside parentheses. *The x-value is always given first.* A pair of numbers such as (2, 13) is called an **ordered pair.** As the name indicates, the order in which the numbers are written is important. The ordered pairs $(2, 13)$ and $(13, 2)$ are not the same. The second pair indicates that $x = 13$ and $y = 2$.

Work Problem 4 at the Side.

3 ▄▄▄ **Decide whether a given ordered pair is a solution of a given equation.** We substitute the x- and y-values of an ordered pair into a linear equation in two variables to see whether the ordered pair is a solution of the equation.

---

4 Write each solution as an ordered pair.

(a) $x = 5$ and $y = 7$

(b) $y = 6$ and $x = -1$

(c) $y = 4$ and $x = -3$

(d) $x = 3$ and $y = -12$

5 Decide whether each ordered pair is a solution of the equation $5x + 2y = 20$.

(a) $(0, 10)$

$$5x + 2y = 20$$
$$5(\quad) + 2(\quad) = 20$$
$$\underline{\qquad} + 20 = 20$$
$$\underline{\qquad} = 20$$

Is $(0, 10)$ a solution?

(b) $(2, -5)$

(c) $(3, 2)$

(d) $(-4, 20)$

6 Complete each ordered pair for the equation $y = 2x - 9$.

(a) $(5, \quad)$

$$y = 2(\quad) - 9$$
$$y = \underline{\qquad} - 9$$
$$y = \underline{\qquad}$$

The ordered pair is _____.

(b) $(2, \quad)$

(c) $(\quad, 7)$

(d) $(\quad, -13)$

Example 4 **Deciding Whether Ordered Pairs Are Solutions of an Equation**

Decide whether each ordered pair is a solution of the equation $2x + 3y = 12$.

(a) $(3, 2)$

To see whether $(3, 2)$ is a solution of the equation $2x + 3y = 12$, substitute 3 for x and 2 for y in the equation.

$$2x + 3y = 12$$
$$2(\mathbf{3}) + 3(\mathbf{2}) = 12 \quad ? \quad \text{Let } x = 3; \text{ let } y = 2.$$
$$6 + 6 = 12 \quad ?$$
$$12 = 12 \quad \text{True}$$

This result is true, so $(3, 2)$ is a solution of $2x + 3y = 12$.

(b) $(-2, -7)$

$$2x + 3y = 12$$
$$2(\mathbf{-2}) + 3(\mathbf{-7}) = 12 \quad ? \quad \text{Let } x = -2; \text{ let } y = -7.$$
$$-4 + (-21) = 12 \quad ?$$
$$-25 = 12 \quad \text{False}$$

This result is false, so $(-2, -7)$ is *not* a solution of $2x + 3y = 12$.

Work Problem 5 at the Side.

4 Complete ordered pairs for a given equation. Choosing a number for one variable in a linear equation makes it possible to find the value of the other variable.

Example 5 **Completing Ordered Pairs**

Complete each ordered pair for the equation $y = 4x + 5$.

(a) $(7, \quad)$

In this ordered pair, $x = 7$. (Remember that x always comes first.) To find the corresponding value of y, replace x with 7 in the equation.

$$y = 4\mathbf{x} + 5$$
$$y = 4(\mathbf{7}) + 5 \quad \text{Let } x = 7.$$
$$y = 28 + 5$$
$$\mathbf{y = 33}$$

The ordered pair is $(\mathbf{7}, \mathbf{33})$.

(b) $(\quad, -3)$

In this ordered pair, $y = -3$. Find the value of x by replacing y with -3 in the equation; then solve for x.

$$\mathbf{y} = 4x + 5$$
$$\mathbf{-3} = 4x + 5 \quad \text{Let } y = -3.$$
$$-8 = 4x \quad \text{Subtract 5 from each side.}$$
$$-2 = x \quad \text{Divide each side by 4.}$$

The ordered pair is $(-2, -3)$.

Work Problem 6 at the Side.

5 Complete a table of values. Ordered pairs are often displayed in a **table of values.** The table may be written either vertically or horizontally.

Example 6 Completing Tables of Values

Complete the table of values for each equation.

(a) $x - 2y = 8$

x	y
2	
10	
	0
	-2

To complete the first two ordered pairs, let $x = 2$ and $x = 10$, respectively.

$$\text{If} \quad x = 2,$$
$$\text{then} \quad x - 2y = 8$$
$$\text{becomes} \quad 2 - 2y = 8$$
$$-2y = 6$$
$$y = -3.$$

$$\text{If} \quad x = 10,$$
$$\text{then} \quad x - 2y = 8$$
$$\text{becomes} \quad 10 - 2y = 8$$
$$-2y = -2$$
$$y = 1.$$

Now complete the last two ordered pairs by letting $y = 0$ and $y = -2$, respectively.

$$\text{If} \quad y = 0,$$
$$\text{then} \quad x - 2y = 8$$
$$\text{becomes} \quad x - 2(0) = 8$$
$$x - 0 = 8$$
$$x = 8.$$

$$\text{If} \quad y = -2,$$
$$\text{then} \quad x - 2y = 8$$
$$\text{becomes} \quad x - 2(-2) = 8$$
$$x + 4 = 8$$
$$x = 4.$$

The completed table of values follows.

x	y
2	-3
10	1
8	0
4	-2

The corresponding ordered pairs are

$$(2, -3), (10, 1), (8, 0), \text{ and } (4, -2).$$

Notice that each ordered pair is a solution of the given equation.

(b) $x = 5$

x	y
	-2
	6
	3

Continued on Next Page

7 Complete the table of values for each equation.

(a) $2x - 3y = 12$

x	y
0	
	0
3	
	−3

(b) $y = 4$

x	y
−3	
2	
5	

8 Name the quadrant in which each point in the figure is located.

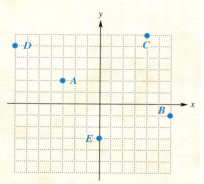

The given equation is $x = 5$. No matter which value of y is chosen, the value of x is always the same, 5.

x	y
5	−2
5	6
5	3

The corresponding ordered pairs are $(5, -2)$, $(5, 6)$, and $(5, 3)$.

NOTE

We can think of $x = 5$ in Example 6(b) as an equation in two variables by rewriting $x = 5$ as $x + 0y = 5$. This form of the equation shows that for any value of y, the value of x is 5. Similarly, $y = 4$ is the same as $0x + y = 4$.

Work Problem 7 at the Side.

6 **Plot ordered pairs.** In Chapter 2, we saw that linear equations in *one* variable had either one, zero, or an infinite number of real number solutions. These solutions could be graphed on *one* number line. Every linear equation in *two* variables has an infinite number of ordered pairs as solutions. Each choice of a number for one variable leads to a particular real number for the other variable.

To graph these solutions, represented as the ordered pairs (x, y), we need *two* number lines, one for each variable. These two number lines are drawn as shown in Figure 4. The horizontal number line is called the **x-axis,** and the vertical line is called the **y-axis.** Together, the x-axis and y-axis form a **rectangular coordinate system.** It is also called the **Cartesian coordinate system,** in honor of René Descartes (1596–1650), the French mathematician who is credited with its invention.

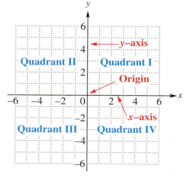

Figure 4

The coordinate system is divided into four regions, called **quadrants.** These quadrants are numbered counterclockwise, as shown in Figure 4. Points on the axes themselves are not in any quadrant. The point at which the x-axis and y-axis meet is called the **origin.** The origin, labeled 0 in Figure 4, is the point corresponding to $(0, 0)$.

Work Problem 8 at the Side.

ANSWERS
7. (a)

x	y
0	−4
6	0
3	−2
$\frac{3}{2}$	−3

(b)

x	y
−3	4
2	4
5	4

8. *A*, II; *B*, IV; *C*, I; *D*, II; *E*, no quadrant

The *x*-axis and *y*-axis determine a **plane,** a flat surface similar to a sheet of paper. By referring to the two axes, every point in the plane can be associated with an ordered pair. The numbers in the ordered pair are called the **coordinates** of the point. For example, locate the point associated with the ordered pair (2, 3) by starting at the origin. Since the *x*-coordinate is 2, go 2 units to the right along the *x*-axis. Then, since the *y*-coordinate is 3, turn and go up 3 units on a line parallel to the *y*-axis. The point (2, 3) is **plotted** in Figure 5. From now on, we will refer to the point with *x*-coordinate 2 and *y*-coordinate 3 as the point (2, 3).

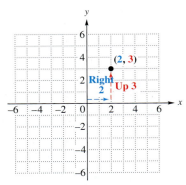

Figure 5

NOTE

When we graph on a number line, one number corresponds to each point. On a plane, however, both numbers in the ordered pair are needed to locate a point. The ordered pair is a name for the point.

Example 7 **Plotting Ordered Pairs**

Plot each ordered pair on a coordinate system.

(a) (1, 5)   **(b)** (−2, 3)   **(c)** (−1, −4)   **(d)** (7, −2)

(e) $\left(\frac{3}{2}, 2\right)$ **(f)** (5, 0)   **(g)** (0, −3)

See Figure 6. In part (c), locate the point (−1, −4) by first going 1 unit to the left along the *x*-axis. Then turn and go 4 units down, parallel to the *y*-axis. Plot the point $\left(\frac{3}{2}, 2\right)$ in part (e) by first going $\frac{3}{2}$ (or $1\frac{1}{2}$) units to the right along the *x*-axis. Then turn and go 2 units up, parallel to the *y*-axis.

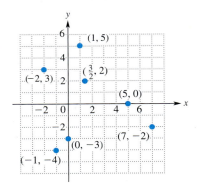

Figure 6

Work Problem ❾ at the Side.

❾ Plot each ordered pair on a coordinate system.

(a) (3, 5)

(b) (−2, 6)

(c) (−4, 0)

(d) (−5, −2)

(e) (5, −2)

(f) (0, −6)

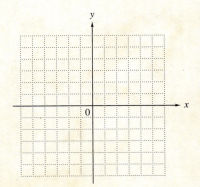

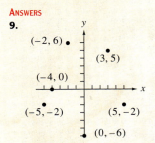

⑩ Refer to the linear equation in Example 8.

(a) Find the y-value for $x = 1996$. Round to the nearest whole number.

Sometimes we can use a linear equation to mathematically describe, or *model,* a real-life situation, as shown in the next example.

Example 8 **Completing Ordered Pairs to Estimate Annual Costs of Doctors' Visits**

The amount Americans pay annually for doctors' visits has increased steadily from 1990 through 2000. This amount can be closely approximated by the linear equation

Cost ⟶   ⟵ Year

$$y = 34.3x - 67{,}693,$$

which relates x, the year, and y, the cost in dollars. (*Source:* U.S. Health Care Financing Administration.)

(a) Complete the table of values for this linear equation.

x (Year)	y (Cost)
1990	
1996	
2000	

To find y when $x = 1990$, substitute into the equation.

$$y = 34.3(\mathbf{1990}) - 67{,}693 \qquad \text{Let } x = 1990.$$
$$y = 564 \qquad\qquad\qquad \text{Use a calculator.}$$

This means that in 1990, Americans each spent about \$564 on doctors' visits.

Work Problem ⑩ at the Side.

Including the results from Problem 10 at the side gives the completed table that follows.

x (Year)	y (Cost)
1990	564
1996	770
2000	907

(b) Find the y-value for $x = 2000$. Interpret your result.

We can write the results from the table of values as ordered pairs (x, y). Each year x is paired with its cost y:

$$(1990, 564), \quad (1996, 770), \quad \text{and} \quad (2000, 907).$$

Continued on Next Page

(b) Graph the ordered pairs found in part (a).

The ordered pairs are graphed in Figure 7. This graph of ordered pairs of data is called a **scatter diagram.** Notice how the axes are labeled: *x* represents the year, and *y* represents the cost in dollars. Different scales are used on the two axes. Here, each square represents two units in the horizontal direction and 100 units in the vertical direction. Because the numbers in the first ordered pair are so large, we show a break in the axes near the origin.

COSTS OF DOCTORS' VISITS

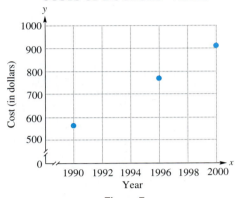

Figure 7

A scatter diagram enables us to tell whether two quantities are related to each other. In Figure 7, the plotted points could be connected to form a straight *line,* so the variables *x* (year) and *y* (cost) have a *line*ar relationship. The increase in costs is also reflected.

CAUTION

The equation in Example 8 is valid only for the years 1990 through 2000 because it was based on data for those years. Do not assume that this equation would provide reliable data for other years since the data for those years may not follow the same pattern.

Real-Data Applications

Linear or Nonlinear? That Is the Question about Windchill

The **windchill factor** measures the cooling effect that the wind has on one's skin. The table gives the windchill factor for various wind speeds and temperatures.

WINDCHILL FACTOR

		Air Temperature (°Fahrenheit)														
		35	30	25	20	15	10	5	0	−5	−10	−15	−20	−25	−30	−35
Wind Speed (mph)	4	35	30	25	20	15	10	5	0	−5	−10	−15	−20	−25	−30	−35
	5	32	27	22	16	11	6	0	−5	−10	−15	−21	−26	−31	−36	−42
	10	22	16	10	3	−3	−9	−15	−22	−27	−34	−40	−46	−52	−58	−64
	15	16	9	2	−5	−11	−18	−25	−31	−38	−45	−51	−58	−65	−72	−78
	20	12	4	−3	−10	−17	−24	−31	−39	−46	−53	−60	−67	−74	−81	−88
	25	8	1	−7	−15	−22	−29	−36	−44	−51	−59	−66	−74	−81	−88	−96
	30	6	−2	−10	−18	−25	−33	−41	−49	−56	−64	−71	−79	−86	−93	−101
	35	4	−4	−12	−20	−27	−35	−43	−52	−58	−67	−74	−82	−89	−97	−105
	40	3	−5	−13	−21	−29	−37	−45	−53	−60	−69	−76	−84	−92	−100	−107
	45	2	−6	−14	−22	−30	−38	−46	−54	−62	−70	−78	−85	−93	−102	−109

Source: USA Today.

The data in the table represents the relationships between two different sets of variable quantities: Windchill versus Air Temperature and Windchill versus Wind Speed. The question is whether either of these relationships is linear, that is, whether the data points, when graphed, could be connected to form a straight line.

Example 1 Windchill versus Air Temperature
Choose one measure of wind speed to keep constant, such as 15 mph. Complete the table with Air Temperature (AT) as the *input* and Windchill (WC) as the *output*. Both variables are measured in degrees Fahrenheit.

AT	35	30	25	20	15	10	5	0	−5	−10	−15				
WC	16	9	2	−5	−11	−18	−25	−31	−38	__	__	__	__	__	__

Example 2 Windchill versus Wind Speed
Choose one measure of air temperature to keep constant, such as 10°F. Complete the table with Wind Speed (WS) as the *input* and Windchill (WC) as the *output*. Wind speed is measured in mph.

WS	4	5	10	15	20	25	30			
WC	10	6	−9	−18	−24	__	__	__	__	__

For Group Discussion

1. Refer to the Windchill versus Air Temperature data (Example 1).

   (a) Write the data as ordered pairs.

   (b) On a sheet of graph paper, draw and label a rectangular coordinate system. Use a scale of 5 on the *x*-axis and the *y*-axis. Make a scatter diagram of the data. Does the graph represent a linear relationship?

2. Repeat Problem 1 using the Windchill versus Wind Speed data (Example 2).

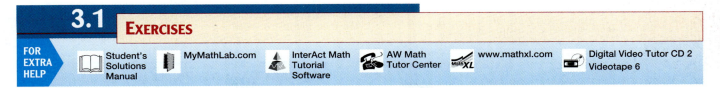

3.1 EXERCISES

FOR
EXTRA
HELP

📖 Student's
Solutions
Manual

🚪 MyMathLab.com

🔺 InterAct Math
Tutorial
Software

📞 AW Math
Tutor Center

Math**XL**

www.mathxl.com

💿 Digital Video Tutor CD 2
Videotape 6

*On February 13, 2000, Peanuts fans bid farewell to this popular comic strip. The circle
graph shows the results of a survey of adults to determine their favorite* Peanuts *characters.
Use the circle graph to work Exercises 1–4. See Example 1.*

1. Which *Peanuts* character was most popular? What
percent of those surveyed named this character as
their favorite?

2. A random sample of 2500 adults is surveyed.

   (a) Estimate how many would be expected to name
   Lucy as their favorite *Peanuts* character. (*Hint*:
   To estimate, round 8% to 10%.)

   (b) Use the actual figure from the graph to determine how many adults would name Lucy as their
   favorite character. Is this answer reasonable based
   on your estimate in part (a)?

3. Regardless of the number of adults surveyed, how
many would we expect to name Charlie Brown as
their favorite *Peanuts* character compared to Linus?

4. Using the group of 2500 adults, confirm your
answer to Exercise 3.

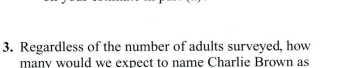

FAVORITE *PEANUTS* CHARACTERS

No opinion 15%
Charlie Brown 26%
Snoopy 31%
Lucy 8%
Linus 13%
Pig Pen 3%
Other 4%

Source: *USA Today*/CNN/Gallop Poll conducted
nationwide February 4–6, 2000.

*The bar graph compares egg production in millions of eggs for six states in June 1999.
Use the bar graph to work Exercises 5–8. See Example 2.*

5. Name the top two egg-producing states in June 1999.
Estimate their production.

6. Which states had egg production less than 400 million
eggs?

7. Which states appear to have had equal production?
Estimate this production.

8. How does egg production in Ohio compare to egg
production in North Carolina?

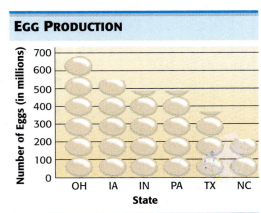

EGG PRODUCTION

Number of Eggs (in millions)

700
600
500
400
300
200
100
0

OH   IA   IN   PA   TX   NC

State

Source: Iowa Agricultural Statistics.

The line graph shows the average price, adjusted for inflation, that Americans have paid for a gallon of gasoline for selected years since 1970. Use the line graph to work Exercises 9–12. See Example 3.

9. Over which period of years did the greatest increase in the price of a gallon of gas occur? About how much was this increase?

10. Estimate the price of a gallon of gas during 1985, 1990, 1995, and 2000.

11. Describe the trend in gas prices from 1980 to 1995.

12. During which year(s) did a gallon of gas cost approximately $1.50?

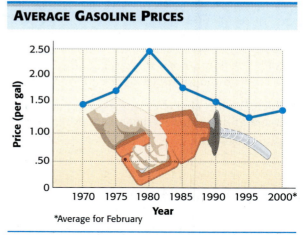

AVERAGE GASOLINE PRICES

*Average for February

Source: American Petroleum Institute; AP research.

Use the concepts of this section to fill in each blank with the correct response.

13. The symbol (x, y) _____ represent an ordered pair, while the
(does/does not)

symbols $[x, y]$ and $\{x, y\}$ _____ represent ordered pairs.
(do/do not)

14. The point whose graph has coordinates $(-4, 2)$ is in quadrant _____.

15. The point whose graph has coordinates $(0, 5)$ lies on the _____-axis.

16. The ordered pair $(4, \_\_\_\_\_)$ is a solution of the equation $y = 3$.

17. The ordered pair $(\_\_\_\_\_, -2)$ is a solution of the equation $x = 6$.

18. The ordered pair $(3, 2)$ is a solution of the equation $2x - 5y = \_\_\_\_\_$.

Decide whether each ordered pair is a solution of the given equation. See Example 4.

19. $x + y = 9$; $(0, 9)$

20. $x + y = 8$; $(0, 8)$

21. $2x - y = 6$; $(4, 2)$

22. $2x + y = 5$; $(3, -1)$

23. $4x - 3y = 6$; $(2, 1)$

24. $5x - 3y = 15$; $(5, 2)$

25. $y = \dfrac{2}{3}x$; $(-6, -4)$

26. $y = -\dfrac{1}{4}x$; $(-8, 2)$

27. $x = -6$; $(5, -6)$

28. $y = 2$; $(2, 4)$

29. Do $(4, -1)$ and $(-1, 4)$ represent the same ordered pair? Explain.

30. Explain why it would be easier to find the corresponding y-value for $x = \dfrac{1}{3}$ in the equation $y = 6x + 2$ than it would be for $x = \dfrac{1}{7}$.

Complete each ordered pair for the equation $y = 2x + 7$. See Example 5.

31. (2,) **32.** (0,) **33.** (, 0) **34.** (, −3)

Complete each ordered pair for the equation $y = -4x - 4$. See Example 5.

35. (0,) **36.** (, 0) **37.** (, 16) **38.** (, 24)

Complete each table of values. In Exercises 39–42, write the results as ordered pairs. See Example 6.

39. $2x + 3y = 12$

x	y
0	
	0
	8

40. $4x + 3y = 24$

x	y
0	
	0
	4

41. $3x - 5y = -15$

x	y
0	
	0
	-6

42. $4x - 9y = -36$

x	y
	0
0	
	8

43. $x = -9$

x	y
	6
	2
	-3

44. $x = 12$

x	y
	3
	8
	0

45. $y = -6$

x	y
8	
4	
-2	

46. $y = -10$

x	y
4	
0	
-4	

47. $x - 8 = 0$

x	y
	8
	3
	0

48. $y + 2 = 0$

x	y
9	
2	
0	

Give the ordered pairs for the points labeled A–F in the figure.

49. *A* **50.** *B* **51.** *C*

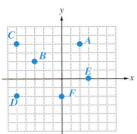

52. *D* **53.** *E* **54.** *F*

Fill in each blank with the word positive *or the word* negative.

The point with coordinates (x, y) is in

55. quadrant III if x is _____ and y is _____.

56. quadrant II if x is _____ and y is _____.

57. quadrant IV if x is _____ and y is _____.

58. quadrant I if x is _____ and y is _____.

59. A point (x, y) has the property that $xy < 0$. In which quadrant(s) must the point lie? Explain.

60. A point (x, y) has the property that $xy > 0$. In which quadrant(s) must the point lie? Explain.

Plot each ordered pair on the rectangular coordinate system provided. See Example 7.

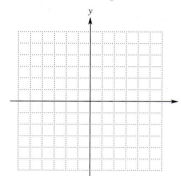

61. $(6, 2)$ **62.** $(5, 3)$ **63.** $(-4, 2)$

64. $(-3, 5)$ **65.** $\left(-\dfrac{4}{5}, -1\right)$ **66.** $\left(-\dfrac{3}{2}, -4\right)$

67. $(0, 4)$ **68.** $(0, -3)$ **69.** $(4, 0)$ **70.** $(-3, 0)$

Complete each table of values, and then plot the ordered pairs. See Examples 6 and 7.

71. $x - 2y = 6$

x	y
0	
	0
2	
	-1

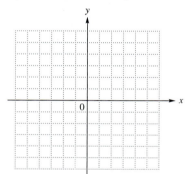

72. $2x - y = 4$

x	y
0	
	0
1	
	-6

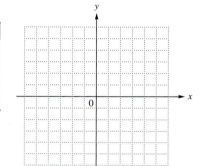

73. $3x - 4y = 12$

x	y
0	
	0
-4	
	-4

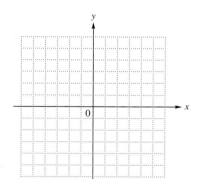

74. $2x - 5y = 10$

x	y
0	
	0
-5	
	-3

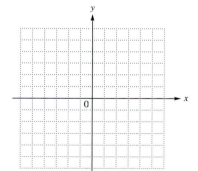

75. $y + 4 = 0$

x	y
0	
5	
-2	
-3	

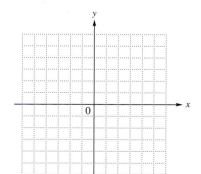

76. $x - 5 = 0$

x	y
	1
	0
	6
	-4

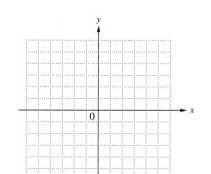

77. Look at the graphs of the ordered pairs in Exercises 71–76. Describe the pattern indicated by the plotted points.

Work each problem. See Example 8.

78. The table shows on-line retail spending in billions of dollars.

Year	Spending (in billions)
1998	7.8
1999	14.9
2000*	23.1
2001*	34.6
2002*	53.0

*Projected
Source: Jupiter Communications.

(a) Write the data from the table as ordered pairs (x, y), where x represents the year and y represents on-line spending in billions of dollars.

(b) What does the ordered pair (2003, 78.0) mean in the context of this problem?

(c) Make a scatter diagram of the data using the ordered pairs from part (a).

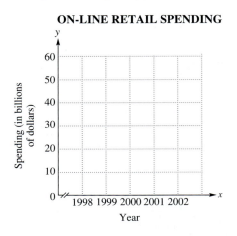

ON-LINE RETAIL SPENDING

(d) Describe the pattern indicated by the points on the scatter diagram. What is the trend in on-line spending?

79. The table shows the rate (in percent) at which 4-year college students graduate within 5 years.

Year	Rate (%)
1996	53.3
1997	52.8
1998	52.1
1999	51.6

Source: ACT.

(a) Write the data from the table as ordered pairs (x, y), where x represents the year and y represents graduation rate.

(b) What does the ordered pair (1995, 54.0) mean in the context of this problem?

(c) Make a scatter diagram of the data using the ordered pairs from part (a).

**4-YEAR COLLEGE STUDENTS
GRADUATING WITHIN 5 YEARS**

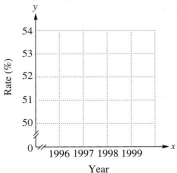

(d) Describe the pattern indicated by the points on the scatter diagram. What is happening to graduation rates for 4-year college students within 5 years?

80. The maximum benefit for the heart from exercising occurs if the heart rate is in the target heart rate zone. The lower limit of this target zone can be approximated by the linear equation

$$y = -.7x + 154,$$

where x represents age and y represents heartbeats per minute. (*Source:* Hockey, R. V., *Physical Fitness: The Pathway to Healthy Living*, Times Mirror/Mosby College Publishing, 1989.)

(a) Complete the table of values for this linear equation.

Age	Heartbeats (per minute)
20	
40	
60	
80	

(b) Write the data from the table of values as ordered pairs.

(c) Make a scatter diagram of the data. Do the points lie in an approximately linear pattern?

TARGET HEART RATE ZONE

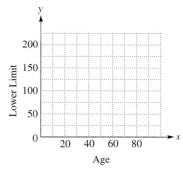

81. (See Exercise 80.) The upper limit of the target heart rate zone can be approximated by the linear equation

$$y = -.8x + 186,$$

where x represents age and y represents heartbeats per minute. (*Source:* Hockey, R. V., *Physical Fitness: The Pathway to Healthy Living*, Times Mirror/Mosby College Publishing, 1989.)

(a) Complete the table of values for this linear equation.

Age	Heartbeats (per minute)
20	
40	
60	
80	

(b) Write the data from the table of values as ordered pairs.

(c) Make a scatter diagram of the data. Describe the pattern indicated by the data.

TARGET HEART RATE ZONE

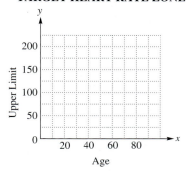

82. Refer to Exercises 80 and 81. What is the target heart rate zone for age 20? age 40?

3.2 GRAPHING LINEAR EQUATIONS IN TWO VARIABLES

1 Graph linear equations by plotting ordered pairs. There are infinitely many ordered pairs that satisfy an equation in two variables. We find ordered pairs that are solutions of the equation $x + 2y = 7$ by choosing as many values of x (or y) as we wish and then completing each ordered pair.

For example, if we choose $x = 1$, then $y = 3$, so the ordered pair $(1, 3)$ is a solution of the equation $x + 2y = 7$.

$$1 + 2(3) = 1 + 6 = 7$$

Work Problem ❶ at the Side.

Figure 8 shows a graph of all the ordered pairs found for $x + 2y = 7$ above and in Problem 1 at the side.

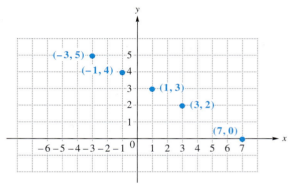

Figure 8

Notice that the points plotted in this figure all appear to lie on a straight line. The line that goes through these points is shown in Figure 9. In fact, all ordered pairs satisfying the equation $x + 2y = 7$ correspond to points that lie on this same straight line. This line gives a "picture" of all the solutions of the equation $x + 2y = 7$. Only a portion of the line is shown here, but it extends indefinitely in both directions, as suggested by the arrowhead on each end of the line.

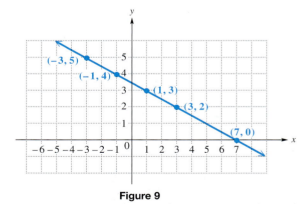

Figure 9

The line in Figure 9 is called the **graph** of the equation $x + 2y = 7$, and the process of plotting the ordered pairs and drawing the line through the corresponding points is called **graphing.** The preceding discussion can be generalized.

OBJECTIVES

1 Graph linear equations by plotting ordered pairs.

2 Find intercepts.

3 Graph linear equations where the intercepts coincide.

4 Graph linear equations of the form $y = k$ or $x = k$.

5 Use a linear equation to model data.

❶ Complete each ordered pair for the equation $x + 2y = 7$.

(a) $(-3, \ \ )$

(b) $(3, \ \ )$

(c) $(-1, \ \ )$

(d) $(7, \ \ )$

> The graph of any linear equation in two variables is a straight line.

(Notice that the word *line* appears in the term "*line*ar equation.")

Because two distinct points determine a line, a straight line can be graphed by finding any two different points on the line. However, it is a good idea to plot a third point as a check.

Example 1 Graphing a Linear Equation

Graph the linear equation $y = -\frac{3}{2}x + 3$.

Although this equation is not in the form $Ax + By = C$, it *could* be put in that form, so it is a linear equation. For most linear equations, two different points on the graph can be found by first letting $x = 0$ and then letting $y = 0$.

If $x = 0$, then

$$y = -\frac{3}{2}x + 3$$

$$y = -\frac{3}{2}(0) + 3 \quad \text{Let } x = 0.$$

$$y = 0 + 3$$

$$y = 3.$$

If $y = 0$, then

$$y = -\frac{3}{2}x + 3$$

$$0 = -\frac{3}{2}x + 3 \quad \text{Let } y = 0.$$

$$\frac{3}{2}x = 3$$

$$\frac{2}{3} \cdot \frac{3}{2}x = \frac{2}{3} \cdot 3$$

$$x = 2.$$

This gives the ordered pairs $(0, 3)$ and $(2, 0)$. Get a third point (as a check) by letting x or y equal some other number. For example, let $x = -2$. (Any number could be used, but a multiple of 2 makes multiplying by $-\frac{3}{2}$ easier.) Replace x with -2 in the given equation.

$$y = -\frac{3}{2}x + 3$$

$$y = -\frac{3}{2}(-2) + 3 \quad \text{Let } x = -2.$$

$$y = 3 + 3$$

$$y = 6$$

These three ordered pairs are shown in the table with Figure 10. Plot the corresponding points, then draw a line through them. This line, shown in Figure 10, is the graph of $y = -\frac{3}{2}x + 3$.

x	y
0	3
2	0
-2	6

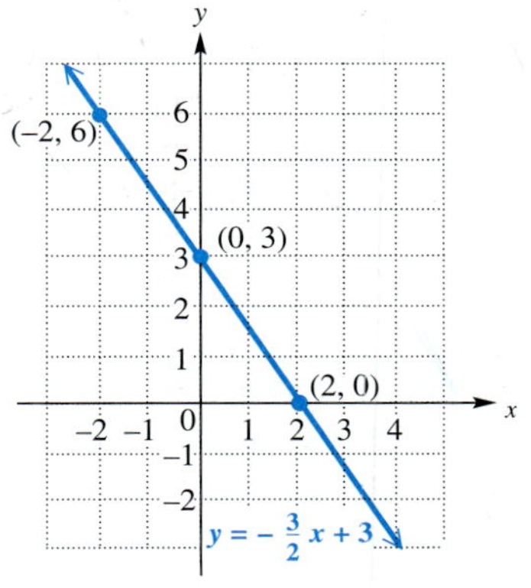

Figure 10

CAUTION

When graphing a linear equation as in Example 1, all three points should lie on the same straight line. If they don't, double-check the ordered pairs you found.

Work Problem ❷ at the Side.

Example 2 **Graphing a Linear Equation**

Graph the linear equation $4x = 5y + 20$.

As before, at least two different points are needed to draw the graph. First let $x = 0$ and then let $y = 0$ to complete two ordered pairs.

$$4x = 5y + 20 \qquad\qquad 4x = 5y + 20$$
$$4(0) = 5y + 20 \quad \text{Let } x = 0. \qquad 4x = 5(0) + 20 \quad \text{Let } y = 0.$$
$$0 = 5y + 20 \qquad\qquad 4x = 20$$
$$-5y = 20 \qquad\qquad x = 5$$
$$y = -4$$

The ordered pairs are $(0, -4)$ and $(5, 0)$. Get a third ordered pair (as a check) by choosing some number other than 0 for x or y. We choose $y = 2$. Replacing y with 2 in the equation $4x = 5y + 20$ leads to the ordered pair $\left(\frac{15}{2}, 2\right)$, or $\left(7\frac{1}{2}, 2\right)$.

Plot the three ordered pairs $(0, -4)$, $(5, 0)$, and $\left(\frac{15}{2}, 2\right)$, and draw a line through them. This line, shown in Figure 11, is the graph of $4x = 5y + 20$.

x	y
0	-4
5	0
$7\frac{1}{2}$	2

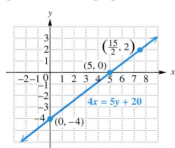

Figure 11

Work Problem ❸ at the Side.

2⬛ **Find intercepts.** In Figure 11, the graph crosses or intersects the y-axis at $(0, -4)$ and the x-axis at $(5, 0)$. For this reason, $(0, -4)$ is called the **y-intercept,** and $(5, 0)$ is called the **x-intercept** of the graph. The intercepts are particularly useful for graphing linear equations, as in Examples 1 and 2. (In general, any point on the y-axis has x-coordinate 0, and any point on the x-axis has y-coordinate 0.) The intercepts are found by replacing, in turn, each variable with 0 in the equation and solving for the value of the other variable.

Finding Intercepts

To find the **x-intercept,** let $y = 0$ in the given equation and solve for x. Then $(x, 0)$ is the x-intercept.

To find the **y-intercept,** let $x = 0$ in the given equation and solve for y. Then $(0, y)$ is the y-intercept.

❷ Complete the table of values, and graph the linear equation.

$x + y = 6$

x	y
0	
	0
2	

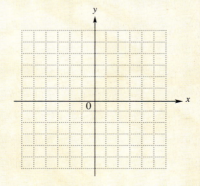

❸ Make a table of values, and graph the linear equation.

$2x = 3y + 6$

x	y

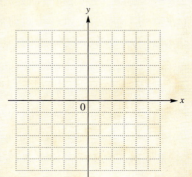

ANSWERS

2.

x	y
0	6
6	0
2	4

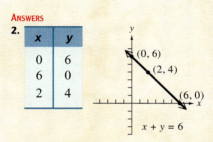

3.

❹ Find the intercepts for the graph of $5x + 2y = 10$. Then draw the graph. (Be sure to get a third point as a check.)

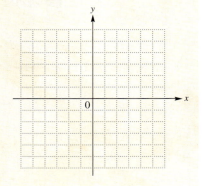

4. x-intercept $(2, 0)$; y-intercept $(0, 5)$

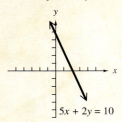

$5x + 2y = 10$

Example 3   **Finding Intercepts**

Find the intercepts for the graph of $2x + y = 4$. Then draw the graph.

Find the y-intercept by letting $x = 0$; find the x-intercept by letting $y = 0$.

$$2x + y = 4 \qquad\qquad 2x + y = 4$$
$$2(0) + y = 4 \quad \text{Let } x = 0. \qquad 2x + 0 = 4 \quad \text{Let } y = 0.$$
$$0 + y = 4 \qquad\qquad 2x = 4$$
$$y = 4 \qquad\qquad\quad x = 2$$

The y-intercept is $(0, 4)$. The x-intercept is $(2, 0)$. The graph with the two intercepts shown in color is given in Figure 12. Get a third point as a check. For example, choosing $x = 1$ gives $y = 2$. Plot $(0, 4)$, $(2, 0)$, and $(1, 2)$ and draw a line through them. This line, shown in Figure 12, is the graph of $2x + y = 4$.

x	y
0	4
2	0
1	2

y-intercept $(0, 4)$
$(1, 2)$ is used as a check.
x-intercept $(2, 0)$
$2x + y = 4$

Figure 12

Work Problem ❹ at the Side.

---

3   **Graph linear equations where the intercepts coincide.** In the preceding examples, the x- and y-intercepts were used to help draw the graphs. This is not always possible. Example 4 shows what to do when the x- and y-intercepts are the same point (that is, coincide).

Example 4   **Graphing an Equation of the Form $Ax + By = 0$**

Graph the linear equation $x - 3y = 0$.

If we let $x = 0$, then $y = 0$, giving the ordered pair $(0, 0)$. Letting $y = 0$ also gives $(0, 0)$. This is the same ordered pair, so choose two *other* values for x or y. Choosing 2 for y gives $x - 3 \cdot 2 = 0$, or $x = 6$, so another ordered pair is $(6, 2)$. For a check point, choose -6 for x to get -2 for y. The ordered pairs $(-6, -2)$, $(0, 0)$, and $(6, 2)$ are used to sketch the graph in Figure 13.

x	y
0	0
6	2
-6	-2

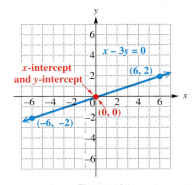

x-intercept and y-intercept
$x - 3y = 0$
$(6, 2)$
$(0, 0)$
$(-6, -2)$

Figure 13

Example 4 can be generalized as follows.

If A and B are nonzero real numbers, the graph of a linear equation of the form

$$Ax + By = 0$$

goes through the origin $(0, 0)$.

Work Problem ❺ at the Side.

4 ▭ **Graph linear equations of the form $y = k$ or $x = k$.**   The equation $y = -4$ is a linear equation in which the coefficient of x is 0. (To see this, write $y = -4$ as $0x + y = -4$.) Also, $x = 3$ is a linear equation in which the coefficient of y is 0. These equations lead to horizontal or vertical straight lines, as the next example shows.

Example 5   **Graphing Equations of the Form $y = k$ and $x = k$**

(a) Graph the linear equation $y = -4$.

As the equation states, for any value of x, y is always equal to -4. Three ordered pairs that satisfy the equation are shown in the table of values. Drawing a line through these points gives the horizontal line in Figure 14. The y-intercept is $(0, -4)$; there is no x-intercept.

x	y
−2	−4
0	−4
3	−4

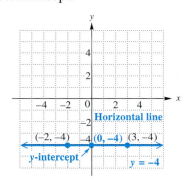

Figure 14

(b) Graph $x - 3 = 0$.

First add 3 to each side of $x - 3 = 0$ to get $x = 3$. All the ordered pairs that satisfy this equation have x-coordinate 3. Any number can be used for y. See Figure 15 for the graph of this vertical line, along with a table of values. The x-intercept is $(3, 0)$; there is no y-intercept.

x	y
3	3
3	0
3	−2

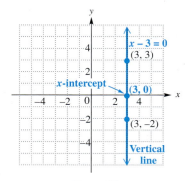

Figure 15

❺ Graph each equation.

(a) $2x - y = 0$

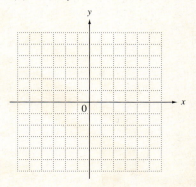

(b) $x = -4y$

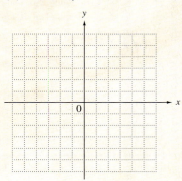

ANSWERS

5. (a)

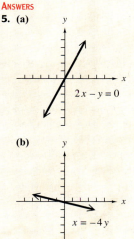

6 Graph each equation.

(a) $y = -5$

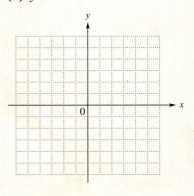

(b) $x + 4 = 6$

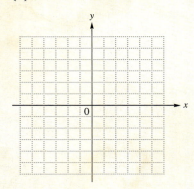

From the results in Example 5, we make the following observations.

Horizontal and Vertical Lines

The graph of the linear equation $y = k$, where k is a real number, is the horizontal line with y-intercept $(0, k)$ and no x-intercept.

The graph of the linear equation $x = k$, where k is a real number, is the vertical line with x-intercept $(k, 0)$ and no y-intercept.

Work Problem 6 at the Side.

The different forms of linear equations from this section and the methods of graphing them are summarized below.

Graphing Linear Equations

Equation	Graphing Method	Example
$y = k$	Draw a horizontal line through $(0, k)$.	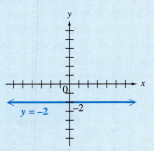
$x = k$	Draw a vertical line through $(k, 0)$.	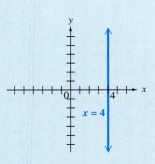
$Ax + By = 0$	Graph goes through $(0, 0)$. To get additional points that lie on the graph, choose any values for x or y, except 0.	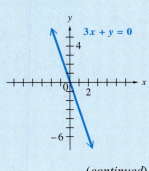

(continued)

ANSWERS

6. (a) **(b)**

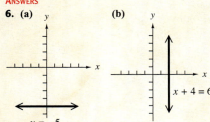

$Ax + By = C$
$A, B,$ and $C \neq 0$

Find any two points on the line. A good choice is to find the intercepts. Let $x = 0$, and find the corresponding value of y; then let $y = 0$, and find x. As a check, get a third point by choosing a value of x or y that has not yet been used.

$3x - 2y = 6$

$(2, 0)$

$(0, -3)$

Work Problem ⑦ at the Side.

⑦ Match the information about the graphs with the linear equations in A–D.

A. $x = 5$
B. $2x - 5y = 8$
C. $y - 2 = 3$
D. $x + 4y = 0$

(a) The graph of the equation is a horizontal line.

5 ▭ Use a linear equation to model data.

Example 6 **Using a Linear Equation to Model Credit Card Debt**

Credit card debt in the United States increased steadily from 1992 through 1999. The amount of debt y in billions of dollars can be modeled by the linear equation

$$y = 47.3x + 281,$$

where $x = 0$ represents 1992, $x = 1$ represents 1993, and so on. (*Source:* Board of Governors of the Federal Reserve System.)

(b) The graph of the equation goes through the origin.

(c) The graph of the equation is a vertical line.

(a) Use the equation to approximate credit card debt in the years 1992, 1993, and 1999.

For 1992:	$y = 47.3(0) + 281$	Replace x with 0.
	$y = 281$ billion dollars	
For 1993:	$y = 47.3(1) + 281$	Replace x with 1.
	$y = 328.3$ billion dollars	
For 1999:	$y = 47.3(7) + 281$	$1999 - 1992 = 7$;
	$y = 612.1$ billion dollars	replace x with 7.

(d) The graph of the equation goes through $(9, 2)$.

Continued on Next Page

⑧ Use the graph and the equation in Example 6 to approximate credit card debt in 1997.

(b) Write the information from part (a) as three ordered pairs, and use them to graph the given linear equation.

Since x represents the year and y represents the debt in billions of dollars, the ordered pairs are (0, 281), (1, 328.3), and (7, 612.1). Figure 16 shows a graph of these ordered pairs and the line through them. (Note that arrowheads are not included with the graphed line since the data are for the years 1992 to 1999 only, that is, from $x = 0$ to $x = 7$.)

U.S. CREDIT CARD DEBT

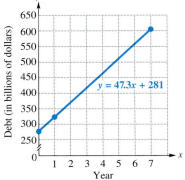

Figure 16

(c) Use the graph and then the equation to approximate credit card debt in 1996.

For 1996, $x = 4$. On the graph, find 4 on the horizontal axis and move up to the graphed line, then across to the vertical axis. It appears that credit card debt in 1996 was about 470 billion dollars.

To use the equation, substitute 4 for x.

$$y = 47.3(4) + 281 \qquad \text{Let } x = 4.$$
$$y = 470.2 \text{ billion dollars}$$

This result is quite similar to our estimate using the graph.

Work Problem ⑧ at the Side.

3.2 **EXERCISES**

Complete the given ordered pairs for each equation. Then graph each equation by plotting the points and drawing a line through them. See Examples 1 and 2.

1. $y = -x + 5$

$(0,\ ),\ (\ ,0),\ (2,\ )$

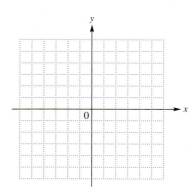

2. $y = x - 2$

$(0,\ ),\ (\ ,0),\ (5,\ )$

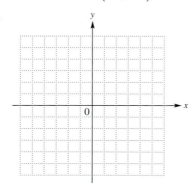

3. $y = \dfrac{2}{3}x + 1$

$(0, 1),\ (3, 3),\ (-3, -1)$

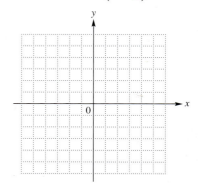

4. $y = -\dfrac{3}{4}x + 2$

$(0,\ ),\ (4,\ ),\ (-4,\ )$

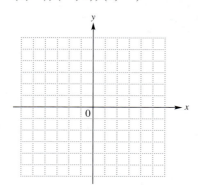

5. $3x = -y - 6$

$(0,\ ),\ (\ ,0),\ \left(-\dfrac{1}{3},\ \right)$

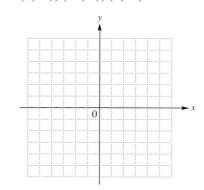

6. $x = 2y + 3$

$(\ ,0),\ (0,\ ),\ \left(\ ,\dfrac{1}{2}\right)$

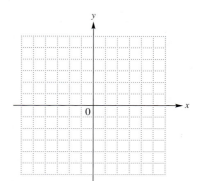

Match the information about each graph in Column I with the correct linear equation in Column II.

	I		**II**

7. The graph of the equation has *y*-intercept $(0, -4)$.

A. $3x + y = -4$

8. The graph of the equation has $(0, 0)$ as *x*-intercept and *y*-intercept.

B. $x - 4 = 0$

9. The graph of the equation does not have an *x*-intercept.

C. $y = 4x$

10. The graph of the equation has *x*-intercept $(4, 0)$.

D. $y = 4$

Find the intercepts for the graph of each equation. See Example 3.

11. $2x - 3y = 24$

    x-intercept:

    y-intercept:

12. $-3x + 8y = 48$

    x-intercept:

    y-intercept:

13. $x + 6y = 0$

    x-intercept:

    y-intercept:

14. $3x - y = 0$

    x-intercept:

    y-intercept:

15. A student attempted to graph $4x + 5y = 0$ by finding intercepts. She first let $x = 0$ and found y; then she let $y = 0$ and found x. In both cases, the resulting point was $(0, 0)$. She knew that she needed at least two points to graph the line, but was unsure what to do next because finding intercepts gave her only one point. How would you explain to her what to do next?

16. What is the equation of the x-axis? What is the equation of the y-axis?

Graph each linear equation. See Examples 1–5.

17. $x = y + 2$

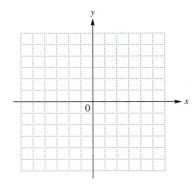

18. $x = -y + 6$

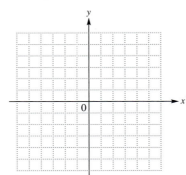

19. $x - y = 4$

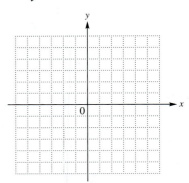

20. $x - y = 5$

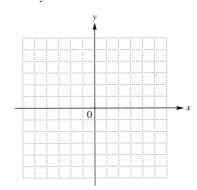

21. $2x + y = 6$

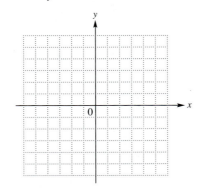

22. $-3x + y = -6$

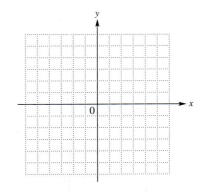

23. $3x + 7y = 14$

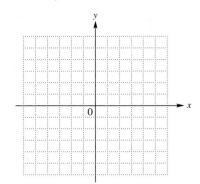

24. $6x - 5y = 18$

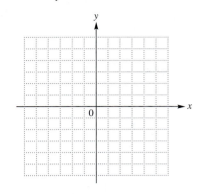

25. $y - 2x = 0$

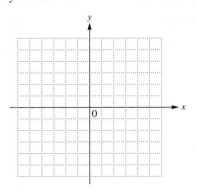

26. $y + 3x = 0$

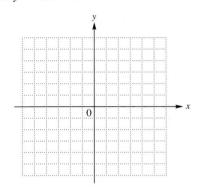

27. $y = -6x$

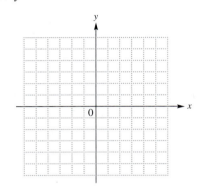

28. $x = 4$

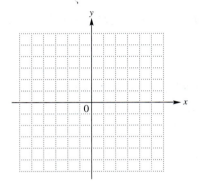

29. $x = -2$

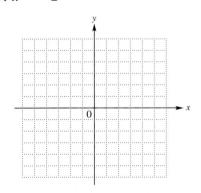

30. $y + 1 = 0$

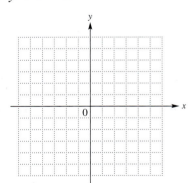

31. $y - 3 = 0$

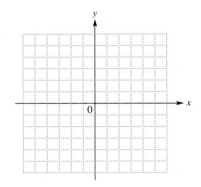

32. Write a few sentences summarizing how to graph a linear equation in two variables.

Solve each problem. See Example 6.

33. The height y (in centimeters) of a woman is related to the length of her radius bone x (from the wrist to the elbow) and is approximated by the linear equation

$$y = 3.9x + 73.5.$$

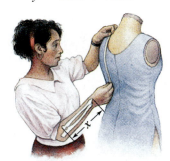

(a) Use the equation to find the approximate heights of women with radius bones of lengths 20 cm, 26 cm, and 22 cm.

(b) Graph the equation using the data from part (a).

HEIGHTS OF WOMEN

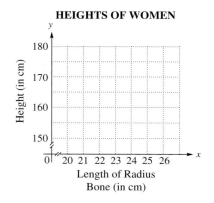

Length of Radius Bone (in cm)

(c) Use the graph to estimate the length of the radius bone in a woman who is 167 cm tall. Then use the equation to find the length of this radius bone to the nearest centimeter. (*Hint:* Substitute for y in the equation.)

34. The weight y (in pounds) of a man taller than 60 in. can be roughly approximated by the linear equation

$$y = 5.5x - 220,$$

where x is the height of the man in inches.

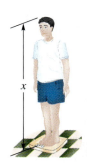

(a) Use the equation to approximate the weights of men whose heights are 62 in., 66 in., and 72 in.

(b) Graph the equation using the data from part (a).

WEIGHTS OF MEN

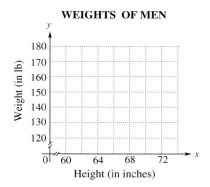

Height (in inches)

(c) Use the graph to estimate the height of a man who weighs 155 lb. Then use the equation to find the height of this man to the nearest inch. (*Hint:* Substitute for y in the equation.)

35. Refer to Section 3.1 Exercise 80. Draw a line through the points you plotted in the scatter diagram there.

(a) Use the graph to estimate the lower limit of the target heart rate zone for age 30.

(b) Use the linear equation given there to approximate the lower limit for age 30.

(c) How does the approximation using the equation compare to the estimate from the graph?

36. Refer to Section 3.1 Exercise 81. Draw a line through the points you plotted in the scatter diagram there.

(a) Use the graph to estimate the upper limit of the target heart rate zone for age 30.

(b) Use the linear equation given there to approximate the upper limit for age 30.

(c) How does the approximation using the equation compare to the estimate from the graph?

37. Use the results of Exercises 35(b) and 36(b) to determine the target heart rate zone for age 30.

38. Should the graphs of the target heart rate zone in the Section 3.1 exercises be used to estimate the target heart rate zone for ages below 20 or above 80? Why or why not?

39. Per capita consumption of carbonated soft drinks increased for the years 1992 through 1997 as shown in the graph. If $x = 0$ represents 1992, $x = 1$ represents 1993, and so on, per capita consumption can be modeled by the linear equation

$$y = .8x + 49,$$

where y is in gallons.

SOFT DRINK CONSUMPTION

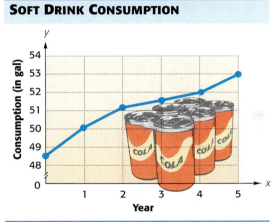

Source: U.S. Department of Agriculture.

(a) Use the equation to approximate consumption in 1993, 1995, and 1997.

(b) Use the graph to estimate consumption for the same years.

(c) How do the approximations using the equation compare to the estimates from the graph?

40. The income generated by the Walgreen Company from 1994 through 1998 is shown in the graph. If $x = 0$ corresponds to 1994, $x = 1$ corresponds to 1995, and so on, the income can be modeled by the linear equation

$$y = 57.3x + 270,$$

where y is in billions of dollars.

WALGREEN COMPANY INCOME

Source: Hoover's Outline.

(a) Use the equation to approximate the income generated in 1994, 1996, and 1998. Round your answers to the nearest billion dollars.

(b) Use the graph to estimate the income for the same years.

(c) How do the approximations using the equation compare to the estimates from the graph?

41. The graph shows the value of a certain sport utility vehicle over the first 5 yr of ownership.

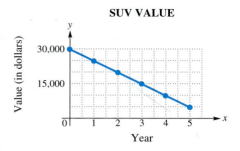

SUV VALUE

Use the graph to do the following.

(a) Determine the initial value of the SUV.

(b) Find the *depreciation* (loss in value) from the original value after the first 3 yr.

(c) What is the annual or yearly depreciation in each of the first 5 yr?

(d) What does the ordered pair (5, 5000) mean in the context of this problem?

42. Demand for an item is often closely related to its price. As price increases, demand decreases, and as price decreases, demand increases. Suppose demand for a video game is 2000 units when the price is $40, and demand is 2500 units when the price is $30.

(a) Let x be the price and y be the demand for the game. Graph the two given pairs of prices and demands.

VIDEO GAME PRICE/DEMAND

(b) Assume the relationship is linear. Draw a line through the two points from part (a). From your graph, estimate the demand if the price drops to $20.

(c) Use the graph to estimate the price if the demand is 3500 units.

43. The graph of the linear equation for credit card debt from Example 6,

$$y = 47.3x + 281,$$

where $x = 0$ represents 1992, and so on, and y is in billions of dollars, is shown in the figure. The actual data for 1992 through 1999 is also plotted.

(a) In general, how well does the linear equation model the actual data?

(b) Use the plotted points to estimate the actual credit card debt for 1996. How does it compare to the answer in Example 6(c)?

(c) Should this equation be used to predict credit card debt for the year 2002? Why or why not?

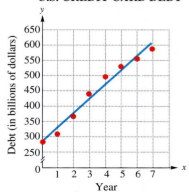

U.S. CREDIT CARD DEBT

Source: Board of Governors of the Federal Reserve System.

3.3 SLOPE

An important characteristic of the lines we graphed in the previous section is their slant or "steepness." See Figure 17.

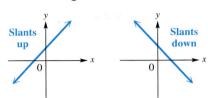

Figure 17

OBJECTIVES

1 Find the slope of a line given two points.

2 Find the slope from the equation of a line.

3 Use slope to determine whether two lines are parallel, perpendicular, or neither.

One way to measure the steepness of a line is to compare the vertical change in the line to the horizontal change while moving along the line from one fixed point to another. This measure of steepness is called the *slope* of the line.

1 **Find the slope of a line given two points.**   Figure 18 shows a line through two nonspecific points (x_1, y_1) and (x_2, y_2). (This notation is called **subscript notation**. Read x_1 as "x-sub-one" and x_2 as "x-sub-two.")

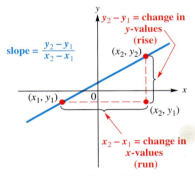

Figure 18

Moving along the line from the point (x_1, y_1) to the point (x_2, y_2) causes y to change by $y_2 - y_1$ units. This is the vertical change or **rise.** Similarly, x changes by $x_2 - x_1$ units, which is the horizontal change or **run.** (In both cases, the change is expressed as a *difference*.) Remember from Section 2.6 that one way to compare two numbers is by using a ratio. Slope is the ratio of the vertical change in y to the horizontal change in x. Traditionally, the letter m represents slope. The slope m of a line is defined as follows.

Slope Formula

The **slope** of the line through the points (x_1, y_1) and (x_2, y_2) is

$$m = \frac{\text{change in } y}{\text{change in } x} = \frac{y_2 - y_1}{x_2 - x_1}, \quad \text{if } x_1 \neq x_2.$$

Work Problem 1 at the Side.

The slope of a line tells how fast y changes for each unit of change in x; that is, the slope gives the rate of change in y for each unit of change in x.

1 Find $\dfrac{y_2 - y_1}{x_2 - x_1}$ for the following values.

(a) $y_2 = 4, y_1 = -1,$
$x_2 = 3, x_1 = 4$

(b) $x_1 = 3, x_2 = -5,$
$y_1 = 7, y_2 = -9$

(c) $x_1 = 2, x_2 = 7,$
$y_1 = 4, y_2 = 9$

ANSWERS

1. **(a)** -5 **(b)** 2 **(c)** 1

The idea of slope is used in many everyday situations. See Figure 19. For example, a highway with a 10% or $\frac{1}{10}$ grade (or slope) rises 1 m for every 10 m horizontally. Architects specify the pitch of a roof using slope; a $\frac{5}{12}$ roof means that the roof rises 5 ft for every 12 ft in the horizontal direction. The slope of a stairwell also indicates the ratio of the vertical rise to the horizontal run. In the figure, the slope of the stairwell is $\frac{8}{10}$ or $\frac{4}{5}$.

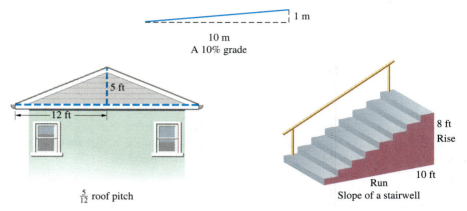

Figure 19

Example 1 Finding Slopes of Lines

Find the slope of each line.

(a) The line through $(1, -2)$ and $(-4, 7)$

Use the slope formula. Let $(-4, 7) = (x_2, y_2)$ and $(1, -2) = (x_1, y_1)$. Then

$$\text{slope } m = \frac{\text{change in } y}{\text{change in } x} = \frac{y_2 - y_1}{x_2 - x_1} = \frac{7 - (-2)}{-4 - 1} = \frac{9}{-5} = -\frac{9}{5}.$$

See Figure 20.

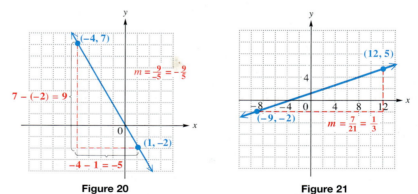

Figure 20   **Figure 21**

(b) The line through $(-9, -2)$ and $(12, 5)$

$$m = \frac{y_2 - y_1}{x_2 - x_1} = \frac{5 - (-2)}{12 - (-9)} = \frac{7}{21} = \frac{1}{3}$$

See Figure 21. The same slope is obtained by subtracting in reverse order.

$$m = \frac{-2 - 5}{-9 - 12} = \frac{-7}{-21} = \frac{1}{3}$$

CAUTION

It makes no difference which point is (x_1, y_1) or (x_2, y_2); however, it is important to be consistent. Start with the x- and y-values of one point (either one) and subtract the corresponding values of the other point. Also, the slope of a line is the same for *any* two points on the line.

Work Problem ❷ at the Side.

In Example 1(a) the slope is negative and the corresponding line in Figure 20 falls from left to right. The slope in Example 1(b) is positive and the corresponding line in Figure 21 rises from left to right. These facts can be generalized.

Positive and Negative Slopes

A line with positive slope rises from left to right.

A line with negative slope falls from left to right.

Example 2   Finding the Slope of a Horizontal Line

Find the slope of the line through $(-8, 4)$ and $(2, 4)$.

$$m = \frac{y_2 - y_1}{x_2 - x_1} = \frac{4 - 4}{-8 - 2} = \frac{0}{-10} = 0 \qquad \text{Zero slope}$$

As shown in Figure 22, the line through the given points is horizontal. *All horizontal lines have slope 0* since the difference in y-values is always 0.

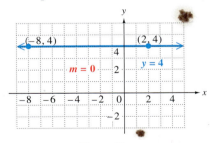

Figure 22

Example 3   Finding the Slope of a Vertical Line

Find the slope of the line through $(6, 2)$ and $(6, -4)$.

$$m = \frac{y_2 - y_1}{x_2 - x_1} = \frac{2 - (-4)}{6 - 6} = \frac{6}{0} \qquad \text{Undefined slope}$$

Because division by 0 is undefined, this line has undefined slope. (This is why the slope formula at the beginning of this section had the restriction $x_1 \neq x_2$.) The graph in Figure 23 on the next page shows that this line is vertical. All points on a vertical line have the same x-value, so *all vertical lines have undefined slope.*

Continued on Next Page

❷ Find the slope of each line.

(a) Through $(6, -2)$ and $(5, 4)$

(b) Through $(-3, 5)$ and $(-4, -7)$

(c) Through $(6, -8)$ and $(-2, 4)$
(Find this slope in two different ways as in Example 1(b).)

ANSWERS

2. (a) -6 **(b)** 12 **(c)** $-\dfrac{3}{2}; -\dfrac{3}{2}$

❸ Find the slope of each line.

(a) Through (2, 5) and (−1, 5)

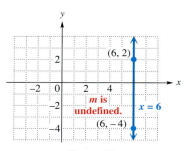

Figure 23

(b) Through (3, 1) and (3, −4)

Slopes of Horizontal and Vertical Lines

Horizontal lines, which have equations of the form $y = k$, have **slope 0.**

Vertical lines, which have equations of the form $x = k$, have **undefined slope.**

Work Problem ❸ at the Side.

2 ▭ **Find the slope from the equation of a line.** The slope of a line can be found directly from its equation. For example, the slope of the line

$$y = -3x + 5$$

can be found using any two points on the line. We get these two points by first choosing two different values of x and then finding the corresponding values of y. Choose $x = -2$ and $x = 4$.

(c) With equation $y = -1$

$y = -3x + 5$	$y = -3x + 5$
$y = -3(-2) + 5$ Let $x = -2$.	$y = -3(4) + 5$ Let $x = 4$.
$y = 6 + 5$	$y = -12 + 5$
$y = 11$	$y = -7$

The ordered pairs are $(-2, 11)$ and $(4, -7)$. Now use the slope formula.

$$m = \frac{11 - (-7)}{-2 - 4} = \frac{18}{-6} = -3$$

The slope, −3, is the same number as the coefficient of x in the equation $y = -3x + 5$. It can be shown that this always happens, *as long as the equation is solved for y.* This fact is used to find the slope of a line from its equation.

(d) With equation $x - 4 = 0$

Finding the Slope of a Line from Its Equation

Step 1   Solve the equation for y.

Step 2   The slope is given by the coefficient of x.

NOTE

We will see in the next section that the equation $y = -3x + 5$ is written using a special form of the equation of a line,

$$y = mx + b,$$

called *slope-intercept form.*

ANSWERS
3. (a) 0   **(b)** undefined
  (c) 0   **(d)** undefined

Example 4 Finding Slopes from Equations

Find the slope of each line.

(a) $2x - 5y = 4$

Step 1 Solve the equation for y.

$$2x - 5y = 4$$
$$-5y = -2x + 4 \qquad \text{Subtract } 2x.$$
$$y = \frac{2}{5}x - \frac{4}{5} \qquad \text{Divide by } -5.$$

Step 2 The slope is given by the coefficient of x, so the slope is $\frac{2}{5}$.

(b) $8x + 4y = 1$
Solve for y.

$$4y = -8x + 1 \qquad \text{Subtract } 8x.$$
$$y = -2x + \frac{1}{4} \qquad \text{Divide by } 4.$$

The slope of this line is given by the coefficient of x, which is -2.

Work Problem ④ at the Side.

③ **Use slope to determine whether two lines are parallel, perpendicular, or neither.** Two lines in a plane that never intersect are **parallel.** We use slopes to tell whether two lines are parallel. For example, Figure 24 shows the graphs of $x + 2y = 4$ and $x + 2y = -6$. These lines appear to be parallel. Solving for y, we find that both $x + 2y = 4$ and $x + 2y = -6$ have slope $-\frac{1}{2}$. Nonvertical parallel lines always have equal slopes.

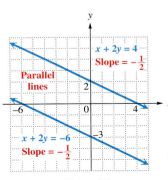

Figure 24

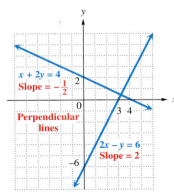

Figure 25

Figure 25 shows the graphs of $x + 2y = 4$ and $2x - y = 6$. These lines appear to be **perpendicular** (intersect at a 90° angle). Solving for y shows that the slope of $x + 2y = 4$ is $-\frac{1}{2}$, while the slope of $2x - y = 6$ is 2. The product of $-\frac{1}{2}$ and 2 is

$$-\frac{1}{2}(2) = -1.$$

This is true in general; the product of the slopes of two perpendicular lines (neither of which is vertical) is always -1.

Slopes of Parallel and Perpendicular Lines

Two lines with the same slope are parallel.

Two lines whose slopes have a product of -1 are perpendicular.

---

④ Find the slope of each line.

(a) $y = -\frac{7}{2}x + 1$

(b) $3x + 2y = 9$

(c) $y + 4 = 0$

(d) $x + 3 = 7$

⑤ Decide whether each pair of lines is *parallel*, *perpendicular*, or *neither*.

(a) $x + y = 6$
$x + y = 1$

(b) $3x - y = 4$
$x + 3y = 9$

(c) $2x - y = 5$
$2x + y = 3$

(d) $3x - 7y = 35$
$7x - 3y = -6$

Example 5 **Deciding Whether Lines Are Parallel, Perpendicular, or Neither**

Decide whether each pair of lines is *parallel, perpendicular,* or *neither.*

(a) $x + 2y = 7$
$-2x + y = 3$
Find the slope of each line by first solving each equation for y.

$x + 2y = 7$	$-2x + y = 3$
$2y = -x + 7$	$y = \mathbf{2}x + 3$
$y = -\dfrac{\mathbf{1}}{\mathbf{2}}x + \dfrac{7}{2}$	
Slope is $-\dfrac{1}{2}$.	Slope is 2.

Because the slopes are not equal, the lines are not parallel. Check the product of the slopes: $-\frac{1}{2}(2) = -1$. The two lines are perpendicular because the product of their slopes is -1. See Figure 26.

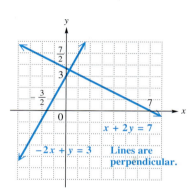

Figure 26

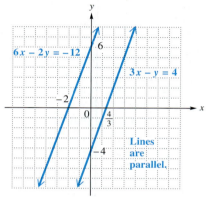

Figure 27

(b) $3x - y = 4$ Solve for y. $y = \mathbf{3}x - 4$
$6x - 2y = -12$ $y = \mathbf{3}x + 6$
Both lines have slope 3, so the lines are parallel. See Figure 27.

(c) $4x + 3y = 6$ Solve for y. $y = -\dfrac{\mathbf{4}}{\mathbf{3}}x + 2$

$2x - y = 5$ $y = \mathbf{2}x - 5$
Here the slopes are $-\frac{4}{3}$ and 2. These lines are neither parallel nor perpendicular.

(d) $5x - y = 1$

$x - 5y = -10$
Solving each equation for y gives

$$y = \mathbf{5}x - 1$$
$$y = \dfrac{\mathbf{1}}{\mathbf{5}}x + 2.$$

The slopes are 5 and $\frac{1}{5}$. The lines are not parallel, nor are they perpendicular. (Be careful! $5\left(\frac{1}{5}\right) = 1$, not -1.)

Work Problem ⑤ at the Side.

3.3 EXERCISES

1. In the context of the graph of a straight line, what is meant by "rise"? What is meant by "run"?

Use the coordinates of the indicated points to find the slope of each line. See Example 1.

2.

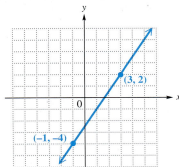

3.

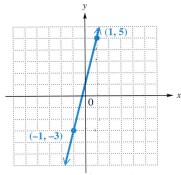

4.

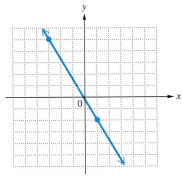

5.

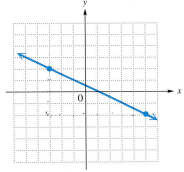

6.

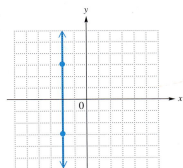

7.

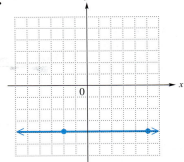

8. Look at the graph in Exercise 2, and answer the following.

   (a) Start at the point $(-1, -4)$ and count vertically up to the horizontal line that goes through the other plotted point. What is this vertical change? (Remember: "up" means positive, "down" means negative.) _____

   (b) From this new position, count horizontally to the other plotted point. What is this horizontal change? (Remember: "right" means positive, "left" means negative.) _____

   (c) What is the quotient of the numbers found in parts (a) and (b)? _____ What do we call this number? _____

9. Refer to Exercise 8. If we were to *start* at the point $(3, 2)$ and *end* at the point $(-1, -4)$, would the answer to part (c) be the same? Explain why or why not.

On the given coordinate system, sketch the graph of a straight line with the indicated slope.

10. Negative

11. Positive

12. Undefined

13. Zero

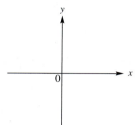

14. Explain in your own words what is meant by the *slope* of a line.

15. A student found the slope of the line through the points $(2, 5)$ and $(-1, 3)$ and got $-\dfrac{2}{3}$ as his answer. He showed his work as

$$\frac{3-5}{2-(-1)} = \frac{-2}{3} = -\frac{2}{3}.$$

Is he correct? If not, find his error and give the correct slope.

Find the slope of the line through each pair of points. See Examples 1–3.

16. $(4, -1)$ and $(-2, -8)$

17. $(1, -2)$ and $(-3, -7)$

18. $(-8, 0)$ and $(0, -5)$

19. $(0, 3)$ and $(-2, 0)$

20. $(-4, -5)$ and $(-5, -8)$

21. $(-2, 4)$ and $(-3, 7)$

22. $(6, -5)$ and $(-12, -5)$

23. $(4, 3)$ and $(-6, 3)$

24. $(-8, 6)$ and $(-8, -1)$

25. $(-12, 3)$ and $(-12, -7)$

26. $(3.1, 2.6)$ and $(1.6, 2.1)$

27. $\left(-\dfrac{7}{5}, \dfrac{3}{10}\right)$ and $\left(\dfrac{1}{5}, -\dfrac{1}{2}\right)$

Find the slope of each line. See Example 4.

28. $y = 2x - 3$

29. $y = 5x + 12$

30. $2y = -x + 4$

31. $4y = x + 1$

32. $-6x + 4y = 4$ **33.** $3x - 2y = 3$ **34.** $y = 4$ **35.** $x = 6$

The figure at the right shows a line that has a positive slope (because it rises from left to right) and a positive y-value for the y-intercept (because it intersects the y-axis above the origin).

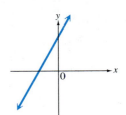

*For each figure in Exercises 36–41, decide whether **(a)** the slope is positive, negative, or 0 and whether **(b)** the y-value of the y-intercept is positive, negative, or 0.*

36. (a) _____
   (b) _____

37. (a) _____
   (b) _____

38. (a) _____
   (b) _____

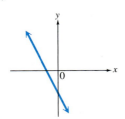
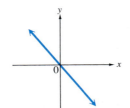
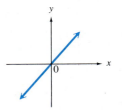

39. (a) _____
   (b) _____

40. (a) _____
   (b) _____

41. (a) _____
   (b) _____

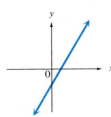
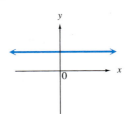
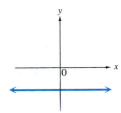

In each pair of equations, give the slope of each line, and then determine whether the two lines are parallel, perpendicular, *or* neither parallel nor perpendicular. *See Example 5.*

42. $2x + 5y = 4$
    $4x + 10y = 1$

43. $-4x + 3y = 4$
    $-8x + 6y = 0$

44. $8x - 9y = 6$
    $8x + 6y = -5$

45. $5x - 3y = -2$
    $3x - 5y = -8$

46. $3x - 2y = 6$
    $2x + 3y = 3$

47. $3x - 5y = -1$
    $5x + 3y = 2$

48. What is the slope (or pitch) of this roof?

49. What is the slope (or grade) of this hill?

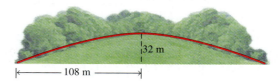

RELATING CONCEPTS (Exercises 50–55) **FOR INDIVIDUAL OR GROUP WORK**

Figure A gives the public school enrollment (in thousands) in grades 9–12 in the United States. Figure B gives the (average) number of public school students per computer.

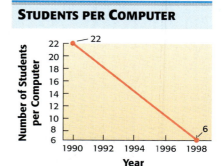

PUBLIC SCHOOL ENROLLMENT

Source: Digest of Educational Statistics, annual, and Projections of Educational Statistics, annual.

Figure A

STUDENTS PER COMPUTER

Source: Quality Education Data.

Figure B

Work Exercises 50–55 in order.

50. Use the ordered pairs (1990, 11,338) and (2005, 14,818) to find the slope of the line in Figure A. Note that enrollment is given in thousands.

51. The slope of the line in Figure A is

_____. This means that during the
  (positive/negative)

period represented, enrollment _____.
                             (increased/decreased)

52. The slope of a line represents its *rate of change.* Based on Figure A, what was the increase in students *per year* during the period shown?

53. Use the given ordered pairs to find the slope of the line in Figure B.

54. The slope of the line in Figure B is

_____. This means that during
  (positive/negative)

the period represented, the number of students per

computer _____.
       (increased/decreased)

55. Based on Figure B, what was the decrease in the number of students per computer *per year* during the period shown?

3.4 EQUATIONS OF LINES

In the previous section, we found the slope (steepness) of a line from the equation of the line by solving the equation for y. In that form, the slope is the coefficient of x. For example, the slope of the line with equation $y = 2x + 3$ is 2, the coefficient of x. What does the number 3 represent? If $x = 0$, the equation becomes

$$y = 2(0) + 3 = 0 + 3 = 3.$$

Since $y = 3$ corresponds to $x = 0$, $(0, 3)$ is the y-intercept of the graph of $y = 2x + 3$. An equation like $y = 2x + 3$ that is solved for y is said to be in **slope-intercept form** because both the slope and the y-intercept of the line can be read directly from the equation.

Slope-Intercept Form

The slope-intercept form of the equation of a line with slope m and y-intercept $(0, b)$ is

$$y = mx + b.$$

NOTE

The slope-intercept form is the most useful form for a linear equation because of the information we can determine from it. It is also the form used by graphing calculators and the one that describes a *linear function*, an important concept in mathematics.

1 **Write an equation of a line given its slope and y-intercept.**   Given the slope and y-intercept of a line, we can use the slope-intercept form to find an equation of the line.

Example 1   **Finding an Equation of a Line**

Find an equation of the line with slope $\frac{2}{3}$ and y-intercept $(0, -1)$.
    Here $m = \frac{2}{3}$ and $b = -1$, so the equation is

Slope ⎯⎯⎯┐   ┌⎯⎯ y-intercept
$$y = mx + b$$
$$y = \frac{2}{3}x - 1.$$

═══ Work Problem **1** at the Side.

2 **Graph a line given its slope and a point on the line.**   We can use the slope and y-intercept to graph a line. For example, to graph $y = \frac{2}{3}x - 1$, first locate the y-intercept $(0, -1)$ on the y-axis. From the definition of slope and the fact that the slope of this line is $\frac{2}{3}$,

$$m = \frac{\text{difference in } y\text{-values}}{\text{difference in } x\text{-values}} = \frac{2}{3}.$$

OBJECTIVES

1 Write an equation of a line given its slope and y-intercept.

2 Graph a line given its slope and a point on the line.

3 Write an equation of a line given its slope and any point on the line.

4 Write an equation of a line given two points on the line.

5 Find an equation of a line that fits a data set.

1 Find an equation of the line with the given slope and y-intercept.

(a) slope $\frac{1}{2}$; y-intercept $(0, -4)$

(b) slope -1; y-intercept $(0, 8)$

(c) slope 3; y-intercept $(0, 0)$

(d) slope 0; y-intercept $(0, 2)$

❷ Graph each line,

(a) Through $(-1, 2)$, with slope $\frac{3}{2}$

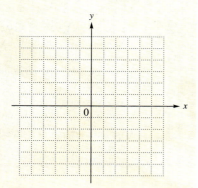

(b) Through $(2, -3)$, with slope $-\frac{1}{3}$

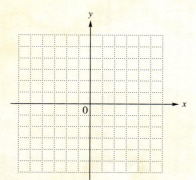

2. (a)

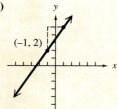

(−1, 2)

(b)

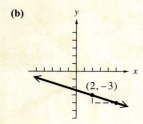

(2, −3)

Recall that slope indicates the change in y (the rise) compared to the change in x (the run) between two points on the line. A slope of $\frac{2}{3}$ indicates the line rises 2 units for a run of 3 units. We can find another point P on the graph by counting from the y-intercept 2 units up and then counting 3 units to the right. We then draw the line through point P and the y-intercept, as shown in Figure 28.

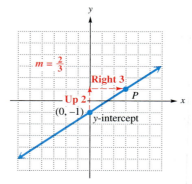

Figure 28

This method can be extended to graph a line given its slope and any point on the line.

Example 2 **Graphing a Line Given a Point and the Slope**

Graph the line through $(-2, 3)$ with slope -4.

First, locate the point $(-2, 3)$. Write the slope as

$$m = \frac{\text{difference in } y\text{-values}}{\text{difference in } x\text{-values}} = -4 = \frac{-4}{1}.$$

Locate another point on the line by counting 4 units down (because of the negative sign) and then 1 unit to the right. Finally, draw the line through this new point P and the given point $(-2, 3)$. See Figure 29.

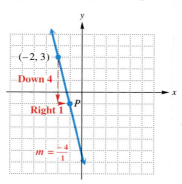

Figure 29

NOTE

In Example 2, we could have written the slope as $\frac{4}{-1}$ instead. In this case, we would move 4 units up from $(-2, 3)$ and then 1 unit to the left (because of the negative sign). Verify that this produces the same line.

Work Problem ❷ at the Side.

3 ▭ **Write an equation of a line given its slope and any point on the line.**   Let m represent the slope of a line and (x_1, y_1) represent a given point on the line. Let (x, y) represent any other point on the line. Then by the slope formula,

$$\frac{y - y_1}{x - x_1} = m \quad \text{or} \quad y - y_1 = m(x - x_1).$$

This result is the **point-slope form** of the equation of a line.

Point-Slope Form

The point-slope form of the equation of a line with slope m going through (x_1, y_1) is

$$y - y_1 = m(x - x_1).$$

This very important result should be memorized.

Example 3   **Using the Point-Slope Form to Write Equations**

Find an equation of each line. Write the equation in slope-intercept form.

(a) Through $(-2, 4)$, with slope -3
   The given point is $(-2, 4)$ so $x_1 = -2$ and $y_1 = 4$. Also, $m = -3$. Substitute these values into the point-slope form.

$$y - y_1 = m(x - x_1) \qquad \text{Point-slope form}$$
$$y - 4 = -3[x - (-2)] \qquad \text{Let } x_1 = -2, y_1 = 4, m = -3.$$
$$y - 4 = -3(x + 2)$$
$$y - 4 = -3x - 6 \qquad \text{Distributive property}$$
$$y = -3x - 2 \qquad \text{Add 4.}$$

The last equation is in slope-intercept form.

(b) Through $(4, 2)$, with slope $\frac{3}{5}$
   Use the point-slope form.

$$y - y_1 = m(x - x_1)$$
$$y - 2 = \frac{3}{5}(x - 4) \qquad \text{Let } x_1 = 4, y_1 = 2, m = \frac{3}{5}.$$
$$y - 2 = \frac{3}{5}x - \frac{12}{5} \qquad \text{Distributive property}$$
$$y = \frac{3}{5}x - \frac{12}{5} + \frac{10}{5} \qquad \text{Add } 2 = \frac{10}{5}.$$
$$y = \frac{3}{5}x - \frac{2}{5} \qquad \text{Combine terms.}$$

We did not clear fractions after the substitution step because we want the equation in slope-intercept form—that is, solved for y.

= **Work Problem ❸ at the Side.**

---

❸ Find an equation for each line. Write answers in slope-intercept form.

(a) Through $(-1, 3)$, with slope -2

$$y - y_1 = m(x - x_1)$$
$$y - \underline{\quad} = \underline{\quad} [x - (\quad)]$$
$$y - 3 = -2(x + \underline{\quad})$$
$$y - 3 = -2x - \underline{\quad}$$
$$y = \underline{\quad}$$

(b) Through $(5, 2)$, with slope $-\frac{1}{3}$

4 Write an equation in slope-intercept form for the line through each pair of points.

(a) $(-3, 1)$ and $(2, 4)$

4 ___ **Write an equation of a line given two points on the line.** We can also use the point-slope form to find an equation of a line when two points on the line are known.

Example 4 Finding the Equation of a Line Given Two Points

Find an equation of the line through the points $(-2, 5)$ and $(3, 4)$. Write the equation in slope-intercept form.

First, find the slope of the line, using the slope formula.

$$\text{slope } m = \frac{y_2 - y_1}{x_2 - x_1} = \frac{5 - 4}{-2 - 3} = \frac{1}{-5} = -\frac{1}{5}$$

Now use either $(-2, 5)$ or $(3, 4)$ and the point-slope form. Using $(3, 4)$ gives

$$y - y_1 = m(x - x_1)$$

$$y - 4 = -\frac{1}{5}(x - 3) \qquad \text{Let } x_1 = 3, y_1 = 4, m = -\tfrac{1}{5}.$$

$$y - 4 = -\frac{1}{5}x + \frac{3}{5} \qquad \text{Distributive property}$$

$$y = -\frac{1}{5}x + \frac{3}{5} + \frac{20}{5} \qquad \text{Add } 4 = \tfrac{20}{5}.$$

$$y = -\frac{1}{5}x + \frac{23}{5}. \qquad \text{Combine terms.}$$

The same result would be found using $(-2, 5)$ for (x_1, y_1).
Work Problem 4 at the Side.

(b) $(2, 5)$ and $(-1, 6)$

Many of the linear equations in Sections 3.1–3.3 were given in the form $Ax + By = C$, called **standard form.** We define the standard form of a linear equation as follows.

Standard Form

A linear equation is in standard form if it is written as

$$Ax + By = C,$$

where A, B, and C are integers and $A > 0$, $B \neq 0$.

NOTE

The above definition of standard form is not the same in all texts. A linear equation can be written in this form in many different, equally correct, ways. For example, $3x + 4y = 12$, $6x + 8y = 24$, and $9x + 12y = 36$ all represent the same set of ordered pairs. Let us agree that $3x + 4y = 12$ is preferable to the other forms because the greatest common factor of 3, 4, and 12 is 1.

A summary of the types of linear equations is given on the next page.

ANSWERS

4. (a) $y = \frac{3}{5}x + \frac{14}{5}$ **(b)** $y = -\frac{1}{3}x + \frac{17}{3}$

Linear Equations

$x = k$	**Vertical line** Slope is undefined; x-intercept is $(k, 0)$.
$y = k$	**Horizontal line** Slope is 0; y-intercept is $(0, k)$.
$y = mx + b$	**Slope-intercept form** Slope is m; y-intercept is $(0, b)$.
$y - y_1 = m(x - x_1)$	**Point-slope form** Slope is m; line goes through (x_1, y_1).
$Ax + By = C$	**Standard form** Slope is $-\frac{A}{B}$; x-intercept is $\left(\frac{C}{A}, 0\right)$; y-intercept is $\left(0, \frac{C}{B}\right)$.

5 ▭ **Find an equation of a line that fits a data set.** Earlier in this chapter, we gave linear equations that modeled real data, such as annual costs of doctors' visits and amounts of credit card debt, and then used these equations to estimate or predict values. Using the information in this section, we can now develop a procedure to find such an equation if the given set of data fits a linear pattern—that is, its graph consists of points lying close to a straight line.

Example 5 Finding an Equation of a Line That Describes Data

The table lists the average annual cost (in dollars) of tuition and fees at public 4-year colleges for selected years. Year 1 represents 1991, year 3 represents 1993, and so on. Plot the data and find an equation that approximates it.

Year	Cost (in dollars)
1	2137
3	2527
5	2860
7	3111
9	3356

rce: The College Board.

.etting y represent the cost in year x, we plot the data as shown in Figure 30.

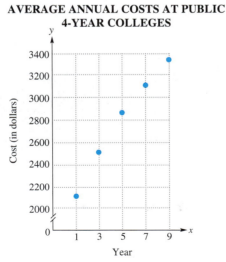

AVERAGE ANNUAL COSTS AT PUBLIC 4-YEAR COLLEGES

Figure 30

ntinued on Next Page

⑤ Use the points (1, 2137) and (7, 3111) to find an equation in slope-intercept form that approximates the data of Example 5. (Round the slope to the nearest tenth.) How well does this equation approximate the cost in 1999?

The points appear to lie approximately in a straight line. We can use two of the data pairs and the point-slope form of the equation of a line to get an equation that describes the relationship between the year and the cost. We choose the ordered pairs (3, 2527) and (7, 3111) from the table and find the slope of the line through these points.

$$m = \frac{y_2 - y_1}{x_2 - x_1} \qquad \text{Slope formula}$$

$$m = \frac{3111 - 2527}{7 - 3} \qquad \begin{array}{l}\text{Let } (7, 3111) = (x_2, y_2) \\ \text{and } (3, 2527) = (x_1, y_1).\end{array}$$

$$m = \mathbf{146}$$

As we might expect, the slope, 146, is positive, indicating that tuition and fees increased \$146 each year. Now use this slope and the point (3, 2527) in the point-slope form to find an equation of the line.

$$y - y_1 = m(x - x_1) \qquad \text{Point-slope form}$$
$$y - \mathbf{2527} = \mathbf{146}(x - \mathbf{3}) \qquad \text{Substitute for } x_1, y_1, \text{ and } m.$$
$$y - 2527 = 146x - 438 \qquad \text{Distributive property}$$
$$y = 146x + 2089 \qquad \text{Add 2527.}$$

To see how well this equation approximates the ordered pairs in the data table, let $x = 9$ (for 1999) and find y.

$$y = 146x + 2089 \qquad \text{Equation of the line}$$
$$y = 146(\mathbf{9}) + 2089 \qquad \text{Substitute 9 for } x.$$
$$y = 3403$$

The corresponding value in the table for $x = 9$ is 3356, so the equation approximates the data reasonably well. With caution, the equation could be used to predict values for years that are not included in the table.

NOTE

In Example 5, if we had chosen two different data points, we would have gotten a slightly different equation.

Work Problem ⑤ at the Side.

Here is a summary of what is needed to find the equation of a line.

Finding the Equation of a Line

To find the equation of a line, you need

1. a point on the line, and
2. the slope of the line.

If two points are known, first find the slope and then use the point-slope form.

3.4 EXERCISES

Match the correct equation in Column II with the description given in Column I.

I	**II**
1. Slope = −2, through the point (4, 1)	**A.** $y = 4x$
2. Slope = −2, *y*-intercept (0, 1)	**B.** $y = \dfrac{1}{4}x$
3. Through the points (0, 0) and (4, 1)	**C.** $y = -2x + 1$
4. Through the points (0, 0) and (1, 4)	**D.** $y - 1 = -2(x - 4)$

Use the geometric interpretation of slope (rise divided by run, from Section 3.3) to find the slope of each line. Then, by identifying the y-intercept from the graph, write the slope-intercept form of the equation of the line.

5.

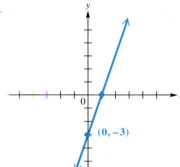

(0, −3)

6.

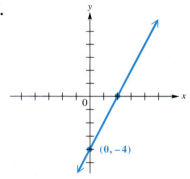

(0, −4)

7.

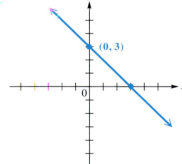

(0, 3)

8.

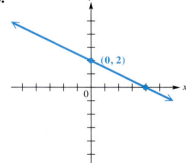

(0, 2)

Write the equation of the line with the given slope and y-intercept. See Example 1.

9. slope 4;
y-intercept (0, −3)

10. slope −5;
y-intercept (0, 6)

11. slope 0;
y-intercept (0, 3)

12. slope 3;
y-intercept (0, 0)

13. Explain why the equation of a vertical line cannot be written in the form $y = mx + b$.

14. Match each equation with the graph that would most closely resemble its graph.

   (a) $y = x + 3$ **A.**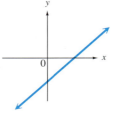

                                       B.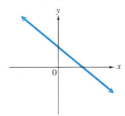

   (b) $y = -x + 3$

   (c) $y = x - 3$ **C.**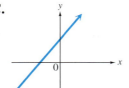

                                         D.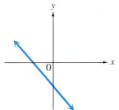

   (d) $y = -x - 3$

Graph the line through the given point with the given slope. (In Exercises 21–24, recall the types of lines having slope 0 and undefined slope.) Give the slope-intercept form of the equation of the line if possible. See Example 2.

15. $(-2, 3)$, $m = \dfrac{1}{2}$ **16.** $(-4, -1)$, $m = \dfrac{3}{4}$ **17.** $(1, -5)$, $m = -\dfrac{2}{5}$

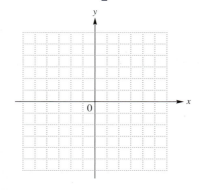   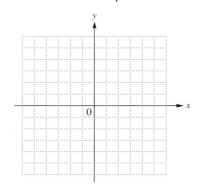   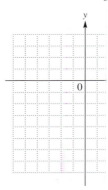

18. $(2, -1)$, $m = -\dfrac{1}{3}$ **19.** $(0, 2)$, $m = 3$ **20.** $(0, -5)$, $m = -2$

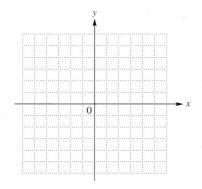   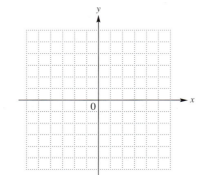   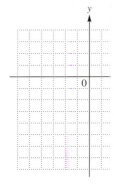

21. $(3, 2)$, $m = 0$

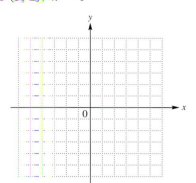

22. $(-2, 3)$, $m = 0$

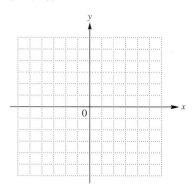

23. $(3, -2)$, undefined slope

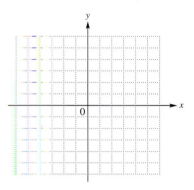

24. $(2, 4)$, undefined slope

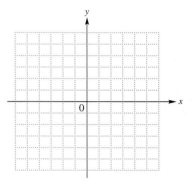

25. What is the common name given to the vertical line whose x-intercept is the origin?

26. What is the common name given to the line with slope 0 whose y-intercept is the origin?

Write an equation of the line through the given point with the given slope. Write the equation in slope-intercept form. See Example 3.

27. $(4, 1)$, $m = 2$

28. $(2, 7)$, $m = 3$

29. $(3, -10)$, $m = -2$

30. $(2, -5)$, $m = -4$

31. $(-2, 5)$, $m = \dfrac{2}{3}$

32. $(-4, 1)$, $m = \dfrac{3}{4}$

33. If a line passes through the origin and a second point whose x- and y-coordinates are equal, what is an equation of the line?

34. What is the only point common to both lines $y = x$ and $y = -x$?

Write an equation, in slope-intercept form if possible, of the line through each pair of points. See Examp

35. (8, 5) and (9, 6)

36. (4, 10) and (6, 12)

37. $(-1, -7)$ and $(-8.$

38. $(-2, -1)$ and $(3, -4)$

39. $(0, -2)$ and $(-3, 0)$

40. $(-4, 0)$ and $(0, 2)$

41. (3, 5) and (3, -2)

42. $(3, -5)$ and $(-1, -5)$

43. $\left(\dfrac{1}{2}, \dfrac{3}{2}\right)$ and $\left(-\dfrac{1}{4}, \dfrac{5}{4}\right)$

44. $\left(-\dfrac{2}{3}, \dfrac{8}{3}\right)$

RELATING CONCEPTS (Exercises 45–52)   **FOR INDIVIDUAL OR GROUP WORK**

If we think of ordered pairs of the form (C, F), then the two most common methods of measuring temperature, Celsius and Fahrenheit, can be related as follows: When C = 0, F = 32, and when C = 100, F = 212. **Work Exercises 45–52 in order.**

45. Write two ordered pairs relating these two temperature scales.

46. Find the slope of the line through the t

47. Use the point-slope form to find an equation of the line. (Your variables should be C and F rather than x and y.)

48. Write an equation for F in terms of C.

49. Use the equation from Exercise 48 to write an equation for C in terms of F.

50. Use the equation from Exercise 48 to f Fahrenheit temperature when $C = 30$.

51. Use the equation from Exercise 49 to find the Celsius temperature when $F = 50$.

52. For what temperature is $F = C$

Write an equation in slope-intercept form of the line satisfying the given conditions.

53. Through (2, -3), parallel to $3x = 4y + 5$

54. Through $(-1, 4)$, perpendicular to $2x -$

55. Perpendicular to $x - 2y = 7$, y-intercept $(0, -3)$

56. Parallel to $5x = 2y + 10$, y-intercept $(($

The cost to produce x items is, in some cases, expressed as y = mx + b. The number b gives the fixed cost (the cost that is the same no matter how many items are produced), and the number m is the variable cost *(the cost to produce an additional item). Use this information to work Exercises 57 and 58.*

57. It costs $400 to start up a business of selling snow cones. Each snow cone costs $.25 to produce.

(a) What is the fixed cost?

(b) What is the variable cost?

(c) Write the cost equation.

(d) What will be the cost to produce 100 snow cones, based on the cost equation?

(e) How many snow cones will be produced if total cost is $775?

58. It costs $2000 to purchase a copier, and each copy costs $.02 to make.

(a) What is the fixed cost?

(b) What is the variable cost?

(c) Write the cost equation.

(d) What will be the cost to produce 10,000 copies, based on the cost equation?

(e) How many copies will be produced if total cost is $2600?

Solve each problem. See Example 5.

59. The table lists the average annual cost (in dollars) of tuition and fees at private 4-year colleges for selected years, where year 1 represents 1991, year 3 represents 1993, and so on.

(a) Write five ordered pairs for the data.

(b) Plot the ordered pairs. Do the points lie approximately in a straight line?

Year	Cost (in dollars)
1	10,017
3	11,025
5	12,432
7	13,785
9	15,380

Source: The College Board.

AVERAGE ANNUAL COSTS AT PRIVATE 4-YEAR COLLEGES

Cost (in dollars): 16,000 / 15,000 / 14,000 / 13,000 / 12,000 / 11,000 / 10,000

Year: 1   3   5   7   9

(c) Use the ordered pairs (3, 11,025) and (9, 15,380) to find the equation of a line that approximates the data. Write the equation in slope-intercept form. (Round the slope to the nearest tenth.)

(d) Use the equation from part (c) to estimate the average annual cost at private 4-year colleges in 2003. (*Hint:* What is the value of x for 2003?)

60. The table gives heavy-metal nuclear waste (in thousands of metric tons) from spent reactor fuel now stored temporarily at reactor sites, awaiting permanent storage. (*Source:* "Burial of Radioactive Nuclear Waste Under the Seabed," *Scientific American,* January 1998.)

Year x	Waste y
1995	32
2000*	42
2010*	61
2020*	76

*Estimates by the U.S. Department of Energy.

Let $x = 0$ represent 1995, $x = 5$ represent 2000 (since $2000 - 1995 = 5$), and so on.

(a) For 1995, the ordered pair is (0, 32). Write ordered pairs for the data for the other years given in the table.

(b) Plot the ordered pairs (x, y). Do the points lie approximately in a straight line?

HEAVY-METAL NUCLEAR WASTE AWAITING STORAGE

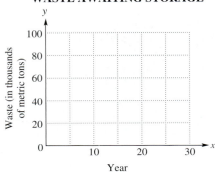

(c) Use the ordered pairs (0, 32) and (25, 76) to find the equation of a line that approximates the other ordered pairs. Write the equation in slope-intercept form.

(d) Use the equation from part (c) to estimate the amount of nuclear waste in 2005. (*Hint:* What is the value of x for 2005?)

61. The graph shows the percent of women civilian labor force for selected years f through 1995.

PERCENT OF WOMEN IN THE CIVILIA LABOR FORCE

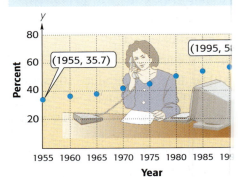

Source: U.S. Bureau of Labor Statistics.

(a) Use the points (1955, 35.7) and (1 find a linear equation in slope-inte that approximates the data points.

(b) Use the equation from part (a) to percent for 1996. How does the re to the actual value of 59.3%?

(c) Is the slope of the equation you w (a) positive or negative? What doe in the context of the problem?

3.5   GRAPHING LINEAR INEQUALITIES IN TWO VARIABLES

In Section 3.2 we graphed linear equations, such as $2x + 3y = 6$. Now this work is extended to **linear inequalities in two variables,** such as

$$2x + 3y \le 6.$$

(Recall that ≤ is read "is less than or equal to.")

1 **Graph ≤ or ≥ linear inequalities.**   The inequality $2x + 3y \le 6$ means that

$$2x + 3y < 6 \quad \text{or} \quad 2x + 3y = 6.$$

As we found earlier, the graph of $2x + 3y = 6$ is a line. This **boundary line** divides the plane into two regions. The graph of the solutions of the inequality $2x - 3y < 6$ will include only *one* of these regions. We find the required region by solving the given inequality for y.

$$2x + 3y \le 6$$

$$3y \le -2x + 6 \quad \text{Subtract } 2x.$$

$$y \le -\frac{2}{3}x + 2 \quad \text{Divide by 3.}$$

By this last statement, ordered pairs in which y is *less than or equal to* $-\frac{2}{3}x + 2$ will be solutions of the inequality. The ordered pairs in which y is equal to $-\frac{2}{3}x + 2$ are on the boundary line, so the ordered pairs in which y is less than $-\frac{2}{3} + 2$ will be *below* that line. (This is because as we move *down* vertically, the y-values become *smaller.*) To indicate the solutions, we shade the region below the line, as in Figure 31. The shaded region, along with the line, is the desired graph.

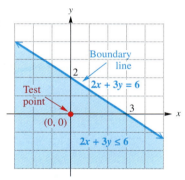

Figure 31

Work Problem ① at the Side.

Alternatively, a test point gives a quick way to find the correct region to shade. We choose any point *not* on the boundary line. Because $(0, 0)$ is easy to substitute into an inequality, it is often a good choice, and we will use it here. We substitute 0 for x and 0 for y in the given inequality to see whether the resulting statement is true or false. In the example above,

$$2x + 3y \le 6$$

$$2(0) + 3(0) \le 6 \quad ? \quad \text{Let } x = 0 \text{ and } y = 0.$$

$$0 + 0 \le 6 \quad ?$$

$$0 \le 6. \quad \text{True}$$

Since the last statement is true, we shade the region that includes the test point $(0, 0)$. This agrees with the result shown in Figure 31.

① Shade the appropriate region for each linear inequality.

(a) $x + 2y \ge 6$

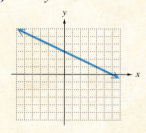

(b) $3x + 4y \le 12$

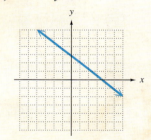

ANSWERS

1. (a)

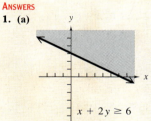

$x + 2y \ge 6$

(b)

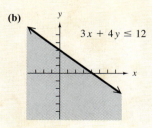

$3x + 4y \le 12$

2 Use $(0, 0)$ as a test point to shade the proper region for the inequality $4x - 5y \le 20$.

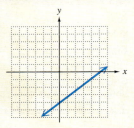

3 Use $(1, 1)$ as a test point to shade the proper region for the inequality $3x + 5y > 15$.

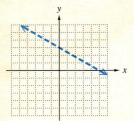

ANSWERS

2.

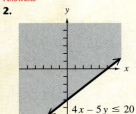

$4x - 5y \le 20$

3.

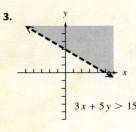

$3x + 5y > 15$

Work Problem 2 at the Side.

2   **Graph < or > linear inequalities.**   An inequality that doe the equals sign is graphed in a similar way.

Example 1   Graphing a Linear Inequality

Graph the inequality $x - y > 5$.

    This inequality does not include the equals sign. Therefore, the line

$$x - y = 5$$

do *not* belong to the graph. However, the line still serves as a two regions, one of which satisfies the inequality. To graph th first graph the equation $x - y = 5$. Use a *dashed line* to show t on the line are *not* solutions of the inequality $x - y > 5$. Then point not on the line to see which side of the line satisfies the i choose $(1, -2)$ this time.

$$x - y > 5$$
$$1 - (-2) > 5 \qquad ? \qquad \text{Let } x = 1 \text{ and } y = -2.$$
$$3 > 5 \qquad\qquad \text{False}$$

    Because $3 > 5$ is false, the graph of the inequality is the re *not* contain $(1, -2)$. We shade the region that does not include $(1, -2)$, as in Figure 32. This shaded region is the desired gra that the proper region is shaded, we select a point in the shad substitute for x and y in the inequality $x - y > 5$. For exa $(4, -3)$ from the shaded region as follows.

$$x - y > 5$$
$$4 - (-3) > 5 \qquad ? \qquad \text{Let } x = 4 \text{ and } y = -3.$$
$$7 > 5 \qquad\qquad \text{True}$$

This verifies that the correct region is shaded in Figure 32.

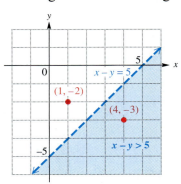

Figure 32

Work Problem 3 at the Side.

    A summary of the steps used to graph a linear inequality in follows.

Graphing a Linear Inequality

Step 1   Graph the boundary. Graph the line that is the boundary of the region. Use the methods of Section 3.2. Draw a solid line if the inequality has ≤ or ≥; draw a dashed line if the inequality has < or >.

Step 2   Shade the appropriate side. Use any point not on the line as a test point. Substitute for *x* and *y* in the *inequality*. If a true statement results, shade the side containing the test point. If a false statement results, shade the other side.

4 Graph $2x - y \geq -4$.

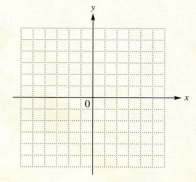

Example 2   Graphing a Linear Inequality

Graph the inequality $2x - 5y \geq 10$.

Start by graphing the equation

$$2x - 5y = 10.$$

Use a solid line to show that the points on the line are solutions of the inequality $2x - 5y \geq 10$. Choose any test point not on the line. Again, we choose $(0, 0)$.

$$2x - 5y \geq 10$$

$$2(0) - 5(0) \geq 10 \qquad ? \qquad \text{Let } x = 0 \text{ and } y = 0.$$

$$0 - 0 \geq 10 \qquad ?$$

$$0 \geq 10 \qquad \qquad \text{False}$$

Because $0 \geq 10$ is false, shade the region *not* containing $(0, 0)$. See Figure 33. Verify that a point in the shaded region satisfies the inequality.

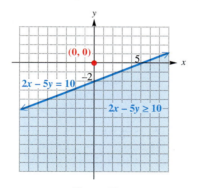

Figure 33

Work Problem **4** at the Side.

Example 3   Graphing a Linear Inequality with a Vertical Boundary Line

Graph the inequality $x \leq 3$.

First graph $x = 3$, a vertical line through the point $(3, 0)$. Use a solid line. Why? Choose $(0, 0)$ as a test point.

$$x \leq 3$$

$$0 \leq 3 \qquad ? \qquad \text{Let } x = 0.$$

$$0 \leq 3 \qquad \qquad \text{True}$$

Continued on Next Page

ANSWERS

4.

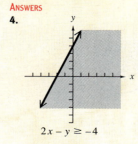

$2x - y \geq -4$

5 Graph $y < 4$.

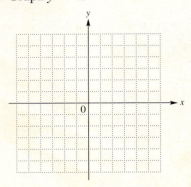

Because $0 \le 3$ is true, shade the region containing $(0, 0)$, as in

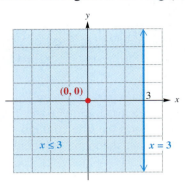

(0, 0)

$x \le 3$ $x = 3$

Figure 34

Work Problem **5** at the Side.

3 ▭ **Graph an inequality with boundary through the origin.** I
an inequality has a boundary line through the origin, $(0, 0)$ can
a test point.

6 Graph $x \ge -3y$.

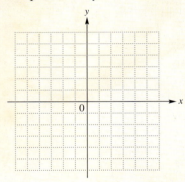

Example 4 **Graphing a Linear Inequality**

Graph the inequality $x \le 2y$.

We begin by graphing $x = 2y$, using a solid line. Some ord
can be used to graph this line are $(0, 0)$, $(6, 3)$, and $(-4, 2)$. \
$(0, 0)$ as a test point because $(0, 0)$ is on the line $x = 2y$. Instea
test point off the line, $(1, 3)$.

$$x \le 2y$$
$$1 \le 2(3) \qquad ? \qquad \text{Let } x = 1 \text{ and } y = 3.$$
$$1 \le 6 \qquad \qquad \text{True}$$

Because $1 \le 6$ is true, we shade the side of the graph cont:
point $(1, 3)$. See Figure 35.

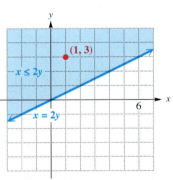

(1, 3)

$x \le 2y$

$x = 2y$

Figure 35

Work Problem **6** at the Side.

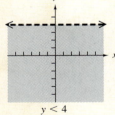
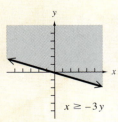

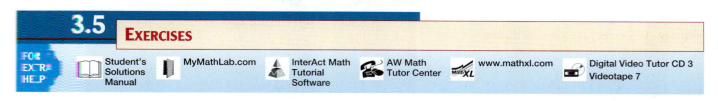

FOR EXTRA HELP
Student's Solutions Manual   MyMathLab.com   InterAct Math Tutorial Software   AW Math Tutor Center   www.mathxl.com   Digital Video Tutor CD 3 Videotape 7

Decide whether each statement is true *or* false.

1. The point $(4, 0)$ lies on the graph of $3x - 4y < 12$.

2. The point $(4, 0)$ lies on the graph of $3x - 4y \leq 12$.

3. Both points $(4, 1)$ and $(0, 0)$ lie on the graph of $3x - 2y \geq 0$.

4. The graph of $y > x$ does not contain points in quadrant IV.

The following statements were taken from recent articles in newspapers or magazines. Each includes a phrase that can be symbolized with one of the inequality symbols $<$, $\leq$, $>$, or $\geq$. Complete each sentence with the appropriate inequality symbol.

5. There are over 6000 plant photos in the Vascular Plant Image Gallery, a Web site created by botanists at Texas A & M University in College Station. (*Source: Science, Vol. 283, January 15, 1999.*)

There are _____ 6000 plant photos.

6. Thus far, more than 11 percent of corporate officers in Fortune 500 companies are women, a proportion that will rise to 17 percent by 2005, according to the New York City–based research organization Catalyst. (*Source: Scientific American, October 1999.*)

Women are _____ 11 percent of the number of Fortune 500 corporate officers.

7. China exported nearly $82 billion worth of footwear, apparel, toys, consumer electronics, and other products to the United States in 1999. (*Source: Sacramento Bee, May 14, 2000.*)

China exported _____ $82 billion worth of products.

8. The committees supervise housing projects with up to 100 families. (*Source: "Long Is It," Fortune, May 25, 1998.*)

The housing projects have _____ 100 families.

In Exercises 9–16, the straight-line boundary has been drawn. Complete each graph by shading the correct region. See Examples 1–4.

9. $x + y \geq 4$

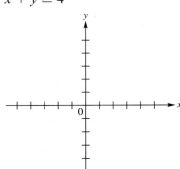

10. $x + y \leq 2$

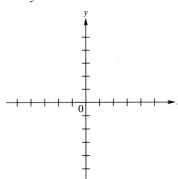

11. $x + 2y \geq 7$

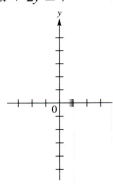

12. $2x + y \geq 5$

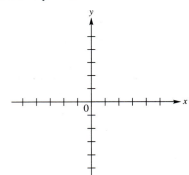

13. $-3x + 4y > 12$

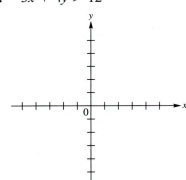

14. $4x - 5y < 20$

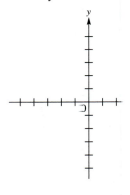

15. $x > 4$

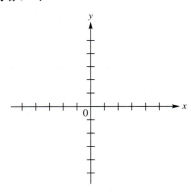

16. $y < -1$

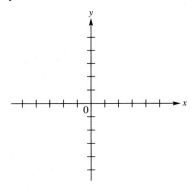

17. Explain how to determine whether to use a dashed line or a solid line when graphing a linear inequality in two variables.

18. Explain why the point $(0, 0)$ is not an appropriate choice for a test point when graphing an inequality whose boundary goes through the origin.

Graph each linear inequality. See Examples 1–4.

19. $x - y \leq 5$

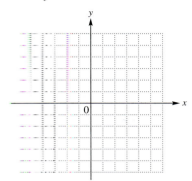

20. $x + y \geq 3$

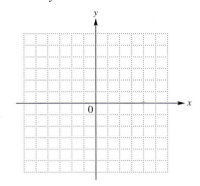

21. $x + 2y < 4$

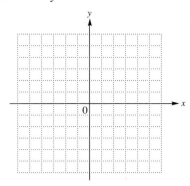

22. $x + 3y > 6$

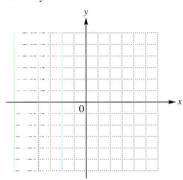

23. $2x + 6 > -3y$

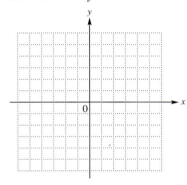

24. $-4y > 3x - 12$

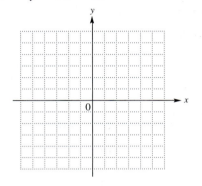

25. $y \geq 2x + 1$

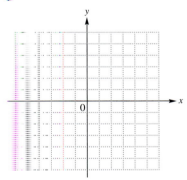

26. $y < -3x + 1$

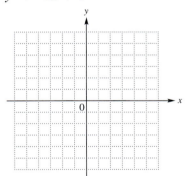

27. $x \leq -2$

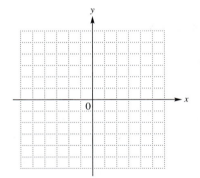

28. $x \geq 1$

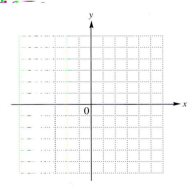

29. $y < 5$

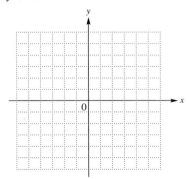

30. $y < -3$

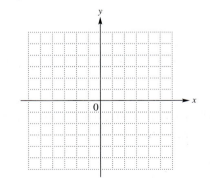

31. $y \geq 4x$

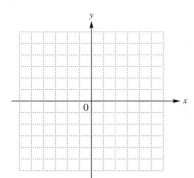

32. $y \leq 2x$

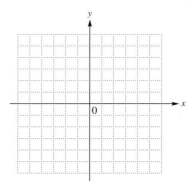

Solve each problem. In part (a), $x \geq 0$ and $y \geq 0$, so graph only the part of the inequality in quadrant I.

33. A company will ship x units of merchandise to outlet I and y units of merchandise to outlet II. The company must ship a total of at least 500 units to these two outlets. This can be expressed by writing

$$x + y \geq 500.$$

(a) Graph the inequality.

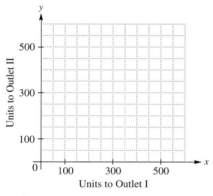

(b) Give two ordered pairs that satisfy the inequality.

34. A toy manufacturer makes stuffed bear It takes 20 min to sew a bear and 30 m goose. There is a total of 480 min of se available to make x bears and y geese. tions lead to the inequality

$$20x + 30y \leq 480.$$

(a) Graph the inequality.

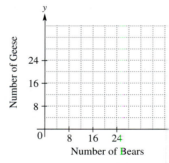

(b) Give two ordered pairs that satisfy inequality.

3

SUMMARY

KEY TERMS

graph — A circle graph is a circle divided into sectors (or wedges) whose sizes show the relative magnitudes of the categories of data represented.

aph — A bar graph is a series of bars used to show comparisons between two categories of data.

aph — A line graph consists of a series of points that are connected with line segments and is used to show changes or trends in data.

equation in variables — An equation that can be written in the form $Ax + By = C$ is a linear equation in two variables. (A and B are real numbers that cannot both be 0.)

d pair — A pair of numbers written between parentheses in which order is important is called an ordered pair.

f values — A table showing selected ordered pairs of numbers that satisfy an equation is called a table of values.

The horizontal axis in a coordinate system is called the x-axis.

The vertical axis in a coordinate system is called the y-axis.

gular (Cartesian) dinate system — An x-axis and y-axis at right angles form a coordinate system.

ants — A coordinate system divides the plane into four regions called quadrants.

The point at which the x-axis and y-axis intersect is called the origin.

A flat surface determined by two intersecting lines is a plane.

nates — The numbers in an ordered pair are called the coordinates of the corresponding point.

To plot an ordered pair is to find the corresponding point on a coordinate system.

diagram — A graph of ordered pairs of data is a scatter diagram.

The graph of an equation is the set of all points that correspond to the ordered pairs that satisfy the equation.

ng — The process of plotting the ordered pairs that satisfy a linear equation and drawing a line through them is called graphing.

cept — If a graph intersects the y-axis at k, then the y-intercept is $(0, k)$.

cept — If a graph intersects the x-axis at k, then the x-intercept is $(k, 0)$.

Rise is the vertical change between two different points on a line.

Run is the horizontal change between two different points on a line.

The slope of a line is the ratio of the change in y compared to the change in x when moving along the line from one point to another.

el lines — Two lines in a plane that never intersect are parallel.

ndicular lines — Perpendicular lines intersect at a 90° angle.

inequality in variables — An inequality that can be written in the form $Ax + By < C$, $Ax + By > C$, $Ax + By \leq C$, or $Ax + By \geq C$ is a linear inequality in two variables.

ary line — In the graph of a linear inequality, the boundary line separates the region that satisfies the inequality from the region that does not satisfy the inequality.

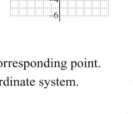

NEW SYMBOLS

(x, y)	ordered pair	(x_1, y_1)	subscript notation; x-sub-one, y-sub-one	m	slope

TEST YOUR WORD POWER

See how well you have learned the vocabulary in this chapter. Answers follow the Quick Review.

1. An **ordered pair** is a pair of numbers written
   (a) in numerical order between brackets
   (b) between parentheses or brackets
   (c) between parentheses in which order is important
   (d) between parentheses in which order does not matter.

2. The **coordinates** of a point are
   (a) the numbers in the corresponding ordered pair
   (b) the solution of an equation
   (c) the values of the x- and y-intercepts
   (d) the graph of the point.

3. An **intercept** is
   (a) the point where the x-axis and y-axis intersect
   (b) a pair of numbers written in parentheses in which order matters
   (c) one of the four regions determined by a rectangular coordinate system
   (d) the point where a graph intersects the x-axis or the y-axis.

4. The **slope** of a line is
   (a) the measure of the run over the rise of the line
   (b) the distance between two points on the line
   (c) the ratio of the change in y to the change in x along the line

   (d) the horizontal change compared to the vertical change of two points on the line

5. Two lines in a plane are **parallel** if
   (a) they represent the same line
   (b) they never intersect
   (c) they intersect at a 90° angle
   (d) one has a positive slope and one has a negative slope

6. Two lines in a plane are **perpendicular** if
   (a) they represent the same line
   (b) they never intersect
   (c) they intersect at a 90° angle
   (d) one has a positive slope and one has a negative slope

QUICK REVIEW

Concepts

3.1 Reading Graphs; Linear Equations in Two Variables
Circle graphs, bar graphs, and line graphs are several ways to represent the relationship between two variables.

Examples

The bar graph indicates that in 2002, worldwide revenue from Internet security software is estimated to be about $7.4 billion.

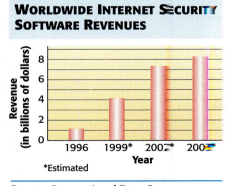

WORLDWIDE INTERNET SECURITY SOFTWARE REVENUES

*Estimated

Source: International Data Corp.

Examples

Graphs; Linear Equations in Two Variables

air is a solution of an equation if it makes the
le statement.

Is $(2, -5)$ or $(0, -6)$ a solution of $4x - 3y = 18$?

$$4(2) - 3(-5) = 23 \neq 18 \quad | \quad 4(0) - 3(-6) = 18$$

$(2, -5)$ is not a solution. | $(0, -6)$ is a solution.

either variable in an equation is given, the
ther variable can be found by substitution.

Complete the ordered pair $(0,\ \ )$ for $3x = y + 4$.

$$3(0) = y + 4 \quad \text{Let } x = 0.$$
$$0 = y + 4$$
$$-4 = y$$

The ordered pair is $(0, -4)$.

dered pair $(-3, 4)$, start at the origin, go 3
ft, and from there go 4 units up.

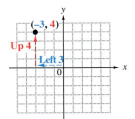

Linear Equations in Two Variables

hear equation:

at least two ordered pairs that are solutions
e equation.

the corresponding points.

v a straight line through the points.

Graph $x - 2y = 4$.

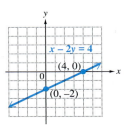

the line through (x_1, y_1) and (x_2, y_2) is

$$= \frac{\text{change in } y}{\text{change in } x} = \frac{y_2 - y_1}{x_2 - x_1} \quad (x_1 \neq x_2).$$

The line through $(-2, 3)$ and $(4, -5)$ has slope

$$m = \frac{-5 - 3}{4 - (-2)} = \frac{-8}{6} = -\frac{4}{3}.$$

nes have slope 0.

have undefined slope.

ope of a line from its equation, solve for y.
the coefficient of x.

The line $y = -2$ has slope 0.

The line $x = 4$ has undefined slope.

Find the slope: $3x - 4y = 12$.

$$-4y = -3x + 12$$
$$y = \frac{3}{4}x - 3$$

The slope is $\frac{3}{4}$.

Concepts

Examples

3.4 Equations of Lines

Slope-Intercept Form

$y = mx + b$

m is the slope.

$(0, b)$ is the y-intercept.

Point-Slope Form

$y - y_1 = m(x - x_1)$

m is the slope.

(x_1, y_1) is a point on the line.

Find an equation of the line with slope **2** and y-intercept $(0, -5)$.

$$y = 2x - 5$$

Find an equation of the line with slope $-\frac{1}{2}$ through $(-4, 5)$.

$$y - 5 = -\frac{1}{2}[x - (-4)]$$

$$y - 5 = -\frac{1}{2}(x + 4)$$

$$y - 5 = -\frac{1}{2}x - 2$$

$$y = -\frac{1}{2}x + 3$$

Standard Form

$Ax + By = C$

$A, B,$ and C are integers and $A > 0, B \neq 0$.

This equation is written in standard form as

$$x + 2y = 6,$$

with $A = 1, B = 2,$ and $C = 6$.

3.5 Graphing Linear Inequalities in Two Variables

Step 1 Graph the line that is the boundary of the region. Make it solid if the inequality is $\leq$ or $\geq$; make it dashed if the inequality is $<$ or $>$.

Step 2 Use any point not on the line as a test point. Substitute for x and y in the inequality. If the result is true, shade the side of the line containing the test point; if the result is false, shade the other side.

Graph $2x + y \leq 5$.

Graph the line $2x + y = 5$. Make it solid because of $\leq$.

Use $(1, 0)$ as a test point.

$$2(1) + 0 \leq 5 \qquad ?$$

$$2 \leq 5 \qquad \text{True}$$

Shade the side of the line containing $(1, 0)$.

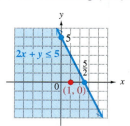

REVIEW EXERCISES

graph shows average prices for a gallon of gasoline in Cedar Rapids,
1999 through June 2000. Use the graph to work Exercises 1–4.

much did a gallon of gas cost in June
June 2000?

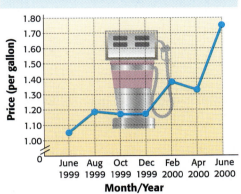

GAS PRICES: JUNE 1999 – JUNE 2000

Source: Iowa DNR.

did the price of a gallon of gas increase
-year period? What percent increase is

nich months did the biggest increase in the
gallon of gas occur? About how much did
ncrease during this time?

which months did the price of a gallon of gas decrease?

given ordered pairs for each equation.

2    $(-1, \ \ ), (0, \ \ ), (\ \ , 5)$ **6.** $4x + 3y = 6$ $(0, \ \ ), (\ \ , 0), (-2, \ \ )$

$(0, \ \ ), (8, \ \ ), (\ \ , -3)$ **8.** $x - 7 = 0$ $(\ \ , -3), (\ \ , 0), (\ \ , 5)$

er each ordered pair is a solution of the given equation.

7; (2, 5)   **10.** $2x + y = 5; (-1, 3)$ **11.** $3x - y = 4; \left(\dfrac{1}{3}, -3\right)$

Plot each ordered pair on the given coordinate system.

12. $(2, 3)$

13. $(-4, 2)$

14. $(3, 0)$

15. $(0, -6)$

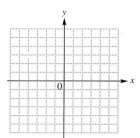

16. If $x > 0$ and $y < 0$, in what quadrant(s) must (x, y) lie? Explain.

17. On what axis does the point $(k, 0)$ lie for any real value of k? the point $(0, k)$? Explain.

Without plotting the given point, name the quadrant in which each point lies.

18. $(-2, 3)$

19. $(-1, -4)$

20. $\left(0, -5\frac{1}{2}\right)$

[3.2] *Find the intercepts for each equation.*

21. $y = 2x + 5$
   x-intercept:
   y-intercept:

22. $2x + y = -7$
   x-intercept:
   y-intercept:

23. $3x + 2y = 8$
   x-intercept:
   y-intercept:

Graph each linear equation.

24. $2x - y = 3$

25. $x + 2y = -4$

26. $x + y = 0$

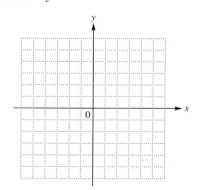

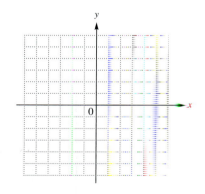

slope of each line.

2, 3) and $(-4, 6)$ **28.** Through $(0, 0)$ and $(-3, 2)$ **29.** Through $(0, 6)$ and $(1, 6)$

2, 5) and $(2, 8)$ **31.** $y = 3x - 4$ **32.** $y = \dfrac{2}{3}x + 1$

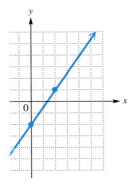

34.

36. $y = 4$

aving these points

x	y
0	1
2	4
6	10

38. (a) A line parallel to the graph of $y = 2x + 3$

(b) A line perpendicular to the graph of $y = -3x + 3$

r each pair of lines is parallel, perpendicular, *or* neither.

6
8   **40.** $x - 3y = 1$
          $3x + y = 4$ **41.** $x - 2y = 8$
          $x + 2y = 8$

e slope of a line perpendicular to a line with undefined slope?

[3.4] *Write an equation in slope-intercept form (if possible) for each line.*

43. $m = -1; b = \dfrac{2}{3}$

44. The line in Exercise 34

45. Through $(4, -3); m = 1$

46. Through $(-1, 4); m = \dfrac{2}{3}$

47. Through $(1, -1); m = -\dfrac{3}{4}$

48. Through $(2, 1)$ and $(-2, 3)$

49. Through $(-4, 1)$ with slope 0

50. Through $\left(\dfrac{1}{3}, -\dfrac{3}{4}\right)$ with undefined slope

51. The Waste Management and Recycling Division of Sacramento County is responsible for managing the disposal of solid waste, including the operation of a landfill. The graph shows the remaining capacity (in millions of tons) at the county landfill during the past few years. These points appear to lie close to a straight line. The equation of this line can be used to project future landfill capacity.

(a) Write an equation in slope-intercept form of the line shown through the points $(0, 5.6)$ and $(2.5, 3.2)$.

(b) Use the equation from part (a) to estimate the remaining capacity of the landfill in 2001.

(c) Based on the graph, when will the capacity of the landfill be used up? Explain.

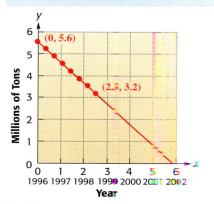

REMAINING LANDFILL CAPACITY

Source: Waste Management and Recycling Division, Sacramento County Public Works Agency, 1998 Report.

[3.5] *Graph each linear inequality.*

52. $3x + 5y > 9$

53. $2x - 3y > -6$

54. $x \geq -4$

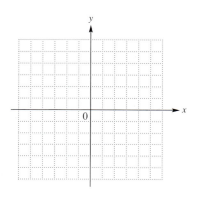

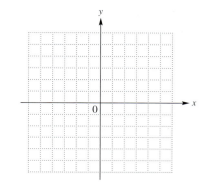

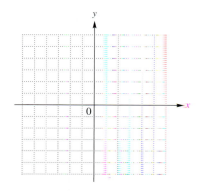

EXERCISES

–60, match each statement to the appropriate graph or graphs in
iay be used more than once.

B.

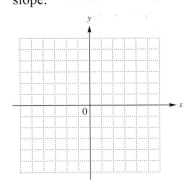

C.

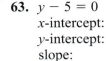

D.

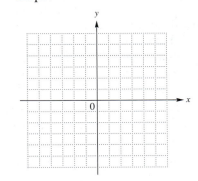

iown in the graph has undefined slope.   **56.** The graph of the equation has y-intercept $(0, -3)$.

of the equation has x-intercept $(-3, 0)$.   **58.** The line shown in the graph has negative slope.

is that of the equation $y = -3$.   **60.** The line shown in the graph has slope 1.

epts and the slope of each line. Then graph the line.

– 5

:

:

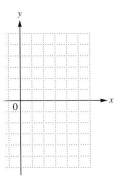

62. $x + 3y = 0$
   x-intercept:
   y-intercept:
   slope:

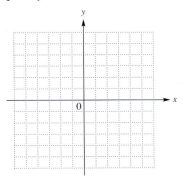

63. $y - 5 = 0$
   x-intercept:
   y-intercept:
   slope:

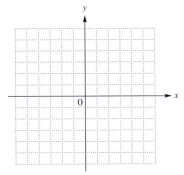

tion in slope-intercept form for each line.

$b = -\dfrac{5}{4}$ **65.** Through $(8, 6)$; $m = -3$ **66.** Through $(3, -5)$ and $(-4, -1)$

equality.

68. $x - 2y \le 6$

RELATING CONCEPTS (Exercises 69–77)   **FOR INDIVIDUAL OR GROUP WORK**

The total amount spent in billions of dollars on video rentals in the United States from 1996 through 2000 is shown in the graph. Use the graph to **work Exercises 69–77 in order.**

69. About how much did the amount spent on video rentals decrease during the years shown in the graph?

70. Since the points of the graph lie approximately in a linear pattern, a straight line can be used to model the data. Will this line have positive or negative slope? Explain.

71. The table gives the actual amounts spent on video rentals in 1996 and 2000. Write two ordered pairs for the data.

VIDEO RENTALS

Source: Blockbuster.

Year x	Amount y (in billions of dollars)
1996	11.1
2000	9.6

72. Use the ordered pairs from Exercise 71 to find the equation of a line that models the data. Write the equation in slope-intercept form.

73. Based on the equation you found in Exercise 72, what is the slope of the line? Does it agree with your answer in Exercise 70?

74. Use the equation from Exercise 72 to approximate the amount spent on video rentals from 1997 through 1999, and complete the table. Round your answers to the nearest tenth.

x	y
1996	11.1
1997	
1998	
1999	
2000	9.6

75. The actual amounts spent on video rentals are given in the following ordered pairs.

(1996, 11.1), (1997, 10.9), (1998, 10.6), (1999, 10.1), (2000, 9.6)

How do the actual amounts compare to those found in Exercise 74 using the linear equation?

76. Since the equation in Exercise 72 models the data fairly well, use it to predict the amount that will be spent on video rentals in 2002.

77. Discuss reasons why the amount spent on video rentals has been decreasing in recent years.

TEST

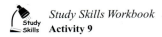 *Study Skills Workbook*
Activity 9

…ese ordered pairs for the equation $3x + 5y = -30$: $(0, \quad)$, …, -3).

…solution of $4x - 7y = 9$?

… find the x-intercept of the graph of a linear equation in two …ow do you find the y-intercept?

…ar equation. Give the x- and y-intercepts.

1. _____

2. _____

3. _____

4. x-intercept: _____
y-intercept: _____

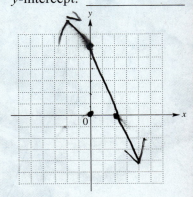

5. x-intercept: _____
y-intercept: _____

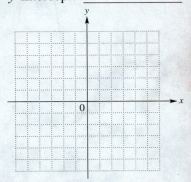

6. x-intercept: _____
y-intercept: _____

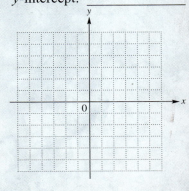

7. *x*-intercept: _____

   y-intercept: _____

7. $y = 1$

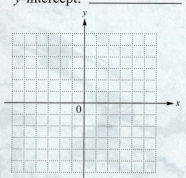

8. *x*-intercept: _____

   y-intercept: _____

8. $x - y = 4$

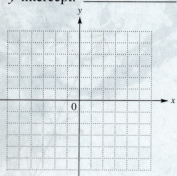

$$M = \frac{y_2 - y_1 \oslash}{x_2 - x_1}$$

Find the slope of each line.

9. _____

9. Through $(-4, 6)$ and $(-1, -2)$

10. _____

10. $2x + y = 10$

11. _____

11. $x + 12 = 0$

12. _____

12.

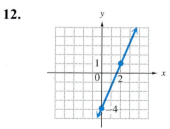

13. _____

13. A line parallel to the graph of $y - 4 = 6$

Write an equation in slope-intercept form for each line.

14. Through $(-1, 4)$; $m = 2$

14. _____

15. The line in Exercise 12

15. _____

16. Through $(2, -6)$ and $(1, 3)$

x y₁   x₂ y₂

16. _____

Graph each linear inequality.

17. $x + y \leq 3$

17.

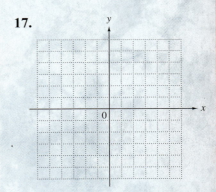

18. $3x - y > 0$

18.

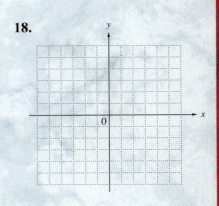

The graph shows total food and drink sales at U.S. restaurants from 1970 through 2000, where 1970 corresponds to x = 0. Use the graph to work Exercises 19–22.

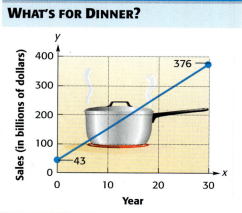

WHAT'S FOR DINNER?

Source: National Restaurant Association.

19. _____

19. Is the slope of the line in the graph positive or negative? Explain.

20. _____

20. Write two ordered pairs for the data points shown in the graph. Use them to find the slope of the line.

21. _____

21. The linear equation

$$y = 11.1x + 43$$

approximates food and drink sales y in billions of dollars, where $x = 0$ again represents 1970. Use the equation to approximate food and drink sales for 1990 and 1995.

22. _____

22. What does the ordered pair (30, 376) mean in the context of this problem?

Perform the indicated operations.

1. $10\dfrac{5}{8} - 3\dfrac{1}{10}$

2. $\dfrac{3}{4} \div \dfrac{1}{8}$

3. $5 - (-4) + (-2)$

4. $\dfrac{(-3)^2 - (-4)(2^4)}{5(2) - (-2)^3}$

5. True or false? $\dfrac{4(3 - 9)}{2 - 6} \geq 6$

6. Find the value of $xz^3 - 5y^2$ when $x = -2$, $y = -3$, and $z = -1$.

7. What property does $3(-2 + x) = -6 + 3x$ illustrate?

8. Simplify $-4p - 6 + 3p + 8$ by combining terms.

Solve.

9. $V = \dfrac{1}{3}\pi r^2 h$ for h

10. $6 - 3(1 + a) = 2(a + 5) - 2$

11. $-(m - 3) = 5 - 2m$

12. $\dfrac{y - 2}{3} = \dfrac{2y + 1}{5}$

Solve each inequality, and graph the solution.

13. $-2.5x < 6.5$

14. $4(x + 3) - 5x < 12$

15. $\dfrac{2}{3}y - \dfrac{1}{6}y \leq -2$

Solve each problem.

16. The gap in average annual earnings by level of education continues to increase. Based on the most recent statistics available, a person with a bachelor's degree can expect to earn $17,583 more each year than someone with a high school diploma. Together the individuals would earn $63,373. How much can a person at each level of education expect to earn? (*Source:* U.S. Bureau of the Census.)

17. Mount Mayon in the Philippines is the most perfectly shaped conical volcano in the world. Its base is a perfect circle with circumference 80 mi, and it has a height of about 8200 ft. (One mile is 5280 ft.) Find the radius of the circular base to the nearest mile. (*Hint:* This problem has some unneeded information.) (*Source: Microsoft Encarta Encyclopedia 2000.*)

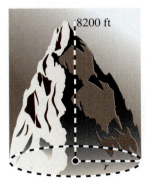

8200 ft

Circumference = 80 mi

18. The winning times in seconds for the women's 1000 m speed skating event in the Winter Olympics for the years 1960 through 1998 can be closely approximated by the linear equation

$$y = -.4685x + 95.07,$$

where x is the number of years since 1960. That is, $x = 4$ represents 1964, $x = 8$ represents 1968, and so on. (*Source: The Universal Almanac, 1998.*)

(a) Use this equation to complete the table of values. Round times to the nearest hundredth of a second.

x	y
12	
28	
36	

(b) What does the ordered pair (20, 85.7) mean in the context of the problem?

19. Baby boomers are expected to inherit $10.4 trillion from their parents over the next 45 yr, an average of $50,000 each. The circle graph shows how they plan to spend their inheritances.

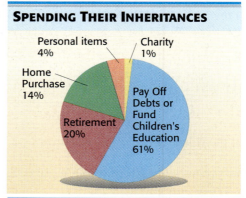

SPENDING THEIR INHERITANCES

Personal items 4%
Charity 1%
Home Purchase 14%
Pay Off Debts or Fund Children's Education 61%
Retirement 20%

Source: First Interstate Bank Trust and Private Banking Group.

(a) How much of the $50,000 is expected to go toward home purchase?

(b) How much is expected to go toward retirement?

(c) Use the answer from part (b) to estimate the amount expected to go toward paying off debts or funding children's education.

Consider the linear equation $-3x + 4y = 12$. *Find the following.*

20. The x- and y-intercepts

21. The slope

22. The graph

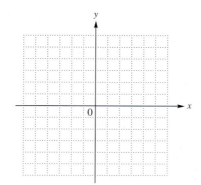

23. Are the lines with equations $x + 5y = -6$ and $y = 5x - 8$ parallel, perpendicular, or neither?

Write an equation in slope-intercept form for each line.

24. Through $(2, -5)$ with slope 3

25. Through $(0, 4)$ and $(2, 4)$

Exponents and Polynomials

4

The expression $100t - 13t^2$ gives the distance in feet a car going approximately 68 mph will skid in t sec. This expression in t is an example of a *polynomial*, the topic of this chapter. Accident investigators use polynomials like this to determine the length of a skid or the elapsed time during a skid. See Exercises 43 and 44 of Section 4.1 where we use this polynomial to approximate skidding distance.

ADDISON - WESLEY
MyMathLab.com
You're Connected

4.1 ADDING AND SUBTRACTING POLYNOMIALS

1 Add like terms.

(a) $5x^4 + 7x^4$

(b) $9pq + 3pq - 2pq$

(c) $r^2 + 3r + 5r^2$

(d) $8t + 6w$

2 Choose all descriptions that apply for each of the expressions in parts (a)–(d).
A. Polynomial
B. Polynomial written in descending powers
C. Not a polynomial

(a) $3m^5 + 5m^2 - 2m + 1$

(b) $2p^4 + p^6$

(c) $\dfrac{1}{x} + 2x^2 + 3$

(d) $x - 3$

Recall from Section 1.8 that in an expression such as

$$4x^3 + 6x^2 + 5x + 8,$$

the quantities that are added, $4x^3$, $6x^2$, $5x$, and 8 are called *terms*. In the term $4x^3$, the number 4 is called the *numerical coefficient*, or simply the *coefficient*, of x^3. In the same way, 6 is the coefficient of x^2 in the term $6x^2$, 5 is the coefficient of x in the term $5x$, and 8 is the constant term.

1 **Review combining like terms.** In Section 1.8, we saw that *like terms* are terms with exactly the same variables, with the same exponents on the variables. Only the coefficients may differ. Like terms are combined, or *added*, by adding their coefficients using the distributive property.

Example 1 Adding Like Terms

Simplify each expression by adding like terms.

(a) $-4x^3 + 6x^3 = (-4 + 6)x^3$ Distributive property
$$= 2x^3$$

(b) $9x^6 - 14x^6 + x^6 = (9 - 14 + 1)x^6 = -4x^6$

(c) $12m^2 + 5m + 4m^2 = (12 + 4)m^2 + 5m$
$$= 16m^2 + 5m$$

(d) $3x^2y + 4x^2y - x^2y = (3 + 4 - 1)x^2y = 6x^2y$

In Example 1(c), we cannot add $16m^2$ and $5m$. These two terms are unlike because the exponents on the variables are different. *Unlike terms* have different variables or different exponents on the same variables.

Work Problem 1 at the Side.

2 **Know the vocabulary for polynomials.** A **polynomial in x** is a term or the sum of a finite number of terms of the form ax^n, for any real number a and any whole number n. For example,

$$16x^8 - 7x^6 + 5x^4 - 3x^2 + 4$$

is a polynomial in x. This polynomial is written in **descending powers**, because the exponents on x decrease from left to right. On the other hand,

$$2x^3 - x^2 + \frac{4}{x}$$

is not a polynomial, since a variable appears in a denominator. Of course, we could define a *polynomial* using any variable, not just x, as in Example 1(c). In fact, polynomials may have terms with more than one variable, as in Example 1(d).

Work Problem 2 at the Side.

The **degree of a term** is the sum of the exponents on the variables. A constant term has degree 0. For example, $3x^4$ has degree 4, while $6x^{17}$ has degree 17. The term $5x$ has degree 1, -7 has degree 0, and $2x^2y$ has degree $2 + 1 = 3$ (y has an exponent of 1). The **degree of a polynomial** is the highest degree of any nonzero term of the polynomial. For example, $3x^4 - 5x^2 + 6$ is of degree 4, the polynomial $5x + 7$ is of degree 1, 3 is of degree 0, and $x^2y + xy - 5xy^2$ is of degree 3.

Three types of polynomials are very common and are given special names. A polynomial with only one term is called a **monomial**. (*Mon(o)*- means "one," as in *mono*rail.) Examples are

$$9m, \quad -6y^5, \quad a^2, \quad \text{and} \quad 6. \quad \text{Monomials}$$

A polynomial with exactly two terms is called a **binomial**. (*Bi*- means "two," as in *bi*cycle.) Examples are

$$-9x^4 + 9x^3, \quad 8m^2 + 6m, \quad \text{and} \quad 3m^5 - 9m^2. \quad \text{Binomials}$$

A polynomial with exactly three terms is called a **trinomial**. (*Tri*- means "three," as in *tri*angle.) Examples are

$$9m^3 - 4m^2 + 6, \quad \frac{19}{3}y^2 + \frac{8}{3}y + 5, \quad \text{and} \quad -3m^5 - 9m^2 + 2. \quad \text{Trinomials}$$

Example 2   Classifying Polynomials

For each polynomial, first simplify if possible by combining like terms. Then give the degree and tell whether the polynomial is a monomial, a binomial, a trinomial, or none of these.

(a) $2x^3 + 5$

The polynomial cannot be simplified. The degree is 3. The polynomial is a binomial.

(b) $4x - 5x + 2x$

Add like terms to simplify: $4x - 5x + 2x = x$. The degree is 1 (since $x = x^1$). The simplified polynomial is a monomial.

=== **Work Problem ❸ at the Side.**

❸ Evaluate polynomials. A polynomial usually represents different numbers for different values of the variable, as shown in the next example.

Example 3   Evaluating a Polynomial

Find the value of $3x^4 + 5x^3 - 4x - 4$ when $x = -2$ and when $x = 3$.
First, substitute -2 for x.

$$3x^4 + 5x^3 - 4x - 4 = 3(-2)^4 + 5(-2)^3 - 4(-2) - 4$$
$$= 3 \cdot 16 + 5(-8) - 4(-2) - 4 \qquad \text{Apply exponents.}$$
$$= 48 - 40 + 8 - 4 \qquad \text{Multiply.}$$
$$= 12 \qquad \text{Add and subtract.}$$

Next, replace x with 3.

$$3x^4 + 5x^3 - 4x - 4 = 3(3)^4 + 5(3)^3 - 4(3) - 4$$
$$= 3 \cdot 81 + 5 \cdot 27 - 4(3) - 4$$
$$= 243 + 135 - 12 - 4$$
$$= 362$$

❸ For each polynomial, first simplify if possible. Then give the degree and tell whether the polynomial is a monomial, binomial, trinomial, or none of these.

(a) $3x^2 + 2x - 4$

(b) $x^3 + 4x^3$

(c) $x^8 - x^7 + 2x^8$

Answers
3. (a) degree 2; trinomial
(b) degree 3; monomial (simplify to $5x^3$)
(c) degree 8; binomial (simplify to $3x^8 - x^7$)

❹ Find the value of $2y^3 + 8y - 6$ in each case.

(a) when $y = -1$

(b) when $y = 4$

❺ Add each pair of polynomials.

(a) $4x^3 - 3x^2 + 2x$
$6x^3 + 2x^2 - 3x$

(b) $x^2 - 2x + 5$
$4x^2 - 2$

CAUTION

Notice the use of parentheses around the numbers that are substituted for the variable in Example 3. This is particularly important when substituting a negative number for a variable that is raised to a power, so the sign of the product is correct.

Work Problem ❹ at the Side.

4 ▬ **Add polynomials.** Polynomials may be added, subtracted, multiplied, and divided.

Adding Polynomials

To add two polynomials, add like terms.

Example 4 Adding Polynomials Vertically

(a) Add $6x^3 - 4x^2 + 3$ and $-2x^3 + 7x^2 - 5$.
Write like terms in columns.

$$6x^3 - 4x^2 + 3$$
$$-2x^3 + 7x^2 - 5$$

Now add, column by column.

$$\begin{array}{ccc} 6x^3 & -4x^2 & 3 \\ -2x^3 & 7x^2 & -5 \\ \hline 4x^3 & 3x^2 & -2 \end{array}$$

Add the three sums together.

$$4x^3 + 3x^2 + (-2) = 4x^3 + 3x^2 - 2$$

(b) Add $2x^2 - 4x + 3$ and $x^3 + 5x$.
Write like terms in columns and add column by column.

$$\begin{array}{l} 2x^2 - 4x + 3 \\ x^3 + 5x \\ \hline x^3 + 2x^2 + x + 3 \end{array}$$

Leave spaces for missing terms.

Work Problem ❺ at the Side.

The polynomials in Example 4 also could be added horizontally by combining like terms, as shown in the next example.

Example 5 Adding Polynomials Horizontally

(a) Add $6x^3 - 4x^2 + 3$ and $-2x^3 + 7x^2 - 5$.
Combine like terms.

$$(6x^3 - 4x^2 + 3) + (-2x^3 + 7x^2 - 5)$$

Continued on Next Page

The sum is

$$4x^3 + 3x^2 - 2,$$

the same answer found in Example 4(a).

(b) Add $2x^2 - 4x + 3$ and $x^3 + 5x$.

$$(2x^2 - 4x + 3) + (x^3 + 5x) = 2x^2 - 4x + 3 + x^3 + 5x$$
$$= x^3 + 2x^2 + x + 3 \quad \text{Combine like terms.}$$

========= Work Problem **6** at the Side.

5 ▭ **Subtract polynomials.** Earlier, the difference $x - y$ was defined as $x + (-y)$. (We find the difference $x - y$ by adding x and the opposite of y.) For example,

$$7 - 2 = 7 + (-2) = 5 \quad \text{and} \quad -8 - (-2) = -8 + 2 = -6.$$

A similar method is used to subtract polynomials.

Subtracting Polynomials

To subtract two polynomials, change all the signs of the second polynomial and add the result to the first polynomial.

Example 6 **Subtracting Polynomials**

(a) Perform the subtraction $(5x - 2) - (3x - 8)$.
   Change the signs in the second polynomial and add like terms.

$$(5x - 2) - (3x - 8) = (5x - 2) + (-3x + 8)$$
$$= 2x + 6$$

(b) Subtract $6x^3 - 4x^2 + 2$ from $11x^3 + 2x^2 - 8$.
   Write the problem.

$$(11x^3 + 2x^2 - 8) - (6x^3 - 4x^2 + 2)$$

Change all the signs in the second polynomial and add the two polynomials.

$$(11x^3 + 2x^2 - 8) + (-6x^3 + 4x^2 - 2) = 5x^3 + 6x^2 - 10$$

To check a subtraction problem, use the fact that if $a - b = c$, then $a = b + c$. For example, $6 - 2 = 4$, so we check by writing $6 = 2 + 4$, which is correct. Check the polynomial subtraction above by adding $6x^3 - 4x^2 + 2$ and $5x^3 + 6x^2 - 10$. Since the sum is $11x^3 + 2x^2 - 8$, the subtraction was performed correctly.

========= Work Problem **7** at the Side.

Subtraction also can be done in columns. We will use vertical subtraction in Section 4.7 when we study polynomial division.

Example 7 **Subtracting Polynomials Vertically**

Use the method of subtracting by columns to find

$$(14y^3 - 6y^2 + 2y - 5) - (2y^3 - 7y^2 - 4y + 6).$$

========= **Continued on Next Page**

6 Find each sum.

(a) $(2x^4 - 6x^2 + 7)$
   $+ (-3x^4 + 5x^2 + 2)$

(b) $(3x^2 + 4x + 2)$
   $+ (6x^3 - 5x - 7)$

7 Subtract, and check your answers by addition.

(a) $(14y^3 - 6y^2 + 2y - 5)$
   $- (2y^3 - 7y^2 - 4y + 6)$

(b) $\left(\frac{7}{2}y^2 - \frac{11}{3}y + 8\right)$
   $- \left(-\frac{3}{2}y^2 + \frac{4}{3}y + 6\right)$

ANSWERS
6. (a) $-x^4 - x^2 + 9$
   (b) $6x^3 + 3x^2 - x - 5$
7. (a) $12y^3 + y^2 + 6y - 11$
   (b) $5y^2 - 5y + 2$

8 Subtract, using the method of subtracting by columns.

$$(4y^3 - 16y^2 + 2y)$$
$$- (12y^3 - 9y^2 + 16)$$

Arrange like terms in columns.

$$14y^3 - 6y^2 + 2y - 5$$
$$\underline{2y^3 - 7y^2 - 4y + 6}$$

Change all signs in the second row, and then add.

$$\begin{aligned}
14y^3 - 6y^2 + 2y - \;\;5 & \\
\underline{-2y^3 + 7y^2 + 4y - \;\;6} & \quad \text{Change signs.} \\
12y^3 + \;\;y^2 + 6y - 11 & \quad \text{Add.}
\end{aligned}$$

Work Problem 8 at the Side.

Either the horizontal or the vertical method may be used for adding or subtracting polynomials.

Example 8 Adding and Subtracting More Than Two Polynomials

Perform the indicated operations to simplify the expression

$$(4 - x + 3x^2) - (2 - 3x + 5x^2) + (8 + 2x - 4x^2).$$

Rewrite, changing the subtraction to adding the opposite.

$$(4 - x + 3x^2) - (2 - 3x + 5x^2) + (8 + 2x - 4x^2)$$
$$= (4 - x + 3x^2) + (-2 + 3x - 5x^2) + (8 + 2x - 4x^2)$$
$$= (2 + 2x - 2x^2) + (8 + 2x - 4x^2) \quad \text{Combine like terms.}$$
$$= 10 + 4x - 6x^2 \quad \text{Combine like terms.}$$

Work Problem 9 at the Side.

9 Perform the indicated operations.

$$(6p^4 - 8p^3 + 2p - 1)$$
$$- (-7p^4 + 6p^2 - 12)$$
$$+ (p^4 - 3p + 8)$$

6 **Add and subtract polynomials with more than one variable.** Polynomials in more than one variable are added and subtracted by combining like terms, just as with single-variable polynomials.

Example 9 Adding and Subtracting Multivariable Polynomials

Add or subtract as indicated.

(a) $(4a + 2ab - b) + (3a - ab + b)$

$$(4a + 2ab - b) + (3a - ab + b)$$
$$= 4a + 2ab - b + 3a - ab + b$$
$$= 7a + ab \quad \text{Combine like terms.}$$

10 Add or subtract.

(a) $(3mn + 2m - 4n)$
$\qquad + (-mn + 4m + n)$

(b) $(2x^2y + 3xy + y^2) - (3x^2y - xy - 2y^2)$

$$(2x^2y + 3xy + y^2) - (3x^2y - xy - 2y^2)$$
$$= 2x^2y + 3xy + y^2 - 3x^2y + xy + 2y^2$$
$$= -x^2y + 4xy + 3y^2$$

Work Problem 10 at the Side.

(b) $(5p^2q^2 - 4p^2 + 2q)$
$\qquad - (2p^2q^2 - p^2 - 3q)$

4.1 EXERCISES

 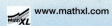

Fill in each blank with the correct response.

1. In the term $7x^5$, the coefficient is _____ and the exponent is _____.

2. The expression $5x^3 - 4x^2$ has _____ term(s).
            (how many?)

3. The degree of the term $-4x^8$ is _____.

4. The polynomial $4x^2 - y^2$ _____ an example of a trinomial.
            (is/is not)

5. When $x^2 + 10$ is evaluated for $x = 4$, the result is _____.

6. _____ is an example of a monomial with coefficient 5, in the variable x, having degree 9.

For each polynomial, determine the number of terms, and name the coefficient of each term.

7. $6x^4$

8. $-9y^5$

9. t^4

10. s^7

11. $-19r^2 - r$

12. $2y^3 - y$

13. $x + 8x^2$

14. $v - 2v^3$

In each polynomial, combine like terms whenever possible. Write the result with descending powers.

15. $-3m^5 + 5m^5$

16. $-4y^3 + 3y^3$

17. $2r^5 + (-3r^5)$

18. $-19y^2 + 9y^2$

19. $.2m^5 - .5m^2$

20. $-.9y + .9y^2$

21. $-3x^5 + 2x^5 - 4x^5$

22. $6x^3 - 8x^3 + 9x^3$

23. $-4p^7 + 8p^7 + 5p^9$

24. $-3a^8 + 4a^8 - 3a^2$

25. $-4y^2 + 3y^2 - 2y^2 + y^2$

26. $3r^5 - 8r^5 + r^5 + 2r^5$

For each polynomial, first simplify, if possible, and write it with descending powers. Then give the degree of the resulting polynomial, and tell whether it is a monomial, a binomial, a trinomial, or none of these. See Example 2.

27. $6x^4 - 9x$

28. $7t^3 - 3t$

29. $5m^4 - 3m^2 + 6m^5 - 7m^3$

30. $6p^5 + 4p^3 - 8p^4 + 10p^2$ **31.** $\dfrac{5}{3}x^4 - \dfrac{2}{3}x^4 + \dfrac{1}{3}x^2 - 4$ **32.** $\dfrac{4}{5}r^6 + \dfrac{1}{5}r^6 - r^4 + \dfrac{2}{5}r$

33. $.8x^4 - .3x^4 - .5x^4 + 7$ **34.** $1.2t^3 - .9t^3 - .3t^3 + 9$

Find the value of each polynomial **(a)** *when* $x = 2$ *and* **(b)** *when* $x = -1$. *See Example 3.*

35. $-2x + 3$ **36.** $5x - 4$ **37.** $2x^2 + 5x + 1$ **38.** $-3x^2 + 14x - 2$

39. $2x^5 - 4x^4 + 5x^3 - x^2$ **40.** $x^4 - 6x^3 + x^2 + 1$ **41.** $-4x^5 + x^2$ **42.** $2x^6 - 4x$

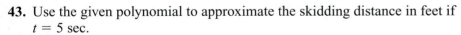

RELATING CONCEPTS (Exercises 43–46)   **FOR INDIVIDUAL OR GROUP WORK**

In the introduction to this chapter, we gave a polynomial that models the distance in feet that a car going approximately 68 mph will skid in t sec. If we let D represent this distance, then

$$D = 100t - 13t^2.$$

Each time we evaluate this polynomial for a value of t, we get one and only one output value D. This idea is basic to the concept of a function, *an important concept in mathematics. Exercises 43–46 illustrate this idea with this polynomial and two others.* **Work them in order.**

43. Use the given polynomial to approximate the skidding distance in feet if $t = 5$ sec.

44. Use the polynomial equation $D = 100t - 13t^2$ to find the distance the car will skid in 1 sec. Write an ordered pair of the form (x, D).

45. If gasoline costs \$1.60 per gal, then the monomial $1.60x$ gives the cost, in dollars, of x gal. How much would 4 gal cost?

46. If it costs \$15 plus \$2 per day to rent a chain saw, the binomial $2x + 15$ gives the cost in dollars to rent the chain saw for x days. How much would it cost to rent the saw for 6 days?

Add or subtract as indicated. See Examples 4 and 7.

47. Add.
$$3m^2 + 5m$$
$$\underline{2m^2 - 2m}$$

48. Add.
$$4a^3 - 4a^2$$
$$\underline{6a^3 + 5a^2}$$

49. Subtract.
$$12x^4 - x^2$$
$$\underline{8x^4 + 3x^2}$$

50. Subtract.
$$13y^5 - y^3$$
$$\underline{7y^5 + 5y^3}$$

51. Add.
$$\frac{2}{3}x^2 + \frac{1}{5}x + \frac{1}{6}$$
$$\underline{\frac{1}{2}x^2 - \frac{1}{3}x + \frac{2}{3}}$$

52. Add.
$$\frac{4}{7}y^2 - \frac{1}{5}y + \frac{7}{9}$$
$$\underline{\frac{1}{3}y^2 - \frac{1}{3}y + \frac{2}{5}}$$

53. Subtract.
$$12m^3 - 8m^2 + 6m + 7$$
$$\underline{ + 5m^2 - 4}$$

54. Subtract.
$$5a^4 - 3a^3 + 2a^2 - a + 6$$
$$\underline{-6a^4 - a^2 + a - 1}$$

Perform the indicated operations. See Examples 5, 6, and 8.

55. $(2r^2 + 3r - 12) + (6r^2 + 2r)$

56. $(3r^2 + 5r - 6) + (2r - 5r^2)$

57. $(8m^2 - 7m) - (3m^2 + 7m - 6)$

58. $(x^2 + x) - (3x^2 + 2x - 1)$

59. $(16x^3 - x^2 + 3x) + (-12x^3 + 3x^2 + 2x)$

60. $(-2b^6 + 3b^4 - b^2) + (b^6 + 2b^4 + 2b^2)$

61. $(7y^4 + 3y^2 + 2y) - (18y^5 - 5y^3 + y)$

62. $(8t^5 + 3t^3 + 5t) - (19t^4 - 6t^2 + t)$

63. $[(8m^2 + 4m - 7) - (2m^3 - 5m + 2)] - (m^2 + m)$

64. $[(9b^3 - 4b^2 + 3b + 2) - (-2b^3 + b)] - (8b^3 + 6b + 4)$

Find the perimeter of each geometric figure.

65.

$4x^2 + 3x + 1$

$x + 2$

66.

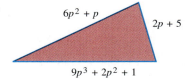

$5y^2 + 3y + 8$

$y + 4$

67.

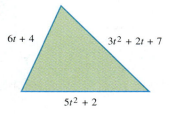

$6t + 4$

$3t^2 + 2t + 7$

$5t^2 + 2$

68.

$6p^2 + p$

$2p + 5$

$9p^3 + 2p^2 + 1$

69. Subtract $9x^2 - 3x + 7$ from $-2x^2 - 6x + 4$.

70. Subtract $-5w^3 + 5w^2 - 7$ from $6w^3 + 8w + 5$.

71. Explain why the degree of the term 3^4 is not 4. What is its degree?

72. Can the sum of two polynomials in x, both of degree 3, be of degree 2? If so, give an example.

Add or subtract as indicated. See Example 9.

73. $(9a^2b - 3a^2 + 2b) + (4a^2b - 4a^2 - 3b)$

74. $(4xy^3 - 3x + y) + (5xy^3 + 13x - 4y)$

75. $(2c^4d + 3c^2d^2 - 4d^2) - (c^4d + 8c^2d^2 - 5d^2)$

76. $(3k^2h^3 + 5kh + 6k^3h^2) - (2k^2h^3 - 9kh + k^3h^2)$

77. Subtract. $\quad 9m^3n - 5m^2n^2 + 4mn^2$
$\underline{-3m^3n + 6m^2n^2 + 8mn^2}$

78. Subtract. $\quad 12r^5t + 11r^4t^2 - 7r^3t^3$
$\underline{-8r^5t + 10r^4t^2 + 3r^3t^3}$

4.2 THE PRODUCT RULE AND POWER RULES FOR EXPONENTS

1 **Use exponents.** In Chapter 1 we used exponents to write repeated products. Recall that in the expression 5^2, the number 5 is called the *base* and 2 is called the *exponent* or *power*. The expression 5^2 is called an *exponential expression*. Usually we do not write a quantity with an exponent of 1, but sometimes it is convenient to do so. In general, for any quantity a, $a = a^1$.

Example 1 Using Exponents

Write $3 \cdot 3 \cdot 3 \cdot 3 \cdot 3$ in exponential form, and evaluate the exponential expression.

Since 3 occurs as a factor five times, the base is 3 and the exponent is 5. The exponential expression is 3^5, read "3 to the fifth power" or simply "3 to the fifth." The value is

$$3^5 = 3 \cdot 3 \cdot 3 \cdot 3 \cdot 3 = 243.$$

Work Problem **1** at the Side.

Example 2 Evaluating Exponential Expressions

Evaluate each exponential expression. Name the base and the exponent.

	Base	Exponent
(a) $5^4 = 5 \cdot 5 \cdot 5 \cdot 5 = 625$	5	4
(b) $-5^4 = -1 \cdot 5^4 = -1 \cdot (5 \cdot 5 \cdot 5 \cdot 5) = -625$	5	4
(c) $(-5)^4 = (-5)(-5)(-5)(-5) = 625$	-5	4

CAUTION

It is important to understand the difference between parts (b) and (c) of Example 2. In -5^4 the lack of parentheses shows that the exponent 4 applies only to the base 5, and not -5; in $(-5)^4$ the parentheses show that the exponent 4 applies to the base -5. In summary, $-a^n$ and $(-a)^n$ are not always the same.

Expression	Base	Exponent	Example
$-a^n$	a	n	$-3^2 = -(3 \cdot 3) = -9$
$(-a)^n$	$-a$	n	$(-3)^2 = (-3)(-3) = 9$

Work Problem **2** at the Side.

2 **Use the product rule for exponents.** To develop the product rule, we use the definition of an exponential expression.

$$2^4 \cdot 2^3 = \overbrace{(2 \cdot 2 \cdot 2 \cdot 2)}^{4 \text{ factors}}\overbrace{(2 \cdot 2 \cdot 2)}^{3 \text{ factors}}$$
$$= \underbrace{2 \cdot 2 \cdot 2 \cdot 2 \cdot 2 \cdot 2 \cdot 2}_{4 + 3 = 7 \text{ factors}}$$
$$= 2^7$$

OBJECTIVES

1 Use exponents.
2 Use the product rule for exponents.
3 Use the rule $(a^m)^n = a^{mn}$.
4 Use the rule $(ab)^m = a^m b^m$.
5 Use the rule $\left(\dfrac{a}{b}\right)^m = \dfrac{a^m}{b^m}$.
6 Use combinations of the rules for exponents.

1 Write $2 \cdot 2 \cdot 2 \cdot 2$ in exponential form, and evaluate.

2 Evaluate each exponential expression. Name the base and the exponent.

(a) $(-2)^5$ **(b)** -2^5

(c) -4^2 **(d)** $(-4)^2$

ANSWERS
1. $2^4 = 16$
2. (a) $-32; -2; 5$ **(b)** $-32; 2; 5$ **(c)** $-16; 4; 2$ **(d)** $16; -4; 2$

❸ Find each product by the product rule, if possible.

(a) $8^2 \cdot 8^5$

(b) $(-7)^5 \cdot (-7)^3$

(c) $y^3 \cdot y$

(d) $4^2 \cdot 3^5$

(e) $6^4 + 6^2$

Also,

$$6^2 \cdot 6^3 = (6 \cdot 6)(6 \cdot 6 \cdot 6)$$
$$= 6 \cdot 6 \cdot 6 \cdot 6 \cdot 6$$
$$= 6^5.$$

Generalizing from these examples, $2^4 \cdot 2^3 = 2^{4+3} = 2^7$ and $6^2 \cdot 6^3 = 6^{2+3} = 6^5$. In each case, adding the exponents gives the exponent of the product, suggesting the **product rule for exponents.**

Product Rule for Exponents

For any positive integers m and n, $\qquad a^m \cdot a^n = a^{m+n}$.
(Keep the same base and add the exponents.)

Example: $6^2 \cdot 6^5 = 6^{2+5} = 6^7$.

CAUTION

Avoid the common error of multiplying the bases when using the product rule.

$$6^2 \cdot 6^5 \neq 36^7$$

Keep the *same* base and add the exponents.

Example 3 **Using the Product Rule**

Use the product rule for exponents to find each product, if possible.

(a) $6^3 \cdot 6^5 = 6^{3+5} = 6^8$ by the product rule.

(b) $(-4)^7(-4)^2 = (-4)^{7+2} = (-4)^9$ by the product rule.

(c) $x^2 \cdot x = x^2 \cdot x^1 = x^{2+1} = x^3$

(d) $m^4 \cdot m^3 = m^{4+3} = m^7$

(e) $2^3 \cdot 3^2$
The product rule does not apply to the product $2^3 \cdot 3^2$ because the bases are different.

$$2^3 \cdot 3^2 = 8 \cdot 9 = 72$$

(f) $2^3 + 2^4$
The product rule does not apply to $2^3 + 2^4$ because it is a *sum*, not a *product*.

$$2^3 + 2^4 = 8 + 16 = 24$$

CAUTION

The bases must be the same before we can apply the product rule for exponents.

Work Problem ❸ at the Side.

Example 4   **Using the Product Rule**

Multiply $2x^3$ and $3x^7$.

Since $2x^3$ means $2 \cdot x^3$ and $3x^7$ means $3 \cdot x^7$, we use the associative and commutative properties and the product rule to get

$$2x^3 \cdot 3x^7 = 2 \cdot 3 \cdot x^3 \cdot x^7 = 6x^{10}.$$

CAUTION

Be sure you understand the difference between *adding* and *multiplying* exponential expressions. For example,

$$8x^3 + 5x^3 = (8+5)x^3 = 13x^3,$$

but $\qquad (8x^3)(5x^3) = (8 \cdot 5)x^{3+3} = 40x^6.$

Work Problem ④ at the Side.

③   **Use the rule $(a^m)^n = a^{mn}$.**   We can simplify an expression such as $(8^3)^2$ with the product rule for exponents, as follows.

$$(8^3)^2 = (8^3)(8^3) = 8^{3+3} = 8^6$$

The product of the exponents in $(8^3)^2$, 3 and 2, gives the exponent in 8^6. As another example,

$$\begin{aligned} (5^2)^4 &= 5^2 \cdot 5^2 \cdot 5^2 \cdot 5^2 & \text{Definition of exponent} \\ &= 5^{2+2+2+2} & \text{Product rule} \\ &= 5^8, \end{aligned}$$

and $2 \cdot 4 = 8$. These examples suggest **power rule (a) for exponents.**

Power Rule (a) for Exponents

For any positive integers m and n, $\quad (a^m)^n = a^{mn}.$
(Raise a power to a power by multiplying exponents.)

Example: $(3^2)^4 = 3^{2 \cdot 4} = 3^8.$

Example 5   **Using Power Rule (a)**

Use power rule (a) for exponents to simplify each expression.

(a) $(2^5)^3 = 2^{5 \cdot 3} = 2^{15}$ **(b)** $(5^7)^2 = 5^{7 \cdot 2} = 5^{14}$

(c) $(x^2)^5 = x^{2 \cdot 5} = x^{10}$ **(d)** $(n^3)^2 = n^{3 \cdot 2} = n^6$

Work Problem ⑤ at the Side.

④   **Use the rule $(ab)^m = a^m b^m$.**   The properties studied in Chapter 1 can be used to develop two more rules for exponents. Using the definition of an exponential expression and the commutative and associative properties, we can rewrite the expression $(4x)^3$ as shown below.

$$\begin{aligned} (4x)^3 &= (4x)(4x)(4x) & \text{Definition of exponent} \\ &= 4 \cdot 4 \cdot 4 \cdot x \cdot x \cdot x & \text{Commutative and associative properties} \\ &= 4^3 x^3 & \text{Definition of exponent} \end{aligned}$$

This example suggests **power rule (b) for exponents.**

④ Multiply.

(a) $5m^2 \cdot 2m^6$

(b) $3p^5 \cdot 9p^4$

(c) $-7p^5 \cdot (3p^8)$

⑤ Simplify each expression.

(a) $(5^3)^4$

(b) $(6^2)^5$

(c) $(3^2)^4$

(d) $(a^6)^5$

ANSWERS
4. (a) $10m^8$ (b) $27p^9$ (c) $-21p^{13}$
5. (a) 5^{12} (b) 6^{10} (c) 3^8 (d) a^{30}

❻ Simplify.

(a) $5(mn)^3$

Power Rule (b) for Exponents

For any positive integer m, $\qquad (ab)^m = a^m b^m$.
(Raise a product to a power by raising each factor to the power.)

Example: $(2p)^5 = 2^5 p^5$.

Example 6 Using Power Rule (b)

Use power rule (b) to simplify each expression.

(a) $(3xy)^2 = 3^2 x^2 y^2$ $\qquad$ Power rule (b)

$\qquad\qquad = 9x^2 y^2$

(b) $9(pq)^2 = 9(p^2 q^2)$ $\qquad$ Power rule (b)

$\qquad\qquad = 9p^2 q^2$

(c) $5(2m^2 p^3)^4 = 5[2^4 (m^2)^4 (p^3)^4]$ $\qquad$ Power rule (b)

$\qquad\qquad = 5(2^4 m^8 p^{12})$ $\qquad$ Power rule (a)

$\qquad\qquad = 5 \cdot 2^4 m^8 p^{12}$

$\qquad\qquad = 80m^8 p^{12}$ $\qquad\qquad 5 \cdot 2^4 = 5 \cdot 16 = 80$

(b) $(3a^2 b^4)^5$

(d) $(-5^6)^3 = (-1 \cdot 5^6)^3$ $\qquad -a = -1 \cdot a$

$\qquad\qquad = (-1)^3 (5^6)^3$ $\qquad$ Power rule (b)

$\qquad\qquad = -1 \cdot 5^{18}$ $\qquad$ Power rule (a)

$\qquad\qquad = -5^{18}$

CAUTION

Power rule (b) *does not* apply to a *sum*.

$$(x + 4)^2 \neq x^2 + 4^2$$

Work Problem ❻ at the Side.

(c) $(-5m^2)^3$

5 ▭ Use the rule $\left(\frac{a}{b}\right)^m = \frac{a^m}{b^m}$. Since the quotient $\frac{a}{b}$ can be written as $a \cdot \frac{1}{b}$, we can use power rule (b), together with some of the properties of real numbers, to get **power rule (c) for exponents.**

Power Rule (c) for Exponents

For any positive integer m, $\qquad \left(\dfrac{a}{b}\right)^m = \dfrac{a^m}{b^m}$ $\quad (b \neq 0)$.

(Raise a quotient to a power by raising both the numerator and the denominator to the power.)

Example: $\left(\dfrac{5}{3}\right)^2 = \dfrac{5^2}{3^2}$.

Example 7   Using Power Rule (c)

Simplify each expression.

(a) $\left(\dfrac{2}{3}\right)^5 = \dfrac{2^5}{3^5}$

(b) $\left(\dfrac{m}{n}\right)^4 = \dfrac{m^4}{n^4}, \; n \neq 0$

Work Problem **7** at the Side.

Next we list the rules for exponents discussed in this section. These rules are basic to the study of algebra and should be *memorized*.

Rules for Exponents

For positive integers m and n: *Examples*

Product rule $a^m \cdot a^n = a^{m+n}$ $6^2 \cdot 6^5 = 6^{2+5} = 6^7$

Power rules (a) $(a^m)^n = a^{mn}$ $(3^2)^4 = 3^{2\cdot4} = 3^8$

(b) $(ab)^m = a^m b^m$ $(2p)^5 = 2^5 p^5$

(c) $\left(\dfrac{a}{b}\right)^m = \dfrac{a^m}{b^m} \; (b \neq 0)$ $\left(\dfrac{5}{3}\right)^2 = \dfrac{5^2}{3^2}$

6▭   Use combinations of the rules for exponents.   As shown in the next example, more than one rule may be needed to simplify an expression.

Example 8   Using Combinations of Rules

Simplify each expression.

(a) $\left(\dfrac{2}{3}\right)^2 \cdot 2^3 = \dfrac{2^2}{3^2} \cdot \dfrac{2^3}{1}$ Power rule (c)

$= \dfrac{2^2 \cdot 2^3}{3^2 \cdot 1}$ Multiply fractions.

$= \dfrac{2^5}{3^2}$ Product rule

(b) $(5x)^3 (5x)^4 = (5x)^7$ Product rule

$= 5^7 x^7$ Power rule (b)

(c) $(2x^2 y^3)^4 (3xy^2)^3 = 2^4 (x^2)^4 (y^3)^4 \cdot 3^3 x^3 (y^2)^3$ Power rule (b)

$= 2^4 \cdot x^8 \cdot y^{12} \cdot 3^3 \cdot x^3 \cdot y^6$ Power rule (a)

$= 2^4 \cdot 3^3 x^8 x^3 y^{12} y^6$ Commutative and associative properties

$= 16 \cdot 27 x^{11} y^{18}$ Product rule

$= 432 x^{11} y^{18}$

Continued on Next Page

7 Simplify. Assume all variables represent nonzero real numbers.

(a) $\left(\dfrac{5}{2}\right)^4$

(b) $\left(\dfrac{p}{q}\right)^2$

(c) $\left(\dfrac{r}{t}\right)^3$

8 Simplify.

(a) $(2m)^3(2m)^4$

(b) $\left(\dfrac{5k^3}{3}\right)^2$

(c) $\left(\dfrac{1}{5}\right)^4(2x)^2$

(d) $(-3xy^2)^3(x^2y)^4$

(d) $(-x^3y)^2(-x^5y^4)^3$

Think of the negative sign in each factor as -1.

$$
\begin{aligned}
(\mathbf{-1}x^3y)^2(\mathbf{-1}x^5y^4)^3 &= (-1)^2(x^3)^2y^2 \cdot (-1)^3(x^5)^3(y^4)^3 & \text{\color{blue}Power rule (b)}\\
&= (-1)^2(x^6)(y^2)(-1)^3(x^{15})(y^{12}) & \text{\color{blue}Power rule (a)}\\
&= (-1)^5(x^{21})(y^{14}) & \text{\color{blue}Product rule}\\
&= -1x^{21}y^{14}\\
&= -x^{21}y^{14}
\end{aligned}
$$

CAUTION

Refer to Example 8(c). Notice that
$$(2x^2y^3)^4 = 2^4x^{2\cdot4}y^{3\cdot4}, \quad \textbf{not} \quad (2\cdot4)x^{2\cdot4}y^{3\cdot4}.$$
Do not multiply the coefficient 2 and the exponent 4.

Work Problem 8 at the Side.

4.2 **EXERCISES**

1. What exponent is understood on the base x in the expression xy^2?

2. How are the expressions 3^2, 5^3, and 7^4 read?

Decide whether each statement is true *or* false.

3. $3^3 = 9$

4. $(-2)^4 = 2^4$

5. $(a^2)^3 = a^5$

6. $\left(\dfrac{1}{4}\right)^2 = \dfrac{1}{4^2}$

Write each expression using exponents. See Example 1.

7. $(-2)(-2)(-2)(-2)(-2)$

8. $w \cdot w \cdot w \cdot w \cdot w \cdot w$

9. $\left(\dfrac{1}{2}\right)\left(\dfrac{1}{2}\right)\left(\dfrac{1}{2}\right)\left(\dfrac{1}{2}\right)\left(\dfrac{1}{2}\right)\left(\dfrac{1}{2}\right)$

10. $\left(-\dfrac{1}{4}\right)\left(-\dfrac{1}{4}\right)\left(-\dfrac{1}{4}\right)\left(-\dfrac{1}{4}\right)\left(-\dfrac{1}{4}\right)$

11. $(-8p)(-8p)$

12. $(-7x)(-7x)(-7x)(-7x)$

13. Explain how the expressions $(-3)^4$ and -3^4 are different.

14. Explain how the expressions $(5x)^3$ and $5x^3$ are different.

Identify the base and the exponent for each exponential expression. In Exercises 15–18, also evaluate the expression. See Example 2.

15. 3^5

16. 2^7

17. $(-3)^5$

18. $(-2)^7$

19. $(-6x)^4$

20. $(-8x)^4$

21. $-6x^4$

22. $-8x^4$

23. Explain why the product rule does not apply to the expression $5^2 + 5^3$. Then evaluate the expression by finding the individual powers and adding the results.

24. Explain why the product rule does not apply to the expression $3^2 \cdot 4^3$. Then evaluate the expression by finding the individual powers and multiplying the results.

Use the product rule to simplify each expression. Write each answer in exponential form. See Examples 3 and 4.

25. $5^2 \cdot 5^6$

26. $3^6 \cdot 3^7$

27. $4^2 \cdot 4^7 \cdot 4^3$

28. $5^3 \cdot 5^8 \cdot 5^2$

29. $(-7)^3(-7)^6$

30. $(-9)^8(-9)^5$

31. $t^3 \cdot t^8 \cdot t^{13}$

32. $n^5 \cdot n^6 \cdot n^9$

33. $(-8r^4)(7r^3)$

34. $(10a^7)(-4a^3)$

35. $(-6p^5)(-7p^5)$

36. $(-5w^8)(-9w^8)$

For each group of terms, first add the given terms. Then start over and multiply them.

37. $5x^4, 9x^4$

38. $8t^5, 3t^5$

39. $-7a^2, 2a^2, 10a^2$

40. $6x^3, 9x^3, -2x^3$

Use the power rules for exponents to simplify each expression. Write each answer in exponential form. See Examples 5–7.

41. $(4^3)^2$

42. $(8^3)^6$

43. $(t^4)^5$

44. $(y^6)^5$

45. $(7r)^3$

46. $(11x)^4$

47. $(5xy)^5$

48. $(9pq)^6$

49. $8(qr)^3$

50. $4(vw)^5$

51. $\left(\dfrac{1}{2}\right)^3$

52. $\left(\dfrac{1}{3}\right)^5$

53. $\left(\dfrac{a}{b}\right)^3 \ (b \neq 0)$

54. $\left(\dfrac{r}{t}\right)^4 \ (t \neq 0)$

55. $\left(\dfrac{9}{5}\right)^8$

56. $\left(\dfrac{12}{7}\right)^3$

57. $(-2x^2y)^3$

58. $(-5m^4p^2)^3$

59. $(3a^3b^2)^2$

60. $(4x^3y^5)^4$

Find the area of each figure. Use the formulas found on the inside covers. (The small squares in the figures indicate 90° right angles.)

61.

$3x^2$

$4x^3$

62.

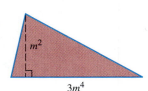

m^2

$3m^4$

Use a combination of the rules for exponents introduced in this section to simplify each expression. See Example 8.

63. $\left(\dfrac{5}{2}\right)^3 \cdot \left(\dfrac{5}{2}\right)^2$

64. $\left(\dfrac{3}{4}\right)^5 \cdot \left(\dfrac{3}{4}\right)^6$

65. $\left(\dfrac{9}{8}\right)^3 \cdot 9^2$

66. $\left(\dfrac{8}{5}\right)^4 \cdot 8^3$

67. $(2x)^9(2x)^3$

68. $(6y)^5(6y)^8$

69. $(-6p)^4(-6p)$

70. $(-13q)^3(-13q)$

71. $(6x^2y^3)^5$

72. $(5r^5t^6)^7$

73. $(x^2)^3(x^3)^5$

74. $(y^4)^5(y^3)^5$

75. $(2w^2x^3y)^2(x^4y)^5$

76. $(3x^4y^2z)^3(yz^4)^5$

77. $(-r^4s)^2(-r^2s^3)^5$

78. $(-ts^6)^4(-t^3s^5)^3$

79. $\left(\dfrac{5a^2b^5}{c^6}\right)^3 \ (c \neq 0)$

80. $\left(\dfrac{6x^3y^9}{z^5}\right)^4 \ (z \neq 0)$

81. $(-5m^3p^4q)^2(p^2q)^3$

82. $(-a^4b^5)(-6a^3b^3)^2$

83. $(2x^2y^3z)^4(xy^2z^3)^2$

4.3 MULTIPLYING POLYNOMIALS

1 **Multiply a monomial and a polynomial.** As shown earlier, we find the product of two monomials by using the rules for exponents and the commutative and associative properties. For example,

$$(-8m^6)(-9n^6) = (-8)(-9)(m^6)(n^6) = 72m^6n^6.$$

CAUTION

> Do not confuse *addition* of terms with *multiplication* of terms. For example,
>
> $$7q^5 + 2q^5 = 9q^5, \quad \text{but} \quad (7q^5)(2q^5) = 7 \cdot 2q^{5+5} = 14q^{10}.$$

To find the product of a monomial and a polynomial with more than one term, we use the distributive property and multiplication of monomials.

Example 1 **Multiplying a Monomial and a Polynomial**

Use the distributive property to find each product.

(a) $4x^2(3x + 5)$

$$4x^2(3x + 5) = 4x^2(3x) + 4x^2(5) \qquad \text{Distributive property}$$
$$= 12x^3 + 20x^2 \qquad \text{Multiply monomials.}$$

(b) $-8m^3(4m^3 + 3m^2 + 2m - 1)$

$$= -8m^3(4m^3) + (-8m^3)(3m^2)$$
$$+ (-8m^3)(2m) + (-8m^3)(-1) \qquad \text{Distributive property}$$
$$= -32m^6 - 24m^5 - 16m^4 + 8m^3 \qquad \text{Multiply monomials.}$$

━━━━━━━━━━━━━ **Work Problem ❶ at the Side.**

2 **Multiply two polynomials.** We can use the distributive property repeatedly to find the product of any two polynomials. For example, to find the product of the polynomials $x^2 + 3x + 5$ and $x - 4$, think of $x - 4$ as a single quantity and use the distributive property as follows.

$$(x^2 + 3x + 5)(x - 4) = x^2(x - 4) + 3x(x - 4) + 5(x - 4)$$

Now use the distributive property three times to find $x^2(x - 4)$, $3x(x - 4)$, and $5(x - 4)$.

$$x^2(x - 4) + 3x(x - 4) + 5(x - 4)$$
$$= x^2(x) + x^2(-4) + 3x(x) + 3x(-4) + 5(x) + 5(-4)$$
$$= x^3 - 4x^2 + 3x^2 - 12x + 5x - 20 \qquad \text{Multiply monomials.}$$
$$= x^3 - x^2 - 7x - 20 \qquad \text{Combine terms.}$$

This example suggests the following rule.

OBJECTIVES

1 Multiply a monomial and a polynomial.

2 Multiply two polynomials.

3 Multiply binomials by the FOIL method.

❶ Find each product.

(a) $5m^3(2m + 7)$

(b) $2x^4(3x^2 + 2x - 5)$

(c)
$-4y^2(3y^3 + 2y^2 - 4y + 8)$

ANSWERS

1. (a) $10m^4 + 35m^3$ **(b)** $6x^6 + 4x^5$
  (c) $-12y^5 - 8y^4 + 16y^3 - 32y^2$

❷ Multiply.

(a) $(m^3 - 2m + 1)$
  $\cdot (2m^2 + 4m + 3)$

Multiplying Polynomials

To multiply two polynomials, multiply each term of the second polynomial by each term of the first polynomial and add the products.

Example 2 — Multiplying Two Polynomials

Multiply $(m^2 + 5)(4m^3 - 2m^2 + 4m)$.

Multiply each term of the second polynomial by each term of the first.

$$(m^2 + 5)(4m^3 - 2m^2 + 4m)$$
$$= m^2(4m^3) + m^2(-2m^2) + m^2(4m) + 5(4m^3) + 5(-2m^2) + 5(4m)$$
$$= 4m^5 - 2m^4 + 4m^3 + 20m^3 - 10m^2 + 20m$$

Now combine like terms.

$$= 4m^5 - 2m^4 + 24m^3 - 10m^2 + 20m$$

Work Problem ❷ at the Side.

When at least one of the factors in a product of polynomials has three or more terms, the multiplication can be simplified by writing one polynomial above the other vertically.

(b) $(6p^2 + 2p - 4)(3p^2 - 5)$

Example 3 — Multiplying Polynomials Vertically

Multiply $(x^3 + 2x^2 + 4x + 1)(3x + 5)$ using the vertical method.

Write the polynomials as follows.

$$\begin{array}{r} x^3 + 2x^2 + 4x + 1 \\ 3x + 5 \\ \hline \end{array}$$

It is not necessary to line up terms in columns, because any terms may be multiplied (not just like terms). Begin by multiplying each of the terms in the top row by 5.

$$\begin{array}{r} x^3 + 2x^2 + 4x + 1 \\ 3x + 5 \\ \hline 5x^3 + 10x^2 + 20x + 5 \quad 5(x^3 + 2x^2 + 4x + 1) \end{array}$$

Notice how this process is similar to multiplication of whole numbers. Now multiply each term in the top row by $3x$. Be careful to place like terms in columns, since the final step will involve addition (as in multiplying two whole numbers).

$$\begin{array}{r} x^3 + 2x^2 + 4x + 1 \\ 3x + 5 \\ \hline 5x^3 + 10x^2 + 20x + 5 \\ 3x^4 + 6x^3 + 12x^2 + \quad 3x \qquad 3x(x^3 + 2x^2 + 4x + 1) \end{array}$$

Continued on Next Page

Add like terms.

$$x^3 + 2x^2 + 4x + 1$$
$$3x + 5$$
$$5x^3 + 10x^2 + 20x + 5$$
$$3x^4 + 6x^3 + 12x^2 + 3x$$
$$3x^4 + 11x^3 + 22x^2 + 23x + 5$$

The product is $3x^4 + 11x^3 + 22x^2 + 23x + 5$.

Work Problem ❸ at the Side.

Example 4 **Multiplying Polynomials Vertically**

Find the product of $4m^3 - 2m^2 + 4m$ and $\frac{1}{2}m^2 + \frac{5}{2}$.

$$4m^3 - 2m^2 + 4m$$
$$\frac{1}{2}m^2 + \frac{5}{2}$$
$$10m^3 - 5m^2 + 10m \qquad \text{Terms of top row multiplied by } \tfrac{5}{2}$$
$$2m^5 - m^4 + 2m^3 \qquad\qquad \text{Terms of top row multiplied by } \tfrac{1}{2}m^2$$
$$2m^5 - m^4 + 12m^3 - 5m^2 + 10m \qquad \text{Add.}$$

Work Problem ❹ at the Side.

We can use a rectangle to model polynomial multiplication. For example, to find the product

$$(2x + 1)(3x + 2),$$

label a rectangle with each term as shown here.

Now put the product of each pair of monomials in the appropriate box.

The product of the original binomials is the sum of these four monomial products.

$$(2x + 1)(3x + 2) = 6x^2 + 4x + 3x + 2$$
$$= 6x^2 + 7x + 2$$

Work Problem ❺ at the Side.

3▭ **Multiply binomials by the FOIL method.** In algebra, many of the polynomials to be multiplied are both binomials (with just two terms). For these products, the **FOIL method** reduces the rectangle method to a systematic approach without the rectangle. To develop the FOIL method, we use the distributive property to find $(x + 3)(x + 5)$.

$$(x + 3)(x + 5) = (x + 3)x + (x + 3)5$$
$$= x(x) + 3(x) + x(5) + 3(5)$$
$$= x^2 + 3x + 5x + 15$$
$$= x^2 + 8x + 15$$

---

❸ Find the product.

$$3x^2 + 4x - 5$$
$$x + 4$$

❹ Find each product.

(a) $k^3 - k^2 + k + 1$
$$\frac{2}{3}k - \frac{1}{3}$$

(b) $a^3 + 3a - 4$
$$2a^2 + 6a + 5$$

❺ Use the rectangle method to find each product.

(a) $(4x + 3)(x + 2)$

(b) $(x + 5)(x^2 + 3x + 1)$

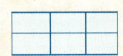

6 For the product $(2p - 5)(3p + 7)$, find the following.

(a) Product of first terms

(b) Outer product

(c) Inner product

(d) Product of last terms

(e) Complete product in simplified form

Here is where the letters of the word FOIL originate.

$(x + 3)(x + 5)$ Multiply the **First** terms: $x(x)$.   **F**

$(x + 3)(x + 5)$ Multiply the **Outer** terms: $x(5)$.   **O**
This is the **outer product**.

$(x + 3)(x + 5)$ Multiply the **Inner** terms: $3(x)$.   **I**
This is the **inner product**.

$(x + 3)(x + 5)$ Multiply the **Last terms**: $3(5)$.   **L**

The inner product and the outer product should be added mentally so that the three terms of the answer can be written without extra steps as

$$(x + 3)(x + 5) = x^2 + 8x + 15.$$

Work Problem 6 at the Side.

A summary of the steps in the FOIL method follows.

Multiplying Binomials by the FOIL Method

Step 1   Multiply the two **F**irst terms of the binomials to get the first term of the answer.

Step 2   Find the **O**uter product and the **I**nner product and add them (when possible) to get the middle term of the answer.

Step 3   Multiply the two **L**ast terms of the binomials to get the last term of the answer.

$$\mathbf{F} = x^2 \qquad \mathbf{L} = 15$$
$$(x + 3)(x + 5)$$
$$\mathbf{I} = 3x$$
$$\mathbf{O} = 5x$$
$$8x \qquad \text{Add.}$$

Example 5   Using the FOIL Method

Use the FOIL method to find the product $(x + 8)(x - 6)$.

Step 1   **F**   Multiply the **first** terms.

$$x(x) = x^2$$

Step 2   **O**   Find the **outer** product.

$$x(-6) = -6x$$

      I   Find the **inner** product.

$$8(x) = 8x$$

Add the outer and inner products mentally.

$$-6x + 8x = 2x$$

Step 3   **L**   Multiply the **last** terms.

$$8(-6) = -48$$

Continued on Next Page

The product of $x + 8$ and $x - 6$ is the sum of the terms found in the three steps above, so

$$(x + 8)(x - 6) = x^2 + 2x - 48.$$

As a shortcut, this product can be found in the following manner.

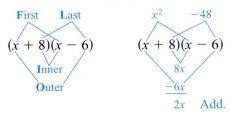

Work Problem **7** at the Side.

It is not possible to add the inner and outer products of the FOIL method if unlike terms result, as shown in the next example.

Example 6 **Using the FOIL Method**

Multiply $(9x - 2)(3y + 1)$.

First	$(\mathbf{9x} - 2)(\mathbf{3y} + 1)$	$\mathbf{27xy}$
Outer	$(\mathbf{9x} - 2)(3y + \mathbf{1})$	$\mathbf{9x}$ ⎤ Unlike terms
Inner	$(9x - \mathbf{2})(\mathbf{3y} + 1)$	$\mathbf{-6y}$ ⎦
Last	$(9x - \mathbf{2})(3y + \mathbf{1})$	-2

$$\mathbf{F}\mathbf{O}\mathbf{I}\mathbf{L}$$
$$(9x - 2)(3y + 1) = 27xy + 9x - 6y - 2$$

Example 7 **Using the FOIL Method**

Find each product.

$$\mathbf{F}\mathbf{O}\mathbf{I}\mathbf{L}$$

(a) $(2k + 5y)(k + 3y) = 2k(k) + 2k(3y) + 5y(k) + 5y(3y)$
$$= 2k^2 + 6ky + 5ky + 15y^2$$
$$= 2k^2 + 11ky + 15y^2$$

(b) $(7p + 2q)(3p - q) = 21p^2 - pq - 2q^2$ FOIL

(c) $2x^2(x - 3)(3x + 4) = 2x^2(3x^2 - 5x - 12)$ FOIL
$$= 6x^4 - 10x^3 - 24x^2 \quad \text{Distributive property}$$

Work Problem **8** at the Side.

NOTE

Example 7(c) showed one way to multiply three polynomials. We could have multiplied $2x^2$ and $x - 3$ first, then multiplied that product and $3x + 4$ as follows.

$$2x^2(x - 3)(3x + 4) = (2x^3 - 6x^2)(3x + 4) = 6x^4 - 10x^3 - 24x^2$$

---

7 Use the FOIL method to find each product.

(a) $(m + 4)(m - 3)$

(b) $(y + 7)(y + 2)$

(c) $(r - 8)(r - 5)$

8 Find each product.

(a) $(4k - 1)(2k + 3)$

(b) $(6m + 5)(m - 4)$

(c) $(3r + 2t)(3r + 4t)$

(d) $y^2(8y + 3)(2y + 1)$

Real-Data Applications

Algebra as Generalized Arithmetic

The rules of algebra are consistent with those for arithmetic. To learn a method for multiplying binomial factors, such as $(3x + 2)(4x + 1)$, we can observe the method for multiplying two-digit numbers, such as $32 \cdot 41$. The number 32 is shorthand for the expanded number $3 \cdot 10 + 2$, and the number 41 is shorthand for $4 \cdot 10 + 1$. So, multiplying $32 \cdot 41$ is the same as multiplying $(3 \cdot 10 + 2)(4 \cdot 10 + 1)$.

1. An expanded version of the usual algorithm is shown in the first column below. Each partial product, such as 1×2 and 40×30, is shown to clarify how it contributes to the process.

2. In the standard algorithm, it is clear that the 32 represents the sum of 1×2 and 1×30, and 1280 represents the sum of 40×2 and 40×30.

3. When FOIL is used to multiply the numbers, the term $11 \cdot 10$ represents the sum of the partial products 1×30 and 40×2. When we simplify $12 \cdot 10^2 + 11 \cdot 10 + 2$, the result is 1312.

4. When FOIL is used to multiply the binomials, each term exactly matches the corresponding term from the multiplication of the numbers.

Partial Products Multiplication Algorithm	Standard Algorithm	FOIL Method for Numbers	FOIL Method for Variables
32	32	$(3 \cdot 10 + 2)(4 \cdot 10 + 1)$	$(3x + 2)(4x + 1)$
$\times\, 41$	$\times\, 41$	$12 \cdot 10^2 + 3 \cdot 10 + 8 \cdot 10 + 2$	$12x^2 + 3x + 8x + 2$
$2 = 1 \times 2$	32	$12 \cdot 10^2 + 11 \cdot 10 + 2$	$12x^2 + 11x + 2$
$30 = 1 \times 30$	1280	$1200 + 110 + 2$	
$80 = 40 \times 2$	1312	1312	
$1200 = 40 \times 30$			
1312			

For Group Discussion

Use FOIL to compute each binomial product and corresponding arithmetic product. Verify that the results of the arithmetic product are valid. For the numerical problems, write the correct *signed* product for each term.

1. $(2x + 1)(2x - 3)$ and $(2 \cdot 10 + 1)(2 \cdot 10 - 3)$, which is $21 \cdot 17$
   Does the arithmetic FOIL result simplify to the correct answer?

2. $(4x - 2)(2x + 5)$ and $(4 \cdot 10 - 2)(2 \cdot 10 + 5)$, which is $38 \cdot 25$
   Does the arithmetic FOIL result simplify to the correct answer?

3. $(7x - 3)(5x - 4)$ and $(7 \cdot 10 - 3)(5 \cdot 10 - 4)$, which is $67 \cdot 46$
   Does the arithmetic FOIL result simplify to the correct answer?

4. A mental trick for multiplying two-digit numbers, such as $27 \cdot 18$, follows: "Multiply the ones $(7 \times 8 = 56)$. Write the 6 in the ones place, and carry the 5. Add the inner product (7×1), the outer product (2×8), and the carried 5 $(7 + 16 + 5 = 28)$. Write the 8 in the tens place, and carry the 2. Multiply the tens (2×1), and add to the carried 2. Write 4 in the hundreds place. The answer is 486." Why does the trick work?

4.3 **EXERCISES**

Find each product using the rectangle method shown in the text.

1. $(x + 3)(x + 4)$

2. $(x + 5)(x + 2)$

3. $(2x + 1)(x^2 + 3x + 2)$

4. $(x + 4)(3x^2 + 2x + 1)$

5. In multiplying a monomial by a polynomial, such as in $4x(3x^2 + 7x^3) = 4x(3x^2) + 4x(7x^3)$, the first property that is used is the _____ property.

6. Match each product in parts (a)–(d) with the correct polynomial in choices A–D.

   (a) $(x - 5)(x + 3)$ **(b)** $(x + 5)(x + 3)$ **(c)** $(x - 5)(x - 3)$ **(d)** $(x + 5)(x - 3)$

   A. $x^2 + 8x + 15$ **B.** $x^2 - 8x + 15$ **C.** $x^2 - 2x - 15$ **D.** $x^2 + 2x - 15$

Find each product. See Example 1.

7. $-2m(3m + 2)$

8. $-5p(6 + 3p)$

9. $\dfrac{3}{4}p(8 - 6p + 12p^3)$

10. $\dfrac{4}{3}x(3 + 2x + 5x^3)$

11. $2y^5(3 + 2y + 5y^4)$

12. $2m^4(3m^2 + 5m + 6)$

Find each product. See Examples 2–4.

13. $(6x + 1)(2x^2 + 4x + 1)$

14. $(9y - 2)(8y^2 - 6y + 1)$

15. $(4m + 3)(5m^3 - 4m^2 + m - 5)$

16. $(y + 4)(3y^3 - 2y^2 + y + 3)$

17. $(2x - 1)(3x^5 - 2x^3 + x^2 - 2x + 3)$

18. $(2a + 3)(a^4 - a^3 + a^2 - a + 1)$

19. $(5x^2 + 2x + 1)(x^2 - 3x + 5)$

20. $(2m^2 + m - 3)(m^2 - 4m + 5)$

Find each binomial product using the FOIL method. See Examples 5–7.

21. $(n - 2)(n + 3)$

22. $(r - 6)(r + 8)$

23. $(4r + 1)(2r - 3)$

24. $(5x + 2)(2x - 7)$

25. $(3x + 2)(3x - 2)$

26. $(7x + 3)(7x - 3)$

27. $(3q + 1)(3q + 1)$ **28.** $(4w + 7)(4w + 7)$ **29.** $(3t + 4s)(2t + 5s)$

30. $(8v + 5w)(2v + 3w)$ **31.** $(-.3t + .4)(t + .6)$ **32.** $(-.5x + .9)(x - .2)$

33. $\left(x - \dfrac{2}{3}\right)\left(x + \dfrac{1}{4}\right)$ **34.** $\left(-\dfrac{8}{3} + 3k\right)\left(-\dfrac{2}{3} - k\right)$ **35.** $\left(-\dfrac{5}{4} + 2r\right)\left(-\dfrac{3}{4} - r\right)$

36. $2m^3(4m - 1)(2m + 3)$ **37.** $3y^3(2y + 3)(y - 5)$ **38.** $5t^4(t + 3)(3t - 1)$

RELATING CONCEPTS (Exercises 39–44)   **FOR INDIVIDUAL OR GROUP WORK**

Work Exercises 39–44 in order. *(All units are in yards.) Refer to the figure as necessary.*

39. Find a polynomial that represents the area of the rectangle.

40. Suppose you know that the area of the rectangle is 600 yd². Use this information and the polynomial from Exercise 39 to write an equation in x, and solve it.

41. What are the dimensions of the rectangle?

42. Suppose the rectangle represents a lawn and it costs $3.50 per square yard to lay sod on the lawn. How much will it cost to sod the entire lawn?

43. Use the result of Exercise 41 to find the perimeter of the lawn.

44. Again, suppose the rectangle represents a lawn and it costs $9.00 per yard to fence the lawn. How much will it cost to fence the lawn?

---

45. Perform the following multiplications: $(x + 4)(x - 4)$; $(y + 2)(y - 2)$; $(r + 7)(r - 7)$. Observe your answers, and explain the pattern that can be found in the answers.

46. Repeat Exercise 45 for the following: $(x + 4)(x + 4)$; $(y - 2)(y - 2)$; $(r + 7)(r + 7)$.

4.4 SPECIAL PRODUCTS

In this section, we develop shortcuts to find certain binomial products that occur frequently.

1   **Square binomials.**   The square of a binomial can be found quickly by using the method shown in Example 1.

OBJECTIVES

1   Square binomials.

2   Find the product of the sum and difference of two terms.

3   Find higher powers of binomials.

Example 1 **Squaring a Binomial**

Find $(m + 3)^2$.
    Squaring $m + 3$ by the FOIL method gives

$$(m + 3)(m + 3) = m^2 + 3m + 3m + 9$$
$$= m^2 + 6m + 9.$$

➊ Consider the binomial $x + 4$.

(a) What is the first term of the binomial? Square it.

The result has the squares of the first and the last terms of the binomial:

$$m^2 = m^2 \quad \text{and} \quad 3^2 = 9.$$

The middle term is twice the product of the two terms of the binomial, since the outer and inner products are $m(3)$ and $3(m)$, and

$$m(3) + 3(m) = 2(m)(3) = 6m.$$

Work Problem ➊ at the Side.

This example suggests the following rules.

(b) What is the last term of the binomial? Square it.

Square of a Binomial

The square of a binomial is a trinomial consisting of the square of the first term, plus twice the product of the two terms, plus the square of the last term of the binomial. For a and b,

$$(a + b)^2 = a^2 + 2ab + b^2.$$

Also, $(a - b)^2 = a^2 - 2ab + b^2.$

(c) Find twice the product of the two terms of the binomial.

Example 2 **Squaring Binomials**

Use the rules to square each binomial.

$$(a - b)^2 = a^2 - 2 \cdot a \cdot b + b^2$$

(a) $(5z - 1)^2 = (5z)^2 - 2(5z)(1) + (1)^2$

$$= 25z^2 - 10z + 1 \qquad (5z)^2 = 5^2 z^2 = 25z^2$$

(b) $(3b + 5r)^2 = (3b)^2 + 2(3b)(5r) + (5r)^2$

$$= 9b^2 + 30br + 25r^2$$

(c) $(2a - 9x)^2 = 4a^2 - 36ax + 81x^2$

(d) $\left(4m + \dfrac{1}{2}\right)^2 = (4m)^2 + 2(4m)\left(\dfrac{1}{2}\right) + \left(\dfrac{1}{2}\right)^2$

$$= 16m^2 + 4m + \dfrac{1}{4}$$

(d) Find $(x + 4)^2$.

② Find each square by using the rules for the square of a binomial.

(a) $(t + u)^2$

Notice that in the square of a sum all of the terms are positive, as in Examples 2(b) and (d). In the square of a difference, the middle term is negative, as in Examples 2(a) and (c).

CAUTION

A common error when squaring a binomial is to forget the middle term of the product. In general,

$$(a + b)^2 \neq a^2 + b^2.$$

Work Problem ② at the Side.

(b) $(2m - p)^2$

2 ▭ **Find the product of the sum and difference of two terms.** Binomial products of the form $(a + b)(a - b)$ also occur frequently. In these products, one binomial is the sum of two terms, and the other is the difference of the same two terms. For example, the product of $x + 2$ and $x - 2$ is

$$(x + 2)(x - 2) = x^2 - 2x + 2x - 4$$
$$= x^2 - 4.$$

As this example suggests, the product of $a + b$ and $a - b$ is the difference between two squares.

(c) $(4p + 3q)^2$

Product of the Sum and Difference of Two Terms

$$(a + b)(a - b) = a^2 - b^2$$

Example 3 · Finding the Product of the Sum and Difference of Two Terms

Find each product.

(d) $(5r - 6s)^2$

(a) $(x + 4)(x - 4)$

Use the rule for the product of the sum and difference of two terms.

$$(x + 4)(x - 4) = x^2 - 4^2 = x^2 - 16$$

(b) $\left(\dfrac{2}{3} - w\right)\left(\dfrac{2}{3} + w\right)$

By the commutative property, this product is the same as $\left(\dfrac{2}{3} + w\right)\left(\dfrac{2}{3} - w\right)$.

(e) $\left(3k - \dfrac{1}{2}\right)^2$

$$\left(\dfrac{2}{3} - w\right)\left(\dfrac{2}{3} + w\right) = \left(\dfrac{2}{3} + w\right)\left(\dfrac{2}{3} - w\right) = \left(\dfrac{2}{3}\right)^2 - w^2 = \dfrac{4}{9} - w^2$$

Example 4 Finding the Product of the Sum and Difference of Two Terms

Find each product.

$$\begin{array}{cccc} (a & + \; b) & (a & - \; b) \\ \downarrow & \downarrow & \downarrow & \downarrow \end{array}$$

(a) $(5m + 3)(5m - 3)$

Use the rule for the product of the sum and difference of two terms.

$$(5m + 3)(5m - 3) = (5m)^2 - 3^2$$
$$= 25m^2 - 9$$

Continued on Next Page

Continued on Next Page

ANSWERS

2. **(a)** $t^2 + 2tu + u^2$
   (b) $4m^2 - 4mp + p^2$
   (c) $16p^2 + 24pq + 9q^2$
   (d) $25r^2 - 60rs + 36s^2$
   (e) $9k^2 - 3k + \dfrac{1}{4}$

(b) $(4x + y)(4x - y) = (4x)^2 - y^2$

$$= 16x^2 - y^2$$

(c) $\left(z - \dfrac{1}{4}\right)\left(z + \dfrac{1}{4}\right) = z^2 - \dfrac{1}{16}$

Work Problem ❸ at the Side.

The product rules of this section will be important later, particularly in Chapters 5 and 6. Therefore, it is important to memorize these rules and practice using them.

3 ▭ **Find higher powers of binomials.** The methods used in the previous section and this section can be combined to find higher powers of binomials.

Example 5 **Finding Higher Powers of Binomials**

Find each product.

(a) $(x + 5)^3 = (x + 5)^2(x + 5)$ $a^3 = a^2 \cdot a$

$$= (x^2 + 10x + 25)(x + 5)$$ Square the binomial.

$$= x^3 + 10x^2 + 25x + 5x^2 + 50x + 125$$ Multiply polynomials.

$$= x^3 + 15x^2 + 75x + 125$$ Combine like terms.

(b) $(2y - 3)^4 = (2y - 3)^2(2y - 3)^2$ $a^4 = a^2 \cdot a^2$

$$= (4y^2 - 12y + 9)(4y^2 - 12y + 9)$$ Square each binomial.

$$= 16y^4 - 48y^3 + 36y^2 - 48y^3 + 144y^2$$ Multiply polynomials.
$$- 108y + 36y^2 - 108y + 81$$

$$= 16y^4 - 96y^3 + 216y^2 - 216y + 81$$ Combine like terms.

Work Problem ❹ at the Side.

❸ Find each product by using the rule for the sum and difference of two terms.

(a) $(6a + 3)(6a - 3)$

(b) $(10m + 7)(10m - 7)$

(c) $(7p + 2q)(7p - 2q)$

(d) $\left(3r - \dfrac{1}{2}\right)\left(3r + \dfrac{1}{2}\right)$

❹ Find each product.

(a) $(m + 1)^3$

(b) $(3k - 2)^4$

Real-Data Applications

Using a Rule of Thumb

A Dynamic Homes neighborhood features variations of three models of houses. Each buyer has an option to purchase a concrete patio extension, constructed to his or her choice of size. Each of the three models is designed so that the patio will be in the shape of a large square that is missing a corner square, similar to the diagram shown on the left. The size of the removed corner and the length of the patio vary from model to model, as does the size of the entire patio. To determine the dimensions of the patio, the rectangle of size $b \times (a - b)$ in the middle patio diagram can be rotated and repositioned to form the rectangle shown on the right, with length $a + b$ and width $a - b$.

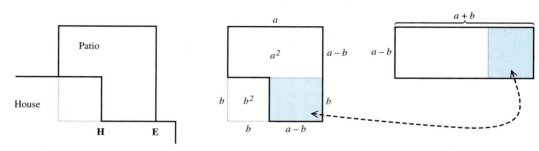

Since every time a house is sold the salesperson must ask the construction supervisor to calculate the cost of the patio extension, the supervisor devises a *rule of thumb* based on the quantity $(a + b)(a - b)$. There are only three possible choices for the value of b, one for each model house. This means that the salesperson needs to know only one additional measurement—the distance from the corner H to the edge of the patio E, which is $a - b$. The thickness of the patio is $\frac{1}{3}$ ft. The cost of concrete is $58.00 per yd^3. (*Source:* Dunham Price, Inc., Westlake, LA.) There is a 100% markup in price to cover profit as well as costs for ground preparation, building forms, and finishing the concrete, giving a cost factor of $116.00 per yd^3.

For Group Discussion

1. Why is knowing the values for b and $a - b$ sufficient for finding the quantity $a + b$?

2. Explain why the area of the patio can be written both as $(a + b)(a - b)$ and $a^2 - b^2$.

3. Apply the construction supervisor's rule of thumb to the three examples given in the table.

   To calculate the total cost for a patio with long side of length a ft, use the table and work from left to right. The second-to-last column is the volume of concrete needed, and the last column is the total cost of the patio. Round to the next tenth.

House Model	Corner Length b ft	Corner Length Doubled $2b$	Distance from Corner H to Patio Edge E $a - b$	Add Previous Two Columns $2b + (a - b)$	Multiply Previous Two Columns $(a + b)(a - b)$	VOLUME Multiply Previous Column by $\frac{1}{3}$ and Divide by 27	COST Multiply Previous Column by Cost Factor of $116.00
A	5		10				
B	6		12				
C	8		14				

4.4 EXERCISES

1. Consider the square $(2x + 3)^2$.

   (a) What is the square of the first term, $(2x)^2$? _____

   (b) What is twice the product of the two terms, $2(2x)(3)$? _____

   (c) What is the square of the last term, 3^2? _____

   (d) Write the final product, which is a trinomial, using your results from parts (a)–(c). _____

2. Repeat Exercise 1 for the square $(3x - 2)^2$.

Find each square. See Examples 1 and 2.

3. $(a - c)^2$

4. $(p - y)^2$

5. $(p + 2)^2$

6. $(r + 5)^2$

7. $(4x - 3)^2$

8. $(5y + 2)^2$

9. $(.8t + .7s)^2$

10. $(.7z - .3w)^2$

11. $\left(5x + \dfrac{2}{5}y\right)^2$

12. $\left(6m - \dfrac{4}{5}n\right)^2$

13. $\left(4a - \dfrac{3}{2}b\right)^2$

14. $x(2x + 5)^2$

15. $-(4r - 2)^2$

16. $-(3y - 8)^2$

17. Consider the product $(7x + 3y)(7x - 3y)$.

   (a) What is the product of the first terms, $7x(7x)$? _____

   (b) Multiply the outer terms, $7x(-3y)$. Then multiply the inner terms, $3y(7x)$. Add the results. What is this sum? _____

   (c) What is the product of the last terms, $3y(-3y)$? _____

   (d) Write the complete product using your answers in parts (a) and (c). _____ Why is the sum found in part (b) omitted here?

18. Repeat Exercise 17 for the product $(5x + 7y)(5x - 7y)$.

Find each product. See Examples 3 and 4.

19. $(q + 2)(q - 2)$

20. $(x + 8)(x - 8)$

21. $(2w + 5)(2w - 5)$

22. $(3z + 8)(3z - 8)$

23. $(10x + 3y)(10x - 3y)$

24. $(13r + 2z)(13r - 2z)$

25. $(2x^2 - 5)(2x^2 + 5)$

26. $(9y^2 - 2)(9y^2 + 2)$

27. $\left(7x + \dfrac{3}{7}\right)\left(7x - \dfrac{3}{7}\right)$

28. $\left(9y + \dfrac{2}{3}\right)\left(9y - \dfrac{2}{3}\right)$

29. $p(3p + 7)(3p - 7)$

30. $q(5q - 1)(5q + 1)$

RELATING CONCEPTS (Exercises 31–40) **FOR INDIVIDUAL OR GROUP WORK**

*Special products can be illustrated by using areas of rectangles. Use the figure and **work Exercises 31–36 in order** to justify the special product $(a + b)^2 = a^2 + 2ab + b^2$.*

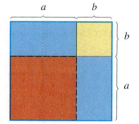

31. Express the area of the large square as the square of a binomial.

32. Give the monomial that represents the area of the red square.

33. Give the monomial that represents the sum of the areas of the blue rectangles.

34. Give the monomial that represents the area of the yellow square.

35. What is the sum of the monomials you obtained in Exercises 32–34?

36. Explain why the binomial square you found in Exercise 31 must equal the polynomial you found in Exercise 35.

*To understand how the special product $(a + b)^2 = a^2 + 2ab + b^2$ can be applied to a purely numerical problem, **work Exercises 37–40 in order.***

37. Evaluate 35^2 using either traditional paper-and-pencil methods or a calculator.

38. The number 35 can be written as $30 + 5$. Therefore, $35^2 = (30 + 5)^2$. Use the special product for squaring a binomial with $a = 30$ and $b = 5$ to write an expression for $(30 + 5)^2$. Do not simplify at this time.

39. Use the order of operations to simplify the expression you found in Exercise 38.

40. How do the answers in Exercises 37 and 39 compare?

Find each product. See Example 5.

41. $(m - 5)^3$

42. $(p + 3)^3$

43. $(2a + 1)^3$

44. $(3m - 1)^3$

45. $(3r - 2t)^4$

46. $(2z + 5y)^4$

47. Explain how the expressions $x^2 + y^2$ and $(x + y)^2$ differ.

48. Does $a^3 + b^3$ equal $(a + b)^3$? Explain your answer.

In Exercises 49 and 50, refer to the figure shown here.

49. Find a polynomial that represents the volume of the cube.

50. If the value of x is 6, what is the volume of the cube?

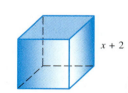

$x + 2$

4.5 INTEGER EXPONENTS AND THE QUOTIENT RULE

In Section 4.2 we studied the product rule for exponents. In all our earlier work, exponents were positive integers. Now we want to develop meaning for exponents that are not positive integers.

Consider the following list of exponential expressions.

$$2^4 = 16$$
$$2^3 = 8$$
$$2^2 = 4$$

Do you see the pattern in the values? Each time we reduce the exponent by 1, the value is divided by 2 (the base). Using this pattern, we can continue the list to smaller and smaller integer exponents.

$$2^1 = 2$$
$$2^0 = 1$$
$$2^{-1} = \frac{1}{2}$$

Work Problem ❶ at the Side.

From the preceding list and the answers to Problem 1, it appears that we should define 2^0 as 1 and negative exponents as reciprocals.

1 ◻◻ **Use 0 as an exponent.**   We want the definitions of 0 and negative exponents to satisfy the rules for exponents from Section 4.2. For example, if $6^0 = 1$,

$$6^0 \cdot 6^2 = 1 \cdot 6^2 = 6^2 \quad \text{and} \quad 6^0 \cdot 6^2 = 6^{0+2} = 6^2,$$

so the product rule is satisfied. Check that the power rules are also valid for a 0 exponent. Thus, we define a 0 exponent as follows.

Zero Exponent

For any nonzero real number a, $a^0 = 1$.

Example: $17^0 = 1$.

Example 1   Using Zero Exponents

Evaluate each exponential expression.

(a) $60^0 = 1$

(b) $(-60)^0 = 1$

(c) $-60^0 = -(1) = -1$

(d) $y^0 = 1$, if $y \neq 0$

(e) $6y^0 = 6(1) = 6$, if $y \neq 0$

(f) $(6y)^0 = 1$, if $y \neq 0$

CAUTION

Notice the difference between parts (b) and (c) of Example 1. In Example 1(b) the base is -60 and the exponent is 0. Any nonzero base raised to the exponent zero is 1. But in Example 1(c), the base is 60. Then $60^0 = 1$, and $-60^0 = -1$.

OBJECTIVES

1 ◻◻ Use 0 as an exponent.

2 ◻◻ Use negative numbers as exponents.

3 ◻◻ Use the quotient rule for exponents.

4 ◻◻ Use combinations of rules.

❶ Continue the list of exponentials using -2, -3, and -4 as exponents.

$$2^{-2} = \underline{\qquad}$$

$$2^{-3} = \underline{\qquad}$$

$$2^{-4} = \underline{\qquad}$$

ANSWERS

1. $2^{-2} = \dfrac{1}{4}$; $2^{-3} = \dfrac{1}{8}$; $2^{-4} = \dfrac{1}{16}$

2 Evaluate.

(a) 28^0

(b) $(-16)^0$

(c) -7^0

(d) $m^0, m \neq 0$

(e) $-p^0, p \neq 0$

Work Problem 2 at the Side.

2 Use negative numbers as exponents. From the lists at the beginning of this section and margin problem 1, since $2^{-2} = \frac{1}{4}$ and $2^{-3} = \frac{1}{8}$, we can deduce that 2^{-n} should equal $\frac{1}{2^n}$. Is the product rule valid in such cases? For example, if we multiply 6^{-2} by 6^2, we get

$$6^{-2} \cdot 6^2 = 6^{-2+2} = 6^0 = 1.$$

The expression 6^{-2} behaves as if it were the reciprocal of 6^2, because their product is 1. The reciprocal of 6^2 may be written $\frac{1}{6^2}$, leading us to define 6^{-2} as $\frac{1}{6^2}$. This is a particular case of the definition of negative exponents.

Negative Exponents

For any nonzero real number a and any integer n, $\quad a^{-n} = \dfrac{1}{a^n}.$

Example: $\quad 3^{-2} = \dfrac{1}{3^2}.$

By definition, a^{-n} and a^n are reciprocals, since

$$a^n \cdot a^{-n} = a^n \cdot \frac{1}{a^n} = 1.$$

Since $1^n = 1$, the definition of a^{-n} can also be written

$$a^{-n} = \frac{1}{a^n} = \frac{1^n}{a^n} = \left(\frac{1}{a}\right)^n.$$

For example,

$$6^{-3} = \left(\frac{1}{6}\right)^3 \quad \text{and} \quad \left(\frac{1}{3}\right)^{-2} = 3^2.$$

Example 2 Using Negative Exponents

Simplify by writing each expression with positive exponents.

(a) $3^{-2} = \dfrac{1}{3^2} = \dfrac{1}{9}$

(b) $5^{-3} = \dfrac{1}{5^3} = \dfrac{1}{125}$

(c) $\left(\dfrac{1}{2}\right)^{-3} = 2^3 = 8 \quad \frac{1}{2}$ and 2 are reciprocals.

Notice that we can change the base to its reciprocal if we also change the sign of the exponent.

(d) $\left(\dfrac{2}{5}\right)^{-4} = \left(\dfrac{5}{2}\right)^4 \quad \frac{2}{5}$ and $\frac{5}{2}$ are reciprocals.

(e) $\left(\dfrac{4}{3}\right)^{-5} = \left(\dfrac{3}{4}\right)^5$

(f) $4^{-1} - 2^{-1} = \dfrac{1}{4} - \dfrac{1}{2} = \dfrac{1}{4} - \dfrac{2}{4} = -\dfrac{1}{4}$

Apply the exponents first, then subtract.

Continued on Next Page

(g) $p^{-2} = \dfrac{1}{p^2}, p \neq 0$

(h) $\dfrac{1}{x^{-4}}, x \neq 0$

$$\dfrac{1}{x^{-4}} = \dfrac{1^{-4}}{x^{-4}} \qquad 1^{-4} = 1$$

$$= \left(\dfrac{1}{x}\right)^{-4} \qquad \text{Power rule (c)}$$

$$= x^4 \qquad \tfrac{1}{x} \text{ and } x \text{ are reciprocals.}$$

CAUTION

A negative exponent does not indicate a negative number; negative exponents lead to reciprocals.

Expression	**Example**	
a^{-n}	$3^{-2} = \dfrac{1}{3^2} = \dfrac{1}{9}$	Not negative
$-a^{-n}$	$-3^{-2} = -\dfrac{1}{3^2} = -\dfrac{1}{9}$	Negative

Work Problem ❸ at the Side.

The definition of negative exponents allows us to move factors in a fraction if we also change the signs of the exponents. For example,

$$\dfrac{2^{-3}}{3^{-4}} = \dfrac{\dfrac{1}{2^3}}{\dfrac{1}{3^4}} = \dfrac{1}{2^3} \cdot \dfrac{3^4}{1} = \dfrac{3^4}{2^3},$$

so that

$$\dfrac{2^{-3}}{3^{-4}} = \dfrac{3^4}{2^3}.$$

Changing from Negative to Positive Exponents

For any nonzero numbers a and b, and any integers m and n,

$$\dfrac{a^{-m}}{b^{-n}} = \dfrac{b^n}{a^m} \quad \text{and} \quad \left(\dfrac{a}{b}\right)^{-m} = \left(\dfrac{b}{a}\right)^m.$$

Examples: $\dfrac{3^{-5}}{2^{-4}} = \dfrac{2^4}{3^5}$ and $\left(\dfrac{4}{5}\right)^{-3} = \left(\dfrac{5}{4}\right)^3.$

---

❸ Write with positive exponents.

(a) 4^{-3}

(b) 6^{-2}

(c) $\left(\dfrac{2}{3}\right)^{-2}$

(d) $2^{-1} + 5^{-1}$

(e) $m^{-5}, m \neq 0$

(f) $\dfrac{1}{z^{-4}}, z \neq 0$

④ Write with only positive exponents. Assume all variables represent nonzero real numbers.

(a) $\dfrac{7^{-1}}{5^{-4}}$

(b) $\dfrac{x^{-3}}{y^{-2}}$

(c) $\dfrac{4h^{-5}}{m^{-2}k}$

(d) $p^2 q^{-5}$

(e) $\left(\dfrac{3m}{p}\right)^{-2}$

Example 3 Changing from Negative to Positive Exponents

Write with only positive exponents. Assume all variables represent nonzero real numbers.

(a) $\dfrac{4^{-2}}{5^{-3}} = \dfrac{5^3}{4^2}$

(b) $\dfrac{m^{-5}}{p^{-1}} = \dfrac{p^1}{m^5} = \dfrac{p}{m^5}$

(c) $\dfrac{a^{-2}b}{3d^{-3}} = \dfrac{bd^3}{3a^2}$

Notice that b in the numerator and 3 in the denominator were not affected.

(d) $x^3 y^{-4} = \dfrac{x^3 y^{-4}}{1} = \dfrac{x^3}{y^4}$

(e) $\left(\dfrac{x}{2y}\right)^{-4} = \left(\dfrac{2y}{x}\right)^4 = \dfrac{2^4 y^4}{x^4}$

Work Problem ④ at the Side.

CAUTION

Be careful. We cannot change negative exponents to positive exponents using this rule if the exponents occur in a sum of terms. For example,

$$\dfrac{5^{-2} + 3^{-1}}{7 - 2^{-3}}$$

cannot be written with positive exponents using the rule given here. We would have to use the definition of a negative exponent to rewrite this expression with positive exponents, as

$$\dfrac{\dfrac{1}{5^2} + \dfrac{1}{3}}{7 - \dfrac{1}{2^3}}.$$

3 ▭ **Use the quotient rule for exponents.** What about the quotient of two exponential expressions with the same base? We know that

$$\dfrac{6^5}{6^3} = \dfrac{6 \cdot 6 \cdot 6 \cdot 6 \cdot 6}{6 \cdot 6 \cdot 6} = 6^2.$$

Notice that the difference between the exponents, $5 - 3 = 2$, is the exponent in the quotient. Also,

$$\dfrac{6^2}{6^4} = \dfrac{6 \cdot 6}{6 \cdot 6 \cdot 6 \cdot 6} = \dfrac{1}{6^2} = 6^{-2}.$$

Here, $2 - 4 = -2$. These examples suggest the quotient rule for exponents.

Quotient Rule for Exponents

For any nonzero real number a and any integers m and n,

$$\frac{a^m}{a^n} = a^{m-n}.$$

(Keep the base and subtract the exponents.)

Example: $\dfrac{5^8}{5^4} = 5^{8-4} = 5^4$.

CAUTION

A common **error** is to write $\frac{5^8}{5^4} = 1^{8-4} = 1^4$. Notice that by the quotient rule, the quotient should have the *same base*, 5. That is,

$$\frac{5^8}{5^4} = 5^{8-4} = 5^4.$$

If you are not sure, use the definition of an exponent to write out the factors:

$$5^8 = 5 \cdot 5 \cdot 5 \cdot 5 \cdot 5 \cdot 5 \cdot 5 \cdot 5 \quad \text{and} \quad 5^4 = 5 \cdot 5 \cdot 5 \cdot 5.$$

Then it is clear that the quotient is 5^4.

Example 4 Using the Quotient Rule for Exponents

Simplify, using the quotient rule for exponents. Write answers with positive exponents.

(a) $\dfrac{5^8}{5^6} = 5^{8-6} = 5^2$

(b) $\dfrac{4^2}{4^9} = 4^{2-9} = 4^{-7} = \dfrac{1}{4^7}$

(c) $\dfrac{5^{-3}}{5^{-7}} = 5^{-3-(-7)} = 5^4$

(d) $\dfrac{q^5}{q^{-3}} = q^{5-(-3)} = q^8, q \neq 0$

(e) $\dfrac{3^2 x^5}{3^4 x^3} = \dfrac{3^2}{3^4} \cdot \dfrac{x^5}{x^3} = 3^{2-4} \cdot x^{5-3} = 3^{-2} x^2 = \dfrac{x^2}{3^2}, x \neq 0$

(f) $\dfrac{(m+n)^{-2}}{(m+n)^{-4}} = (m+n)^{-2-(-4)} = (m+n)^{-2+4} = (m+n)^2, m \neq -n$

(g) $\dfrac{7x^{-3}y^2}{2^{-1}x^2 y^{-5}} = \dfrac{7 \cdot 2^1 y^2 y^5}{x^2 x^3} = \dfrac{14y^7}{x^5}$

Work Problem ⑤ at the Side.

---

⑤ Simplify. Write answers with positive exponents.

(a) $\dfrac{5^{11}}{5^8}$

(b) $\dfrac{4^7}{4^{10}}$

(c) $\dfrac{6^{-5}}{6^{-2}}$

(d) $\dfrac{8^4 m^9}{8^5 m^{10}}, m \neq 0$

(e) $\dfrac{3^{-1}(x+y)^{-3}}{2^{-2}(x+y)^{-4}}, x \neq -y$

ANSWERS

5. (a) 5^3 **(b)** $\dfrac{1}{4^3}$ **(c)** $\dfrac{1}{6^3}$ **(d)** $\dfrac{1}{8m}$

(e) $\dfrac{4}{3}(x+y)$

The definitions and rules for exponents given in this section and Section 4.2 are summarized below.

Definitions and Rules for Exponents

For any integers m and n:

Examples

Product rule $a^m \cdot a^n = a^{m+n}$ $7^4 \cdot 7^5 = 7^9$

Zero exponent $a^0 = 1$ $(a \neq 0)$ $(-3)^0 = 1$

Negative exponent $a^{-n} = \dfrac{1}{a^n}$ $(a \neq 0)$ $5^{-3} = \dfrac{1}{5^3}$

Quotient rule $\dfrac{a^m}{a^n} = a^{m-n}$ $(a \neq 0)$ $\dfrac{2^2}{2^5} = 2^{2-5} = 2^{-3} = \dfrac{1}{2^3}$

Power rules (a) $(a^m)^n = a^{mn}$ $(4^2)^3 = 4^6$

(b) $(ab)^m = a^m b^m$ $(3k)^4 = 3^4 k^4$

(c) $\left(\dfrac{a}{b}\right)^m = \dfrac{a^m}{b^m}$ $(b \neq 0)$ $\left(\dfrac{2}{3}\right)^2 = \dfrac{2^2}{3^2}$

Negative to positive rules $\dfrac{a^{-m}}{b^{-n}} = \dfrac{b^n}{a^m}$ $(a \neq 0, b \neq 0)$ $\dfrac{2^{-4}}{5^{-3}} = \dfrac{5^3}{2^4}$

$\left(\dfrac{a}{b}\right)^{-m} = \left(\dfrac{b}{a}\right)^m$ $\left(\dfrac{4}{7}\right)^{-2} = \left(\dfrac{7}{4}\right)^2$

---

4   **Use combinations of rules.**   As shown in the next example, we may sometimes need to use more than one rule to simplify an expression.

Example 5   **Using a Combination of Rules**

Use a combination of the rules for exponents to simplify each expression. Assume all variables represent nonzero real numbers.

(a) $\dfrac{(4^2)^3}{4^5} = \dfrac{4^6}{4^5}$ Power rule (a)

$= 4^{6-5}$ Quotient rule

$= 4^1 = 4$

(b) $(2x)^3(2x)^2 = (2x)^5$ Product rule

$= 2^5 x^5$ or $32x^5$ Power rule (b)

(c) $\left(\dfrac{2x^3}{5}\right)^{-4} = \left(\dfrac{5}{2x^3}\right)^4$ Negative to positive rule

$= \dfrac{5^4}{2^4 x^{12}}$ Power rules (a)–(c)

(d) $\left(\dfrac{3x^{-2}}{4^{-1}y^3}\right)^{-3} = \dfrac{3^{-3}x^6}{4^3 y^{-9}}$ Power rules (a)–(c)

$= \dfrac{x^6 y^9}{4^3 \cdot 3^3}$ Negative to positive rule

Continued on Next Page

(e) $\dfrac{(4m)^{-3}}{(3m)^{-4}} = \dfrac{4^{-3}m^{-3}}{3^{-4}m^{-4}}$ Power rule (b)

$\qquad\qquad = \dfrac{3^4 m^4}{4^3 m^3}$ Negative to positive rule

$\qquad\qquad = \dfrac{3^4 m^{4-3}}{4^3}$ Quotient rule

$\qquad\qquad = \dfrac{3^4 m}{4^3}$

NOTE

Since the steps can be done in several different orders, there are many equally correct ways to simplify expressions like Examples 5(d) and 5(e).

Work Problem ❻ at the Side.

❻ Simplify. Assume all variables represent nonzero real numbers.

(a) $12^5 \cdot 12^{-7} \cdot 12^6$

(b) $y^{-2} \cdot y^5 \cdot y^{-8}$

(c) $\dfrac{(6x)^{-1}}{(3x^2)^{-2}}$

(d) $\dfrac{3^9 \cdot (x^2 y)^{-2}}{3^3 \cdot x^{-4} y}$

ANSWERS

6. **(a)** 12^4 **(b)** $\dfrac{1}{y^5}$ **(c)** $\dfrac{3x^3}{2}$ **(d)** $\dfrac{3^6}{y^3}$

Real-Data Applications

Numbers BIG and SMALL

The ancient Egyptians used the *astonished man* symbol to represent a million. We now use exponents to write very big and very small numbers efficiently. Columbia University professor Edward Kasner asked his nine-year-old nephew to think of a name for the exceedingly large number, 10^{100}, which is a 1 followed by 100 zeros. His nephew proclaimed it to be a **googol.** Kasner then called the unimaginably large number, 10^{googol}, a **googolplex.**

> A googol is 10^{100} or
>
> 10,000,000,000,000,000,000,000,000,000,000,
> 000,000,000,000,000,000,000,000,000,000,
> 000,000,000,000,000,000,000,000,000,000.

Real-world examples of very big and very small numbers are described in Howard Eves's book, *Mathematical Circles Revisited.* A few such examples are listed here.

- The total number of electrons in the universe is, according to an estimate by Sir Arthur Eddington, about 10^{79}.
- The number of grains of sand on the beach at Coney Island, New York, is about 10^{20}.
- The total number of printed words since the Gutenberg Bible appeared is approximately 10^{16}.
- The temperature at the center of an atomic bomb explosion is 2×10^8 degrees Fahrenheit.
- The diameter of a human hair is about 10^{-4} millimeters.
- The diameter of a nucleus of a cell is about 10^{-6} millimeters (1 micron).
- The probability of winning a lottery by choosing 6 numbers from among 50 is about 6.3×10^{-8}.
- The size of a quark is approximately 10^{-18} millimeters.

Interestingly enough, the Web search engine Google is named after a googol. Sergey Brin, president and co-founder of Google, Inc., was a math major. He chose the name Google to describe the vast reach of this search engine. (*Source: The Gazette*, March 2, 2001.)

For Group Discussion

Suppose that you have been offered a new job with "salary negotiable." You present to your new employer the following offer: You will work for 30 days and will be paid 1¢ on day one, 2¢ on day two, 4¢ on day three, 8¢ on day four, and so on, doubling your pay each day for the month. Your employer accepts your proposal, and you begin work on the first day of the next month.

1. On which day will you have received half of your month's wages?

2. When will you have received one-fourth of your total month's wages?

3. What do you think your approximate monthly salary will be?

4. How many days would it take you to become a "googolaire?"

4.5 **EXERCISES**

Decide whether each expression is positive, negative, or 0.

1. $(-2)^{-3}$

2. $(-3)^{-2}$

3. -2^4

4. -3^6

5. $\left(\dfrac{1}{4}\right)^{-2}$

6. $\left(\dfrac{1}{5}\right)^{-2}$

7. $1 - 5^0$

8. $1 - 7^0$

Each expression is equal to either 0, 1, or −1. Decide which is correct. See Example 1.

9. $(-4)^0$

10. $(-10)^0$

11. -9^0

12. -5^0

13. $(-2)^0 - 2^0$

14. $(-8)^0 - 8^0$

15. $\dfrac{0^{10}}{10^0}$

16. $\dfrac{0^5}{5^0}$

Evaluate each expression. See Examples 1 and 2.

17. $7^0 + 9^0$

18. $8^0 + 6^0$

19. 4^{-3}

20. 5^{-4}

21. $\left(\dfrac{1}{2}\right)^{-4}$

22. $\left(\dfrac{1}{3}\right)^{-3}$

23. $\left(\dfrac{6}{7}\right)^{-2}$

24. $\left(\dfrac{2}{3}\right)^{-3}$

25. $5^{-1} + 3^{-1}$

26. $6^{-1} + 2^{-1}$

27. $-2^{-1} + 3^{-2}$

28. $(-3)^{-2} + (-4)^{-1}$

RELATING CONCEPTS (Exercises 29–32) **FOR INDIVIDUAL OR GROUP WORK**

*In Objective 1, we used the product rule to motivate the definition of a 0 exponent. We can also use the quotient rule. To see this, **work Exercises 29–32 in order.***

29. Consider the expression $\frac{25}{25}$. What is its simplest form?

30. Write the quotient in Exercise 29 using the fact that $25 = 5^2$.

31. Apply the quotient rule for exponents to your answer for Exercise 30. Give the answer as a power of 5.

32. Because your answers for Exercises 29 and 31 both represent $\frac{25}{25}$, they must be equal. Write this equality. What definition does it support?

Use the quotient rule to simplify each expression. Write each expression with positive exponents. Assume that all variables represent nonzero real numbers. See Examples 2–4.

33. $\dfrac{9^4}{9^5}$

34. $\dfrac{7^3}{7^4}$

35. $\dfrac{6^{-3}}{6^2}$

36. $\dfrac{4^{-2}}{4^3}$

37. $\dfrac{1}{6^{-3}}$

38. $\dfrac{1}{5^{-2}}$

39. $\dfrac{2}{r^{-4}}$

40. $\dfrac{3}{s^{-8}}$

41. $\dfrac{4^{-3}}{5^{-2}}$

42. $\dfrac{6^{-2}}{5^{-4}}$

43. $p^5 q^{-8}$

44. $x^{-8} y^4$

45. $\dfrac{r^5}{r^{-4}}$

46. $\dfrac{a^6}{a^{-4}}$

47. $\dfrac{6^4 x^8}{6^5 x^3}$

48. $\dfrac{3^8 y^5}{3^{10} y^2}$

49. $\dfrac{6y^3}{2y}$

50. $\dfrac{5m^2}{m}$

51. $\dfrac{3x^5}{3x^2}$

52. $\dfrac{10p^8}{2p^4}$

Use a combination of the rules for exponents to simplify each expression. Write answers with only positive exponents. Assume that all variables represent nonzero real numbers. See Example 5.

53. $\dfrac{(7^4)^3}{7^9}$

54. $\dfrac{(5^3)^2}{5^2}$

55. $x^{-3} \cdot x^5 \cdot x^{-4}$

56. $y^{-8} \cdot y^5 \cdot y^{-2}$

57. $\dfrac{(3x)^{-2}}{(4x)^{-3}}$

58. $\dfrac{(2y)^{-3}}{(5y)^{-4}}$

59. $\left(\dfrac{x^{-1}y}{z^2}\right)^{-2}$

60. $\left(\dfrac{p^{-4}q}{r^{-3}}\right)^{-3}$

61. $(6x)^4(6x)^{-3}$

62. $(10y)^9(10y)^{-8}$

63. $\dfrac{(m^7 n)^{-2}}{m^{-4} n^3}$

64. $\dfrac{(m^8 n^{-4})^2}{m^{-2} n^5}$

65. $\dfrac{5x^{-3}}{(4x)^2}$

66. $\dfrac{-3k^5}{(2k)^2}$

67. $\left(\dfrac{2p^{-1}q}{3^{-1}m^2}\right)^2$

68. $\left(\dfrac{4xy^2}{x^{-1}y}\right)^{-2}$

4.6 DIVIDING A POLYNOMIAL BY A MONOMIAL

1 **Divide a polynomial by a monomial.** We add two fractions with a common denominator as follows.

$$\frac{a}{c} + \frac{b}{c} = \frac{a+b}{c}$$

Looking at this statement in reverse gives us a rule for dividing a polynomial by a monomial.

Dividing a Polynomial by a Monomial

To divide a polynomial by a monomial, divide each term of the polynomial by the monomial:

$$\frac{a+b}{c} = \frac{a}{c} + \frac{b}{c} \quad (c \neq 0).$$

For example,

$$\frac{2+5}{3} = \frac{2}{3} + \frac{5}{3} \quad \text{and} \quad \frac{x+3z}{2y} = \frac{x}{2y} + \frac{3z}{2y}.$$

The parts of a division problem are named here.

$$\text{Dividend} \rightarrow \frac{12x^2 + 6x}{6x} = 2x + 1 \leftarrow \text{Quotient}$$
$$\text{Divisor} \rightarrow$$

Example 1 Dividing a Polynomial by a Monomial

Divide $5m^5 - 10m^3$ by $5m^2$.
   Use the preceding rule, with $+$ replaced by $-$. Then use the quotient rule.

$$\frac{5m^5 - 10m^3}{5m^2} = \frac{5m^5}{5m^2} - \frac{10m^3}{5m^2} = m^3 - 2m$$

Check by multiplying: $5m^2(m^3 - 2m) = 5m^5 - 10m^3$.

Because division by 0 is undefined, the quotient

$$\frac{5m^5 - 10m^3}{5m^2}$$

is undefined if $m = 0$. From now on, we assume that no denominators are 0.

= **Work Problem 1 at the Side.**

Example 2 Dividing a Polynomial by a Monomial

Divide: $\dfrac{16a^5 - 12a^4 + 8a^2}{4a^3}$.
   Divide each term of $16a^5 - 12a^4 + 8a^2$ by $4a^3$.

$$\frac{16a^5 - 12a^4 + 8a^2}{4a^3} = \frac{16a^5}{4a^3} - \frac{12a^4}{4a^3} + \frac{8a^2}{4a^3}$$

$$= 4a^2 - 3a + \frac{2}{a} \qquad \text{Quotient rule}$$

Continued on Next Page

OBJECTIVE

1 Divide a polynomial by a monomial.

1 Divide.

(a) $\dfrac{6p^4 + 18p^7}{3p^2}$

(b) $\dfrac{12m^6 + 18m^5 + 30m^4}{6m^2}$

(c) $(18r^7 - 9r^2) \div (3r)$

❷ Divide.

(a) $\dfrac{20x^4 - 25x^3 + 5x}{5x^2}$

(b) $\dfrac{50m^4 - 30m^3 + 20m}{10m^3}$

❸ Divide.

(a) $\dfrac{8y^7 - 9y^6 - 11y - 4}{y^2}$

(b) $\dfrac{12p^5 + 8p^4 + 3p^3 - 5p^2}{3p^3}$

(c) $\dfrac{45x^4y^3 + 30x^3y^2 - 60x^2y}{-15x^2y}$

The quotient is not a polynomial because of the expression $\frac{2}{a}$, which has a variable in the denominator. While the sum, difference, and product of two polynomials are always polynomials, the quotient of two polynomials may not be.

Again, check by multiplying.

$$4a^3\left(4a^2 - 3a + \frac{2}{a}\right) = 4a^3(4a^2) - 4a^3(3a) + 4a^3\left(\frac{2}{a}\right)$$
$$= 16a^5 - 12a^4 + 8a^2$$

Work Problem ❷ at the Side.

Example 3 Dividing a Polynomial by a Monomial

Divide.

$$\frac{12x^4 - 7x^3 + 4x}{4x} = \frac{12x^4}{4x} - \frac{7x^3}{4x} + \frac{4x}{4x}$$
$$= 3x^3 - \frac{7x^2}{4} + 1 \qquad \textcolor{blue}{\text{Quotient rule}}$$

Check by multiplying.

CAUTION

In Example 3, notice that the quotient $\frac{4x}{4x} = 1$. It is a common error to leave the 1 out of the answer. Multiplying to check will show that the answer $3x^3 - \frac{7}{4}x^2$ is not correct.

Example 4 Dividing a Polynomial by a Monomial

Divide the polynomial

$$180x^4y^{10} - 150x^3y^8 + 120x^2y^6 - 90xy^4 + 100y$$

by the monomial $-30xy^2$.

$$\frac{180x^4y^{10} - 150x^3y^8 + 120x^2y^6 - 90xy^4 + 100y}{-30xy^2}$$
$$= \frac{180x^4y^{10}}{-30xy^2} - \frac{150x^3y^8}{-30xy^2} + \frac{120x^2y^6}{-30xy^2} - \frac{90xy^4}{-30xy^2} + \frac{100y}{-30xy^2}$$
$$= -6x^3y^8 + 5x^2y^6 - 4xy^4 + 3y^2 - \frac{10}{3xy}$$

Work Problem ❸ at the Side.

ANSWERS

2. (a) $4x^2 - 5x + \dfrac{1}{x}$ **(b)** $5m - 3 + \dfrac{2}{m^2}$

3. (a) $8y^5 - 9y^4 - \dfrac{11}{y} - \dfrac{4}{y^2}$

(b) $4p^2 + \dfrac{8p}{3} + 1 - \dfrac{5}{3p}$

(c) $-3x^2y^2 - 2xy + 4$

4.6 **EXERCISES**

Fill in each blank with the correct response.

1. In the statement $\dfrac{6x^2 + 8}{2} = 3x^2 + 4$, _____ is the dividend, _____ is the divisor, and _____ is the quotient.

2. The expression $\dfrac{3x + 12}{x}$ is undefined if $x =$ _____ .

3. To check the division shown in Exercise 1, multiply _____ by _____ and show that the product is _____ .

4. The expression $5x^2 - 3x + 6 + \dfrac{2}{x}$ _____ a polynomial.
   $\underset{\text{(is/is not)}}{}$

5. Explain why the division problem $\dfrac{16m^3 - 12m^2}{4m}$ can be performed using the method of this section, while the division problem $\dfrac{4m}{16m^3 - 12m^2}$ cannot.

6. Evaluate $\dfrac{5y + 6}{2}$ when $y = 2$. Evaluate $5y + 3$ when $y = 2$. Does $\dfrac{5y + 6}{2}$ equal $5y + 3$?

Perform each division. See Examples 1–4.

7. $\dfrac{60x^4 - 20x^2 + 10x}{2x}$

8. $\dfrac{120x^6 - 60x^3 + 80x^2}{2x}$

9. $\dfrac{20m^5 - 10m^4 + 5m^2}{-5m^2}$

10. $\dfrac{12t^5 - 6t^3 + 6t^2}{-6t^2}$

11. $\dfrac{8t^5 - 4t^3 + 4t^2}{2t}$

12. $\dfrac{8r^4 - 4r^3 + 6r^2}{2r}$

13. $\dfrac{4a^5 - 4a^2 + 8}{4a}$

14. $\dfrac{5t^8 + 5t^7 + 15}{5t}$

15. $\dfrac{12x^5 - 4x^4 + 6x^3}{-6x^2}$

16. $\dfrac{24x^6 - 12x^5 + 30x^4}{-6x^2}$

17. $\dfrac{4x^2 + 20x^3 - 36x^4}{4x^2}$

18. $\dfrac{5x^2 - 30x^4 + 30x^5}{5x^2}$

19. $\dfrac{4x^4 + 3x^3 + 2x}{3x^2}$

20. $\dfrac{5x^4 - 6x^3 + 8x}{3x^2}$

21. $\dfrac{27r^4 - 36r^3 - 6r^2 + 3r - 2}{3r}$

22. $\dfrac{8k^4 - 12k^3 - 2k^2 - 2k - 3}{2k}$

23. $\dfrac{2m^5 - 6m^4 + 8m^2}{-2m^3}$

24. $\dfrac{6r^5 - 8r^4 + 10r^2}{-2r^4}$

25. $(20a^4b^3 - 15a^5b^2 + 25a^3b) \div (-5a^4b)$

26. $(16y^5z - 8y^2z^2 + 12yz^3) \div (-4y^2z^2)$

27. $(120x^{11} - 60x^{10} + 140x^9 - 100x^8) \div (10x^{12})$

28. $(120x^{12} - 84x^9 + 60x^8 - 36x^7) \div (12x^9)$

29. The quotient in Exercise 19 is $\dfrac{4x^2}{3} + x + \dfrac{2}{3x}$. Notice how the third term is written with x in the denominator. Would $\dfrac{2}{3}x$ be an acceptable form for this term? Explain why or why not. Is $\dfrac{4}{3}x^2$ an acceptable form for the first term? Why or why not?

30. What expression represents the length of the rectangle?

$2x$

Area $= 12x^2 - 4x + 2$

31. What polynomial, when divided by $5x^3$, yields $3x^2 - 7x + 7$ as a quotient?

32. The quotient of a certain polynomial and $-12y^3$ is $6y^3 - 5y^2 + 2y - 3 + \dfrac{7}{y}$. Find the polynomial.

RELATING CONCEPTS (Exercises 33–36) **FOR INDIVIDUAL OR GROUP WORK**

Our system of numeration is called a decimal system. It is based on powers of ten. In a whole number such as 2846, each digit is understood to represent the number of powers of ten for its place value. The 2 represents two thousands (2×10^3), the 8 represents eight hundreds (8×10^2), the 4 represents four tens (4×10^1), and the 6 represents six ones (or units) (6×10^0). In expanded form we write

$$2846 = (2 \times 10^3) + (8 \times 10^2) + (4 \times 10^1) + (6 \times 10^0).$$

Keeping this information in mind, **work Exercises 33–36 in order.**

33. Divide 2846 by 2, using paper-and-pencil methods: $2\overline{)2846}$.

34. Write your answer in Exercise 33 in expanded form.

35. Use the methods of this section to divide the polynomial $2x^3 + 8x^2 + 4x + 6$ by 2.

36. Compare your answers in Exercises 34 and 35. How are they similar? How are they different? For what value of x does the answer in Exercise 35 equal the answer in Exercise 34?

4.7   THE QUOTIENT OF TWO POLYNOMIALS

1 ▭   **Divide a polynomial by a polynomial.**   We use a method of "long division" to divide a polynomial by a polynomial (other than a monomial). This method is similar to the method of long division used for two whole numbers. For comparison, the division of whole numbers is shown alongside the division of polynomials. Both polynomials must first be written in descending powers.

OBJECTIVE

1 ▭   Divide a polynomial by a polynomial.

Dividing Whole Numbers	Dividing Polynomials

Step 1

Divide 6696 by 27.

$$27\overline{)6696}$$

Divide $8x^3 - 4x^2 - 14x + 15$ by $2x + 3$.

$$2x + 3\overline{)8x^3 - 4x^2 - 14x + 15}$$

Step 2

66 divided by $27 = 2$;
$2 \cdot 27 = 54$.

$$\begin{array}{r} 2 \\ 27\overline{)6696} \\ 54 \end{array}$$

$8x^3$ divided by $2x = 4x^2$;
$4x^2(2x + 3) = 8x^3 + 12x^2$.

$$\begin{array}{r} 4x^2 \\ 2x + 3\overline{)8x^3 - 4x^2 - 14x + 15} \\ 8x^3 + 12x^2 \end{array}$$

Step 3

Subtract; then bring down the next digit.

$$\begin{array}{r} 2 \\ 27\overline{)6696} \\ 54\downarrow \\ 129 \end{array}$$

Subtract; then bring down the next term.

$$\begin{array}{r} 4x^2 \\ 2x + 3\overline{)8x^3 - 4x^2 - 14x + 15} \\ 8x^3 + 12x^2\downarrow \\ -16x^2 - 14x \end{array}$$

(To subtract two polynomials, change the signs of the second and then add.)

Step 4

129 divided by $27 = 4$;
$4 \cdot 27 = 108$.

$$\begin{array}{r} 24 \\ 27\overline{)6696} \\ 54 \\ 129 \\ 108 \end{array}$$

$-16x^2$ divided by $2x = -8x$;
$-8x(2x + 3) = -16x^2 - 24x$.

$$\begin{array}{r} 4x^2 - 8x \\ 2x + 3\overline{)8x^3 - 4x^2 - 14x + 15} \\ 8x^3 + 12x^2 \\ -16x^2 - 14x \\ -16x^2 - 24x \end{array}$$

Step 5

Subtract; then bring down the next digit.

$$\begin{array}{r} 24 \\ 27\overline{)6696} \\ 54 \\ 129 \\ 108\downarrow \\ 216 \end{array}$$

Subtract; then bring down the next term.

$$\begin{array}{r} 4x^2 - 8x \\ 2x + 3\overline{)8x^3 - 4x^2 - 14x + 15} \\ 8x^3 + 12x^2 \\ -16x^2 - 14x \\ -16x^2 - 24x \\ 10x + 15 \end{array}$$

(continued)

Step 6

216 divided by 27 = **8**;
8 · 27 = **216**.

$$
\begin{array}{r}
248 \\
27)\overline{6696} \\
54 \\
129 \\
108 \\
216 \\
216 \\
\hline
0
\end{array}
$$

10x divided by 2x = **5**;
5(2x + 3) = **10x + 15**.

$$
\begin{array}{r}
4x^2 - \ 8x + \ 5 \\
2x + 3)\overline{8x^3 - \ 4x^2 - 14x + 15} \\
8x^3 + 12x^2 \\
-16x^2 - 14x \\
-16x^2 - 24x \\
10x + 15 \\
10x + 15 \\
\hline
0
\end{array}
$$

6696 divided by 27 is 248.
There is no remainder.

$8x^3 - 4x^2 - 14x + 15$ divided by $2x + 3$ is $4x^2 - 8x + 5$. There is no remainder.

Step 7

Check by multiplying.

$27 \cdot 248 = 6696$

Check by multiplying.

$(2x + 3)(4x^2 - 8x + 5)$
$= 8x^3 - 4x^2 - 14x + 15$

Example 1 Dividing a Polynomial by a Polynomial

Divide $5x + 4x^3 - 8 - 4x^2$ by $2x - 1$.

Both polynomials must be written with the exponents in descending order. Rewrite the first polynomial as $4x^3 - 4x^2 + 5x - 8$. Then begin the division process.

Divide $4x^3 - 4x^2 + 5x - 8$ by $2x - 1$.

$$
\begin{array}{r}
2x^2 - \ x + 2 \\
2x - 1)\overline{4x^3 - 4x^2 + 5x - 8} \\
4x^3 - 2x^2 \\
-2x^2 + 5x \\
-2x^2 + \ x \\
4x - 8 \\
4x - 2 \\
\hline
-6 \ \leftarrow \text{Remainder}
\end{array}
$$

Step 1 $4x^3$ divided by $2x = \mathbf{2x^2}$; $2x^2(2x - 1) = 4x^3 - 2x^2$.

Step 2 Subtract; bring down the next term.

Step 3 $-2x^2$ divided by $2x = \mathbf{-x}$; $-x(2x - 1) = -2x^2 + x$.

Step 4 Subtract; bring down the next term.

Step 5 $4x$ divided by $2x = \mathbf{2}$; $2(2x - 1) = 4x - 2$.

Step 6 Subtract. The remainder is **−6**. Thus $4x^3 - 4x^2 + 5x - 8$ divided by $2x - 1$ has a quotient of $2x^2 - x + 2$ and a remainder of -6. Write the remainder as the numerator of a fraction that has $2x - 1$ as its denominator. The answer is not a polynomial because of the remainder.

$$
\frac{4x^3 - 4x^2 + 5x - 8}{2x - 1} = 2x^2 - x + 2 + \frac{-6}{2x - 1}
$$

Continued on Next Page

Step 7 Check by multiplying.

$$(2x - 1)\left(2x^2 - x + 2 + \frac{-6}{2x - 1}\right)$$

$$= (2x - 1)(2x^2) + (2x - 1)(-x) + (2x - 1)(2)$$

$$+ (2x - 1)\left(\frac{-6}{2x - 1}\right)$$

$$= 4x^3 - 2x^2 - 2x^2 + x + 4x - 2 - 6$$

$$= 4x^3 - 4x^2 + 5x - 8$$

Work Problem ❶ at the Side.

Example 2 Dividing into a Polynomial with Missing Terms

Divide $x^3 - 1$ by $x - 1$.

Here the polynomial $x^3 - 1$ is missing the x^2 term and the x term. When terms are missing, use 0 as the coefficient for each missing term. (Zero acts as a placeholder here, just as it does in our number system.)

$$x^3 - 1 = x^3 + 0x^2 + 0x - 1$$

Now divide.

$$
\begin{array}{r}
x^2 + x + 1 \\
x - 1 \overline{)x^3 + 0x^2 + 0x - 1} \\
\underline{x^3 - x^2} \\
x^2 + 0x \\
\underline{x^2 - x} \\
x - 1 \\
\underline{x - 1} \\
0
\end{array}
$$

The remainder is 0. The quotient is $x^2 + x + 1$. Check by multiplying.

$$(x^2 + x + 1)(x - 1) = x^3 - 1$$

Work Problem ❷ at the Side.

Example 3 Dividing by a Polynomial with Missing Terms

Divide $x^4 + 2x^3 + 2x^2 - x - 1$ by $x^2 + 1$.

Since $x^2 + 1$ has a missing x term, write it as $x^2 + 0x + 1$. Then go through the division process as follows.

$$
\begin{array}{r}
x^2 + 2x + 1 \\
x^2 + 0x + 1 \overline{)x^4 + 2x^3 + 2x^2 - x - 1} \\
\underline{x^4 + 0x^3 + x^2} \\
2x^3 + x^2 - x \\
\underline{2x^3 + 0x^2 + 2x} \\
x^2 - 3x - 1 \\
\underline{x^2 + 0x + 1} \\
-3x - 2 \leftarrow \text{Remainder}
\end{array}
$$

Continued on Next Page

❶ Divide.

(a) $(x^3 + x^2 + 4x - 6) \div (x - 1)$

(b) $\dfrac{p^3 - 2p^2 - 5p + 9}{p + 2}$

❷ Divide.

(a) $\dfrac{r^2 - 5}{r + 4}$

(b) $(x^3 - 8) \div (x - 2)$

❸ Divide.

(a) $(2x^4 + 3x^3 - x^2 + 6x + 5)$
$\div (x^2 - 1)$

When the result of subtracting $(-3x - 2$, in this case) is a polynomial of smaller degree than the divisor $(x^2 + 0x + 1)$, that polynomial is the remainder. Write the answer as

$$x^2 + 2x + 1 + \frac{-3x - 2}{x^2 + 1}.$$

Multiply to check that this is the correct quotient.

Work Problem ❸ at the Side.

Example 4 Dividing a Polynomial with a Quotient That Has Fractional Coefficients

Divide $4x^3 + 2x^2 + 3x + 1$ by $4x - 4$.

$$
\begin{array}{r}
x^2 + \dfrac{3}{2}x + \dfrac{9}{4} \\
4x - 4 \overline{\smash{)}\, 4x^3 + 2x^2 + 3x + 1} \\
\underline{4x^3 - 4x^2} \\
6x^2 + 3x \\
\underline{6x^2 - 6x} \\
9x + 1 \\
\underline{9x - 9} \\
10
\end{array}
$$

(b)
$$\frac{2m^5 + m^4 + 6m^3 - 3m^2 - 18}{m^2 + 3}$$

The quotient is $x^2 + \dfrac{3}{2}x + \dfrac{9}{4} + \dfrac{10}{4x - 4}$.

Work Problem ❹ at the Side.

❹ Divide $3x^3 + 7x^2 + 7x + 10$ by $3x + 6$.

4.7   EXERCISES

 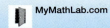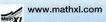

1. In the division problem $(4x^4 + 2x^3 - 14x^2 + 19x + 10) \div (2x + 5) = 2x^3 - 4x^2 + 3x + 2$, which polynomial is the divisor? Which is the quotient?

2. When dividing one polynomial by another, how do you know when to stop dividing?

3. In dividing $12m^2 - 20m + 3$ by $2m - 3$, what is the first step?

4. In the division in Exercise 3, what is the second step?

Perform each division. See Example 1.

5. $\dfrac{x^2 - x - 6}{x - 3}$

6. $\dfrac{m^2 - 2m - 24}{m - 6}$

7. $\dfrac{2y^2 + 9y - 35}{y + 7}$

8. $\dfrac{2y^2 + 9y + 7}{y + 1}$

9. $\dfrac{p^2 + 2p + 20}{p + 6}$

10. $\dfrac{x^2 + 11x + 16}{x + 8}$

11. $(r^2 - 8r + 15) \div (r - 3)$

12. $(t^2 + 2t - 35) \div (t - 5)$

13. $\dfrac{4a^2 - 22a + 32}{2a + 3}$

14. $\dfrac{9w^2 + 6w + 10}{3w - 2}$

15. $\dfrac{8x^3 - 10x^2 - x + 3}{2x + 1}$

16. $\dfrac{12t^3 - 11t^2 + 9t + 18}{4t + 3}$

RELATING CONCEPTS (Exercises 17–20)   FOR INDIVIDUAL OR GROUP WORK

We can find the value of a polynomial in x for a given value of x by substituting that number for x. Surprisingly, we can accomplish the same thing by division. For example, to find the value of $2x^2 - 4x + 3$ for $x = -3$, we would divide $2x^2 - 4x + 3$ by $x - (-3)$. The remainder will give the value of the polynomial for $x = -3$. **Work Exercises 17–20 in order.**

17. Find the value of $2x^2 - 4x + 3$ for $x = -3$ by substitution.

18. Divide $2x^2 - 4x + 3$ by $x + 3$. Give the remainder.

(continued)

19. Compare your answers to Exercises 17 and 18. What do you notice?

20. Choose another polynomial and evaluate it both ways at some value of the variable. Do the answers agree?

Perform each division. See Examples 2–4.

21. $\dfrac{3y^3 + y^2 + 2}{y + 1}$

22. $\dfrac{2r^3 - 6r - 36}{r - 3}$

23. $\dfrac{3k^3 - 4k^2 - 6k + 10}{k^2 - 2}$

24. $\dfrac{5z^3 - z^2 + 10z + 2}{z^2 + 2}$

25. $(x^4 - x^2 - 2) \div (x^2 - 2)$

26. $(r^4 + 2r^2 - 3) \div (r^2 - 1)$

27. $\dfrac{6p^4 - 15p^3 + 14p^2 - 5p + 10}{3p^2 + 1}$

28. $\dfrac{6r^4 - 10r^3 - r^2 + 15r - 8}{2r^2 - 3}$

29. $\dfrac{2x^5 + 9x^4 + 8x^3 + 10x^2 + 14x + 5}{2x^2 + 3x + 1}$

30. $\dfrac{4t^5 - 11t^4 - 6t^3 + 5t^2 - t + 3}{4t^2 + t - 3}$

31. $\dfrac{x^4 - 1}{x^2 - 1}$

32. $\dfrac{y^3 + 1}{y + 1}$

33. $(10x^3 + 13x^2 + 4x + 1) \div (5x + 5)$

34. $(6x^3 - 19x^2 - 19x - 4) \div (2x - 8)$

Work each problem.

35. Give the length of the rectangle.

$5x + 2$

The area is $5x^3 + 7x^2 - 13x - 6$ sq. units.

36. Find the measure of the base of the parallelogram.

$x - 1$

The area is $2x^3 + 2x^2 - 3x - 1$ sq. units.

4.8   AN APPLICATION OF EXPONENTS: SCIENTIFIC NOTATION

1 **Express numbers in scientific notation.** One example of the use of exponents comes from science. The numbers occurring in science are often extremely large (such as the distance from Earth to the sun, 93,000,000 mi) or extremely small (the wavelength of yellow-green light, approximately .0000006 m). Because of the difficulty of working with many zeros, scientists often express such numbers with exponents. Each number is written as $a \times 10^n$, where $1 \le |a| < 10$ and n is an integer. This form is called **scientific notation.** There is always one nonzero digit before the decimal point. For example, 35 is written 3.5×10^1, or 3.5×10; 56,200 is written 5.62×10^4, since

$$56,200 = 5.62 \times \mathbf{10,000} = 5.62 \times \mathbf{10^4},$$

and .09 is written as 9×10^{-2}.

The steps involved in writing a number in scientific notation are given next. For negative numbers, follow these steps using the absolute value of the number; then make the result negative.

OBJECTIVES

1 Express numbers in scientific notation.

2 Convert numbers in scientific notation to numbers without exponents.

3 Use scientific notation in calculations.

Writing a Number in Scientific Notation

Step 1   Move the decimal point to the right of the first nonzero digit.

Step 2   Count the number of places you moved the decimal point.

Step 3   The number of places in Step 2 is the absolute value of the exponent on 10.

Step 4   The exponent on 10 is positive if the original number is larger than the number in Step 1; the exponent is negative if the original number is smaller than the number in Step 1. If the decimal point is not moved, the exponent is 0.

Example 1   Using Scientific Notation

Write each number in scientific notation.

(a) 93,000,000.
   Move the decimal point to follow the first nonzero digit. Count the number of places the decimal point was moved.

$$9.3\ 000\ 000 \qquad \text{7 places}$$

The number will be written in scientific notation as 9.3×10^n. To find the value of n, first compare 9.3 with 93,000,000. Since 93,000,000 is *larger* than 9.3, we must multiply by a *positive* power of 10 so the product 9.3×10^n will equal the larger number.

Since the decimal point was moved 7 places, and since n is positive,

$$93,000,000 = 9.3 \times 10^7.$$

(b) $463,000,000,000,000 = 4.63\ 000\ 000\ 000\ 000.$ \qquad 14 places
$$= 4.63 \times 10^{14}$$

Continued on Next Page

1 Write each number in scientific notation.

(a) 63,000

(b) 5,870,000

(c) .0571

(d) −.000062

2 Write without exponents.

(a) 4.2×10^3

(b) 8.7×10^5

(c) 6.42×10^{-3}

3 Simplify, and write without exponents.

(a) $(2.6 \times 10^4)(2 \times 10^{-6})$

(b) $\dfrac{4.8 \times 10^2}{2.4 \times 10^{-3}}$

(c) $3.021 = 3.021 \times 10^0$

(d) .00462

Move the decimal point to the right of the first nonzero digit and count the number of places the decimal point was moved.

$$004.62 \quad \text{3 places}$$

Because .00462 is *smaller* than 4.62, the exponent must be *negative*.

$$.00462 = 4.62 \times \mathbf{10^{-3}}$$

(e) $-.0000762 = -7.62 \times 10^{-5}$

Work Problem ① at the Side.

2 Convert numbers in scientific notation to numbers without exponents. To convert a number written in scientific notation to a number without exponents, work in reverse. Multiplying a number by a positive power of 10 will make the number larger; multiplying by a negative power of 10 will make the number smaller.

Example 2 Writing Numbers without Exponents

Write each number without exponents.

(a) 6.2×10^3

Since the exponent is positive, make 6.2 larger by moving the decimal point 3 places to the right.

$$6.2 \times \mathbf{10^3} = 6.200 = 6200$$

(b) $4.283 \times 10^5 = 4.28300 = 428,300$ Move 5 places to the right.

(c) $-9.73 \times 10^{-2} = -09.73 = -.0973$ Move 2 places to the left.

As these examples show, the exponent tells the number of places and the direction that the decimal point is moved.

Work Problem ② at the Side.

3 Use scientific notation in calculations. The next example shows how scientific notation can be used with products and quotients.

Example 3 Multiplying and Dividing with Scientific Notation

Write each product or quotient without exponents.

(a) $(6 \times 10^3)(5 \times 10^{-4})$

$= (6 \times 5)(10^3 \times 10^{-4})$ Commutative and associative properties

$= 30 \times 10^{-1}$ Product rule for exponents

$= 30. = 3$ Write without exponents.

(b) $\dfrac{6 \times 10^{-5}}{2 \times 10^3} = \dfrac{6}{2} \times \dfrac{10^{-5}}{10^3} = 3 \times 10^{-8} = .00000003$

Work Problem ③ at the Side.

Calculator Tip Calculators usually have a key labeled EE or EXP for scientific notation. See An Introduction to Calculators at the front of this book for more information.

Example 4 Applying Scientific Notation

Convert to scientific notation, calculate each computation, then give the result without scientific notation.

(a) In determining helium usage at Kennedy Space Center, the product 70,000(.0283)(1000) must be calculated. (*Source: NASA-AMATYC-NSF Mathematics Explorations II,* Capital Community College, 2000.)

$$70{,}000(.0283)(1000) = (7 \times 10^4)(2.83 \times 10^{-2})(1 \times 10^3)$$
$$= (7 \times 2.83 \times 1)(10^{4-2+3})$$
$$= 19.81 \times 10^5$$
$$= 1{,}981{,}000$$

(b) The ratio of the tidal force exerted by the moon compared to that exerted by the sun is given by

$$\frac{73.5 \times 10^{21} \times (1.5 \times 10^8)^3}{1.99 \times 10^{30} \times (3.84 \times 10^5)^3}.$$

(*Source:* Kastner, Bernice, *Space Mathematics,* NASA.)

$$\frac{7.35 \times 10^1 \times 10^{21} \times 1.5^3 \times 10^{24}}{1.99 \times 10^{30} \times 3.84^3 \times 10^{15}} \approx .22 \times 10^{1+21+24-30-15}$$
$$= .22 \times 10^1$$
$$= 2.2$$

Work Problem ❹ at the Side.

❹ The speed of light is approximately 3.0×10^5 km per sec. (*Source: World Almanac and Book of Facts,* 2000.) Write answers without exponents.

(a) How far does light travel in 6.0×10^1 sec?

(b) How many seconds does it take light to travel approximately 1.5×10^8 km from the sun to Earth?

Real-Data Applications

Earthquake Intensities Measured by the Richter Scale

Charles F. Richter devised a scale in 1935 to compare the intensities, or relative power, of earthquakes. The **intensity** of an earthquake is measured relative to the intensity of a standard **zero-level** earthquake of intensity I_0. The relationship is equivalent to $I = I_0 \times 10^R$, where R is the **Richter scale** measure. For example, if an earthquake has magnitude 5.0 on the Richter scale, then its intensity is calculated as $I = I_0 \times 10^{5.0} = I_0 \times 100{,}000$, which is 100,000 times as intense as a zero-level earthquake. The following diagram illustrates the intensities of earthquakes and their Richter scale magnitudes.

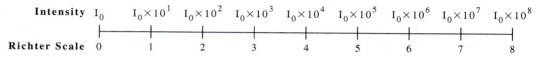

To compare two earthquakes to each other, a ratio of the intensities is calculated. For example, to compare an earthquake that measures 8.0 on the Richter scale to one that measures 5.0, simply find the ratio of the intensities:

$$\frac{\text{intensity 8.0}}{\text{intensity 5.0}} = \frac{I_0 \times 10^{8.0}}{I_0 \times 10^{5.0}} = \frac{10^8}{10^5} = 10^{8-5} = 10^3 = 1000.$$

Therefore an earthquake that measures 8.0 on the Richter Scale is 1000 times as intense as one that measures 5.0.

For Group Discussion

The table gives Richter scale measurements for several earthquakes.

Earthquake		Richter Scale Measurement
1960	Concepción, Chile	9.5
1906	San Francisco, California	8.3
1939	Erzincan, Turkey	8.0
1998	Sumatra, Indonesia	7.0
1998	Adana, Turkey	6.3

Source: World Almanac and Book of Facts, 2000.

1. Compare the intensity of the 1939 Erzincan earthquake to the 1998 Sumatra earthquake.

2. Compare the intensity of the 1998 Adana earthquake to the 1906 San Francisco earthquake.

3. Compare the intensity of the 1939 Erzincan earthquake to the 1998 Adana earthquake.

4. Suppose an earthquake measures 7.2 on the Richter scale. How would the intensity of a second earthquake compare if its Richter scale measure differed by $+3.0$? By -1.0?

4.8 EXERCISES

Write the numbers (other than dates) mentioned in the following statements in scientific notation.

1. NASA has budgeted $13,750,400,000 in each of the years 2003 and 2004 for the international space station. (*Source:* U.S. National Aeronautics and Space Administration.)

2. The mass of Pluto, the smallest planet, is .0021 times that of Earth; the mass of Jupiter, the largest planet, is 317.83 times that of Earth. (*Source: World Almanac and Book of Facts,* 2000.)

3. In 1998, the federal government spent $66,636,000,000 on research and development. Industry spent $143,714,000,000 in that same year. (*Source:* U.S. National Science Foundation.)

4. The risk to industrial workers at the Hansom Landfill at the Kennedy Space Center depends on the reference doses of materials dumped there. For thallium, the reference dose is 700,000 mg/kg per day, and the reference dose for beryllium is 5000 mg/kg per day. (*Source: NASA-AMATYC-NSF Math Explorations I*, Capital Community College, 1999.)

Determine whether or not the given number is written in scientific notation as defined in Objective 1. If it is not, write it as such.

5. 4.56×10^3 6. 7.34×10^5 7. 5,600,000   8. 34,000

9. .004   10. .0007   11. $.8 \times 10^2$ 12. $.9 \times 10^3$

13. Explain in your own words what it means for a number to be written in scientific notation.

14. Explain how to multiply a number by a positive power of ten. Then explain how to multiply a number by a negative power of ten.

Write each number in scientific notation. See Example 1.

15. 5,876,000,000   16. 9,994,000,000   17. 82,350   18. 78,330

19. .000007   20. .0000004   21. −.00203   22. −.0000578

Write each number without exponents. See Example 2.

23. 7.5×10^5 24. 8.8×10^6 25. 5.677×10^{12} 26. 8.766×10^9

27. -6.21×10^0

28. -8.56×10^0

29. 7.8×10^{-4}

30. 8.9×10^{-5}

31. 5.134×10^{-9}

32. 7.123×10^{-10}

Perform the indicated operations, and write the answers in scientific notation and then without exponents. See Example 3.

33. $(2 \times 10^8) \times (3 \times 10^3)$

34. $(4 \times 10^7) \times (3 \times 10^3)$

35. $(5 \times 10^4) \times (3 \times 10^2)$

36. $(8 \times 10^5) \times (2 \times 10^3)$

37. $(3.15 \times 10^{-4}) \times (2.04 \times 10^8)$

38. $(4.92 \times 10^{-3}) \times (2.25 \times 10^7)$

Perform the indicated operations, and write the answers in scientific notation. See Example 3.

39. $\dfrac{9 \times 10^{-5}}{3 \times 10^{-1}}$

40. $\dfrac{12 \times 10^{-4}}{4 \times 10^{-3}}$

41. $\dfrac{8 \times 10^3}{2 \times 10^2}$

42. $\dfrac{5 \times 10^4}{1 \times 10^3}$

43. $\dfrac{2.6 \times 10^{-3} \times 7.0 \times 10^{-1}}{2 \times 10^2 \times 3.5 \times 10^{-3}}$

44. $\dfrac{9.5 \times 10^{-1} \times 2.4 \times 10^4}{5 \times 10^3 \times 1.2 \times 10^{-2}}$

Work each problem. Give answers without exponents. See Example 4.

45. There are 10^9 social security numbers. The population of the U.S. is about 3×10^8. How many social security numbers are available for each person? (*Source:* U.S. Bureau of the Census.)

46. The number of possible hands in contract bridge is about 6.35×10^{11}. The probability of being dealt one particular hand is $\dfrac{1}{6.35 \times 10^{11}}$. Express this number without scientific notation.

47. The top-grossing movie of 1997 was *Titanic*, with box office receipts of about 6×10^8 dollars. That amount represented a fraction of about 9.5×10^{-3} of the total receipts for motion pictures in that year. (*Source:* U.S. Bureau of the Census.) What were the total receipts?

48. There were 6.3×10^{10} dollars spent to attend motion pictures in a recent year. Approximately 1.3×10^8 adults attended a motion picture theatre at least once. (*Source:* U.S. National Endowment for the Arts.) What was the average amount spent per person that year?

49. The body of a 150-lb person contains about 2.3×10^{-4} lb of copper. How much copper is contained in the bodies of 1200 such people?

50. It takes about 3.6×10^1 sec at a speed of 3.0×10^5 km per sec for light from the sun to reach Venus. (*Source: World Almanac and Book of Facts,* 2000.) How far is Venus from the sun?

SUMMARY

KEY TERMS

4.1	**polynomial**	A polynomial is a term or the sum of a finite number of terms with whole number exponents.
	descending powers	A polynomial in x is written in descending powers if the exponents on x in its terms are in decreasing order.
	degree of a term	The degree of a term is the sum of the exponents on the variables.
	degree of a polynomial	The degree of a polynomial is the highest degree of any term of the polynomial.
	monomial	A monomial is a polynomial with one term.
	binomial	A binomial is a polynomial with two terms.
	trinomial	A trinomial is a polynomial with three terms.
4.3	**FOIL**	FOIL is a shortcut method for finding the product of two binomials.
	outer product	The outer product of $(2x + 3)(x - 5)$ is $2x(-5)$.
	inner product	The inner product of $(2x + 3)(x - 5)$ is $3x$.
4.7	**scientific notation**	A number written as $a \times 10^n$, where $1 \le |a| < 10$ and n is an integer, is in scientific notation.

NEW SYMBOLS

x^{-n}	x to the negative n power

TEST YOUR WORD POWER

See how well you have learned the vocabulary in this chapter. Answers follow the Quick Review.

1. A **polynomial** is an algebraic expression made up of
   (a) a term or a finite product of terms with positive coefficients and exponents
   (b) a term or a finite sum of terms with real coefficients and whole number exponents
   (c) the product of two or more terms with positive exponents
   (d) the sum of two or more terms with whole number coefficients and exponents.

2. The **degree of a term** is
   (a) the number of variables in the term

   (b) the product of the exponents on the variables
   (c) the smallest exponent on the variables
   (d) the sum of the exponents on the variables.

3. A **trinomial** is a polynomial with
   (a) only one term
   (b) exactly two terms
   (c) exactly three terms
   (d) more than three terms.

4. A **binomial** is a polynomial with
   (a) only one term
   (b) exactly two terms
   (c) exactly three terms
   (d) more than three terms.

5. A **monomial** is a polynomial with
   (a) only one term
   (b) exactly two terms
   (c) exactly three terms
   (d) more than three terms.

6. **FOIL** is a method for
   (a) adding two binomials
   (b) adding two trinomials
   (c) multiplying two binomials
   (d) multiplying two trinomials.

Concepts	Examples

4.1 Adding and Subtracting Polynomials

Addition: Add like terms.

Add: $2x^2 + 5x - 3$
$\underline{5x^2 - 2x + 7}$
$7x^2 + 3x + 4$

Subtraction: Change the signs of the terms in the second polynomial and add to the first polynomial.

$(2x^2 + 5x - 3) - (5x^2 - 2x + 7)$
$= (2x^2 + 5x - 3) + (-5x^2 + 2x - 7)$
$= -3x^2 + 7x - 10$

4.2 The Product Rule and Power Rules for Exponents

For any integers m and n:

Product rule $a^m \cdot a^n = a^{m+n}$

$2^4 \cdot 2^5 = 2^9$

Power rules (a) $(a^m)^n = a^{mn}$

$(3^4)^2 = 3^8$

$\quad\quad$ **(b)** $(ab)^m = a^m b^m$

$(6a)^5 = 6^5 a^5$

$\quad\quad$ **(c)** $\left(\dfrac{a}{b}\right)^m = \dfrac{a^m}{b^m} \; (b \neq 0).$

$\left(\dfrac{2}{3}\right)^4 = \dfrac{2^4}{3^4}$

4.3 Multiplying Polynomials

Multiply each term of the first polynomial by each term of the second polynomial. Then add like terms.

Multiply: $\quad\quad 3x^3 - 4x^2 + 2x - 7$
$\underline{\quad\quad\quad\quad\quad\quad\quad 4x + 3}$
$\quad\quad\quad 9x^3 - 12x^2 + 6x - 21$
$\underline{12x^4 - 16x^3 + 8x^2 - 28x}$
$12x^4 - 7x^3 - 4x^2 - 22x - 21$

FOIL Method

Step 1 Multiply the two first terms to get the first term of the answer.

Multiply: $(2x + 3)(5x - 4)$.
$2x(5x) = \mathbf{10x^2}$

Step 2 Find the outer product and the inner product and mentally add them, when possible, to get the middle term of the answer.

$2x(-4) + 3(5x) = \mathbf{7x}$

Step 3 Multiply the two last terms to get the last term of the answer.

$3(-4) = \mathbf{-12}$
$(2x + 3)(5x - 4) = \mathbf{10x^2 + 7x - 12}$

4.4 Special Products

Square of a Binomial

$(a + b)^2 = a^2 + 2ab + b^2$

$(a - b)^2 = a^2 - 2ab + b^2$

$(3x + 1)^2 = 9x^2 + 6x + 1$
$(2m - 5n)^2 = 4m^2 - 20mn + 25n^2$

Product of the Sum and Difference of Two Terms

$(a + b)(a - b) = a^2 - b^2$

$(4a + 3)(4a - 3) = 16a^2 - 9$

4.5 Integer Exponents and the Quotient Rule

If $a \neq 0$, for integers m and n:

Zero exponent $a^0 = 1$

$15^0 = 1$

Negative exponent $a^{-n} = \dfrac{1}{a^n}$

$5^{-2} = \dfrac{1}{5^2} = \dfrac{1}{25}$

Quotient rule $\dfrac{a^m}{a^n} = a^{m-n}$

$\dfrac{4^8}{4^3} = 4^5$

Negative to positive rules $\dfrac{a^{-m}}{b^{-n}} = \dfrac{b^n}{a^m} \quad \left(\dfrac{a}{b}\right)^{-m} = \left(\dfrac{b}{a}\right)^m.$

$\dfrac{6^{-2}}{7^{-3}} = \dfrac{7^3}{6^2} \quad \left(\dfrac{5}{3}\right)^{-4} = \left(\dfrac{3}{5}\right)^4$

Concepts	Examples
4.6 *Dividing a Polynomial by a Monomial* Divide each term of the polynomial by the monomial: $$\dfrac{a + b}{c} = \dfrac{a}{c} + \dfrac{b}{c}.$$	Divide: $\dfrac{4x^3 - 2x^2 + 6x - 8}{2x} = \dfrac{4x^3}{2x} - \dfrac{2x^2}{2x} + \dfrac{6x}{2x} - \dfrac{8}{2x}$ $= 2x^2 - x + 3 - \dfrac{4}{x}$

4.7 *The Quotient of Two Polynomials* Use "long division."	Divide: $\begin{array}{r} 2x - 5 + \dfrac{-1}{3x+4} \\ 3x + 4 \overline{)6x^2 - 7x - 21} \\ \underline{6x^2 + 8x} \\ -15x - 21 \\ \underline{-15x - 20} \\ -1 \leftarrow \text{Remainder} \end{array}$

4.8 *An Application of Exponents: Scientific Notation* To write a number in scientific notation (as $a \times 10^n$), move the decimal point to the right of the first nonzero digit. If the decimal point is moved n places, and this makes the number smaller, n is positive; otherwise, n is negative. If the decimal point is not moved, n is 0.	$247 = 2.47 \times 10^2$ $.0051 = 5.1 \times 10^{-3}$ $4.8 = 4.8 \times 10^0$ $3.25 \times 10^5 = 325{,}000$ $8.44 \times 10^{-6} = .00000844$

ANSWERS TO TEST YOUR WORD POWER

1. (b) *Example:* $5x^3 + 2x^2 - 7$ **2. (d)** *Examples:* The term 6 has degree 0, $3x$ has degree 1, $-2x^8$ has degree 8, and $5x^2y^4$ has degree 6.   **3. (c)** *Example:* $2a^2 - 3ab + b^2$ **4. (b)** *Example:* $3t^3 + 5t$
5. (a) *Examples:* -5 and $4xy^5$

$\qquad\qquad\qquad\qquad\qquad$ F $\qquad$ O $\quad$ I $\qquad$ L
6. (c) *Example:* $(m + 4)(m - 3) = m(m) - 3m + 4m + 4(-3) = m^2 + m - 12$

Real-Data Applications

Algebra in Euclid's *Elements*

The word *algebra* is derived from *Al-jabr wa'l muqabalah*, a ninth-century treatise written by the Arabic mathematician, al-Khwarizmi. The notation that we use today, including the use of letters to represent variables and the symbols $+$ for addition and $-$ for subtraction, was introduced in the sixteenth century by François Viète. In Book II of Euclid's *Elements*, algebraic relationships were written in terms of geometric figures. The following proposition is an example.

Proposition 4: "If a straight line is cut at random, the square on the whole equals the squares on the segments plus twice the rectangle contained by the segments."

This proposition can be viewed geometrically. The straight line segment has length $a + b$. The "square on the whole" is the large outer square, $(a + b)^2$. The "squares on the segments" are a^2 and b^2, and the "rectangle contained by the segments" is ab, as shown in the figure. The algebraic statement equivalent to Proposition 4 is $(a + b)^2 = a^2 + b^2 + 2ab$. You should recognize this as the formula for computing the square of a binomial. We usually write it in the form $(a + b)^2 = a^2 + 2ab + b^2$.

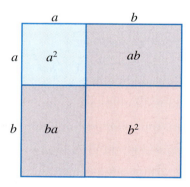

For Group Discussion

Consider the following propositions from Book II of Euclid's *Elements*. An equivalent algebraic statement accompanies each proposition.

Proposition 1: "If there are two straight lines and one of them is cut into any number of segments whatever [2 in the example], then the rectangle contained by the two straight lines equals the sum of the rectangles contained by the uncut straight line and each of the segments."

The equivalent algebraic form is $a(b + c) = ab + ac$.

1. By what name do we know this property?

Proposition 7: "If a straight line is cut at random, then the sum of the square on the whole [a] and that on one of the segments [b] equals twice the rectangle contained by the whole and the said segment plus the square on the remaining segment [a − b]."

The equivalent algebraic form is $a^2 + b^2 = 2ab + (a - b)^2$.

2. How is this property written in your textbook?

Proposition 8: An equivalent algebraic form of this proposition is $4ab + (a - b)^2 = (a + b)^2$.

3. Expand both the left and right sides of the formula. Are they the same expression?

[4.1] *Combine terms where possible in each polynomial. Write the answer in descending powers of the variable. Give the degree of the answer. Identify the polynomial as a monomial, binomial, trinomial, or none of these.*

1. $9m^2 + 11m^2 + 2m^2$

2. $-4p + p^3 - p^2 + 8p + 2$

3. $12a^5 - 9a^4 + 8a^3 + 2a^2 - a + 3$

4. $-7y^5 - 8y^4 - y^5 + y^4 + 9y$

Add or subtract as indicated.

5. Add.
$$-2a^3 + 5a^2$$
$$-3a^3 - a^2$$

6. Add.
$$4r^3 - 8r^2 + 6r$$
$$-2r^3 + 5r^2 + 3r$$

7. Subtract.
$$6y^2 - 8y + 2$$
$$-5y^2 + 2y - 7$$

8. Subtract.
$$-12k^4 - 8k^2 + 7k - 5$$
$$k^4 + 7k^2 + 11k + 1$$

9. $(2m^3 - 8m^2 + 4) + (8m^3 + 2m^2 - 7)$

10. $(-5y^2 + 3y + 11) + (4y^2 - 7y + 15)$

11. $(6p^2 - p - 8) - (-4p^2 + 2p + 3)$

12. $(12r^4 - 7r^3 + 2r^2) - (5r^4 - 3r^3 + 2r^2 + 1)$

[4.2] *Use the product rule or power rules to simplify each expression. Write the answer in exponential form.*

13. $4^3 \cdot 4^8$

14. $(-5)^6(-5)^5$

15. $(-8x^4)(9x^3)$

16. $(2x^2)(5x^3)(x^9)$

17. $(19x)^5$

18. $(-4y)^7$

19. $5(pt)^4$

20. $\left(\dfrac{7}{5}\right)^6$

21. $(3x^2y^3)^3$

22. $(t^4)^8(t^2)^5$

23. $(6x^2z^4)^2(x^3yz^2)^4$

24. Explain why the product rule for exponents does not apply to the expression $7^2 + 7^4$.

[4.3] *Find each product.*

25. $5x(2x + 14)$

26. $-3p^3(2p^2 - 5p)$

27. $(3r - 2)(2r^2 + 4r - 3)$

28. $(2y + 3)(4y^2 - 6y + 9)$

29. $(5p^2 + 3p)(p^3 - p^2 + 5)$

30. $(3k - 6)(2k + 1)$

31. $(6p - 3q)(2p - 7q)$

32. $(m^2 + m - 9)(2m^2 + 3m - 1)$

[4.4] *Find each product.*

33. $(a + 4)^2$

34. $(3p - 2)^2$

35. $(2r + 5s)^2$

36. $(r + 2)^3$

37. $(2x - 1)^3$

38. $(6m - 5)(6m + 5)$

39. $(2z + 7)(2z - 7)$

40. $(5a + 6b)(5a - 6b)$

41. $(2x^2 + 5)(2x^2 - 5)$

42. Explain why $(a + b)^2$ is not equal to $a^2 + b^2$.

[4.5] *Evaluate each expression.*

43. $5^0 + 8^0$

44. 2^{-5}

45. $\left(\dfrac{6}{5}\right)^{-2}$

46. $4^{-2} - 4^{-1}$

Simplify. Write each answer in exponential form, using only positive exponents. Assume all variables are nonzero.

47. $\dfrac{6^{-3}}{6^{-5}}$

48. $\dfrac{x^{-7}}{x^{-9}}$

49. $\dfrac{p^{-8}}{p^4}$

50. $\dfrac{r^{-2}}{r^{-6}}$

51. $(2^4)^2$

52. $(9^3)^{-2}$

53. $(5^{-2})^{-4}$

54. $(8^{-3})^4$

55. $\dfrac{(m^2)^3}{(m^4)^2}$

56. $\dfrac{y^4 \cdot y^{-2}}{y^{-5}}$

57. $\dfrac{r^9 \cdot r^{-5}}{r^{-2} \cdot r^{-7}}$

58. $(-5m^3)^2$

59. $(2y^{-4})^{-3}$

60. $\dfrac{ab^{-3}}{a^4b^2}$

61. $\dfrac{(6r^{-1})^2 \cdot (2r^{-4})}{r^{-5}(r^2)^{-3}}$

62. $\dfrac{(2m^{-5}n^2)^3(3m^2)^{-1}}{m^{-2}n^{-4}(m^{-1})^2}$

[4.6] *Perform each division.*

63. $\dfrac{-15y^4}{-9y^2}$

64. $\dfrac{-12x^3y^2}{6xy}$

65. $\dfrac{6y^4 - 12y^2 + 18y}{-6y}$

66. $\dfrac{2p^3 - 6p^2 + 5p}{2p^2}$

67. $(5x^{13} - 10x^{12} + 20x^7 - 35x^5) \div (-5x^4)$

68. $(-10m^4n^2 + 5m^3n^3 + 6m^2n^4) \div (5m^2n)$

[4.7] *Perform each division.*

69. $(2r^2 + 3r - 14) \div (r - 2)$

70. $\dfrac{12m^2 - 11m - 10}{3m - 5}$

71. $\dfrac{10a^3 + 5a^2 - 14a + 9}{5a^2 - 3}$

72. $\dfrac{2k^4 + 4k^3 + 9k^2 - 8}{2k^2 + 1}$

[4.8] *Write each number in scientific notation.*

73. 48,000,000

74. 28,988,000,000

75. .000065

76. .0000000824

Write each number without exponents.

77. 2.4×10^4

78. 7.83×10^7

79. 8.97×10^{-7}

80. 9.95×10^{-12}

Perform the indicated operations and write the answers without exponents.

81. $(2 \times 10^{-3}) \times (4 \times 10^5)$

82. $\dfrac{8 \times 10^4}{2 \times 10^{-2}}$

83. $\dfrac{12 \times 10^{-5} \times 5 \times 10^4}{4 \times 10^3 \times 6 \times 10^{-2}}$

84. $\dfrac{2.5 \times 10^5 \times 4.8 \times 10^{-4}}{7.5 \times 10^8 \times 1.6 \times 10^{-5}}$

85. There are 13 red balls and 39 black balls in a box. Mix them up and draw 13 out one at a time without returning any ball. The probability that the 13 drawings each will produce a red ball is 1.6×10^{-12}. Write the number given in scientific notation without exponents. (*Source:* Warren Weaver, *Lady Luck*, Doubleday & Company, 1963.)

86. A Boeing 747 is too big to maneuver with the lively agility required of an Air Force fighter. It performs best when it flies straight and level, as does the worldwide airline industry. That industry has annual revenues that approach $200 billion. Write this dollar amount in scientific notation. (*Source:* Heppenheimer, T. A., *Turbulent Skies: The History of Commercial Aviation*, John Wiley & Sons, 1995.)

MIXED REVIEW EXERCISES

Perform the indicated operations. Write with positive exponents. Assume that no denominators are equal to 0.

87. $19^0 - 3^0$

88. $(3p)^4(3p^{-7})$

89. 7^{-2}

90. $(-7 + 2k)^2$

91. $\dfrac{2y^3 + 17y^2 + 37y + 7}{2y + 7}$

92. $\left(\dfrac{6r^2s}{5}\right)^4$

93. $-m^5(8m^2 + 10m + 6)$

94. $\left(\dfrac{1}{2}\right)^{-5}$

95. $(25x^2y^3 - 8xy^2 + 15x^3y) \div (5x)$

96. $(6r^{-2})^{-1}$

97. $(2x + y)^3$

98. $2^{-1} + 4^{-1}$

99. $(a + 2)(a^2 - 4a + 1)$

100. $(5y^3 - 8y^2 + 7) - (-3y^3 + y^2 + 2)$

101. $(2r + 5)(5r - 2)$

102. $(12a + 1)(12a - 1)$

103. Find a polynomial that represents the area of the rectangle shown.

$2x - 3$

$x + 2$

104. If the side of a square has a measure represented by $5x^4 + 2x^2$, what polynomial represents its area?

$5x^4 + 2x^2$

Chapter 4   **TEST**

 Study Skills Workbook
Activity 10

Perform the indicated operations.

1. $(5t^4 - 3t^2 + 7t + 3) - (t^4 - t^3 + 3t^2 + 8t + 3)$

    1. _____

2. $(2y^2 - 8y + 8) + (-3y^2 + 2y + 3) - (y^2 + 3y - 6)$

    2. _____

3. Subtract.
$$9t^3 - 4t^2 + 2t + 2$$
$$9t^3 + 8t^2 - 3t - 6$$

    3. _____

Simplify, and write each answer with only positive exponents.

4. $(-2)^3(-2)^2$

    4. _____

5. $\left(\dfrac{6}{m^2}\right)^3, \quad m \neq 0$

    5. _____

6. $3x^2(-9x^3 + 6x^2 - 2x + 1)$

    6. _____

7. $(2r - 3)(r^2 + 2r - 5)$

    7. _____

8. $(t - 8)(t + 3)$

    8. _____

9. $(4x + 3y)(2x - y)$

    9. _____

10. $(5x - 2y)^2$

    10. _____

11. $(10v + 3w)(10v - 3w)$

    11. _____

12. $(x + 1)^3$

    12. _____

Evaluate each expression.

13. 5^{-4}

    13. _____

14. $(-3)^0 + 4^0$

    14. _____

15. $4^{-1} + 3^{-1}$

    15. _____

Perform the indicated operations. In Exercises 16 and 17, write each answer using only positive exponents. Assume that variables represent nonzero numbers.

16. _____

16. $\dfrac{8^{-1} \cdot 8^4}{8^{-2}}$

17. _____

17. $\dfrac{(x^{-3})^{-2}(x^{-1}y)^2}{(xy^{-2})^2}$

18. _____

18. $\dfrac{8y^3 - 6y^2 + 4y + 10}{2y}$

19. _____

19. $(-9x^2y^3 + 6x^4y^3 + 12xy^3) \div (3xy)$

20. _____

20. $\dfrac{2x^2 + x - 36}{x - 4}$ *long*

21. _____

21. $(3x^3 - x + 4) \div (x - 2)$ *long*

Write each number in scientific notation.

22. (a)_____

   (b)_____

22. (a) 344,000,000,000
    (b) .00000557

Write each number without exponents.

23. (a)_____

   (b)_____

23. (a) 2.96×10^7
    (b) 6.07×10^{-8}

24. _____

24. What polynomial expression represents the area of this square?

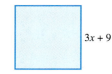

$3x + 9$

25. _____

25. Give an example of this situation: the sum of two fourth-degree polynomials in x is a third-degree polynomial in x.

1. Write $\frac{28}{16}$ in lowest terms.

Work each problem.

2. $\dfrac{2}{3} + \dfrac{1}{8}$

3. $\dfrac{7}{4} - \dfrac{9}{5}$

4. $8.32 - 4.6$

5. 7.21×8.6

6. A retailer has \$34,000 invested in her business. She finds that last year she earned 5.4% on this investment. How much did she earn?

Find the value of each expression if $x = -2$ and $y = 4$.

7. $\dfrac{4x - 2y}{x + y}$

8. $x^3 - 4xy$

Perform the indicated operations.

9. $\dfrac{(-13 + 15) - (3 + 2)}{6 - 12}$

10. $-7 - 3[2 + (5 - 8)]$

Decide what property justifies each statement.

11. $(9 + 2) + 3 = 9 + (2 + 3)$

12. $-7 + 7 = 0$

13. $6(4 + 2) = 6(4) + 6(2)$

Solve each equation.

14. $2x - 7x + 8x = 30$

15. $2 - 3(t - 5) = 4 + t$

16. $2(5h + 1) = 10h + 4$

17. $d = rt$ for r

18. $\dfrac{x}{5} = \dfrac{x - 2}{7}$

19. $\dfrac{1}{3}p - \dfrac{1}{6}p = -2$

20. $.05x + .15(50 - x) = 5.50$

21. $4 - (3x + 12) = (2x - 9) - (5x - 1)$

Solve each problem.

22. Each month, Janet's allowance is six times as much as Louis's allowance. Together their allowances total \$56. What is the monthly allowance for each?

23. A 1-oz mouse takes about 16 times as many breaths as does a 3-ton elephant. (*Source: Dinosaurs, Spitfires, and Sea Dragons,* McGowan, C., Harvard University Press, 1991.) If the two animals take a combined total of 170 breaths per minute, how many breaths does each take during that time period?

24. If a number is subtracted from 8 and this difference is tripled, the result is three times the number. Find this number, and you will learn how many times a dolphin rests during a 24-hr period.

Solve each inequality.

25. $-8x \leq -80$

26. $-2(x + 4) > 3x + 6$

27. $-3 \leq 2x + 5 < 9$

Given $2x - 3y = -6$, find the following.

28. The intercepts of the graph

29. The graph

30. The slope of the line

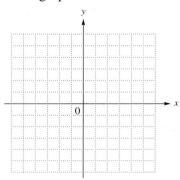

Evaluate each expression.

31. $4^{-1} + 3^0$

32. $2^{-4} \cdot 2^5$

33. $\dfrac{8^{-5} \cdot 8^7}{8^2}$

34. Write with positive exponents only: $\dfrac{(a^{-3}b^2)^2}{(2a^{-4}b^{-3})^{-1}}.$

35. Write in scientific notation: 34,500.

Perform the indicated operations.

36. $(7x^3 - 12x^2 - 3x + 8) + (6x^2 + 4) - (-4x^3 + 8x^2 - 2x - 2)$

37. $6x^5(3x^2 - 9x + 10)$

38. $(7x + 4)(9x + 3)$

39. $(5x + 8)^2$

40. $\dfrac{y^3 - 3y^2 + 8y - 6}{y - 1}$

Factoring and Applications

5

Galileo Galilei, born near Pisa, Italy, in 1564, became a professor of mathematics at the University of Pisa at age 25. He and his students conducted experiments involving the famous Leaning Tower to investigate the relationship between an object's speed of fall and its weight. (*Source: Microsoft Encarta Encyclopedia 2000.*) We will use the concepts of this chapter and the formula Galileo developed from his experiments in Section 5.6.

5.1   FACTORS; THE GREATEST COMMON FACTOR

Recall from Section R.1 that to **factor** a number means to write it as the product of two or more numbers. The product is called the **factored form** of the number. For example,

Factors

$$12 = \underbrace{6 \cdot 2}.$$

Factored form

Factoring is a process that "undoes" multiplying. We multiply $6 \cdot 2$ to get 12, but we factor 12 by writing it as $6 \cdot 2$. In this chapter, we extend these ideas to polynomials.

1 **Find the greatest common factor of a list of numbers.** An integer that is a factor of two or more integers is a **common factor** of those integers. For example, 6 is a common factor of 18 and 24 because 6 is a factor of both 18 and 24. Other common factors of 18 and 24 are 1, 2, and 3. The **greatest common factor (GCF)** of a list of integers is the largest common factor of those integers. This means 6 is the greatest common factor of 18 and 24, since it is the largest of their common factors.

NOTE

Factors of a number are also divisors of the number. The greatest common factor is the same as the greatest common divisor.

Example 1   **Finding the Greatest Common Factor for Numbers**

Find the greatest common factor for each list of numbers.

(a) 30, 45

First write each number in prime factored form.

$$30 = 2 \cdot \mathbf{3} \cdot \mathbf{5}$$
$$45 = \mathbf{3} \cdot 3 \cdot \mathbf{5}$$

Use each prime the *least* number of times it appears in *all* the factored forms. There is no 2 in the prime factored form of 45, so there will be no 2 in the greatest common factor. The least number of times 3 appears in all the factored forms is 1; the least number of times 5 appears is also 1. From this, the

$$GCF = \mathbf{3^1} \cdot \mathbf{5^1} = 3 \cdot 5 = 15.$$

(b) 72, 120, 432

Find the prime factored form of each number.

$$72 = \mathbf{2} \cdot \mathbf{2} \cdot \mathbf{2} \cdot \mathbf{3} \cdot 3$$
$$120 = \mathbf{2} \cdot \mathbf{2} \cdot \mathbf{2} \cdot \mathbf{3} \cdot 5$$
$$432 = \mathbf{2} \cdot \mathbf{2} \cdot \mathbf{2} \cdot 2 \cdot \mathbf{3} \cdot 3 \cdot 3$$

The least number of times 2 appears in all the factored forms is 3, and the least number of times 3 appears is 1. There is no 5 in the prime factored form of either 72 or 432, so the

$$GCF = \mathbf{2^3} \cdot \mathbf{3^1} = 24.$$

Continued on Next Page

(c) 10, 11, 14

Write the prime factored form of each number.

$$10 = 2 \cdot 5$$
$$11 = 11$$
$$14 = 2 \cdot 7$$

There are no primes common to all three numbers, so the GCF is 1.

================ Work Problem **1** at the Side.

2▭ **Find the greatest common factor of a list of variable terms.**   The greatest common factor can also be found for a list of variable terms. For example, the terms x^4, x^5, x^6, and x^7 have x^4 as the greatest common factor because the smallest exponent on the variable x is 4.

$$x^4 = 1 \cdot x^4, \quad x^5 = x \cdot x^4, \quad x^6 = x^2 \cdot x^4, \quad x^7 = x^3 \cdot x^4$$

NOTE

The exponent on a variable in the GCF is the *smallest* exponent that appears on that variable in *all* the terms.

Example 2 **Finding the Greatest Common Factor for Variable Terms**

Find the greatest common factor for each list of terms.

(a) $21m^7$, $-18m^6$, $45m^8$

$$21m^7 = 3 \cdot 7 \cdot m^7$$
$$-18m^6 = -1 \cdot 2 \cdot 3 \cdot 3 \cdot m^6$$
$$45m^8 = 3 \cdot 3 \cdot 5 \cdot m^8$$

First, 3 is the greatest common factor of the coefficients 21, -18, and 45. The smallest exponent on m is 6, so the

$$GCF = 3m^6.$$

(b) x^4y^2, x^7y^5, x^3y^7, y^{15}

$$x^4y^2, \quad x^7y^5, \quad x^3y^7, \quad y^{15}$$

There is no x in the last term, y^{15}, so x will not appear in the greatest common factor. There is a y in each term, however, and 2 is the smallest exponent on y. The GCF is y^2.

(c) $-a^2b$, $-ab^2$

$$-a^2b = -1a^2b = -1 \cdot 1 \cdot a^2b$$
$$-ab^2 = -1ab^2 = -1 \cdot 1 \cdot ab^2$$

The factors of -1 are -1 and 1. Since $1 > -1$, the GCF is $1ab$ or ab.

NOTE

In a list of negative terms, sometimes a negative common factor is preferable (even though it is not the greatest common factor). In Example 2(c), for instance, we might prefer $-ab$ as the common factor. In factoring exercises, either answer will be acceptable.

1 Find the greatest common factor for each list of numbers.

(a) 30, 20, 15

$$30 = 2 \cdot 3 \cdot 5$$
$$20 = 2 \cdot \underline{\quad} \cdot \underline{\quad}$$
$$15 = 3 \cdot \underline{\quad}$$
$$GCF = \underline{\quad}$$

(b) 42, 28, 35

(c) 12, 18, 26, 32

(d) 10, 15, 21

❷ Find the greatest common factor for each list of terms.

(a) $6m^4, 9m^2, 12m^5$

$$6m^4 = 2 \cdot \underline{} \cdot m^4$$

$$9m^2 = 3 \cdot \underline{} \cdot \underline{}$$

$$12m^5 = 2 \cdot 2 \cdot \underline{} \cdot \underline{}$$

$$\text{GCF} = \underline{}$$

(b) $-12p^5, -18q^4$

(c) y^4z^2, y^6z^8, z^9

(d) $12p^{11}, 17q^5$

In summary, we find the greatest common factor of a list of terms as follows.

> ### Finding the Greatest Common Factor (GCF)
>
> *Step 1* **Factor.** Write each number in prime factored form.
>
> *Step 2* **List common factors.** List each prime number or each variable that is a factor of every term in the list. (If a prime does not appear in one of the prime factored forms, it cannot appear in the greatest common factor.)
>
> *Step 3* **Choose smallest exponents.** Use as exponents on the common prime factors the *smallest* exponents from the prime factored forms.
>
> *Step 4* **Multiply.** Multiply the primes from Step 3. If there are no primes left after Step 3, the greatest common factor is 1.

Work Problem ❷ at the Side.

3 Factor out the greatest common factor. The idea of a greatest common factor can be used to write a polynomial in factored form. For example, the polynomial

$$3m + 12$$

has two terms, $3m$ and 12. The greatest common factor of these two terms is 3. We can write $3m + 12$ so that each term is a product with 3 as one factor.

$$3m + 12 = \mathbf{3} \cdot m + \mathbf{3} \cdot 4$$

Using the distributive property,

$$3m + 12 = \mathbf{3} \cdot m + \mathbf{3} \cdot 4 = \mathbf{3}(m + 4).$$

The factored form of $3m + 12$ is $3(m + 4)$. This process is called **factoring out the greatest common factor.**

> **CAUTION**
>
> The polynomial $3m + 12$ is *not* in factored form when written as the *sum*
>
> $$3 \cdot m + 3 \cdot 4.$$
>
> The *terms* are factored, but the polynomial is not. The factored form of $3m + 12$ is the *product*
>
> $$3(m + 4).$$

Writing a polynomial as a product, that is, in factored form, is called **factoring** the polynomial.

Example 3 Factoring Out the Greatest Common Factor

Factor out the greatest common factor.

(a) $5y^2 + 10y = \mathbf{5y}(y) + \mathbf{5y}(2)$ GCF $= 5y$

$ = \mathbf{5y}(y + 2)$ Distributive property

Continued on Next Page

Answers
2. **(a)** $3; 3; m^2; 3; m^5; 3m^2$ **(b)** 6
   (c) z^2 **(d)** 1

Check by multiplying: $\quad 5y(y + 2) = 5y(y) + 5y(2)$
$$= 5y^2 + 10y. \qquad \text{Original polynomial}$$

(b) $20m^5 + 10m^4 - 15m^3$

The GCF for the terms of this polynomial is $5m^3$.

$20m^5 + 10m^4 - 15m^3$

$\quad = \mathbf{5m^3}(4m^2) + \mathbf{5m^3}(2m) - \mathbf{5m^3}(3) \qquad$ Factor each term.

$\quad = \mathbf{5m^3}(4m^2 + 2m - 3) \qquad$ Factor out $5m^3$.

Check: $\quad 5m^3(4m^2 + 2m - 3) = 20m^5 + 10m^4 - 15m^3$, which is the original polynomial.

(c) $x^5 + x^3 = x^3(x^2) + x^3(\mathbf{1}) = x^3(x^2 + \mathbf{1}) \qquad$ Don't forget the 1.

(d) $20m^7p^2 - 36m^3p^4 = 4m^3p^2(5m^4) - 4m^3p^2(9p^2) \qquad$ GCF $= 4m^3p^2$

$\qquad\qquad\qquad\quad = 4m^3p^2(5m^4 - 9p^2)$

(e) $\dfrac{1}{6}n^2 + \dfrac{5}{6}n = \dfrac{1}{6}n(n) + \dfrac{1}{6}n(5) = \dfrac{1}{6}n(n + 5)$

CAUTION

Be sure to include the 1 in a problem like Example 3(c). *Always* check that the factored form can be multiplied out to give the original polynomial.

Work Problem ❸ at the Side.

Example 4 Factoring Out the Greatest Common Factor

Factor out the greatest common factor.

(a) $a(a + 3) + 4(a + 3)$

The binomial $a + 3$ is the greatest common factor here.

$$\overset{\text{Same}}{a(\mathbf{a + 3}) + 4(\mathbf{a + 3}) = (a + 3)(a + 4)}$$

(b) $x^2(x + 1) - 5(x + 1) = (x + 1)(x^2 - 5) \qquad$ Factor out $x + 1$.

Work Problem ❹ at the Side.

4 Factor by grouping. When a polynomial has four terms, common factors can sometimes be used to **factor by grouping.**

Example 5 Factoring by Grouping

Factor by grouping.

(a) $2x + 6 + ax + 3a$

Group the first two terms and the last two terms, since the first two terms have a common factor of 2 and the last two terms have a common factor of a.

$$2x + 6 + ax + 3a = (2x + 6) + (ax + 3a)$$
$$= \mathbf{2}(x + 3) + \mathbf{a}(x + 3)$$

The expression is still not in factored form because it is the *sum* of two terms. Now, however, $x + 3$ is a common factor and can be factored out.

$$2x + 6 + ax + 3a = 2(\mathbf{x + 3}) + a(\mathbf{x + 3})$$
$$= (\mathbf{x + 3})(2 + a)$$

Continued on Next Page

---

❸ Factor out the greatest common factor.

(a) $4x^2 + 6x$

(b) $10y^5 - 8y^4 + 6y^2$

(c) $m^7 + m^9$

(d) $8p^5q^2 + 16p^6q^3 - 12p^4q^7$

(e) $\dfrac{1}{3}b^2 - \dfrac{2}{3}b$

(f) $13x^2 - 27$

❹ Factor out the greatest common factor.

(a) $r(t - 4) + 5(t - 4)$

(b) $y^2(y + 2) - 3(y + 2)$

(c) $x(x - 1) - 5(x - 1)$

ANSWERS
3. **(a)** $2x(2x + 3)$
(b) $2y^2(5y^3 - 4y^2 + 3)$
(c) $m^7(1 + m^2)$
(d) $4p^4q^2(2p + 4p^2q - 3q^5)$
(e) $\dfrac{1}{3}b(b - 2)$

(f) no common factor (except 1)
4. **(a)** $(t - 4)(r + 5)$ **(b)** $(y + 2)(y^2 - 3)$
(c) $(x - 1)(x - 5)$

5 Factor by grouping.

(a) $pq + 5q + 2p + 10$

The final result is in factored form because it is a *product.* Note that the goal in factoring by grouping is to get a common factor, $x + 3$ here, so that the last step is possible. Check by multiplying the binomials using the FOIL method from the previous chapter.

Check: $(x + 3)(2 + a) = 2x + ax + 6 + 3a$
$$= 2x + 6 + ax + 3a, \quad \text{Rearrange terms.}$$

which is the original polynomial.

(b) $2x^2 - 10x + 3xy - 15y = (2x^2 - 10x) + (3xy - 15y)$ Group terms.
$$= 2x(x - 5) + 3y(x - 5) \quad \begin{array}{l}\text{Factor each}\\\text{group.}\end{array}$$
$$= (x - 5)(2x + 3y) \quad \begin{array}{l}\text{Factor out the}\\\text{common factor,}\\x - 5.\end{array}$$

Check: $(x - 5)(2x + 3y) = 2x^2 + 3xy - 10x - 15y$ FOIL
$$= 2x^2 - 10x + 3xy - 15y \quad \text{Original polynomial}$$

(b) $2a^2 - 4a + 3ab - 6b$

(c) $t^3 + 2t^2 - 3t - 6 = (t^3 + 2t^2) + (-3t - 6)$ Group terms.
$$= t^2(t + 2) - 3(t + 2) \quad \begin{array}{l}\text{Factor out } -3 \text{ so there is a}\\\text{common factor, } t + 2;\\-3(t + 2) = -3t - 6.\end{array}$$
$$= (t + 2)(t^2 - 3) \quad \text{Factor out } t + 2.$$

Check by multiplying.

CAUTION

Be careful with signs when grouping in a problem like Example 5(c). It is wise to check the factoring in the second step, as shown in the example side comment, before continuing.

Work Problem 5 at the Side.

Use these steps to factor a polynomial with four terms by grouping.

(c) $x^3 + 3x^2 - 5x - 15$

Factoring by Grouping

Step 1   **Group terms.** Collect the terms into two groups so that each group has a common factor.

Step 2   **Factor within groups.** Factor out the greatest common factor from each group.

Step 3   **Factor the entire polynomial.** Factor a common binomial factor from the results of Step 2.

Step 4   **If necessary, rearrange terms.** If Step 2 does not result in a common binomial factor, try a different grouping.

Example 6 Rearranging Terms Before Factoring by Grouping

Factor by grouping.

(a) $10x^2 - 12y + 15x - 8xy$

Factoring out the common factor of 2 from the first two terms and the common factor of x from the last two terms gives

$$10x^2 - 12y + 15x - 8xy = 2(5x^2 - 6y) + x(15 - 8y).$$

This did not lead to a common factor, so we try rearranging the terms. There is usually more than one way to do this. Let's try

$$10x^2 - 8xy - 12y + 15x,$$

and group the first two terms and the last two terms as follows.

$$10x^2 - 8xy - 12y + 15x = 2x(5x - 4y) + 3(-4y + 5x)$$
$$= 2x(\mathbf{5x - 4y}) + 3(\mathbf{5x - 4y})$$
$$= (\mathbf{5x - 4y})(2x + 3)$$

Check: $(5x - 4y)(2x + 3) = 10x^2 + 15x - 8xy - 12y$ FOIL
$$= 10x^2 - 12y + 15x - 8xy$$ Original polynomial

(b) $2xy + 12 - 3y - 8x$

We need to rearrange these terms to get two groups that each have a common factor. Trial and error suggests the following grouping.

$$2xy + 12 - 3y - 8x = (2xy - 3y) + (-8x + 12)$$ Group terms.
$$= y(2x - 3) - 4(2x - 3)$$ Factor each group. Be careful with signs.
$$= (2x - 3)(y - 4)$$ Factor out the common factor.

Since the quantities in parentheses in the second step must be the same, we factored out -4 rather than 4. Check by multiplying.

CAUTION

Use negative signs carefully when grouping, as in Example 6(b), or a sign error will occur. *Always* check by multiplying.

Work Problem 6 at the Side.

6 Factor by grouping.

(a) $6y^2 - 20w + 15y - 8yw$

(b) $9mn - 4 + 12m - 3n$

Real-Data Applications

Idle Prime Time

A positive integer greater than 1 is a prime number if its only factors are 1 and itself. Every positive integer can be written as a product of prime numbers in a unique way, except for the order of the factors. Finding new primes has intrigued people from ancient Greece to modern times. The *Great Internet Mersenne Prime Search* is a consortium headed by George Woltman and Scott Kurowski that has discovered four of the ten largest primes. The prime number $2^{6972593} - 1$ was found during 111 days of idle time on Nayan Hajratwala's home computer. It would take over $4\frac{1}{2}$ miles to actually write this number without commas using a 10-point font. (*Source*: www.utm. edu/research/primes/largest.html)

Prime numbers are essential in the development of unbreakable codes that, in an era of Internet commerce, ensure security in transmitting and storing computer data.

The oldest known method for finding prime numbers is the Sieve of Eratosthenes, similar to the version shown below. Numbers that are not prime (composite numbers) are eliminated and only the prime numbers are left. Begin with 2. Two is prime but multiples of 2 are not, so delete the remaining numbers in Column 2 and all of Columns 4 and 6. Three is prime, but multiples of 3 are not, so delete the remaining numbers in Column 3. Examine the remaining numbers and eliminate any that are composite (such as 25 or 91). The prime numbers are highlighted.

For Group Discussion

1. **Twin primes** occur in pairs that differ by 2. List all the twin primes from the table.

2. Observe that all prime numbers larger than 3 are in Columns 1 and 5. Each number in Column 5 is 1 less than a multiple of 6, and therefore has the form $6n - 1$. Each number in Column 1 has a similar structure, $6n + 1$. Show that each of these twin primes, found in the year 2000, has the form $6n + 1$:

$$1693965 \times 2^{66443} \pm 1$$

and $4648619711505 \times 2^{60000} \pm 1$. (*Hint:* Show that the leading term is divisible by both 2 and 3.)

SIEVE OF ERATOSTHENES

Col 1	Col 2	Col 3	Col 4	Col 5	Col 6
1	2	3	4	5	6
7	8	9	10	11	12
13	14	15	16	17	18
19	20	21	22	23	24
25	26	27	28	29	30
31	32	33	34	35	36
37	38	39	40	41	42
43	44	45	46	47	48
49	50	51	52	53	54
55	56	57	58	59	60
61	62	63	64	65	66
67	68	69	70	71	72
73	74	75	76	77	78
79	80	81	82	83	84
85	86	87	88	89	90
91	92	93	94	95	96
97	98	99	100	101	102

3. **Mersenne primes,** named for the 17th century French monk Marin Mersenne, have the form $2^p - 1$, where p is a prime number. Not all such numbers are prime. Show that $2^{11} - 1$ is composite and $2^5 - 1$ is prime.

4. A **Sophie Germain prime,** named for an 18th century French mathematician, is an odd prime p for which $2p + 1$ is also prime. For example, 5 is a Sophie Germain prime since 11 ($2 \cdot 5 + 1$) is prime, but 13 is not since 27 ($2 \cdot 13 + 1$) is composite. List the Sophie Germain primes from the table.

5.1 EXERCISES

 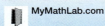

Find the greatest common factor for each list of numbers. See Example 1.

1. 12, 16

2. 18, 24

3. 40, 20, 4

4. 50, 30, 5

5. 18, 24, 36, 48

6. 15, 30, 45, 75

7. 4, 9, 12

8. 9, 16, 24

Find the greatest common factor for each list of terms. See Example 2.

9. $16y$, 24

10. $18w$, 27

11. $30x^3$, $40x^6$, $50x^7$

12. $60z^4$, $70z^8$, $90z^9$

13. $-x^4y^3$, $-xy^2$

14. $-a^4b^5$, $-a^3b$

15. $42ab^3$, $-36a$, $90b$, $-48ab$

16. $45c^3d$, $75c$, $90d$, $-105cd$

Complete each factoring.

17. $9m^4 = 3m^2(\quad)$

18. $12p^5 = 6p^3(\quad)$

19. $-8z^9 = -4z^5(\quad)$

20. $-15k^{11} = -5k^8(\quad)$

21. $6m^4n^5 = 3m^3n(\quad)$

22. $27a^3b^2 = 9a^2b(\quad)$

23. $12y + 24 = 12(\quad)$

24. $18p + 36 = 18(\quad)$

25. $10a^2 - 20a = 10a(\quad)$

26. $15x^2 - 30x = 15x(\quad)$

27. $8x^2y + 12x^3y^2 = 4x^2y(\quad)$

28. $18s^3t^2 + 10st = 2st(\quad)$

Factor out the greatest common factor. See Examples 3 and 4.

29. $x^2 - 4x$

30. $m^2 - 7m$

31. $6t^2 + 15t$

32. $8x^2 + 6x$

33. $\dfrac{1}{4}d^2 - \dfrac{3}{4}d$

34. $\dfrac{1}{5}z^2 + \dfrac{3}{5}z$

35. $12x^3 + 6x^2$

36. $21b^3 - 7b^2$

37. $65y^{10} + 35y^6$

38. $100a^5 + 16a^3$

39. $11w^3 - 100$

40. $13z^5 - 80$

41. $8m^2n^3 + 24m^2n^2$

42. $19p^2y - 38p^2y^3$

43. $4x^3 - 10x^2 + 6x$

44. $9z^3 - 6z^2 + 12z$

45. $13y^8 + 26y^4 - 39y^2$

46. $5x^5 + 25x^4 - 20x^3$

47. $45q^4p^5 + 36qp^6 + 81q^2p^3$

48. $125a^3z^5 + 60a^4z^4 - 85a^5z^2$

49. $c(x + 2) + d(x + 2)$

50. $r(5 - x) + t(5 - x)$

51. $a^2(2a + b) - b(2a + b)$

52. $3x(x^2 + 5) - y(x^2 + 5)$

Factor by grouping. See Examples 5 and 6.

53. $5m + mn + 20 + 4n$

54. $ts + 5t + 2s + 10$

55. $6xy - 21x + 8y - 28$

56. $2mn - 8n + 3m - 12$

57. $7z^2 + 14z - az - 2a$

58. $2b^2 + 3b - 8ab - 12a$

59. $18r^2 + 12ry - 3xr - 2xy$

60. $5m^2 + 15mp - 2mp - 6p^2$

61. $w^3 + w^2 + 9w + 9$

62. $y^3 + y^2 + 6y + 6$

63. $3a^3 + 6a^2 - 2a - 4$

64. $10x^3 + 15x^2 - 8x - 12$

65. $16m^3 - 4m^2p^2 - 4mp + p^3$

66. $10t^3 - 2t^2s^2 - 5ts + s^3$

67. $y^2 + 3x + 3y + xy$

68. $m^2 + 14p + 7m + 2mp$

69. $2z^2 + 6w - 4z - 3wz$

70. $2a^2 + 20b - 8a - 5ab$

RELATING CONCEPTS (Exercises 71–74) **FOR INDIVIDUAL OR GROUP WORK**

In many cases, the choice of which pairs of terms to group when factoring by grouping can be made in different ways. To see this for Example 6(b), **work Exercises 71–74 in order.**

71. Start with the polynomial from Example 6(b), $2xy + 12 - 3y - 8x$, and rearrange the terms as follows: $2xy - 8x - 3y + 12$. What property from Section 1.7 allows this?

72. Group the first two terms and the last two terms of the rearranged polynomial in Exercise 71. Then factor each group.

73. Is your result from Exercise 72 in factored form? Explain your answer.

74. If your answer to Exercise 73 is *no*, factor the polynomial. Is the result the same as the one shown for Example 6(b)?

5.2   FACTORING TRINOMIALS

Using FOIL, the product of the binomials $k - 3$ and $k + 1$ is

$$(k - 3)(k + 1) = k^2 - 2k - 3. \quad \text{Multiplying}$$

Suppose instead that we are given the polynomial $k^2 - 2k - 3$ and want to rewrite it as the product $(k - 3)(k + 1)$. That is,

$$k^2 - 2k - 3 = (k - 3)(k + 1). \quad \text{Factoring}$$

Recall from the previous section that this process is called factoring the polynomial. Factoring reverses or "undoes" multiplying.

1   **Factor trinomials with a coefficient of 1 for the squared term.**   When factoring polynomials with integer coefficients, we use only integers in the factors. For example, we can factor $x^2 + 5x + 6$ by finding integers m and n such that

$$x^2 + 5x + 6 = (x + m)(x + n).$$

To find these integers m and n, we first use FOIL to multiply the two binomials on the right side of the equation:

$$(x + m)(x + n) = x^2 + nx + mx + mn.$$

By the distributive property,

$$x^2 + nx + mx + mn = x^2 + (n + m)x + mn.$$

Comparing this result with $x^2 + 5x + 6$ shows that we must find integers m and n having a sum of 5 and a product of 6.

Product of m and n is 6.

$$x^2 + 5x + 6 = x^2 + (n + m)x + mn$$

Sum of m and n is 5.

Because many pairs of integers have a sum of 5, it is best to begin by listing those pairs of integers whose product is 6. Both 5 and 6 are positive, so we consider only pairs in which both integers are positive.

Work Problem 1 at the Side.

From Problem 1 at the side, we see that the numbers 1 and 6 and the numbers 2 and 3 both have a product of 6, but only the pair 2 and 3 has a sum of 5. So 2 and 3 are the required integers, and

$$x^2 + 5x + 6 = (x + 2)(x + 3).$$

Check by multiplying the binomials using FOIL. *Make sure that the sum of the outer and inner products produces the correct middle term.*

Check: $(x + 2)(x + 3) = x^2 + 5x + 6$

$$\begin{array}{c} 2x \\ 3x \\ \hline 5x \quad \text{Add.} \end{array}$$

This method of factoring can be used only for trinomials that have 1 as the coefficient of the squared term. Methods for factoring other trinomials will be given in the next two sections.

OBJECTIVES

1 Factor trinomials with a coefficient of 1 for the squared term.

2 Factor trinomials after factoring out the greatest common factor.

---

1 **(a)** List all pairs of positive integers whose product is 6.

(b) Find the pair from part (a) whose sum is 5.

---

❷ Factor each trinomial.

(a) $y^2 + 12y + 20$

First complete the given list of numbers.

Factors of 20	Sums of Factors
20, 1	20 + 1 = 21
10, ___	10 + ___ = ___
5, ___	5 + ___ = ___

(b) $x^2 + 9x + 18$

❸ Factor each trinomial.

(a) $t^2 - 12t + 32$

First complete the given list of numbers.

Factors of 32	Sums of Factors
−32, −1	−32 + (−1) = −33
−16, ___	−16 + (___) = ___
−8, ___	−8 + (___) = ___

(b) $y^2 - 10y + 24$

Example 1   **Factoring a Trinomial with All Positive Terms**

Factor $m^2 + 9m + 14$.

Look for two integers whose product is 14 and whose sum is 9. List the pairs of integers whose products are 14. Then examine the sums. Only positive integers are needed since all signs in $m^2 + 9m + 14$ are positive.

Factors of 14	Sums of Factors	
14, 1	14 + 1 = 15	
7, 2	7 + 2 = **9**	Sum is 9.

From the list, 7 and 2 are the required integers, since $7 \cdot 2 = 14$ and $7 + 2 = 9$. Thus,

$$m^2 + 9m + 14 = (m + 2)(m + 7).$$

Check: $(m + 2)(m + 7) = m^2 + 7m + 2m + 14$
$$= m^2 + 9m + 14$$

NOTE

In Example 1, the answer also could have been written $(m + 7)(m + 2)$. Because of the commutative property of multiplication, the order of the factors does not matter. *Always* check by multiplying.

Work Problem ❷ at the Side.

Example 2   **Factoring a Trinomial with a Negative Middle Term**

Factor $x^2 - 9x + 20$.

Find two integers whose product is 20 and whose sum is −9. Since the numbers we are looking for have a positive product and a negative sum, we consider only pairs of negative integers.

Factors of 20	Sums of Factors	
−20, −1	−20 + (−1) = −21	
−10, −2	−10 + (−2) = −12	
−5, −4	−5 + (−4) = **−9**	Sum is −9.

The required integers are −5 and −4, so
$$x^2 - 9x + 20 = (x - 5)(x - 4).$$

Check: $(x - 5)(x - 4) = x^2 - 4x - 5x + 20$
$$= x^2 - 9x + 20$$

Work Problem ❸ at the Side.

ANSWERS
2. (a) 2; 2; 12; 4; 4; 9; $(y + 10)(y + 2)$
   (b) $(x + 3)(x + 6)$
3. (a) −2; −2; −18; −4; −4; −12;
      $(t - 8)(t - 4)$
   (b) $(y - 6)(y - 4)$

Example 3 Factoring a Trinomial with Two Negative Terms

Factor $p^2 - 2p - 15$.

Find two integers whose product is -15 and whose sum is -2. If these numbers do not come to mind right away, find them (if they exist) by listing all the pairs of integers whose product is -15. Because the last term, -15, is negative, we need pairs of integers with different signs.

Factors of -15	Sums of Factors
15, -1	$15 + (-1) = 14$
-15, 1	$-15 + 1 = -14$
5, -3	$5 + (-3) = 2$
-5, 3	$-5 + 3 = \mathbf{-2}$ Sum is -2.

The required integers are -5 and 3, so

$$p^2 - 2p - 15 = (p - 5)(p + 3).$$

Check: Multiply $(p - 5)(p + 3)$.

NOTE

In Examples 1–3, notice that we listed factors in descending order (disregarding sign) when we were looking for the required pair of integers. This helps avoid skipping the correct combination.

Work Problem ④ at the Side.

As shown in the next example, some trinomials cannot be factored using only integers. We call such trinomials **prime polynomials.**

Example 4 Deciding whether Polynomials Are Prime

Factor each trinomial.

(a) $x^2 - 5x + 12$

As in Example 2, both factors must be negative to give a positive product and a negative sum. First, list all pairs of negative integers whose product is 12. Then examine the sums.

Factors of 12	Sums of Factors
-12, -1	$-12 + (-1) = -13$
-6, -2	$-6 + (-2) = -8$
-4, -3	$-4 + (-3) = -7$

None of the pairs of integers has a sum of -5. Therefore, the trinomial $x^2 - 5x + 12$ *cannot be factored using only integers; it is a prime polynomial.*

(b) $k^2 - 8k + 11$

There is no pair of integers whose product is 11 and whose sum is -8, so $k^2 - 8k + 11$ is a prime polynomial.

Work Problem ⑤ at the Side.

④ Factor each trinomial.

(a) $a^2 - 9a - 22$

(b) $r^2 - 6r - 16$

⑤ Factor each trinomial, if possible.

(a) $r^2 - 3r - 4$

(b) $m^2 - 2m + 5$

ANSWERS
4. (a) $(a - 11)(a + 2)$ **(b)** $(r - 8)(r + 2)$
5. (a) $(r - 4)(r + 1)$ **(b)** prime

❻ Factor each trinomial.

(a) $b^2 - 3ab - 4a^2$

(b) $r^2 - 6rs + 8s^2$

The procedure for factoring a trinomial of the form $x^2 + bx + c$ follows.

Factoring $x^2 + bx + c$

Find two integers whose product is c and whose sum is b.

1. Both integers must be positive if b and c are positive.

2. Both integers must be negative if c is positive and b is negative.

3. One integer must be positive and one must be negative if c is negative.

Example 5 **Factoring a Trinomial with Two Variables**

Factor $z^2 - 2bz - 3b^2$.

Here, the coefficient of the middle term is $-2b$, so we need to find two expressions whose product is $-3b^2$ and whose sum is $-2b$. The expressions are $-3b$ and b, so

$$z^2 - 2bz - 3b^2 = (z - 3b)(z + b).$$

Check: $(z - 3b)(z + b) = z^2 + zb - 3bz - 3b^2$

$$= z^2 + 1bz - 3bz - 3b^2$$

$$= z^2 - 2bz - 3b^2$$

Work Problem ❻ at the Side.

---

❼ Factor each trinomial completely.

(a) $2p^3 + 6p^2 - 8p$

2 **Factor trinomials after factoring out the greatest common factor.** The trinomial in the next example does not have a coefficient of 1 for the squared term. (In fact, there is no squared term.) However, there may be a common factor.

Example 6 **Factoring a Trinomial with a Common Factor**

Factor $4x^5 - 28x^4 + 40x^3$.

First, factor out the greatest common factor, $4x^3$.

$$4x^5 - 28x^4 + 40x^3 = \mathbf{4x^3}(x^2 - 7x + 10)$$

Now factor $x^2 - 7x + 10$. The integers -5 and -2 have a product of 10 and a sum of -7. The complete factored form is

$$4x^5 - 28x^4 + 40x^3 = 4x^3(x - 5)(x - 2). \quad \text{Include } 4x^3.$$

(b) $3x^4 - 15x^3 + 18x^2$

Check: $4x^3(x - 5)(x - 2) = 4x^3(x^2 - 7x + 10)$

$$= 4x^5 - 28x^4 + 40x^3$$

CAUTION

When factoring, always look for a common factor first. Remember to include the common factor as part of the answer. As a check, multiplying out the complete factored form should give the original polynomial.

Work Problem ❼ at the Side.

5.2   EXERCISES

1. When factoring a trinomial in x as $(x + a)(x + b)$, what must be true of a and b, if the last term of the trinomial is negative?

2. In Exercise 1, what must be true of a and b if the last term is positive?

3. What is meant by a *prime polynomial*?

4. How can you check your work when factoring a trinomial? Does the check ensure that the trinomial is completely factored?

In Exercises 5–8, list all pairs of integers with the given product. Then find the pair whose sum is given. See the tables in Examples 1–4.

5. Product: 12   Sum: 7

6. Product: 18   Sum: 9

7. Product: -24 Sum: -5

8. Product: -36 Sum: -16

9. Which one of the following is the correct factored form of $x^2 - 12x + 32$?

   A. $(x - 8)(x + 4)$ **B.** $(x + 8)(x - 4)$
   C. $(x - 8)(x - 4)$ **D.** $(x + 8)(x + 4)$

10. What would be the first step in factoring $2x^3 + 8x^2 - 10x$?

Complete each factoring.

11. $x^2 + 15x + 44 = (x + 4)(\qquad)$

12. $r^2 + 15r + 56 = (r + 7)(\qquad)$

13. $x^2 - 9x + 8 = (x - 1)(\qquad)$

14. $t^2 - 14t + 24 = (t - 2)(\qquad)$

15. $y^2 - 2y - 15 = (y + 3)(\qquad)$

16. $t^2 - t - 42 = (t + 6)(\qquad)$

17. $x^2 + 9x - 22 = (x - 2)(\qquad)$

18. $x^2 + 6x - 27 = (x - 3)(\qquad)$

19. $y^2 - 7y - 18 = (y + 2)(\qquad)$

20. $y^2 - 2y - 24 = (y + 4)(\qquad)$

Factor completely. If a polynomial cannot be factored, write prime. *See Examples 1–4.*

21. $y^2 + 9y + 8$

22. $a^2 + 9a + 20$

23. $b^2 + 8b + 15$

24. $x^2 + 6x + 8$

25. $m^2 + m - 20$

26. $p^2 + 4p - 5$

27. $x^2 + 3x - 40$

28. $d^2 + 4d - 45$

29. $y^2 - 8y + 15$

30. $y^2 - 6y + 8$

31. $z^2 - 15z + 56$

32. $x^2 - 13x + 36$

33. $r^2 - r - 30$

34. $q^2 - q - 42$

35. $a^2 - 8a - 48$

36. $m^2 - 10m - 24$

37. $x^2 + 4x + 5$

38. $t^2 + 11t + 12$

Factor completely. See Examples 5 and 6.

39. $r^2 + 3ra + 2a^2$

40. $x^2 + 5xa + 4a^2$

41. $x^2 + 4xy + 3y^2$

42. $p^2 + 9pq + 8q^2$

43. $t^2 - tz - 6z^2$

44. $a^2 - ab - 12b^2$

45. $v^2 - 11vw + 30w^2$

46. $v^2 - 11vx + 24x^2$

47. $4x^2 + 12x - 40$

48. $5y^2 - 5y - 30$

49. $2t^3 + 8t^2 + 6t$

50. $3t^3 + 27t^2 + 24t$

51. $2x^6 + 8x^5 - 42x^4$

52. $4y^5 + 12y^4 - 40y^3$

53. $a^5 + 3a^4b - 4a^3b^2$

54. $z^{10} - 4z^9y - 21z^8y^2$

55. $m^3n - 10m^2n^2 + 24mn^3$

56. $y^3z + 3y^2z^2 - 54yz^3$

57. Use the FOIL method from Section 4.3 to show that $(2x + 4)(x - 3) = 2x^2 - 2x - 12$. Why, then, is it incorrect to completely factor $2x^2 - 2x - 12$ as $(2x + 4)(x - 3)$?

58. Why is it incorrect to completely factor $3x^2 + 9x - 12$ as the product $(x - 1)(3x + 12)$?

5.3 FACTORING TRINOMIALS BY GROUPING

Trinomials like $2x^2 + 7x + 6$, in which the coefficient of the squared term is *not* 1, are factored with extensions of the methods from the previous sections. One such method uses factoring by grouping from Section 5.1.

1 **Factor trinomials by grouping when the coefficient of the squared term is not 1.** Recall that a trinomial such as $m^2 + 3m + 2$ is factored by finding two numbers whose product is 2 and whose sum is 3. To factor $2x^2 + 7x + 6$, we look for two integers whose product is $2 \cdot 6 = 12$ and whose sum is 7.

$$2x^2 + 7x + 6$$

Sum is 7. Product is $2 \cdot 6 = 12$.

By considering pairs of positive integers whose product is 12, the necessary integers are found to be 3 and 4. We use these integers to write the middle term, $7x$, as $7x = 3x + 4x$. The trinomial $2x^2 + 7x + 6$ becomes

$$2x^2 + 7x + 6 = 2x^2 + \underbrace{3x + 4x}_{7x} + 6.$$

$$= (2x^2 + 3x) + (4x + 6) \quad \text{Group terms.}$$
$$= x(2x + 3) + 2(2x + 3) \quad \text{Factor each group.}$$

Must be same

$$2x^2 + 7x + 6 = (2x + 3)(x + 2) \quad \text{Factor out } 2x + 3.$$

Check: $(2x + 3)(x + 2) = 2x^2 + 7x + 6$

In the example above, we could have written $7x$ as $4x + 3x$. Factoring by grouping this way would give the same answer.

Work Problem 1 at the Side.

Example 1 Factoring Trinomials by Grouping

Factor each trinomial.

(a) $6r^2 + r - 1$
We must find two integers with a product of $6(-1) = -6$ and a sum of 1.

Sum is 1.

$$6r^2 + r - 1 = 6r^2 + 1r - 1$$

Product is $6(-1) = -6$.

The integers are -2 and 3. We write the middle term, r, as $-2r + 3r$.

$$6r^2 + r - 1 = 6r^2 - 2r + 3r - 1 \quad r = -2r + 3r$$
$$= (6r^2 - 2r) + (3r - 1) \quad \text{Group terms.}$$
$$= 2r(3r - 1) + 1(3r - 1) \quad \text{The binomials must be the same.}$$
$$= (3r - 1)(2r + 1) \quad \text{Factor out } 3r - 1.$$

Check: $(3r - 1)(2r + 1) = 6r^2 + r - 1$

Continued on Next Page

OBJECTIVE

1 Factor trinomials by grouping when the coefficient of the squared term is not 1.

1 **(a)** Factor $2x^2 + 7x + 6$ by writing $7x$ as $4x + 3x$. Complete the following.

$2x^2 + 7x + 6$
$= 2x^2 + 4x + 3x + 6$
$= (2x^2 + \underline{\quad}) + (3x + \underline{\quad})$
$= 2x(x + \underline{\quad}) + 3(x + \underline{\quad})$
$= (\underline{\quad})(2x + 3)$

(b) Is the answer the same? (Remember that the order of the factors does not matter.)

2 Factor each trinomial by grouping.

(a) $2m^2 + 7m + 3$

(b) $5p^2 - 2p - 3$

(c) $15k^2 - km - 2m^2$

3 Factor each trinomial completely.

(a) $4x^2 - 2x - 30$

(b) $18p^4 + 63p^3 + 27p^2$

(c) $6a^2 + 3ab - 18b^2$

(b) $12z^2 - 5z - 2$

Look for two integers whose product is $12(-2) = -24$ and whose sum is -5. The required integers are 3 and -8, so

$$12z^2 - 5z - 2 = 12z^2 + 3z - 8z - 2 \qquad -5z = 3z - 8z$$
$$= (12z^2 + 3z) + (-8z - 2) \qquad \text{Group terms.}$$
$$= 3z(4z + 1) - 2(4z + 1) \qquad \text{Factor each group; be careful with signs.}$$
$$= (4z + 1)(3z - 2). \qquad \text{Factor out } 4z + 1.$$

Check: $(4z + 1)(3z - 2) = 12z^2 - 5z - 2$

(c) $10m^2 + mn - 3n^2$

Two integers whose product is $10(-3) = -30$ and whose sum is 1 are -5 and 6. Rewrite the trinomial with four terms.

$$10m^2 + mn - 3n^2 = 10m^2 - 5mn + 6mn - 3n^2 \qquad mn = -5mn + 6mn$$
$$= 5m(2m - n) + 3n(2m - n) \qquad \text{Group terms; factor each group.}$$
$$= (2m - n)(5m + 3n) \qquad \text{Factor out } 2m - n.$$

Check by multiplying.

Work Problem ② at the Side.

Example 2   **Factoring a Trinomial with a Common Factor by Grouping**

Factor $28x^5 - 58x^4 - 30x^3$.

First factor out the greatest common factor, $2x^3$.

$$28x^5 - 58x^4 - 30x^3 = 2x^3(14x^2 - 29x - 15)$$

To factor $14x^2 - 29x - 15$, find two integers whose product is $14(-15) = -210$ and whose sum is -29. Factoring 210 into prime factors gives

$$210 = 2 \cdot 3 \cdot 5 \cdot 7.$$

Combine these prime factors in pairs in different ways, using one positive and one negative (to get -210). The factors 6 and -35 have the correct sum. Now rewrite the given trinomial and factor it.

$$28x^5 - 58x^4 - 30x^3 = 2x^3(14x^2 + 6x - 35x - 15)$$
$$= 2x^3[(14x^2 + 6x) + (-35x - 15)]$$
$$= 2x^3[2x(7x + 3) - 5(7x + 3)]$$
$$= 2x^3[(7x + 3)(2x - 5)]$$
$$= 2x^3(7x + 3)(2x - 5)$$

Check by multiplying.

CAUTION

Remember to include the common factor in the final result.

Work Problem ③ at the Side.

5.3 **EXERCISES**

FOR EXTRA HELP

 Student's Solutions Manual

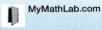 MyMathLab.com

 InterAct Math Tutorial Software

AW Math Tutor Center

 www.mathxl.com MathXL

 Digital Video Tutor CD 4 Videotape 10

Factor each polynomial by grouping. (The middle term of an equivalent trinomial has already been rewritten.) See Example 1.

1. $m^2 + 6m + 2m + 12$

2. $x^2 + 7x + 2x + 14$

3. $a^2 + 5a - 2a - 10$

4. $y^2 + 4y - 6y - 24$

5. $10t^2 + 5t + 4t + 2$

6. $6x^2 + 9x + 4x + 6$

7. $15z^2 - 10z - 9z + 6$

8. $12p^2 - 9p - 8p + 6$

9. $8s^2 - 4st + 6st - 3t^2$

10. $3x^2 - 7xy + 6xy - 14y^2$

11. Which pair of integers would be used to rewrite the middle term when factoring $12y^2 + 5y - 2$ by grouping?

A. $-8, 3$ **B.** $8, -3$ **C.** $-6, 4$ **D.** $6, -4$

12. Which pair of integers would be used to rewrite the middle term when factoring $20b^2 - 13b + 2$ by grouping?

A. $10, 3$ **B.** $-10, -3$ **C.** $8, 5$ **D.** $-8, -5$

Complete the steps to factor each trinomial by grouping.

13. $2m^2 + 11m + 12$

(a) Find two integers whose product is

_____ · _____ = _____ and whose

sum is _____.

(b) The required integers are _____ and

_____.

(c) Write the middle term $11m$ as _____ +

_____.

(d) Rewrite the given trinomial as

_____.

(e) Factor the polynomial in part (d) by grouping.

(f) Check by multiplying.

14. $6y^2 - 19y + 10$

(a) Find two integers whose product is

_____ · _____ = _____ and whose

sum is _____.

(b) The required integers are _____ and

_____.

(c) Write the middle term $-19y$ as _____ +

_____.

(d) Rewrite the given trinomial as

_____.

(e) Factor the polynomial in part (d) by grouping.

(f) Check by multiplying.

Factor each trinomial by grouping. See Examples 1 and 2.

15. $2x^2 + 7x + 3$

16. $3y^2 + 13y + 4$

17. $4r^2 + r - 3$

18. $4r^2 + 3r - 10$

19. $8m^2 - 10m - 3$

20. $20x^2 - 28x - 3$

21. $21m^2 + 13m + 2$

22. $38x^2 + 23x + 2$

23. $6b^2 + 7b + 2$

24. $6w^2 + 19w + 10$

25. $12y^2 - 13y + 3$

26. $15a^2 - 16a + 4$

27. $24x^2 - 42x + 9$

28. $48b^2 - 74b - 10$

29. $2m^3 + 2m^2 - 40m$

30. $3x^3 + 12x^2 - 36x$

31. $32z^5 - 20z^4 - 12z^3$

32. $18x^5 + 15x^4 - 75x^3$

33. $12p^2 + 7pq - 12q^2$

34. $6m^2 - 5mn - 6n^2$

35. $6a^2 - 7ab - 5b^2$

36. $25g^2 - 5gh - 2h^2$

37. $5 - 6x + x^2$

38. $7 + 8x + x^2$

39. On a quiz, a student factored $16x^2 - 24x + 5$ by grouping as follows.

$$16x^2 - 24x + 5$$
$$= 16x^2 - 4x - 20x + 5$$
$$= 4x(4x - 1) - 5(4x - 1) \quad \text{His answer}$$

He thought his answer was correct since it checked by multiplying. Why was the answer marked wrong? What is the correct factored form?

40. On the same quiz, another student factored $3k^3 - 12k^2 - 15k$ by first factoring out the common factor $3k$ to get $3k(k^2 - 4k - 5)$. Then she wrote

$$k^2 - 4k - 5 = k^2 - 5k + k - 5$$
$$= k(k - 5) + 1(k - 5)$$
$$= (k - 5)(k + 1). \quad \text{Her answer}$$

Why was the answer marked wrong? What is the correct factored form?

5.4   FACTORING TRINOMIALS USING FOIL

1   **Factor trinomials using FOIL.**   This section shows an alternative method of factoring trinomials in which the coefficient of the squared term is not 1. This method uses trial and error.

To factor $2x^2 + 7x + 6$ (the same trinomial factored at the beginning of Section 5.3) by trial and error, we use FOIL backwards. We want to write $2x^2 + 7x + 6$ as the product of two binomials.

$$2x^2 + 7x + 6 = (\qquad)(\qquad)$$

The product of the two first terms of the binomials is $2x^2$. The possible factors of $2x^2$ are $2x$ and x or $-2x$ and $-x$. Since all terms of the trinomial are positive, only positive factors should be considered. Thus, we have

$$2x^2 + 7x + 6 = (2x \qquad)(x \qquad).$$

The product of the two last terms, 6, can be factored as $6 \cdot 1$, $1 \cdot 6$, $2 \cdot 3$, or $3 \cdot 2$. Try each pair to find the pair that gives the correct middle term.

Work Problem ❶ at the Side.

In part (b) at the side, since $2x + 6 = 2(x + 3)$, the binomial $2x + 6$ has a common factor of 2, while $2x^2 + 7x + 6$ has no common factor other than 1. The product $(2x + 6)(x + 1)$ cannot be correct. (Part (c) also has one binomial factor with a common factor.)

> **NOTE**
>
> If the original polynomial has no common factor, then none of its binomial factors will either.

Now try the numbers 2 and 3 as factors of 6. Because of the common factor of 2 in $2x + 2$, $(2x + 2)(x + 3)$ will not work. Try $(2x + 3)(x + 2)$.

$$\begin{array}{c}(2x + 3)(x + 2) = 2x^2 + \textcolor{red}{7x} + 6 \quad \textcolor{red}{\text{Correct}}\\ \underset{3x}{\diagdown}\ \underset{4x}{\diagup}\\ \overline{7x} \quad \text{Add.}\end{array}$$

Finally, we see that $2x^2 + 7x + 6$ factors as

$$2x^2 + 7x + 6 = (2x + 3)(x + 2).$$

Check by multiplying: $(2x + 3)(x + 2) = 2x^2 + 7x + 6$.

Example 1   **Factoring a Trinomial with All Positive Terms Using FOIL**

Factor $8p^2 + 14p + 5$.

The number 8 has several possible pairs of factors, but 5 has only 1 and 5 or -1 and -5. For this reason, it is easier to begin by considering the factors of 5. Ignore the negative factors since all coefficients in the trinomial are positive. If $8p^2 + 14p + 5$ can be factored, the factors will have the form

$$(\qquad + 5)(\qquad + 1).$$

Continued on Next Page

OBJECTIVE

1   **Factor trinomials using FOIL.**

Study Skills Workbook
Activity 5

❶ Multiply to decide whether each factored form is correct or incorrect for $2x^2 + 7x + 6$.

  (a) $(2x + 1)(x + 6)$

  (b) $(2x + 6)(x + 1)$

  (c) $(2x + 2)(x + 3)$

ANSWERS

1. **(a)** incorrect   **(b)** incorrect   **(c)** incorrect

❷ Factor each trinomial.

(a) $2p^2 + 9p + 9$

(b) $6p^2 + 19p + 10$

(c) $8x^2 + 14x + 3$

The possible pairs of factors of $8p^2$ are $8p$ and p, or $4p$ and $2p$. Try various combinations, checking to see if the middle term is $14p$ in each case.

$$(8p + 5)(p + 1) \qquad \text{Incorrect}$$

$$\begin{array}{c} 5p \\ 8p \\ \hline 13p \quad \text{Add.} \end{array}$$

$$(p + 5)(8p + 1) \qquad \text{Incorrect}$$

$$\begin{array}{c} 40p \\ p \\ \hline 41p \quad \text{Add.} \end{array}$$

$$(4p + 5)(2p + 1) \qquad \text{Correct}$$

$$\begin{array}{c} 10p \\ 4p \\ \hline 14p \quad \text{Add.} \end{array}$$

Since $14p$ is the correct middle term,

$$8p^2 + 14p + 5 = (4p + 5)(2p + 1).$$

Check: $(4p + 5)(2p + 1) = 8p^2 + 14p + 5$

Work Problem ❷ at the Side.

Example 2 **Factoring a Trinomial with a Negative Middle Term Using FOIL**

Factor $6x^2 - 11x + 3$.

Since 3 has only 1 and 3 or -1 and -3 as factors, it is better here to begin by factoring 3. The last term of the trinomial $6x^2 - 11x + 3$ is positive and the middle term has a negative coefficient, so only negative factors should be considered. Try -3 and -1 as factors of 3:

$$(\quad - 3)(\quad - 1).$$

❸ Factor each trinomial.

(a) $4y^2 - 11y + 6$

The factors of $6x^2$ may be either $6x$ and x, or $2x$ and $3x$. Try $2x$ and $3x$.

$$(2x - 3)(3x - 1) \qquad \text{Correct}$$

$$\begin{array}{c} -9x \\ -2x \\ \hline -11x \quad \text{Add.} \end{array}$$

These factors give the correct middle term, so

$$6x^2 - 11x + 3 = (2x - 3)(3x - 1).$$

Check by multiplying.

Work Problem ❸ at the Side.

(b) $9x^2 - 21x + 10$

Example 3 ▸ Factoring a Trinomial with a Negative Last Term Using FOIL

Factor $8x^2 + 6x - 9$.

The integer 8 has several possible pairs of factors, as does -9. Since the last term is negative, one positive factor and one negative factor of -9 are needed. Since the coefficient of the middle term is small, it is wise to avoid large factors such as 8 or 9. We try 4 and 2 as factors of 8, and 3 and -3 as factors of -9, and check the middle term.

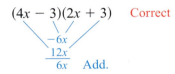

$(4x + 3)(2x - 3)$ Incorrect

$6x$
$-12x$
$-6x$ Add.

Now, let's try exchanging 3 and -3, since only the sign of the middle term is incorrect.

$(4x - 3)(2x + 3)$ Correct

$-6x$
$12x$
$6x$ Add.

This combination produces the correct middle term, so

$$8x^2 + 6x - 9 = (4x - 3)(2x + 3).$$

━━━━━ **Work Problem ④ at the Side.**

Example 4 ▸ Factoring a Trinomial with Two Variables

Factor $12a^2 - ab - 20b^2$.

There are several pairs of factors of $12a^2$, including $12a$ and a, $6a$ and $2a$, and $4a$ and $3a$, just as there are many possible pairs of factors of $-20b^2$, including $20b$ and $-b$, $-20b$ and b, $10b$ and $-2b$, $-10b$ and $2b$, $4b$ and $-5b$, and $-4b$ and $5b$. Once again, since the desired middle term is small, avoid the larger factors. Try the factors $6a$ and $2a$ and $4b$ and $-5b$.

$$(6a + 4b)(2a - 5b)$$

This cannot be correct, as mentioned before, since $6a + 4b$ has a common factor while the given trinomial has none. Try $3a$ and $4a$ with $4b$ and $-5b$.

$$(3a + 4b)(4a - 5b) = 12a^2 + ab - 20b^2 \quad \text{Incorrect}$$

Here the middle term has the wrong sign, so change the signs in the factors.

$$(3a - 4b)(4a + 5b) = 12a^2 - ab - 20b^2 \quad \text{Correct}$$

━━━━━ **Work Problem ⑤ at the Side.**

④ Factor each trinomial, if possible.

(a) $6x^2 + 5x - 4$

(b) $6m^2 - 11m - 10$

(c) $4x^2 - 3x - 7$

(d) $3y^2 + 8y - 6$

⑤ Factor each trinomial.

(a) $2x^2 - 5xy - 3y^2$

(b) $8a^2 + 2ab - 3b^2$

ANSWERS
4. (a) $(3x + 4)(2x - 1)$
   (b) $(2m - 5)(3m + 2)$
   (c) $(4x - 7)(x + 1)$
   (d) prime
5. (a) $(2x + y)(x - 3y)$
   (b) $(4a + 3b)(2a - b)$

6 Factor each trinomial.

(a) $36z^3 - 6z^2 - 72z$

Example 5 Factoring Trinomials with Common Factors

Factor each trinomial.

(a) $15y^3 + 55y^2 + 30y$

First factor out the greatest common factor, $5y$.

$$15y^3 + 55y^2 + 30y = \mathbf{5y}(3y^2 + 11y + 6)$$

Now factor $3y^2 + 11y + 6$. Try $3y$ and y as factors of $3y^2$ and 2 and 3 as factors of 6.

$$(3y + 2)(y + 3) = 3y^2 + 11y + 6 \qquad \text{Correct}$$

The complete factored form of $15y^3 + 55y^2 + 30y$ is

$$15y^3 + 55y^2 + 30y = 5y(3y + 2)(y + 3).$$

Check by multiplying.

(b) $-24a^3 - 42a^2 + 45a$

The common factor could be $3a$ or $-3a$. If we factor out $-3a$, the first term of the trinomial will be positive, which makes it easier to factor.

$$
\begin{aligned}
-24a^3 - 42a^2 + 45a &= \mathbf{-3a}(8a^2 + 14a - 15) &&\text{Factor out } -3a. \\
&= -3a(4a - 3)(2a + 5) &&\text{Use trial and error.}
\end{aligned}
$$

Check by multiplying.

(b) $-12x^3 + 16x^2y + 3xy^2$

CAUTION

This caution bears repeating: Remember to include the common factor in the final factored form.

Work Problem 6 at the Side.

5.4   **EXERCISES**

| FOR EXTRA HELP | Student's Solutions Manual | MyMathLab.com | 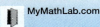 InterAct Math Tutorial Software | AW Math Tutor Center | www.mathxl.com |  Digital Video Tutor CD 4 Videotape 10 |

Decide which is the correct factored form of the given polynomial.

1. $2x^2 - x - 1$

   A. $(2x - 1)(x + 1)$ **B.** $(2x + 1)(x - 1)$

2. $3a^2 - 5a - 2$

   A. $(3a + 1)(a - 2)$ **B.** $(3a - 1)(a + 2)$

3. $4y^2 + 17y - 15$

   A. $(y + 5)(4y - 3)$ **B.** $(2y - 5)(2y + 3)$

4. $12c^2 - 7c - 12$

   A. $(6c - 2)(2c + 6)$ **B.** $(4c + 3)(3c - 4)$

5. $4k^2 + 13mk + 3m^2$

   A. $(4k + m)(k + 3m)$ **B.** $(4k + 3m)(k + m)$

6. $2x^2 + 11x + 12$

   A. $(2x + 3)(x + 4)$ **B.** $(2x + 4)(x + 3)$

Complete each factoring.

7. $6a^2 + 7ab - 20b^2 = (3a - 4b)(\qquad)$

8. $9m^2 - 3mn - 2n^2 = (3m + n)(\qquad)$

9. $2x^2 + 6x - 8 = 2(\qquad)$
$= 2(\qquad)(\qquad)$

10. $3x^2 - 9x - 30 = 3(\qquad)$
$= 3(\qquad)(\qquad)$

11. $4z^3 - 10z^2 - 6z = 2z(\qquad)$
$= 2z(\qquad)(\qquad)$

12. $15r^3 - 39r^2 - 18r = 3r(\qquad)$
$= 3r(\qquad)(\qquad)$

13. For the polynomial $12x^2 + 7x - 12$, 2 is not a common factor. Explain why the binomial $2x - 6$, then, cannot be a factor of the polynomial.

14. Explain how the signs of the last terms of the two binomial factors of a trinomial are determined.

Factor each trinomial completely. See Examples 1–5.

15. $3a^2 + 10a + 7$

16. $7r^2 + 8r + 1$

17. $2y^2 + 7y + 6$

18. $5z^2 + 12z + 4$

19. $15m^2 + m - 2$

20. $6x^2 + x - 1$

21. $12s^2 + 11s - 5$

22. $20x^2 + 11x - 3$

23. $10m^2 - 23m + 12$

24. $6x^2 - 17x + 12$

25. $8w^2 - 14w + 3$

26. $9p^2 - 18p + 8$

27. $20y^2 - 39y - 11$

28. $10x^2 - 11x - 6$

29. $3x^2 - 15x + 16$

30. $2t^2 + 13t - 18$

31. $20x^2 + 22x + 6$

32. $36y^2 + 81y + 45$

33. $40m^2q + mq - 6q$

34. $15a^2b + 22ab + 8b$

35. $15n^4 - 39n^3 + 18n^2$

36. $24a^4 + 10a^3 - 4a^2$

37. $15x^2y^2 - 7xy^2 - 4y^2$

38. $14a^2b^3 + 15ab^3 - 9b^3$

39. $5a^2 - 7ab - 6b^2$

40. $6x^2 - 5xy - y^2$

41. $12s^2 + 11st - 5t^2$

42. $25a^2 + 25ab + 6b^2$

43. $6m^6n + 7m^5n^2 + 2m^4n^3$

44. $12k^3q^4 - 4k^2q^5 - kq^6$

If a trinomial has a negative coefficient for the squared term, such as $-2x^2 + 11x - 12$, *it may be easier to factor by first factoring out the common factor* -1:

$$-2x^2 + 11x - 12 = -1(2x^2 - 11x + 12)$$
$$= -1(2x - 3)(x - 4).$$

Use this method to factor the trinomials in Exercises 45–50.

45. $-x^2 - 4x + 21$

46. $-x^2 + x + 72$

47. $-3x^2 - x + 4$

48. $-5x^2 + 2x + 16$

49. $-2a^2 - 5ab - 2b^2$

50. $-3p^2 + 13pq - 4q^2$

RELATING CONCEPTS (Exercises 51–56) **FOR INDIVIDUAL OR GROUP WORK**

One of the most common problems that beginning algebra students face is this: If an answer obtained doesn't look exactly like the one given in the back of the book, is it necessarily incorrect? Often there are several different equivalent forms of an answer that are all correct. **Work Exercises 51–56 in order,** *to see how and why this is possible for factoring problems.*

51. Factor the integer 35 as the product of two prime numbers.

52. Factor the integer 35 as the product of the negatives of two prime numbers.

53. Verify the following factored form: $6x^2 - 11x + 4 = (3x - 4)(2x - 1)$.

54. Verify the following factored form: $6x^2 - 11x + 4 = (4 - 3x)(1 - 2x)$.

55. Compare the two valid factored forms in Exercises 53 and 54. How do the factors in each case compare?

56. Suppose you know that the correct factored form of a particular trinomial is $(7t - 3)(2t - 5)$. Based on your observations in Exercises 51–55, what is another valid factored form?

5.5 SPECIAL FACTORING TECHNIQUES

By reversing the rules for multiplication of binomials from the last chapter, we get rules for factoring polynomials in certain forms.

OBJECTIVES

1 Factor a difference of squares.

2 Factor a perfect square trinomial.

1 **Factor a difference of squares.** The formula for the product of the sum and difference of the same two terms is

$$(a + b)(a - b) = a^2 - b^2.$$

Reversing this rule leads to the following special factoring rule.

Factoring a Difference of Squares

$$a^2 - b^2 = (a + b)(a - b)$$

For example,

$$m^2 - 16 = m^2 - 4^2 = (m + 4)(m - 4).$$

As the next examples show, the following conditions must be true for a binomial to be a difference of squares.

1. Both terms of the binomial must be squares, such as

$$x^2, \quad 9y^2, \quad 25, \quad 1, \quad m^4.$$

2. The terms must have different signs (one positive and one negative).

Example 1 Factoring Differences of Squares

Factor each binomial, if possible.

$$a^2 - b^2 = (a + b)(a - b)$$

(a) $x^2 - 49 = x^2 - 7^2 = (x + 7)(x - 7)$

(b) $y^2 - m^2 = (y + m)(y - m)$

(c) $z^2 - \dfrac{9}{16} = z^2 - \left(\dfrac{3}{4}\right)^2 = \left(z + \dfrac{3}{4}\right)\left(z - \dfrac{3}{4}\right)$

(d) $x^2 - 8$

Because 8 is not the square of an integer, this binomial is not a difference of squares. It is a prime polynomial.

(e) $p^2 + 16$

Since $p^2 + 16$ is a *sum* of squares, it is not equal to $(p + 4)(p - 4)$. Also, using FOIL,

$$(p - 4)(p - 4) = p^2 - 8p + 16 \neq p^2 + 16$$

and

$$(p + 4)(p + 4) = p^2 + 8p + 16 \neq p^2 + 16,$$

so $p^2 + 16$ is a prime polynomial.

❶ Factor, if possible.

(a) $p^2 - 100$

(b) $x^2 - \dfrac{25}{36}$

(c) $x^2 + y^2$

(d) $9m^2 - 49$

(e) $64a^2 - 25$

❷ Factor completely.

(a) $50r^2 - 32$

(b) $27y^2 - 75$

(c) $25a^2 - 64b^2$

(d) $k^4 - 49$

(e) $81r^4 - 16$

ANSWERS
1. (a) $(p + 10)(p - 10)$
(b) $\left(x + \dfrac{5}{6}\right)\left(x - \dfrac{5}{6}\right)$
(c) prime
(d) $(3m + 7)(3m - 7)$
(e) $(8a + 5)(8a - 5)$
2. (a) $2(5r + 4)(5r - 4)$
(b) $3(3y + 5)(3y - 5)$
(c) $(5a + 8b)(5a - 8b)$
(d) $(k^2 + 7)(k^2 - 7)$
(e) $(9r^2 + 4)(3r + 2)(3r - 2)$

CAUTION

As Example 1(e) suggests, after any common factor is removed, a *sum* of squares cannot be factored.

Example 2 **Factoring Differences of Squares**

Factor each difference of squares.

$$a^2 \quad - \ b^2 \ = \ (a \ + \ b) \ (a \ - \ b)$$

(a) $25m^2 - 16 = (5m)^2 - 4^2 = (5m + 4)(5m - 4)$

(b) $49z^2 - 64 = (7z)^2 - 8^2 = (7z + 8)(7z - 8)$

NOTE

As in previous sections, you should always check a factored form by multiplying.

Work Problem ❶ at the Side.

Example 3 **Factoring More Complex Differences of Squares**

Factor completely.

(a) $81y^2 - 36$
First factor out the common factor, 9.

$$\begin{aligned} 81y^2 - 36 &= \mathbf{9}(9y^2 - 4) &&\text{Factor out 9.} \\ &= 9[(3y)^2 - 2^2] \\ &= 9(3y + 2)(3y - 2) &&\text{Difference of squares} \end{aligned}$$

(b) $9x^2 - 4z^2 = (\mathbf{3x})^2 - (\mathbf{2z})^2 = (\mathbf{3x + 2z})(\mathbf{3x - 2z})$

(c) $\mathbf{p^4} - 36 = (\mathbf{p^2})^2 - 6^2 = (p^2 + 6)(p^2 - 6)$
Neither $p^2 + 6$ nor $p^2 - 6$ can be factored further.

(d) $m^4 - 16 = (m^2)^2 - 4^2$
$$\begin{aligned} &= (m^2 + 4)(\mathbf{m^2 - 4}) &&\text{Difference of squares} \\ &= (m^2 + 4)(\mathbf{m + 2})(\mathbf{m - 2}) &&\text{Difference of squares again} \end{aligned}$$

CAUTION

Remember to factor again when any of the factors is a difference of squares, as in Example 3(d). Check by multiplying.

Work Problem ❷ at the Side.

2 ▬▬ **Factor a perfect square trinomial.** The expressions 144, $4x^2$, and $81m^6$ are called *perfect squares* because

$$144 = \mathbf{12^2}, \quad 4x^2 = (\mathbf{2x})^2, \quad \text{and} \quad 81m^6 = (\mathbf{9m^3})^2.$$

A **perfect square trinomial** is a trinomial that is the square of a binomial. For example, $x^2 + 8x + 16$ is a perfect square trinomial because it is the square of the binomial $x + 4$:

$$x^2 + 8x + 16 = (x + 4)(x + 4) = \mathbf{(x + 4)^2}.$$

For a trinomial to be a perfect square, *two of its terms must be perfect squares*. For this reason, $16x^2 + 4x + 15$ is not a perfect square trinomial because only the term $16x^2$ is a perfect square.

On the other hand, even if two of the terms are perfect squares, the trinomial may not be a perfect square trinomial. For example, $x^2 + 6x + 36$ has two perfect square terms, but it is not a perfect square trinomial. (Try to find a binomial that can be squared to give $x^2 + 6x + 36$.)

We can multiply to see that the square of a binomial gives one of the following perfect square trinomials.

Factoring Perfect Square Trinomials

$$a^2 + 2ab + b^2 = (a + b)^2$$
$$a^2 - 2ab + b^2 = (a - b)^2$$

The middle term of a perfect square trinomial is always twice the product of the two terms in the squared binomial. (This was shown in Section 4.4.) Use this to check any attempt to factor a trinomial that appears to be a perfect square.

> **Example 4**   Factoring a Perfect Square Trinomial

Factor $x^2 + 10x + 25$.

The term x^2 is a perfect square, and so is 25. Try to factor the trinomial as

$$x^2 + 10x + 25 = (x + 5)^2.$$

To check, take twice the product of the two terms in the squared binomial.

$$2 \cdot x \cdot 5 = 10x$$

Twice   First term ⎯   ⎯ Last term
of binomial   of binomial

Since $10x$ is the middle term of the trinomial, the trinomial is a perfect square and can be factored as $(x + 5)^2$. Thus,

$$x^2 + 10x + 25 = (x + 5)^2.$$

━━━━━━━━━ **Work Problem ❸ at the Side.**

> **Example 5**   Factoring Perfect Square Trinomials

Factor each trinomial.

(a) $x^2 - 22x + 121$

The first and last terms are perfect squares ($121 = 11^2$ or $(-11)^2$). Check to see whether the middle term of $x^2 - 22x + 121$ is twice the product of the first and last terms of the binomial $x - 11$.

━━━━ **Continued on Next Page**

❸ Factor each trinomial.

(a) $p^2 + 14p + 49$

(b) $m^2 + 8m + 16$

(c) $x^2 + 2x + 1$

ANSWERS
3. **(a)** $(p + 7)^2$ **(b)** $(m + 4)^2$ **(c)** $(x + 1)^2$

4 Factor each trinomial.

(a) $p^2 - 18p + 81$

(b) $16a^2 + 56a + 49$

(c) $121p^2 + 110p + 100$

(d) $64x^2 - 48x + 9$

(e) $27y^3 + 72y^2 + 48y$

$$2 \cdot x \cdot (-11) = -22x$$

Twice — First term — Last term

Since twice the product of the first and last terms of the binomial is the middle term, $x^2 - 22x + 121$ is a perfect square trinomial and

$$x^2 - 22x + 121 = (x - 11)^2.$$

Notice that the sign of the second term in the squared binomial is the same as the sign of the middle term in the trinomial.

(b) $9m^2 - 24m + 16 = (3m)^2 + 2(3m)(-4) + (-4)^2 = (3m - 4)^2$

Twice — First term — Last term

(c) $25y^2 + 20y + 16$

The first and last terms are perfect squares.

$$25y^2 = (5y)^2 \quad \text{and} \quad 16 = 4^2$$

Twice the product of the first and last terms of the binomial $5y + 4$ is

$$2 \cdot 5y \cdot 4 = 40y,$$

which is not the middle term of $25y^2 + 20y + 16$. This trinomial is not a perfect square. In fact, the trinomial cannot be factored even with the methods of the previous sections; it is a prime polynomial.

(d) $12z^3 + 60z^2 + 75z$

Factor out the common factor, $3z$, first.

$$12z^3 + 60z^2 + 75z = 3z(4z^2 + 20z + 25)$$
$$= 3z[(2z)^2 + 2(2z)(5) + 5^2]$$
$$= 3z(2z + 5)^2$$

NOTE

As noted in Example 5(a), the sign of the second term in the squared binomial is always the same as the sign of the middle term in the trinomial. Also, the first and last terms of a perfect square trinomial must be *positive*, because they are squares. For example, the polynomial $x^2 - 2x - 1$ cannot be a perfect square because the last term is negative.

Perfect square trinomials can also be factored using grouping or FOIL, although using the method of this section is often easier.

Work Problem **4** at the Side.

The methods of factoring discussed in this section are summarized here.

Special Factoring Rules

Difference of squares	$a^2 - b^2 = (a + b)(a - b)$
Perfect square trinomials	$a^2 + 2ab + b^2 = (a + b)^2$
	$a^2 - 2ab + b^2 = (a - b)^2$

ANSWERS
4. (a) $(p - 9)^2$ **(b)** $(4a + 7)^2$ **(c)** prime
(d) $(8x - 3)^2$ **(e)** $3y(3y + 4)^2$

5.5 **EXERCISES**

FOR
EXTRA
HELP

 Student's Solutions Manual   MyMathLab.com   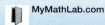 InterAct Math Tutorial Software   AW Math Tutor Center   www.mathxl.com    Digital Video Tutor CD 4 Videotape 11

1. To help you factor a difference of squares, complete the following list of squares.

$1^2 =$ _____ $2^2 =$ _____ $3^2 =$ _____ $4^2 =$ _____ $5^2 =$ _____

$6^2 =$ _____ $7^2 =$ _____ $8^2 =$ _____ $9^2 =$ _____ $10^2 =$ _____

$11^2 =$ _____ $12^2 =$ _____ $13^2 =$ _____ $14^2 =$ _____ $15^2 =$ _____

$16^2 =$ _____ $17^2 =$ _____ $18^2 =$ _____ $19^2 =$ _____ $20^2 =$ _____

2. To use the factoring techniques described in this section, you will sometimes need to recognize fourth powers of integers. Complete the following list of fourth powers.

$1^4 =$ _____ $2^4 =$ _____ $3^4 =$ _____ $4^4 =$ _____ $5^4 =$ _____

3. The following powers of x are all perfect squares: x^2, x^4, x^6, x^8, x^{10}. Based on this observation, we may make a conjecture (an educated guess) that if the power of a variable is divisible by _____ (with 0 remainder), then it is a perfect square.

4. Which of the following are differences of squares?

   A. $x^2 - 4$ **B.** $y^2 + 9$ **C.** $2a^2 - 25$ **D.** $9m^2 - 1$

Factor each binomial completely. Use your answers in Exercises 1 and 2 as necessary. See Examples 1–3.

5. $y^2 - 25$

6. $t^2 - 16$

7. $9r^2 - 4$

8. $4x^2 - 9$

9. $36m^2 - \dfrac{16}{25}$

10. $100b^2 - \dfrac{4}{49}$

11. $36x^2 - 16$

12. $32a^2 - 8$

13. $196p^2 - 225$

14. $361q^2 - 400$

15. $16r^2 - 25a^2$

16. $49m^2 - 100p^2$

17. $100x^2 + 49$

18. $81w^2 + 16$

19. $p^4 - 49$

20. $r^4 - 25$

21. $x^4 - 1$

22. $y^4 - 16$

23. $p^4 - 256$

24. $16k^4 - 1$

25. When a student was directed to factor $x^4 - 81$ completely, his teacher did not give him full credit when he answered $(x^2 + 9)(x^2 - 9)$. The student argued that because his answer does indeed give $x^4 - 81$ when multiplied out, he should be given full credit. Was the teacher justified in her grading of this item? Why or why not?

26. The binomial $4x^2 + 16$ is a sum of squares that *can* be factored. How is this binomial factored? When can a sum of squares be factored?

27. In the polynomial $9y^2 + 14y + 25$, the first and last terms are perfect squares. Can the polynomial be factored? If it can, factor it. If it cannot, explain why it is not a perfect square trinomial.

28. Which of the following are perfect square trinomials?

    A. $y^2 - 13y + 36$ **B.** $x^2 + 6x + 9$ **C.** $4z^2 - 4z + 1$ **D.** $16m^2 + 10m + 1$

Factor each trinomial completely. It may be necessary to factor out the greatest common factor first. See Examples 4 and 5.

29. $w^2 + 2w + 1$

30. $p^2 + 4p + 4$

31. $x^2 - 8x + 16$

32. $x^2 - 10x + 25$

33. $t^2 + t + \dfrac{1}{4}$

34. $m^2 + \dfrac{2}{3}m + \dfrac{1}{9}$

35. $x^2 - 1.0x + .25$

36. $y^2 - 1.4y + .49$

37. $2x^2 + 24x + 72$

38. $3y^2 - 48y + 192$

39. $16x^2 - 40x + 25$

40. $36y^2 - 60y + 25$

41. $49x^2 - 28xy + 4y^2$

42. $4z^2 - 12zw + 9w^2$

43. $64x^2 + 48xy + 9y^2$

44. $9t^2 + 24tr + 16r^2$

45. $50h^3 - 40h^2y + 8hy^2$

46. $18x^3 + 48x^2y + 32xy^2$

RELATING CONCEPTS (Exercises 47–50)   **FOR INDIVIDUAL OR GROUP WORK**

We have seen that multiplication and factoring are reverse processes. We know that multiplication and division are also related. To check a division problem, we multiply the quotient by the divisor to get the dividend. To see how factoring and division are related, **work Exercises 47–50 in order.**

47. Factor $10x^2 + 11x - 6$.

48. Use long division to divide $10x^2 + 11x - 6$ by $2x + 3$.

49. Could we have predicted the result in Exercise 48 from the result in Exercise 47? Explain.

50. Divide $x^3 - 1$ by $x - 1$. Use your answer to factor $x^3 - 1$.

Summary Exercises on **FACTORING**

As you factor a polynomial, ask yourself these questions to decide on a suitable factoring technique.

Factoring a Polynomial

1. Is there a common factor? If so, factor it out.

2. How many terms are in the polynomial?
Two terms: Check to see whether it is a difference of squares.
Three terms: Is it a perfect square trinomial? If the trinomial is not a perfect square, check to see whether the coefficient of the squared term is 1. If so, use the method of Section 5.2. If the coefficient of the squared term of the trinomial is not 1, use the general factoring methods of Sections 5.3 and 5.4.
Four terms: Try to factor the polynomial by grouping.

3. Can any factors be factored further? If so, factor them.

Factor each polynomial completely. Remember to check by multiplying.

1. $32m^9 + 16m^5 + 24m^3$

2. $2m^2 - 10m - 48$

3. $14k^3 + 7k^2 - 70k$

4. $9z^2 + 64$

5. $6z^2 + 31z + 5$

6. $m^2 - 3mn - 4n^2$

7. $49z^2 - 16y^2$

8. $100n^2r^2 + 30nr^3 - 50n^2r$

9. $16x^2 + 20x$

10. $20 + 5m + 12n + 3mn$

11. $10y^2 - 7yz - 6z^2$

12. $y^4 - 81$

13. $m^2 + 2m - 15$

14. $6y^2 - 5y - 4$

15. $32z^3 + 56z^2 - 16z$

16. $15y^2 + 5y$

17. $z^2 - 12z + 36$

18. $9m^2 - 64$

19. $y^2 - 4yk - 12k^2$

20. $16z^2 - 8z + 1$

21. $6y^2 - 6y - 12$

22. $x^2 + \frac{1}{2}x + \frac{1}{16}$

23. $p^2 - 17p + 66$

24. $a^2 + 17a + 72$

25. $k^2 + 9$

26. $108m^2 - 36m + 3$

27. $z^2 - 3za - 10a^2$

28. $2a^3 + a^2 - 14a - 7$

29. $4k^2 - 12k + 9$

30. $a^2 - 3ab - 28b^2$

31. $16r^2 + 24rm + 9m^2$

32. $3k^2 + 4k - 4$

33. $n^2 - 12n - 35$

34. $a^4 - 625$

35. $16k^2 - 48k + 36$

36. $8k^2 - 10k - 3$

37. $36y^6 - 42y^5 - 120y^4$

38. $5z^3 - 45z^2 + 70z$

39. $8p^2 + 23p - 3$

40. $8k^2 - 2kh - 3h^2$

41. $54m^2 - 24z^2$

42. $4k^2 - 20kz + 25z^2$

43. $6a^2 + 10a - 4$

44. $15h^2 + 11hg - 14g^2$

45. $m^2 - 81$

46. $10z^2 - 7z - 6$

47. $125m^4 - 400m^3n + 195m^2n^2$

48. $9y^2 + 12y - 5$

49. $m^2 - 4m + 4$

50. $36x^2 + 32x + 9$

51. $27p^{10} - 45p^9 - 252p^8$

52. $10m^2 + 25m - 60$

53. $4 - 2q - 6p + 3pq$

54. $k^2 - \dfrac{64}{121}$

55. $64p^2 - 100m^2$

56. $m^3 + 4m^2 - 6m - 24$

57. $100a^2 - 81y^2$

58. $8a^2 + 23ab - 3b^2$

59. $a^2 + 8a + 16$

60. $4y^2 - 25$

5.6   SOLVING QUADRATIC EQUATIONS BY FACTORING

Galileo Galilei (1564–1642) developed theories to explain physical phenomena and set up experiments to test his ideas. According to legend, Galileo dropped objects of different weights from the Leaning Tower of Pisa to disprove the belief that heavier objects fall faster than lighter objects. He developed a formula for freely falling objects described by

$$d = 16t^2,$$

where d is the distance in feet that an object falls (disregarding air resistance) in t seconds, regardless of weight. (*Source:* Miller, Charles D., Heeren, Vern E., and Hornsby, John, *Mathematical Ideas,* Ninth Edition, Addison-Wesley Publishing Company, 2001.)

The equation $d = 16t^2$ is a *quadratic equation,* the subject of this section. A quadratic equation contains a squared term and no terms of higher degree.

Quadratic Equation

A **quadratic equation** is an equation that can be written in the form

$$ax^2 + bx + c = 0,$$

where a, b, and c are real numbers, with $a \neq 0$.

The form $ax^2 + bx + c = 0$ is the **standard form** of a quadratic equation. For example,

$$x^2 + 5x + 6 = 0, \quad 2a^2 - 5a = 3, \quad \text{and} \quad y^2 = 4$$

are all quadratic equations, but only $x^2 + 5x + 6 = 0$ is in standard form.

Work Problems ❶ and ❷ at the Side.

1 **Solve quadratic equations by factoring.** We use the **zero-factor property** to solve a quadratic equation by factoring.

Zero-Factor Property

If a and b are real numbers and $ab = 0$, then $a = 0$ or $b = 0$.

In words, if the product of two numbers is 0, then at least one of the numbers must be 0. One number *must* be 0, but both *may* be 0.

Example 1   **Using the Zero-Factor Property**

Solve each equation.

(a) $(x + 3)(2x - 1) = 0$

The product $(x + 3)(2x - 1)$ is equal to 0. By the zero-factor property, the only way that the product of these two factors can be 0 is if at least one of the factors equals 0. Therefore, either $x + 3 = 0$ or $2x - 1 = 0$. Solve each of these two linear equations as in Chapter 2.

$$x + 3 = 0 \quad \text{or} \quad 2x - 1 = 0 \quad \text{Zero-factor property}$$
$$x = -3 \qquad\qquad 2x = 1 \quad \text{Add 1 to each side.}$$
$$x = \frac{1}{2} \quad \text{Divide each side by 2.}$$

Continued on Next Page

❶ Which of the following equations are quadratic equations?

 A. $y^2 - 4y - 5 = 0$

 B. $x^3 - x^2 + 16 = 0$

 C. $2z^2 + 7z = -3$

 D. $x + 2y = -4$

❷ Write each quadratic equation in standard form.

 (a) $x^2 - 3x = 4$

 (b) $y^2 = 9y - 8$

❸ Solve each equation. Check your solutions.

(a) $(x - 5)(x + 2) = 0$

(b) $(3x - 2)(x + 6) = 0$

(c) $z(2z + 5) = 0$

The given equation, $(x + 3)(2x - 1) = 0$, has two solutions, -3 and $\frac{1}{2}$. Check these solutions by substituting -3 for x in the original equation, $(x + 3)(2x - 1) = 0$. Then start over and substitute $\frac{1}{2}$ for x.

If $x = -3$, then

$$(x + 3)(2x - 1) = 0$$

$$(-3 + 3)[2(-3) - 1] = 0 \quad ?$$

$$0(-7) = 0. \quad \text{True}$$

If $x = \frac{1}{2}$, then

$$(x + 3)(2x - 1) = 0$$

$$\left(\frac{1}{2} + 3\right)\left(2 \cdot \frac{1}{2} - 1\right) = 0 \quad ?$$

$$\frac{7}{2}(1 - 1) = 0 \quad ?$$

$$\frac{7}{2} \cdot 0 = 0. \quad \text{True}$$

Both -3 and $\frac{1}{2}$ result in true equations, so they are solutions to the original equation.

(b) $y(3y - 4) = 0$

$$y(3y - 4) = 0$$

$$y = 0 \quad \text{or} \quad 3y - 4 = 0 \qquad \text{Zero-factor property}$$

$$3y = 4$$

$$y = \frac{4}{3}$$

Check these solutions by substituting each one in the original equation. The solutions are 0 and $\frac{4}{3}$.

NOTE

The word *or* as used in Example 1 means "one or the other or both."

Work Problem ❸ at the Side.

In Example 1, each equation to be solved was given with the polynomial in factored form. If the polynomial in an equation is not already factored, first make sure that the equation is in standard form. Then factor.

Example 2 Solving Quadratic Equations

Solve each equation.

(a) $x^2 - 5x = -6$

First, rewrite the equation in standard form by adding 6 to each side.

$$x^2 - 5x = -6$$

$$x^2 - 5x + 6 = 0 \qquad \text{Add 6.}$$

Now factor $x^2 - 5x + 6$. Find two numbers whose product is 6 and whose sum is -5. These two numbers are -2 and -3, so the equation becomes

$$(x - 2)(x - 3) = 0. \qquad \text{Factor.}$$

$$x - 2 = 0 \quad \text{or} \quad x - 3 = 0 \qquad \text{Zero-factor property}$$

$$x = 2 \quad \text{or} \qquad x = 3 \qquad \text{Solve each equation.}$$

Continued on Next Page

ANSWERS

3. **(a)** $-2, 5$ **(b)** $-6, \frac{2}{3}$ **(c)** $-\frac{5}{2}, 0$

Check: If $x = 2$, then   If $x = 3$, then

$$2^2 - 5(2) = -6 \quad ?$$ $$3^2 - 5(3) = -6 \quad ?$$

$$4 - 10 = -6 \quad ?$$ $$9 - 15 = -6 \quad ?$$

$$-6 = -6. \quad \text{True}$$ $$-6 = -6. \quad \text{True}$$

Both solutions check, so the solutions are 2 and 3.

(b) $y^2 = y + 20$

Rewrite the equation in standard form.

$$y^2 = y + 20$$

$$y^2 - y - 20 = 0 \qquad \text{Subtract } y \text{ and 20.}$$

$$(y - 5)(y + 4) = 0 \qquad \text{Factor.}$$

$$y - 5 = 0 \quad \text{or} \quad y + 4 = 0 \qquad \text{Zero-factor property}$$

$$y = 5 \quad \text{or} \qquad y = -4 \qquad \text{Solve each equation.}$$

Check these solutions by substituting each one in the original equation. The solutions are 5 and -4.

================================== **Work Problem ❹ at the Side.**

In summary, follow these steps to solve quadratic equations by factoring.

Solving a Quadratic Equation by Factoring

Step 1   **Write in standard form.** Write the equation so that all terms are on one side of the equals sign in descending powers of the variable, with 0 on the other side.

Step 2   **Factor.** Factor completely.

Step 3   **Use the zero-factor property.** Set each factor with a variable equal to 0, and solve the resulting equations.

Step 4   **Check.** Check each solution in the original equation.

NOTE

Not all quadratic equations can be solved by factoring. A more general method for solving such equations is given in Chapter 9.

Example 3   Solving a Quadratic Equation with a Common Factor

Solve $4p^2 + 40 = 26p$.

Subtract $26p$ from each side and write the equation in standard form to get

$$4p^2 - 26p + 40 = 0.$$

$$2(2p^2 - 13p + 20) = 0 \qquad \text{Factor out 2.}$$

$$2p^2 - 13p + 20 = 0 \qquad \text{Divide each side by 2.}$$

$$(2p - 5)(p - 4) = 0 \qquad \text{Factor.}$$

$$2p - 5 = 0 \quad \text{or} \quad p - 4 = 0 \qquad \text{Zero-factor property}$$

$$2p = 5 \qquad\qquad p = 4$$

$$p = \frac{5}{2}$$

Check that the solutions are $\frac{5}{2}$ and 4 by substituting each one in the original equation.

---

❹ Solve each equation. Check your solutions.

(a) $m^2 - 3m - 10 = 0$

(b) $r^2 + 2r = 8$

5 Solve each equation. Check your solutions.

(a) $10a^2 - 5a - 15 = 0$

(b) $4x^2 - 2x = 42$

Work Problem 5 at the Side.

Example 4 Solving Quadratic Equations

Solve each equation.

(a) $16m^2 - 25 = 0$

$$16m^2 - 25 = 0$$
$$(4m + 5)(4m - 5) = 0 \qquad \text{Factor.}$$
$$4m + 5 = 0 \quad \text{or} \quad 4m - 5 = 0 \qquad \text{Zero-factor property}$$
$$4m = -5 \quad \text{or} \qquad 4m = 5$$
$$m = -\frac{5}{4} \quad \text{or} \qquad m = \frac{5}{4}$$

Check the solutions $-\frac{5}{4}$ and $\frac{5}{4}$ in the original equation.

(b) $k(2k + 5) = 3$

We need to write this equation in standard form.

$$k(2k + 5) = 3$$
$$2k^2 + 5k = 3 \qquad \text{Multiply.}$$
$$2k^2 + 5k - 3 = 0 \qquad \text{Standard form}$$
$$(2k - 1)(k + 3) = 0 \qquad \text{Factor.}$$
$$2k - 1 = 0 \quad \text{or} \quad k + 3 = 0 \qquad \text{Zero-factor property}$$
$$2k = 1 \qquad\qquad k = -3$$
$$k = \frac{1}{2}$$

Check that the solutions are $\frac{1}{2}$ and -3.

(c) $y^2 = 2y$

First write the equation in standard form.

$$y^2 - 2y = 0 \qquad \text{Standard form}$$
$$y(y - 2) = 0 \qquad \text{Factor.}$$
$$y = 0 \quad \text{or} \quad y - 2 = 0 \qquad \text{Zero-factor property}$$
$$y = 2$$

Check that the solutions are 0 and 2.

CAUTION

In Example 4(b), the zero-factor property could not be used to solve the equation as given because of the 3 on the right. Remember that the zero-factor property applies only to a product that equals 0.

In Example 4(c), it is tempting to begin by dividing each side of the equation by y to get $y = 2$. Note that we do not get the other solution, 0, if we divide by a variable. (We *may* divide each side of an equation by a *nonzero* real number, however. For instance, in Example 3 we divided each side by 2.)

Work Problem ➏ at the Side.

2   **Solve other equations by factoring.**   We can also use the zero-factor property to solve equations that involve more than two factors with variables, as shown in Examples 5 and 6. (These equations are *not* quadratic equations. Why not?)

Example 5   **Solving an Equation with More Than Two Factors**

Solve $6z^3 - 6z = 0$.

$$6z^3 - 6z = 0$$
$$6z(z^2 - 1) = 0 \quad \text{Factor out } 6z.$$
$$6z(z + 1)(z - 1) = 0 \quad \text{Factor } z^2 - 1.$$

By an extension of the zero-factor property, this product can equal 0 only if at least one of the factors equals 0. Write and solve three equations, one for each factor with a variable.

$$6z = 0 \quad \text{or} \quad z + 1 = 0 \quad \text{or} \quad z - 1 = 0$$
$$z = 0 \quad \text{or} \quad z = -1 \quad \text{or} \quad z = 1$$

Check by substituting, in turn, 0, -1, and 1 in the original equation. The solutions are 0, -1, and 1.

Work Problem ➐ at the Side.

Example 6   **Solving an Equation with a Quadratic Factor**

Solve $(2x - 1)(x^2 - 9x + 20) = 0$.

$$(2x - 1)(x^2 - 9x + 20) = 0$$
$$(2x - 1)(x - 5)(x - 4) = 0 \quad \text{Factor } x^2 - 9x + 20.$$
$$2x - 1 = 0 \quad \text{or} \quad x - 5 = 0 \quad \text{or} \quad x - 4 = 0 \quad \text{Zero-factor property}$$
$$x = \frac{1}{2} \quad \text{or} \quad x = 5 \quad \text{or} \quad x = 4$$

Check. The solutions are $\frac{1}{2}$, 5, and 4.

Work Problem ➑ at the Side.

CAUTION

In Example 6, it would be unproductive to begin by multiplying the two factors together. Keep in mind that the zero-factor property requires the product of two or more factors to equal 0. Always consider first whether an equation is given in the appropriate form to apply the zero-factor property.

➏ Solve each equation. Check your solutions.

(a) $49m^2 - 9 = 0$

(b) $p(4p + 7) = 2$

(c) $m^2 = 3m$

➐ Solve each equation. Check your solutions.

(a) $r^3 - 16r = 0$

(b) $x^3 - 3x^2 - 18x = 0$

➑ Solve each equation. Check your solutions.

(a) $(m + 3)(m^2 - 11m + 10) = 0$

(b) $(2x + 5)(4x^2 - 9) = 0$

ANSWERS

6. (a) $-\frac{3}{7}, \frac{3}{7}$ (b) $-2, \frac{1}{4}$ (c) 0, 3

7. (a) $-4, 0, 4$ (b) $-3, 0, 6$

8. (a) $-3, 1, 10$ (b) $-\frac{5}{2}, -\frac{3}{2}, \frac{3}{2}$

Real-Data Applications

Factoring Trinomials Made Easy

FOIL is a memory aid that stands for *First, Outer, Inner, Last* and explains how to multiply binomials such as $(3x - 2)(2x + 1)$. The result of multiplying two binomials is typically a trinomial, $6x^2 - x - 2$ in this case. The ***first*** term of the trinomial, $6x^2$, is the *First* product in FOIL; the ***middle*** term, $-x$, is the sum of the *Outer* and *Inner* products in FOIL; and the ***last*** term, -2, is the *Last* product in FOIL. To factor a trinomial, all we have to do is find the *Outer* and *Inner* coefficients that sum to give the coefficient of the middle term, and then use grouping.

Our approach begins with a **key number,** which is found by multiplying the coefficients of the first and last terms of the trinomial. In our example, the key number is -12 since $6(-2) = -12$. We can display the factors of -12 by entering $Y_1 = -12/X$ in a graphing calculator (Screen 1), and using an automatic table (Screen 2). Factors of -12 are automatically displayed in pairs as $1, -12$; $2, -6$; $3, -4$; $4, -3$; and $6, -2$ (Screen 3). You could scroll up or down to find other factors. Note that $5, -2.4$ and $7, -1.714$ are not factor pairs since -2.4 and -1.714 are not integers.

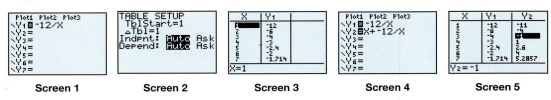

| Screen 1 | Screen 2 | Screen 3 | Screen 4 | Screen 5 |

We want to find the pair of factors that sum to the *middle* term coefficient, -1. We can let the calculator do this, too. Enter $Y_2 = X + -12/X$. In this case, X is one of the factors, and $-12/X$ is the other, so Y_2 will give the sum (Screen 4). Look for -1 in the Y_2 column in Screen 5. (You may have to scroll up or down to find it.)

Now we know that the coefficients of the Outer and Inner products are 3 and -4. So we can write $6x^2 - x - 2$ as $6x^2 + \mathbf{3x - 4x} - 2$. Using the grouping method,

$$(6x^2 + 3x) + (-4x - 2) = 3x\mathbf{(2x + 1)} - 2\mathbf{(2x + 1)} = (3x - 2)\mathbf{(2x + 1)}.$$

For Group Discussion

Factor each trinomial given in the column heads of the table. First find the key number, and then use a calculator to help you find the coefficients of the Outer and Inner products of FOIL. Then apply the grouping method. (If you do not have a graphing calculator, simply use a regular calculator and create a table similar to that shown in Screen 3. Then add across each row to create a table similar to Screen 5.)

Trinomial	$3x^2 - 2x - 8$	$2x^2 - 11x + 15$	$10x^2 + 11x - 6$	$4x^2 + 5x + 3$
Key Number	-24 (Why?)	(*Hint:* Scroll up table.)	(*Hint:* Scroll down table.)	
Outer, Inner Coefficients				
Grouping Method				(*Hint:* What does it mean if the middle term coefficient is *not* listed in the Y_2 column?)

5.6 EXERCISES

 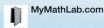
Solve each equation, and check your solutions. See Example 1.

1. $(x + 5)(x - 2) = 0$

2. $(x - 1)(x + 8) = 0$

3. $(2m - 7)(m - 3) = 0$

4. $(6k + 5)(k + 4) = 0$

5. $t(6t + 5) = 0$

6. $w(4w + 1) = 0$

7. $2x(3x - 4) = 0$

8. $6y(4y + 9) = 0$

9. $\left(x + \dfrac{1}{2}\right)\left(2x - \dfrac{1}{3}\right) = 0$

10. $\left(a + \dfrac{2}{3}\right)\left(5a - \dfrac{1}{2}\right) = 0$

11. $(.5z - 1)(2.5z + 2) = 0$

12. $(.25x + 1)(x - .5) = 0$

13. $(x - 9)(x - 9) = 0$

14. $(2y + 1)(2y + 1) = 0$

15. What is wrong with this "solution"?

$$2x(3x - 4) = 0$$
$$x = 2 \quad \text{or} \quad x = 0 \quad \text{or} \quad 3x - 4 = 0$$
$$x = \dfrac{4}{3}$$

The solutions are 2, 0, and $\dfrac{4}{3}$.

16. What is wrong with this "solution"?

$$x(7x - 1) = 0$$
$$7x - 1 = 0 \quad \text{Zero-factor property}$$
$$x = \dfrac{1}{7}$$

The solution is $\dfrac{1}{7}$.

Solve each equation, and check your solutions. See Examples 2–6.

17. $y^2 + 3y + 2 = 0$

18. $p^2 + 8p + 7 = 0$

19. $y^2 - 3y + 2 = 0$

20. $r^2 - 4r + 3 = 0$

21. $x^2 = 24 - 5x$

22. $t^2 = 2t + 15$

23. $x^2 = 3 + 2x$

24. $m^2 = 4 + 3m$

25. $z^2 + 3z = -2$

26. $p^2 - 2p = 3$

27. $m^2 + 8m + 16 = 0$

28. $b^2 - 6b + 9 = 0$

29. $3x^2 + 5x - 2 = 0$

30. $6r^2 - r - 2 = 0$

31. $6p^2 = 4 - 5p$

32. $6x^2 = 4 + 5x$

33. $9s^2 + 12s = -4$

34. $36x^2 + 60x = -25$

35. $y^2 - 9 = 0$

36. $m^2 - 100 = 0$

37. $16k^2 - 49 = 0$

38. $4w^2 - 9 = 0$

39. $n^2 = 121$

40. $x^2 = 400$

41. $x^2 = 7x$

42. $t^2 = 9t$

43. $6r^2 = 3r$

44. $10y^2 = -5y$

45. $g(g - 7) = -10$

46. $r(r - 5) = -6$

47. $z(2z + 7) = 4$

48. $b(2b + 3) = 9$

49. $2(y^2 - 66) = -13y$

50. $3(t^2 + 4) = 20t$

51. $5x^3 - 20x = 0$

52. $3x^3 - 48x = 0$

53. $9y^3 - 49y = 0$

54. $16r^3 - 9r = 0$

55. $(2r + 5)(3r^2 - 16r + 5) = 0$

56. $(3m + 4)(6m^2 + m - 2) = 0$

57. $(2x + 7)(x^2 + 2x - 3) = 0$

58. $(x + 1)(6x^2 + x - 12) = 0$

59. Galileo's formula for freely falling objects, $d = 16t^2$, was given at the beginning of this section. The distance d in feet an object falls depends on the time elapsed t in seconds. (This is an example of an important mathematical concept, the *function.*)

(a) Use Galileo's formula and complete the following table. (*Hint:* Substitute each given value into the formula and solve for the unknown value.)

t in seconds	0	1	2	3	__	__
d in feet	0	16	__	__	256	576

(b) When $t = 0$, $d = 0$. Explain this in the context of the problem.

(c) When you substituted 256 for d and solved for t, you should have found two solutions, 4 and -4. Why doesn't -4 make sense as an answer?

APPLICATIONS OF QUADRATIC EQUATIONS

We can now use factoring to solve quadratic equations that arise in application problems. We follow the same six problem-solving steps given in Section 2.4.

Solving an Applied Problem

Step 1 **Read** the problem carefully until you understand what is given and what is to be found.

Step 2 **Assign a variable** to represent the unknown value, using diagrams or tables as needed. Write down what the variable represents. If necessary, express any other unknown values in terms of the variable.

Step 3 **Write an equation** using the variable expression(s).

Step 4 **Solve** the equation.

Step 5 **State the answer.** Does it seem reasonable?

Step 6 **Check** the answer in the words of the original problem.

1 **Solve problems about geometric figures.** Some of the applied problems in this section require one of the formulas given on the inside covers of the text.

Example 1 **Solving an Area Problem**

The Moens want to plant a rectangular garden in their yard. The width of the garden will be 4 ft less than its length, and they want it to have an area of 96 ft^2. (Recall that ft^2 means square feet.) Find the length and width of the garden.

Step 1 **Read** the problem carefully. We need to find the dimensions of a garden with area 96 ft^2.

Step 2 **Assign a variable.**

Let x = the length of the garden.

Then $x - 4$ = the width. (The width is 4 ft less than the length.)

See Figure 1.

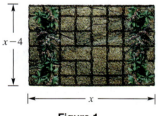

$x - 4$

x

Figure 1

Step 3 **Write an equation.** The area of a rectangle is given by the formula

$$\text{Area} = LW = \text{Length} \times \text{Width}.$$

Substitute 96 for area, x for length, and $x - 4$ for width in the formula.

$$A = LW$$

$$96 = x(x - 4) \qquad \text{Let } A = 96, L = x, W = x - 4.$$

Continued on Next Page

❶ Solve each problem.

(a) The length of a rectangular room is 2 m more than the width. The area of the floor is 48 m². Find the length and width of the room.

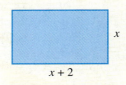

x + 2

(b) The length of each side of a square is increased by 4 in. The sum of the areas of the original square and the larger square is 106 in². What is the length of a side of the original square?

Figure 2

Step 4 **Solve.**

$$96 = x^2 - 4x \qquad \text{Distributive property}$$
$$0 = x^2 - 4x - 96 \qquad \text{Standard form}$$
$$0 = (x - 12)(x + 8) \qquad \text{Factor.}$$
$$x - 12 = 0 \quad \text{or} \quad x + 8 = 0 \qquad \text{Zero-factor property}$$
$$x = 12 \quad \text{or} \qquad x = -8$$

Step 5 **State the answer.** The solutions are 12 and −8. Because a rectangle cannot have a side of negative length, discard the solution −8. Then the length of the garden will be 12 ft, and the width will be 12 − 4 = 8 ft.

Step 6 **Check.** The width is 4 ft less than the length, and the area is 12 · 8 = 96 ft².

CAUTION

When solving applied problems, *always* check solutions against physical facts and discard any answers that are not appropriate.

Work Problem ❶ at the Side.

2 ▭ **Solve problems about consecutive integers.** Recall from our work in Section 2.4 that consecutive integers are integers that are next to each other on a number line, such as 5 and 6, or −11 and −10. Consecutive odd integers are *odd* integers that are next to each other, such as 5 and 7, or −13 and −11. Consecutive even integers are defined similarly; for example, 4 and 6 are consecutive even integers, as are −10 and −8. The following list may be helpful.

Consecutive Integers

Let *x* represent the first of the integers.

Two consecutive integers	$x, x + 1$
Three consecutive integers	$x, x + 1, x + 2$
Two consecutive even or odd integers	$x, x + 2$
Three consecutive even or odd integers	$x, x + 2, x + 4$

Example 2 **Solving a Consecutive Integer Problem**

The product of the numbers on two consecutive post-office boxes is 210. Find the box numbers.

Step 1 **Read** the problem. Note that the boxes are consecutive.

Step 2 **Assign a variable.**

Let x = the first box number.

Then $x + 1$ = the next consecutive box number.

See Figure 2.

Step 3 **Write an equation.** The product of the box numbers is 210, so

$$x(x + 1) = 210.$$

Continued on Next Page

Step 4 **Solve.**

$$x^2 + x = 210$$
$$x^2 + x - 210 = 0 \qquad \text{Standard form}$$
$$(x + 15)(x - 14) = 0 \qquad \text{Factor.}$$
$$x + 15 = 0 \quad \text{or} \quad x - 14 = 0 \qquad \text{Zero-factor property}$$
$$x = -15 \quad \text{or} \qquad x = 14$$

Step 5 **State the answer.** The solutions are -15 and 14. Discard the solution -15 since a box number cannot be negative. When $x = 14$, then $x + 1 = 15$, so the post office boxes have the numbers 14 and 15.

Step 6 **Check.** The numbers 14 and 15 are consecutive and $14 \cdot 15 = 210$, as required.

======================== **Work Problem ❷ at the Side.**

❷ Solve the problem.
   The product of the numbers on two consecutive lockers at a health club is 132. Find the locker numbers.

Example 3 Solving a Consecutive Integer Problem

The product of two consecutive odd integers is 1 less than five times their sum. Find the integers.

Step 1 **Read** carefully. This problem is a little more complicated.

Step 2 **Assign a variable.**

Let $s = $ the smaller integer.

Because the problem mentions consecutive *odd* integers,

$$s + 2 = \text{the next larger odd integer.}$$

Step 3 **Write an equation.** According to the problem, the product is 1 less than five times the sum.

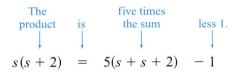

$$s(s + 2) = 5(s + s + 2) - 1$$

Step 4 **Solve.**

$$s^2 + 2s = 5s + 5s + 10 - 1 \qquad \text{Distributive property}$$
$$s^2 + 2s = 10s + 9 \qquad \text{Combine like terms.}$$
$$s^2 - 8s - 9 = 0 \qquad \text{Standard form}$$
$$(s - 9)(s + 1) = 0 \qquad \text{Factor.}$$
$$s - 9 = 0 \quad \text{or} \quad s + 1 = 0 \qquad \text{Zero-factor property}$$
$$s = 9 \quad \text{or} \qquad s = -1$$

Step 5 **State the answer.** We need to find two consecutive odd integers.

If $s = 9$ is the first, then $s + 2 = 9 + 2 = 11$ is the second.

If $s = -1$ is the first, then $s + 2 = -1 + 2 = 1$ is the second.

There are two sets of answers here since integers can be positive or negative.

Step 6 **Check.** The product of the first pair of integers is $9 \cdot 11 = 99$. One less than five times their sum is $5(9 + 11) - 1 = 99$. Thus 9 and 11 satisfy the problem. Repeat the check with -1 and 1.

======================== **Work Problem ❸ at the Side.**

❸ Solve each problem.

   (a) The product of two consecutive even integers is 4 more than two times their sum. Find the integers.

   (b) Find three consecutive odd integers such that the product of the smallest and largest is 16 more than the middle integer.

CAUTION

Do *not* use x, $x + 1$, $x + 3$, and so on to represent consecutive odd integers. To see why, let $x = 3$. Then $x + 1 = 3 + 1 = 4$ and $x + 3 = 3 + 3 = 6$, and 3, 4, and 6 are not consecutive odd integers.

3 **Solve problems using the Pythagorean formula.** The next example requires the Pythagorean formula from geometry.

Pythagorean Formula

If a right triangle (a triangle with a 90° angle) has longest side of length c and two other sides of lengths a and b, then

$$a^2 + b^2 = c^2.$$

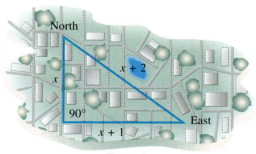

The longest side, the **hypotenuse,** is opposite the right angle. The two shorter sides are the **legs** of the triangle.

Example 4 **Using the Pythagorean Formula**

Ed and Mark leave their office, with Ed traveling north and Mark traveling east. When Mark is 1 mi farther than Ed from the office, the distance between them is 2 mi more than Ed's distance from the office. Find their distances from the office and the distance between them.

Step 1 **Read** the problem again. There will be three answers to this problem.

Step 2 **Assign a variable.** Let x represent Ed's distance from the office, $x + 1$ represent Mark's distance from the office, and $x + 2$ represent the distance between them. Place these on a right triangle, as in Figure 3.

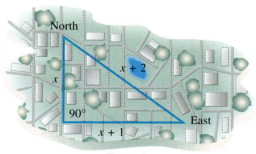

Figure 3

Step 3 **Write an equation.** Substitute into the Pythagorean formula.

$$a^2 + b^2 = c^2$$
$$x^2 + (x + 1)^2 = (x + 2)^2$$

Continued on Next Page

Step 4 **Solve.** $x^2 + x^2 + 2x + 1 = x^2 + 4x + 4$

$$x^2 - 2x - 3 = 0 \quad \text{Standard form}$$

$$(x - 3)(x + 1) = 0 \quad \text{Factor.}$$

$$x - 3 = 0 \quad \text{or} \quad x + 1 = 0 \quad \text{Zero-factor property}$$

$$x = 3 \quad \text{or} \qquad x = -1$$

Step 5 **State the answer.** Since -1 cannot represent a distance, 3 is the only possible answer. Ed's distance is 3 mi, Mark's distance is $3 + 1 = 4$ mi, and the distance between them is $3 + 2 = 5$ mi.

Step 6 **Check.** Since $3^2 + 4^2 = 5^2$, the answers are correct.

CAUTION

When solving a problem involving the Pythagorean formula, be sure that the expressions for the sides are properly placed.

$$\textbf{leg}^2 + \textbf{leg}^2 = \textbf{hypotenuse}^2$$

Work Problem ④ at the Side.

4▭ **Solve problems using given quadratic models.** In Examples 1–4, we wrote quadratic equations to model, or mathematically describe, various situations and then solved the equations. In the final examples, you are given the quadratic models and must use them to determine data.

Example 5 **Finding the Height of a Ball**

A tennis player's serve travels 180 ft per sec (125 mph). If she serves upward, the height h of the ball in feet at time t in seconds is modeled by the quadratic equation

$$h = -16t^2 + 180t + 6.$$

How long will it take for the ball to reach a height of 206 ft?

A height of 206 ft means $h = 206$, so we substitute 206 for h in the equation.

$$\mathbf{206} = -16t^2 + 180t + 6 \quad \text{Let } h = 206.$$

To solve the equation, we first write it in standard form. For convenience, we reverse the sides of the equation.

$$-16t^2 + 180t + 6 = 206$$

$$-16t^2 + 180t - 200 = 0 \quad \text{Standard form}$$

$$4t^2 - 45t + 50 = 0 \quad \text{Divide by } -4.$$

$$(4t - 5)(t - 10) = 0 \quad \text{Factor.}$$

$$4t - 5 = 0 \quad \text{or} \quad t - 10 = 0 \quad \text{Zero-factor property}$$

$$t = \frac{5}{4} \quad \text{or} \qquad t = 10$$

Since we found two acceptable answers, the ball will be 206 ft above the ground twice (once on its way up and once on its way down)—at $\frac{5}{4}$ sec and at 10 sec. See Figure 4.

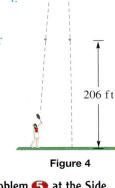

206 ft

Figure 4

Work Problem ⑤ at the Side.

---

④ Solve the problem.

The hypotenuse of a right triangle is 3 in. longer than the longer leg. The shorter leg is 3 in. shorter than the longer leg. Find the lengths of the sides of the triangle.

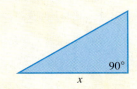

90°

x

⑤ Solve the problem.

The number of impulses fired after a nerve has been stimulated is modeled by

$$I = -x^2 + 2x + 60,$$

where x is in milliseconds (ms) after the stimulation. When will 45 impulses occur? Do you get two solutions? Why is only one answer given?

6 Solve the problem.

Use the model in Example 6 to find the annual percent increase in the amount pharmacies paid for drugs in 1995. Give your answer to the nearest tenth. How does it compare to the actual data from the table?

Example 6   Modeling Increases in Drug Prices

The annual percent increase y in the amount pharmacies paid wholesalers for drugs in the years 1990–1999 can be modeled by the quadratic equation

$$y = .23x^2 - 2.6x + 9,$$

where $x = 0$ represents 1990, $x = 1$ represents 1991, and so on. (*Source: IMS Health,* Retail and Provider Perspective.)

(a) Use the model to find the annual percent increase to the nearest tenth in 1997.

In 1997, $x = 1997 - 1990 = 7$. Substitute 7 for x in the equation.

$$y = .23(7)^2 - 2.6(7) + 9 \quad \text{Let } x = 7.$$
$$y = 2.07 \quad \text{Use a calculator.}$$

To the nearest tenth, pharmacies paid about 2.1% more for drugs in 1997.

(b) Repeat part (a) for 1999.

For 1999, $x = 9$.

$$y = .23(9)^2 - 2.6(9) + 9 \quad \text{Let } x = 1999 - 1990 = 9.$$
$$y = 4.23$$

In 1999, pharmacies paid about 4.2% more for drugs.

(c) The model used in parts (a) and (b) was developed using the data in the table below. How do the results in parts (a) and (b) compare to the actual data from the table?

Year	Percent Increase
1990	8.4
1991	7.2
1992	5.5
1993	3.0
1994	1.7
1995	1.9
1996	1.6
1997	2.5
1998	3.2
1999	4.2

From the table, the actual data for 1997 is 2.5%. Our answer, 2.1%, is a little low. For 1999, the actual data is 4.2%, which is the same as our answer in part (b).

Work Problem 6 at the Side.

NOTE

A graph of the quadratic equation from Example 6 is shown in Figure 5. Notice the basic shape of this graph, which follows the general pattern of the data in the table—it decreases from 1990 to 1996 (with the exception of the data for 1995) and then increases from 1997 to 1999. We will consider such graphs of quadratic equations, called *parabolas*, in more detail in Chapter 9.

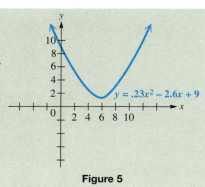

$y = .23x^2 - 2.6x + 9$

Figure 5

5.7   **EXERCISES**

 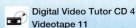

1. To review the six problem-solving steps first introduced in Section 2.4, complete each statement.

Step 1: _____ the problem carefully until you understand what is given and what must be found.

Step 2: Assign a _____ to represent the unknown value.

Step 3: Write a(n) _____ using the variable expression(s).

Step 4: _____ the equation.

Step 5: State the _____.

Step 6: _____ the answer in the words of the _____ problem.

2. A student solves an applied problem and gets 6 or −3 for the length of the side of a square. Which of these answers is reasonable? Explain.

In Exercises 3–6, a figure and a corresponding geometric formula are given. Using x as the variable, complete Steps 3–6 for each problem. (Refer to the steps in Exercise 1 as needed.)

3.

$x + 1$

$2x + 1$

Area of a parallelogram: $A = bh$

The area of this parallelogram is 45 sq. units. Find its base and height.

4.

$x + 5$

$3x + 6$

Area of a triangle: $A = \dfrac{1}{2} bh$

The area of this triangle is 60 sq. units. Find its base and height.

5.

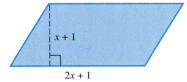

$x - 8$

$x + 8$

Area of a rectangular rug: $A = LW$

The area of this rug is 80 sq. units. Find its length and width.

6.

4

x $x + 2$

Volume of a rectangular Chinese box: $V = LWH$

The volume of this box is 192 cu. units. Find its length and width.

Solve each problem. Check your answers to be sure they are reasonable. Refer to the formulas on the inside covers. See Example 1.

7. The length of a VHS videocassette shell is 3 in. more than its width. The area of the rectangular top side of the shell is 28 in.². Find the length and width of the videocassette shell.

8. A plastic box that holds a standard audiocassette has length 4 cm longer than its width. The area of the rectangular top of the box is 77 cm². Find the length and width of the box.

9. The dimensions of a Gateway EV700 computer monitor screen are such that its length is 3 in. more than its width. If the length is increased by 1 in. while the width remains the same, the area is increased by 10 in.². What are the dimensions of the screen? (*Source:* Author's computer.)

10. The keyboard of the computer in Exercise 9 is 11 in. longer than it is wide. If both its length and width are increased by 2 in., the area of the top of the keyboard is increased by 54 in.². Find the length and width of the keyboard. (*Source:* Author's computer.)

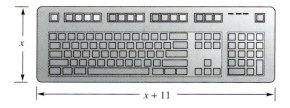

11. A ten-gallon aquarium is 3 in. higher than it is wide. Its length is 21 in., and its volume is 2730 in.³. What are the height and width of the aquarium?

12. A toolbox is 2 ft high, and its width is 3 ft less than its length. If its volume is 80 ft³, find the length and width of the box.

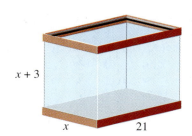

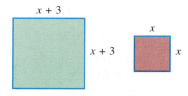

13. A square mirror has sides measuring 2 ft less than the sides of a square painting. If the difference between their areas is 32 ft², find the lengths of the sides of the mirror and the painting.

14. The sides of one square have length 3 m more than the sides of a second square. If the area of the larger square is subtracted from 4 times the area of the smaller square, the result is 36 m². What are the lengths of the sides of each square?

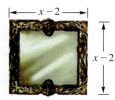

Solve each problem about consecutive integers. See Examples 2 and 3.

15. The product of the numbers on two consecutive volumes of research data is 420. Find the volume numbers.

16. The product of the page numbers on two facing pages of a book is 600. Find the page numbers.

17. The product of two consecutive integers is 11 more than their sum. Find the integers.

18. The product of two consecutive integers is 4 less than four times their sum. Find the integers.

19. Find two consecutive odd integers such that their product is 15 more than three times their sum.

20. Find two consecutive odd integers such that five times their sum is 23 less than their product.

21. Find three consecutive even integers such that the sum of the squares of the smaller two is equal to the square of the largest.

22. Find three consecutive even integers such that the square of the sum of the smaller two is equal to twice the largest.

Use the Pythagorean formula to solve each problem. See Example 4.

23. The hypotenuse of a right triangle is 1 cm longer than the longer leg. The shorter leg is 7 cm shorter than the longer leg. Find the length of the longer leg of the triangle.

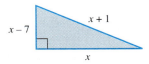

24. The longer leg of a right triangle is 1 m longer than the shorter leg. The hypotenuse is 1 m shorter than twice the shorter leg. Find the length of the shorter leg of the triangle.

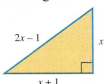

25. Wei-Jen works due north of home. Her husband Alan works due east. They leave for work at the same time. By the time Wei-Jen is 5 mi from home, the distance between them is 1 mi more than Alan's distance from home. How far from home is Alan?

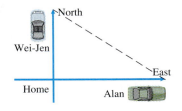

26. Two cars left an intersection at the same time. One traveled north. The other traveled 14 mi farther, but to the east. How far apart were they then, if the distance between them was 4 mi more than the distance traveled east?

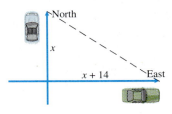

27. A ladder is leaning against a building. The distance from the bottom of the ladder to the building is 4 ft less than the length of the ladder. How high up the side of the building is the top of the ladder if that distance is 2 ft less than the length of the ladder?

28. A lot has the shape of a right triangle with one leg 2 m longer than the other. The hypotenuse is 2 m less than twice the length of the shorter leg. Find the length of the shorter leg.

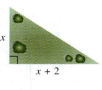

Solve each problem. See Examples 5 and 6.

29. An object propelled from a height of 48 ft with an initial velocity of 32 ft per sec after *t* seconds has height

$$h = -16t^2 + 32t + 48.$$

(a) After how many seconds is the height 64 ft? (*Hint:* Let $h = 64$ and solve.)

(b) After how many seconds is the height 60 ft?

(c) After how many seconds does the object hit the ground? (*Hint:* When the object hits the ground, $h = 0$.)

(d) The quadratic equation from part (c) has two solutions, yet only one of them is appropriate for answering the question. Why is this so?

30. If an object is propelled upward from ground level with an initial velocity of 64 ft per sec, its height *h* in feet *t* seconds later is

$$h = -16t^2 + 64t.$$

(a) After how many seconds is the height 48 ft?

(b) The object reaches its maximum height 2 sec after it is propelled. What is this maximum height?

(c) After how many seconds does the object hit the ground?

(d) The quadratic equation from part (c) has two solutions, yet only one of them is appropriate for answering the question. Why is this so?

31. The table shows the number of cellular phones (in millions) owned by Americans.

Year	Cellular Phones (in millions)
1988	2
1990	5
1992	11
1994	24
1996	44
1998	62

Source: Cellular Telecommunications Industry Association.

We used the data to develop the quadratic equation

$$y = .585x^2 + .295x + 1.75,$$

which models the number of cellular phones y (in millions) in the year x, where $x = 0$ represents 1988, $x = 2$ represents 1990, and so on.

(a) Use the model to find the number of cellular phones in 1990. How does the result compare to the actual data in the table?

(b) What value of x corresponds to 1998?

(c) Use the model to find the number of cellular phones in 1998. How does the result compare to the actual data in the table?

(d) Assuming that the trend in the data continues, use the quadratic equation to predict the number of cellular phones in 2002.

RELATING CONCEPTS (Exercises 32–40)   **FOR INDIVIDUAL OR GROUP WORK**

The U.S. trade deficit represents the amount by which exports are less than imports. It provides not only a sign of economic prosperity but also a warning of potential decline. The data in the table shows the U.S. trade deficit for 1995 through 1999.

Year	Deficit (in billions of dollars)
1995	97.5
1996	104.3
1997	104.7
1998	164.3
1999	271.3

Source: U.S. Department of Commerce.

*Use the data to **work Exercises 32–40 in order.***

32. How much did the trade deficit increase from 1998 to 1999? What percent increase is this (to the nearest percent)?

(continued)

33. The U.S. trade deficit for the years shown in the table can be approximated by the linear equation

$$y = 40.8x + 66.9,$$

where y is the deficit in billions of dollars. Here $x = 0$ represents 1995, $x = 1$ represents 1996, and so on. Use this equation to approximate the trade deficits in 1995, 1997, and 1999.

34. How do your answers from Exercise 33 compare to the actual data in the table?

35. The trade deficit y (in billions of dollars) can also be approximated by the quadratic equation

$$y = 18.5x^2 - 33.4x + 104,$$

where $x = 0$ again represents 1995, $x = 1$ represents 1996, and so on. Use this equation to approximate the trade deficits in 1995, 1997, and 1999.

36. Compare your answers from Exercise 35 to the actual data in the table. Which equation, the linear or quadratic one, models the data better?

37. We can also see graphically why the linear equation is not a very good model for the data. To do so, write the data from the table as a set of ordered pairs (x, y), where x represents the year since 1995 and y represents the trade deficit in billions of dollars.

38. Plot the ordered pairs from Exercise 37 on the graph.

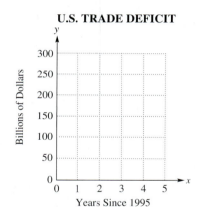

U.S. TRADE DEFICIT

Recall from Chapter 3 that a linear equation has a straight line for its graph. Do the ordered pairs you plotted lie in a linear pattern?

39. Assuming that the trend in the data continues and since the quadratic equation models the data fairly well, use the quadratic equation to predict the trade deficit for the year 2000.

40. The actual trade deficit for 2000 was 369.7 billion dollars. (*Source*: www.census.gov)

   (a) How does the actual deficit for 2000 compare to your prediction from Exercise 39?

   (b) Should the quadratic equation be used to predict the U.S. trade deficit for years after 2000? Explain.

SUMMARY

KEY TERMS

5.1 **factor** An expression A is a factor of an expression B if B can be divided by A with 0 remainder.

factored form An expression is in factored form when it is written as a product.

greatest common factor (GCF) The greatest common factor is the largest quantity that is a factor of each of a group of quantities.

factoring The process of writing a polynomial as a product is called factoring.

5.2 **prime polynomial** A prime polynomial is a polynomial that cannot be factored using only integers.

5.5 **perfect square trinomial** A perfect square trinomial is a trinomial that can be factored as the square of a binomial.

5.6 **quadratic equation** A quadratic equation is an equation that can be written in the form $ax^2 + bx + c = 0$, with $a \neq 0$.

standard form The form $ax^2 + bx + c = 0$ is the standard form of a quadratic equation.

5.7 **hypotenuse** The longest side of a right triangle, opposite the right angle, is the hypotenuse.

legs The two shorter sides of a right triangle are the legs.

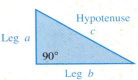

TEST YOUR WORD POWER

See how well you have learned the vocabulary in this chapter. Answers follow the Quick Review.

1. Factoring is
  (a) a method of multiplying polynomials
  (b) the process of writing a polynomial as a product
  (c) the answer in a multiplication problem
  (d) a way to add the terms of a polynomial.

2. A polynomial is in **factored form** when
  (a) it is prime
  (b) it is written as a sum
  (c) the squared term has a coefficient of 1
  (d) it is written as a product.

3. A **perfect square trinomial** is a trinomial
  (a) that can be factored as the square of a binomial
  (b) that cannot be factored
  (c) that is multiplied by a binomial
  (d) where all terms are perfect squares.

4. A **quadratic equation** is an equation that can be written in the form
  (a) $y = mx + b$
  (b) $ax^2 + bx + c = 0 \ (a \neq 0)$
  (c) $Ax + By = C$
  (d) $x = k$.

5. A **hypotenuse** is
  (a) either of the two shorter sides of a triangle
  (b) the shortest side of a right triangle
  (c) the side opposite the right angle in a right triangle
  (d) the longest side in any triangle.

Concepts	*Examples*

5.1 Factors; The Greatest Common Factor

Finding the Greatest Common Factor (GCF)

Step 1   Write each number in prime factored form.

Step 2   List each prime number or each variable that is a factor of every term in the list.

Step 3   Use as exponents on the common prime factors the smallest exponents from the prime factored forms.

Step 4   Multiply the primes from Step 3.

Find the greatest common factor of $4x^2y$, $-6x^2y^3$, and $2xy^2$.

$$4x^2y = \mathbf{2} \cdot 2 \cdot \mathbf{x^2} \cdot \mathbf{y}$$
$$-6x^2y^3 = -1 \cdot \mathbf{2} \cdot 3 \cdot \mathbf{x^2} \cdot \mathbf{y^3}$$
$$2xy^2 = \mathbf{2} \cdot \mathbf{x} \cdot \mathbf{y^2}$$

The greatest common factor is $2xy$.

Factoring by Grouping

Step 1   Group the terms.

Step 2   Factor out the greatest common factor from each group.

Step 3   Factor a common factor from the results of Step 2.

Step 4   If necessary, rearrange terms.

Factor by grouping.

$$2a^2 + 2ab + a + b = (2a^2 + 2ab) + (a + b)$$
$$= 2a(\mathbf{a + b}) + 1(\mathbf{a + b})$$
$$= \mathbf{(a + b)}(2a + 1)$$

5.2 Factoring Trinomials

To factor $x^2 + bx + c$, find m and n such that $mn = c$ and $m + n = b$.

$$mn = c$$
$$x^2 + bx + c$$
$$m + n = b$$

Then $x^2 + bx + c = (x + m)(x + n)$.

Check by multiplying.

Factor $x^2 + \mathbf{6}x + \mathbf{8}$.

$$mn = 8$$
$$x^2 + 6x + 8$$
$$m + n = 6$$

$m = 2$ and $n = 4$

$x^2 + 6x + 8 = (x + 2)(x + 4)$

Check: $(x + 2)(x + 4) = x^2 + 4x + 2x + 8$
$$= x^2 + 6x + 8$$

5.3 Factoring Trinomials by Grouping

To factor $ax^2 + bx + c$ by grouping:
Find m and n.

$$m + n = b$$
$$ax^2 + bx + c$$
$$mn = ac$$

Then factor $ax^2 + mx + nx + b$ by grouping.

Factor $\mathbf{3}x^2 + \mathbf{14}x - \mathbf{5}$.

$$-15$$

Find two integers with a product of $3(-5) = -15$ and a sum of 14. The integers are -1 and 15.

$$3x^2 + \mathbf{14}x - 5 = 3x^2 - x + 15x - 5$$
$$= (3x^2 - x) + (15x - 5)$$
$$= x(3x - 1) + 5(3x - 1)$$
$$= (3x - 1)(x + 5)$$

Concepts	Examples
5.4 *Factoring Trinomials Using FOIL*	By trial and error,
To factor $ax^2 + bx + c$ by trial and error:	
Use FOIL backwards.	$$3x^2 + 14x - 5 = (3x - 1)(x + 5).$$

Concepts	Examples
5.5 *Special Factoring Techniques*	Factor.
Difference of Squares	
$$a^2 - b^2 = (a + b)(a - b)$$	$$4x^2 - 9 = (2x + 3)(2x - 3)$$
Perfect Square Trinomials	
$$a^2 + 2ab + b^2 = (a + b)^2$$	$$9x^2 + 6x + 1 = (3x + 1)^2$$
$$a^2 - 2ab + b^2 = (a - b)^2$$	$$4x^2 - 20x + 25 = (2x - 5)^2$$

Concepts	Examples
5.6 *Solving Quadratic Equations by Factoring*	
Zero-Factor Property	
If a and b are real numbers and $ab = 0$, then $a = 0$ or $b = 0$.	If $(x - 2)(x + 3) = 0$, then $x - 2 = 0$ or $x + 3 = 0$.
Solving a Quadratic Equation by Factoring	Solve $2x^2 = 7x + 15$.
Step 1 Write the equation in standard form.	$$2x^2 - 7x - 15 = 0$$
Step 2 Factor.	$$(2x + 3)(x - 5) = 0$$
Step 3 Use the zero-factor property.	$$2x + 3 = 0 \quad \text{or} \quad x - 5 = 0$$
	$$2x = -3 \qquad\qquad x = 5$$
	$$x = -\frac{3}{2}$$
Step 4 Check.	The solutions $-\frac{3}{2}$ and 5 satisfy the original equation.

Concepts	Examples
5.7 *Applications of Quadratic Equations*	In a right triangle, one leg measures 2 ft longer than the other. The hypotenuse measures 4 ft longer than the shorter leg. Find the lengths of the three sides of the triangle.
Pythagorean Formula	
In a right triangle, the square of the hypotenuse equals the sum of the squares of the legs.	
$$a^2 + b^2 = c^2$$	Let $x =$ the length of the shorter leg. Then
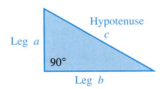	$$x^2 + (x + 2)^2 = (x + 4)^2.$$
Hypotenuse c, Leg a, $90°$, Leg b	Solve this equation to get $x = 6$ or $x = -2$. Discard -2 as a solution. Check that the sides measure 6 ft, $6 + 2 = 8$ ft, and $6 + 4 = 10$ ft.

ANSWERS TO TEST YOUR WORD POWER

1. (b) *Example:* $x^2 - 5x - 14 = (x - 7)(x + 2)$ **2. (d)** *Example:* The factored form of $x^2 - 5x - 14$ is $(x - 7)(x + 2)$.   **3. (a)** *Example:* $a^2 + 2a + 1$ is a perfect square trinomial; its factored form is $(a + 1)^2$.
4. (b) *Examples:* $y^2 - 3y + 2 = 0$, $x^2 - 9 = 0$, $2m^2 = 6m + 8$ **5. (c)** *Example:* See the triangle included in the Quick Review above for Section 5.7.

Real-Data Applications

Stopping Distance

The overall *stopping distance* is the sum of the *thinking distance* (how far the car travels once you realize you have to brake) and the *braking distance* (how far the car travels after you apply the brakes).

The data in the table represents three distinct relationships. The *input* is speed in miles per hour for all three relationships. The *output* is thinking distance in feet for the first relationship, braking distance in feet for the second relationship, and overall stopping distance in feet for the third relationship.

Speed	Thinking Distance	Braking Distance	Overall Stopping Distance
20 mph	20 ft	20 ft	40 ft
30 mph	30 ft	45 ft	75 ft
40 mph	40 ft	80 ft	120 ft
50 mph	50 ft	125 ft	175 ft
60 mph	60 ft	180 ft	240 ft
70 mph	70 ft	245 ft	315 ft

Source: Pass Your Driving Theory Test, British School of Motoring (1996).

For Group Discussion

1. In the relationship between thinking distance and speed, the *output* (y) is numerically the same as the *input* (x).

   (a) Write the equation that expresses this relationship.

   (b) When the speed is doubled from 20 mph to 40 mph, how does the thinking distance change?

   (c) Does the same pattern hold true if a 30 mph speed is doubled?

   (d) Is the equation linear or quadratic? Explain.

2. The relationship between braking distance and speed is given by the equation $y = \frac{1}{20}x^2$.

   (a) Show that this equation corresponds to the table values for speeds of 20 mph and 40 mph.

   (b) When the speed is doubled from 20 mph to 40 mph, how does the braking distance change?

   (c) Does the same pattern hold true if a 30 mph speed is doubled?

   (d) Is the equation linear or quadratic? Explain.

3. The relationship between overall stopping distance and speed is based on the equations in Problems 1 and 2.

   (a) Use those results to write the equation that expresses the relationship between overall stopping distance and speed.

   (b) Is the equation linear or quadratic? Explain.

   (c) A *rule of thumb* for calculating overall stopping distance is to take speed in *tens* of miles per hour, divide by 2, add 1, and multiply the result by the speed in miles per hour. For example, at 40 mph, the rule says: $4 \div 2 + 1 = 3$; $3 \times 40 = 120$ ft. Show why this rule of thumb works. [*Hint:* Factor the right side of the equation in part (a).]

Chapter 5   **REVIEW EXERCISES**

[5.1] *Factor out the greatest common factor or factor by grouping.*

1. $7t + 14$

2. $60z^3 + 30z$

3. $35x^3 + 70x^2$

4. $100m^2n^3 - 50m^3n^4 + 150m^2n^2$

5. $2xy - 8y + 3x - 12$

6. $6y^2 + 9y + 4xy + 6x$

[5.2] *Factor completely.*

7. $x^2 + 5x + 6$

8. $y^2 - 13y + 40$

9. $q^2 + 6q - 27$

10. $r^2 - r - 56$

11. $r^2 - 4rs - 96s^2$

12. $p^2 + 2pq - 120q^2$

13. $8p^3 - 24p^2 - 80p$

14. $3x^4 + 30x^3 + 48x^2$

15. $m^2 - 3mn - 18n^2$

16. $y^2 - 8yz + 15z^2$

17. $p^7 - p^6q - 2p^5q^2$

18. $3r^5 - 6r^4s - 45r^3s^2$

19. $x^2 + x + 1$

20. $3x^2 + 6x + 6$

[5.3–5.4]

21. To begin factoring $6r^2 - 5r - 6$, what are the possible first terms of the two binomial factors, if we consider only positive integer coefficients?

22. What is the first step you would use to factor $2z^3 + 9z^2 - 5z$?

Factor completely.

23. $2k^2 - 5k + 2$

24. $3r^2 + 11r - 4$

25. $6r^2 - 5r - 6$

26. $10z^2 - 3z - 1$

27. $5t^2 - 11t + 12$

28. $24x^5 - 20x^4 + 4x^3$

29. $-6x^2 + 3x + 30$

30. $10r^3s + 17r^2s^2 + 6rs^3$

[5.5]

31. Which one of the following is a difference of squares?

    A. $32x^2 - 1$ **B.** $4x^2y^2 - 25z^2$

    C. $x^2 + 36$ **D.** $25y^3 - 1$

32. Which one of the following is a perfect square trinomial?

    A. $x^2 + x + 1$ **B.** $y^2 - 4y + 9$

    C. $4x^2 + 10x + 25$ **D.** $x^2 - 20x + 100$

Factor completely.

33. $n^2 - 64$

34. $25b^2 - 121$

35. $49y^2 - 25w^2$

36. $144p^2 - 36q^2$

37. $x^2 + 100$

38. $z^2 + 10z + 25$

39. $r^2 - 12r + 36$

40. $9t^2 - 42t + 49$

41. $16m^2 + 40mn + 25n^2$

42. $54x^3 - 72x^2 + 24x$

[5.6] *Solve each equation, and check the solutions.*

43. $(4t + 3)(t - 1) = 0$

44. $(x + 7)(x - 4)(x + 3) = 0$

45. $x(2x - 5) = 0$

46. $z^2 + 4z + 3 = 0$

47. $m^2 - 5m + 4 = 0$

48. $x^2 = -15 + 8x$

49. $3z^2 - 11z - 20 = 0$

50. $81t^2 - 64 = 0$

51. $y^2 = 8y$

52. $n(n - 5) = 6$

53. $t^2 - 14t + 49 = 0$

54. $t^2 = 12(t - 3)$

55. $(5z + 2)(z^2 + 3z + 2) = 0$

56. $x^2 = 9$

[5.7] *Solve each problem.*

57. The length of a rug is 6 ft more than the width. The area is 40 ft². Find the length and width of the rug.

x

$x + 6$

58. The surface area S of a box is given by

$$S = 2WH + 2WL + 2LH.$$

A treasure chest from a sunken galleon has dimensions as shown in the figure. Its surface area is 650 ft². Find its width.

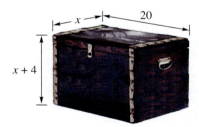

20

x

$x + 4$

59. The length of a rectangle is three times the width. If the width were increased by 3 m while the length remained the same, the new rectangle would have an area of 30 m². Find the length and width of the original rectangle.

60. The volume of a rectangular box is 120 m³. The width of the box is 4 m, and the height is 1 m less than the length. Find the length and height of the box.

61. The product of two consecutive integers is 29 more than their sum. What are the integers?

62. Two cars left an intersection at the same time. One traveled west, and the other traveled 14 mi less, but to the south. How far apart were they then, if the distance between them was 16 mi more than the distance traveled south?

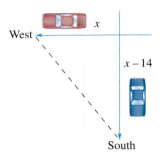

If an object is propelled upward with an initial velocity of 128 ft per sec, its height h after t seconds is

$$h = 128t - 16t^2.$$

Find the height of the object after each period of time.

63. 1 sec

64. 2 sec

65. 4 sec

66. For the object described above, when does it return to the ground?

67. Annual revenue in millions of dollars for eBay is shown in the table.

Year	Annual Revenue (in millions of dollars)
1997	5.1
1998	47.4
1999	224.7

Source: eBay.

Using the data, we developed the quadratic equation

$$y = 67.5x^2 - 25.2x + 5.1$$

to model eBay revenues *y* in year *x,* where $x = 0$ represents 1997, $x = 1$ represents 1998, and so on. Because only three years of data were used to determine the model, we must be careful about using it to predict revenue for years beyond 1999.

(a) Use the model to predict annual revenue for eBay in 2000.

(b) The revenue for eBay through the first half of 2000 was $183.2 million. Given this information, do you think your prediction in part (a) is reliable? Explain.

MIXED REVIEW EXERCISES

68. Which of the following is *not* factored completely?

    A. $3(7t)$ **B.** $3x(7t + 4)$ **C.** $(3 + x)(7t + 4)$ **D.** $3(7t + 4) + x(7t + 4)$

69. Although $(2x + 8)(3x - 4) = 6x^2 + 16x - 32$ is a true statement, the polynomial is not factored completely. Explain why and give the complete factored form.

Factor completely.

70. $z^2 - 11zx + 10x^2$ **71.** $3k^2 + 11k + 10$

72. $15m^2 + 20mp - 12m - 16p$ **73.** $y^4 - 625$

74. $6m^3 - 21m^2 - 45m$ **75.** $24ab^3c^2 - 56a^2bc^3 + 72a^2b^2c$

76. $25a^2 + 15ab + 9b^2$ **77.** $12x^2yz^3 + 12xy^2z - 30x^3y^2z^4$

78. $2a^5 - 8a^4 - 24a^3$ **79.** $12r^2 + 8rq - 15q^2$

80. $100a^2 - 9$ **81.** $49t^2 + 56t + 16$

Solve.

82. $t(t - 7) = 0$ **83.** $x^2 + 3x = 10$ **84.** $25x^2 + 20x + 4 = 0$

Solve each problem.

85. A lot is shaped like a right triangle. The hypotenuse is 3 m longer than the longer leg. The longer leg is 6 m longer than twice the length of the shorter leg. Find the lengths of the sides of the lot.

86. A pyramid has a rectangular base with a length that is 2 m more than the width. The height of the pyramid is 6 m, and its volume is 48 m^3. Find the length and width of the base.

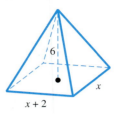

87. The product of the smaller two of three consecutive integers is equal to 23 plus the largest. Find the integers.

88. If an object is dropped, the distance d in feet it falls in t seconds (disregarding air resistance) is given by the quadratic equation

$$d = 16t^2.$$

Find the distance an object would fall in the following times.

(a) 4 sec   **(b)** 8 sec

89. The floor plan for a house is a rectangle with length 7 m more than its width. The area is 170 m^2. Find the width and length of the house.

90. The triangular sail of a schooner has an area of 30 m^2. The height of the sail is 4 m more than the base. Find the base of the sail.

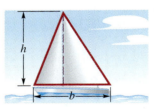

Chapter 5 TEST

1. Which one of the following is the correct, completely factored form of $2x^2 - 2x - 24$?

   A. $(2x + 6)(x - 4)$ **B.** $(x + 3)(2x - 8)$

   C. $2(x + 4)(x - 3)$ **D.** $2(x + 3)(x - 4)$

1. _____

Factor each polynomial completely.

2. $12x^2 - 30x$

2. _____

3. $2m^3n^2 + 3m^3n - 5m^2n^2$

3. _____

4. $2ax - 2bx + ay - by$

4. _____

5. $x^2 - 9x + 14$

5. _____

6. $2x^2 + x - 3$

6. _____

7. $6x^2 - 19x - 7$

7. _____

8. $3x^2 - 12x - 15$

8. _____

9. $10z^2 - 17z + 3$

9. _____

10. $t^2 + 2t + 3$

10. _____

11. $x^2 + 36$

11. _____

12. $y^2 - 49$

12. _____

13. $9y^2 - 64$

13. _____

14. $x^2 + 16x + 64$

14. _____

15. $4x^2 - 28xy + 49y^2$

15. _____

16. $-2x^2 - 4x - 2$

16. _____

17. $6t^4 + 3t^3 - 108t^2$

17. _____

18. $4r^2 + 10rt + 25t^2$

18. _____

19. _____

19. $4t^3 + 32t^2 + 64t$

20. _____

20. $x^4 - 81$

21. _____

21. Why is $(p + 3)(p + 3)$ *not* the correct factored form of $p^2 + 9$?

Solve each equation.

22. _____

22. $(x + 3)(x - 9) = 0$

23. _____

23. $2r^2 - 13r + 6 = 0$

24. _____

24. $25x^2 - 4 = 0$

25. _____

25. $x(x - 20) = -100$

26. _____

26. $t^2 = 3t$

Solve each problem.

27. _____

27. The length of a rectangular flower bed is 3 ft less than twice its width. The area of the bed is 54 ft². Find the dimensions of the flower bed.

28. _____

28. Find two consecutive integers such that the square of the sum of the two integers is 11 more than the smaller integer.

29. _____

29. A carpenter needs to cut a brace to support a wall stud, as shown in the figure. The brace should be 7 ft less than three times the length of the stud. If the brace will be anchored on the floor 15 ft away from the stud, how long should the brace be?

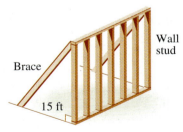

Brace

Wall
stud

15 ft

30. _____

30. TV viewers have more choices than ever. The number of cable TV channels y from 1984 through 1999 can be approximated by the quadratic equation

$$y = .57x^2 + .31x + 48,$$

where $x = 0$ represents 1984, $x = 1$ represents 1985, and so on. (*Source:* National Cable Television Association.) Use the model to estimate the number of cable TV channels in 1999. Round your answer to the nearest whole number.

Solve each equation.

1. $3x + 2(x - 4) = 4(x - 2)$

2. $.3x + .9x = .06$

3. $\frac{2}{3}y - \frac{1}{2}(y - 4) = 3$

4. Solve for P: $A = P + Prt$

5. From a list of "everyday items" often taken for granted, adults were recently surveyed as to those items they wouldn't want to live without. Complete the results shown in the table if 500 adults were surveyed.

Item	Percent That Wouldn't Want to Live Without	Number That Wouldn't Want to Live Without
Toilet paper	69%	____
Zipper	42%	____
Frozen foods	____	190
Self-stick note pads	____	75

(Other items included tape, hairspray, pantyhose, paper clips, and Velcro.)
Source: Market Facts for Kleenex Cottonelle.

Solve each problem.

6. At the 1998 Winter Olympics in Nagano, Japan, the top medal winner was Germany with 29. Germany won 1 more silver medal than bronze and 3 more gold medals than silver. Find the number of each type of medal won. (*Source: The World Almanac and Book of Facts,* 2000.)

7. In July 2000, roughly 144 million people surfed the Web from home. This was a 35% increase from the same month the previous year. How many people, to the nearest million, surfed the Web from home in July 1999? (*Source: The Gazette,* September 3, 2000.)

8. Find the measures of the marked angles.

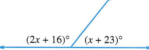

$(2x + 16)°$ $(x + 23)°$

9. Fill in each blank with *positive* or *negative*. The point with coordinates (a, b) is in
   (a) quadrant II if a is _____ and b is _____.
   (b) quadrant III if a is _____ and b is _____.

Consider the equation $y = 12x + 3$. Find the following.

10. The x- and y-intercepts

11. The slope

12. The graph

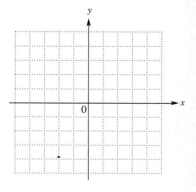

13. The points on the graph show the number of U.S. radio stations in the years 1993–1999, along with the graph of a linear equation that models the data. Use the ordered pairs shown on the graph to find the slope of the line to the nearest whole number. Interpret the slope.

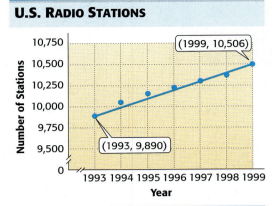

U.S. RADIO STATIONS

(1999, 10,506)

(1993, 9,890)

Source: M Street Corporation.

Evaluate each expression.

14. $2^{-3} \cdot 2^5$

15. $\left(\dfrac{3}{4}\right)^{-2}$

16. $\dfrac{6^5 \cdot 6^{-2}}{6^3}$

17. $\left(\dfrac{4^{-3} \cdot 4^4}{4^5}\right)^{-1}$

Simplify each expression and write the answer using only positive exponents. Assume no denominators are 0.

18. $\dfrac{(p^2)^3 p^{-4}}{(p^{-3})^{-1} p}$

19. $\dfrac{(m^{-2})^3 m}{m^5 m^{-4}}$

Perform the indicated operations.

20. $(2k^2 + 4k) - (5k^2 - 2) - (k^2 + 8k - 6)$

21. $(9x + 6)(5x - 3)$

22. $(3p + 2)^2$

23. $\dfrac{8x^4 + 12x^3 - 6x^2 + 20x}{2x}$

24. To make a pound of honey, bees may travel 55,000 mi and visit more than 2,000,000 flowers. (*Source: Home & Garden* magazine.) Write the two given numbers in scientific notation.

Factor completely.

25. $2a^2 + 7a - 4$

26. $10m^2 + 19m + 6$

27. $8t^2 + 10tv + 3v^2$

28. $4p^2 - 12p + 9$

29. $25r^2 - 81t^2$

30. $2pq + 6p^3q + 8p^2q$

Solve each equation.

31. $6m^2 + m - 2 = 0$

32. $8x^2 = 64x$

33. The length of the hypotenuse of a right triangle is twice the length of the shorter leg, plus 3 m. The longer leg is 7 m longer than the shorter leg. Find the lengths of the sides.

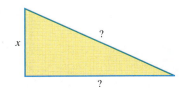

Rational Expressions and Applications

6

At the 2000 Olympic games in Sydney, Australia, Dutch swimmer Inge de Bruijn won three events, the last of which was the women's 50-m freestyle. She completed the race in 24.32 sec. In Example 2(c) of Section 6.7, we use a rational expression to find de Bruijn's rate.

You're Connected

6.1 THE FUNDAMENTAL PROPERTY OF RATIONAL EXPRESSIONS

OBJECTIVES

1 Find the values of the variable for which a rational expression is undefined.

2 Find the numerical value of a rational expression.

3 Write rational expressions in lowest terms.

4 Recognize equivalent forms of rational expressions.

The quotient of two integers (with denominator not 0) is called a rational number. In the same way, the quotient of two polynomials with denominator not equal to 0 is called a *rational expression*. Our work with rational expressions will require much of what we learned in Chapters 4 and 5 on polynomials and factoring, as well as the rules for fractions from Chapter R.

Rational Expression

A **rational expression** is an expression of the form

$$\frac{P}{Q},$$

where P and Q are polynomials, with $Q \neq 0$.

Examples of rational expressions include

$$\frac{-6x}{x^3 + 8}, \quad \frac{9x}{y + 3}, \quad \text{and} \quad \frac{2m^3}{8}.$$

1 **Find the values of the variable for which a rational expression is undefined.** A fraction with denominator 0 is *not* a rational expression because division by 0 is undefined. For this reason, be careful when substituting a number in the denominator of a rational expression. For example, in

$$\frac{8x^2}{x - 3},$$

the variable x can take on any value except 3. When $x = 3$, the denominator becomes $3 - 3 = 0$, making the expression undefined.

To determine the values for which a rational expression is undefined, use the following procedure.

Determining When a Rational Expression Is Undefined

Step 1 Set the denominator of the rational expression equal to 0.

Step 2 Solve this equation.

Step 3 The solutions of the equation are the values that make the rational expression undefined.

Example 1 **Finding Values That Make Rational Expressions Undefined**

Find any values of the variable for which each rational expression is undefined.

(a) $\dfrac{p + 5}{3p + 2}$

Remember that the *numerator* may be any number; we must find any value of p that makes the *denominator* equal to 0 since division by 0 is undefined.

Continued on Next Page

Step 1 Set the denominator equal to 0.

$$3p + 2 = 0$$

Step 2 Solve this equation.

$$3p = -2$$

$$p = -\frac{2}{3}$$

Step 3 Since $p = -\frac{2}{3}$ will make the denominator 0, the given expression is undefined for $-\frac{2}{3}$.

(b) $\dfrac{9m^2}{m^2 - 5m + 6}$

Set the denominator equal to 0, and then find the solutions of the equation $m^2 - 5m + 6 = 0$.

$$(m - 2)(m - 3) = 0 \qquad \text{Factor.}$$

$$m - 2 = 0 \quad \text{or} \quad m - 3 = 0 \qquad \text{Zero-factor property}$$

$$m = 2 \quad \text{or} \qquad m = 3$$

The original expression is undefined for $m = 2$ and for $m = 3$.

(c) $\dfrac{2r}{r^2 + 1}$

This denominator cannot equal 0 for any value of r because r^2 is always greater than or equal to 0, and adding 1 makes the sum greater than 0. Thus, there are no values for which this rational expression is undefined.

=== **Work Problem ❶ at the Side.**

2 **Find the numerical value of a rational expression.** We use substitution to evaluate a rational expression for a given value of the variable.

Example 2 **Evaluating Rational Expressions**

Find the numerical value of $\dfrac{3x + 6}{2x - 4}$ for each value of x.

(a) $x = 1$

$$\frac{3x + 6}{2x - 4} = \frac{3(1) + 6}{2(1) - 4} \qquad \text{Let } x = 1.$$

$$= \frac{9}{-2} = -\frac{9}{2}$$

(b) $x = 2$

$$\frac{3x + 6}{2x - 4} = \frac{3(2) + 6}{2(2) - 4} = \frac{12}{0} \qquad \text{Let } x = 2.$$

Substituting 2 for x makes the denominator 0, so the expression is undefined when $x = 2$.

=== **Work Problem ❷ at the Side.**

---

❶ Find all values for which each rational expression is undefined.

(a) $\dfrac{x + 2}{x - 5}$

(b) $\dfrac{3r}{r^2 + 6r + 8}$

(c) $\dfrac{-5m}{m^2 + 4}$

❷ Find the value of each rational expression when $x = 3$.

(a) $\dfrac{x}{2x + 1}$

(b) $\dfrac{2x + 6}{x - 3}$

❸ Use the fundamental property of rational expressions to write each rational expression in lowest terms.

(a) $\dfrac{5x^4}{15x^2}$

(b) $\dfrac{6p^3}{2p^2}$

3 ▢ **Write rational expressions in lowest terms.** A fraction such as $\frac{2}{3}$ is said to be in *lowest terms*. How can "lowest terms" be defined? We use the idea of greatest common factor for this definition, which applies to all rational expressions.

Lowest Terms

A rational expression $\dfrac{P}{Q}$ ($Q \neq 0$) is in **lowest terms** if the greatest common factor of its numerator and denominator is 1.

The properties of rational numbers also apply to rational expressions. We use the fundamental property of rational expressions to write a rational expression in lowest terms.

Fundamental Property of Rational Expressions

If $\dfrac{P}{Q}$ ($Q \neq 0$) is a rational expression and if K represents any polynomial, where $K \neq 0$, then

$$\frac{PK}{QK} = \frac{P}{Q}.$$

This property is based on the identity property of multiplication, since

$$\frac{PK}{QK} = \frac{P}{Q} \cdot \frac{K}{K} = \frac{P}{Q} \cdot 1 = \frac{P}{Q}.$$

The next example shows how to write both a rational number and a rational expression in lowest terms. Notice the similarity in the procedures. In both cases, we factor and then divide out the greatest common factor.

Example 3 **Writing in Lowest Terms**

Write each expression in lowest terms.

(a) $\dfrac{30}{72}$
Begin by factoring.

$$\frac{30}{72} = \frac{2 \cdot 3 \cdot 5}{2 \cdot 2 \cdot 2 \cdot 3 \cdot 3}$$

(b) $\dfrac{14k^2}{2k^3}$
Write k^2 as $k \cdot k$ and k^3 as $k \cdot k \cdot k$.

$$\frac{14k^2}{2k^3} = \frac{2 \cdot 7 \cdot k \cdot k}{2 \cdot k \cdot k \cdot k}$$

Group any factors common to the numerator and denominator.

$$\frac{30}{72} = \frac{5 \cdot (2 \cdot 3)}{2 \cdot 2 \cdot 3 \cdot (2 \cdot 3)}$$

$$\frac{14k^2}{2k^3} = \frac{7(2 \cdot k \cdot k)}{k(2 \cdot k \cdot k)}$$

Use the fundamental property.

$$\frac{30}{72} = \frac{5}{2 \cdot 2 \cdot 3} = \frac{5}{12}$$

$$\frac{14k^2}{2k^3} = \frac{7}{k}$$

Work Problem ❸ at the Side.

Example 4 **Writing in Lowest Terms**

Write each rational expression in lowest terms.

(a) $\dfrac{3x - 12}{5x - 20}$

Begin by factoring both numerator and denominator. Then use the fundamental property of rational expressions.

$$\frac{3x - 12}{5x - 20} = \frac{3(x - 4)}{5(x - 4)} = \frac{3}{5}$$

The rational expression $\dfrac{3x - 12}{5x - 20}$ is equal to $\dfrac{3}{5}$ for all values of x, where $x \neq 4$ (since the denominator of the original rational expression is 0 when x is 4).

(b) $\dfrac{m^2 + 2m - 8}{2m^2 - m - 6}$

$$\frac{m^2 + 2m - 8}{2m^2 - m - 6} = \frac{(m + 4)(m - 2)}{(2m + 3)(m - 2)} \quad \text{Factor.}$$

$$= \frac{m + 4}{2m + 3} \quad \text{Fundamental property}$$

Thus, $\dfrac{m^2 + 2m - 8}{2m^2 - m - 6} = \dfrac{m + 4}{2m + 3}$ for $m \neq -\dfrac{3}{2}$ and $m \neq 2$, since the denominator of the original expression is 0 for these values of m.

From now on, we will write statements of equality of rational expressions with the understanding that they apply only to those real numbers that make neither denominator equal to 0.

CAUTION

One of the most common errors in algebra occurs when students attempt to write rational expressions in lowest terms *before factoring*. The fundamental property is applied only *after* the numerator and denominator are expressed in factored form. For example, although x appears in both the numerator and denominator in Example 4(a), and 12 and 20 have a common factor of 4, the fundamental property cannot be used before factoring because $3x$, $5x$, 12, and 20 are *terms*, not *factors*. Terms are *added* or *subtracted*; factors are *multiplied* or *divided*.

Correct: $\dfrac{6 + 2}{3 + 2} = \dfrac{8}{5}$

Incorrect: $\dfrac{6 + 2}{3 + 2} = \dfrac{6}{3} + \dfrac{2}{2}$
$= 2 + 1$
$= 3$

Correct: $\dfrac{2x + 3}{4x + 6} = \dfrac{2x + 3}{2(2x + 3)} = \dfrac{1}{2}$

Incorrect: $\dfrac{2x + 3}{4x + 6} = \dfrac{1 + 1}{2 + 2}$
$= \dfrac{2}{4}$
$= \dfrac{1}{2}$

Work Problem ④ at the Side.

④ Write each rational expression in lowest terms.

(a) $\dfrac{4y + 2}{6y + 3}$

(b) $\dfrac{8p + 8q}{5p + 5q}$

(c) $\dfrac{x^2 + 4x + 4}{4x + 8}$

(d) $\dfrac{a^2 - b^2}{a^2 + 2ab + b^2}$

5 Write each rational expression in lowest terms.

(a) $\dfrac{5 - y}{y - 5}$

Example 5 Writing in Lowest Terms (Factors Are Opposites)

Write $\dfrac{x - y}{y - x}$ in lowest terms.

At first glance, there does not seem to be any way in which $x - y$ and $y - x$ can be factored to get a common factor. However, $y - x$ can be factored as

$$y - x = \mathbf{-1}(-y + x) = -1(x - y).$$

Now, use the fundamental property to simplify.

$$\frac{x - y}{y - x} = \frac{1(x - y)}{-1(x - y)} = \frac{1}{-1} = -1$$

In Example 5, notice that $y - x$ is the opposite of $x - y$. A general rule for this situation follows.

> If the numerator and the denominator of a rational expression are opposites, such as in $\dfrac{x - y}{y - x}$, the rational expression is equal to -1.

(b) $\dfrac{m - n}{n - m}$

CAUTION

> Although x and y appear in both the numerator and denominator in Example 5, it is not possible to use the fundamental property right away because they are *terms*, not *factors*. Terms are *added*, while factors are *multiplied*.

Example 6 Writing in Lowest Terms (Factors Are Opposites)

Write each rational expression in lowest terms.

(a) $\dfrac{2 - m}{m - 2}$

Since $2 - m$ and $m - 2$ (or $-2 + m$) are opposites,

$$\frac{2 - m}{m - 2} = -1.$$

(c) $\dfrac{9 - k}{9 + k}$

(b) $\dfrac{3 + r}{3 - r}$

The quantity $3 - r$ *is not* the opposite of $3 + r$. This rational expression is already in lowest terms.

Work Problem 5 at the Side.

4 ▭ **Recognize equivalent forms of rational expressions.**   When working with rational expressions, it is important to be able to recognize equivalent forms of an expression. For example, the common fraction $-\frac{5}{6}$ can also be written as $\frac{-5}{6}$ and as $\frac{5}{-6}$. Consider also the fraction

$$-\frac{2x + 3}{2}.$$

The $-$ sign representing the -1 factor is in front of the fraction, on the same line as the fraction bar. The -1 factor may be placed in front of the fraction, in the numerator, or in the denominator. Some other acceptable forms of this fraction are

$$\frac{-(2x + 3)}{2},$$

$$\frac{-2x - 3}{2},$$

and
$$\frac{2x + 3}{-2}.$$

However, $\frac{-2x + 3}{2}$ is *not* an acceptable form, because the sign preceding 3 in the numerator should be $-$ rather than $+$.

Example 7   **Writing Equivalent Forms of a Rational Expression**

Write four equivalent forms of the rational expression

$$-\frac{3x + 2}{x - 6}.$$

   If we apply the negative sign to the numerator, we have the equivalent form

$$\frac{-(3x + 2)}{x - 6}.$$

By distributing the negative sign in this expression, we have another equivalent form,

$$\frac{-3x - 2}{x - 6}.$$

If we apply the negative sign to the denominator of the fraction, we get

$$\frac{3x + 2}{-(x - 6)}$$

or, distributing once again,

$$\frac{3x + 2}{-x + 6}.$$

6 Decide whether each rational expression is equivalent to

$$-\frac{2x - 6}{x + 3}.$$

(a) $\dfrac{-(2x - 6)}{x + 3}$

(b) $\dfrac{-2x + 6}{x + 3}$

(c) $\dfrac{-2x - 6}{x + 3}$

(d) $\dfrac{2x - 6}{-(x + 3)}$

(e) $\dfrac{2x - 6}{-x - 3}$

(f) $\dfrac{2x - 6}{x - 3}$

CAUTION

In Example 7, it would be incorrect to distribute the negative sign to *both* the numerator *and* the denominator. This would lead to the *opposite* of the original expression.

Work Problem 6 at the Side.

6.1 EXERCISES

1. Fill in each blank with the correct response.

 (a) The rational expression $\dfrac{x + 5}{x - 3}$ is undefined when $x = $ _____, and is equal to 0 when $x = $ _____.

 (b) The rational expression $\dfrac{p - q}{q - p}$ is undefined when $p = $ _____, and in all other cases when written in lowest terms is equal to _____.

2. Make the correct choice for each blank.

 (a) $\dfrac{4 - r^2}{4 + r^2}$ _____ equal to -1.
     (is/is not)

 (b) $\dfrac{5 + 2x}{3 - x}$ and $\dfrac{-5 - 2x}{x - 3}$ _____ equivalent rational expressions.
     (are/are not)

3. Define *rational expression* in your own words, and give an example.

4. Give an example of a rational expression that is not in lowest terms, and then show the steps required to write it in lowest terms.

Find any value(s) for which each rational expression is undefined. See Example 1.

5. $\dfrac{2}{5y}$

6. $\dfrac{7}{3z}$

7. $\dfrac{4x^2}{3x - 5}$

8. $\dfrac{2x^3}{3x - 4}$

9. $\dfrac{m + 2}{m^2 + m - 6}$

10. $\dfrac{r - 5}{r^2 - 5r + 4}$

11. $\dfrac{3x}{x^2 + 2}$

12. $\dfrac{4q}{q^2 + 9}$

Find the numerical value of each rational expression when (a) $x = 2$ and (b) $x = -3$. See Example 2.

13. $\dfrac{5x - 2}{4x}$

14. $\dfrac{3x + 1}{5x}$

15. $\dfrac{2x^2 - 4x}{3x}$

16. $\dfrac{4x^2 - 1}{5x}$

17. $\dfrac{(-3x)^2}{4x + 12}$

18. $\dfrac{(-2x)^3}{3x + 9}$

19. $\dfrac{5x + 2}{2x^2 + 11x + 12}$

20. $\dfrac{7 - 3x}{3x^2 - 7x + 2}$

21. If 2 is substituted for x in the rational expression $\dfrac{x - 2}{x^2 - 4}$, the result is $\dfrac{0}{0}$. We often hear the statement "Any number divided by itself is 1." Does this mean that this expression is equal to 1 for $x = 2$? If not, explain.

22. For $x \neq 2$, the rational expression $\dfrac{2(x - 2)}{x - 2}$ is equal to 2. Can $\dfrac{2x - 2}{x - 2}$ also be simplified to 2? Explain.

Write each rational expression in lowest terms. See Examples 3 and 4.

23. $\dfrac{18r^3}{6r}$

24. $\dfrac{27p^2}{3p}$

25. $\dfrac{4(y-2)}{10(y-2)}$

26. $\dfrac{15(m-1)}{9(m-1)}$

27. $\dfrac{(x+1)(x-1)}{(x+1)^2}$

28. $\dfrac{(t+5)(t-3)}{(t-1)(t+5)}$

29. $\dfrac{7m+14}{5m+10}$

30. $\dfrac{8z-24}{4z-12}$

31. $\dfrac{m^2-n^2}{m+n}$

32. $\dfrac{a^2-b^2}{a-b}$

33. $\dfrac{12m^2-3}{8m-4}$

34. $\dfrac{20p^2-45}{6p-9}$

35. $\dfrac{3m^2-3m}{5m-5}$

36. $\dfrac{6t^2-6t}{2t-2}$

37. $\dfrac{9r^2-4s^2}{9r+6s}$

38. $\dfrac{16x^2-9y^2}{12x-9y}$

39. $\dfrac{zw+4z-3w-12}{zw+4z+5w+20}$

40. $\dfrac{km+4k+4m+16}{km+4k+5m+20}$

41. $\dfrac{2x^2-3x-5}{2x^2-7x+5}$

42. $\dfrac{3x^2+8x+4}{3x^2-4x-4}$

Write each rational expression in lowest terms. See Examples 5 and 6.

43. $\dfrac{6-t}{t-6}$

44. $\dfrac{2-k}{k-2}$

45. $\dfrac{m^2-1}{1-m}$

46. $\dfrac{a^2-b^2}{b-a}$

47. $\dfrac{q^2-4q}{4q-q^2}$

48. $\dfrac{z^2-5z}{5z-z^2}$

Write four equivalent expressions for each expression. See Example 7.

49. $-\dfrac{x+4}{x-3}$

50. $-\dfrac{x+6}{x-1}$

51. $-\dfrac{2x-3}{x+3}$

52. $-\dfrac{5x-6}{x+4}$

53. $\dfrac{-3x+1}{5x-6}$

54. $\dfrac{-2x-9}{3x+1}$

55. The area of the rectangle is represented by $x^4 + 10x^2 + 21$. What is the width?

$\left(Hint: \text{Use } W = \dfrac{A}{L}.\right)$

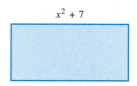

$x^2 + 7$

56. The volume of the box is represented by

$$(x^2 + 8x + 15)(x + 4).$$

Find the polynomial that represents the area of the bottom of the box.

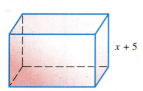

$x + 5$

6.2 MULTIPLYING AND DIVIDING RATIONAL EXPRESSIONS

1 **Multiply rational expressions.**   The product of two fractions is found by multiplying the numerators and multiplying the denominators. Rational expressions are multiplied in the same way.

Multiplying Rational Expressions

The product of the rational expressions $\frac{P}{Q}$ and $\frac{R}{S}$ is

$$\frac{P}{Q} \cdot \frac{R}{S} = \frac{PR}{QS}.$$

In words: To multiply rational expressions, multiply the numerators and multiply the denominators.

In the following example, the parallel discussion with rational numbers and rational expressions lets you compare the steps.

Example 1   **Multiplying Rational Expressions**

Multiply. Write answers in lowest terms.

(a) $\dfrac{3}{10} \cdot \dfrac{5}{9}$ **(b)** $\dfrac{6}{x} \cdot \dfrac{x^2}{12}$

Indicate the product of the numerators and the product of the denominators.

$$\frac{3}{10} \cdot \frac{5}{9} = \frac{3 \cdot 5}{10 \cdot 9} \qquad \frac{6}{x} \cdot \frac{x^2}{12} = \frac{6 \cdot x^2}{x \cdot 12}$$

Leave the products in factored form because common factors are needed to write the product in lowest terms. Factor the numerator and denominator to further identify any common factors. Then use the fundamental property to write each product in lowest terms.

$$\frac{3}{10} \cdot \frac{5}{9} = \frac{3 \cdot 5}{2 \cdot 5 \cdot 3 \cdot 3} = \frac{1}{6} \qquad \frac{6}{x} \cdot \frac{x^2}{12} = \frac{6 \cdot x \cdot x}{2 \cdot 6 \cdot x} = \frac{x}{2}$$

Work Problem 1 at the Side.

Example 2   **Multiplying Rational Expressions**

Multiply $\dfrac{x + y}{2x} \cdot \dfrac{x^2}{(x + y)^2}$. Write the answer in lowest terms.

Use the definition of multiplication. Indicate the products in the first step so that common factors are easily identified.

$$\frac{x + y}{2x} \cdot \frac{x^2}{(x + y)^2} = \frac{(x + y)x^2}{2x(x + y)^2} \qquad \text{Multiply numerators.}$$
$$\text{Multiply denominators.}$$
$$= \frac{(x + y)x \cdot x}{2x(x + y)(x + y)} \qquad \text{Factor; identify common factors.}$$
$$= \frac{x}{2(x + y)} \qquad \text{Lowest terms}$$

Continued on Next Page

OBJECTIVES

1 Multiply rational expressions.

2 Find reciprocals.

3 Divide rational expressions.

1 Multiply. Write each answer in lowest terms.

(a) $\dfrac{3m^2}{2} \cdot \dfrac{10}{m}$

(b) $\dfrac{8p^2q}{3} \cdot \dfrac{9}{q^2p}$

ANSWERS

1. (a) $15m$ **(b)** $\dfrac{24p}{q}$

❷ Multiply. Write each answer in lowest terms.

(a) $\dfrac{a + b}{5} \cdot \dfrac{30}{2(a + b)}$

(b) $\dfrac{3(p - q)}{p} \cdot \dfrac{q}{2(p - q)}$

❸ Multiply. Write each answer in lowest terms.

(a)

$\dfrac{x^2 + 7x + 10}{3x + 6} \cdot \dfrac{6x - 6}{x^2 + 2x - 15}$

(b)

$\dfrac{m^2 + 4m - 5}{m + 5} \cdot \dfrac{m^2 + 8m + 15}{m - 1}$

❹ Find each reciprocal.

(a) $\dfrac{6b^5}{3r^2 b}$

(b) $\dfrac{t^2 - 4t}{t^2 + 2t - 3}$

Notice the quotient of factors $\dfrac{(x + y)x}{x(x + y)}$ in the second line of the solution.

Since it is equal to 1, the final product is $\dfrac{x}{2(x + y)}$.

Work Problem ❷ at the Side.

Example 3 Multiplying Rational Expressions

Multiply. Write the answer in lowest terms.

$$\dfrac{x^2 + 3x}{x^2 - 3x - 4} \cdot \dfrac{x^2 - 5x + 4}{x^2 + 2x - 3} = \dfrac{(x^2 + 3x)(x^2 - 5x + 4)}{(x^2 - 3x - 4)(x^2 + 2x - 3)} \quad \begin{array}{l}\text{Definition of}\\ \text{multiplication}\end{array}$$

$$= \dfrac{x(x + 3)(x - 4)(x - 1)}{(x - 4)(x + 1)(x + 3)(x - 1)} \quad \text{Factor.}$$

$$= \dfrac{x}{x + 1} \quad \text{Lowest terms}$$

The quotients

$$\dfrac{x + 3}{x + 3}, \quad \dfrac{x - 4}{x - 4}, \quad \text{and} \quad \dfrac{x - 1}{x - 1}$$

are all equal to 1, justifying the final product $\dfrac{x}{x + 1}$.

Work Problem ❸ at the Side.

❷ **Find reciprocals.** If the product of two rational expressions is 1, the rational expressions are called **reciprocals** (or **multiplicative inverses**) of each other. The reciprocal of a rational expression is found by inverting it. For example,

$$\dfrac{2x - 1}{x - 5} \quad \text{has reciprocal} \quad \dfrac{x - 5}{2x - 1}.$$

Example 4 Finding Reciprocals of Rational Expressions

Find the reciprocal of each rational expression.

(a) $\dfrac{4p^3}{9q}$

Invert the rational expression. The reciprocal is $\dfrac{9q}{4p^3}$.

(b) $\dfrac{k^2 - 9}{k^2 - k - 20}$

The reciprocal is $\dfrac{k^2 - k - 20}{k^2 - 9}$.

Work Problem ❹ at the Side.

ANSWERS

2. (a) 3 **(b)** $\dfrac{3q}{2p}$

3. (a) $\dfrac{2(x - 1)}{x - 3}$ **(b)** $(m + 5)(m + 3)$

4. (a) $\dfrac{3r^2 b}{6b^5}$ **(b)** $\dfrac{t^2 + 2t - 3}{t^2 - 4t}$

3 Divide rational expressions.   To develop a method for dividing rational numbers and rational expressions, consider the following problem. Suppose that you have $\frac{7}{8}$ gal of milk and you wish to find how many quarts you have. Since 1 qt is $\frac{1}{4}$ gal, you must ask yourself, "How many $\frac{1}{4}$s are there in $\frac{7}{8}$?" This would be interpreted as

$$\frac{7}{8} \div \frac{1}{4} \quad \text{or} \quad \frac{\frac{7}{8}}{\frac{1}{4}}$$

since the fraction bar means division.

The fundamental property of rational expressions discussed earlier can be applied to rational number values of P, Q, and K. With $P = \frac{7}{8}$, $Q = \frac{1}{4}$, and $K = 4$ (K is the reciprocal of $Q = \frac{1}{4}$),

$$\frac{P}{Q} = \frac{P \cdot K}{Q \cdot K} = \frac{\frac{7}{8} \cdot 4}{\frac{1}{4} \cdot 4} = \frac{\frac{7}{8} \cdot 4}{1} = \frac{7}{8} \cdot \frac{4}{1}.$$

So, to divide $\frac{7}{8}$ by $\frac{1}{4}$, we must multiply $\frac{7}{8}$ by the reciprocal of $\frac{1}{4}$, namely 4. Since $\frac{7}{8}(4) = \frac{7}{2}$, there are $\frac{7}{2}$ or $3\frac{1}{2}$ qt in $\frac{7}{8}$ gal.

The preceding discussion illustrates the rule for dividing common fractions. To divide $\frac{a}{b}$ by $\frac{c}{d}$, multiply $\frac{a}{b}$ by the reciprocal of $\frac{c}{d}$. Division of rational expressions is defined in the same way.

Dividing Rational Expressions

If $\frac{P}{Q}$ and $\frac{R}{S}$ are any two rational expressions, with $\frac{R}{S} \neq 0$, then

$$\frac{P}{Q} \div \frac{R}{S} = \frac{P}{Q} \cdot \frac{S}{R} = \frac{PS}{QR}.$$

In words: To divide one rational expression by another rational expression, multiply the first rational expression by the reciprocal of the second rational expression.

The next example shows the division of two rational numbers and the division of two rational expressions.

Example 5   Dividing Rational Expressions

Divide. Write each answer in lowest terms.

(a) $\dfrac{5}{8} \div \dfrac{7}{16}$

(b) $\dfrac{y}{y + 3} \div \dfrac{4y}{y + 5}$

Continued on Next Page

5 Find each quotient. Write each answer in lowest terms.

(a) $\dfrac{r}{r-1} \div \dfrac{3r}{r+4}$

(b) $\dfrac{6x-4}{3} \div \dfrac{15x-10}{9}$

6 Find each quotient. Write each answer in lowest terms.

(a) $\dfrac{5a^2b}{2} \div \dfrac{10ab^2}{8}$

(b) $\dfrac{(3t)^2}{w} \div \dfrac{3t^2}{5w^4}$

7 Divide. Write each answer in lowest terms.

(a)

$\dfrac{y^2+4y+3}{y+3} \div \dfrac{y^2-4y-5}{y-3}$

(b) $\dfrac{4x(x+3)}{2x+1} \div \dfrac{-x^2(x+3)}{4x^2-1}$

Multiply the first expression by the reciprocal of the second.

$$\frac{5}{8} \div \frac{7}{16} = \frac{5}{8} \cdot \frac{\mathbf{16}}{\mathbf{7}} \qquad \text{Reciprocal of } \tfrac{7}{16}$$

$$= \frac{5 \cdot 16}{8 \cdot 7}$$

$$= \frac{5 \cdot \mathbf{8} \cdot 2}{\mathbf{8} \cdot 7}$$

$$= \frac{10}{7}$$

$$\frac{y}{y+3} \div \frac{4y}{y+5}$$

$$= \frac{y}{y+3} \cdot \frac{\mathbf{y+5}}{\mathbf{4y}} \qquad \text{Reciprocal of } \tfrac{4y}{y+5}$$

$$= \frac{y(y+5)}{(y+3)(4y)}$$

$$= \frac{y+5}{4(y+3)}$$

Work Problem 5 at the Side.

Example 6 Dividing Rational Expressions

Find the quotient $\dfrac{(3m)^2}{(2p)^3} \div \dfrac{6m^3}{16p^2}$. Write the answer in lowest terms.

Use the properties of exponents as necessary.

$$\frac{(3m)^2}{(2p)^3} \div \frac{6m^3}{16p^2} = \frac{(3m)^2}{(2p)^3} \cdot \frac{16p^2}{6m^3} \qquad \text{Multiply by the reciprocal.}$$

$$= \frac{9m^2}{8p^3} \cdot \frac{16p^2}{6m^3} \qquad \text{Power rule for exponents}$$

$$= \frac{9 \cdot 16m^2p^2}{8 \cdot 6p^3m^3} \qquad \begin{array}{l}\text{Multiply numerators.}\\ \text{Multiply denominators.}\end{array}$$

$$= \frac{3}{mp} \qquad \text{Lowest terms}$$

Work Problem 6 at the Side.

Example 7 Dividing Rational Expressions

Find the quotient $\dfrac{x^2-4}{(x+3)(x-2)} \div \dfrac{(x+2)(x+3)}{-2x}$. Write the answer in lowest terms.

First, use the definition of division.

$$\frac{x^2-4}{(x+3)(x-2)} \div \frac{(x+2)(x+3)}{-2x}$$

$$= \frac{x^2-4}{(x+3)(x-2)} \cdot \frac{-2x}{(x+2)(x+3)} \qquad \text{Multiply by the reciprocal.}$$

$$= \frac{(x+2)(x-2)}{(x+3)(x-2)} \cdot \frac{-2x}{(x+2)(x+3)} \qquad \text{Factor.}$$

$$= \frac{-2x(x+2)(x-2)}{(x+3)(x-2)(x+2)(x+3)} \qquad \begin{array}{l}\text{Multiply numerators.}\\ \text{Multiply denominators.}\end{array}$$

$$= \frac{-2x}{(x+3)^2} = -\frac{2x}{(x+3)^2} \qquad \text{Lowest terms}$$

Work Problem 7 at the Side.

Example 8 Dividing Rational Expressions (Factors Are Opposites)

Divide $\dfrac{m^2 - 4}{m^2 - 1} \div \dfrac{2m^2 + 4m}{1 - m}$. Write the answer in lowest terms.

$\dfrac{m^2 - 4}{m^2 - 1} \div \dfrac{2m^2 + 4m}{1 - m}$

$= \dfrac{m^2 - 4}{m^2 - 1} \cdot \dfrac{1 - m}{2m^2 + 4m}$ Multiply by the reciprocal.

$= \dfrac{(m + 2)(m - 2)}{(m + 1)(m - 1)} \cdot \dfrac{1 - m}{2m(m + 2)}$ Factor; $1 - m$ and $m - 1$ differ only in sign.

$= \dfrac{-1(m - 2)}{2m(m + 1)}$ From Section 6.1, $\frac{1 - m}{m - 1} = -1$.

$= \dfrac{2 - m}{2m(m + 1)}$ Distribute the negative sign in the numerator.

Work Problem ❽ at the Side.

In summary, follow these steps to multiply or divide rational expressions.

Multiplying or Dividing Rational Expressions

Step 1   **Note the operation.** If the operation is division, use the definition of division to rewrite as multiplication.

Note: Steps 2 and 3 may be interchanged. It is a matter of personal preference.

Step 2   **Factor.** Factor all numerators and denominators completely.

Step 3   **Multiply.** Multiply numerators and multiply denominators.

Step 4   **Write in lowest terms.** Use the fundamental property to write the answer in lowest terms.

---

❽ Divide. Write each answer in lowest terms.

(a) $\dfrac{ab - a^2}{a^2 - 1} \div \dfrac{a - b}{a - 1}$

(b) $\dfrac{x^2 - 9}{2x + 6} \div \dfrac{9 - x^2}{4x - 12}$

Answers

8. (a) $\dfrac{-a}{a + 1}$ **(b)** $\dfrac{-2x + 6}{3 + x}$

Real-Data Applications

Is 5 or 10 Minutes Really Worth the Risk?

A poignant e-mail entitled *The Drive Home* recounted the story of the habitual speeder named Jack, who was stopped by a policeman. The policeman happened to be his friend. Rather than writing a speeding ticket, the policeman handed Jack a note that described the loss of his daughter to a speeding driver. Jack was deeply affected and changed his attitude toward life on the road. The e-mail then gave some facts about speed.

- For a trip of 20 miles in a 40 mile-per-hour (mph) zone, speeding by 10 mph will save 6 minutes.

- For a trip of 50 miles in a 55 mph zone, speeding by 10 mph will save 8.3 minutes and speeding by 15 mph will save 11.6 minutes.

The question is, "Is 5 or 10 minutes really worth the risk?"

The relationship between time, distance, and rate is given by $t = \dfrac{d}{r}$. In the example, the time it takes to travel 20 miles in a 40 mph zone is $\dfrac{20 \text{ mi}}{40 \text{ mph}} = \dfrac{1}{2}$ hr, or 30 minutes. At a speed of 50 mph, the time is $\dfrac{20 \text{ mi}}{50 \text{ mph}} = \dfrac{2}{5}$ hr, or 24 minutes, since $\dfrac{2}{5}(60 \text{ min}) = 24$ min. The author's claim that only 6 minutes were saved is valid.

For Group Discussion

1. Verify the claims that traveling 50 miles at 65 mph in a 55 mph zone saves only 8.3 minutes and traveling at 70 mph saves only 11.6 minutes.

2. To save 10 minutes in a trip of 20 miles in a 40 mph zone, by how many miles per hour would Jack have to exceed the speed limit? (*Hint:* To save 10 minutes, the time for the trip would have to be only 20 minutes, which is $\dfrac{20}{60} = \dfrac{1}{3}$ hr. Determine the fraction of the form $\dfrac{20}{40 + x}$ that is equivalent to $\dfrac{1}{3}$ and then determine the value for *x*.)

3. Suppose you have to drive 10 miles to school in a 30 mph zone and you are running late. To save 5 minutes on your trip, at what speed would you have to travel? If stopped by a policeman, would you likely be given a ticket?

6.2   EXERCISES

1. Match each multiplication problem in Column I with the correct product in Column II.

I		II
(a) $\dfrac{5x^3}{10x^4} \cdot \dfrac{10x^7}{2x}$		**A.** $\dfrac{2}{5x^5}$
(b) $\dfrac{10x^4}{5x^3} \cdot \dfrac{10x^7}{2x}$		**B.** $\dfrac{5x^5}{2}$
(c) $\dfrac{5x^3}{10x^4} \cdot \dfrac{2x}{10x^7}$		**C.** $\dfrac{1}{10x^7}$
(d) $\dfrac{10x^4}{5x^3} \cdot \dfrac{2x}{10x^7}$		**D.** $10x^7$

2. Match each division problem in Column I with the correct quotient in Column II.

I		II
(a) $\dfrac{5x^3}{10x^4} \div \dfrac{10x^7}{2x}$		**A.** $\dfrac{5x^5}{2}$
(b) $\dfrac{10x^4}{5x^3} \div \dfrac{10x^7}{2x}$		**B.** $10x^7$
(c) $\dfrac{5x^3}{10x^4} \div \dfrac{2x}{10x^7}$		**C.** $\dfrac{2}{5x^5}$
(d) $\dfrac{10x^4}{5x^3} \div \dfrac{2x}{10x^7}$		**D.** $\dfrac{1}{10x^7}$

Multiply. Write each answer in lowest terms. See Examples 1 and 2.

3. $\dfrac{10m^2}{7} \cdot \dfrac{14}{15m}$

4. $\dfrac{36z^3}{6z} \cdot \dfrac{28}{z^2}$

5. $\dfrac{16y^4}{18y^5} \cdot \dfrac{15y^5}{y^2}$

6. $\dfrac{20x^5}{-2x^2} \cdot \dfrac{8x^4}{35x^3}$

7. $\dfrac{2(c+d)}{3} \cdot \dfrac{18}{6(c+d)^2}$

8. $\dfrac{4(y-2)}{x} \cdot \dfrac{3x}{6(y-2)^2}$

Find the reciprocal of each rational expression. See Example 4.

9. $\dfrac{3p^3}{16q}$

10. $\dfrac{6x^4}{9y^2}$

11. $\dfrac{r^2+rp}{7}$

12. $\dfrac{16}{9a^2+36a}$

13. $\dfrac{z^2+7z+12}{z^2-9}$

14. $\dfrac{p^2-4p+3}{p^2-3p}$

Divide. Write each answer in lowest terms. See Examples 5, 6, and 7.

15. $\dfrac{9z^4}{3z^5} \div \dfrac{3z^2}{5z^3}$

16. $\dfrac{35q^8}{9q^5} \div \dfrac{25q^6}{10q^5}$

17. $\dfrac{4t^4}{2t^5} \div \dfrac{(2t)^3}{-6}$

18. $\dfrac{-12a^6}{3a^2} \div \dfrac{(2a)^3}{27a}$

19. $\dfrac{3}{2y-6} \div \dfrac{6}{y-3}$

20. $\dfrac{4m+16}{10} \div \dfrac{3m+12}{18}$

21. Explain in your own words how to multiply rational expressions.

22. Explain in your own words how to divide rational expressions.

Multiply or divide. Write each answer in lowest terms. See Examples 3, 7, and 8.

23. $\dfrac{5x - 15}{3x + 9} \cdot \dfrac{4x + 12}{6x - 18}$

24. $\dfrac{8r + 16}{24r - 24} \cdot \dfrac{6r - 6}{3r + 6}$

25. $\dfrac{2 - t}{8} \div \dfrac{t - 2}{6}$

26. $\dfrac{4}{m - 2} \div \dfrac{16}{2 - m}$

27. $\dfrac{27 - 3z}{4} \cdot \dfrac{12}{2z - 18}$

28. $\dfrac{5 - x}{5 + x} \cdot \dfrac{x + 5}{x - 5}$

29. $\dfrac{6(m - 2)^2}{5(m + 4)^2} \cdot \dfrac{15(m + 4)}{2(2 - m)}$

30. $\dfrac{7(q - 1)}{3(q + 1)^2} \cdot \dfrac{6(q + 1)}{3(1 - q)^2}$

31. $\dfrac{p^2 + 4p - 5}{p^2 + 7p + 10} \div \dfrac{p - 1}{p + 4}$

32. $\dfrac{z^2 - 3z + 2}{z^2 + 4z + 3} \div \dfrac{z - 1}{z + 1}$

33. $\dfrac{2k^2 - k - 1}{2k^2 + 5k + 3} \div \dfrac{4k^2 - 1}{2k^2 + k - 3}$

34. $\dfrac{2m^2 - 5m - 12}{m^2 + m - 20} \div \dfrac{4m^2 - 9}{m^2 + 4m - 5}$

35. $\dfrac{2k^2 + 3k - 2}{6k^2 - 7k + 2} \cdot \dfrac{4k^2 - 5k + 1}{k^2 + k - 2}$

36. $\dfrac{2m^2 - 5m - 12}{m^2 - 10m + 24} \div \dfrac{4m^2 - 9}{m^2 - 9m + 18}$

37. $\dfrac{m^2 + 2mp - 3p^2}{m^2 - 3mp + 2p^2} \div \dfrac{m^2 + 4mp + 3p^2}{m^2 + 2mp - 8p^2}$

38. $\dfrac{r^2 + rs - 12s^2}{r^2 - rs - 20s^2} \div \dfrac{r^2 - 2rs - 3s^2}{r^2 + rs - 30s^2}$

39. $\left(\dfrac{x^2 + 10x + 25}{x^2 + 10x} \cdot \dfrac{10x}{x^2 + 15x + 50} \right) \div \dfrac{x + 5}{x + 10}$

40. $\left(\dfrac{m^2 - 12m + 32}{8m} \cdot \dfrac{m^2 - 8m}{m^2 - 8m + 16} \right) \div \dfrac{m - 8}{m - 4}$

41. Consider the division problem $\dfrac{x - 6}{x + 4} \div \dfrac{x + 7}{x + 5}$.
We know that division by 0 is undefined, so the restrictions on x are $x \ne -4$, $x \ne -5$, and $x \ne -7$. Why is the last restriction needed?

42. If the rational expression $\dfrac{5x^2y^3}{2pq}$ represents the area of a rectangle and $\dfrac{2xy}{p}$ represents the length, what rational expression represents the width?

Width

Length $= \dfrac{2xy}{p}$

The area is $\dfrac{5x^2y^3}{2pq}$.

6.3   LEAST COMMON DENOMINATORS

1 **Find the least common denominator for a group of fractions.**   Just as with common fractions, adding or subtracting rational expressions (to be discussed in the next section) often requires a **least common denominator (LCD),** the simplest expression that is divisible by all denominators. For example, the least common denominator for $\frac{2}{9}$ and $\frac{5}{12}$ is 36 because 36 is the smallest positive number divisible by both 9 and 12.

Least common denominators can often be found by inspection. For example, the LCD for $\frac{1}{6}$ and $\frac{2}{3m}$ is $6m$. In other cases, the LCD can be found by a procedure similar to that used in Chapter 5 for finding the greatest common factor.

OBJECTIVES

1 Find the least common denominator for a group of fractions.

2 Rewrite rational expressions with given denominators.

Finding the Least Common Denominator (LCD)

Step 1 **Factor.** Factor each denominator into prime factors.

Step 2 **List the factors.** List each different denominator factor the *greatest* number of times it appears in any of the denominators.

Step 3 **Multiply.** Multiply the denominator factors from Step 2 to get the LCD.

When each denominator is factored into prime factors, every prime factor must be a factor of the least common denominator.

In Example 1, we find the LCD for both numerical and algebraic denominators.

1 Find the LCD for each pair of fractions.

(a) $\dfrac{7}{20p}, \dfrac{11}{30p}$

> **Example 1**   Finding the LCD

Find the LCD for each pair of fractions.

(a) $\dfrac{1}{24}, \dfrac{7}{15}$ (b) $\dfrac{1}{8x}, \dfrac{3}{10x}$

Step 1 Write each denominator in factored form with numerical coefficients in prime factored form.

$$24 = 2^3 \cdot 3 \qquad\qquad 8x = 2^3 \cdot x$$
$$15 = 3 \cdot 5 \qquad\qquad 10x = 2 \cdot 5 \cdot x$$

Step 2 We find the LCD by taking each different factor the *greatest* number of times it appears as a factor in any of the denominators.

The factor 2 appears three times in one product and not at all in the other, so the greatest number of times 2 appears is three. The greatest number of times both 3 and 5 appear is one.

Here 2 appears three times in one product and once in the other, so the greatest number of times 2 appears is three. The greatest number of times 5 appears is one, and the greatest number of times x appears in either product is one.

(b) $\dfrac{4}{5x}, \dfrac{12}{10x}$

Step 3 LCD $= 2 \cdot 2 \cdot 2 \cdot 3 \cdot 5$
         $= 2^3 \cdot 3 \cdot 5$
         $= 120$

LCD $= 2 \cdot 2 \cdot 2 \cdot 5 \cdot x$
    $= 2^3 \cdot 5 \cdot x$
    $= 40x$

━━━━ Work Problem **1** at the Side.

ANSWERS
1. (a) $60p$ **(b)** $10x$

2 Find the LCD.

(a) $\dfrac{4}{16m^3n}, \dfrac{5}{9m^5}$

(b) $\dfrac{3}{25a^2}, \dfrac{2}{10a^3b}$

3 Find the LCD.

(a) $\dfrac{7}{3a}, \dfrac{11}{a^2 - 4a}$

(b)

$\dfrac{2m}{m^2 - 3m + 2}, \dfrac{5m - 3}{m^2 + 3m - 10},$

$\dfrac{4m + 7}{m^2 + 4m - 5}$

(c) $\dfrac{6}{x - 4}, \dfrac{3x - 1}{4 - x}$

Example 2 Finding the LCD

Find the LCD for $\dfrac{5}{6r^2}$ and $\dfrac{3}{4r^3}$.

Step 1 Factor each denominator.

$$6r^2 = 2 \cdot \mathbf{3} \cdot r^2$$
$$4r^3 = \mathbf{2^2} \cdot \mathbf{r^3}$$

Step 2 The greatest number of times 2 appears is two, the greatest number of times 3 appears is one, and the greatest number of times r appears is three; therefore,

Step 3 $\qquad\qquad$ LCD $= 2^2 \cdot 3 \cdot r^3 = 12r^3$.

Work Problem 2 at the Side.

Example 3 Finding the LCD

Find the LCD.

(a) $\dfrac{6}{5m}, \dfrac{4}{m^2 - 3m}$

Factor each denominator.

$$5m = \mathbf{5} \cdot \mathbf{m}$$
$$m^2 - 3m = m(\mathbf{m - 3})$$

Use each different factor the greatest number of times it appears.

$$\text{LCD} = 5 \cdot m \cdot (m - 3) = 5m(m - 3)$$

Because m is not a *factor* of $m - 3$, both factors, m and $m - 3$, must appear in the LCD.

(b) $\dfrac{1}{r^2 - 4r - 5}, \dfrac{3}{r^2 - r - 20}, \dfrac{1}{r^2 - 10r + 25}$

Factor each denominator.

$$r^2 - 4r - 5 = (r - 5)(\mathbf{r + 1})$$
$$r^2 - r - 20 = (r - 5)(\mathbf{r + 4})$$
$$r^2 - 10r + 25 = (\mathbf{r - 5})^2$$

Use each different factor the greatest number of times it appears as a factor. The LCD is

$$(r - 5)^2(r + 1)(r + 4).$$

(c) $\dfrac{1}{q - 5}, \dfrac{3}{5 - q}$

The expressions $q - 5$ and $5 - q$ are opposites of each other because

$$-(q - 5) = -q + 5 = 5 - q.$$

Therefore, either $q - 5$ or $5 - q$ can be used as the LCD.

Work Problem 3 at the Side.

2 ▭ **Rewrite rational expressions with given denominators.** Once the LCD has been found, the next step in preparing to add or subtract two rational expressions is to use the fundamental property to write equivalent rational expressions. The next example shows how to do this with both numerical and algebraic fractions.

ANSWERS
2. (a) $144m^5n$ (b) $50a^3b$
3. (a) $3a(a - 4)$ (b) $(m - 1)(m - 2)(m + 5)$
(c) either $x - 4$ or $4 - x$

Example 4   Writing Rational Expressions with Given Denominators

Rewrite each rational expression with the indicated denominator.

(a) $\dfrac{3}{8} = \dfrac{}{40}$ **(b)** $\dfrac{9k}{25} = \dfrac{}{50k}$

For each example, first factor the denominator on the right. Then compare the denominator on the left with the one on the right to decide what factors are missing. (It may be necessary to factor both denominators.)

$$\dfrac{3}{8} = \dfrac{}{5 \cdot 8}$$ $$\dfrac{9k}{25} = \dfrac{}{25 \cdot 2k}$$

A factor of 5 is missing. Using the fundamental property, multiply $\frac{3}{8}$ by $\frac{5}{5}$.

Factors of 2 and k are missing. Multiply by $\frac{2k}{2k}$.

$$\dfrac{3}{8} = \dfrac{3}{8} \cdot \dfrac{5}{5} = \dfrac{15}{40}$$ $$\dfrac{9k}{25} = \dfrac{9k}{25} \cdot \dfrac{2k}{2k} = \dfrac{18k^2}{50k}$$

$\dfrac{5}{5} = 1$ $\dfrac{2k}{2k} = 1$

Example 5   Writing Rational Expressions with Given Denominators

Rewrite each rational expression with the indicated denominator.

(a) $\dfrac{8}{3x + 1} = \dfrac{}{12x + 4}$

Factor the denominator on the right.

$$\dfrac{8}{3x + 1} = \dfrac{}{4(3x + 1)} \quad \text{Factor.}$$

The missing factor is 4, so multiply the fraction on the left by $\dfrac{4}{4}$.

$$\dfrac{8}{3x + 1} \cdot \dfrac{4}{4} = \dfrac{32}{12x + 4} \quad \text{Fundamental property}$$

(b) $\dfrac{12p}{p^2 + 8p} = \dfrac{}{p^3 + 4p^2 - 32p}$

Factor $p^2 + 8p$ as $p(p + 8)$. Compare with the denominator on the right, which factors as $p(p + 8)(p - 4)$. The factor $p - 4$ is missing, so multiply $\dfrac{12p}{p(p + 8)}$ by $\dfrac{p - 4}{p - 4}$.

$$\dfrac{12p}{p^2 + 8p} = \dfrac{12p}{p(p + 8)} \cdot \dfrac{p - 4}{p - 4} \quad \text{Fundamental property}$$

$$= \dfrac{12p(p - 4)}{p(p + 8)(p - 4)} \quad \text{Multiplication of rational expressions}$$

$$= \dfrac{12p^2 - 48p}{p^3 + 4p^2 - 32p} \quad \text{Multiply the factors.}$$

④ Rewrite each rational expression with the indicated denominator.

(a) $\dfrac{7k}{5} = \dfrac{}{30p}$

Work Problem ④ at the Side.

(b) $\dfrac{9}{2a + 5} = \dfrac{}{6a + 15}$

(c)

$\dfrac{5k + 1}{k^2 + 2k} = \dfrac{}{k(k + 2)(k - 1)}$

Answers

4. (a) $\dfrac{42kp}{30p}$ (b) $\dfrac{27}{6a + 15}$

(c) $\dfrac{(5k + 1)(k - 1)}{k(k + 2)(k - 1)}$

6.3 EXERCISES

 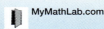
Choose the correct response in Exercises 1–4.

1. Suppose that the greatest common factor of a and b is 1. Then the least common denominator for $\frac{1}{a}$ and $\frac{1}{b}$ is

 A. a **B.** b **C.** ab **D.** 1.

2. If a is a factor of b, then the least common denominator for $\frac{1}{a}$ and $\frac{1}{b}$ is

 A. a **B.** b **C.** ab **D.** 1.

3. The least common denominator for $\frac{11}{20}$ and $\frac{1}{2}$ is

 A. 40   **B.** 2   **C.** 20   **D.** none of these.

4. Suppose that we wish to write the fraction $\dfrac{1}{(x-4)^2(y-3)}$ with denominator $(x-4)^3(y-3)^2$. We must multiply both the numerator and the denominator by

 A. $(x-4)(y-3)$ **B.** $(x-4)^2$ **C.** $x-4$
 D. $(x-4)^2(y-3)$.

Find the least common denominator for the fractions in each list. See Examples 1–3.

5. $\dfrac{2}{15}, \dfrac{3}{10}, \dfrac{7}{30}$

6. $\dfrac{5}{24}, \dfrac{7}{12}, \dfrac{9}{28}$

7. $\dfrac{3}{x^4}, \dfrac{5}{x^7}$

8. $\dfrac{2}{y^5}, \dfrac{3}{y^6}$

9. $\dfrac{5}{36q}, \dfrac{17}{24q}$

10. $\dfrac{4}{30p}, \dfrac{9}{50p}$

11. $\dfrac{6}{21r^3}, \dfrac{8}{12r^5}$

12. $\dfrac{9}{35t^2}, \dfrac{5}{49t^6}$

13. If the denominators of two fractions in prime factored form are $2^3 \cdot 3$ and $2^2 \cdot 5$, what is the factored form of their LCD?

14. Suppose two rational expressions have denominators $(t+4)^3(t-3)$ and $(t+4)^2(t+8)$. Find the factored form of their LCD. What is the similarity between the answers for this problem and for Exercise 13?

15. If two denominators have greatest common factor equal to 1, how can you easily find their least common denominator?

16. Suppose two fractions have denominators a^k and a^r, where k and r are natural numbers, with $k > r$. What is their least common denominator?

Find the least common denominator for each pair of fractions. See Examples 1–3.

17. $\dfrac{9}{28m^2}, \dfrac{3}{12m-20}$

18. $\dfrac{15}{27a^3}, \dfrac{8}{9a-45}$

19. $\dfrac{7}{5b-10}, \dfrac{11}{6b-12}$

20. $\dfrac{3}{7x^2+21x}, \dfrac{1}{5x^2+15x}$

21. $\dfrac{5}{c-d}, \dfrac{8}{d-c}$

22. $\dfrac{4}{y-x}, \dfrac{7}{x-y}$

23. $\dfrac{3}{k^2 + 5k}, \dfrac{2}{k^2 + 3k - 10}$

24. $\dfrac{1}{z^2 - 4z}, \dfrac{4}{z^2 - 3z - 4}$

25. $\dfrac{5}{p^2 + 8p + 15}, \dfrac{3}{p^2 - 3p - 18}, \dfrac{2}{p^2 - p - 30}$

26. $\dfrac{10}{y^2 - 10y + 21}, \dfrac{2}{y^2 - 2y - 3}, \dfrac{5}{y^2 - 6y - 7}$

Rewrite each rational expression with the given denominator. See Examples 4 and 5.

27. $\dfrac{4}{11} = \dfrac{}{55}$

28. $\dfrac{6}{7} = \dfrac{}{42}$

29. $\dfrac{-5}{k} = \dfrac{}{9k}$

30. $\dfrac{-3}{q} = \dfrac{}{6q}$

31. $\dfrac{13}{40y} = \dfrac{}{80y^3}$

32. $\dfrac{5}{27p} = \dfrac{}{108p^4}$

33. $\dfrac{5t^2}{6r} = \dfrac{}{42r^4}$

34. $\dfrac{8y^2}{3x} = \dfrac{}{30x^3}$

35. $\dfrac{5}{2(m + 3)} = \dfrac{}{8(m + 3)}$

36. $\dfrac{7}{4(y - 1)} = \dfrac{}{16(y - 1)}$

37. $\dfrac{-4t}{3t - 6} = \dfrac{}{12 - 6t}$

38. $\dfrac{-7k}{5k - 20} = \dfrac{}{40 - 10k}$

39. $\dfrac{14}{z^2 - 3z} = \dfrac{}{z(z - 3)(z - 2)}$

40. $\dfrac{12}{x(x + 4)} = \dfrac{}{x(x + 4)(x - 9)}$

41. $\dfrac{2(b - 1)}{b^2 + b} = \dfrac{}{b^3 + 3b^2 + 2b}$

42. $\dfrac{3(c + 2)}{c(c - 1)} = \dfrac{}{c^3 - 5c^2 + 4c}$

6.4 ADDING AND SUBTRACTING RATIONAL EXPRESSIONS

To add and subtract rational expressions, we will need the skills developed in the previous section to find least common denominators and to write equivalent fractions with the LCD.

1 **Add rational expressions having the same denominator.** We find the sum of two rational expressions with the same procedure that we used for adding two fractions in Chapter R.

Adding Rational Expressions

If $\frac{P}{Q}$ and $\frac{R}{Q}$ ($Q \neq 0$) are rational expressions, then

$$\frac{P}{Q} + \frac{R}{Q} = \frac{P+R}{Q}.$$

In words: To add rational expressions with the same denominator, add the numerators and keep the same denominator.

The first example shows how the addition of rational expressions compares with that of rational numbers.

Example 1 Adding Rational Expressions with the Same Denominator

Add. Write each answer in lowest terms.

(a) $\frac{4}{7} + \frac{2}{7}$ (b) $\frac{3x}{x+1} + \frac{3}{x+1}$

The denominators are the same, so the sum is found by adding the two numerators and keeping the same (common) denominator.

$$\frac{4}{7} + \frac{2}{7} = \frac{4+2}{7}$$
$$= \frac{6}{7}$$

$$\frac{3x}{x+1} + \frac{3}{x+1} = \frac{3x+3}{x+1}$$
$$= \frac{3(x+1)}{x+1}$$
$$= 3$$

Work Problem **1** at the Side.

2 **Add rational expressions having different denominators.** We use the following steps to add two rational expressions with different denominators. These are the same steps we used to add fractions with different denominators in Chapter R.

Adding with Different Denominators

Step 1 **Find the LCD.** Find the least common denominator (LCD).

Step 2 **Rewrite fractions.** Rewrite each rational expression as an equivalent rational expression with the LCD as the denominator.

Step 3 **Add.** Add the numerators to get the numerator of the sum. The LCD is the denominator of the sum.

Step 4 **Write in lowest terms.** Use the fundamental property to write the answer in lowest terms.

OBJECTIVES

1 Add rational expressions having the same denominator.

2 Add rational expressions having different denominators.

3 Subtract rational expressions.

1 Add. Write each answer in lowest terms.

(a) $\frac{3}{y+4} + \frac{2}{y+4}$

(b) $\frac{x}{x+y} + \frac{1}{x+y}$

(c) $\frac{a}{a+b} + \frac{b}{a+b}$

(d) $\frac{x^2}{x+1} + \frac{x}{x+1}$

ANSWERS

1. (a) $\frac{5}{y+4}$ (b) $\frac{x+1}{x+y}$ (c) 1 (d) x

② Add. Write each answer in lowest terms.

(a) $\dfrac{6}{5x} + \dfrac{9}{2x}$

Example 2 Adding Rational Expressions with Different Denominators

Add. Write each answer in lowest terms.

(a) $\dfrac{1}{12} + \dfrac{7}{15}$

(b) $\dfrac{2}{3y} + \dfrac{1}{4y}$

Step 1 First find the LCD using the methods of the previous section.

$$\text{LCD} = 2^2 \cdot 3 \cdot 5 = 60 \qquad\qquad \text{LCD} = 2^2 \cdot 3 \cdot y = 12y$$

Step 2 Now rewrite each rational expression as a fraction with the LCD (either 60 or $12y$) as the denominator.

$$\dfrac{1}{12} + \dfrac{7}{15} = \dfrac{1(5)}{12(5)} + \dfrac{7(4)}{15(4)} \qquad\qquad \dfrac{2}{3y} + \dfrac{1}{4y} = \dfrac{2(4)}{3y(4)} + \dfrac{1(3)}{4y(3)}$$

$$= \dfrac{5}{60} + \dfrac{28}{60} \qquad\qquad\qquad\qquad = \dfrac{8}{12y} + \dfrac{3}{12y}$$

Step 3 Since the fractions now have common denominators, add the numerators.

Step 4 Write in lowest terms if necessary.

$$\dfrac{5}{60} + \dfrac{28}{60} = \dfrac{5 + 28}{60} \qquad\qquad \dfrac{8}{12y} + \dfrac{3}{12y} = \dfrac{8 + 3}{12y}$$

$$= \dfrac{33}{60} = \dfrac{11}{20} \qquad\qquad\qquad = \dfrac{11}{12y}$$

Work Problem ② at the Side.

(b) $\dfrac{m}{3n} + \dfrac{2}{7n}$

Example 3 Adding Rational Expressions

Add. Write the answer in lowest terms.

$$\dfrac{2x}{x^2 - 1} + \dfrac{-1}{x + 1}$$

Step 1 Since the denominators are different, find the LCD.

$$x^2 - 1 = (x + 1)(x - 1)$$

$$x + 1 \text{ is prime.}$$

The LCD is $(x + 1)(x - 1)$.

Step 2 Rewrite each rational expression as a fraction with common denominator $(x + 1)(x - 1)$.

$$\dfrac{2x}{x^2 - 1} + \dfrac{-1}{x + 1} = \dfrac{2x}{(x + 1)(x - 1)} + \dfrac{-1(x - 1)}{(x + 1)(x - 1)} \qquad \text{Multiply the second fraction by } \tfrac{x - 1}{x - 1}.$$

$$= \dfrac{2x}{(x + 1)(x - 1)} + \dfrac{-x + 1}{(x + 1)(x - 1)} \qquad \text{Distributive property}$$

Step 3
$$= \dfrac{2x - x + 1}{(x + 1)(x - 1)} \qquad \text{Add numerators; keep the same denominator.}$$

$$= \dfrac{x + 1}{(x + 1)(x - 1)} \qquad \text{Combine like terms in the numerator.}$$

Continued on Next Page

ANSWERS

2. (a) $\dfrac{57}{10x}$ (b) $\dfrac{7m + 6}{21n}$

Step 4 $= \dfrac{1(x + 1)}{(x + 1)(x - 1)}$ Identity property for multiplication

$= \dfrac{1}{x - 1}$ Fundamental property

==== **Work Problem ❸ at the Side.**

Example 4 Adding Rational Expressions

Add. Write the answer in lowest terms.

$\dfrac{2x}{x^2 + 5x + 6} + \dfrac{x + 1}{x^2 + 2x - 3}$

$= \dfrac{2x}{(x + 2)(x + 3)} + \dfrac{x + 1}{(x + 3)(x - 1)}$ Factor the denominators.

The LCD is $(x + 2)(x + 3)(x - 1)$. Use the fundamental property.

$= \dfrac{2x(x - 1)}{(x + 2)(x + 3)(x - 1)} + \dfrac{(x + 1)(x + 2)}{(x + 2)(x + 3)(x - 1)}$

$= \dfrac{2x(x - 1) + (x + 1)(x + 2)}{(x + 2)(x + 3)(x - 1)}$ Add numerators; keep the same denominator.

$= \dfrac{2x^2 - 2x + x^2 + 3x + 2}{(x + 2)(x + 3)(x - 1)}$ Multiply.

$= \dfrac{3x^2 + x + 2}{(x + 2)(x + 3)(x - 1)}$ Combine like terms.

It is usually more convenient to leave the denominator in factored form. The numerator cannot be factored here, so the expression is in lowest terms.

==== **Work Problem ❹ at the Side.**

Rational expressions to be added or subtracted may have denominators that are opposites of each other. The next example illustrates this.

Example 5 Adding Rational Expressions with Denominators That Are Opposites

Add. Write the answer in lowest terms.

$$\dfrac{y}{y - 2} + \dfrac{8}{2 - y}$$

One way to get a common denominator is to multiply the second expression by -1 in both the numerator and the denominator, giving $y - 2$ as a common denominator.

$\dfrac{y}{y - 2} + \dfrac{8}{2 - y} = \dfrac{y}{y - 2} + \dfrac{8(-1)}{(2 - y)(-1)}$ Fundamental property

$= \dfrac{y}{y - 2} + \dfrac{-8}{y - 2}$ Distributive property

$= \dfrac{y - 8}{y - 2}$ Add numerators; keep the same denominator.

If we had chosen to use $2 - y$ as the common denominator, the final answer would be in the form $\dfrac{8 - y}{2 - y}$, which is equivalent to $\dfrac{y - 8}{y - 2}$.

==== **Work Problem ❺ at the Side.**

---

❸ Add. Write each answer in lowest terms.

(a) $\dfrac{2p}{3p + 3} + \dfrac{5p}{2p + 2}$

(b) $\dfrac{4}{y^2 - 1} + \dfrac{6}{y + 1}$

(c) $\dfrac{-2}{p + 1} + \dfrac{4p}{p^2 - 1}$

❹ Add. Write each answer in lowest terms.

(a) $\dfrac{2k}{k^2 - 5k + 4} + \dfrac{3}{k^2 - 1}$

(b)

$\dfrac{4m}{m^2 + 3m + 2} + \dfrac{2m - 1}{m^2 + 6m + 5}$

❺ Add. Write the answer in lowest terms.

$\dfrac{m}{2m - 3n} + \dfrac{n}{3n - 2m}$

6 Subtract. Write each answer in lowest terms.

(a) $\dfrac{3}{m^2} - \dfrac{2}{m^2}$

(b) $\dfrac{x}{2x+3} - \dfrac{3x+4}{2x+3}$

3 ▮ **Subtract rational expressions.** To subtract rational expressions, use the following rule.

Subtracting Rational Expressions

If $\dfrac{P}{Q}$ and $\dfrac{R}{Q}$ $(Q \neq 0)$ are rational expressions, then

$$\frac{P}{Q} - \frac{R}{Q} = \frac{P-R}{Q}.$$

In words: To subtract rational expressions with the same denominator, subtract the numerators and keep the same denominator.

We will not show a parallel subtraction problem from arithmetic because the steps for subtraction are essentially the same as for addition.

Example 6 Subtracting Rational Expressions with the Same Denominator

Subtract. Write the answer in lowest terms.

$$\frac{2m}{m-1} - \frac{m+3}{m-1}$$

By the definition of subtraction,

$$\frac{2m}{m-1} - \frac{m+3}{m-1} = \frac{2m - (m+3)}{m-1} \quad \text{Subtract numerators; keep the same denominator.}$$

$$= \frac{2m - m - 3}{m-1} \quad \text{Distributive property}$$

$$= \frac{m-3}{m-1}. \quad \text{Combine like terms.}$$

CAUTION

Sign errors often occur in subtraction problems like the one in Example 6. Remember that the numerator of the fraction being subtracted must be treated as a single quantity. Be sure to use parentheses after the subtraction sign to avoid this common error.

Work Problem 6 at the Side.

Example 7 Subtracting Rational Expressions with Different Denominators

Subtract. Write the answer in lowest terms.

$$\frac{9}{x-2} - \frac{3}{x}$$

The LCD is $x(x-2)$.

$$\frac{9}{x-2} - \frac{3}{x} = \frac{9x}{x(x-2)} - \frac{3(x-2)}{x(x-2)} \quad \text{Get the least common denominator.}$$

$$= \frac{9x - 3(x-2)}{x(x-2)} \quad \text{Subtract numerators; keep the same denominator.}$$

Continued on Next Page

$$= \frac{9x - 3x + 6}{x(x - 2)}$$ Distributive property

$$= \frac{6x + 6}{x(x - 2)}$$ Combine like terms.

$$= \frac{6(x + 1)}{x(x - 2)}$$ Factor.

NOTE

We factor the final numerator in Example 7 to get an answer in the form $\frac{6(x + 1)}{x(x - 2)}$; however, the fundamental property does not apply, since there are no common factors that would allow us to write the answer in lower terms.

Work Problem ❼ at the Side.

Example 8 **Subtracting Rational Expressions with Denominators That Are Opposites**

Subtract. Write the answer in lowest terms.

$$\frac{3x}{x - 5} - \frac{2x - 25}{5 - x}$$

The denominators are opposites, so either may be used as the common denominator. We will choose $x - 5$.

$$\frac{3x}{x - 5} - \frac{2x - 25}{5 - x} = \frac{3x}{x - 5} - \frac{2x - 25}{5 - x} \cdot \frac{-1}{-1}$$ Fundamental property

$$= \frac{3x}{x - 5} - \frac{-2x + 25}{x - 5}$$ Multiply.

$$= \frac{3x - (-2x + 25)}{x - 5}$$ Subtract numerators.

$$= \frac{3x + 2x - 25}{x - 5}$$ Distributive property

$$= \frac{5x - 25}{x - 5}$$ Combine like terms.

$$= \frac{5(x - 5)}{x - 5}$$ Factor.

$$= 5$$ Lowest terms

Work Problem ❽ at the Side.

Example 9 **Subtracting Rational Expressions**

Subtract. Write the answer in lowest terms.

$$\frac{6x}{x^2 - 2x + 1} - \frac{1}{x^2 - 1}$$

Begin by factoring the denominators.

$$x^2 - 2x + 1 = (x - 1)^2 \quad \text{and} \quad x^2 - 1 = (x - 1)(x + 1)$$

Continued on Next Page

❼ Subtract. Write each answer in lowest terms.

(a) $\dfrac{1}{k + 4} - \dfrac{2}{k}$

(b) $\dfrac{6}{a + 2} - \dfrac{1}{a - 3}$

❽ Subtract. Write each answer in lowest terms.

(a) $\dfrac{5}{x - 1} - \dfrac{3x}{1 - x}$

(b) $\dfrac{2y}{y - 2} - \dfrac{1 + y}{2 - y}$

ANSWERS

7. (a) $\dfrac{-k - 8}{k(k + 4)}$ (b) $\dfrac{5(a - 4)}{(a + 2)(a - 3)}$

8. (a) $\dfrac{5 + 3x}{x - 1}$ (b) $\dfrac{3y + 1}{y - 2}$

9 Subtract. Write each answer in lowest terms.

(a) $\dfrac{4y}{y^2 - 1} - \dfrac{5}{y^2 + 2y + 1}$

(b) $\dfrac{3r}{r^2 - 5r} - \dfrac{4}{r^2 - 10r + 25}$

10 Subtract. Write each answer in lowest terms.

(a) $\dfrac{2}{p^2 - 5p + 4} - \dfrac{3}{p^2 - 1}$

(b)

$\dfrac{q}{2q^2 + 5q - 3} - \dfrac{3q + 4}{3q^2 + 10q + 3}$

From the factored denominators, identify the LCD,

$$(x - 1)^2(x + 1).$$

Use the factor $x - 1$ twice because it appears twice in the first denominator.

$$\frac{6x}{(x - 1)^2} - \frac{1}{(x - 1)(x + 1)}$$

$$= \frac{6x(x + 1)}{(x - 1)^2(x + 1)} - \frac{1(x - 1)}{(x - 1)(x - 1)(x + 1)} \qquad \text{Fundamental property}$$

$$= \frac{6x(x + 1) - 1(x - 1)}{(x - 1)^2(x + 1)} \qquad \text{Subtract numerators.}$$

$$= \frac{6x^2 + 6x - x + 1}{(x - 1)^2(x + 1)} \qquad \text{Distributive property}$$

$$= \frac{6x^2 + 5x + 1}{(x - 1)^2(x + 1)} \quad \text{or} \quad \frac{(2x + 1)(3x + 1)}{(x - 1)^2(x + 1)} \qquad \text{Combine like terms.}$$

Verify that the final expression is in lowest terms.

Work Problem 9 at the Side.

Example 10 Subtracting Rational Expressions

Subtract. Write the answer in lowest terms.

$$\frac{q}{q^2 - 4q - 5} - \frac{3}{2q^2 - 13q + 15}$$

To find the LCD, factor each denominator.

$$q^2 - 4q - 5 = (q + 1)(q - 5)$$

$$2q^2 - 13q + 15 = (q - 5)(2q - 3)$$

The LCD is $(q + 1)(q - 5)(2q - 3)$. Rewrite each rational expression with the LCD, using the fundamental property.

$$\frac{q}{(q + 1)(q - 5)} - \frac{3}{(q - 5)(2q - 3)}$$

$$= \frac{q(2q - 3)}{(q + 1)(q - 5)(2q - 3)} - \frac{3(q + 1)}{(q + 1)(q - 5)(2q - 3)}$$

$$= \frac{q(2q - 3) - 3(q + 1)}{(q + 1)(q - 5)(2q - 3)} \qquad \text{Subtract numerators.}$$

$$= \frac{2q^2 - 3q - 3q - 3}{(q + 1)(q - 5)(2q - 3)} \qquad \text{Distributive property}$$

$$= \frac{2q^2 - 6q - 3}{(q + 1)(q - 5)(2q - 3)} \qquad \text{Combine like terms.}$$

Verify that the final expression is in lowest terms.

Work Problem 10 at the Side.

NOTE

When adding and subtracting rational expressions, several different equivalent forms of the answer often exist. If your answer does not look exactly like the one given in the back of the book, check to see whether you have written an equivalent form.

ANSWERS

9. (a) $\dfrac{4y^2 - y + 5}{(y + 1)^2(y - 1)}$

(b) $\dfrac{3r - 19}{(r - 5)^2}$

10. (a) $\dfrac{14 - p}{(p - 4)(p - 1)(p + 1)}$

(b) $\dfrac{-3q^2 - 4q + 4}{(2q - 1)(q + 3)(3q + 1)}$

6.4 EXERCISES

Match the problem in Column I with the correct sum or difference in Column II.

Study Skills Workbook Activity 11

I

1. $\dfrac{x}{x+6} + \dfrac{6}{x+6}$

2. $\dfrac{2x}{x-6} - \dfrac{12}{x-6}$

3. $\dfrac{6}{x-6} - \dfrac{x}{x-6}$

4. $\dfrac{6}{x+6} - \dfrac{x}{x+6}$

5. $\dfrac{x}{x+6} - \dfrac{6}{x+6}$

6. $\dfrac{1}{x} + \dfrac{1}{6}$

7. $\dfrac{1}{6} - \dfrac{1}{x}$

8. $\dfrac{1}{6x} - \dfrac{1}{6x}$

II

A. 2

B. $\dfrac{x-6}{x+6}$

C. -1

D. $\dfrac{6+x}{6x}$

E. 1

F. 0

G. $\dfrac{x-6}{6x}$

H. $\dfrac{6-x}{x+6}$

Add or subtract. Write each answer in lowest terms. See Examples 1 and 6.

9. $\dfrac{4}{m} + \dfrac{7}{m}$

10. $\dfrac{5}{p} + \dfrac{11}{p}$

11. $\dfrac{a+b}{2} - \dfrac{a-b}{2}$

12. $\dfrac{x-y}{2} - \dfrac{x+y}{2}$

13. $\dfrac{x^2}{x+5} + \dfrac{5x}{x+5}$

14. $\dfrac{t^2}{t-3} + \dfrac{-3t}{t-3}$

15. $\dfrac{y^2-3y}{y+3} + \dfrac{-18}{y+3}$

16. $\dfrac{r^2-8r}{r-5} + \dfrac{15}{r-5}$

17. Explain with an example how to add or subtract rational expressions with the same denominator.

18. Explain with an example how to add or subtract rational expressions with different denominators.

Add or subtract. Write each answer in lowest terms. See Examples 2, 3, 4, and 7.

19. $\dfrac{z}{5} + \dfrac{1}{3}$

20. $\dfrac{p}{8} + \dfrac{3}{5}$

21. $\dfrac{5}{7} - \dfrac{r}{2}$

22. $\dfrac{10}{9} - \dfrac{z}{3}$

23. $-\dfrac{3}{4} - \dfrac{1}{2x}$

24. $-\dfrac{5}{8} - \dfrac{3}{2a}$

25. $\dfrac{x+1}{6} + \dfrac{3x+3}{9}$

26. $\dfrac{2x-6}{4} + \dfrac{x+5}{6}$

27. $\dfrac{x+3}{3x} + \dfrac{2x+2}{4x}$

28. $\dfrac{x+2}{5x} + \dfrac{6x+3}{3x}$

29. $\dfrac{2}{x+3} + \dfrac{1}{x}$

30. $\dfrac{3}{x-4} + \dfrac{2}{x}$

31. $\dfrac{x}{x-2} + \dfrac{4}{x+2}$

32. $\dfrac{2x}{x-1} + \dfrac{3}{x+1}$

33. $\dfrac{t}{t+2} + \dfrac{5-t}{t} - \dfrac{4}{t^2+2t}$

34. $\dfrac{2p}{p-3} + \dfrac{2+p}{p} - \dfrac{-6}{p^2-3p}$

35. What are the two possible LCDs that could be used for the sum

$$\frac{10}{m-2} + \frac{5}{2-m}?$$

36. If one form of the correct answer to a sum or difference of rational expressions is $\frac{4}{k-3}$, what would be an alternate form of the answer if the denominator is $3 - k$?

Add or subtract. Write each answer in lowest terms. See Examples 5 and 8.

37. $\dfrac{4}{x-5} + \dfrac{6}{5-x}$

38. $\dfrac{10}{m-2} + \dfrac{5}{2-m}$

39. $\dfrac{-1}{1-y} + \dfrac{3-4y}{y-1}$

40. $\dfrac{-4}{p-3} - \dfrac{p+1}{3-p}$

41. $\dfrac{2}{x-y^2} + \dfrac{7}{y^2-x}$

42. $\dfrac{-8}{p-q^2} + \dfrac{3}{q^2-p}$

43. $\dfrac{x}{5x-3y} - \dfrac{y}{3y-5x}$

44. $\dfrac{t}{8t-9s} - \dfrac{s}{9s-8t}$

45. $\dfrac{3}{4p-5} + \dfrac{9}{5-4p}$

46. $\dfrac{8}{3-7y} - \dfrac{2}{7y-3}$

In each subtraction problem, the rational expression that follows the subtraction sign has a numerator with more than one term. Be very careful with signs and find each difference. See Examples 6–10.

47. $\dfrac{2m}{m-n} - \dfrac{5m+n}{2m-2n}$

48. $\dfrac{5p}{p-q} - \dfrac{3p+1}{4p-4q}$

49. $\dfrac{5}{x^2 - 9} - \dfrac{x + 2}{x^2 + 4x + 3}$

50. $\dfrac{1}{a^2 - 1} - \dfrac{a - 1}{a^2 + 3a - 4}$

51. $\dfrac{2q + 1}{3q^2 + 10q - 8} - \dfrac{3q + 5}{2q^2 + 5q - 12}$

52. $\dfrac{4y - 1}{2y^2 + 5y - 3} - \dfrac{y + 3}{6y^2 + y - 2}$

Perform the indicated operations. See Examples 1–10.

53. $\dfrac{4}{r^2 - r} + \dfrac{6}{r^2 + 2r} - \dfrac{1}{r^2 + r - 2}$

54. $\dfrac{6}{k^2 + 3k} - \dfrac{1}{k^2 - k} + \dfrac{2}{k^2 + 2k - 3}$

55. $\dfrac{x + 3y}{x^2 + 2xy + y^2} + \dfrac{x - y}{x^2 + 4xy + 3y^2}$

56. $\dfrac{m}{m^2 - 1} + \dfrac{m - 1}{m^2 + 2m + 1}$

57. $\dfrac{r + y}{18r^2 + 12ry - 3ry - 2y^2} + \dfrac{3r - y}{36r^2 - y^2}$

58. $\dfrac{2x - z}{2x^2 - 4xz + 5xz - 10z^2} - \dfrac{x + z}{x^2 - 4z^2}$

59. Refer to the rectangle in the figure.
  (a) Find an expression that represents its perimeter. Give the simplified form.
  (b) Find an expression that represents its area. Give the simplified form.

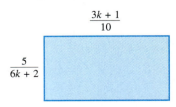

60. Refer to the triangle in the figure. Find an expression that represents its perimeter.

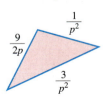

6.5 COMPLEX FRACTIONS

The quotient of two mixed numbers in arithmetic, such as $2\frac{1}{2} \div 3\frac{1}{4}$, can be written as a fraction:

$$2\frac{1}{2} \div 3\frac{1}{4} = \frac{2\frac{1}{2}}{3\frac{1}{4}} = \frac{2 + \frac{1}{2}}{3 + \frac{1}{4}}.$$

The last expression is the quotient of expressions that involve fractions. In algebra, some rational expressions also have fractions in the numerator, or denominator, or both.

Complex Fraction

A rational expression with one or more fractions in the numerator, denominator, or both, is called a **complex fraction.**

Examples of complex fractions include

$$\frac{2 + \frac{1}{2}}{3 + \frac{1}{4}}, \quad \frac{\frac{3x^2 - 5x}{6x^2}}{2x - \frac{1}{x}}, \quad \text{and} \quad \frac{3 + x}{5 - \frac{2}{x}}.$$

The parts of a complex fraction are named as follows.

$$\begin{array}{l} \dfrac{2}{p} - \dfrac{1}{q} \quad \leftarrow \text{Numerator of complex fraction} \\ \hline \qquad\qquad \leftarrow \text{Main fraction bar} \\ \dfrac{3}{p} + \dfrac{5}{q} \quad \leftarrow \text{Denominator of complex fraction} \end{array}$$

1   **Simplify a complex fraction by writing it as a division problem (Method 1).** Since the main fraction bar represents division in a complex fraction, one method of simplifying a complex fraction involves division.

Method 1

To simplify a complex fraction:

Step 1   Write both the numerator and denominator as single fractions.

Step 2   Change the complex fraction to a division problem.

Step 3   Perform the indicated division.

Once again, the first example shows complex fractions from both arithmetic and algebra.

① Simplify each complex fraction using Method 1.

(a) $\dfrac{6 + \dfrac{1}{x}}{5 - \dfrac{2}{x}}$

(b) $\dfrac{9 - \dfrac{4}{p}}{\dfrac{2}{p} + 1}$

② Simplify each complex fraction using Method 1.

(a) $\dfrac{\dfrac{rs^2}{t}}{\dfrac{r^2 s}{t^2}}$

(b) $\dfrac{\dfrac{m^2 n^3}{p}}{\dfrac{m^4 n}{p^2}}$

Example 1 Simplifying Complex Fractions (Method 1)

Simplify each complex fraction.

(a) $\dfrac{\dfrac{2}{3} + \dfrac{5}{9}}{\dfrac{1}{4} + \dfrac{1}{12}}$

(b) $\dfrac{6 + \dfrac{3}{x}}{\dfrac{x}{4} + \dfrac{1}{8}}$

Step 1 First, write each numerator as a single fraction.

$$\frac{2}{3} + \frac{5}{9} = \frac{2(3)}{3(3)} + \frac{5}{9} \qquad\qquad 6 + \frac{3}{x} = \frac{6}{1} + \frac{3}{x}$$
$$= \frac{6}{9} + \frac{5}{9} = \frac{11}{9} \qquad\qquad = \frac{6x}{x} + \frac{3}{x} = \frac{6x + 3}{x}$$

Do the same thing with each denominator.

$$\frac{1}{4} + \frac{1}{12} = \frac{1(3)}{4(3)} + \frac{1}{12} \qquad\qquad \frac{x}{4} + \frac{1}{8} = \frac{x(2)}{4(2)} + \frac{1}{8}$$
$$= \frac{3}{12} + \frac{1}{12} = \frac{4}{12} \qquad\qquad = \frac{2x}{8} + \frac{1}{8} = \frac{2x + 1}{8}$$

Step 2 The original complex fraction can now be written as follows.

$$\dfrac{\dfrac{11}{9}}{\dfrac{4}{12}} \qquad\qquad\qquad \dfrac{\dfrac{6x + 3}{x}}{\dfrac{2x + 1}{8}}$$

Step 3 Now use the definition of division and the fundamental property.

$$\frac{11}{9} \div \frac{4}{12} = \frac{11}{9} \cdot \frac{12}{4} \qquad\qquad \frac{6x + 3}{x} \div \frac{2x + 1}{8} = \frac{6x + 3}{x} \cdot \frac{8}{2x + 1}$$
$$= \frac{11 \cdot 3 \cdot 4}{3 \cdot 3 \cdot 4} \qquad\qquad\qquad = \frac{3(2x + 1)}{x} \cdot \frac{8}{2x + 1}$$
$$= \frac{11}{3} \qquad\qquad\qquad\qquad = \frac{24}{x}$$

Work Problem ① at the Side.

Example 2 Simplifying a Complex Fraction (Method 1)

Simplify the complex fraction.

$$\dfrac{\dfrac{xp}{q^3}}{\dfrac{p^2}{qx^2}}$$

Here the numerator and denominator are already single fractions, so use the definition of division and then the fundamental property.

$$\frac{xp}{q^3} \div \frac{p^2}{qx^2} = \frac{xp}{q^3} \cdot \frac{qx^2}{p^2} = \frac{x^3}{q^2 p}$$

Work Problem ② at the Side.

Example 3   **Simplifying a Complex Fraction (Method 1)**

Simplify the complex fraction.

$$\dfrac{\dfrac{3}{x+2}-4}{\dfrac{2}{x+2}+1} = \dfrac{\dfrac{3}{x+2}-\dfrac{4(x+2)}{x+2}}{\dfrac{2}{x+2}+\dfrac{1(x+2)}{x+2}}$$

Write both second terms with a denominator of $x+2$.

$$= \dfrac{\dfrac{3-4(x+2)}{x+2}}{\dfrac{2+1(x+2)}{x+2}}$$

Subtract in the numerator.

Add in the denominator.

$$= \dfrac{\dfrac{3-4x-8}{x+2}}{\dfrac{2+x+2}{x+2}}$$

Distributive property

$$= \dfrac{\dfrac{-5-4x}{x+2}}{\dfrac{4+x}{x+2}}$$

Combine like terms.

$$= \dfrac{-5-4x}{x+2}\cdot\dfrac{x+2}{4+x}$$

Multiply by the reciprocal.

$$= \dfrac{-5-4x}{4+x}$$

Lowest terms

CAUTION

Be aware that

$$\dfrac{\dfrac{a}{b}+\dfrac{c}{d}}{\dfrac{e}{f}+\dfrac{g}{h}} \neq \left(\dfrac{a}{b}+\dfrac{c}{d}\right)\cdot\left(\dfrac{f}{e}+\dfrac{h}{g}\right).$$

Work Problem ❸ at the Side.

2   **Simplify a complex fraction by multiplying by the least common denominator (Method 2).**   Since any expression can be multiplied by a form of 1 to get an equivalent expression, we may multiply both the numerator and the denominator of a complex fraction by the same nonzero expression to get an equivalent complex fraction. If we choose the expression to be the LCD of all the fractions within the complex fraction, the complex fraction will be simplified. This is Method 2.

❸ Simplify by Method 1.

$$\dfrac{\dfrac{2}{x-1}+\dfrac{1}{x+1}}{\dfrac{3}{x-1}-\dfrac{4}{x+1}}$$

④ Simplify by Method 2.

(a) $\dfrac{2 - \dfrac{6}{a}}{3 + \dfrac{4}{a}}$

(b) $\dfrac{\dfrac{p}{5 - p}}{\dfrac{4p}{2p + 1}}$

Method 2

To simplify a complex fraction:

Step 1 Find the LCD of all fractions within the complex fraction.

Step 2 Multiply both the numerator and the denominator of the complex fraction by this LCD using the distributive property as necessary. Write in lowest terms.

In the next example, Method 2 is used to simplify the complex fractions from Example 1.

Example 4 Simplifying Complex Fractions (Method 2)

Simplify each complex fraction.

(a) $\dfrac{\dfrac{2}{3} + \dfrac{5}{9}}{\dfrac{1}{4} + \dfrac{1}{12}}$

(b) $\dfrac{6 + \dfrac{3}{x}}{\dfrac{x}{4} + \dfrac{1}{8}}$

Step 1 Find the LCD for all denominators in the complex fraction.

The LCD for 3, 9, 4, and 12 is 36. | The LCD for x, 4, and 8 is $8x$.

Step 2 Multiply the numerator and denominator of the complex fraction by the LCD.

$$\dfrac{\dfrac{2}{3} + \dfrac{5}{9}}{\dfrac{1}{4} + \dfrac{1}{12}} = \dfrac{36\left(\dfrac{2}{3} + \dfrac{5}{9}\right)}{36\left(\dfrac{1}{4} + \dfrac{1}{12}\right)}$$

$$= \dfrac{36\left(\dfrac{2}{3}\right) + 36\left(\dfrac{5}{9}\right)}{36\left(\dfrac{1}{4}\right) + 36\left(\dfrac{1}{12}\right)}$$

$$= \dfrac{24 + 20}{9 + 3}$$

$$= \dfrac{44}{12} = \dfrac{4 \cdot 11}{4 \cdot 3}$$

$$= \dfrac{11}{3}$$

$$\dfrac{6 + \dfrac{3}{x}}{\dfrac{x}{4} + \dfrac{1}{8}} = \dfrac{8x\left(6 + \dfrac{3}{x}\right)}{8x\left(\dfrac{x}{4} + \dfrac{1}{8}\right)}$$

$$= \dfrac{8x(6) + 8x\left(\dfrac{3}{x}\right)}{8x\left(\dfrac{x}{4}\right) + 8x\left(\dfrac{1}{8}\right)}$$

$$= \dfrac{48x + 24}{2x^2 + x}$$

$$= \dfrac{24(2x + 1)}{x(2x + 1)}$$

$$= \dfrac{24}{x}$$

Work Problem **④** at the Side.

Example 5 Simplifying a Complex Fraction (Method 2)

Simplify the complex fraction.

$$\dfrac{\dfrac{3}{5m} - \dfrac{2}{m^2}}{\dfrac{9}{2m} + \dfrac{3}{4m^2}}$$

Continued on Next Page

ANSWERS

4. (a) $\dfrac{2a - 6}{3a + 4}$ (b) $\dfrac{2p + 1}{4(5 - p)}$

The LCD for $5m$, m^2, $2m$, and $4m^2$ is $20m^2$. Multiply the numerator and denominator by $20m^2$.

$$\frac{\dfrac{3}{5m} - \dfrac{2}{m^2}}{\dfrac{9}{2m} + \dfrac{3}{4m^2}} = \frac{20m^2\left(\dfrac{3}{5m} - \dfrac{2}{m^2}\right)}{20m^2\left(\dfrac{9}{2m} + \dfrac{3}{4m^2}\right)}$$

$$= \frac{20m^2\left(\dfrac{3}{5m}\right) - 20m^2\left(\dfrac{2}{m^2}\right)}{20m^2\left(\dfrac{9}{2m}\right) + 20m^2\left(\dfrac{3}{4m^2}\right)} \qquad \text{Distributive property}$$

$$= \frac{12m - 40}{90m + 15}$$

═══ Work Problem ❺ at the Side.

Either of the two methods shown in this section can be used to simplify a complex fraction. You may want to choose one method and stick with it to eliminate confusion. However, some students prefer to use Method 1 for problems like Example 2, which is the quotient of two fractions. They prefer Method 2 for problems like Examples 1, 3, 4, and 5, which have sums or differences in the numerators or denominators or both.

Example 6 **Deciding on a Method and Simplifying a Complex Fraction**

Simplify $\dfrac{\dfrac{1}{y} + \dfrac{2}{y + 2}}{\dfrac{4}{y} - \dfrac{3}{y + 2}}$.

Although either method will work, we will use Method 2 since there are sums and differences in the numerator and denominator. The LCD is $y(y + 2)$. Multiply the numerator and denominator by the LCD.

$$\frac{\dfrac{1}{y} + \dfrac{2}{y + 2}}{\dfrac{4}{y} - \dfrac{3}{y + 2}} \cdot \frac{y(y + 2)}{y(y + 2)} = \frac{1(y + 2) + 2y}{4(y + 2) - 3y} \qquad \text{Fundamental property}$$

$$= \frac{y + 2 + 2y}{4y + 8 - 3y} \qquad \text{Distributive property}$$

$$= \frac{3y + 2}{y + 8} \qquad \text{Combine like terms.}$$

═══ Work Problem ❻ at the Side.

Example 7 **Deciding on a Method and Simplifying a Complex Fraction**

Simplify $\dfrac{\dfrac{x + 2}{x - 3}}{\dfrac{x^2 - 4}{x^2 - 9}}$.

═══ Continued on Next Page

---

❺ Simplify by Method 2.

$$\frac{\dfrac{2}{5x} - \dfrac{3}{x^2}}{\dfrac{7}{4x} + \dfrac{1}{2x^2}}$$

❻ Simplify. Use either method.

$$\frac{\dfrac{1}{x} + \dfrac{2}{x - 1}}{\dfrac{4}{x - 1}}$$

7 Simplify. Use either method.

$$\frac{\dfrac{a + b}{a - b}}{\dfrac{a^2 + 2ab + b^2}{a^2 - b^2}}$$

Since this is simply a quotient of two rational expressions, we will use Method 1.

$$\frac{\dfrac{x + 2}{x - 3}}{\dfrac{x^2 - 4}{x^2 - 9}} = \frac{x + 2}{x - 3} \div \frac{x^2 - 4}{x^2 - 9}$$

$$= \frac{x + 2}{x - 3} \cdot \frac{x^2 - 9}{x^2 - 4} \qquad \text{Definition of division}$$

$$= \frac{x + 2}{x - 3} \cdot \frac{(x + 3)(x - 3)}{(x + 2)(x - 2)} \qquad \text{Factor.}$$

$$= \frac{x + 3}{x - 2} \qquad \text{Multiply.}$$

Work Problem 7 at the Side.

6.5 EXERCISES

FOR EXTRA HELP

Student's Solutions Manual | MyMathLab.com | InterAct Math Tutorial Software | AW Math Tutor Center | www.mathxl.com | Digital Video Tutor CD 5 Videotape 13

Note: In many problems involving complex fractions, several different equivalent forms of the answers exist. If your answer does not look exactly like the one given in the back of the book, check to see whether your answer is an equivalent form.

1. Consider the complex fraction $\dfrac{\frac{1}{2}-\frac{1}{3}}{\frac{5}{6}-\frac{1}{12}}$. Answer each part, outlining Method 1 for simplifying this complex fraction.

   (a) To combine the terms in the numerator, we must find the LCD of $\frac{1}{2}$ and $\frac{1}{3}$. What is this LCD? Determine the simplified form of the numerator of the complex fraction.

   (b) To combine the terms in the denominator, we must find the LCD of $\frac{5}{6}$ and $\frac{1}{12}$. What is this LCD? Determine the simplified form of the denominator of the complex fraction.

   (c) Now use the results from parts (a) and (b) to write the complex fraction as a division problem using the symbol $\div$.

   (d) Perform the operation from part (c) to obtain the final simplification.

2. Consider the same complex fraction given in Exercise 1, $\dfrac{\frac{1}{2}-\frac{1}{3}}{\frac{5}{6}-\frac{1}{12}}$. Answer each part, outlining Method 2 for simplifying this complex fraction.

   (a) We must determine the LCD of all the fractions within the complex fraction. What is this LCD?

   (b) Multiply every term in the complex fraction by the LCD found in part (a), but do not combine the terms in the numerator and the denominator yet.

   (c) Combine the terms from part (b) to obtain the simplified form of the complex fraction.

Simplify each complex fraction. Use either method. See Examples 1–7.

3. $\dfrac{-\frac{4}{3}}{\frac{2}{9}}$

4. $\dfrac{-\frac{5}{6}}{\frac{5}{4}}$

5. $\dfrac{\frac{p}{q^2}}{\frac{p^2}{q}}$

6. $\dfrac{\frac{a}{x}}{\frac{a^2}{2x}}$

7. $\dfrac{\frac{x}{y^2}}{\frac{x^2}{y}}$

8. $\dfrac{\frac{p^4}{r}}{\frac{p^2}{r^2}}$

9. $\dfrac{\frac{4a^4b^3}{3a}}{\frac{2ab^4}{b^2}}$

10. $\dfrac{\frac{2r^4t^2}{3t}}{\frac{5r^2t^5}{3r}}$

11. $\dfrac{\dfrac{m+2}{3}}{\dfrac{m-4}{m}}$

12. $\dfrac{\dfrac{q-5}{q}}{\dfrac{q+5}{3}}$

13. $\dfrac{\dfrac{2}{x}-3}{\dfrac{2-3x}{2}}$

14. $\dfrac{6+\dfrac{2}{r}}{\dfrac{3r+1}{4}}$

15. $\dfrac{\dfrac{1}{x}+x}{\dfrac{x^2+1}{8}}$

16. $\dfrac{\dfrac{3}{m}-m}{\dfrac{3-m^2}{4}}$

17. $\dfrac{a-\dfrac{5}{a}}{a+\dfrac{1}{a}}$

18. $\dfrac{q+\dfrac{1}{q}}{q+\dfrac{4}{q}}$

19. $\dfrac{\dfrac{1}{2}+\dfrac{1}{p}}{\dfrac{2}{3}+\dfrac{1}{p}}$

20. $\dfrac{\dfrac{3}{4}-\dfrac{1}{r}}{\dfrac{1}{5}+\dfrac{1}{r}}$

21. $\dfrac{\dfrac{t}{t+2}}{\dfrac{4}{t^2-4}}$

22. $\dfrac{\dfrac{m}{m+1}}{\dfrac{3}{m^2-1}}$

23. $\dfrac{\dfrac{1}{k+1}-1}{\dfrac{1}{k+1}+1}$

24. $\dfrac{\dfrac{2}{p-1}+2}{\dfrac{3}{p-1}-2}$

25. $\dfrac{\dfrac{1}{m-1}+\dfrac{2}{m+2}}{\dfrac{2}{m+2}-\dfrac{1}{m-3}}$

26. $\dfrac{\dfrac{5}{r+3}-\dfrac{1}{r-1}}{\dfrac{2}{r+2}+\dfrac{3}{r+3}}$

27. $2-\dfrac{2}{2+\dfrac{2}{2+2}}$

28. $3-\dfrac{2}{4+\dfrac{2}{4-2}}$

RELATING CONCEPTS (Exercises 29–32)   **FOR INDIVIDUAL OR GROUP WORK**

To find the average of two numbers, we add them and divide by 2. Suppose that we wish to find the average of $\frac{3}{8}$ and $\frac{5}{6}$. **Work Exercises 29–32 in order,** *to see how a complex fraction occurs in a problem like this.*

29. Write in symbols: the sum of $\frac{3}{8}$ and $\frac{5}{6}$, divided by 2. Your result should be written as a complex fraction.

30. Simplify the complex fraction from Exercise 29 using Method 1.

31. Simplify the complex fraction from Exercise 29 using Method 2.

32. Your answers in Exercises 30 and 31 should be the same. Which method did you prefer? Why?

6.6   SOLVING EQUATIONS WITH RATIONAL EXPRESSIONS

In Section 2.3 we solved equations with fractions as coefficients. By using the multiplication property of equality, we cleared the fractions by multiplying by the LCD. We continue this work here.

OBJECTIVES

1 Distinguish between operations with rational expressions and equations with terms that are rational expressions.

2 Solve equations with rational expressions.

3 Solve a formula for a specified variable.

1 **Distinguish between operations with rational expressions and equations with terms that are rational expressions.**   Before solving equations with rational expressions, you must understand the difference between *sums* and *differences* of terms with rational coefficients, and *equations* with terms that are rational expressions. Sums and differences are operations to be *performed,* while equations are *solved.*

Example 1   **Distinguishing between Operations and Equations**

Identify each of the following as an operation or an equation. Then perform the operation or solve the equation.

(a) $\dfrac{3}{4}x - \dfrac{2}{3}x$

This is a difference of two terms, so it is an operation. (There is no equals sign.) Find the LCD, write each coefficient with this LCD, and combine like terms.

$$\frac{3}{4}x - \frac{2}{3}x = \frac{9}{12}x - \frac{8}{12}x \qquad \text{Get a common denominator.}$$

$$= \frac{1}{12}x \qquad \text{Combine like terms.}$$

(b) $\dfrac{3}{4}x - \dfrac{2}{3}x = \dfrac{1}{2}$

Because of the equals sign, this is an equation to be solved. We proceed as in Section 2.3, using the multiplication property of equality to clear fractions. The LCD is 12.

$$\frac{3}{4}x - \frac{2}{3}x = \frac{1}{2}$$

$$12\left(\frac{3}{4}x - \frac{2}{3}x\right) = 12\left(\frac{1}{2}\right) \qquad \text{Multiply by 12.}$$

$$12\left(\frac{3}{4}x\right) - 12\left(\frac{2}{3}x\right) = 12\left(\frac{1}{2}\right) \qquad \text{Distributive property}$$

$$9x - 8x = 6 \qquad \text{Multiply.}$$

$$x = 6 \qquad \text{Combine like terms.}$$

Continued on Next Page

1 Identify each as an operation or an equation. Then perform the operation or solve the equation.

(a) $\dfrac{x}{3} + \dfrac{x}{5} = 7 + x$

(b) $\dfrac{2x}{3} - \dfrac{4x}{9}$

2 Solve each equation, and check your answers.

(a) $\dfrac{x}{5} + 3 = \dfrac{3}{5}$

(b) $\dfrac{x}{2} - \dfrac{x}{3} = \dfrac{5}{6}$

Check:

$$\frac{3}{4}x - \frac{2}{3}x = \frac{1}{2} \qquad \text{Original equation}$$

$$\frac{3}{4}(6) - \frac{2}{3}(6) = \frac{1}{2} \quad ? \qquad \text{Let } x = 6.$$

$$\frac{9}{2} - 4 = \frac{1}{2} \quad ? \qquad \text{Multiply.}$$

$$\frac{1}{2} = \frac{1}{2} \qquad \text{True}$$

The check shows that 6 is the solution of the equation.

The ideas of Example 1 can be summarized as follows.

> When adding or subtracting, the LCD must be kept throughout the simplification. When solving an equation, the LCD is used to multiply each side so that denominators are eliminated.

Work Problem ❶ at the Side.

2 Solve equations with rational expressions. When an equation involves fractions as in Example 1(b), we use the multiplication property of equality to clear it of fractions. Choose as multiplier the LCD of all denominators in the fractions of the equation.

Example 2 Solving an Equation with Rational Expressions

Solve $\dfrac{x}{3} + \dfrac{x}{4} = 10 + x$. Check the solution.

Because the LCD of the two fractions is 12, we begin by multiplying each side of the equation by 12.

$$12\left(\frac{x}{3} + \frac{x}{4}\right) = 12(10 + x)$$

$$12\left(\frac{x}{3}\right) + 12\left(\frac{x}{4}\right) = 12(10) + 12x \qquad \text{Distributive property}$$

$$4x + 3x = 120 + 12x$$

$$7x = 120 + 12x \qquad \text{Combine like terms.}$$

$$-5x = 120 \qquad \text{Subtract } 12x.$$

$$x = -24 \qquad \text{Divide by } -5.$$

Check:

$$\frac{x}{3} + \frac{x}{4} = 10 + x \qquad \text{Original equation}$$

$$\frac{-24}{3} + \frac{-24}{4} = 10 - 24 \quad ? \qquad \text{Let } x = -24.$$

$$-8 - 6 = -14 \quad ?$$

$$-14 = -14 \qquad \text{True}$$

The solution is -24.

Work Problem ❷ at the Side.

CAUTION

Note that the use of the LCD here is different from its use in the previous section. Here, we use the multiplication property of equality to multiply each side of an *equation* by the LCD. Earlier, we used the fundamental property to multiply a *fraction* by another fraction that had the LCD as both its numerator and denominator. Be careful not to confuse these two methods.

Example 3 Solving an Equation with Rational Expressions

Solve $\dfrac{p}{2} - \dfrac{p-1}{3} = 1$.

$$6\left(\frac{p}{2} - \frac{p-1}{3}\right) = 6 \cdot 1 \quad \text{Multiply by the LCD, 6.}$$

$$6\left(\frac{p}{2}\right) - 6\left(\frac{p-1}{3}\right) = 6 \quad \text{Distributive property}$$

$$3p - 2(p-1) = 6$$

Be very careful to put parentheses around $p - 1$; otherwise, you may find an incorrect solution. Continue simplifying and solve.

$$3p - 2p + 2 = 6 \quad \text{Distributive property}$$
$$p + 2 = 6 \quad \text{Combine like terms.}$$
$$p = 4 \quad \text{Subtract 2.}$$

Check to see that 4 is correct by replacing p with 4 in the original equation.

Work Problem ❸ at the Side.

When solving an equation that has a variable in the denominator, remember that the number 0 cannot be used as a denominator. Therefore, the solution cannot be a number that will make the denominator equal 0.

Example 4 Solving an Equation with Rational Expressions

Solve $\dfrac{x}{x-2} = \dfrac{2}{x-2} + 2$. Check the proposed solution.

The common denominator is $x - 2$. (*Note:* Because $x = 2$ makes a denominator 0, x cannot equal 2.) Solve the equation by multiplying each side of the equation by $x - 2$.

$$(x-2)\left(\frac{x}{x-2}\right) = (x-2)\left(\frac{2}{x-2} + 2\right)$$

$$(x-2)\left(\frac{x}{x-2}\right) = (x-2)\left(\frac{2}{x-2}\right) + (x-2)(2)$$

$$x = 2 + 2x - 4$$
$$x = -2 + 2x \quad \text{Combine like terms.}$$
$$-x = -2 \quad \text{Subtract } 2x.$$
$$x = 2 \quad \text{Divide by } -1.$$

Continued on Next Page

❸ Solve each equation, and check your answers.

(a) $\dfrac{k}{6} - \dfrac{k+1}{4} = -\dfrac{1}{2}$

(b) $\dfrac{2m-3}{5} - \dfrac{m}{3} = -\dfrac{6}{5}$

④ Solve the equation, and check your answer.

$$1 - \frac{2}{x + 1} = \frac{2x}{x + 1}$$

Check: The proposed solution is 2. If we substitute 2 in the original equation, we get

$$\frac{2}{2 - 2} = \frac{2}{2 - 2} + 2 \quad ?$$

$$\frac{2}{0} = \frac{2}{0} + 2. \quad ?$$

Notice that 2 makes both denominators equal 0. Because 0 cannot be the denominator, there is no solution.

While it is always a good idea to check solutions to guard against arithmetic and algebraic errors, it is *essential* to check proposed solutions when variables appear in denominators in the original equation. Some students like to determine which numbers cannot be solutions *before* solving the equation.

Work Problem ④ at the Side.

The steps used to solve an equation with rational expressions follow.

> ### Solving Equations with Rational Expressions
>
> *Step 1* **Multiply by the LCD.** Multiply each side of the equation by the least common denominator. (This clears the equation of fractions.)
>
> *Step 2* **Solve.** Solve the resulting equation.
>
> *Step 3* **Check.** Check each proposed solution by substituting it in the original equation. Reject any that cause a denominator to equal 0.

Example 5 Solving an Equation with Rational Expressions

Solve $\dfrac{2}{x^2 - x} = \dfrac{1}{x^2 - 1}$. Check the proposed solution.

Step 1 Begin by finding the LCD.

$$\frac{2}{x(x - 1)} = \frac{1}{(x + 1)(x - 1)} \qquad \text{Factor the denominators to find the LCD.}$$

Since $x^2 - x$ can be factored as $x(x - 1)$, and $x^2 - 1$ can be factored as $(x + 1)(x - 1)$, the LCD is $x(x + 1)(x - 1)$.

Step 2 Notice that 0, -1, and 1 cannot be solutions of this equation. Multiply each side of the equation by $x(x + 1)(x - 1)$.

$$x(x + 1)(x - 1)\frac{2}{x(x - 1)} = x(x + 1)(x - 1)\frac{1}{(x + 1)(x - 1)}$$

$$2(x + 1) = x$$

$$2x + 2 = x \qquad \text{Distributive property}$$

$$2 = -x \qquad \text{Subtract } 2x.$$

$$x = -2 \qquad \text{Multiply by } -1.$$

Continued on Next Page

ANSWERS

4. no solution (When the equation is solved, -1 is found. However, because $x = -1$ leads to a 0 denominator in the original equation, there is no solution.)

Step 3 The proposed solution is -2, which does not make any denominator equal 0.

Check:

$$\frac{2}{x^2 - x} = \frac{1}{x^2 - 1}$$ Original equation

$$\frac{2}{(-2)^2 - (-2)} = \frac{1}{(-2)^2 - 1}$$? Let $x = -2$.

$$\frac{2}{4 + 2} = \frac{1}{4 - 1}$$?

$$\frac{1}{3} = \frac{1}{3}$$ True

The solution is indeed -2.

Work Problem ⑤ at the Side.

⑤ Solve each equation, and check your answers.

(a) $\dfrac{4}{x^2 - 3x} = \dfrac{1}{x^2 - 9}$

(b) $\dfrac{2}{p^2 - 2p} = \dfrac{3}{p^2 - p}$

Example 6 Solving an Equation with Rational Expressions

Solve $\dfrac{2m}{m^2 - 4} + \dfrac{1}{m - 2} = \dfrac{2}{m + 2}$.

Multiply by the LCD, $(m + 2)(m - 2)$. (Notice that -2 and 2 cannot be solutions.)

$$(m + 2)(m - 2)\left(\frac{2m}{m^2 - 4} + \frac{1}{m - 2}\right)$$

$$= (m + 2)(m - 2)\frac{2}{m + 2}$$

$$(m + 2)(m - 2)\frac{2m}{m^2 - 4} + (m + 2)(m - 2)\frac{1}{m - 2}$$

$$= (m + 2)(m - 2)\frac{2}{m + 2}$$

$$2m + m + 2 = 2(m - 2)$$

$$3m + 2 = 2m - 4$$ Distributive property; combine like terms.

$$m + 2 = -4$$ Subtract $2m$.

$$m = -6$$ Subtract 2.

Check to see that -6 is a valid solution for the given equation.

Work Problem ⑥ at the Side.

⑥ Solve each equation, and check your answers.

(a)

$$\frac{2p}{p^2 - 1} = \frac{2}{p + 1} - \frac{1}{p - 1}$$

(b)

$$\frac{8r}{4r^2 - 1} = \frac{3}{2r + 1} + \frac{3}{2r - 1}$$

Example 7 Solving an Equation with Rational Expressions

Solve $\dfrac{1}{x - 1} + \dfrac{1}{2} = \dfrac{2}{x^2 - 1}$.

Multiply each side of the equation by the LCD, $2(x + 1)(x - 1)$. (Notice that -1 and 1 cannot be solutions.)

Continued on Next Page

7 Solve the equation, and check your solution.

$$\frac{2}{3x + 1} - \frac{1}{x} = \frac{-6x}{3x + 1}$$

$$2(x + 1)(x - 1)\left(\frac{1}{x - 1} + \frac{1}{2}\right) = 2(x + 1)(x - 1)\frac{2}{(x + 1)(x - 1)}$$

$$2(x + 1)(x - 1)\frac{1}{x - 1} + 2(x + 1)(x - 1)\frac{1}{2}$$

$$= 2(x + 1)(x - 1)\frac{2}{(x + 1)(x - 1)}$$

$$2(x + 1) + (x + 1)(x - 1) = 4$$

$2x + 2 + x^2 - 1 = 4$	Distributive property
$x^2 + 2x + 1 = 4$	Combine like terms.
$x^2 + 2x - 3 = 0$	Subtract 4.
$(x + 3)(x - 1) = 0$	Factor.

Solving this equation suggests that $x = -3$ or $x = 1$. But 1 makes a denominator of the original equation equal 0, so 1 is not a solution. However, -3 is a solution, as shown by substituting -3 for x in the original equation.

Check:

$$\frac{1}{x - 1} + \frac{1}{2} = \frac{2}{x^2 - 1} \qquad \text{Original equation}$$

$$\frac{1}{-3 - 1} + \frac{1}{2} = \frac{2}{(-3)^2 - 1} \quad ? \quad \text{Let } x = -3.$$

$$\frac{1}{-4} + \frac{1}{2} = \frac{2}{9 - 1} \quad ? \quad \text{Simplify.}$$

$$\frac{1}{4} = \frac{1}{4} \qquad \text{True}$$

The check shows that -3 is a solution.

Work Problem 7 at the Side.

8 Solve each equation, and check your proposed solutions.

(a) $\dfrac{1}{x - 2} + \dfrac{1}{5} = \dfrac{2}{5(x^2 - 4)}$

(b) $\dfrac{6}{5a + 10} - \dfrac{1}{a - 5}$

$\quad = \dfrac{4}{a^2 - 3a - 10}$

Example 8 Solving an Equation with Rational Expressions

Solve $\dfrac{1}{k^2 + 4k + 3} + \dfrac{1}{2k + 2} = \dfrac{3}{4k + 12}$.

Factor the three denominators to get the common denominator, $4(k + 1)(k + 3)$. Multiply each side by this product. (Notice that -1 and -3 cannot be solutions.)

$$4(k + 1)(k + 3)\left(\frac{1}{(k + 1)(k + 3)} + \frac{1}{2(k + 1)}\right)$$

$$= 4(k + 1)(k + 3)\frac{3}{4(k + 3)}$$

$$4(k + 1)(k + 3)\frac{1}{(k + 1)(k + 3)} + 2 \cdot 2(k + 1)(k + 3)\frac{1}{2(k + 1)}$$

$$= 4(k + 1)(k + 3)\frac{3}{4(k + 3)}$$

$4 + 2(k + 3) = 3(k + 1)$	Simplify.
$4 + 2k + 6 = 3k + 3$	Distributive property
$2k + 10 = 3k + 3$	Combine like terms.
$7 = k$	Subtract $2k$ and 3.

Check to see that 7 is a solution of the given equation.

Work Problem 8 at the Side.

3 Solve a formula for a specified variable. Solving a formula for a specified variable was first discussed in Chapter 2. Remember to treat the variable for which you are solving as if it were the only variable, and all others as if they were constants.

Example 9 Solving for a Specified Variable

Solve $a = \dfrac{v - w}{t}$ for v.

$$a = \frac{v - w}{t} \qquad \text{Given equation}$$

$$at = v - w \qquad \text{Multiply by } t.$$

$$at + w = v \qquad \text{Add } w.$$

or $$v = at + w$$

To check this, substitute $at + w$ for v in the original equation. The final result will be the identity $a = a$, indicating that the result obtained is correct.

════════════════ **Work Problem 9 at the Side.**

Example 10 Solving for a Specified Variable

Solve the formula $\dfrac{1}{a} = \dfrac{1}{b} + \dfrac{1}{c}$ for c.

The LCD of all the fractions in the equation is abc, so multiply each side by abc.

$$abc\left(\frac{1}{a}\right) = abc\left(\frac{1}{b} + \frac{1}{c}\right)$$

$$abc\left(\frac{1}{a}\right) = abc\left(\frac{1}{b}\right) + abc\left(\frac{1}{c}\right) \qquad \text{Distributive property}$$

$$bc = ac + ab$$

Since we are solving for c, get all terms with c on one side of the equation. Do this by subtracting ac from each side.

$$bc - ac = ab \qquad \text{Subtract } ac.$$

Factor out the common factor c on the left.

$$c(b - a) = ab \qquad \text{Factor out } c.$$

Finally, divide each side by the coefficient of c, which is $b - a$.

$$c = \frac{ab}{b - a}$$

9 Solve $z = \dfrac{x}{x + y}$ for y.

10 Solve $\dfrac{2}{x} = \dfrac{1}{y} + \dfrac{1}{z}$ for z.

CAUTION

Students often have trouble in the step that involves factoring out the variable for which they are solving. In Example 10, we had to factor out c on the left side so that we could divide both sides by $b - a$.

When solving an equation for a specified variable, be sure that the specified variable appears alone on only one side of the equals sign in the final equation.

Work Problem 10 at the Side.

6.6 EXERCISES

 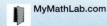

Identify each as an operation or an equation. Then perform the operation or solve the equation. See Example 1.

1. $\dfrac{7}{8}x + \dfrac{1}{5}x$

2. $\dfrac{4}{7}x + \dfrac{3}{5}x$

3. $\dfrac{7}{8}x + \dfrac{1}{5}x = 1$

4. $\dfrac{4}{7}x + \dfrac{3}{5}x = 1$

5. $\dfrac{3}{5}y - \dfrac{7}{10}y$

6. $\dfrac{3}{5}y - \dfrac{7}{10}y = 1$

7. Explain how the LCD is used in a different way when adding and subtracting rational expressions compared to solving equations with rational expressions.

8. If we multiply each side of the equation $\dfrac{6}{x+5} = \dfrac{6}{x+5}$ by $x + 5$, we get $6 = 6$. Are all real numbers solutions of this equation? Explain.

Solve each equation, and check your answers. See Examples 2 and 3.

9. $\dfrac{2}{3}x + \dfrac{1}{2}x = -7$

10. $\dfrac{1}{4}x - \dfrac{1}{3}x = 1$

11. $\dfrac{p}{3} - \dfrac{p}{6} = 4$

12. $\dfrac{x}{15} + \dfrac{x}{5} = 4$

13. $\dfrac{3x}{5} - 6 = x$

14. $\dfrac{5t}{4} + t = 9$

15. $\dfrac{4m}{7} + m = 11$

16. $a - \dfrac{3a}{2} = 1$

17. $\dfrac{z-1}{4} = \dfrac{z+3}{3}$

18. $\dfrac{r-5}{2} = \dfrac{r+2}{3}$

19. $\dfrac{3p+6}{8} = \dfrac{3p-3}{16}$

20. $\dfrac{2z+1}{5} = \dfrac{7z+5}{15}$

21. $\dfrac{2x+3}{-6} = \dfrac{3}{2}$

22. $\dfrac{4y+3}{6} = \dfrac{5}{2}$

23. $\dfrac{q+2}{3} + \dfrac{q-5}{5} = \dfrac{7}{3}$

24. $\dfrac{b+7}{8} - \dfrac{b-2}{3} = \dfrac{4}{3}$

25. $\dfrac{t}{6} + \dfrac{4}{3} = \dfrac{t-2}{3}$

26. $\dfrac{x}{2} = \dfrac{5}{4} + \dfrac{x-1}{4}$

27. $\dfrac{3m}{5} - \dfrac{3m-2}{4} = \dfrac{1}{5}$

28. $\dfrac{8p}{5} = \dfrac{3p-4}{2} + \dfrac{5}{2}$

29. What values of x cannot possibly be solutions of the equation $\dfrac{1}{x-4} = \dfrac{3}{2x}$?

30. What is wrong with the following problem? "Solve $\dfrac{2}{3x} + \dfrac{1}{5x}$."

Solve each equation, and check your answers. See Examples 4–8.

31. $\dfrac{2x+3}{x} = \dfrac{3}{2}$

32. $\dfrac{5-2y}{y} = \dfrac{1}{4}$

33. $\dfrac{k}{k-4} - 5 = \dfrac{4}{k-4}$

34. $\dfrac{-5}{a+5} = \dfrac{a}{a+5} + 2$

35. $\dfrac{3}{x-1} + \dfrac{2}{4x-4} = \dfrac{7}{4}$

36. $\dfrac{2}{p+3} + \dfrac{3}{8} = \dfrac{5}{4p+12}$

37. $\dfrac{y}{3y+3} = \dfrac{2y-3}{y+1} - \dfrac{2y}{3y+3}$

38. $\dfrac{2k+3}{k+1} - \dfrac{3k}{2k+2} = \dfrac{-2k}{2k+2}$

39. $\dfrac{2}{m} = \dfrac{m}{5m+12}$

40. $\dfrac{x}{4-x} = \dfrac{2}{x}$

41. $\dfrac{-2}{z+5} + \dfrac{3}{z-5} = \dfrac{20}{z^2-25}$

42. $\dfrac{3}{r+3} - \dfrac{2}{r-3} = \dfrac{-12}{r^2-9}$

43. $\dfrac{3y}{y^2+5y+6} = \dfrac{5y}{y^2+2y-3} - \dfrac{2}{y^2+y-2}$

44. $\dfrac{x+4}{x^2-3x+2} - \dfrac{5}{x^2-4x+3} = \dfrac{x-4}{x^2-5x+6}$

45. $\dfrac{5x}{14x+3} = \dfrac{1}{x}$

46. $\dfrac{m}{8m+3} = \dfrac{1}{3m}$

47. $\dfrac{2}{z-1} - \dfrac{5}{4} = \dfrac{-1}{z+1}$

48. $\dfrac{5}{p-2} = 7 - \dfrac{10}{p+2}$

Solve each formula for the specified variable. See Example 9.

49. $m = \dfrac{kF}{a}$ for F

50. $I = \dfrac{kE}{R}$ for E

51. $m = \dfrac{kF}{a}$ for a

52. $I = \dfrac{kE}{R}$ for R

53. $I = \dfrac{E}{R+r}$ for R

54. $I = \dfrac{E}{R+r}$ for r

55. $h = \dfrac{2A}{B+b}$ for A

56. $d = \dfrac{2S}{n(a+L)}$ for S

57. $d = \dfrac{2S}{n(a+L)}$ for a

58. $h = \dfrac{2A}{B+b}$ for B

Solve each equation for the specified variable. See Example 10.

59. $\dfrac{2}{r} + \dfrac{3}{s} + \dfrac{1}{t} = 1$ for t

60. $\dfrac{5}{p} + \dfrac{2}{q} + \dfrac{3}{r} = 1$ for r

61. $\dfrac{1}{a} - \dfrac{1}{b} - \dfrac{1}{c} = 2$ for c

62. $\dfrac{-1}{x} + \dfrac{1}{y} + \dfrac{1}{z} = 4$ for y

Summary Exercises on OPERATIONS AND EQUATIONS WITH RATIONAL EXPRESSIONS

We have performed the four operations of arithmetic with rational expressions and solved equations with rational expressions. The exercises in this summary include a mixed variety of problems of these types. To work them, recall the procedures explained in the earlier sections of this chapter. They are summarized here.

Multiplication of Rational Expressions	Multiply numerators and multiply denominators. Use the fundamental property to express the answer in lowest terms.
Division of Rational Expressions	Change the second rational expression to its reciprocal; then multiply as just described.
Addition of Rational Expressions	Find the least common denominator (LCD) if necessary. Write all rational expressions with this LCD. Add numerators, and keep the same denominator. Express the answer in lowest terms.
Subtraction of Rational Expressions	Find the LCD if necessary. Write all rational expressions with this LCD. Subtract numerators (use parentheses as required), and keep the same denominator. Express the answer in lowest terms.
Solving Equations with Rational Expressions	Multiply each side of the equation by the LCD of all the rational expressions in the equation. Solve, using methods described in earlier chapters. Be sure to check all proposed solutions and reject any that cause a denominator to equal 0.

Students often confuse *operations* on rational expressions with the *solution of equations* with rational expressions. For example, the four possible operations on the rational expressions $\frac{1}{x}$ and $\frac{1}{x-2}$ are performed as follows.

Add:

$$\frac{1}{x} + \frac{1}{x-2} = \frac{x-2}{x(x-2)} + \frac{x}{x(x-2)} \quad \text{Write with a common denominator.}$$

$$= \frac{x-2+x}{x(x-2)} \quad \text{Add numerators; keep the same denominator.}$$

$$= \frac{2x-2}{x(x-2)} \quad \text{Combine like terms.}$$

Subtract:

$$\frac{1}{x} - \frac{1}{x-2} = \frac{x-2}{x(x-2)} - \frac{x}{x(x-2)} \quad \text{Write with a common denominator.}$$

$$= \frac{x-2-x}{x(x-2)} \quad \text{Subtract numerators; keep the same denominator.}$$

$$= \frac{-2}{x(x-2)} \quad \text{Combine like terms.}$$

Multiply:

$$\frac{1}{x} \cdot \frac{1}{x-2} = \frac{1}{x(x-2)} \quad \text{Multiply numerators and multiply denominators.}$$

Divide:

$$\frac{1}{x} \div \frac{1}{x-2} = \frac{1}{x} \cdot \frac{x-2}{1} = \frac{x-2}{x} \quad \text{Change to multiplication by the reciprocal of the second fraction.}$$

On the other hand, consider the *equation*

$$\frac{1}{x} + \frac{1}{x-2} = \frac{3}{4}.$$

Neither 0 nor 2 can be a solution of this equation, since each will cause a denominator to equal 0. We use the multiplication property of equality to multiply each side by the LCD, $4x(x-2)$, leading to an equation with no denominators.

$$4x(x-2)\frac{1}{x} + 4x(x-2)\frac{1}{x-2} = 4x(x-2)\frac{3}{4}$$

$$4x - 8 + 4x = 3x^2 - 6x \qquad \text{Distributive property}$$

$$0 = 3x^2 - 14x + 8 \qquad \text{Get 0 on one side.}$$

$$0 = (3x-2)(x-4) \qquad \text{Factor.}$$

$$3x - 2 = 0 \quad \text{or} \quad x - 4 = 0 \qquad \text{Zero-factor property}$$

$$x = \frac{2}{3} \quad \text{or} \qquad x = 4$$

Both $\frac{2}{3}$ and 4 are solutions since neither makes a denominator equal 0.

In conclusion, remember the following points when working exercises involving rational expressions.

Points to Remember When Working with Rational Expressions

1. The fundamental property is applied only after numerators and denominators have been *factored*.
2. When adding and subtracting rational expressions, the common denominator must be kept throughout the problem and in the final result.
3. Always look to see if the answer is in lowest terms; if it is not, use the fundamental property.
4. When solving equations, the LCD is used to clear the equation of fractions. Multiply each side by the LCD. (Notice how this differs from the use of the LCD in Point 2.)
5. When solving equations with rational expressions, reject any proposed solution that causes an original denominator to equal 0.

For each exercise, indicate "operation" if an operation is to be performed or "equation" if an equation is to be solved. Then perform the operation or solve the equation.

1. $\dfrac{4}{p} + \dfrac{6}{p}$

2. $\dfrac{x^3 y^2}{x^2 y^4} \cdot \dfrac{y^5}{x^4}$

3. $\dfrac{1}{x^2 + x - 2} \div \dfrac{4x^2}{2x - 2}$

4. $\dfrac{8}{m-5} = 2$

5. $\dfrac{2y^2 + y - 6}{2y^2 - 9y + 9} \cdot \dfrac{y^2 - 2y - 3}{y^2 - 1}$

6. $\dfrac{2}{k^2 - 4k} + \dfrac{3}{k^2 - 16}$

7. $\dfrac{x-4}{5} = \dfrac{x+3}{6}$

8. $\dfrac{3t^2 - t}{6t^2 + 15t} \div \dfrac{6t^2 + t - 1}{2t^2 - 5t - 25}$

9. $\dfrac{4}{p+2} + \dfrac{1}{3p+6}$

10. $\dfrac{1}{y} + \dfrac{1}{y-3} = -\dfrac{5}{4}$

11. $\dfrac{3}{t-1} + \dfrac{1}{t} = \dfrac{7}{2}$

12. $\dfrac{6}{y} - \dfrac{2}{3y}$

13. $\dfrac{5}{4z} - \dfrac{2}{3z}$

14. $\dfrac{k+2}{3} = \dfrac{2k-1}{5}$

15. $\dfrac{1}{m^2 + 5m + 6} + \dfrac{2}{m^2 + 4m + 3}$

16. $\dfrac{2k^2 - 3k}{20k^2 - 5k} \div \dfrac{2k^2 - 5k + 3}{4k^2 + 11k - 3}$

6.7 APPLICATIONS OF RATIONAL EXPRESSIONS

When we learn how to solve a new type of equation, we are able to apply our knowledge to solving new types of applications. In Section 6.6 we solved equations with rational expressions; now we can solve applications that involve this type of equation. The six-step problem solving method of Chapter 2 still applies.

1 **Solve problems about numbers.**   We begin with an example about an unknown number.

Example 1 **Solving a Problem about an Unknown Number**

If the same number is added to both the numerator and the denominator of the fraction $\frac{2}{5}$, the result is equivalent to $\frac{2}{3}$. Find the number.

Step 1   **Read** the problem carefully. We are trying to find a number.

Step 2   **Assign a variable.** Here, let $x =$ the number added to the numerator and the denominator.

Step 3   **Write an equation.** The fraction

$$\frac{2 + x}{5 + x}$$

represents the result of adding the same number to both the numerator and the denominator. Since this result is equivalent to $\frac{2}{3}$, the equation is

$$\frac{2 + x}{5 + x} = \frac{2}{3}.$$

Step 4   **Solve** this equation by multiplying each side by the LCD, $3(5 + x)$.

$$3(5 + x)\frac{2 + x}{5 + x} = 3(5 + x)\frac{2}{3}$$

$$3(2 + x) = 2(5 + x)$$

$$6 + 3x = 10 + 2x \qquad \text{Distributive property}$$

$$x = 4 \qquad \text{Subtract } 2x; \text{ subtract } 6.$$

Step 5   **State the answer.** The number is 4.

Step 6   **Check** the solution in the words of the original problem. If 4 is added to both the numerator and the denominator of $\frac{2}{5}$, the result is $\frac{6}{9} = \frac{2}{3}$, as required.

1 Solve each problem.

(a) A certain number is added to the numerator and subtracted from the denominator of $\frac{5}{8}$. The new fraction equals the reciprocal of $\frac{5}{8}$. Find the number.

(b) The denominator of a fraction is 1 more than the numerator. If 6 is added to the numerator and subtracted from the denominator, the result is $\frac{15}{4}$. Find the original fraction.

===== **Work Problem 1 at the Side.**

2 **Solve problems about distance, rate, and time.**   If an automobile travels at an average rate of 50 mph for 2 hr, then it travels $50 \times 2 = 100$ mi. This is an example of the basic relationship between distance, rate, and time:

$$\text{distance} = \text{rate} \times \text{time}.$$

This relationship is given by the formula $d = rt$. By solving, in turn, for r and t in the formula, we obtain two other equivalent forms of the formula. The three forms are given below.

Distance, Rate, and Time Relationship

$$d = rt \qquad r = \frac{d}{t} \qquad t = \frac{d}{r}$$

② Solve each problem.

(a) The world record in the men's 100-m dash was set in 1999 by Maurice Green, who ran it in 9.79 sec. What was his speed in meters per second? (*Source:* http://english.sydneylink.com)

(b) The world record for the women's 3000-m run was set by Junxia Wang in 1993. Her speed was 6.173 m per sec. What was her time in seconds?

(c) A small plane flew from Warsaw to Rome averaging 164 mph. The trip took 2 hr. What is the distance between Warsaw and Rome?

The next example illustrates the uses of these formulas.

Example 2 Finding Distance, Rate, or Time

(a) The speed of sound is 1088 ft per sec at sea level at 32°F. In 5 sec under these conditions, sound travels

$$1088 \times 5 = 5440 \text{ ft.}$$

$$\text{Rate} \times \text{Time} = \text{Distance}$$

Here, we found distance given rate and time, using $d = rt$.

(b) The winner of the first Indianapolis 500 race (in 1911) was Ray Harroun, driving a Marmon Wasp at an average speed of 74.59 mph. (*Source: The Universal Almanac,* 1997.) To complete the 500 mi, it took him

$$\text{Distance} \rightarrow \frac{500}{74.59} = 6.70 \text{ hr} \quad \text{(rounded).} \leftarrow \text{Time}$$
$$\text{Rate} \rightarrow$$

Here, we found time given rate and distance using $t = \frac{d}{r}$. To convert .70 hr to minutes, multiply by 60 to get .70(60) = 42. It took Harroun about 6 hr, 42 min to complete the race.

(c) At the 2000 Olympic Games in Sydney, Australia, Dutch swimmer Inge de Bruijn won the women's 50-m freestyle swimming event in 24.32 sec. (*Source:* www.olympics.com) Her rate was

$$\text{Rate} = \frac{\text{Distance} \rightarrow 50}{\text{Time} \rightarrow 24.32} = 2.06 \text{ m per sec (rounded).}$$

Work Problem ② at the Side.

Problem Solving

Many applied problems use the formulas just discussed. The next two examples show how to solve typical applications of the formula $d = rt$. A helpful strategy for solving such problems is to first make a sketch showing what is happening in the problem. Then make a table using the information given, along with the unknown quantities. The table will help you organize the information, and the sketch will help you set up the equation.

Example 3 Solving a Motion Problem about Distance, Rate, and Time

Two cars leave Baton Rouge, Louisiana, at the same time and travel east on Interstate 12. One travels at a constant speed of 55 mph and the other travels at a constant speed of 63 mph. In how many hours will the distance between them be 24 mi?

Continued on Next Page

Step 1 **Read** the problem. We are trying to find the time when the distance between the cars will be 24 mi.

Step 2 **Assign a variable.** Since we are looking for time, let t = the number of hours until the distance between them is 24 mi. The sketch in Figure 1 shows what is happening in the problem.

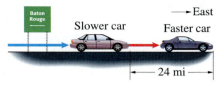

East

Slower car Faster car

Baton Rouge

|←— 24 mi —→|

Figure 1

Now, construct a table like the one that follows. Fill in the information given in the problem, and use t for the time traveled by each car. Multiply rate by time to get the expressions for distances traveled.

	Rate	× Time =	Distance
Faster Car	63	t	$63t$
Slower Car	55	t	$55t$

⎤ Difference is 24 mi.

The quantities $63t$ and $55t$ represent the two distances. Refer to Figure 1, and notice that the *difference* between the larger distance and the smaller distance is 24 mi.

Step 3 **Write an equation.**

$$63t - 55t = 24$$

Step 4 **Solve.**

$$63t - 55t = 24$$
$$8t = 24 \qquad \text{Combine like terms.}$$
$$t = 3 \qquad \text{Divide by 8.}$$

Step 5 **State the answer.** It will take the cars 3 hr to be 24 mi apart.

Step 6 **Check.** After 3 hr the faster car will have traveled $63 \times 3 = 189$ mi, and the slower car will have traveled $55 \times 3 = 165$ mi. Since $189 - 165 = 24$, the conditions of the problem are satisfied.

NOTE

In motion problems like the one in Example 3, once you have filled in two pieces of information in each row of the table, you should automatically fill in the third piece of information, using the appropriate form of the formula relating distance, rate, and time. Set up the equation based on your sketch and the information in the table.

Work Problem ❸ at the Side.

❸ Solve each problem.

(a) From a point on a straight road, Lupe and Maria ride bicycles in opposite directions. Lupe rides 10 mph and Maria rides 12 mph. In how many hours will they be 55 mi apart?

(b) At a given hour, two steamboats leave a city in the same direction on a straight canal. One travels at 18 mph, and the other travels at 25 mph. In how many hours will the boats be 35 mi apart?

ANSWERS

3. (a) $2\frac{1}{2}$ hr **(b)** 5 hr

Example 4 Solving a Problem about Distance, Rate, and Time

The Tickfaw River has a current of 3 mph. A motorboat takes as long to go 12 mi downstream as to go 8 mi upstream. What is the speed of the boat in still water?

Step 1 **Read** the problem again. We are looking for the speed of the boat in still water.

Step 2 **Assign a variable.** Let x = the speed of the boat in still water. Because the current pushes the boat when the boat is going downstream, the speed of the boat downstream will be the sum of the speed of the boat and the speed of the current, $x + 3$ mph. Also, the boat's speed going upstream is given by $x - 3$ mph. See Figure 2.

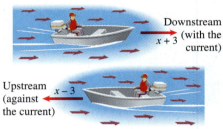

Downstream (with the current) $x + 3$

Upstream (against the current) $x - 3$

Figure 2

This information is summarized in the following table.

	d	r	t
Downstream	12	$x + 3$	
Upstream	8	$x - 3$	

Fill in the column representing time by using the formula $t = \frac{d}{r}$. Then the time upstream is the distance divided by the rate, or

$$t = \frac{d}{r} = \frac{8}{x - 3},$$

and the time downstream is also the distance divided by the rate, or

$$t = \frac{d}{r} = \frac{12}{x + 3}.$$

Now complete the table.

	d	r	t
Downstream	12	$x + 3$	$\dfrac{12}{x + 3}$
Upstream	8	$x - 3$	$\dfrac{8}{x - 3}$

Times are equal.

Step 3 **Write an equation.** According to the original problem, the time upstream equals the time downstream. The two times from the chart must therefore be equal, giving the equation

$$\frac{12}{x + 3} = \frac{8}{x - 3}.$$

Continued on Next Page

Step 4 **Solve.** Begin by multiplying each side by $(x + 3)(x - 3)$.

$$(x + 3)(x - 3)\frac{12}{x + 3} = (x + 3)(x - 3)\frac{8}{x - 3}$$

$$12(x - 3) = 8(x + 3)$$

$$12x - 36 = 8x + 24 \qquad \text{Distributive property}$$

$$4x = 60 \qquad \text{Subtract } 8x; \text{ add } 36.$$

$$x = 15 \qquad \text{Divide by } 4.$$

Step 5 **State the answer.** The speed of the boat in still water is 15 mph.

Step 6 **Check.** First find the speed of the boat downstream, which is $15 + 3 = 18$ mph. Traveling 12 mi would take

$$t = \frac{d}{r} = \frac{12}{18} = \frac{2}{3} \text{ hr.}$$

On the other hand, the speed of the boat upstream is $15 - 3 = 12$ mph, and traveling 8 mi would take

$$t = \frac{d}{r} = \frac{8}{12} = \frac{2}{3} \text{ hr.}$$

The time upstream equals the time downstream, as required.

Work Problem 4 at the Side.

3   **Solve problems about work.** Suppose that you can mow your lawn in 4 hr. Then after 1 hr, you will have mowed $\frac{1}{4}$ of the lawn. After 2 hr, you will have mowed $\frac{2}{4}$ or $\frac{1}{2}$ of the lawn, and so on. This idea is generalized as follows.

Rate of Work

If a job can be completed in t units of time, then the rate of work is

$$\frac{1}{t} \text{ job per unit of time.}$$

Problem Solving

The relationship between problems involving work and problems involving distance is a very close one. Recall that the formula $d = rt$ says that distance traveled is equal to rate of travel multiplied by time traveled. Similarly, the fractional part of a job accomplished is equal to the rate of work multiplied by the time worked. In the lawn mowing example, after 3 hr, the fractional part of the job done is

$$\underbrace{\frac{1}{4}}_{\substack{\text{Rate of} \\ \text{work}}} \cdot \underbrace{3}_{\substack{\text{Time} \\ \text{worked}}} = \underbrace{\frac{3}{4}}_{\substack{\text{Fractional part} \\ \text{of job done}}}.$$

After 4 hr, $\frac{1}{4}(4) = 1$ whole job has been done.

4 Solve each problem.

(a) A boat can go 20 mi against the current in the same time it can go 60 mi with the current. The current is flowing at 4 mph. Find the speed of the boat with no current.

(b) An airplane, maintaining a constant airspeed, takes as long to go 450 mi with the wind as it does to go 375 mi against the wind. If the wind is blowing at 15 mph, what is the speed of the plane?

Example 5 Solving a Problem about Work Rates

With spraying equipment, Mateo can paint the woodwork in a small house in 8 hr. His assistant, Chet, needs 14 hr to complete the same job painting by hand. If both Mateo and Chet work together, how long will it take them to paint the woodwork?

Step 1 **Read** the problem again. We are looking for time working together.

Step 2 **Assign a variable.** Let x = the number of hours it will take for Mateo and Chet to paint the woodwork, working together.

Certainly, x will be less than 8, since Mateo alone can complete the job in 8 hr. Begin by making a table as shown. Remember that based on the previous discussion, Mateo's rate alone is $\frac{1}{8}$ job per hour, and Chet's rate is $\frac{1}{14}$ job per hour.

	Rate	Time Working Together	Fractional Part of the Job Done When Working Together	
Mateo	$\frac{1}{8}$	x	$\frac{1}{8}x$	⎤ Sum is 1
Chet	$\frac{1}{14}$	x	$\frac{1}{14}x$	⎦ whole job.

Step 3 **Write an equation.** Since together Mateo and Chet complete 1 whole job, we must add their individual fractional parts and set the sum equal to 1.

$$\underbrace{\text{Fractional part}}_{\text{done by Mateo}} + \underbrace{\text{Fractional part}}_{\text{done by Chet}} = \underbrace{\text{1 whole job}}$$

$$\frac{1}{8}x \quad + \quad \frac{1}{14}x \quad = \quad 1$$

Step 4 **Solve.**

$$56\left(\frac{1}{8}x + \frac{1}{14}x\right) = 56(1) \qquad \text{Multiply by the LCD, 56.}$$

$$56\left(\frac{1}{8}x\right) + 56\left(\frac{1}{14}x\right) = 56(1) \qquad \text{Distributive property}$$

$$7x + 4x = 56$$

$$11x = 56 \qquad \text{Combine like terms.}$$

$$x = \frac{56}{11} \qquad \text{Divide by 11.}$$

Step 5 **State the answer.** Working together, Mateo and Chet can paint the woodwork in $\frac{56}{11}$ hr, or $5\frac{1}{11}$ hr.

Step 6 **Check** to be sure the answer is correct.

NOTE

An alternative approach in work problems is to consider the part of the job that can be done in 1 hr. For instance, in Example 5 Mateo can do the entire job in 8 hr, and Chet can do it in 14 hr. Thus, their work rates, as we saw in Example 5, are $\frac{1}{8}$ and $\frac{1}{14}$, respectively. Since it takes them x hr to complete the job when working together, in 1 hr they can paint $\frac{1}{x}$ of the woodwork. The amount painted by Mateo in 1 hr plus the amount painted by Chet in 1 hr must equal the amount they can do together. This leads to the equation

Amount by Chet

Amount by Mateo → $\dfrac{1}{8} + \dfrac{1}{14} = \dfrac{1}{x}.$ ← Amount together

Compare this with the equation in Example 5. Multiplying each side by $56x$ leads to

$$7x + 4x = 56,$$

the same equation found in the third line of Step 4 in the example. The same solution results.

Work Problem ⑤ at the Side.

⑤ Solve each problem.

(a) Michael can paint a room, working alone, in 8 hr. Lindsay can paint the same room, working alone, in 6 hr. How long will it take them if they work together?

(b) Roberto can detail his Camaro in 2 hr working alone. His brother Marco can do the job in 3 hr working alone. How long would it take them if they worked together?

Real-Data Applications

Upward Mobility*

As a struggling college student you have been driving the "Wimp," a 1989 Honda CRX, which has one saving grace—it gets 30 miles per gallon in the city. Now that you are graduating and are being recruited for your dream job, your first major purchase will be a new truck or sport utility vehicle. In addition to car payments, you also must consider increased gasoline costs.

   Based on past experience, you anticipate driving 15,000 miles a year. The new vehicle requires premium unleaded gasoline at $1.40 per gallon, instead of the $1.35 cost for regular unleaded that you use now. Currently, you use 500 gallons of gasoline per year, and you spend $675 per year on gasoline.

Gasoline usage $\quad \dfrac{15{,}000 \text{ mi}}{1 \text{ yr}} \div \dfrac{30 \text{ mi}}{1 \text{ gal}} = \dfrac{15{,}000 \text{ mi}}{1 \text{ yr}} \cdot \dfrac{1 \text{ gal}}{30 \text{ mi}} = 500 \text{ gal per yr}$

Current gasoline costs $\quad \dfrac{500 \text{ gal}}{1 \text{ yr}} \cdot \dfrac{\$1.35 \text{ gal}}{1 \text{ gal}} = \$675 \text{ per yr}$

You test drove a 4.8 Liter, V8, Chevy Tahoe that is rated to get 15 miles to the gallon (mpg). To calculate the additional costs for gasoline, you must first compute the cost for gasoline usage in the Tahoe and then subtract your current gasoline costs. Observe the *process* so that you can copy it to devise a **cost equation** that you can use to evaluate costs for all the other vehicles that you want to test drive. The *variable* quantity (the number of miles per gallon) is shown in blue.

Chevy Tahoe gasoline usage $\quad \dfrac{15{,}000 \text{ mi}}{1 \text{ yr}} \div \dfrac{\textcolor{blue}{15 \text{ mi}}}{\textcolor{blue}{1 \text{ gal}}} = \dfrac{15{,}000 \text{ mi}}{1 \text{ yr}} \cdot \dfrac{1 \text{ gal}}{\textcolor{blue}{15 \text{ mi}}}$

Projected gasoline cost increase $\quad \dfrac{15{,}000 \text{ gal}}{\textcolor{blue}{15} \text{ yr}} \cdot \dfrac{\$1.40}{1 \text{ gal}} - \$675 = \dfrac{(15{,}000)(1.40)}{\textcolor{blue}{15}} - 675$

For Group Discussion

1. Write a cost equation that computes the additional costs, *y*, for gasoline for a truck or SUV that gets *x* mpg. Assume that the new vehicle requires premium gasoline. (*Hint:* Replace the variable quantity in the process above with *x*.)

2. Evaluate the cost equation for the following vehicles to predict increased gasoline costs. (*Source*: *Consumer's Report*, on-line.)

   (a) Honda CR-V, 4 L, 23 mpg   **(b)** Ford Expedition, 5.4 L, V8, 14 mpg

   (c) Chevy Yukon, 5.3 L, V8, 12 mpg

3. A calculator graph of the cost equation is shown here.

   (a) What can you conclude about the effect of *decreasing* mileage rating on the additional gasoline costs?

   (b) The graph crosses the *x* axis at approximately *x* = 31. Explain why.

*Based on *Driving Rationally* by Patricia Stone, Tomball College.

6.7 **EXERCISES**

 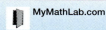

Use Steps 2 and 3 of the six-step method to set up the equation you would use to solve each problem. (Remember that Step 1 is to read the problem carefully.) Do not actually solve the equation. See Example 1.

1. The numerator of the fraction $\frac{5}{6}$ is increased by an amount so that the value of the resulting fraction is equivalent to $\frac{13}{3}$. By what amount was the numerator increased?

(a) Let $x =$ _____. (*Step 2*)

(b) Write an expression for "the numerator of the fraction $\frac{5}{6}$ is increased by an amount."

(c) Set up an equation to solve the problem.
(*Step 3*)

2. If the same number is added to the numerator and subtracted from the denominator of $\frac{23}{12}$, the resulting fraction is equivalent to $\frac{3}{2}$. What is the number?

(a) Let $x =$ _____. (*Step 2*)

(b) Write an expression for "a number is added to the numerator of $\frac{23}{12}$." Then write an expression for "the same number is subtracted from the denominator of $\frac{23}{12}$."

(c) Set up an equation to solve the problem.
(*Step 3*)

Use the six-step method to solve each problem. See Example 1.

3. In a certain fraction, the denominator is 6 more than the numerator. If 3 is added to both the numerator and the denominator, the resulting fraction is equivalent to $\frac{5}{7}$. What was the original fraction?

4. In a certain fraction, the denominator is 4 less than the numerator. If 3 is added to both the numerator and the denominator, the resulting fraction is equivalent to $\frac{3}{2}$. What was the original fraction?

5. The numerator of a certain fraction is four times the denominator. If 6 is added to both the numerator and the denominator, the resulting fraction is equivalent to 2. What was the original fraction?

6. The denominator of a certain fraction is three times the numerator. If 2 is added to the numerator and subtracted from the denominator, the resulting fraction is equivalent to 1. What was the original fraction?

7. One-third of a number is 2 more than one-sixth of the same number. What is the number?

8. One-sixth of a number is 5 more than the same number. What is the number?

9. A quantity, its $\frac{2}{3}$, its $\frac{1}{2}$, and its $\frac{1}{7}$, added together, become 33. What is the quantity? (*Source: Rhind Mathematical Papyrus.*)

10. A quantity, its $\frac{3}{4}$, its $\frac{1}{2}$, and its $\frac{1}{3}$, added together, become 93. What is the quantity? (*Source: Rhind Mathematical Papyrus.*)

Solve each problem. See Example 2.

11. In the 1998 World Championships, Amy Van Dyken of the United States won the 50-m freestyle swimming event for women in 25.15 sec. What was her rate? (*Source: Sports Illustrated 2000 Sports Almanac.*)

12. In the 1998 Winter Games, Catriona LeMay Doan of Canada won the 500-m speed skating event for women. Her rate was 13.0856 m per sec. What was her time (to the nearest hundredth of a second)? (*Source: Sports Illustrated 2000 Sports Almanac.*)

13. The winner of the 1998 Charlotte 500 (mile) race was Mark Martin, who drove his Ford to victory with a rate of 123.188 mph. What was his time? (*Source: Sports Illustrated 2000 Sports Almanac.*)

14. In 1998, Jeff Gordon drove his Chevrolet to victory in the North Carolina 400 (mile) race. His rate was 128.423 mph. What was his time? (*Source: Sports Illustrated 2000 Sports Almanac.*)

15. The winner of the women's 1500-m race in the 2000 Olympics was Nouria Merah-Benida of Algeria with a time of 4.085 min. What was her rate? (*Source:* www.olympics.com)

16. Gabriela Szabo of Romania won the women's 5000-m race in the 2000 Olympics with a time of 14.680 min. What was her rate? (*Source:* www.olympics.com)

Set up the equation you would use to solve each problem. Do not actually solve the equation. See Examples 3 and 4.

17. Julio flew his airplane 500 mi against the wind in the same time it took him to fly it 600 mi with the wind. If the speed of the wind was 10 mph, what was the average speed of his plane? (Let x = speed of the plane in still air.)

	d	r	t
Against the wind	500	$x - 10$	
With the wind	600	$x + 10$	

18. Luvenia can row 4 mph in still water. It takes as long to row 8 mi upstream as 24 mi downstream. How fast is the current? (Let x = speed of the current.)

	d	r	t
Upstream	8	$4 - x$	
Downstream	24	$4 + x$	

Solve each problem. See Examples 3 and 4.

19. Suppose Stephanie walks D mi at R mph in the same time that Wally walks d mi at r mph. Give an equation relating D, R, d, and r.

20. If a migrating hawk travels m mph in still air, what is its rate when it flies into a steady headwind of 5 mph? What is its rate with a tailwind of 5 mph?

21. A boat can go 20 mi against a current in the same time that it can go 60 mi with the current. The current is 4 mph. Find the speed of the boat in still water.

22. A plane flies 350 mi with the wind in the same time that it can fly 310 mi against the wind. The plane has a still-air speed of 165 mph. Find the speed of the wind.

23. Sandi Goldstein flew from Dallas to Indianapolis at 180 mph and then flew back at 150 mph. The trip at the slower speed took 1 hr longer than the trip at the higher speed. Find the distance between the two cities.

24. The distance from Seattle, Washington, to Victoria, British Columbia, is about 148 mi by ferry. It takes about 4 hr less to travel by the same ferry to Vancouver, British Columbia, a distance of about 74 mi. What is the average speed of the ferry?

In Exercises 25 and 26, set up the equation you would use to solve each problem. Do not actually solve the equation. See Example 5.

25. Working alone, Jorge can paint a room in 8 hr. Caterina can paint the same room working alone in 6 hr. How long will it take them if they work together? (Let x represent the time working together.)

	r	t	w
Jorge		x	
Caterina		x	

26. Edwin Bedford can tune up his Chevy in 2 hr working alone. His son, Beau, can do the job in 3 hr working alone. How long would it take them if they worked together? (Let t represent the time working together.)

	r	t	w
Edwin		t	
Beau		t	

Solve each problem. See Example 5.

27. Geraldo and Luisa Hernandez operate a small laundry. Luisa, working alone, can clean a day's laundry in 9 hr. Geraldo can clean a day's laundry in 8 hr. How long would it take them if they work together?

28. Lea can groom the horses in her boarding stable in 5 hr, while Tran needs 4 hr. How long will it take them to groom the horses if they work together?

29. A pump can pump the water out of a flooded basement in 10 hr. A smaller pump takes 12 hr. How long would it take to pump the water from the basement using both pumps?

30. Doug Todd's copier can do a printing job in 7 hr. Scott's copier can do the same job in 12 hr. How long would it take to do the job using both copiers?

31. One pipe can fill a swimming pool in 6 hr, and another pipe can do it in 9 hr. How long will it take the two pipes working together to fill the pool $\frac{3}{4}$ full?

32. An inlet pipe can fill a swimming pool in 9 hr, and an outlet pipe can empty the pool in 12 hr. Through an error, both pipes are left open. How long will it take to fill the pool?

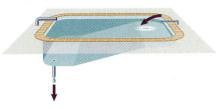

33. A cold water faucet can fill a sink in 12 min, and a hot water faucet can fill it in 15 min. The drain can empty the sink in 25 min. If both faucets are on and the drain is open, how long will it take to fill the sink?

34. Refer to Exercise 32. Assume the error was discovered after both pipes had been running for 3 hr, and the outlet pipe was then closed. How much more time would then be required to fill the pool? (*Hint:* Consider how much of the job had been done when the error was discovered.)

35. Students often wonder how teachers and textbook authors make up problems for them to solve. One way to do this is to start with the answer and work backwards.

For example, suppose that we start with the fraction $\frac{7}{3}$ and add 3 to both the numerator and the denominator. We get $\frac{10}{6}$, which simplifies to $\frac{5}{3}$. Based on this observation, we can write the following problem.

If a number is added to both the numerator and the denominator of $\frac{7}{3}$, the resulting fraction is equivalent to $\frac{5}{3}$. What is the number?

Because of how we constructed the problem, the answer must be 3. Make up your own problem similar to this one, and then solve it using an equation with rational expressions.

36. Refer to the table in Exercise 18. Suppose that a student made the error of interchanging the positions of the expressions $4 - x$ and $4 + x$, but used the correct method of setting up the equation, applying the formula $t = \frac{d}{r}$. Solve the equation the student used, and explain how the student should immediately know that there is something wrong in the setup.

6.8 VARIATION

1 **Solve problems about direct variation.** Suppose that gasoline costs $1.50 per gal. Then 1 gal costs $1.50, 2 gal cost 2($1.50) = $3.00, 3 gal cost 3($1.50) = $4.50, and so on. Each time, the total cost is obtained by multiplying the number of gallons by the price per gallon. In general, if k equals the price per gallon and x equals the number of gallons, then the total cost y is equal to kx. Notice that as the number of gallons increases, the total cost increases.

The preceding discussion is an example of variation. Equations with fractions often result when discussing variation. As in the gasoline example, two variables **vary directly** if one is a constant multiple of the other.

Direct Variation

y **varies directly as** x if there exists a constant k such that

$$y = kx.$$

The constant k in the equation for direct variation is a numerical value, such as 1.50 in the gasoline price discussion.

Example 1 **Using Direct Variation**

Suppose y varies directly as x, and $y = 20$ when $x = 4$. Find y when $x = 9$.

Since y varies directly as x, there is a constant k such that $y = kx$. We know that $y = 20$ when $x = 4$. Substituting these values into $y = kx$ and solving for k gives

$$y = kx$$
$$20 = k \cdot 4$$
$$k = 5.$$

Since $y = kx$ and $k = 5$,

$$y = 5x. \qquad \text{Let } k = 5.$$

When $x = 9$,

$$y = 5x = 5 \cdot 9 = 45. \qquad \text{Let } x = 9.$$

Thus, $y = 45$ when $x = 9$.

───── Work Problem **1** at the Side.

2 **Solve problems about inverse variation.** In another common type of variation, the value of one variable increases while the value of another decreases. For example, an increase in the supply of an item causes a decrease in the price of the item.

1 Solve each problem.

(a) If z varies directly as t, and $z = 11$ when $t = 4$, find z when $t = 32$.

(b) The circumference of a circle varies directly as the radius. A circle with a radius of 7 cm has a circumference of 43.96 cm. Find the circumference if the radius is 11 cm.

2 Solve the problem.

Suppose z varies inversely as t, and $z = 8$ when $t = 2$. Find z when $t = 32$.

Inverse Variation

y **varies inversely as** *x* if there exists a constant k such that

$$y = \frac{k}{x}.$$

Example 2   Using Inverse Variation

Suppose y varies inversely as x, and $y = 3$ when $x = 8$. Find y when $x = 6$.

Since y varies inversely as x, there is a constant k such that $y = \frac{k}{x}$. We know that $y = 3$ when $x = 8$, so we can find k.

$$y = \frac{k}{x}$$

$$3 = \frac{k}{8}$$

$$k = 24$$

Since $y = \frac{24}{x}$, we let $x = 6$ and solve for y.

$$y = \frac{24}{x} = \frac{24}{6} = 4$$

Therefore, when $x = 6$, $y = 4$.

Work Problem 2 at the Side.

3 Solve the problem.

The current in a simple electrical circuit varies inversely as the resistance. If the current is 80 amps when the resistance is 10 ohms, find the current if the resistance is 16 ohms.

Example 3   Using Inverse Variation

In the manufacturing of a certain medical syringe, the cost of producing the syringe varies inversely as the number produced. If 10,000 syringes are produced, the cost is $2 per unit. Find the cost per unit to produce 25,000 syringes.

Let $x =$ the number of syringes produced

and $c =$ the cost per unit.

Since c varies inversely as x, there is a constant k such that

$$c = \frac{k}{x}.$$

Find k by replacing c with 2 and x with 10,000.

$$2 = \frac{k}{10,000}$$

$$20,000 = k \qquad \text{Multiply by 10,000.}$$

Since $c = \frac{k}{x}$,

$$c = \frac{20,000}{25,000} = .80. \qquad \text{Let } k = 20,000 \text{ and } x = 25,000.$$

The cost per unit to make 25,000 syringes is $.80.

Work Problem 3 at the Side.

ANSWERS
2. $\frac{1}{2}$
3. 50 amps

6.8 EXERCISES

 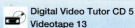

In Exercises 1 and 2, fill in each blank with the correct response.

1. If the constant of variation is positive and y varies directly as x, then as x increases,

y _____ .
(increases/decreases)

2. If the constant of variation is positive and y varies inversely as x, then as x increases,

y _____ .
(increases/decreases)

Solve each problem involving direct variation. See Example 1.

3. If y varies directly as x, and $x = 27$ when $y = 6$, find x when $y = 2$.

4. If z varies directly as x, and $z = 30$ when $x = 8$, find z when $x = 4$.

5. If d varies directly as t, and $d = 150$ when $t = 3$, find d when $t = 5$.

6. If d varies directly as r, and $d = 200$ when $r = 40$, find d when $r = 60$.

Solve each problem involving inverse variation. See Example 2.

7. If x varies inversely as y, and $x = 3$ when $y = 8$, find y when $x = 4$.

8. If z varies inversely as x, and $z = 50$ when $x = 2$, find z when $x = 25$.

9. If p varies inversely as q, and $p = 7$ when $q = 6$, find p when $q = 2$.

10. If m varies directly as r, and $m = 12$ when $r = 8$, find m when $r = 16$.

Solve each variation problem. See Examples 1–3.

11. The interest on an investment varies directly as the rate of interest. If the interest is $48 when the interest rate is 5%, find the interest when the rate is 4.2%.

12. For a given base, the area of a triangle varies directly as its height. Find the area of a triangle with a height of 6 in., if the area is 10 in.² when the height is 4 in.

13. The pressure exerted by water at a given point varies directly with the depth of the point beneath the surface of the water. Water exerts 4.34 lb per in.² for every 10 ft traveled below the water's surface. What is the pressure exerted on a scuba diver at 20 ft?

14. Hooke's law for an elastic spring states that the distance a spring stretches varies directly with the force applied. If a force of 75 lb stretches a certain spring 16 in., how much will a force of 200 lb stretch the spring?

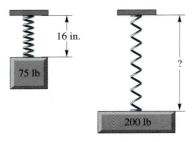

15. Over a specified distance, speed varies inversely with time. If a Dodge Viper on a test track goes a certain distance in one-half minute at 160 mph, what speed is needed to go the same distance in three-fourths minute?

16. For a constant area, the length of a rectangle varies inversely as the width. The length of a rectangle is 27 ft when the width is 10 ft. Find the width of a rectangle with the same area if the length is 18 ft.

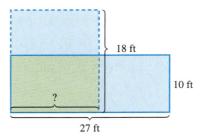

17. The current in a simple electrical circuit varies inversely as the resistance. If the current is 20 amps when the resistance is 5 ohms, find the current when the resistance is 8 ohms.

18. If the temperature is constant, the pressure of a gas in a container varies inversely as the volume of the container. If the pressure is 10 lb per ft² in a container with volume 3 ft³, what is the pressure in a container with volume 1.5 ft³?

19. The force required to compress a spring varies directly as the change in the length of the spring. If a force of 12 lb is required to compress a certain spring 3 in., how much force is required to compress the spring 5 in.?

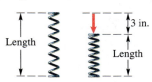

20. In the inversion of raw sugar, the rate of change of the amount of raw sugar varies directly as the amount of raw sugar remaining. The rate is 200 kg per hr when there are 800 kg left. What is the rate of change per hour when only 100 kg are left?

Use personal experience or intuition to determine whether the situation suggests direct or inverse variation.

21. The number of different lottery tickets you buy and your probability of winning that lottery

22. The rate and the distance traveled by a pickup truck in 3 hr

23. The amount of pressure put on the accelerator of a car and the speed of the car

24. The number of days from now until December 25 and the magnitude of the frenzy of Christmas shopping

25. The surface area of a balloon and its diameter

26. Your age and the probability that you believe in Santa Claus

27. The number of days until the end of the baseball season and the number of home runs that Sammy Sosa has

28. The amount of gasoline you pump and the amount you will pay

*Two triangles are **similar** if they have the same shape (but not necessarily the same size). Similar triangles have sides that vary directly. The figure shows two similar triangles. Notice that the ratios of the corresponding sides are all equal to $\frac{3}{2}$:*

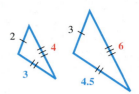

$$\frac{3}{2} = \frac{3}{2} \qquad \frac{4.5}{3} = \frac{3}{2} \qquad \frac{6}{4} = \frac{3}{2}.$$

If we know that two triangles are similar, we can set up a direct variation equation to solve for the length of an unknown side.

Find the length x, given that the pair of triangles are similar.

29.

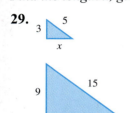

30.

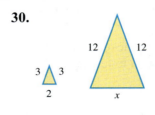

31.

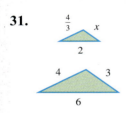

32.

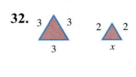

Use similar triangles and direct variation to solve each problem. (Source: The Guinness Book of World Records.)

33. An enlarged version of the chair used by George Washington at the Constitutional Convention casts a shadow 18 ft long at the same time a vertical pole 12 ft high casts a shadow 4 ft long. How tall is the chair?

34. One of the tallest candles ever constructed was exhibited at the 1897 Stockholm Exhibition. If it cast a shadow 5 ft long at the same time a vertical pole 32 ft high cast a shadow 2 ft long, how tall was the candle?

Chapter 6

KEY TERMS

6.1 **rational expression** The quotient of two polynomials with denominator not 0 is called a rational expression.

lowest terms A rational expression is written in lowest terms if the greatest common factor of its numerator and denominator is 1.

6.3 **least common denominator (LCD)** The simplest expression that is divisible by all denominators is called the least common denominator.

6.5 **complex fraction** A rational expression with one or more fractions in the numerator, denominator, or both, is called a complex fraction.

6.8 **direct variation** y varies directly as x if there is a constant k such that $y = kx$.

inverse variation y varies inversely as x if there is a constant k such that $y = \frac{k}{x}$.

TEST YOUR WORD POWER

See how well you have learned the vocabulary in this chapter. Answers follow the Quick Review.

1. A **rational expression** is
   (a) an algebraic expression made up of a term or the sum of a finite number of terms with real coefficients and whole number exponents
   (b) a polynomial equation of degree 2
   (c) an expression with one or more fractions in the numerator, denominator, or both
   (d) the quotient of two polynomials with denominator not 0.

2. A **complex fraction** is
   (a) an algebraic expression made up of a term or the sum of a finite number of terms with real coefficients and whole number exponents
   (b) a polynomial equation of degree 2
   (c) a rational expression with one or more fractions in the numerator, denominator, or both
   (d) the quotient of two polynomials with denominator not 0.

QUICK REVIEW

| *Concepts* | *Examples* |

6.1 *The Fundamental Property of Rational Expressions*

To find the value(s) for which a rational expression is not defined, set the denominator equal to 0 and solve the equation.

Find the values for which the expression

$$\frac{x - 4}{x^2 - 16}$$

is not defined.

$$x^2 - 16 = 0$$
$$(x - 4)(x + 4) = 0$$
$$x - 4 = 0 \quad \text{or} \quad x + 4 = 0$$
$$x = 4 \quad \text{or} \quad x = -4$$

The rational expression is not defined for 4 or -4.

To write a rational expression in lowest terms, (1) factor, and (2) use the fundamental property to remove common factors from the numerator and denominator.

Write $\dfrac{x^2 - 1}{(x - 1)^2}$ in lowest terms.

$$\frac{x^2 - 1}{(x - 1)^2} = \frac{(x - 1)(x + 1)}{(x - 1)(x - 1)} = \frac{x + 1}{x - 1}$$

There are often several different equivalent forms of a rational expression.

Give two equivalent forms of $-\dfrac{x - 1}{x + 2}$.

Distribute the $-$ sign in the numerator to get $\dfrac{-x + 1}{x + 2}$; do

so in the denominator to get $\dfrac{x - 1}{-x - 2}$. (There are other

forms as well.)

6.2 *Multiplying and Dividing Rational Expressions*

Multiplying Rational Expressions

Multiply. $\dfrac{3x + 9}{x - 5} \cdot \dfrac{x^2 - 3x - 10}{x^2 - 9}$

Step 1 Factor.

$$= \frac{3(x + 3)}{x - 5} \cdot \frac{(x - 5)(x + 2)}{(x + 3)(x - 3)}$$

Step 2 Multiply numerators and multiply denominators.

$$= \frac{3(x + 3)(x - 5)(x + 2)}{(x - 5)(x + 3)(x - 3)}$$

Step 3 Write in lowest terms.

$$= \frac{3(x + 2)}{x - 3}$$

Dividing Rational Expressions

Divide. $\dfrac{2x + 1}{x + 5} \div \dfrac{6x^2 - x - 2}{x^2 - 25}$

Step 1 Factor.

$$= \frac{2x + 1}{x + 5} \div \frac{(2x + 1)(3x - 2)}{(x + 5)(x - 5)}$$

Step 2 Multiply the first rational expression by the reciprocal of the second.

$$= \frac{2x + 1}{x + 5} \cdot \frac{(x + 5)(x - 5)}{(2x + 1)(3x - 2)}$$

Step 3 Write in lowest terms.

$$= \frac{x - 5}{3x - 2}$$

Concepts	Examples

6.3 *Least Common Denominators*

Finding the LCD

Find the LCD for $\dfrac{3}{k^2 - 8k + 16}$ and $\dfrac{1}{4k^2 - 16k}$.

Step 1 Factor each denominator into prime factors.

$$k^2 - 8k + 16 = (k - 4)^2$$
$$4k^2 - 16k = 4k(k - 4)$$

Step 2 List each different factor the greatest number of times it appears.

$$\text{LCD} = (k - 4)^2 \cdot 4 \cdot k$$

Step 3 Multiply the factors from Step 2 to get the LCD.

$$= 4k(k - 4)^2$$

Writing a Rational Expression with a Specified Denominator

Find the numerator: $\dfrac{5}{2z^2 - 6z} = \dfrac{}{4z^3 - 12z^2}$.

Step 1 Factor both denominators.

$$\dfrac{5}{2z(z - 3)} = \dfrac{}{4z^2(z - 3)}$$

Step 2 Decide what factors the denominator must be multiplied by to equal the specified denominator.

$2z(z - 3)$ must be multiplied by $2z$.

Step 3 Multiply the rational expression by that factor divided by itself (multiply by 1).

$$\dfrac{5}{2z(z - 3)} \cdot \dfrac{2z}{2z} = \dfrac{10z}{4z^2(z - 3)} = \dfrac{10z}{4z^3 - 12z^2}$$

6.4 *Adding and Subtracting Rational Expressions*

Adding Rational Expressions

Add. $\dfrac{2}{3m + 6} + \dfrac{m}{m^2 - 4}$

Step 1 Find the LCD.

$$3m + 6 = 3(m + 2)$$
$$m^2 - 4 = (m + 2)(m - 2)$$

The LCD is $3(m + 2)(m - 2)$.

Step 2 Rewrite each rational expression with the LCD as denominator.

$$= \dfrac{2(m - 2)}{3(m + 2)(m - 2)} + \dfrac{3m}{3(m + 2)(m - 2)}$$

Step 3 Add the numerators to get the numerator of the sum. The LCD is the denominator of the sum.

$$= \dfrac{2m - 4 + 3m}{3(m + 2)(m - 2)}$$

Step 4 Write in lowest terms.

$$= \dfrac{5m - 4}{3(m + 2)(m - 2)}$$

Subtracting Rational Expressions

Subtract. $\dfrac{6}{k + 4} - \dfrac{2}{k}$

Follow the same steps as for addition, but subtract in Step 3.

The LCD is $k(k + 4)$.

$$\dfrac{6k}{(k + 4)k} - \dfrac{2(k + 4)}{k(k + 4)} = \dfrac{6k - 2(k + 4)}{k(k + 4)}$$

$$= \dfrac{6k - 2k - 8}{k(k + 4)}$$

$$= \dfrac{4k - 8}{k(k + 4)} \quad \text{or} \quad \dfrac{4(k - 2)}{k(k + 4)}$$

Concepts	Examples

6.5 Complex Fractions

Simplifying Complex Fractions

Simplify.

Method 1 Simplify the numerator and denominator separately. Then divide the simplified numerator by the simplified denominator.

Method 1

$$\frac{\dfrac{1}{a} - a}{1 - a} = \frac{\dfrac{1}{a} - \dfrac{a^2}{a}}{1 - a} = \frac{\dfrac{1 - a^2}{a}}{1 - a}$$

$$= \frac{1 - a^2}{a} \cdot \frac{1}{1 - a}$$

$$= \frac{(1 - a)(1 + a)}{a(1 - a)} = \frac{1 + a}{a}$$

Method 2 Multiply the numerator and denominator of the complex fraction by the LCD of all the denominators in the complex fraction. Write in lowest terms.

Method 2

$$\frac{\dfrac{1}{a} - a}{1 - a} = \frac{\dfrac{1}{a} - a}{1 - a} \cdot \frac{a}{a} = \frac{\dfrac{a}{a} - a^2}{(1 - a)a}$$

$$= \frac{1 - a^2}{(1 - a)a} = \frac{(1 + a)(1 - a)}{(1 - a)a}$$

$$= \frac{1 + a}{a}$$

6.6 Solving Equations with Rational Expressions

Solving Equations with Rational Expressions

Solve $\dfrac{2}{x - 1} + \dfrac{3}{4} = \dfrac{5}{x - 1}$.

Step 1 Find the LCD of all denominators in the equation.

The LCD is $4(x - 1)$. Note that 1 cannot be a solution.

Step 2 Multiply each side of the equation by the LCD.

$$4(x - 1)\left(\frac{2}{x - 1} + \frac{3}{4}\right) = 4(x - 1)\left(\frac{5}{x - 1}\right)$$

$$4(x - 1)\left(\frac{2}{x - 1}\right) + 4(x - 1)\left(\frac{3}{4}\right) = 4(x - 1)\left(\frac{5}{x - 1}\right)$$

Step 3 Solve the resulting equation, which should have no fractions.

$$8 + 3(x - 1) = 20$$
$$8 + 3x - 3 = 20$$
$$3x = 15$$
$$x = 5$$

Step 4 Check each proposed solution.

The proposed solution, 5, checks.

Concepts	Examples

6.7 *Applications of Rational Expressions*

Solving Problems about Distance
Use the six-step method.

Step 1 **Read** the problem carefully.

Step 2 **Assign a variable.** Use a table to identify distance, rate, and time. Solve $d = rt$ for the unknown quantity in the table.

On a trip from Sacramento to Monterey, Marge traveled at an average speed of 60 mph. The return trip, at an average speed of 64 mph, took $\frac{1}{4}$ hr less. How far did she travel between the two cities?

Let x = the unknown distance.

	d	r	$t = \dfrac{d}{r}$
Going	x	60	$\dfrac{x}{60}$
Returning	x	64	$\dfrac{x}{64}$

Step 3 **Write an equation.** From the wording in the problem, decide the relationship between the quantities. Use those expressions to write an equation.

Since the time for the return trip was $\frac{1}{4}$ hr less, the time going equals the time returning plus $\frac{1}{4}$.

$$\frac{x}{60} = \frac{x}{64} + \frac{1}{4}$$

Step 4 **Solve** the equation.

$$16x = 15x + 240 \qquad \text{Multiply by 960.}$$
$$x = 240 \qquad \text{Subtract } 15x.$$

Step 5 **State the answer.**

She traveled 240 mi.

Step 6 **Check** the solution.

The trip there took $\frac{240}{60} = 4$ hr, while the return trip took $\frac{240}{64} = 3\frac{3}{4}$ hr, which is $\frac{1}{4}$ hr less time. The solution checks.

Solving Problems about Work

Step 1 **Read** the problem carefully.

It takes the regular mail carrier 6 hr to cover her route. A substitute takes 8 hr to cover the same route. How long would it take them to cover the route together?

Let x = the number of hours to cover the route together.

Step 2 **Assign a variable.** State what the variable represents. Put the information from the problem in a table. If a job is done in t units of time, the rate is $\frac{1}{t}$.

The rate of the regular carrier is $\frac{1}{6}$ job per hour; the rate of the substitute is $\frac{1}{8}$ job per hour. Multiply rate by time to get the fractional part of the job done.

	Rate	Time	Part of the Job Done
Regular	$\dfrac{1}{6}$	x	$\dfrac{1}{6}x$
Substitute	$\dfrac{1}{8}$	x	$\dfrac{1}{8}x$

Step 3 **Write an equation.** The sum of the fractional parts should equal 1 (whole job).

The equation is $\dfrac{1}{6}x + \dfrac{1}{8}x = 1$.

Step 4 **Solve** the equation.

The solution of the equation is $\frac{24}{7}$. The solution checks, because $\frac{1}{6}\left(\frac{24}{7}\right) + \frac{1}{8}\left(\frac{24}{7}\right) = 1$ is true.

Steps 5 **State the answer** and **check** the solution.
and 6

It would take them $\frac{24}{7}$ or $3\frac{3}{7}$ hr to cover the route together.

Concepts	Examples
6.8 *Variation*	If y varies inversely as x, and $y = 4$ when $x = 9$, find y when $x = 6$.
Solving Variation Problems	
Step 1 Write the variation equation. Use	The equation for inverse variation is
$$y = kx \quad \text{for direct variation,}$$	
$$y = \frac{k}{x} \quad \text{for inverse variation.}$$	$$y = \frac{k}{x}.$$
Step 2 Find k by substituting the given values of x and y into the equation.	$$4 = \frac{k}{9}$$ $$k = 36$$
Step 3 Write the equation with the value of k from Step 2 and the given value of x or y. Solve for the remaining variable.	$$y = \frac{36}{x} \quad k = 36$$ $$y = \frac{36}{6} \quad \text{Let } x = 6.$$ $$y = 6$$

ANSWERS TO TEST YOUR WORD POWER

1. (d) *Examples:* $-\dfrac{3}{4y}, \dfrac{5x^3}{x+2}, \dfrac{a+3}{a^2-4a-5}$ **2. (c)** *Examples:* $\dfrac{\frac{2}{3}}{\frac{4}{7}}, \dfrac{x-\frac{1}{y}}{x+\frac{1}{y}}, \dfrac{2}{a^2-1}$

Chapter 6 **REVIEW EXERCISES**

[6.1] *Find the value(s) of the variable for which each rational expression is undefined.*

1. $\dfrac{4}{x-3}$

2. $\dfrac{y+3}{2y}$

3. $\dfrac{m-2}{m^2-2m-3}$

4. $\dfrac{2k+1}{3k^2+17k+10}$

*Find the numerical value of each rational expression when **(a)** $x = -2$ and **(b)** $x = 4$.*

5. $\dfrac{x^2}{x-5}$

6. $\dfrac{4x-3}{5x+2}$

7. $\dfrac{3x}{x^2-4}$

8. $\dfrac{x-1}{x+2}$

Write each rational expression in lowest terms.

9. $\dfrac{5a^3b^3}{15a^4b^2}$

10. $\dfrac{m-4}{4-m}$

11. $\dfrac{4x^2-9}{6-4x}$

12. $\dfrac{4p^2+8pq-5q^2}{10p^2-3pq-q^2}$

Write four equivalent expressions for each fraction.

13. $-\dfrac{4x-9}{2x+3}$

14. $\dfrac{8-3x}{3+6x}$

[6.2] *Find each product or quotient. Write each answer in lowest terms.*

15. $\dfrac{8x^2}{12x^5} \cdot \dfrac{6x^4}{2x}$

16. $\dfrac{9m^2}{(3m)^4} \div \dfrac{6m^5}{36m}$

17. $\dfrac{x-3}{4} \cdot \dfrac{5}{2x-6}$

18. $\dfrac{3q+3}{5-6q} \div \dfrac{4q+4}{2(5-6q)}$

19. $\dfrac{2r+3}{r-4} \cdot \dfrac{r^2-16}{6r+9}$

20. $\dfrac{y^2-6y+8}{y^2+3y-18} \div \dfrac{y-4}{y+6}$

21. $\dfrac{2p^2+13p+20}{p^2+p-12} \cdot \dfrac{p^2+2p-15}{2p^2+7p+5}$

22. $\dfrac{3z^2+5z-2}{9z^2-1} \cdot \dfrac{9z^2+6z+1}{z^2+5z+6}$

[6.3] *Find the least common denominator for each list of fractions.*

23. $\dfrac{1}{8}, \dfrac{5}{12}, \dfrac{7}{32}$

24. $\dfrac{4}{9y}, \dfrac{7}{12y^2}, \dfrac{5}{27y^4}$

25. $\dfrac{1}{m^2 + 2m}, \dfrac{4}{m^2 + 7m + 10}$

26. $\dfrac{3}{x^2 + 4x + 3}, \dfrac{5}{x^2 + 5x + 4}, \dfrac{2}{x^2 + 7x + 12}$

Rewrite each rational expression with the given denominator.

27. $\dfrac{5}{8} = \dfrac{}{56}$

28. $\dfrac{10}{k} = \dfrac{}{4k}$

29. $\dfrac{3}{2a^3} = \dfrac{}{10a^4}$

30. $\dfrac{9}{x - 3} = \dfrac{}{18 - 6x}$

31. $\dfrac{-3y}{2y - 10} = \dfrac{}{50 - 10y}$

32. $\dfrac{4b}{b^2 + 2b - 3} = \dfrac{}{(b + 3)(b - 1)(b + 2)}$

[6.4] *Add or subtract as indicated. Write each answer in lowest terms.*

33. $\dfrac{10}{x} + \dfrac{5}{x}$

34. $\dfrac{6}{3p} - \dfrac{12}{3p}$

35. $\dfrac{9}{k} - \dfrac{5}{k - 5}$

36. $\dfrac{4}{y} + \dfrac{7}{7 + y}$

37. $\dfrac{m}{3} - \dfrac{2 + 5m}{6}$

38. $\dfrac{12}{x^2} - \dfrac{3}{4x}$

39. $\dfrac{5}{a - 2b} + \dfrac{2}{a + 2b}$

40. $\dfrac{4}{k^2 - 9} - \dfrac{k + 3}{3k - 9}$

41. $\dfrac{8}{z^2 + 6z} - \dfrac{3}{z^2 + 4z - 12}$

42. $\dfrac{11}{2p - p^2} - \dfrac{2}{p^2 - 5p + 6}$

[6.5] *Simplify each complex fraction.*

43. $\dfrac{\dfrac{a^4}{b^2}}{\dfrac{a^3}{b}}$

44. $\dfrac{\dfrac{y - 3}{y}}{\dfrac{y + 3}{4y}}$

45. $\dfrac{\dfrac{3m + 2}{m}}{\dfrac{2m - 5}{6m}}$

46. $\dfrac{\dfrac{1}{p} - \dfrac{1}{q}}{\dfrac{1}{q - p}}$

47. $\dfrac{x + \dfrac{1}{w}}{x - \dfrac{1}{w}}$

48. $\dfrac{\dfrac{1}{r + t} - 1}{\dfrac{1}{r + t} + 1}$

[6.6] *Solve each equation. Check your answers.*

49. $\dfrac{k}{5} - \dfrac{2}{3} = \dfrac{1}{2}$

50. $\dfrac{4 - z}{z} + \dfrac{3}{2} = \dfrac{-4}{z}$

51. $\dfrac{x}{2} - \dfrac{x - 3}{7} = -1$

52. $\dfrac{3y - 1}{y - 2} = \dfrac{5}{y - 2} + 1$

53. $\dfrac{3}{m - 2} + \dfrac{1}{m - 1} = \dfrac{7}{m^2 - 3m + 2}$

Solve for the specified variable.

54. $m = \dfrac{Ry}{t}$ for t

55. $x = \dfrac{3y - 5}{4}$ for y

56. $\dfrac{1}{r} - \dfrac{1}{s} = \dfrac{1}{t}$ for t

[6.7] *Solve each problem. Use the six-step method.*

57. In a certain fraction, the denominator is 5 less than the numerator. If 5 is added to both the numerator and the denominator, the resulting fraction is equivalent to $\dfrac{5}{4}$. Find the original fraction.

58. The denominator of a certain fraction is six times the numerator. If 3 is added to the numerator and subtracted from the denominator, the resulting fraction is equivalent to $\dfrac{2}{5}$. Find the original fraction.

59. On August 18, 1996, Scott Sharp won the True Value 200-mi Indy race driving a Ford with an average speed of 130.934 mph. What was his time? (*Source: Sports Illustrated 1998 Sports Almanac.*)

60. A man can plant his garden in 5 hr, working alone. His daughter can do the same job in 8 hr. How long would it take them if they worked together?

61. At a given hour, two steamboats leave a city in the same direction on a straight canal. One travels at 18 mph, and the other travels at 25 mph. In how many hours will the boats be 70 mi apart?

18 mph

25 mph

[6.8] *Solve each problem.*

62. If a parallelogram has a fixed area, the height varies inversely as the base. A parallelogram has a height of 8 cm and a base of 12 cm. Find the height if the base is changed to 24 cm.

63. If y varies directly as x, and $x = 12$ when $y = 5$, find x when $y = 3$.

MIXED REVIEW EXERCISES

Perform the indicated operations.

64. $\dfrac{4}{m-1} - \dfrac{3}{m+1}$

65. $\dfrac{8p^5}{5} \div \dfrac{2p^3}{10}$

66. $\dfrac{r-3}{8} \div \dfrac{3r-9}{4}$

67. $\dfrac{\dfrac{5}{x} - 1}{\dfrac{5-x}{3x}}$

68. $\dfrac{4}{z^2 - 2z + 1} - \dfrac{3}{z^2 - 1}$

Solve.

69. $F = \dfrac{k}{d - D}$ for d

70. $\dfrac{2}{z} - \dfrac{z}{z+3} = \dfrac{1}{z+3}$

71. Anne Kelly flew her plane 400 km with the wind in the same time it took her to go 200 km against the wind. The speed of the wind is 50 km per hr. Find the speed of the plane in still air.

72. "If Joe can paint a house in 3 hours, and Sam can paint the same house in 5 hours, how long does it take for them to do it together?" (From the movie *Little Big League*.)

73. In rectangles of constant area, length and width vary inversely. When the length is 24, the width is 2. What is the width when the length is 12?

74. If w varies inversely as z, and $w = 16$ when $z = 3$, find w when $z = 2$.

Chapter 6 TEST

 Study Skills Workbook
Activity 12

1. Find any values for which $\dfrac{3x - 1}{x^2 - 2x - 8}$ is undefined.

        1. _____

2. Find the numerical value of $\dfrac{6r + 1}{2r^2 - 3r - 20}$ when

    (a) $r = -2$ and **(b)** $r = 4$.

        2. **(a)** _____

            (b) _____

3. Write four rational expressions equivalent to $-\dfrac{6x - 5}{2x + 3}$.

        3. _____

Write each rational expression in lowest terms.

4. $\dfrac{-15x^6 y^4}{5x^4 y}$ 5. $\dfrac{6a^2 + a - 2}{2a^2 - 3a + 1}$

        4. _____

        5. _____

Multiply or divide. Write each answer in lowest terms.

6. $\dfrac{5(d - 2)}{9} \div \dfrac{3(d - 2)}{5}$ 7. $\dfrac{6k^2 - k - 2}{8k^2 + 10k + 3} \cdot \dfrac{4k^2 + 7k + 3}{3k^2 + 5k + 2}$

        6. _____

        7. _____

8. $\dfrac{4a^2 + 9a + 2}{3a^2 + 11a + 10} \div \dfrac{4a^2 + 17a + 4}{3a^2 + 2a - 5}$

        8. _____

Find the least common denominator for each list of fractions.

9. $\dfrac{-3}{10p^2}, \dfrac{21}{25p^3}, \dfrac{-7}{30p^5}$ 10. $\dfrac{r + 1}{2r^2 + 7r + 6}, \dfrac{-2r + 1}{2r^2 - 7r - 15}$

        9. _____

       10. _____

Rewrite each rational expression with the given denominator.

11. $\dfrac{15}{4p} = \dfrac{}{64p^3}$ 12. $\dfrac{3}{6m - 12} = \dfrac{}{42m - 84}$

       11. _____

       12. _____

Add or subtract. Write each answer in lowest terms.

13. $\dfrac{4x + 2}{x + 5} + \dfrac{-2x + 8}{x + 5}$ 14. $\dfrac{-4}{y + 2} + \dfrac{6}{5y + 10}$

       13. _____

       14. _____

15. $\dfrac{x + 1}{3 - x} - \dfrac{x^2}{x - 3}$ 16. $\dfrac{3}{2m^2 - 9m - 5} - \dfrac{m + 1}{2m^2 - m - 1}$

       15. _____

       16. _____

Simplify each complex fraction.

17. _____

$$17. \quad \frac{\dfrac{2p}{k^2}}{\dfrac{3p^2}{k^3}}$$

18. _____

$$18. \quad \frac{\dfrac{1}{x+3} - 1}{1 + \dfrac{1}{x+3}}$$

19. _____

19. Solve the equation $\dfrac{2x}{x-3} + \dfrac{1}{x+3} = \dfrac{-6}{x^2-9}$. Be sure to check your answer(s).

20. _____

20. Solve the formula $F = \dfrac{k}{d-D}$ for D.

Solve each problem.

21. _____

21. If the same number is added to the numerator and subtracted from the denominator of $\dfrac{5}{6}$, the resulting fraction is equivalent to $\dfrac{1}{10}$. What is the number?

22. _____

22. A boat goes 7 mph in still water. It takes as long to go 20 mi upstream as 50 mi downstream. Find the speed of the current.

23. _____

23. A man can paint a room in his house, working alone, in 5 hr. His wife can do the job in 4 hr. How long will it take them to paint the room if they work together?

24. _____

24. If x varies directly as y, and $x = 12$ when $y = 4$, find x when $y = 9$.

25. _____

25. Under certain conditions, the length of time that it takes for fruit to ripen during the growing season varies inversely as the average maximum temperature during the season. If it takes 25 days for fruit to ripen with an average maximum temperature of 80°, find the number of days it would take at 75°. Round your answer to the nearest whole number.

1. Evaluate $3 + 4\left(\dfrac{1}{2} - \dfrac{3}{4}\right)$.

Solve. Graph the solutions in Exercises 5 and 6.

2. $3(2y - 5) = 2 + 5y$

3. $A = \dfrac{1}{2}bh$ for b

4. $\dfrac{2 + m}{2 - m} = \dfrac{3}{4}$

5. $5y \le 6y + 8$

6. $5m - 9 > 2m + 3$

7. For the graph of $4x + 3y = -12$,

    (a) what is the x-intercept and **(b)** what is the y-intercept?

Sketch each graph.

8. $y = -3x + 2$

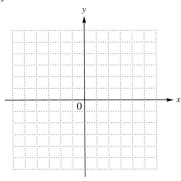

9. $y \ge 2x + 3$

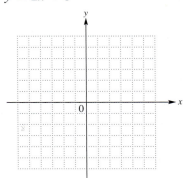

Simplify each expression. Write with only positive exponents.

10. $\dfrac{(2x^3)^{-1} \cdot x}{2^3 x^5}$

11. $\dfrac{(m^{-2})^3 m}{m^5 m^{-4}}$

12. $\dfrac{2p^3 q^4}{8p^5 q^3}$

Perform the indicated operations.

13. $(2k^2 + 3k) - (k^2 + k - 1)$

14. $8x^2 y^2 (9x^4 y^5)$

15. $(2a - b)^2$

16. $(y^2 + 3y + 5)(3y - 1)$

17. $\dfrac{12p^3 + 2p^2 - 12p + 4}{2p - 2}$

18. A computer can do one operation in 1.4×10^{-7} sec. How long would it take for the computer to do one trillion (10^{12}) operations?

Factor completely.

19. $8t^2 + 10tv + 3v^2$

20. $8r^2 - 9rs + 12s^2$

21. $16x^4 - 1$

Solve each equation.

22. $r^2 = 2r + 15$

23. $(r - 5)(2r + 1)(3r - 2) = 0$

Solve each problem.

24. One number is 4 more than another. The product of the numbers is 2 less than the smaller number. Find the smaller number.

25. The length of a rectangle is 2 m less than twice the width. The area is 60 m². Find the width of the rectangle.

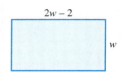

26. For what value(s) of t is $\dfrac{2 + t}{t^2 - 4}$ undefined?

27. One of the following is equal to 1 for *all* real numbers. Which one is it?

 A. $\dfrac{k^2 + 2}{k^2 + 2}$ **B.** $\dfrac{4 - m}{4 - m}$ **C.** $\dfrac{2x + 9}{2x + 9}$ **D.** $\dfrac{x^2 - 1}{x^2 - 1}$

28. Which one of the following rational expressions is *not* equivalent to $\dfrac{4 - 3x}{7}$?

 A. $-\dfrac{-4 + 3x}{7}$ **B.** $-\dfrac{4 - 3x}{-7}$ **C.** $\dfrac{-4 + 3x}{-7}$ **D.** $\dfrac{-(3x + 4)}{7}$

Perform each operation, and write the answer in lowest terms.

29. $\dfrac{5}{q} - \dfrac{1}{q}$

30. $\dfrac{3}{7} + \dfrac{4}{r}$

31. $\dfrac{4}{5q - 20} - \dfrac{1}{3q - 12}$

32. $\dfrac{2}{k^2 + k} - \dfrac{3}{k^2 - k}$

33. $\dfrac{7z^2 + 49z + 70}{16z^2 + 72z - 40} \div \dfrac{3z + 6}{4z^2 - 1}$

34. Simplify the complex fraction $\dfrac{\dfrac{4}{a} + \dfrac{5}{2a}}{\dfrac{7}{6a} - \dfrac{1}{5a}}$.

Solve each equation. Check your answers.

35. $\dfrac{r + 2}{5} = \dfrac{r - 3}{3}$

36. $\dfrac{1}{x} = \dfrac{1}{x + 1} + \dfrac{1}{2}$

Solve each problem.

37. On a business trip, Arlene traveled to her destination at an average speed of 60 mph. Coming home, her average speed was 50 mph, and the trip took $\dfrac{1}{2}$ hr longer. How far did she travel each way?

38. Juanita can weed the yard in 3 hr. Benito can weed the yard in 2 hr. How long would it take them if they worked together?

Systems of Equations and Inequalities

7

Systems of equations are used to solve many applications of mathematics. An important example is their use in modeling the cost to produce an item and the revenue received from that item. The owner or manager of a business needs to know the *break-even point*, where the number of items produced and sold leads to equal cost and revenue. If more items are produced and sold, the business will make a profit. In the Exercises for Sections 7.2 and 7.4, we explore how the concepts of this and earlier chapters are used to find the break-even point.

ADDISON · WESLEY
MyMathLab.com
You're Connected

7.1 SOLVING SYSTEMS OF LINEAR EQUATIONS BY GRAPHING

1 Fill in the blanks, and decide whether the given ordered pair is a solution of the system.

(a) $(2, 5)$
$$3x - 2y = -4$$
$$5x + y = 15$$

$$3x - 2y = -4$$
$$3(\underline{\quad}) - 2(\underline{\quad}) = -4$$

$$5x + y = 15$$
$$5(2) + \underline{\quad} = \underline{\quad}$$

$(2, 5)$ _____ a solution.
(is/is not)

(b) $(1, -2)$
$$x - 3y = 7$$
$$4x + y = 5$$

$(1, -2)$ _____ a solution.
(is/is not)

A **system of linear equations** consists of two or more linear equations with the same variables. Examples of systems of two linear equations include

$$2x + 3y = 4 \qquad x + 3y = 1 \qquad x - y = 1$$
$$3x - y = -5 \qquad -y = 4 - 2x \qquad y = 3.$$

In the system on the right, think of $y = 3$ as an equation in two variables by writing it as $0x + y = 3$.

1 **Decide whether a given ordered pair is a solution of a system.** A **solution of a system** of linear equations is an ordered pair that makes both equations true at the same time. A solution of an equation is said to *satisfy* the equation.

Example 1 Determining Whether an Ordered Pair Is a Solution

Is $(4, -3)$ a solution of each system?

(a) $x + 4y = -8$
 $3x + 2y = 6$

To decide whether or not $(4, -3)$ is a solution of the system, substitute 4 for x and -3 for y in each equation.

$x + 4y = -8$		$3x + 2y = 6$	
$4 + 4(-3) = -8$	?	$3(4) + 2(-3) = 6$	?
$4 + (-12) = -8$	? Multiply.	$12 + (-6) = 6$	? Multiply.
$-8 = -8$	True	$6 = 6$	True

Because $(4, -3)$ satisfies both equations, it is a solution of the system.

(b) $2x + 5y = -7$
 $3x + 4y = 2$

Again, substitute 4 for x and -3 for y in both equations.

$2x + 5y = -7$		$3x + 4y = 2$	
$2(4) + 5(-3) = -7$	?	$3(4) + 4(-3) = 2$	?
$8 + (-15) = -7$	? Multiply.	$12 + (-12) = 2$	? Multiply.
$-7 = -7$	True	$0 = 2$	False

The ordered pair $(4, -3)$ is not a solution of this system because it does not satisfy the second equation.

Work Problem 1 at the Side.

We discuss several methods of solving a system of two linear equations in two variables in this chapter.

2 **Solve linear systems by graphing.** One way to find the solution of a system of two linear equations is to graph both equations on the same axes. The graph of each line shows points whose coordinates satisfy the equation of that line. Any intersection point would be on both lines and would therefore be a solution of both equations. Thus, the coordinates of any point where the lines intersect give a solution of the system. Because two different straight lines can intersect at no more than one point, there can never be more than one solution for such a system. The graph in Figure 1 shows that the solution of the system in Example 1(a) is the intersection point $(4, -3)$.

ANSWERS
1. (a) 2; 5; 5; 15; is **(b)** is not

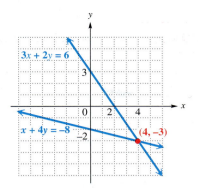

Figure 1

Example 2 **Solving a System by Graphing**

Solve the system of equations by graphing both equations on the same axes.

$$2x + 3y = 4$$
$$3x - y = -5$$

As shown in Chapter 3, we graph these two equations by plotting several points for each line. Recall from Section 3.2 that we can choose *any* number for either x or y to get an ordered pair. The intercepts are often convenient choices. It is a good idea to use a third ordered pair as a check.

$2x + 3y = 4$			$3x - y = -5$	
x	**y**		**x**	**y**
0	$\frac{4}{3}$		0	5
2	0		$-\frac{5}{3}$	0
-2	$\frac{8}{3}$		-2	-1

The lines in Figure 2 suggest that the graphs intersect at the point $(-1, 2)$. We check this by substituting -1 for x and 2 for y in both equations. Because $(-1, 2)$ satisfies both equations, the solution of this system is $(-1, 2)$.

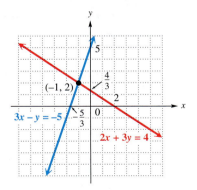

Figure 2

CAUTION

A difficulty with the graphing method of solution is that it may not be possible to determine from the graph the exact coordinates of the point that represents the solution, particularly if these coordinates are not integers. For this reason, algebraic methods of solution are explained later in this chapter. The graphing method does, however, show geometrically how solutions are found and is useful when approximate answers will do.

Work Problem ❷ at the Side.

---

❷ Solve each system of equations by graphing both equations on the same axes. Check your answers.

(a) $5x - 3y = 9$
$x + 2y = 7$
(One of the lines is already graphed.)

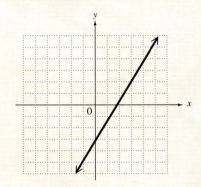

(b) $x + y = 4$
$2x - y = -1$

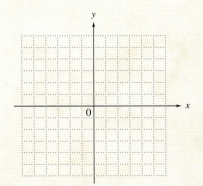

3 ▭ **Solve special systems by graphing.** Sometimes the graphs of the two equations in a system either do not intersect at all or are the same line, as in the systems in Example 3.

Example 3 Solving Special Systems

Solve each system by graphing.

(a) $2x + y = 2$

$2x + y = 8$

The graphs of these lines are shown in Figure 3. The two lines are parallel and have no points in common. For such a system, we will write "no solution."

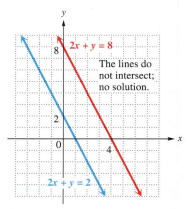

Figure 3

(b) $2x + 5y = 1$

$6x + 15y = 3$

The graphs of these two equations are the same line. See Figure 4. The second equation can be obtained by multiplying each side of the first equation by 3. In this case, every point on the line is a solution of the system, and the solutions are the infinite number of ordered pairs that satisfy the equations. We will write "infinite number of solutions" or "same line" to indicate this case.

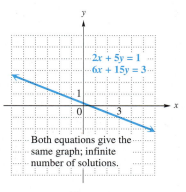

Figure 4

The system in Example 2 has exactly one solution. A system with at least one solution is called a **consistent system.** A system of equations with no solutions, such as the one in Example 3(a), is called an **inconsistent system.** The equations in Example 2 are **independent equations** with different graphs. The equations of the system in Example 3(b) have the same graph and are equivalent. Because they are different forms of the same equation, these equations are called **dependent equations.**

Work Problem ❸ at the Side.

Examples 2 and 3 show the three cases that may occur when solving a system of two equations with two variables.

Possible Types of Solutions

1. The graphs intersect at exactly one point, which gives the (single) solution of the system. The **system is consistent,** and the **equations are independent.**

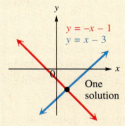

2. The graphs are parallel lines, so there is no solution. The **system is inconsistent.**

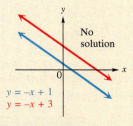

3. The graphs are the same line. The solution is an infinite number of ordered pairs. The **equations are dependent.**

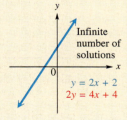

❸ Solve each system of equations by graphing both equations on the same axes.

(a) $3x - y = 4$
$6x - 2y = 12$
(One of the lines is already graphed.)

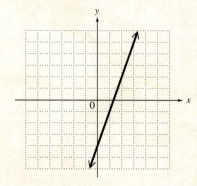

(b) $-x + 3y = 2$
$2x - 6y = -4$

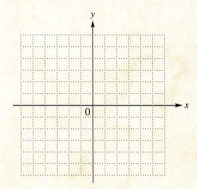

Real-Data Applications

Estimating Fahrenheit Temperature

When traveling in countries other than the United States, you will hear the daily high and low temperatures reported in degrees Celsius, instead of degrees Fahrenheit. The following information may help you.

- The linear equation to convert degrees Celsius to degrees Fahrenheit is $F = \frac{9}{5}C + 32$.

- Travel books advise you to use a *rule of thumb* to estimate Fahrenheit temperature that says "*Double the temperature (degrees Celsius) and add 30.*" This rule of thumb is written mathematically as $F = 2C + 30$.

For Group Discussion

Suppose you are interested in knowing for what temperature the rule of thumb and the actual formulas give the same result. You also want to know if the rule of thumb formula is predicting temperatures that are lower or higher than the actual temperature.

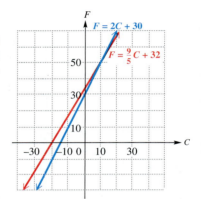

1. The two formulas can be written as the system of equations

$$F = \frac{9}{5}C + 32$$
$$F = 2C + 30.$$

 (a) Use the graph of the system of equations to find the point of intersection of the two formulas. (*Hint:* To check your answer, use substitution to see if it satisfies both formulas.)

 (b) For what temperature in degrees Celsius do the two formulas agree?

 (c) For what temperature in degrees Fahrenheit do the two formulas agree?

2. **(a)** Complete the table of values to compare the *actual* and the *rule of thumb* formulas for temperature conversion.

°C	°F (*Actual*)	°F (*Rule of Thumb*)
0		
5		
10		
15		
20		
30		

 (b) If the daily low is predicted to be 5°C, then is the rule of thumb estimate too high or too low?

 (c) If the daily high is predicted to be 20°C, then is the rule of thumb estimate too high or too low?

 (d) If the daily high is predicted to be 30°C, then is the rule of thumb estimate too high or too low?

 (e) How many degrees "off" is the rule of thumb estimate for the boiling point of water?

 (f) Comment on the accuracy of using the rule of thumb as an estimate of the actual Fahrenheit temperature.

7.1 **EXERCISES**

 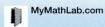

1. Which ordered pair could be a solution of the system graphed? Why is it the only valid choice?

   A. $(2, 2)$

   B. $(-2, 2)$

   C. $(-2, -2)$

   D. $(2, -2)$

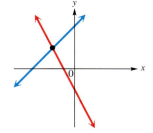

2. Which ordered pair could be a solution of the system graphed? Why is it the only valid choice?

   A. $(2, 0)$

   B. $(0, 2)$

   C. $(-2, 0)$

   D. $(0, -2)$

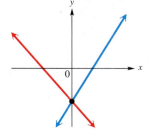

3. How can you tell without graphing that this system has no solution?

$$x + y = 2$$
$$x + y = 4$$

4. Explain why a system of two linear equations cannot have exactly two solutions.

Decide whether the given ordered pair is a solution of the given system. See Example 1.

5. $(2, -3)$
   $x + y = -1$
   $2x + 5y = 19$

6. $(4, 3)$
   $x + 2y = 10$
   $3x + 5y = 3$

7. $(-1, -3)$
   $3x + 5y = -18$
   $4x + 2y = -10$

8. $(-9, -2)$
   $2x - 5y = -8$
   $3x + 6y = -39$

9. $(7, -2)$
   $4x = 26 - y$
   $3x = 29 + 4y$

10. $(9, 1)$
   $2x = 23 - 5y$
   $3x = 24 + 3y$

11. $(6, -8)$
   $-2y = x + 10$
   $3y = 2x + 30$

12. $(-5, 2)$
   $5y = 3x + 20$
   $3y = -2x - 4$

Solve each system of equations by graphing both equations on the same axes. See Example 2.

13. $x - y = 2$
$x + y = 6$

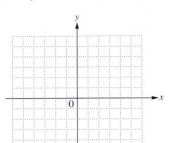

14. $x - y = 3$
$x + y = -1$

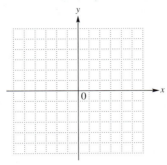

15. $x + y = 4$
$y - x = 4$

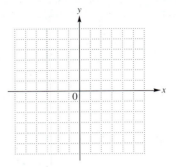

16. $x + y = -5$
$x - y = 5$

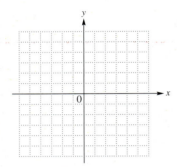

17. $x - 2y = 6$
$x + 2y = 2$

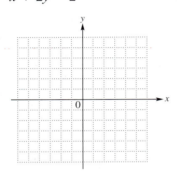

18. $2x - y = 4$
$4x + y = 2$

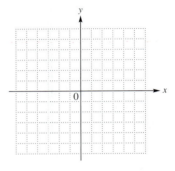

19. $3x - 2y = -3$
$-3x - y = -6$

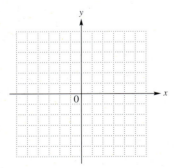

20. $2x - y = 4$
$2x + 3y = 12$

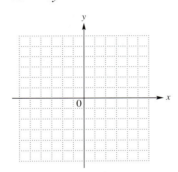

21. $2x - 3y = -6$
$y = -3x + 2$

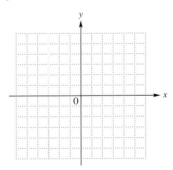

22. $-3x + y = -3$
$y = x - 3$

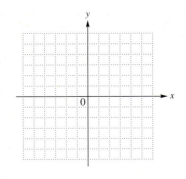

23. $3x - 4y = 24$
$y = -\dfrac{3}{2}x + 3$

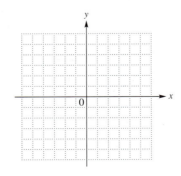

24. $3x - 2y = 12$
$y = -4x + 5$

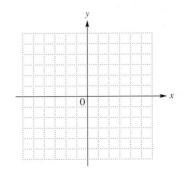

RELATING CONCEPTS (Exercises 25–28) **FOR INDIVIDUAL OR GROUP WORK**

In Exercises 25–27, first write each equation in slope-intercept form. Then use what you learned in Chapter 3 about slope and the y-intercept to describe the graphs of each system of equations. **Work these exercises in order.**

25. $3x + 2y = 6$
$-2y = 3x - 5$

26. $2x - y = 4$
$x = .5y + 2$

27. $x - 3y = 5$
$2x + y = 8$

28. Use the results of Exercises 25–27 to determine the number of solutions of each system.

Solve each system by graphing. If the two equations produce parallel lines, write no solution. *If the two equations produce the same line, write* infinite number of solutions. *See Example 3.*

29. $x + 2y = 6$
$2x + 4y = 8$

30. $2x - y = 6$
$6x - 3y = 12$

31. $-2x + y = -4$
$4x = 2y + 8$

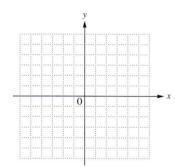
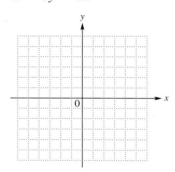
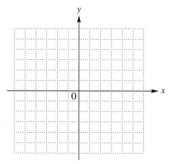

32. $3x + y = 5$
$6x = 10 - 2y$

33. $3x = y + 5$
$6x - 5 = 2y$

34. $2x = y - 4$
$4x - 2y = -4$

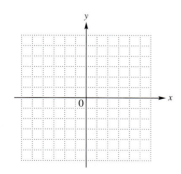
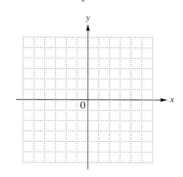
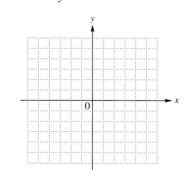

The graph shows how production levels of vinyl LPs, cassettes, and CDs changed from 1986 to 1998. Use the graph to respond to the questions in Exercises 35–38.

35. In what year did cassette production and CD production reach an equal level? What was that production level?

36. For which years was the production of CDs less than the production of LPs?

37. Between which two nonconsecutive years was the production of cassettes approximately constant?

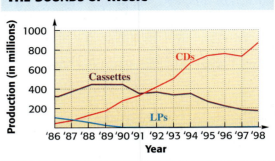

THE SOUNDS OF MUSIC

Source: Recording Industry Association of America.

38. If a straight line were used to approximate the graph of CD production from 1986 to 1998, would its slope be positive, negative, or 0?

39. Explain one of the drawbacks of solving a system of equations graphically.

40. If the two lines that are the graphs of the equations in a system are parallel, how many solutions does the system have? If the two lines coincide, how many solutions does the system have?

41. Find a system of equations with the solution $(-2, 3)$, and show the graph.

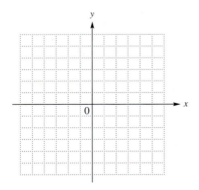

42. Solve the system

$$2x + 3y = 6$$
$$x - 3y = 5$$

by graphing. Can you check your answer? Why or why not?

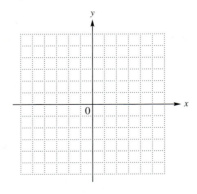

7.2 SOLVING SYSTEMS OF LINEAR EQUATIONS BY SUBSTITUTION

1 **Solve linear systems by substitution.** Graphing to solve a system of equations has a serious drawback: It is difficult to accurately find a solution such as $\left(\frac{1}{3}, -\frac{5}{6}\right)$ from a graph. One algebraic method for solving a system of equations is the **substitution method.** This method is particularly useful for solving systems where one equation is already solved, or can be solved quickly, for one of the variables.

OBJECTIVES

1 Solve linear systems by substitution.

2 Solve special systems.

3 Solve linear systems with fractions.

Example 1 Using the Substitution Method

Solve the system

$$3x + 5y = 26$$
$$y = 2x.$$

The second equation is already solved for y. This equation says that $y = 2x$. Substituting $2x$ for y in the first equation gives

$$3x + 5y = 26$$
$$3x + 5(2x) = 26 \qquad \text{Let } y = 2x.$$
$$3x + 10x = 26 \qquad \text{Multiply.}$$
$$13x = 26 \qquad \text{Combine terms.}$$
$$x = 2. \qquad \text{Divide by 13.}$$

Because $x = 2$, we find y from the equation $y = 2x$ by substituting 2 for x.

$$y = 2(2) = 4 \qquad \text{Let } x = 2.$$

Check that the solution of the given system is $(2, 4)$ by substituting 2 for x and 4 for y in *both* equations.

Work Problem ❶ at the Side.

❶ Fill in the blanks to solve by the substitution method. Check your solution.

$$3x + 5y = 69$$
$$y = 4x$$
$$3x + 5(\underline{\quad}) = 69$$
$$\underline{\quad} = 69$$
$$x = \underline{\quad}$$
$$y = 4(\underline{\quad}) = \underline{\quad}$$

The solution is _____.

Example 2 Using the Substitution Method

Solve the system

$$2x + 5y = 7$$
$$x = -1 - y.$$

The second equation gives x in terms of y. Substitute $-1 - y$ for x in the first equation.

$$2x + 5y = 7$$
$$2(-1 - y) + 5y = 7 \qquad \text{Let } x = -1 - y.$$
$$-2 - 2y + 5y = 7 \qquad \text{Distributive property}$$
$$-2 + 3y = 7 \qquad \text{Combine terms.}$$
$$3y = 9 \qquad \text{Add 2.}$$
$$y = 3 \qquad \text{Divide by 3.}$$

To find x, substitute 3 for y in the equation $x = -1 - y$ to get $x = -1 - 3 = -4$. Check that the solution of the given system is $(-4, 3)$.

Work Problem ❷ at the Side.

❷ Solve by the substitution method. Check your solution.

$$2x + 7y = -12$$
$$x = 3 - 2y$$

ANSWERS
1. $4x$; $23x$; 3; 3; 12; $(3, 12)$
2. $(15, -6)$

③ Solve each system by substitution. Check each solution.

(a) Fill in the blanks to solve

$$x + 4y = -1$$
$$2x - 5y = 11.$$

Solve the first equation for x.

$$x = -1 - \underline{\quad}$$

Substitute into the second equation.

$$2(\underline{\quad}) - 5y = 11$$
$$\underline{\quad}y = \underline{\quad}$$
$$y = \underline{\quad}$$
$$x = -1 - \underline{\quad}$$
$$x = \underline{\quad}$$

The solution is ___.

(b) $2x + 5y = 4$
      $x + y = -1$

Example 3 Using the Substitution Method

Use substitution to solve the system

$$2x + 3y = 10$$
$$-3x - 2y = 0.$$

To use the substitution method, one of the equations must be solved for one of the variables. We choose the first equation of the system, $2x + 3y = 10$, and solve for x.

$$2x + 3y = 10$$
$$2x = 10 - 3y \qquad \text{Subtract } 3y.$$
$$x = 5 - \frac{3}{2}y \qquad \text{Divide by 2.}$$

Substitute this expression for x in the second equation of the system.

$$-3x - 2y = 0$$
$$-3\left(5 - \frac{3}{2}y\right) - 2y = 0 \qquad \text{Let } x = 5 - \frac{3}{2}y.$$
$$-15 + \frac{9}{2}y - 2y = 0 \qquad \text{Distributive property}$$
$$-15 + \frac{5}{2}y = 0 \qquad \text{Combine terms.}$$
$$\frac{5}{2}y = 15 \qquad \text{Add 15.}$$
$$y = \frac{30}{5} = 6 \qquad \text{Multiply by } \frac{2}{5}.$$

Find x by substituting **6** for y in $x = 5 - \frac{3}{2}y$.

$$x = 5 - \frac{3}{2}(6) = -4$$

Check:

$$2x + 3y = 10 \qquad\qquad -3x - 2y = 0$$
$$2(-4) + 3(6) = 10 \quad ? \qquad -3(-4) - 2(6) = 0 \quad ?$$
$$-8 + 18 = 10 \quad ? \qquad\qquad 12 - 12 = 0 \quad ?$$
$$10 = 10 \quad \text{True} \qquad\qquad\qquad 0 = 0 \quad \text{True}$$

The solution of the system is $(-4, 6)$.

NOTE

In Example 3, we could have started the solution by solving the second equation for either x or y and then substituting the result into the first equation. The solution would be the same.

Work Problem ③ at the Side.

Example 4 **Using the Substitution Method**

Use substitution to solve the system

$$2x = 4 - y \qquad (1)$$
$$6 + 3y + 4x = 16 - x. \qquad (2)$$

Start by simplifying the second equation by adding x and subtracting 6 on each side. This gives the simplified system

$$2x = 4 - y \qquad (1)$$
$$5x + 3y = 10. \qquad (3)$$

For the substitution method, one of the equations must be solved for either x or y. Because the coefficient of y in equation (1) is -1, we avoid fractions by solving this equation for y.

$$2x = 4 - y \qquad (1)$$
$$2x - 4 = -y \qquad \text{Subtract 4.}$$
$$-2x + 4 = y \qquad \text{Multiply by } -1.$$

Now substitute $-2x + 4$ for y in equation (3).

$$5x + 3y = 10$$
$$5x + 3(\mathbf{-2x + 4}) = 10 \qquad \text{Let } y = -2x + 4.$$
$$5x - 6x + 12 = 10 \qquad \text{Distributive property}$$
$$-x + 12 = 10 \qquad \text{Combine like terms.}$$
$$-x = -2 \qquad \text{Subtract 12.}$$
$$x = 2 \qquad \text{Multiply by } -1.$$

Since $y = -2x + 4$ and $x = 2$,

$$y = -2(\mathbf{2}) + 4 = 0,$$

and the solution is $(2, 0)$.

Check:

$$2\mathbf{x} = 4 - \mathbf{y} \qquad (1) \qquad\qquad 6 + 3\mathbf{y} + 4\mathbf{x} = 16 - \mathbf{x} \qquad (2)$$
$$2(\mathbf{2}) = 4 - \mathbf{0} \quad ? \qquad\qquad 6 + 3(\mathbf{0}) + 4(\mathbf{2}) = 16 - \mathbf{2} \quad ?$$
$$4 = 4 \qquad \text{True} \qquad\qquad 6 + 0 + 8 = 14 \qquad ?$$
$$14 = 14 \qquad \text{True}$$

=========== **Work Problem ④ at the Side.**

2 ▭ **Solve special systems.** In the previous section we solved inconsistent systems with graphs that are parallel lines and systems of dependent equations with graphs that are the same line. We can also solve these special systems with the substitution method.

④ Solve each system by substitution. First simplify where necessary.

(a) $$x = 5 - 3y$$
$$2x + 3 = 5x - 4y + 14$$

(b) $$5x - y = -14 + 2x + y$$
$$7x + 9y + 4 = 3x + 8y$$

Example 5 Solving an Inconsistent System by Substitution

Use substitution to solve the system

$$x = 5 - 2y \quad (1)$$
$$2x + 4y = 6. \quad (2)$$

Substitute $5 - 2y$ for x in equation (2).

$$2x + 4y = 6$$
$$2(5 - 2y) + 4y = 6 \quad \text{Let } x = 5 - 2y.$$
$$10 - 4y + 4y = 6 \quad \text{Distributive property}$$
$$10 = 6 \quad \text{False}$$

This false result means that the equations in the system have graphs that are parallel lines. The system is inconsistent and has no solution. See Figure 5.

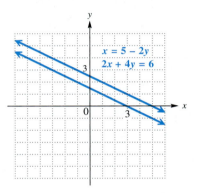

Figure 5

CAUTION

It is a common error to give "false" as the answer to an inconsistent system. The correct response is "no solution."

Example 6 Solving a System with Dependent Equations by Substitution

Solve the system by the substitution method.

$$3x - y = 4 \quad (1)$$
$$-9x + 3y = -12 \quad (2)$$

Begin by solving equation (1) for y to get $y = 3x - 4$. Substitute $3x - 4$ for y in equation (2) and solve the resulting equation.

$$-9x + 3y = -12$$
$$-9x + 3(3x - 4) = -12 \quad \text{Let } y = 3x - 4.$$
$$-9x + 9x - 12 = -12 \quad \text{Distributive property}$$
$$0 = 0 \quad \text{Add 12; combine terms.}$$

Continued on Next Page

This true result means that every solution of one equation is also a solution of the other, so the system has an infinite number of solutions: all the ordered pairs corresponding to points that lie on the common graph. A graph of the equations of this system is shown in Figure 6.

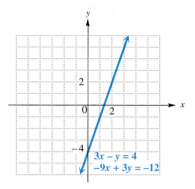

Figure 6

CAUTION

It is a common error to give "true" as the answer to a system of dependent equations. The correct response is "infinite number of solutions."

Work Problem ❺ at the Side.

3   **Solve linear systems with fractions.**   When a system includes an equation with fractions as coefficients, eliminate the fractions by multiplying each side of the equation by a common denominator. Then solve the resulting system.

Example 7   **Using the Substitution Method with Fractions as Coefficients**

Solve the system by the substitution method.

$$3x + \frac{1}{4}y = 2 \qquad (1)$$

$$\frac{1}{2}x + \frac{3}{4}y = -\frac{5}{2} \qquad (2)$$

Clear equation (1) of fractions by multiplying each side by 4.

$$4\left(3x + \frac{1}{4}y\right) = 4(2) \qquad \text{Multiply by 4.}$$

$$4(3x) + 4\left(\frac{1}{4}y\right) = 4(2) \qquad \text{Distributive property}$$

$$12x + y = 8 \qquad (3)$$

Now clear equation (2) of fractions by multiplying each side by the common denominator 4.

Continued on Next Page

---

❺ Solve each system by substitution.

  (a) $8x - y = 4$
$$y = 8x + 4$$

  (b) $\quad 7x - 6y = 10$
$$-14x + 20 = -12y$$

6 Solve the system by the substitution method. First clear all fractions.

$$\frac{2}{3}x + \frac{1}{2}y = 6$$

$$\frac{1}{2}x - \frac{3}{4}y = 0$$

$$4\left(\frac{1}{2}x + \frac{3}{4}y\right) = 4\left(-\frac{5}{2}\right) \quad \text{Multiply by 4.}$$

$$4\left(\frac{1}{2}x\right) + 4\left(\frac{3}{4}y\right) = 4\left(-\frac{5}{2}\right) \quad \text{Distributive property}$$

$$2x + 3y = -10 \qquad (4)$$

The given system of equations has been simplified as

$$12x + y = 8 \qquad (3)$$

$$2x + 3y = -10. \qquad (4)$$

Solve this system by the substitution method. Equation (3) can be solved for y by subtracting $12x$ from each side.

$$12x + y = 8$$

$$y = -12x + 8 \quad \text{Subtract } 12x.$$

Now substitute the result for y in equation (4).

$$2x + 3(\mathbf{-12x + 8}) = -10 \quad \text{Let } y = -12x + 8.$$

$$2x - 36x + 24 = -10 \quad \text{Distributive property}$$

$$-34x = -34 \quad \text{Combine terms; subtract 24.}$$

$$x = 1 \quad \text{Divide by } -34.$$

Substitute 1 for x in $y = -12x + 8$ to get $y = -12(1) + 8 = -4$. The solution is $(1, -4)$. Check by substituting 1 for x and -4 for y in both of the original equations.

Work Problem 6 at the Side.

7.2 EXERCISES

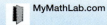

1. A student solves the system

$$5x - y = 15$$
$$7x + y = 21$$

and finds that $x = 3$, which is the correct value for x. The student gives the solution as "$x = 3$." Is this correct? Explain.

2. A student solves the system

$$x + y = 4$$
$$2x + 2y = 8$$

and obtains the equation $0 = 0$. The student gives the solution as $(0, 0)$. Is this correct? Explain.

3. Professor Brandsma gave the following item on a test in algebra:
Use the substitution method to solve the system

$$3x - y = 13$$
$$2x + 5y = 20.$$

One student worked the problem by solving first for y in the first equation. Another student worked it by solving first for x in the second equation. Both students got the correct solution, $(5, 2)$. Which student, do you think, had less work to do? Explain.

4. When you use the substitution method, how can you tell that a system has

(a) no solution?

(b) an infinite number of solutions?

Solve each system by the substitution method. Check each solution. See Examples 1–3, 5, and 6.

5. $3x + 2y = 27$
     $x = y + 4$

6. $4x + 3y = -5$
     $x = y - 3$

7. $3x + 5y = 14$
     $x - 2y = -10$

8. $5x + 2y = -1$
$2x - y = -13$

9. $3x + 4 = -y$
$2x + y = 0$

10. $2x - 5 = -y$
$x + 3y = 0$

11. $7x + 4y = 13$
$x + y = 1$

12. $3x - 2y = 19$
$x + y = 8$

13. $3x - y = 5$
$y = 3x - 5$

14. $4x - y = -3$
$y = 4x + 3$

15. $6x - 8y = 6$
$2y = -2 + 3x$

16. $3x + 2y = 6$
$6x = 8 + 4y$

17. $2x + 8y = 3$
$x = 8 - 4y$

18. $2x + 10y = 3$
$x = 1 - 5y$

19. $12x - 16y = 8$
$3x = 4y + 2$

20. $6x + 9y = 6$
$2x = 2 - 3y$

Solve each system by the substitution method. First simplify equations where necessary. Check each solution. See Example 4.

21. $4 + 4x - 3y = 34 + x$
$4x = -y - 2 + 3x$

22. $5x - 4y = 42 - 8y - 2$
$2x + y = x + 1$

23. $2x - 8y + 3y + 2 = 5y + 16$
$8x - 2y = 4x + 28$

24. $7x - 9 + 2y - 8 = -3y + 4x + 13$
$4y - 8x = -8 + 9x + 32$

25. $-2x + 3y = 12 + 2y$
$2x - 5y + 4 = -8 - 4y$

26. $2x + 5y = 7 + 4y - x$
$5x + 3y + 8 = 22 - x + y$

27. $5x + y = 12 - x - 7y$
$3x + 2y = 10 - 6x - 10y$

28. $-2x + 3y = 7 - 5x - y$
$-4x + 2y = 1 - 10x - 6y$

29. Solve each system.

(a) $5x - 4y = 7$
$x = 3$

(b) $5x - 4y = 7$
$y = -3$

Why are these systems easier to solve than the examples in this section?

30. One student solved the system

$$\frac{1}{3}x - \frac{1}{2}y = 7$$

$$\frac{1}{6}x + \frac{1}{3}y = 0$$

and wrote as his answer "$x = 12$," while another solved it and wrote as her answer "$y = -6$." Who, if either, was correct? Why?

Solve each system by the substitution method. First clear all fractions. Check each solution. See Example 7.

31. $x + \frac{1}{3}y = y - 2$
$\frac{1}{4}x + y = x + y$

32. $\frac{5}{3}x + 2y = \frac{1}{3} + y$
$3x - 3 + \frac{y}{3} = -2 + 2x$

33. $\frac{x}{6} + \frac{y}{6} = 2$
$-\frac{1}{2}x - \frac{1}{3}y = -8$

34. $\frac{x}{2} - \frac{y}{3} = 9$
$\frac{x}{5} - \frac{y}{4} = 5$

35. $\frac{x}{3} - \frac{3y}{4} = -\frac{1}{2}$
$\frac{x}{6} + \frac{y}{8} = \frac{3}{4}$

36. $\frac{x}{5} + 2y = \frac{16}{5}$
$\frac{3x}{5} + \frac{y}{2} = -\frac{7}{5}$

RELATING CONCEPTS (Exercises 37–40) **FOR INDIVIDUAL OR GROUP WORK**

A system of linear equations can be used to model the cost and the revenue of a business.
Work Exercises 37–40 in order.

37. Suppose that you start a business manufacturing and selling bicycles, and it costs you $5000 to get started. You determine that each bicycle will cost $400 to manufacture. Explain why the linear equation $y_1 = 400x + 5000$ gives your *total* cost to manufacture x bicycles (y_1 in dollars).

38. You decide to sell each bike for $600. What expression in x represents the revenue you will take in if you sell x bikes? Write an equation using y_2 to express your revenue when you sell x bikes (y_2 in dollars).

39. Form a system from the two equations in Exercises 37 and 38, and then solve the system, assuming $y_1 = y_2$, that is, cost = revenue.

40. The value of x from Exercise 39 is the number of bikes it takes to *break even*. Fill in the blanks:

When _____ bikes are sold, the break-even point is reached. At that point, you have spent _____ dollars and taken in _____ dollars.

 *Work each problem.*

41. During the period from 1991 to 1996, average ticket prices rose in the National Football League from $25.21 to $35.74. If we let $x = 1$ represent 1991, $x = 2$ represent 1992, and so on, the linear equation $y = 2.1x + 22.8$ gives a good approximation for this average price, where y is in dollars. To determine the year in which the average ticket price was $28.68, solve the system

$$y = 2.1x + 22.8$$
$$y = 28.68.$$

The rounded x-value will then give us the year. Solve this system by substitution to determine the year. (*Hint:* After finding the value of x, you must then determine which year it represents.) (*Source:* Team Marketing Report, Chicago.)

42. During the period from 1991 to 1996, the average price of a ticket to a National Basketball Association game rose from $23.24 to $34.08. If we let $x = 1$ represent the 1991–1992 season, $x = 2$ represent 1992–1993, and so on, the linear equation $y = 2.14x + 20.81$ gives a good approximation for this average price, where y is in dollars. To determine the period in which the average ticket price was $28.63, solve the system

$$y = 2.14x + 20.81$$
$$y = 28.63.$$

The rounded x-value will then give us the first year of the period. Solve this system by substitution to determine the period. See the hint in Exercise 41. (*Source:* Team Marketing Report, Chicago.)

7.3 SOLVING SYSTEMS OF LINEAR EQUATIONS BY ELIMINATION

1 **Solve linear systems by elimination.** An algebraic method that depends on the addition property of equality can be used to solve systems. As mentioned earlier, adding the same quantity to each side of an equation results in equal sums.

$$\text{If}\quad A = B, \quad\text{then}\quad A + C = B + C.$$

This addition can be taken a step further. Adding *equal* quantities, rather than the *same* quantity, to both sides of an equation also results in equal sums.

$$\text{If}\quad A = B \quad\text{and}\quad C = D, \quad\text{then}\quad A + C = B + D.$$

Using the addition property to solve systems is called the **elimination method.** When using this method, the idea is to *eliminate* one of the variables. To do this, one of the variables in the two equations must have coefficients that are opposites.

OBJECTIVES

1 Solve linear systems by elimination.

2 Multiply when using the elimination method.

3 Use an alternative method to find the second value in a solution.

4 Use the elimination method to solve special systems.

Example 1 **Using the Elimination Method**

Use the elimination method to solve the system

$$x + y = 5$$
$$x - y = 3.$$

Each equation in this system is a statement of equality, so the sum of the right sides equals the sum of the left sides. Adding in this way gives

$$(x + y) + (x - y) = 5 + 3.$$

Combine terms and simplify to get

$$2x = 8$$
$$x = 4. \quad \text{Divide by 2.}$$

Notice that y has been eliminated. The result, $x = 4$, gives the x-value of the solution of the given system. To find the y-value of the solution, substitute 4 for x in either of the two equations of the system.

Work Problem ❶ at the Side.

The solution found at the side, (4, 1), can be checked by substituting 4 for x and 1 for y in both equations of the given system.

Check: $x + y = 5$ $x - y = 3$

         $4 + 1 = 5$? $4 - 1 = 3$?

          $5 = 5$ True $3 = 3$ True

Since both results are true, the solution of the system is (4, 1).

❶ **(a)** Substitute 4 for x in the equation $x + y = 5$ to find the value of y.

(b) Give the solution of the system.

CAUTION

A system is not completely solved until values for *both* x and y are found. Do not stop after finding the value of only one variable. Remember to write the solution as an ordered pair.

ANSWERS
1. (a) $y = 1$ **(b)** (4, 1)

② Solve each system by the elimination method. Check each solution.

(a) Fill in the blanks to find the solution.

$$x + y = 8$$
$$\underline{x - y = 2}$$

_____ + _____ = 10   Add.

$$x = \underline{\quad}$$

$$\underline{\quad} - y = 2$$

$$-y = \underline{\quad}$$

$$y = \underline{\quad}$$

The solution is _____.

(b) $3x - y = 7$
$2x + y = 3$

Work Problem ② at the Side.

In general, we use the following steps to solve a linear system of equations by the elimination method.

> ### Solving Linear Systems by Elimination
>
> *Step 1* **Write in standard form.** Write both equations of the system in standard form $Ax + By = C$.
>
> *Step 2* **Multiply.** Multiply one or both equations by appropriate numbers (if necessary) so that the coefficients of x (or y) are opposites of each other.
>
> *Step 3* **Add.** Add the two equations to get an equation with only one variable (or no variable).
>
> *Step 4* **Solve.** Solve the equation from Step 3.
>
> *Step 5* **Substitute.** Substitute the solution from Step 4 into either of the original equations to find the value of the remaining variable.
>
> *Step 6* **Check.** Check the solution in both of the original equations. Write the solution as an ordered pair.

It does not matter which variable is eliminated first. Usually we choose the one that is more convenient to work with.

Example 2 **Using the Elimination Method**

Solve the system

$$y + 11 = 2x$$
$$4 + 5x + y = 2y + 30.$$

Step 1 Rewrite both equations in the form $Ax + By = C$ to get the system

$$-2x + y = -11 \qquad \text{Subtract } 2x \text{ and } 11.$$
$$5x - y = 26. \qquad \text{Subtract } 4 \text{ and } 2y.$$

Step 2 Because the coefficients of y are 1 and -1, adding will eliminate y. It is not necessary to multiply either equation by a number.

Step 3 Add the two equations. This time we use vertical addition.

$$-2x \mathbf{+ y} = -11$$
$$\underline{5x \mathbf{- y} = 26}$$
$$3x = 15 \qquad \text{Add in columns.}$$

Step 4 Solve the equation.

$$3x = 15$$
$$x = 5 \qquad \text{Divide by 3.}$$

Step 5 Find the value of y by substituting 5 for x in either of the original equations. Choosing the first gives

$$y + 11 = 2\mathbf{x}$$
$$y + 11 = 2(\mathbf{5}) \qquad \text{Let } x = 5.$$
$$y = 10 - 11 \qquad \text{Subtract 11.}$$
$$y = -1.$$

Continued on Next Page

Step 6 Check the solution by substitution into both of the original equations. Let $x = 5$ and $y = -1$.

$$y + 11 = 2x \qquad\qquad 4 + 5x + y = 2y + 30$$
$$(-1) + 11 = 2(5) \quad ? \qquad 4 + 5(5) + (-1) = 2(-1) + 30 \quad ?$$
$$10 = 10 \qquad \text{True} \qquad 28 = 28 \qquad\qquad \text{True}$$

The solution $(5, -1)$ is correct.

=========== **Work Problem ❸ at the Side.**

❸ Solve each system by the elimination method. Check each solution.

(a) $2x - y = 2$
$\quad 4x + y = 10$

2 ▭ **Multiply when using the elimination method.** In both of the preceding examples, a variable was eliminated by adding the equations. Sometimes we need to multiply each side of one or both equations in a system by some number before adding the equations will eliminate a variable.

Example 3 **Multiplying Both Equations When Using the Elimination Method**

Solve the system

$$2x + 3y = -15 \qquad (1)$$
$$5x + 2y = 1. \qquad (2)$$

Adding the two equations gives $7x + 5y = -14$, which does not eliminate either variable. However, we can multiply each equation by a suitable number so that the coefficients of one of the two variables are opposites. For example, to eliminate x, multiply each side of equation (1) by 5, and each side of equation (2) by -2.

$$\begin{array}{ll} 10x + 15y = -75 & \text{Multiply equation (1) by 5.} \\ \underline{-10x - 4y = -2} & \text{Multiply equation (2) by } -2. \\ 11y = -77 & \text{Add.} \\ y = -7 & \end{array}$$

Substituting -7 for y in either equation (1) or (2) gives $x = 3$. Check that the solution of the system is $(3, -7)$.

=========== **Work Problem ❹ at the Side.**

(b) $8x - 5y = 32$
$\quad 4x + 5y = 4$

❹ (a) Solve the system in Example 3 by first eliminating the variable y. Check your solution.

3 ▭ **Use an alternative method to find the second value in a solution.** Sometimes it is easier to find the value of the second variable in a solution by using the elimination method twice. The next example shows this approach.

Example 4 **Finding the Second Value Using an Alternative Method**

Solve the system

$$4x = 9 - 3y \qquad (1)$$
$$5x - 2y = 8. \qquad (2)$$

Rearrange the terms in equation (1) so that like terms are aligned in columns. Add $3y$ to each side to get the following system.

$$4x + 3y = 9 \qquad (3)$$
$$5x - 2y = 8 \qquad (2)$$

=========== **Continued on Next Page**

(b) Solve

$6x + 7y = 4$
$5x + 8y = -1,$

and check your solution.

5 Solve each system of equations.

(a) $5x = 7 + 2y$
$5y = 5 - 3x$

One way to proceed is to eliminate y by multiplying each side of equation (3) by 2 and each side of equation (2) by 3, and then adding.

$$
\begin{aligned}
8x + 6y &= 18 && \text{Multiply equation (3) by 2.} \\
15x - 6y &= 24 && \text{Multiply equation (2) by 3.} \\
\hline
23x &= 42 && \text{Add.} \\
x &= \frac{42}{23} && \text{Divide by 23.}
\end{aligned}
$$

Substituting $\frac{42}{23}$ for x in one of the given equations would give y, but the arithmetic involved would be messy. Instead, solve for y by starting again with the original equations and eliminating x. Multiply each side of equation (3) by 5 and each side of equation (2) by -4, and then add.

$$
\begin{aligned}
20x + 15y &= 45 && \text{Multiply equation (3) by 5.} \\
-20x + 8y &= -32 && \text{Multiply equation (2) by } -4. \\
\hline
23y &= 13 && \text{Add.} \\
y &= \frac{13}{23} && \text{Divide by 23.}
\end{aligned}
$$

Check that the solution is $\left(\frac{42}{23}, \frac{13}{23}\right)$.

When the value of the first variable is a fraction, the method used in Example 4 helps avoid arithmetic errors. Of course, this method could be used to solve any system of equations.

Work Problem 5 at the Side.

(b) $3y = 8 + 4x$
$6x = 9 - 2y$

4 **Use the elimination method to solve special systems.** The next example shows the elimination method when a system is inconsistent or the equations of the system are dependent. To contrast the elimination method with the substitution method, in part (b) we use the same system solved in Example 6 of the previous section.

Example 5 **Using the Elimination Method for an Inconsistent System or Dependent Equations**

Solve each system by the elimination method.

(a) $2x + 4y = 5$

$4x + 8y = -9$

Multiply each side of $2x + 4y = 5$ by -2; then add to $4x + 8y = -9$.

$$
\begin{aligned}
-4x - 8y &= -10 \\
4x + 8y &= -9 \\
\hline
0 &= -19 && \text{False}
\end{aligned}
$$

The false statement $0 = -19$ shows that the given system has no solution.

Continued on Next Page

(b) $3x - y = 4$

$-9x + 3y = -12$

Multiply each side of the first equation by 3; then add the two equations.

$$9x - 3y = 12$$
$$\underline{-9x + 3y = -12}$$
$$0 = 0 \quad \text{True}$$

A true statement occurs when the equations are equivalent. As before, this result indicates that every solution of one equation is also a solution of the other; there are an infinite number of solutions.

━━━━━━━━━━━ **Work Problem ❻ at the Side.**

NOTE

A good way to decide whether two linear equations are equivalent is to write them both in slope-intercept form (solved for y). The resulting equations should be the same.

Summary of Situations That May Occur

One of three situations may occur when the elimination method is used to solve a linear system of equations.

1. The result of the addition step is a statement such as $x = 2$ or $y = -3$. The solution will be exactly one ordered pair. The graphs of the equations of the system will intersect at exactly one point. The system is *consistent*, and the equations are *independent*. See Examples 1–4.

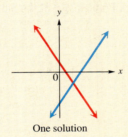

One solution

2. The result of the addition step is a false statement, such as $0 = 4$. In this case, the graphs are parallel lines, and there is no solution for the system. The system is *inconsistent*. See Example 5(a).

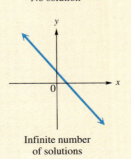

No solution

3. The result of the addition step is a true statement, such as $0 = 0$. The graphs of the equations of the system are the same line, and an infinite number of ordered pairs are solutions. These ordered pairs must satisfy the equation of the line. The equations are *dependent*. See Example 5(b).

Infinite number
of solutions

❻ Solve each system by the elimination method.

(a) $4x + 3y = 10$

$2x + \dfrac{3}{2}y = 12$

(b) $4x - 6y = 10$

$-10x + 15y = -25$

7 Use the guidelines to decide whether to use substitution or elimination to solve each system. (Do not actually solve the system.)

(a) $6x - y = 5$
$y = 11x$

When no method of solution of a system is specified and a choice of substitution or elimination is allowed, use the following guidelines.

1. If one of the equations of the system is already solved for one of the variables, such as

$$3x + 4y = 9 \qquad -5x + 3y = 9$$
$$\text{or}$$
$$y = 2x - 6 \qquad x = 3y - 7,$$

the substitution method is the better choice.

2. If both equations are in standard $Ax + By = C$ form, such as

$$4x - 11y = 3$$
$$-2x + 3y = 4,$$

and none of the variables has coefficient -1 or 1, the elimination method is the better choice.

3. If one or both of the equations are in standard form and the coefficient of one of the variables is -1 or 1, such as

$$3x + y = -2 \qquad -x + 3y = -4$$
$$\text{or}$$
$$-5x + 2y = 4 \qquad 3x - 2y = 8,$$

choose the elimination method, or solve for the variable with coefficient -1 or 1 and then use the substitution method.

(b) $3x - 5y = 7$
$2x + 3y = 30$

Work Problem 7 at the Side.

(c) $2x + 3y = 10$
$3x + y = 18$

7.3   **EXERCISES**

| FOR EXTRA HELP | | Student's Solutions Manual | | MyMathLab.com | | InterAct Math Tutorial Software | | AW Math Tutor Center | | www.mathxl.com | | Digital Video Tutor CD 5 Videotape 15 |

In Exercises 1–4, answer true *or* false *for each statement. If* false, *tell why.*

1. The ordered pair (0, 0) *must* be a solution of a system of the form
$$Ax + By = 0$$
$$Cx + Dy = 0.$$

2. To eliminate the *y*-terms in the system
$$2x + 12y = 7$$
$$3x + 4y = 1,$$
we should multiply the bottom equation by 3 and then add.

3. The system
$$x + y = 1$$
$$x + y = 2$$
has no solution.

4. Which one of the following systems would be easier to solve using the substitution method? Why?
$$5x - 3y = 7 \qquad 7x + 2y = 4$$
$$2x + 8y = 3 \qquad y = -3x$$

Solve each system by the elimination method. Check each solution. See Examples 1 and 2.

5. $x + y = 2$
  $2x - y = -5$

6. $3x - y = -12$
  $x + y = 4$

7. $2x + y = -5$
  $x - y = 2$

8. $2x + y = -15$
  $-x - y = 10$

9. $3x + 2y = 0$
  $-3x - y = 3$

10. $5x - y = 5$
  $-5x + 2y = 0$

11. $6x - y = -1$
  $-8x + 6y = 17 - 2x + y$

12. $3x + 2y = 9 - 3x + y$
  $-6x + 3y = 15$

Solve each system by the elimination method. Check each solution. See Example 3.

13. $2x - y = 12$
$3x + 2y = -3$

14. $x + y = 3$
$-3x + 2y = -19$

15. $x + 3y = 19$
$2x - y = 10$

16. $4x - 3y = -19$
$2x + y = 13$

17. $x + 4y = 16$
$3x + 5y = 20$

18. $2x + y = 8$
$5x - 2y = -16$

19. $5x - 3y = -20$
$-3x + 6y = 12$

20. $4x + 3y = -28$
$5x - 6y = -35$

21. $2x - 8y = 0$
$4x + 5y = 0$

22. $3x - 15y = 0$
$6x + 10y = 0$

Solve each system by the elimination method. Check each solution. See Example 4.

23. $3x - 7 = -5y$
$5x + 4y = -10$

24. $2x + 3y = 13$
$6 + 2y = -5x$

25. $2x + 3y = 0$
$4x + 12 = 9y$

26. $-4x + 3y = 2$
        $5x + 3 = -2y$

27. $24x + 12y = -7$
        $16x - 17 = 18y$

28. $9x + 4y = -3$
        $6x + 7 = -6y$

29. $3x = 3 + 2y$
        $-\dfrac{4}{3}x + y = \dfrac{1}{3}$

30. $3x = 27 + 2y$
        $x - \dfrac{7}{2}y = -25$

Use the elimination method to solve each system. See Example 5.

31. $x + y = 7$
        $x + y = -3$

32. $x - y = 4$
        $x - y = -3$

33. $-x + 3y = 4$
        $-2x + 6y = 8$

34. $6x - 2y = 24$
        $-3x + y = -12$

35. $5x - 2y = 3$
        $10x - 4y = 5$

36. $3x - 5y = 1$
        $6x - 10y = 4$

37. $6x + 3y = 0$
        $-18x - 9y = 0$

38. $3x - 5y = 0$
        $9x - 15y = 0$

RELATING CONCEPTS (Exercises 39–44) **FOR INDIVIDUAL OR GROUP WORK**

Attending the movies is one of America's favorite forms of entertainment. The graph shows how attendance gradually increased from 1991 to 1996. In 1991, attendance was 1141 million, as represented by the point P(1991, 1141). In 1996, attendance was 1339 million, as represented by the point Q(1996, 1339). We can find an equation of line segment PQ using a system of equations, and then we can use the equation to approximate the attendance in any of the years between 1991 and 1996. **Work Exercises 39–44 in order.**

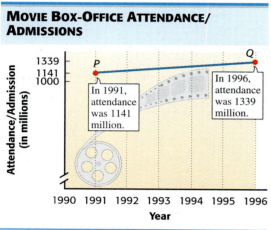

MOVIE BOX-OFFICE ATTENDANCE/ ADMISSIONS

Source: Motion Picture Association of America.

39. The line segment has an equation that can be written in the form $y = ax + b$. Using the coordinates of point P with $x = 1991$ and $y = 1141$, write an equation in the variables a and b.

40. Using the coordinates of point Q with $x = 1996$ and $y = 1339$, write a second equation in the variables a and b.

41. Write the system of equations formed from the two equations in Exercises 39 and 40, and solve the system using the elimination method.

42. What is the equation of the segment PQ?

43. Let $x = 1993$ in the equation of Exercise 42, and solve for y. How does the result compare with the actual figure of 1244 million?

44. The data points for the years 1991 through 1996 do not lie in a perfectly straight line. Explain the pitfalls of relying too heavily on using the equation in Exercise 42 to predict attendance.

7.4   APPLICATIONS OF LINEAR SYSTEMS

Recall from Chapter 2 the six-step method for solving applied problems. We modify those steps slightly to allow for two variables and two equations.

Solving Applied Problems with Two Variables

Step 1   **Read** the problem carefully until you understand what is given and what is to be found.

Step 2   **Assign variables** to represent the unknown values, using diagrams or tables as needed. Write down what each variable represents.

Step 3   **Write two equations** using both variables.

Step 4   **Solve** the system of two equations.

Step 5   **State the answer** to the problem. Is the answer reasonable?

Step 6   **Check** the answer in the words of the original problem.

❶ Solve the system

$$x = 179 + y$$
$$x + y = 26{,}601.$$

1 **Solve problems about unknown numbers.**   Use the modified six-step method to solve problems about unknown numbers.

Example 1   Solving a Problem about Two Unknown Numbers

In 1999, sales of sports clothing were $179 million more than sales of sports footwear. Together, the total sales for these items amounted to $26,601 million. (*Source:* National Sporting Goods Association.) What were the sales for each?

Step 1   **Read** the problem carefully. We must find the 1999 sales (in millions of dollars) for sports clothing and sports footwear. We know how much more sports clothing sales were than sports footwear sales. Also, we know the total sales.

Step 2   **Assign variables.** Let x represent the sales of sports clothing in millions of dollars and y represent the sales of sports footwear in millions of dollars.

Step 3   **Write two equations.**

$x = 179 + y$ Sales of sports clothing were $179 million more than sales of sports footwear.

$x + y = 26{,}601$ Total sales were $26,601 million.

Step 4   **Solve** the system from Step 3. The substitution method works well here.

Work Problem ❶ at the Side.

Step 5   **State the answer.** Clothing sales were $13,390 million and footwear sales were $13,211 million.

Step 6   **Check** the answer in the original problem.

Since

$$13{,}390 - 13{,}211 = 179 \quad \text{and} \quad 13{,}390 + 13{,}211 = 26{,}601,$$

the answer satisfies the information in the problem.

❷ Set up a system of equations for the following problem. Do not solve the system.

The two top-selling Disney videos of 1996 were *Toy Story* and *Pocahontas*. Together they sold 38 million copies. *Pocahontas* sold 4 million fewer copies than *Toy Story*. (*Source:* Paul Kagan Associates, Inc.) How many copies of each title were sold?

Let x = the number of copies of *Toy Story* sold (in millions)

and y = the number of copies of _____ sold.

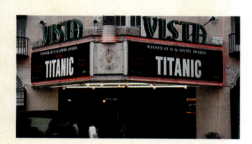

CAUTION

If an applied problem asks for *two* values as in Example 1, be sure to give both of them in your answer.

Work Problem ❷ at the Side.

2 ▭ **Solve problems about quantities and their costs.** We can also use a linear system to solve a common type of applied problem involving two quantities and their costs.

Example 2 Solving a Problem about Quantities and Costs

The 1997 box-office smash *Titanic* earned more in Europe than in the United States. This may be because average movie prices in Europe exceed those of the United States. (*Source: Parade* magazine, September 13, 1998.) For example, while the average movie ticket (to the nearest dollar) in 1997–1998 cost $5 in the United States, it cost an equivalent of $11 in London. Suppose that a group of 41 Americans and Londoners who paid these average prices spent a total of $307 for tickets. How many from each country were in the group?

Step 1 **Read** the problem again.

Step 2 **Assign variables.**

Let x = the number of Americans in the group;

 y = the number of Londoners in the group.

Summarize the information given in the problem in a table. The entries in the first two rows of the Total Value column were found by multiplying the number of tickets sold by the price per ticket.

	Number of Tickets	Price per Ticket (in dollars)	Total Value
American	x	5	$5x$
Londoner	y	11	$11y$
Total	41		307

Step 3 **Write two equations.** The total number of tickets was 41, so one equation is

$$x + y = 41.$$

Since the total value was $307, the final column leads to

$$5x + 11y = 307.$$

These two equations form a system.

$$x + \quad y = 41 \qquad (1)$$
$$5x + 11y = 307 \qquad (2)$$

Step 4 **Solve** the system of equations using the elimination method. First, eliminate the *x*-terms. Multiply each side of equation (1) by −5 to get

$$-5x - 5y = -205.$$

Continued on Next Page

Then add this result to equation (2).

$$-5x - 5y = -205$$
$$\underline{5x + 11y = 307}$$
$$6y = 102$$
$$y = 17$$

Substitute 17 for y in equation (1) to get

$$x + y = 41$$
$$x + 17 = 41$$
$$x = 24.$$

Step 5   **State the answer.** There were 24 Americans and 17 Londoners in the group.

Step 6   **Check.** The sum of 24 and 17 is 41, so the number of movie-goers is correct. Since 24 Americans paid $5 each and 17 Londoners paid $11 each, the total of the admission prices is $5(24) + $11(17) = $307, which agrees with the total amount stated in the problem.

========== Work Problem ❸ at the Side.

3 ▭ **Solve problems about mixtures.**   Earlier we solved percent problems using one variable. Many problems about mixtures that involve percent can be solved using a system of two equations in two variables.

Example 3   Solving a Mixture Problem Involving Percent

A pharmacist needs 100 L of 50% alcohol solution. She has on hand 30% alcohol solution and 80% alcohol solution, which she can mix. How many liters of each will be required to make the 100 L of 50% alcohol solution?

Step 1   **Read** the problem. Note the percent of each solution and of the mixture.

Step 2   **Assign variables.**

Let x = the number of liters of 30% alcohol needed;

y = the number of liters of 80% alcohol needed.

Summarize the information in a table. Percents are represented in decimal form.

Percent	Liters of Mixture	Liters of Pure Alcohol
.30	x	.30x
.80	y	.80y
.50	100	.50(100)

========== Continued on Next Page

---

❸ The average movie ticket (to the nearest U.S. dollar) costs $10 in Geneva and $8 in Paris. (*Source: Parade* magazine, September 13, 1998.) If a group of 36 people from these two cities paid $298 for tickets to see *As Good as It Gets*, how many people from each city were there?

(a) Complete the table.

	Number of Tickets Sold	Price (in dollars)	Total Value
Genevan	x		
Parisian	y		
Total			

(b) Write a system of equations.

(c) Solve the system and check your answer in the words of the original problem.

❹ Solve the system

$$x + y = 100$$
$$.30x + .80y = 50.$$

❺ How many liters of 25% alcohol solution must be mixed with 12% solution to get 13 L of 15% solution?

(a) Complete the table.

Percent	Liters	Liters of Pure Alcohol
.25	x	.25x
.12	y	
.15	13	

(b) Write a system of equations, and solve it.

❻ Joe needs 100 cc (cubic centimeters) of 20% acid solution for a chemistry experiment. The lab has on hand only 10% and 25% solutions. How much of each should he mix to get the desired amount of 20% solution?

Figure 7 gives an idea of what is actually happening in this problem.

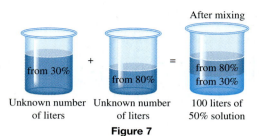

Figure 7

Step 3 **Write two equations.** Since the total number of liters in the final mixture will be 100, the first equation is

$$x + y = 100.$$

To find the amount of pure alcohol in each mixture, multiply the number of liters by the concentration. The amount of pure alcohol in the 30% solution added to the amount of pure alcohol in the 80% solution will equal the amount of pure alcohol in the final 50% solution. This gives the second equation,

$$.30x + .80y = .50(100).$$

These two equations form the system

$$x + y = 100$$
$$.30x + .80y = 50. \qquad \text{.50(100)} = 50$$

Step 4 **Solve** this system.

Work Problem ❹ at the Side.

Step 5 **State the answer.** From Problem 4 at the side, the pharmacist should use 60 L of the 30% solution and 40 L of the 80% solution.

Step 6 Since $60 + 40 = 100$ and $.30(60) + .80(40) = 50$, this mixture will give the 100 L of 50% solution, as required in the original problem.

Work Problems ❺ and ❻ at the Side.

4 ▮▮▮ **Solve problems about distance, rate (or speed), and time.** Problems that use the distance formula $d = rt$ were first introduced in Section 6.7. In many cases, these problems can be solved with systems of two linear equations. Keep in mind that setting up a table and drawing a sketch will help you solve such problems.

Example 4 Solving a Problem about Distance, Rate, and Time

Two executives in cities 400 mi apart drive to a business meeting at a location on the line between their cities. They meet after 4 hr. Find the speed of each car if one car travels 20 mph faster than the other.

Step 1 **Read** the problem carefully.

Step 2 **Assign variables.** Let $x =$ the speed of the faster car; $y =$ the speed of the slower car.

We use the formula $d = rt$. Since each car travels for 4 hr, the time, t, for each car is 4. This information is shown in the table. The distance is found by using the formula $d = rt$ and the expressions already entered in the table.

ANSWERS

4. (60, 40)
5. (a)

Percent	Liters	Liters of Pure Alcohol
.25	x	.25x
.12	y	.12y
.15	13	.15(13)

(b) $x + y = 13$
$.25x + .12y = .15(13)$
3 L of 25%, 10 L of 12%

6. $33\frac{1}{3}$ cc of 10%, $66\frac{2}{3}$ cc of 25%

	r	t	d
Faster Car	x	4	$4x$
Slower Car	y	4	$4y$

Find d from $d = rt$.

Draw a sketch showing what is happening in the problem. See Figure 8.

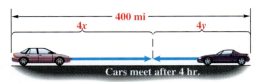

400 mi

$4x$ $4y$

Cars meet after 4 hr.

Figure 8

Step 3 **Write two equations.** As shown in the figure, since the total distance traveled by both cars is 400 mi, one equation is

$$4x + 4y = 400.$$

Because the faster car goes 20 mph faster than the slower car, the second equation is

$$x = 20 + y.$$

Step 4 **Solve.** This system of equations,

$$4x + 4y = 400 \qquad (1)$$
$$x = 20 + y, \qquad (2)$$

can be solved by substitution. Replace x with $20 + y$ in equation (1) and solve for y.

$4(\mathbf{20 + y}) + 4y = 400$	Let $x = 20 + y$.
$80 + 4y + 4y = 400$	Distributive property
$80 + 8y = 400$	Combine like terms.
$8y = 320$	Subtract 80.
$y = 40$	Divide by 8.

Since $x = 20 + y$, and $y = \mathbf{40}$,

$$x = 20 + \mathbf{40} = 60.$$

Step 5 **State the answer.** The speeds of the two cars are 40 mph and 60 mph.

Step 6 **Check** the answer. Since each car travels for 4 hr, the total distance traveled is

$$4(60) + 4(40) = 240 + 160 = 400 \text{ mi,}$$

as required.

=========== **Work Problem ❼ at the Side.**

The problems in Examples 1–4 also could be solved using only one variable. Many students find that the solution is simpler with two variables.

CAUTION

Be careful! When you use two variables to solve a problem, you must write two equations.

---

❼ Solve the problem.

(a) Two cars that were 450 mi apart traveled toward each other. They met after 5 hr. If one car traveled twice as fast as the other, what were their speeds? Complete this table.

	r	t	d
Faster Car	x	5	
Slower Car	y	5	

Write a system, and solve it.

(b) In 1 hr, Ann can row 2 mi against the current or 10 mi with the current. Find the speed of the current and Ann's speed in still water. (*Hint:* Let x = the speed of the current and y = Ann's speed in still water. Then her rate against the current is $y - x$, and her rate with the current is $y + x$.)

Real-Data Applications

Sales of Compact Discs versus Cassettes, Part 1

Since 1990, sales of compact discs have increased while sales of cassettes have declined. The data is given in the table. Even though the *input* is the year, we will rescale x to represent the number of years since 1990. The advantage of rescaling x is that the numbers in the linear models of the sales performance of CDs and cassettes will be smaller. Also, note that the slope represents the increase or decline in sales per year.

Year	x	CD Sales (in millions)	Cassette Sales (in millions)
1990	0	286.5	442.2
1991	1	333.4	360.1
1992	2	407.5	366.4
1993	3	495.4	339.5
1994	4	662.1	345.4
1995	5	722.9	272.6
1996	6	778.9	225.3
1997	7	753.1	172.6
1998	8	847.0	158.5

Source: World Almanac and Book of Facts, 2000.

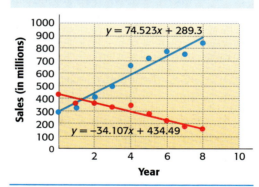

SALES OF CDs VERSUS CASSETTES

$y = 74.523x + 289.3$

$y = -34.107x + 434.49$

The graph depicts the actual data with the linear models for sales of CDs and cassettes superimposed.
- The linear model for the sales of CDs is $y = 74.523x + 289.3$.
- The linear model for the sales of cassettes is $y = -34.107x + 434.49$.

For Group Discussion

1. Using the linear models, when did sales of CDs overtake sales of cassettes? Give the month and year.

2. It appears from the model that sales of cassettes will eventually be 0. Do you believe that will really occur? Explain your answer.

3. It appears from the model that sales of CDs will continue to increase *linearly* well beyond one billion. Do you believe that will really occur? Explain your answer.

7.4   EXERCISES

Choose the correct response in Exercises 1–5.

1. Which expression represents the monetary value of *x* 20-dollar bills?

   A. $\dfrac{x}{20}$ dollars   **B.** $\dfrac{20}{x}$ dollars   **C.** $20 + x$ dollars   **D.** $20x$ dollars

2. Which expression represents the cost of *t* pounds of candy that sells for $1.95 per lb?

   A. $\$1.95t$ **B.** $\dfrac{\$1.95}{t}$ **C.** $\dfrac{t}{\$1.95}$ **D.** $\$1.95 + t$

3. Suppose that *x* liters of a 40% acid solution are mixed with *y* liters of a 35% solution to obtain 100 L of a 38% solution. One equation in a system for solving this problem is $x + y = 100$. Which one of the following is the other equation?

   A. $.35x + .40y = .38(100)$ **B.** $.40x + .35y = .38(100)$

   C. $35x + 40y = 38$ **D.** $40x + 35y = .38(100)$

4. What is the speed of a plane that travels at a rate of 560 mph *into* a wind of *r* mph?

   A. $560 + r$ mph   **B.** $\dfrac{560}{r}$ mph   **C.** $560 - r$ mph   **D.** $r - 560$ mph

5. What is the speed of a plane that travels at a rate of 560 mph *with* a wind of *r* mph?

   A. $\dfrac{r}{560}$ mph   **B.** $560 - r$ mph   **C.** $560 + r$ mph   **D.** $r - 560$ mph

6. Using the list of steps for solving an applied problem with two variables, write a short paragraph describing the general procedure you will use to solve the problems that follow in this exercise set.

Write a system of equations for each problem, and then solve the problem. Use the modified six-step method. See Example 1.

7. During 1995, two of the top-grossing concert tours were by Boyz II Men and Bruce Springsteen & the E Street Band. Together the two tours visited 174 cities. Boyz II Men visited 94 cities more than Bruce Springsteen. How many cities did each group visit? (*Source:* Pollstar.)

8. In 1997, the two top formats of U.S. commercial radio stations were country and adult contemporary. There were 981 fewer adult contemporary stations than country stations, and together they comprised a total of 4023 stations. How many stations of each format were there? (*Source:* M Street Corporation.)

9. In 1996, a total of 2683 thousand people lived in the metropolitan areas of Las Vegas, Nevada and Sacramento, California. Sacramento had 281 thousand more residents than Las Vegas. What was the population of each metropolitan area? (*Source: The World Almanac and Book of Facts*, 2000.)

10. The Terminal Tower in Cleveland, Ohio, is 242 ft shorter than the Key Tower, also in Cleveland. The total of the heights of the two buildings is 1658 ft. Find the heights of the buildings. (*Source: The World Almanac and Book of Facts*, 2000.)

242 ft

Terminal Tower   Key Tower

If x units of a product cost C dollars to manufacture and earn revenue of R dollars, the value of x where the expressions for C and R are equal is called the break-even quantity, *the number of units that produce 0 profit. In Exercises 11 and 12,* **(a)** *find the break-even quantity, and* **(b)** *decide whether the product should be produced based on whether it will earn a profit. (Profit equals revenue minus cost.)*

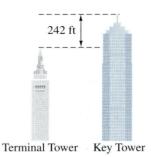

11. $C = 85x + 900$; $R = 105x$; no more than 38 units can be sold.

12. $C = 105x + 6000$; $R = 255x$; no more than 400 units can be sold.

Write a system of equations for each problem, and then solve the system. See Example 2.

13. A motel clerk counts his $1 and $10 bills at the end of a day. He finds that he has a total of 74 bills having a combined monetary value of $326. Find the number of bills of each denomination that he has.

14. Letarsha is a bank teller. At the end of a day, she has a total of 69 $5 and $10 bills. The total value of the money is $590. How many of each denomination does she have?

Denomination of Bill	Number of Bills	Total Value
$1	x	
$10	y	
Totals	74	$326

Denomination of Bill	Number of Bills	Total Value
$5	x	$5x$
$10	y	
Totals		

15. On November 1, 1998, K-Mart advertised videocassettes and CDs in its Sunday newspaper insert. Christine May went shopping and bought each of her seven nephews a gift, either a copy of the movie *The Grinch* or the latest N'Sync CD. The movie cost $14.95 and the CD cost $16.88, and she spent a total of $114.30. How many movies and how many CDs did she buy? (*Source: Times Picayune.*)

16. Terry Wong saw the K-Mart ad (see Exercise 15) and he, too, went shopping. He bought each of his five nieces a gift, either a copy of *Rudolph: The Movie* or the CD soundtrack to *Touched by an Angel*. The movie cost $14.99 and the soundtrack cost $13.88, and he spent a total of $70.51. How many movies and soundtracks did he buy? (*Source: Times Picayune.*)

17. Maria Lopez has twice as much money invested at 5% simple annual interest as she does at 4%. If her yearly income from these two investments is $350, how much does she have invested at each rate?

18. Charles Miller invested his textbook royalty income in two accounts, one paying 3% annual simple interest and the other paying 2% interest. He earned a total of $11 interest. If he invested three times as much in the 3% account as he did in the 2% account, how much did he invest at each rate?

19. Average movie ticket prices in the United States are, in general, lower than in other countries. It would cost $77.87 to buy three tickets in Japan plus two tickets in Switzerland. Three tickets in Switzerland plus two tickets in Japan would cost $73.83. How much does an average movie ticket cost in each of these countries? (*Source:* Business Traveler International.)

20. (See Exercise 19.) Four movie tickets in Germany plus three movie tickets in France would cost $62.27. Three tickets in Germany plus four tickets in France would cost $62.19. How much does an average movie ticket cost in each of these countries? (*Source:* Business Traveler International.)

Write a system of equations for each problem, and then solve the system. See Example 3.

21. A 40% dye solution is to be mixed with a 70% dye solution to get 120 L of a 50% solution. How many liters of the 40% and 70% solutions will be needed?

Percent (as a Decimal)	Liters of Solution	Liters of Pure Dye
.40	x	
.70	y	
.50	120	

22. A 90% antifreeze solution is to be mixed with a 75% solution to make 120 L of a 78% solution. How many liters of the 90% and 75% solutions will be used?

Percent (as a Decimal)	Liters of Solution	Liters of Pure Antifreeze
.90	x	
.75	y	
.78	120	

23. A merchant wishes to mix coffee worth $6 per lb with coffee worth $3 per lb to get 90 lb of a mixture worth $4 per lb. How many pounds of the $6 and the $3 coffees will be needed?

Dollars per Pound	Pounds	Cost
6	x	
	y	
	90	

24. A grocer wishes to blend candy selling for $1.20 per lb with candy selling for $1.80 per lb to get a mixture that will be sold for $1.40 per lb. How many pounds of the $1.20 and the $1.80 candies should be used to get 45 lb of the mixture?

Dollars per Pound	Pounds	Cost
	x	
1.80	y	
	45	

25. How many barrels of pickles worth $40 per barrel and pickles worth $60 per barrel must be mixed to obtain 50 barrels of a mixture worth $48 per barrel?

26. The owner of a nursery wants to mix some fertilizer worth $70 per bag with some worth $90 per bag to obtain 40 bags of mixture worth $77.50 per bag. How many bags of each type should she use?

Write a system of equations for each problem, and then solve the system. See Example 4.

27. A boat takes 3 hr to go 24 mi upstream. It can go 36 mi downstream in the same time. Find the speed of the current and the speed of the boat in still water if x = the speed of the boat in still water and y = the speed of the current.

	d	r	t
Downstream	36	$x + y$	
Upstream	24	$x - y$	

28. It takes a boat $1\frac{1}{2}$ hr to go 12 mi downstream, and 6 hr to return. Find the speed of the boat in still water and the speed of the current. Let x = the speed of the boat in still water and y = the speed of the current.

	d	r	t
Downstream	12	$x + y$	$\frac{3}{2}$
Upstream			6

Downstream Upstream

29. If a plane can travel 440 mph into the wind and 500 mph with the wind, find the speed of the wind and the speed of the plane in still air.

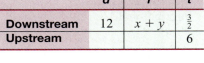

440 mph into wind

500 mph with wind

30. A small plane travels 200 mph with the wind and 120 mph against it. Find the speed of the wind and the speed of the plane in still air.

31. Toledo and Cincinnati are 200 mi apart. A car leaves Toledo traveling toward Cincinnati, and another car leaves Cincinnati at the same time, traveling toward Toledo. The car leaving Toledo averages 15 mph faster than the other, and they meet after 1 hr and 36 min. What are the rates of the cars?

Michigan
Lake Erie
Toledo
Pennsylvania
Ohio
200 mi
Indiana
Cincinnati
West Virginia
Kentucky

32. Kansas City and Denver are 600 mi apart. Two cars start from these cities, traveling toward each other. They meet after 6 hr. Find the rate of each car if one travels 30 mph slower than the other.

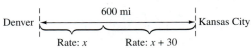

600 mi
Denver ⊢——————————⊣ Kansas City
Rate: x Rate: $x + 30$

33. At the beginning of a bicycle ride for charity, Roberto and Juana are 30 mi apart. If they leave at the same time and ride in the same direction, Roberto overtakes Juana in 6 hr. If they ride toward each other, they meet in 1 hr. What are their speeds?

34. Mr. Abbot left Farmersville in a plane at noon to travel to Exeter. Mr. Baker left Exeter in his automobile at 2 P.M. to travel to Farmersville. It is 400 mi from Exeter to Farmersville. If the sum of their speeds was 120 mph, and if they crossed paths at 4 P.M., find the speed of each.

7.5 SOLVING SYSTEMS OF LINEAR INEQUALITIES

OBJECTIVE

1 Solve systems of linear inequalities by graphing.

We graphed the solutions of a linear inequality in Section 3.5. Recall that to graph the solutions of $x + 3y > 12$, for example, we first graph $x + 3y = 12$ by finding and plotting a few ordered pairs that satisfy the equation. Because the points on the line do *not* satisfy the inequality, we use a dashed line. To decide which side of the line includes the points that are solutions, we choose a test point not on the line, such as $(0, 0)$. Substituting these values for x and y in the inequality gives

$$x + 3y > 12$$
$$0 + 3(0) > 12$$
$$0 > 12,$$

a false result. This indicates that the solutions are those points on the side of the line that does not include $(0, 0)$, as shown in Figure 9.

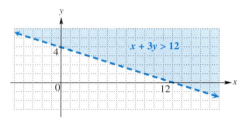

Figure 9

Now we use the same techniques to solve systems of linear inequalities.

1 Solve systems of linear inequalities by graphing. A **system of linear inequalities** consists of two or more linear inequalities. The **solution of a system of linear inequalities** includes all points that make all inequalities of the system true at the same time. To solve a system of linear inequalities, use the following steps.

Solving a System of Linear Inequalities

Step 1 **Graph the inequalities.** Graph each inequality using the method of Section 3.5.

Step 2 **Choose the intersection.** Indicate the solution of the system by shading the intersection of the graphs (the region where the graphs overlap).

Example 1 Solving a System of Two Linear Inequalities

Graph the solution of the system

$$3x + 2y \leq 6$$
$$2x - 5y \geq 10.$$

Begin by graphing each inequality on the same axes. To graph $3x + 2y \leq 6$, graph the solid boundary line $3x + 2y = 6$ and shade the region containing $(0, 0)$, as shown in Figure 10(a). Then graph $2x - 5y \geq 10$ with the solid boundary line $2x - 5y = 10$. The test point $(0, 0)$ makes this inequality false, so shade the region on the other side of the boundary line. See Figure 10(b).

Continued on Next Page

❶ Graph the solution of the system

$$x - 2y \le 8$$
$$3x + y \ge 6.$$

To get you started, the graphs of $x - 2y = 8$ and $3x + y = 6$ are shown.

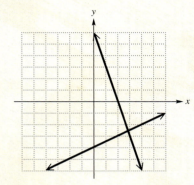

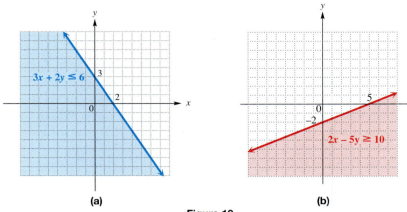

(a)   (b)

Figure 10

The solution of this system includes all points in the intersection (overlap) of the graphs of the two inequalities. It includes the shaded region and portions of the two boundary lines shown in Figure 11.

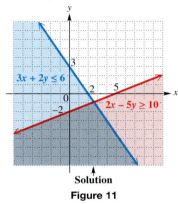

Solution

Figure 11

Work Problem ❶ at the Side.

NOTE

We usually do all the work on one set of axes. In the following examples, only one graph is shown. Be sure that the region of the final solution is clearly indicated.

Example 2 Solving a System of Linear Inequalities

Graph the solution of the system

$$x - y > 5$$
$$2x + y < 2$$

Figure 12 shows the graphs of both $x - y > 5$ and $2x + y < 2$. Dashed lines show that the graphs of the inequalities do not include their boundary lines. The solution of the system is the region with the darkest shading. The solution does not include either boundary line.

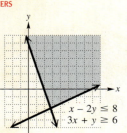

Continued on Next Page

Continued on Next Page

ANSWERS
1.

$$x - 2y \le 8$$
$$3x + y \ge 6$$

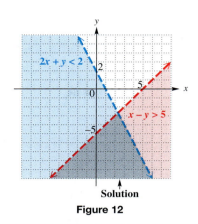

Figure 12

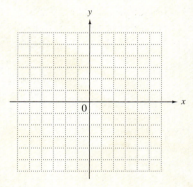

② Graph the solution of each system.

(a) $x + 2y < 0$
$3x - 4y < 12$

┌─ **Example 3**   **Solving a System of Three Linear Inequalities**

Graph the solution of the system

$$4x - 3y \leq 8$$
$$x \geq 2$$
$$y \leq 4.$$

Recall that $x = 2$ is a vertical line through the point $(2, 0)$, and $y = 4$ is the horizontal line through $(0, 4)$. The graph of the solution is the shaded region in Figure 13, including all boundary lines.

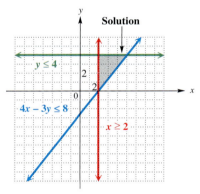

Figure 13

Work Problem ② at the Side.

(b) $3x + 2y \leq 12$
$x \leq 2$
$y \leq 4$

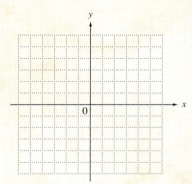

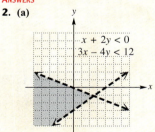
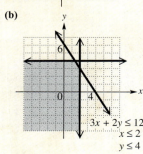

Real-Data Applications

Sales of Compact Discs versus Cassettes, Part 2

Sales since 1990 of compact discs and cassettes are given in the table. The number of years since 1990 is represented by x. Sales of compact discs and cassettes can be modeled by linear equations.

Year	x	CD Sales (in millions)	Cassette Sales (in millions)
1990	0	286.5	442.2
1991	1	333.4	360.1
1992	2	407.5	366.4
1993	3	495.4	339.5
1994	4	662.1	345.4
1995	5	722.9	272.6
1996	6	778.9	225.3
1997	7	753.1	172.6
1998	8	847.0	158.5

Source: World Almanac and Book of Facts, 2000.

- The linear model for the sales of compact discs is $y = 74.523x + 289.3$.
- The linear model for the sales of cassettes is $y = -34.107x + 434.49$.

Recall that y is sales in millions. The actual data and linear models are graphed below.

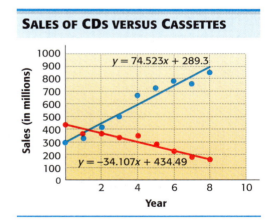

SALES OF CDS VERSUS CASSETTES

For Group Discussion

1. Shade the region on the graph that corresponds to the solution of the following system.

$$y \leq 74.523x + 289.3$$
$$y \geq -34.107x + 434.49$$

Interpret the solution in the context of sales of compact discs and cassettes.

2. Solve the linear inequality $74.523x + 289.3 > -34.107x + 434.49$. Round your answer to the nearest hundredth.

3. What does the solution to the inequality in Problem 2 represent in the context of sales of CDs and cassettes?

7.5 EXERCISES

 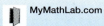
Match each system of inequalities with the correct graph from choices A–D.

1. $x \geq 5$
$y \leq -3$

2. $x \leq 5$
$y \geq -3$

3. $x > 5$
$y < -3$

4. $x < 5$
$y > -3$

A.

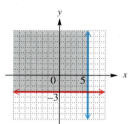

B.

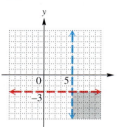

C.

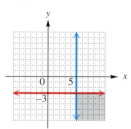

D.

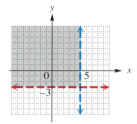

Graph the solution of each system of linear inequalities. See Examples 1–3.

5. $x + y \leq 6$
$x - y \geq 1$

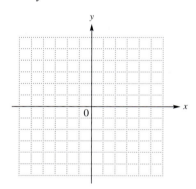

6. $x + y \leq 2$
$x - y \geq 3$

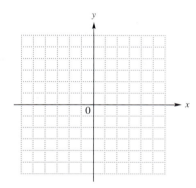

7. $4x + 5y \geq 20$
$x - 2y \leq 5$

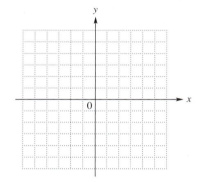

8. $x + 4y \leq 8$
$2x - y \geq 4$

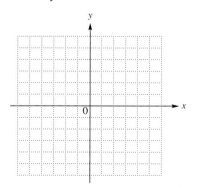

9. $2x + 3y < 6$
$x - y < 5$

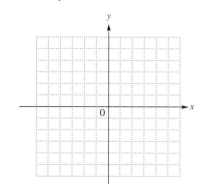

10. $x + 2y < 4$
$x - y < -1$

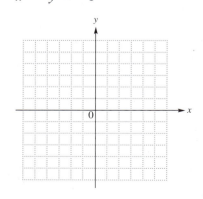

11. $y \leq 2x - 5$
$x < 3y + 2$

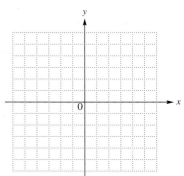

12. $x \geq 2y + 6$
$y > -2x + 4$

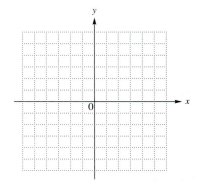

13. $4x + 3y < 6$
$x - 2y > 4$

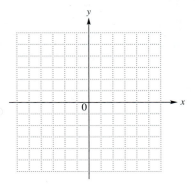

14. $3x + y > 4$
$x + 2y < 2$

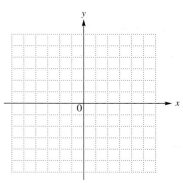

15. $x \leq 2y + 3$
$x + y < 0$

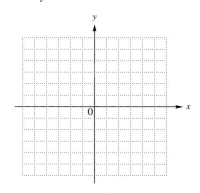

16. $x \leq 4y + 3$
$x + y > 0$

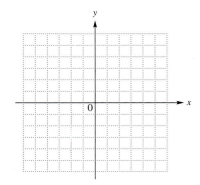

17. $4x + 5y < 8$
$y > -2$
$x > -4$

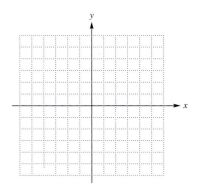

18. $x + y \geq -3$
$x - y \leq 3$
$y \leq 3$

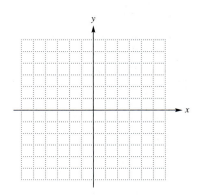

19. $3x - 2y \geq 6$
$x + y \leq 4$
$x \geq 0$
$y \geq -4$

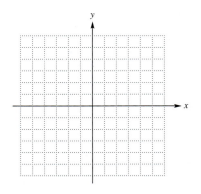

20. Every system of inequalities illustrated in the examples of this section has infinitely many solutions. Explain why this is so. Does this mean that *any* ordered pair is a solution?

SUMMARY

7.1	**system of linear equations**	A system of linear equations consists of two or more linear equations with the same variables.
	solution of a system	The solution of a system of linear equations includes all the ordered pairs that make all the equations of the system true at the same time.
	consistent system	A system of equations with at least one solution is a consistent system.
	inconsistent system	An inconsistent system of equations is a system with no solution.
	independent equations	Equations of a system that have different graphs are called independent equations.
	dependent equations	Equations of a system that have the same graph (because they are different forms of the same equation) are called dependent equations.
7.5	**system of linear inequalities**	A system of linear inequalities contains two or more linear inequalities (and no other kinds of inequalities).
	solution of a system of linear inequalities	The solution of a system of linear inequalities includes all points that make all inequalities of the system true at the same time.

TEST YOUR WORD POWER

See how well you have learned the vocabulary in this chapter. Answers follow the Quick Review.

1. A **system of linear equations** consists of
   (a) at least two linear equations with different variables
   (b) two or more linear equations that have an infinite number of solutions
   (c) two or more linear equations with the same variables
   (d) two or more linear inequalities.

2. A **solution of a system** of linear equations is
   (a) an ordered pair that makes one equation of the system true

   (b) an ordered pair that makes all the equations of the system true at the same time
   (c) any ordered pair that makes one or the other or both equations of the system true
   (d) the set of values that make all the equations of the system false.

3. A **consistent system** is a system of equations
   (a) with one solution
   (b) with no solution
   (c) with two solutions
   (d) that has parallel lines at its graph.

4. An **inconsistent system** is a system of equations
   (a) with one solution
   (b) with no solution
   (c) with an infinite number of solutions
   (d) that have the same graph.

5. **Dependent equations**
   (a) have different graphs
   (b) have no solution
   (c) have one solution
   (d) are different forms of the same equation.

Concepts	*Examples*

7.1 *Solving Systems of Linear Equations by Graphing*

An ordered pair is a solution of a system if it makes all equations of the system true at the same time.

Is $(4, -1)$ a solution of the following system?

$$x + y = 3$$
$$2x - y = 9$$

Yes, because $4 + (-1) = 3$ and $2(4) - (-1) = 9$ are both true.

If the graphs of the equations of a system are both sketched on the same axes, the points of intersection, if any, are solutions of the system.

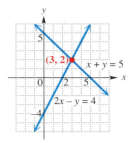

$(3, 2)$ is the solution of the system

$$x + y = 5$$
$$2x - y = 4.$$

7.2 *Solving Systems of Linear Equations by Substitution*

Solve one equation for one variable, and substitute the expression into the other equation to get an equation in one variable. Solve the equation, and then substitute the solution into either of the original equations to obtain the value of the other variable. Check the solution of the system.

Solve by substitution.

$$x + 2y = -5 \qquad (1)$$
$$y = \mathbf{-2x - 1} \qquad (2)$$

Substitute $-2x - 1$ for y in equation (1).

$$x + 2(\mathbf{-2x - 1}) = -5$$
$$x - 4x - 2 = -5$$
$$-3x - 2 = -5$$
$$-3x = -3$$
$$x = 1$$

To find y, let $x = 1$ in equation (2):

$$y = -2(\mathbf{1}) - 1 = -3.$$

The solution is $(1, -3)$.

7.3 *Solving Systems of Linear Equations by Elimination*

Step 1 Write both equations in standard form $Ax + By = C$.

Step 2 If necessary, multiply one or both equations by appropriate numbers so that the coefficients of x (or y) are opposites of each other.

Step 3 Add the equations to get an equation with only one variable (or no variable).

Step 4 Solve the equation from Step 3.

Solve by elimination.

$$x + 3y = 7 \qquad (1)$$
$$3x - y = 1 \qquad (2)$$

Multiply equation (1) by -3 to eliminate the x-terms.

$$\mathbf{-3x} - 9y = -21$$
$$\underline{\mathbf{3x} - y = 1}$$
$$-10y = -20 \qquad \text{Add.}$$

$$y = 2 \qquad \text{Divide by } -10.$$

Concepts	Examples

7.3 Solving Systems of Linear Equations by Elimination (*continued*)

Step 5 Substitute the solution from Step 4 into either of the original equations to find the value of the remaining variable.

$$x + 3(\mathbf{2}) = 7 \quad (1)$$
$$x + 6 = 7$$
$$x = 1$$

Step 6 Check the solution in both of the original equations. Write the solution as an ordered pair.

Since $1 + 3(2) = 7$ and $3(1) - 2 = 1$, the solution $(1, 2)$ checks.

7.4 Applications of Linear Systems

Use the modified six-step method.

The sum of two numbers is 30. Their difference is 6. Find the numbers.

Step 1 **Read** the problem carefully.

Step 2 **Assign variables** for each unknown value. Use diagrams or tables as needed.

Let x represent one number.
Let y represent the other number.

Step 3 **Write two equations** using both variables.

$$x + y = 30$$
$$\underline{x - y = \;\; 6}$$
$$2x \qquad = 36 \quad \text{Add.}$$

Step 4 **Solve** the system.

$$x = 18 \quad \text{Divide by 2.}$$

Step 5 **State the answer.**

Let $x = 18$ in the first equation: $18 + y = 30$. Solve to get $y = 12$. The numbers are 18 and 12.

Step 6 **Check** the answer in the words of the original problem.

The sum of 18 and 12 is 30, and the difference between 18 and 12 is 6, so the answer checks.

7.5 Solving Systems of Linear Inequalities

To solve a system of two or more linear inequalities, graph the inequalities on the same axes. (This was explained in Section 3.5.) The solution of the system is the overlap of the regions of the graphs. The portions of the boundary lines that bound the region of solutions are included for a $\leq$ or $\geq$ inequality and excluded for a $<$ or $>$ inequality.

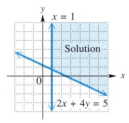

The shaded region is the solution of the system

$$2x + 4y \geq 5$$
$$x \geq 1.$$

ANSWERS TO TEST YOUR WORD POWER

1. **(c)** *Example:* $2x + y = 7$, $3x - y = 3$ **2.** **(b)** *Example:* The ordered pair $(2, 3)$ satisfies both equations of the system in the Item 1 example, so it is a solution of the system. **3.** **(a)** *Example:* The system in the Item 1 example is consistent. The graphs of the equations intersect at exactly one point, in this case the solution $(2, 3)$. **4.** **(b)** *Example:* The equations of two parallel lines make up an inconsistent system; their graphs never intersect, so there is no solution to the system. **5.** **(d)** *Example:* The equations $4x - y = 8$ and $8x - 2y = 16$ are dependent because their graphs are the same line.

Real-Data Applications

Systems of Linear Equations and Modeling

A system of linear equations is an efficient tool for finding a linear *model*, or the equation for data that is known to be linear. Recall that the slope-intercept form of the equation of a line is $y = mx + b$. Once we know the values of the slope, m, and the y-intercept, b, we can then write the exact model. If we know two ordered pairs, then we can write a system of linear equations in which the unknown quantities are m and b.

For example, to find the formula to convert from Kelvin (K), the most commonly used thermodynamic temperature scale, as the *input* to degrees Fahrenheit (°F) as the *output*, we only need to know the data presented in the table. The linear equation has the form $F = mK + b$, and ordered pairs have the format (K, F).

	K	°F
Water Freezes	273	32
Water Boils	373	212

The system of linear equations to be solved is

$$32 = m(273) + b$$
$$212 = m(373) + b.$$

For Group Discussion

Use systems of linear equations to find the model for each problem.

1. Solve the system of linear equations given above to find the conversion formula from K to °F.

2. Water freezes at 0°C and boils at 100°C. Write the system of linear equations to find the model for converting from degrees Celsius as the input to degrees Fahrenheit as the output.

3. Suppose you begin a carefully managed weight-loss program in which you expect your weight to decline steadily. (A constant weight loss means that it is reasonable to assume that the relationship between number of weeks on the program and weight is linear.) After two weeks you weigh 179 lb, and after six weeks you weigh 169 lb. Use x to represent the number of weeks on the program and y to represent your weight in pounds at the end of x weeks. Write the linear model for your weight-loss program.

Extension: Suppose a line contains the distinct points (x_1, y_1) and (x_2, y_2), where $x_1 \neq x_2$. Use the method of this activity to derive the slope formula.

Chapter 7

REVIEW EXERCISES

[7.1] *Decide whether the given ordered pair is a solution of the given system.*

1. $(3, 4)$
$4x - 2y = 4$
$5x + \ y = 19$

2. $(-5, 2)$
$x - 4y = -13$
$2x + 3y = 4$

Solve each system by graphing.

3. $x + y = 4$
$2x - y = 5$

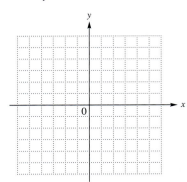

4. $x - 2y = 4$
$2x + \ y = -2$

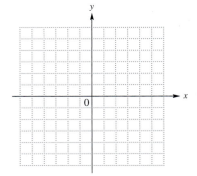

5. $x - 2 \ = 2y$
$2x - 4y = 4$

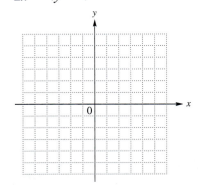

6. $2x + 4 = 2y$
$y - x = -3$

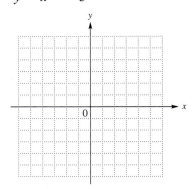

7. When a student was asked to determine whether the ordered pair $(1, -2)$ is a solution of the system

$$x + y = -1$$
$$2x + y = 4,$$

he answered "yes." His reasoning was that the ordered pair satisfies the equation $x + y = -1$; that is, $1 + (-2) = -1$ is true. Why is the student's answer wrong?

[7.2] *Solve each system by the substitution method.*

8. $3x + y = 7$
$\quad\ x = 2y$

9. $2x - 5y = -19$
$\quad\ \ y = x + 2$

10. $4x + 5y = 44$
$\quad\ \ x + 2 = 2y$

11. $5x + 15y = 3$
$\quad\ \ x + \ 3y = 2$

[7.3] *Solve each system by the elimination method.*

12. $2x - y = 13$
$\quad\ \ x + y = 8$

13. $3x - \ y = -13$
$\quad\ \ x - 2y = -1$

14. $-4x + 3y = 25$
$\quad\ \ 6x - 5y = -39$

15. $3x - 4y = 9$
$\quad\ \ 6x - 8y = 18$

16. For the system

$$2x + 12y = 7$$
$$3x + \ \ 4y = 1,$$

if we were to multiply the first (top) equation by -3, by what number would we have to multiply the second (bottom) equation in order to

(a) eliminate the x-terms when solving by the elimination method?

(b) eliminate the y-terms when solving by the elimination method?

Solve each system by any method. First simplify equations, and clear them of fractions where necessary.

17. $2x + y - x = 3y + 5$
$\quad\quad\ \ y + 2 = x - 5$

18. $5x - 3 + y = 4y + 8$
$\quad\quad\ \ 2y + 1 = x - 3$

19. $\dfrac{x}{2} + \dfrac{y}{3} = 7$

$\dfrac{x}{4} + \dfrac{2y}{3} = 8$

20. $\dfrac{3x}{4} - \dfrac{y}{3} = \dfrac{7}{6}$

$\dfrac{x}{2} + \dfrac{2y}{3} = \dfrac{5}{3}$

[7.4] *Solve each problem by using a system of equations. Use the modified six-step method.*

21. A popular leisure activity of Americans is reading. In 1996, two popular fiction titles were *How Stella Got Her Groove Back* by Terry McMillan and *The Deep End of the Ocean* by Jacquelyn Mitchard. Together, these two titles sold 1,622,962 copies. The Mitchard book sold 57,564 more copies than the McMillan book. How many copies of each title were sold? (*Source: Publishers Weekly.*)

22. When people are not reading fiction during their leisure time, they are often reading magazines. Two of the most popular magazines in the United States are *Modern Maturity* and *Reader's Digest.* Together, the average total circulation for these two magazines during July–December 1996 was 35.6 million. *Reader's Digest* circulation was 5.4 million less than that of *Modern Maturity.* What were the circulation figures for each magazine? (*Source:* Audit Bureau of Circulations and Magazine Publishers of America.)

23. The perimeter of a rectangle is 90 m. Its length is $1\frac{1}{2}$ times its width. Find the length and width of the rectangle.

24. A cashier has 20 bills, all of which are $10 or $20 bills. The total value of the money is $330. How many of each type does the cashier have?

Denomination of Bill	Number of Bills	Total Value
$10	x	$10x$
$20		
Totals		$330

25. Candy that sells for $1.30 per lb is to be mixed with candy selling for $.90 per lb to get 100 lb of a mix that will sell for $1 per lb. How much of each type should be used?

26. A certain plane flying with the wind travels 540 mi in 2 hr. Later, flying against the same wind, the plane travels 690 mi in 3 hr. Find the speed of the plane in still air and the speed of the wind.

27. After taxes, Ms. Cesar's game show winnings were $18,000. She invested part of it at 3% annual simple interest and the rest at 4%. Her interest income for the first year was $650. How much did she invest at each rate?

Percent	Amount of Principal	Interest
.03	x	
.04	y	
Totals	$18,000	

28. A 40% antifreeze solution is to be mixed with a 70% solution to get 90 L of a 50% solution. How many liters of the 40% and 70% solutions will be needed?

Percent	Number of Liters	Amount of Pure Antifreeze
.40	x	
.70	y	
.50	90	

[7.5] *Graph the solution for each system of linear inequalities.*

29. $x + y \geq 2$
$x - y \leq 4$

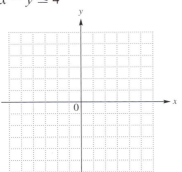

30. $y \geq 2x$
$2x + 3y \leq 6$

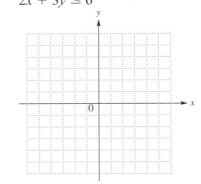

31. $x + y < 3$
$2x > y$

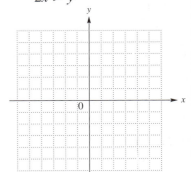

32. Which system of linear inequalities is graphed in the figure?

A. $x \le 3$
$\quad\; y \le 1$

B. $x \le 3$
$\quad\; y \ge 1$

C. $x \ge 3$
$\quad\; y \le 1$

D. $x \ge 3$
$\quad\; y \ge 1$

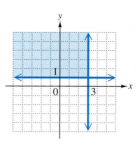

33. Without actually graphing, determine which system of inequalities has no solution.

A. $x \ge 4$
$\quad\; y \le 3$

B. $x + y > 4$
$\quad\; x + y < 3$

C. $x > 2$
$\quad\; y < 1$

D. $x + y > 4$
$\quad\; x - y < 3$

MIXED REVIEW EXERCISES

Solve each system.

34. $3x + 4y = 6$
$\quad\; 4x - 5y = 8$

35. $\dfrac{3x}{2} + \dfrac{y}{5} = -3$

$\quad\; 4x + \dfrac{y}{3} = -11$

36. $x + y < 5$
$\quad\; x - y \ge 2$

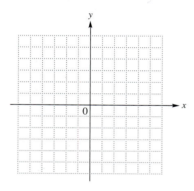

37. $y \le 2x$
$\quad\; x + 2y > 4$

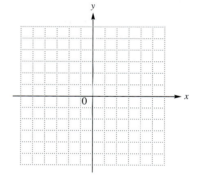

38. The perimeter of an isosceles triangle is 29 in. One side of the triangle is 5 in. longer than each of the two equal sides. Find the lengths of the sides of the triangle.

$y = x + 5$

39. In the 2000 National Football League, the Baltimore Ravens and the Philadelphia Eagles both advanced to the second round of the playoffs, winning by identical scores on their home fields. Each team beat its opponent by 18 points, and the winning score was seven times the losing score. What was the final score of each game? (*Source:* NFL.)

40. $x + 6y = 3$
$\quad\; 2x + 12y = 2$

41. $y < -4x$
$\quad\; y < -2$

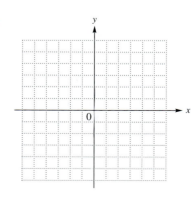

Chapter 7 TEST

 Study Skills Workbook
Activity 12

1. Solve the system by graphing.

$$2x + y = 1$$
$$3x - y = 9$$

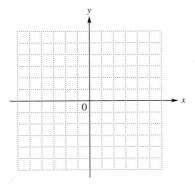

1. _____

2. _____

3. _____

2. Suppose that the graph of a system of two linear equations consists of lines that have the same slope but different *y*-intercepts. How many solutions does the system have?

4. _____

Solve each system by substitution.

3. $2x + y = -4$
　　$x = y + 7$

4. $4x + 3y = -35$
　　$x + y = 0$

5. _____

Solve each system by the elimination method.

6. _____

5. $2x - y = 4$
　　$3x + y = 21$

6. $4x + 2y = 2$
　　$5x + 4y = 7$

7. _____

7. $6x - 5y = 0$
　　$-2x + 3y = 0$

8. $4x + 5y = 2$
　　$-8x - 10y = 6$

8. _____

9. _____

Solve each system by any method.

9. $8 + 3x - 4y = 14 - 3y$
　　$3x + y + 12 = 9x - y$

10. $\dfrac{x}{2} - \dfrac{y}{4} = 7$

　　　$\dfrac{2x}{3} + \dfrac{5y}{4} = 3$

10. _____

Write a system of two equations for each problem, and then solve the system. Use the modified six-step method.

11. _____

11. The distance between Memphis and Atlanta is 300 mi less than the distance between Minneapolis and Houston. Together, the two distances total 1042 mi. How far is it between Memphis and Atlanta? How far is it between Minneapolis and Houston?

12. _____

12. In 1996, the two most popular amusement parks in the United States were Disneyland and the Magic Kingdom at Walt Disney World. Disneyland had 1.2 million more visitors than the Magic Kingdom, and together they had 28.8 million visitors. How many visitors did each park have? (*Source: The Wall Street Journal Almanac,* 1998.)

13. _____

13. A 25% solution of alcohol is to be mixed with a 40% solution to get 50 L of a final mixture that is 30% alcohol. How much of each of the original solutions should be used?

14. _____

14. Two cars leave from Perham, Minnesota, and travel in the same direction. One car travels $1\frac{1}{3}$ times as fast as the other. After 3 hr they are 45 mi apart. What are the speeds of the cars?

15.

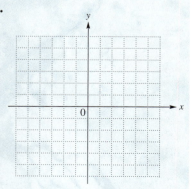

Graph the solution of each system of inequalities.

15. $2x + 7y \le 14$
$x - y \ge 1$

16. $2x - y > 6$
$4y + 12 \ge -3x$

16.

1. List all integer factors of 40.

2. Find the value of the expression if $x = 1$ and $y = 5$.

$$\frac{3x^2 + 2y^2}{10y + 3}$$

Name the property that justifies each statement.

3. $5 + (-4) = (-4) + 5$

4. $r(s - k) = rs - rk$

5. $-\dfrac{2}{3} + \dfrac{2}{3} = 0$

6. Evaluate $-2 + 6[3 - (4 - 9)]$.

Solve each linear equation.

7. $2 - 3(6x + 2) = 4(x + 1) + 18$

8. $\dfrac{3}{2}\left(\dfrac{1}{3}x + 4\right) = 6\left(\dfrac{1}{4} + x\right)$

Solve each linear inequality.

9. $-\dfrac{5}{6}x < 15$

10. $-8 < 2x + 3$

11. No baseball fan should be without a copy of *The Sports Encyclopedia: Baseball 2000* by David S. Neft and Richard M. Cohen. Now in its 20th edition, it provides exhaustive statistics for professional baseball dating back to 1876. This book has a perimeter of 38 in., and its width measures 2.5 in. less than its length. What are the dimensions of the book?

Graph each linear equation.

12. $x - y = 4$

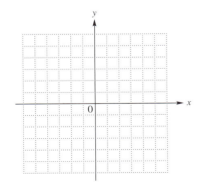

13. $3x + y = 6$

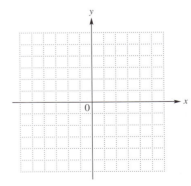

Find the slope of each line.

14. Through $(-5, 6)$ and $(1, -2)$

15. Perpendicular to the line $y = 4x - 3$

Write an equation for each line in slope-intercept form.

16. Through $(-4, 1)$ with slope $\frac{1}{2}$

17. Through the points $(1, 3)$ and $(-2, -3)$

18. (a) Write an equation of the vertical line through $(9, -2)$.

   (b) Write an equation of the horizontal line through $(4, -1)$.

19. The graph shows the growth of investment clubs from 1994 to 1997. If 1994 is represented by $x = 4$ and 1997 is represented by $x = 7$, what is the equation of the line joining the tops of the bars for these years? How is the slope interpreted in the context of these data?

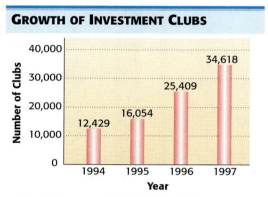

GROWTH OF INVESTMENT CLUBS

Source: National Association of Investment Clubs.

Perform the indicated operations.

20. $-3(-5x^2 + 3x - 10) - (x^2 - 4x + 7)$

21. $(3x - 7)(2y + 4)$

22. $\dfrac{3k^3 + 17k^2 - 27k + 7}{k + 7}$

23. The current population of the world is about 6,000,000,000. Some estimate world population will reach 10,000,000,000 in 2050, while others predict this population will be reached in 2183. (*Source: Scientific American*, October 2000.)

   (a) Write these population figures in scientific notation.

   (b) In 1968, the population was about 3.5×10^9. Write this number without exponents.

24. Simplify, and write the answer using only positive exponents: $\left(\dfrac{x^{-4}y^3}{x^2y^4} \right)^{-1}$.

Factor completely.

25. $10m^2 + 7mp - 12p^2$

26. $64t^2 - 48t + 9$

Solve each quadratic equation.

27. $6x^2 - 7x - 3 = 0$

28. $r^2 - 121 = 0$

Perform each operation, and express answers in lowest terms.

29. $\dfrac{-3x + 6}{2x + 4} - \dfrac{-3x - 8}{2x + 4}$

30. $\dfrac{16k^2 - 9}{8k + 6} \div \dfrac{16k^2 - 24k + 9}{6}$

31. Solve the equation $\dfrac{4}{x + 1} + \dfrac{3}{x - 2} = 4.$

32. Solve the formula $P = \dfrac{kT}{V}$ for T.

Solve each system by any method.

33. $2x - \ y = -8$
  $x + 2y = 11$

34. $4x + 5y = -8$
  $3x + 4y = -7$

35. $3x + 5y = 1$
  $x = y + 3$

36. $3x + 4y = 2$
  $6x + 8y = 1$

Use a system of equations to solve each problem.

37. Admission prices at a football game were $6 for adults and $2 for children. The total value of the tickets sold was $2528, and 454 tickets were sold. How many adults and how many children attended the game?

Kind of Ticket	Number Sold	Cost of Each (in dollars)	Total Value (in dollars)
Adult	x	6	$6x$
Child	y		
Total	454		

38. The perimeter of a triangle is 53 in. If two sides are of equal length, and the third side measures 4 in. less than each of the equal sides, what are the lengths of the three sides?

39. The Smith family is coming to visit, and no one knows how many children they have. Janet, one of the girls, says she has as many brothers as sisters; her brother Steve says he has twice as many sisters as brothers. How many boys and how many girls are in the family?

40. Graph the solution of the system

$$x + 2y \le 12$$
$$2x - y \le 8.$$

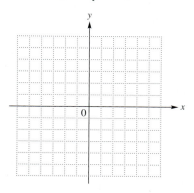

Roots and Radicals

8

The United States ranks among the world's leading trading countries. In December 2000 alone, U.S. exports of goods and services totaled $89.8 billion, while imports totaled $122.8 billion, a trade deficit of $33.0 billion. Our top trading partners include Canada, Mexico, Japan, China, and Germany. (*Source:* U.S. Bureau of the Census.) We will use radical equations to investigate U.S. exports and imports of electronics in Section 8.6 of this chapter.

ADDISON · WESLEY
MyMathLab.com
You're Connected

8.1 EVALUATING ROOTS

❶ Find all square roots.

(a) 100

(b) 25

(c) 36

(d) $\dfrac{25}{36}$

Early radical symbol

In Section 1.1, we discussed the idea of the *square* of a number. Recall that squaring a number means multiplying the number by itself.

$$\text{If } a = 8, \quad \text{then} \quad a^2 = \mathbf{8} \cdot \mathbf{8} = 64.$$
$$\text{If } a = -4, \quad \text{then} \quad a^2 = (-4)(-4) = 16.$$
$$\text{If } a = -\frac{1}{2}, \quad \text{then} \quad a^2 = \left(-\frac{1}{2}\right)\left(-\frac{1}{2}\right) = \frac{1}{4}.$$

In this chapter, the opposite process is considered.

$$\text{If } a^2 = 49, \quad \text{then} \quad a = \text{?}$$
$$\text{If } a^2 = 100, \quad \text{then} \quad a = \text{?}$$
$$\text{If } a^2 = 25, \quad \text{then} \quad a = \text{?}$$

1 **Find square roots.** To find a in the three preceding statements, we must find a number that when multiplied by itself results in the given number. The number a is called a **square root** of the number a^2.

Example 1 Finding All Square Roots of a Number

Find all square roots of 49.

To find a square root of 49, think of a number that when multiplied by itself gives 49. One square root is 7 because $7 \cdot 7 = 49$. Another square root of 49 is -7 because $(-7)(-7) = 49$. The number 49 has two square roots, 7 and -7; one is positive, and one is negative.

Work Problem ❶ at the Side.

The **positive** or **principal square root** of a number is written with the symbol $\sqrt{\ }$. For example, the positive square root of 121 is 11, written

$$\sqrt{121} = 11.$$

The symbol $-\sqrt{\ }$ is used for the **negative square root** of a number. For example, the negative square root of 121 is -11, written

$$-\sqrt{121} = -11.$$

The symbol $\sqrt{\ }$, called a **radical sign,** always represents the positive square root (except that $\sqrt{0} = 0$). The number inside the radical sign is called the **radicand,** and the entire expression, radical sign and radicand, is called a **radical.**

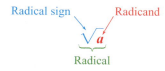

An algebraic expression containing a radical is called a **radical expression.**

Radicals have a long mathematical history. The radical sign $\sqrt{\ }$ has been used since sixteenth-century Germany and was probably derived from the letter R. The radical symbol in the margin comes from the Latin word for root, *radix.* It was first used by Leonardo da Pisa (Fibonnaci) in 1220. (*Source:* Miller, Charles D., Heeren, Vern E., and Hornsby, John, *Mathematical Ideas*, Ninth Edition, Addison-Wesley, 2001.)

Our discussion of square roots is summarized as follows.

Square Roots of a

If a is a positive real number,

$\sqrt{a}$ is the positive or principal square root of a,

and $-\sqrt{a}$ is the negative square root of a.

For nonnegative a,

$$\sqrt{a} \cdot \sqrt{a} = (\sqrt{a})^2 = a \quad \text{and} \quad -\sqrt{a} \cdot -\sqrt{a} = (-\sqrt{a})^2 = a.$$

Also, $\sqrt{0} = 0$.

Calculator Tip Most calculators have a square root key, usually labeled $\boxed{\sqrt{x}}$, that allows us to find the square root of a number. On some models, the square root key must be used in conjunction with the key marked $\boxed{INV}$ or $\boxed{2nd}$.

Example 2 Finding Square Roots

Find each square root.

(a) $\sqrt{144}$
 The radical $\sqrt{144}$ represents the positive or principal square root of 144. Think of a positive number whose square is 144.

$$12^2 = 144, \quad \text{so} \quad \sqrt{144} = 12.$$

(b) $-\sqrt{1024}$
 This symbol represents the negative square root of 1024. A calculator with a square root key can be used to find $\sqrt{1024} = 32$. Then, $-\sqrt{1024} = -32$.

(c) $\sqrt{\dfrac{4}{9}} = \dfrac{2}{3}$ **(d)** $-\sqrt{\dfrac{16}{49}} = -\dfrac{4}{7}$

=== **Work Problem ❷ at the Side.**

As shown in the preceding definition, when the square root of a positive real number is squared, the result is that positive real number. $\left(\text{Also}, (\sqrt{0})^2 = 0.\right)$

Example 3 Squaring Radical Expressions

Find the *square* of each radical expression.

(a) $\sqrt{13}$
 $(\sqrt{13})^2 = 13$ Definition of square root

(b) $-\sqrt{29}$
 $(-\sqrt{29})^2 = 29$ The square of a *negative* number is positive.

(c) $\sqrt{p^2 + 1}$
 $(\sqrt{p^2 + 1})^2 = p^2 + 1$

=== **Work Problem ❸ at the Side.**

❷ Find each square root.

(a) $\sqrt{16}$

(b) $-\sqrt{169}$

(c) $-\sqrt{225}$

(d) $\sqrt{729}$

(e) $\sqrt{\dfrac{36}{25}}$

❸ Find the *square* of each radical expression.

(a) $\sqrt{41}$

(b) $-\sqrt{39}$

(c) $\sqrt{2x^2 + 3}$

ANSWERS
2. **(a)** 4 **(b)** -13 **(c)** -15
  (d) 27 **(e)** $\dfrac{6}{5}$
3. **(a)** 41 **(b)** 39 **(c)** $2x^2 + 3$

4 Tell whether each square root is *rational*, *irrational*, or *not a real number*.

(a) $\sqrt{9}$

(b) $\sqrt{7}$

(c) $\sqrt{\dfrac{4}{9}}$

(d) $\sqrt{72}$

(e) $\sqrt{-43}$

2 **Decide whether a given root is rational, irrational, or not a real number.** All numbers with square roots that are rational are called **perfect squares.** For example, 144 and $\frac{4}{9}$ are perfect squares since their respective square roots, 12 and $\frac{2}{3}$, are rational numbers.

A number that is not a perfect square has a square root that is not a rational number. For example, $\sqrt{5}$ is not a rational number because it cannot be written as the ratio of two integers. Its decimal equivalent (or approximation) neither terminates nor repeats. However, $\sqrt{5}$ is a real number and corresponds to a point on the number line. As mentioned in Chapter 1, a real number that is not rational is called an *irrational number.* The number $\sqrt{5}$ is irrational. Many square roots of integers are irrational.

> If a is a positive real number that is not a perfect square, then $\sqrt{a}$ is irrational.

Not every number has a *real number* square root. For example, there is no real number that can be squared to get -36. (The square of a real number can never be negative.) Because of this, $\sqrt{-36}$ is not a real number.

> If a is a negative real number, $\sqrt{a}$ is not a real number.

CAUTION

> Be careful not to confuse $\sqrt{-36}$ and $-\sqrt{36}$. $\sqrt{-36}$ is not a real number since there is no real number that can be squared to get -36. However, $-\sqrt{36}$ is the negative square root of 36, which is -6.

Example 4 Identifying Types of Square Roots

Tell whether each square root is *rational*, *irrational*, or *not a real number*.

(a) $\sqrt{17}$
Because 17 is not a perfect square, $\sqrt{17}$ is irrational.

(b) $\sqrt{64}$
The number 64 is a perfect square, 8^2, so $\sqrt{64} = 8$, a rational number.

(c) $\sqrt{-25}$
There is no real number whose square is -25. Therefore, $\sqrt{-25}$ is not a real number.

Work Problem 4 at the Side.

NOTE

> Not all irrational numbers are square roots of integers. For example, π (approximately 3.14159) is an irrational number that is not a square root of any integer.

3 Find decimal approximations for irrational square roots. Even if a number is irrational, a decimal that approximates the number can be found using a calculator. For example, if we use a calculator to find $\sqrt{10}$, the display will show 3.16227766, which is only an *approximation* of $\sqrt{10}$, not an exact rational value.

Example 5 Approximating Irrational Square Roots

Find a decimal approximation for each square root. Round answers to the nearest thousandth.

(a) $\sqrt{11}$
 Using the square root key of a calculator gives $3.31662479 \approx 3.317$, where $\approx$ means "is approximately equal to."

(b) $\sqrt{39} \approx 6.245$ Use a calculator.

(c) $-\sqrt{740} \approx -27.203$

== Work Problem **5** at the Side.

4 Use the Pythagorean formula. Many applications of square roots use the Pythagorean formula. Recall from Section 5.7 that by this formula if c is the length of the hypotenuse of a right triangle, and a and b are the lengths of the two legs, then

$$a^2 + b^2 = c^2.$$

See Figure 1.

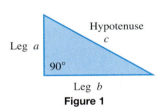

Leg a

Hypotenuse
c

90°

Leg b
Figure 1

Example 6 Using the Pythagorean Formula

Find the length of the unknown side of each right triangle with sides a, b, and c, where c is the hypotenuse.

(a) $a = 3, b = 4$
 Use the Pythagorean formula to find c^2 first.

$$c^2 = a^2 + b^2$$
$$c^2 = 3^2 + 4^2 \quad \text{Let } a = 3 \text{ and } b = 4.$$
$$c^2 = 9 + 16 \quad \text{Square.}$$
$$c^2 = 25 \quad \text{Add.}$$

Since the length of a side of a triangle must be a positive number, find the positive square root of 25 to get c.

$$c = \sqrt{25} = 5$$

======== Continued on Next Page

5 Find a decimal approximation for each square root. Round answers to the nearest thousandth.

(a) $\sqrt{28}$

(b) $\sqrt{63}$

(c) $-\sqrt{190}$

(d) $\sqrt{1000}$

6 Find the length of the unknown side in each right triangle. Give any decimal approximations to the nearest thousandth.

(a) $a = 7, b = 24$

(b) $c = 9, b = 5$

Substitute the given values in the Pythagorean formula. Then solve for a^2.

$$c^2 = a^2 + b^2$$
$$9^2 = a^2 + 5^2 \quad \text{Let } c = 9 \text{ and } b = 5.$$
$$81 = a^2 + 25 \quad \text{Square.}$$
$$56 = a^2 \quad \text{Subtract 25.}$$

Use a calculator to find $a = \sqrt{56} \approx 7.483$.

CAUTION

Be careful not to make the common mistake of thinking that $\sqrt{a^2 + b^2}$ equals $a + b$. As Example 6(a) shows, $\sqrt{9 + 16} = \sqrt{25} = 5$. However, $\sqrt{9} + \sqrt{16} = 3 + 4 = 7$. Since $5 \neq 7$, in general,

$$\sqrt{a^2 + b^2} \neq a + b.$$

(b) $c = 15, b = 13$

Work Problem 6 at the Side.

The Pythagorean formula can be used to solve applied problems that involve right triangles. Use the same six problem-solving steps that we have been using throughout the text.

Example 7 Using the Pythagorean Formula to Solve an Application

A ladder 10 ft long leans against a wall. The foot of the ladder is 6 ft from the base of the wall. How high up the wall does the top of the ladder rest?

Step 1 **Read** the problem again.

Step 2 **Assign a variable.** As shown in Figure 2, a right triangle is formed with the ladder as the hypotenuse. Let a represent the height of the top of the ladder when measured straight down to the ground.

(c)

[triangle with sides 8, 11, and ? with a right angle at the bottom]

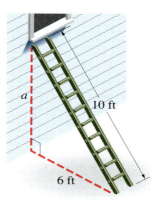

6 ft

10 ft

a

Figure 2

Continued on Next Page

ANSWERS

6. (a) 25 **(b)** $\sqrt{56} \approx 7.483$
(c) $\sqrt{57} \approx 7.550$

Step 3 **Write an equation** using the Pythagorean formula.

$$c^2 = a^2 + b^2$$
$$\mathbf{10}^2 = a^2 + \mathbf{6}^2 \qquad \text{Let } c = 10 \text{ and } b = 6.$$

Step 4 **Solve.**
$$100 = a^2 + 36 \qquad \text{Square.}$$
$$64 = a^2 \qquad \text{Subtract 36.}$$
$$\sqrt{64} = a$$
$$a = 8 \qquad \sqrt{64} = 8$$

Choose the positive square root of 64 because *a* represents a length.

Step 5 **State the answer.** The top of the ladder rests 8 ft up the wall.

Step 6 **Check.** From Figure 2, we see that we must have

$$8^2 + 6^2 = 10^2 \qquad ?$$
$$64 + 36 = 100. \qquad \text{True}$$

The check confirms that the top of the ladder rests 8 ft up the wall.

= **Work Problem ❼ at the Side.**

5▭ **Find higher roots.**   Finding the square root of a number is the inverse (reverse) of squaring a number. In a similar way, there are inverses to finding the cube of a number, or finding the fourth or higher power of a number. These inverses are the **cube root,** written $\sqrt[3]{a}$, and the **fourth root,** written $\sqrt[4]{a}$. Similar symbols are used for higher roots. In general, we have the following.

$\sqrt[n]{a}$

The *n*th root of *a* is written $\sqrt[n]{a}$.

In $\sqrt[n]{a}$, the number *n* is the **index** or **order** of the radical. It is possible to write $\sqrt[2]{a}$ instead of $\sqrt{a}$, but the simpler symbol $\sqrt{a}$ is customary since the square root is the most commonly used root.

 Calculator Tip   A calculator that has a key marked $\sqrt[x]{y}$, $\boxed{x^y}$, or $\boxed{y^x}$ (again perhaps in conjunction with the $\boxed{\text{INV}}$ or $\boxed{\text{2nd}}$ key) can be used to find higher roots.

When working with cube roots or fourth roots, it is helpful to memorize the first few *perfect cubes* ($2^3 = 8$, $3^3 = 27$, and so on) and the first few perfect fourth powers ($2^4 = 16$, $3^4 = 81$, and so on).

┌─ **Example 8**   **Finding Cube Roots**

Find each cube root.

(a) $\sqrt[3]{8}$
    Look for a number that can be cubed to give 8. Because $2^3 = 8$, $\sqrt[3]{8} = 2$.

(b) $\sqrt[3]{-8} = -2$ because $(-2)^3 = -8$.

(c) $\sqrt[3]{216} = 6$ because $6^3 = 216$.

---

❼ A rectangle has dimensions 5 ft by 12 ft. Find the length of its diagonal.

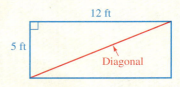

(Note that the diagonal divides the rectangle into two right triangles with itself as the hypotenuse.)

8 Find each cube root.

(a) $\sqrt[3]{27}$

(b) $\sqrt[3]{64}$

(c) $\sqrt[3]{-125}$

Notice in Example 8(b) that we can find the cube root of a negative number. (Contrast this with the square root of a negative number, which is not real.) In fact, the cube root of a positive number is positive, and the cube root of a negative number is negative. *There is only one real number cube root for each real number.*

Work Problem 8 at the Side.

When the index of the radical is even (square root, fourth root, and so on), *the radicand must be nonnegative* to get a real number root. Also, for even indexes, the symbols $\sqrt{}$, $\sqrt[4]{}$, $\sqrt[6]{}$, and so on are used for the positive or principal roots. The symbols $-\sqrt{}$, $-\sqrt[4]{}$, $-\sqrt[6]{}$, and so on are used for the negative roots.

Example 9 Finding Higher Roots

Find each root.

(a) $\sqrt[4]{16}$

$\sqrt[4]{16} = 2$ because 2 is positive and $2^4 = 16$.

(b) $-\sqrt[4]{16}$

From part (a), $\sqrt[4]{16} = 2$, so the negative root $-\sqrt[4]{16} = -2$.

(c) $\sqrt[4]{-16}$

For a real number fourth root, the radicand must be nonnegative. There is no real number that equals $\sqrt[4]{-16}$.

(d) $-\sqrt[5]{32}$

First find $\sqrt[5]{32}$. Because 2 is the number whose fifth power is 32, $\sqrt[5]{32} = 2$. If $\sqrt[5]{32} = 2$, then

$$-\sqrt[5]{32} = -2.$$

(e) $\sqrt[5]{-32}$

Because $(-2)^5 = -32$, $\sqrt[5]{-32} = -2$.

Work Problem 9 at the Side.

9 Find each root.

(a) $\sqrt[4]{81}$

(b) $\sqrt[4]{-81}$

(c) $-\sqrt[4]{625}$

(d) $\sqrt[5]{243}$

(e) $\sqrt[5]{-243}$

ANSWERS
8. (a) 3 (b) 4 (c) −5
9. (a) 3 (b) not a real number
(c) −5 (d) 3 (e) −3

8.1 **EXERCISES**

**FOR
EXTRA
HELP**

 Student's
Solutions
Manual

 MyMathLab.com

 InterAct Math
Tutorial
Software

 AW Math
Tutor Center

 www.mathxl.com
MathXL

 Digital Video Tutor CD 6
Videotape 16

Decide whether each statement is true *or* false. *If* false, *tell why.*

1. Every positive number has two real square roots.

2. A negative number has negative square roots.

3. Every nonnegative number has two real square roots.

4. The positive square root of a positive number is its principal square root.

5. The cube root of every real number has the same sign as the number itself.

6. Every positive number has three real cube roots.

Find all square roots of each number. See Example 1.

7. 9

8. 16

9. 64

10. 100

11. 144

12. 225

13. $\dfrac{25}{196}$

14. $\dfrac{81}{400}$

15. 900

16. 1600

Find each square root. See Examples 2 and 4(c).

17. $\sqrt{1}$

18. $\sqrt{4}$

19. $\sqrt{49}$

20. $\sqrt{81}$

21. $-\sqrt{121}$

22. $-\sqrt{196}$

23. $-\sqrt{\dfrac{144}{121}}$

24. $-\sqrt{\dfrac{49}{36}}$

25. $\sqrt{-121}$

26. $\sqrt{-64}$

Find the square of each radical expression. See Example 3.

27. $\sqrt{100}$

28. $\sqrt{36}$

29. $-\sqrt{19}$

30. $-\sqrt{99}$

31. $\sqrt{\dfrac{2}{3}}$

32. $\sqrt{\dfrac{5}{7}}$

33. $\sqrt{3x^2 + 4}$

34. $\sqrt{9y^2 + 3}$

What must be true about a for each statement in Exercises 35–38 to be true?

35. $\sqrt{a}$ represents a positive number.

36. $-\sqrt{a}$ represents a negative number.

37. $\sqrt{a}$ is not a real number.

38. $-\sqrt{a}$ is not a real number.

Write rational, irrational, or not a real number for each number. If a number is rational, give its exact value. If a number is irrational, give a decimal approximation to the nearest thousandth. Use a calculator as necessary. See Examples 4 and 5.

39. $\sqrt{25}$

40. $\sqrt{169}$

41. $\sqrt{29}$

42. $\sqrt{33}$

43. $-\sqrt{64}$

44. $-\sqrt{81}$

45. $-\sqrt{300}$

46. $-\sqrt{500}$

47. $\sqrt{-29}$

48. $\sqrt{-47}$

49. $\sqrt{1200}$

50. $\sqrt{1500}$

Work Exercises 51 and 52 without using a calculator.

51. Choose the best estimate for the length and width (in meters) of this rectangle.

   A. 11 by 6 **B.** 11 by 7 **C.** 10 by 7 **D.** 10 by 6

52. Choose the best estimate for the base and height (in feet) of this triangle.

   A. $b = 8, h = 5$ **B.** $b = 8, h = 4$

   C. $b = 9, h = 5$ **D.** $b = 9, h = 4$

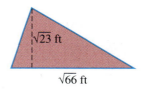

Find the length of the unknown side of each right triangle with sides a, b, and c, where c is the hypotenuse. See Figure 1 and Example 6. Give any decimal approximations to the nearest thousandth.

53. $a = 8, b = 15$

54. $a = 24, b = 10$

55. $a = 6, c = 10$

56. $b = 12, c = 13$

57. $a = 11, b = 4$

58. $a = 13, b = 9$

 *Use the Pythagorean formula to solve each problem. See Example 7.*

59. The diagonal of a rectangle measures 25 cm. The width of the rectangle is 7 cm. Find the length of the rectangle.

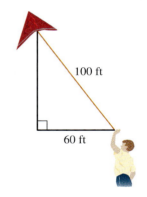

25 cm

7 cm

60. The length of a rectangle is 40 m, and the width is 9 m. Find the measure of the diagonal of the rectangle.

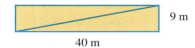

9 m

40 m

61. Tyler is flying a kite on 100 ft of string. How high is it above his hand (vertically) if the horizontal distance between Tyler and the kite is 60 ft?

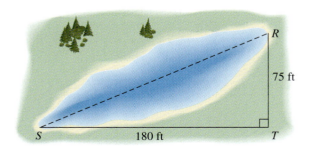

100 ft

60 ft

62. A guy wire is attached to the mast of a short-wave transmitting antenna. It is attached 96 ft above ground level. If the wire is staked to the ground 72 ft from the base of the mast, how long is the wire?

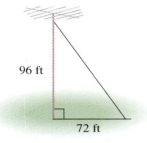

96 ft

72 ft

63. A surveyor measured the distances shown in the figure. Find the distance across the lake between points *R* and *S*.

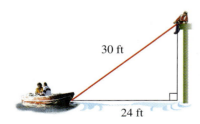

R

75 ft

S   180 ft   T

64. A boat is being pulled toward a dock with a rope attached at water level. When the boat is 24 ft from the dock, 30 ft of rope is extended. What is the height of the dock above the water?

30 ft

24 ft

65. One of the authors of this text took this photo of a broken tree in a field near his home. The vertical distance from the base of the broken tree to the point of the break is 4.5 ft. The length of the broken part is 12 ft. How far along the ground (to the nearest tenth) is it from the base of the tree to the point where the broken part touches the ground?

66. Another of the authors recently purchased a new television set. A television set is "sized" according to the diagonal measurement of the viewing screen. The author purchased a 19-in. TV, so the TV measures 19 in. from one corner of the viewing screen diagonally to the other corner. The viewing screen is 15.5 in. wide. Find the height of the viewing screen (to the nearest tenth). (*Source:* Phillips Magnavox color television 19PR21C1.)

19 in.

67. What is the value of x (to the nearest thousandth) in the figure?

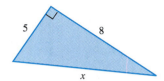

5   8

x

68. What is the value of y (to the nearest thousandth) in the figure?

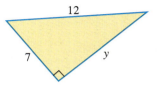

12

7   y

69. Use specific values for a and b different from those given in the "Caution" following Example 6 to show that $\sqrt{a^2 + b^2} \neq a + b$.

70. Why would the values $a = 0$ and $b = 1$ *not* be satisfactory in Exercise 69?

Find each root. See Examples 8 and 9.

71. $\sqrt[3]{1}$

72. $\sqrt[3]{8}$

73. $\sqrt[3]{125}$

74. $\sqrt[3]{1000}$

75. $\sqrt[3]{-27}$

76. $\sqrt[3]{-64}$

77. $-\sqrt[3]{8}$

78. $-\sqrt[3]{216}$

79. $\sqrt[4]{625}$

80. $\sqrt[4]{10,000}$

81. $\sqrt[4]{-1}$

82. $\sqrt[4]{-625}$

83. $-\sqrt[4]{81}$

84. $-\sqrt[4]{256}$

85. $\sqrt[5]{-1024}$

86. $\sqrt[5]{-100,000}$

8.2   MULTIPLYING, DIVIDING, AND SIMPLIFYING RADICALS

1 ▭ **Multiply radicals.**   We develop several rules for finding products and quotients of radicals in this section. To illustrate the rule for products, notice that

$$\sqrt{4} \cdot \sqrt{9} = 2 \cdot 3 = 6 \quad \text{and} \quad \sqrt{4 \cdot 9} = \sqrt{36} = 6,$$

showing that

$$\sqrt{4} \cdot \sqrt{9} = \sqrt{4 \cdot 9}.$$

This result is a particular case of the more general product rule for radicals.

Product Rule for Radicals

For nonnegative real numbers a and b,

$$\sqrt{a} \cdot \sqrt{b} = \sqrt{a \cdot b} \quad \text{and} \quad \sqrt{a \cdot b} = \sqrt{a} \cdot \sqrt{b}.$$

In words, the product of two radicals is the radical of the product, and the radical of a product is the product of the radicals.

Example 1   **Using the Product Rule to Multiply Radicals**

Use the product rule for radicals to find each product.

(a) $\sqrt{2} \cdot \sqrt{3} = \sqrt{2 \cdot 3} = \sqrt{6}$ Product rule

(b) $\sqrt{7} \cdot \sqrt{5} = \sqrt{35}$ Product rule

(c) $\sqrt{11} \cdot \sqrt{a} = \sqrt{11a}$ Assume $a \geq 0$.

Work Problem **1** at the Side.

---

2 ▭ **Simplify radicals using the product rule.**   A square root radical is *simplified* when no perfect square factor remains under the radical sign. This is accomplished by using the product rule in the form $\sqrt{a \cdot b} = \sqrt{a} \cdot \sqrt{b}$, as in Example 2.

Example 2   **Using the Product Rule to Simplify Radicals**

Simplify each radical.

(a) $\sqrt{20}$

Because 20 has a perfect square factor of 4, we can write

$$\sqrt{20} = \sqrt{4 \cdot 5} \qquad \text{4 is a perfect square.}$$

$$= \sqrt{4} \cdot \sqrt{5} \qquad \text{Product rule}$$

$$= 2\sqrt{5}. \qquad \sqrt{4} = 2$$

Thus, $\sqrt{20} = 2\sqrt{5}$. Because 5 has no perfect square factor (other than 1), $2\sqrt{5}$ is called the *simplified form* of $\sqrt{20}$. Note that $2\sqrt{5}$ represents a product, where the factors are 2 and $\sqrt{5}$.

We could also factor 20 into prime factors and look for pairs of like factors. Each pair of like factors produces one factor outside the radical in the simplified form. Therefore,

$$\sqrt{20} = \sqrt{2 \cdot 2 \cdot 5} = 2\sqrt{5}.$$

Continued on Next Page

---

1 Use the product rule for radicals to find each product.

(a) $\sqrt{6} \cdot \sqrt{11}$

(b) $\sqrt{2} \cdot \sqrt{5}$

(c) $\sqrt{10} \cdot \sqrt{r}, r \geq 0$

2 Simplify each radical.

(a) $\sqrt{8}$

(b) $\sqrt{27}$

(c) $\sqrt{50}$

(d) $\sqrt{60}$

(e) $\sqrt{30}$

(b) $\sqrt{72}$

Begin by looking for the *largest* perfect square factor of 72. This number is 36, so

$$\sqrt{72} = \sqrt{36 \cdot 2} \qquad \text{36 is a perfect square.}$$
$$= \sqrt{36} \cdot \sqrt{2} \qquad \text{Product rule}$$
$$= 6\sqrt{2}. \qquad \sqrt{36} = 6$$

We could also factor 72 into its prime factors and look for pairs of like factors.

$$\sqrt{72} = \sqrt{2 \cdot 2 \cdot 2 \cdot 3 \cdot 3} = 2 \cdot 3 \cdot \sqrt{2} = 6\sqrt{2}$$

In either case, we obtain $6\sqrt{2}$ as the simplified form of $\sqrt{72}$. However, our work is simpler if we begin with the largest perfect square factor.

(c) $\sqrt{300} = \sqrt{100 \cdot 3} \qquad \text{100 is a perfect square.}$
$$= \sqrt{100} \cdot \sqrt{3} \qquad \text{Product rule}$$
$$= 10\sqrt{3} \qquad \sqrt{100} = 10$$

(d) $\sqrt{15}$

The number 15 has no perfect square factors (except 1), so $\sqrt{15}$ cannot be simplified further.

Work Problem 2 at the Side.

Sometimes the product rule can be used to simplify a product, as Example 3 shows.

Example 3 Multiplying and Simplifying Radicals

Find each product and simplify.

(a) $\sqrt{9} \cdot \sqrt{75} = 3\sqrt{75} \qquad \sqrt{9} = 3$
$$= 3\sqrt{25 \cdot 3} \qquad \text{25 is a perfect square.}$$
$$= 3\sqrt{25} \cdot \sqrt{3} \qquad \text{Product rule}$$
$$= 3 \cdot 5\sqrt{3} \qquad \sqrt{25} = 5$$
$$= 15\sqrt{3} \qquad \text{Multiply.}$$

Notice that we could have used the product rule to get $\sqrt{9} \cdot \sqrt{75} = \sqrt{675}$, and then simplified. However, the product rule as used here allows us to obtain the final answer without using a large number like 675.

(b) $\sqrt{8} \cdot \sqrt{12} = \sqrt{8 \cdot 12} \qquad \text{Product rule}$
$$= \sqrt{4 \cdot 2 \cdot 4 \cdot 3} \qquad \text{Factor; 4 is a perfect square.}$$
$$= \sqrt{4} \cdot \sqrt{4} \cdot \sqrt{2 \cdot 3} \qquad \text{Product rule}$$
$$= 2 \cdot 2 \cdot \sqrt{6} \qquad \sqrt{4} = 2$$
$$= 4\sqrt{6} \qquad \text{Multiply.}$$

NOTE

We could also simplify Example 3(b) as follows.

$$\sqrt{8} \cdot \sqrt{12} = \sqrt{4 \cdot 2} \cdot \sqrt{4 \cdot 3}$$
$$= 2\sqrt{2} \cdot 2\sqrt{3}$$
$$= 2 \cdot 2 \cdot \sqrt{2} \cdot \sqrt{3}$$
$$= 4\sqrt{6}$$

Both approaches are correct. There is often more than one way to find such a product.

Work Problem ❸ at the Side.

3   **Simplify radicals using the quotient rule.**   The quotient rule for radicals is very similar to the product rule. It, too, can be used either way.

Quotient Rule for Radicals

If a and b are nonnegative real numbers and $b \neq 0$, then

$$\sqrt{\frac{a}{b}} = \frac{\sqrt{a}}{\sqrt{b}} \quad \text{and} \quad \frac{\sqrt{a}}{\sqrt{b}} = \sqrt{\frac{a}{b}}.$$

In words, the radical of a quotient is the quotient of the radicals, and the quotient of two radicals is the radical of the quotient.

Example 4   **Using the Quotient Rule to Simplify Radicals**

Use the quotient rule to simplify each radical.

(a) $\sqrt{\dfrac{25}{9}} = \dfrac{\sqrt{25}}{\sqrt{9}} = \dfrac{5}{3}$ *Quotient rule*

(b) $\dfrac{\sqrt{288}}{\sqrt{2}} = \sqrt{\dfrac{288}{2}} = \sqrt{144} = 12$ *Quotient rule*

(c) $\sqrt{\dfrac{3}{4}} = \dfrac{\sqrt{3}}{\sqrt{4}} = \dfrac{\sqrt{3}}{2}$ *Quotient rule*

Example 5   **Using the Quotient Rule to Divide Radicals**

Divide $27\sqrt{15}$ by $9\sqrt{3}$.

Use multiplication of fractions and the quotient rule as follows.

$$\frac{27\sqrt{15}}{9\sqrt{3}} = \frac{27}{9} \cdot \frac{\sqrt{15}}{\sqrt{3}} = \frac{27}{9} \cdot \sqrt{\frac{15}{3}} = 3\sqrt{5}$$

Work Problem ❹ at the Side.

---

3   Find each product and simplify.

(a) $\sqrt{3} \cdot \sqrt{15}$

(b) $\sqrt{10} \cdot \sqrt{50}$

(c) $\sqrt{12} \cdot \sqrt{2}$

(d) $\sqrt{7} \cdot \sqrt{14}$

4   Use the quotient rule to simplify each radical.

(a) $\sqrt{\dfrac{81}{16}}$

(b) $\dfrac{\sqrt{192}}{\sqrt{3}}$

(c) $\sqrt{\dfrac{10}{49}}$

(d) $\dfrac{8\sqrt{50}}{4\sqrt{5}}$

Answers

3. (a) $3\sqrt{5}$ **(b)** $10\sqrt{5}$ **(c)** $2\sqrt{6}$
   (d) $7\sqrt{2}$

4. (a) $\dfrac{9}{4}$ **(b)** 8 **(c)** $\dfrac{\sqrt{10}}{7}$ **(d)** $2\sqrt{10}$

Some problems require both the product and quotient rules.

5 Multiply and then simplify each product.

(a) $\sqrt{\dfrac{5}{6}} \cdot \sqrt{120}$

> **Example 6** Using Both the Product and Quotient Rules
>
> Simplify $\sqrt{\dfrac{3}{5}} \cdot \sqrt{\dfrac{1}{5}}$.
>
> $$\sqrt{\frac{3}{5}} \cdot \sqrt{\frac{1}{5}} = \sqrt{\frac{3}{5} \cdot \frac{1}{5}} \qquad \text{Product rule}$$
>
> $$= \sqrt{\frac{3}{25}} \qquad \text{Multiply fractions.}$$
>
> $$= \frac{\sqrt{3}}{\sqrt{25}} \qquad \text{Quotient rule}$$
>
> $$= \frac{\sqrt{3}}{5} \qquad \sqrt{25} = 5$$

Work Problem 5 at the Side.

4 **Simplify radicals involving variables.** Radicals can also involve variables, such as $\sqrt{x^2}$. Simplifying such radicals can get a little tricky. If x represents a nonnegative number, then $\sqrt{x^2} = x$. If x represents a negative number, then $\sqrt{x^2} = -x$, the opposite of x (which is positive). For example,

$$\sqrt{5^2} = 5, \qquad \text{but} \qquad \sqrt{(-5)^2} = \sqrt{25} = 5, \quad \text{the opposite of } -5.$$

This means that the square root of a squared number is always nonnegative. We can use absolute value to express this.

(b) $\sqrt{\dfrac{3}{8}} \cdot \sqrt{\dfrac{7}{2}}$

> For any real number a,
>
> $$\sqrt{a^2} = |a|.$$

The product and quotient rules apply when variables appear under the radical sign, as long as the variables represent only *nonnegative* real numbers. *To avoid negative radicands, variables under radical signs are assumed to be nonnegative in this text.* Therefore, absolute value bars are not necessary, since for $x \geq 0$, $|x| = x$.

> **Example 7** Simplifying Radicals Involving Variables
>
> Simplify each radical. Remember that we assume all variables represent nonnegative real numbers.
>
> (a) $\sqrt{x^4} = x^2$ since $(x^2)^2 = x^4$.
>
> (b) $\sqrt{25m^6} = \sqrt{25} \cdot \sqrt{m^6} \qquad \text{Product rule}$
> $\phantom{\sqrt{25m^6}} = 5m^3 \qquad (m^3)^2 = m^6$
>
> **Continued on Next Page**

(c) $\sqrt{8p^{10}} = \sqrt{4 \cdot 2 \cdot p^{10}}$ 4 is a perfect square.

$\qquad\qquad = \sqrt{4} \cdot \sqrt{2} \cdot \sqrt{p^{10}}$ Product rule

$\qquad\qquad = 2 \cdot \sqrt{2} \cdot p^5$ $(p^5)^2 = p^{10}$

$\qquad\qquad = 2p^5\sqrt{2}$

(d) $\sqrt{r^9} = \sqrt{r^8 \cdot r}$

$\qquad\quad = \sqrt{r^8} \cdot \sqrt{r}$ Product rule

$\qquad\quad = r^4\sqrt{r}$ $(r^4)^2 = r^8$

(e) $\sqrt{\dfrac{5}{x^2}} = \dfrac{\sqrt{5}}{\sqrt{x^2}}$ Quotient rule

$\qquad\quad = \dfrac{\sqrt{5}}{x}$ $x \neq 0$

NOTE

A quick way to find the square root of a variable raised to an even power is to divide the exponent by the index, 2. For example,

$$\sqrt{x^6} = x^3 \quad \text{and} \quad \sqrt{x^{10}} = x^5.$$

$\qquad\qquad 6 \div 2 = 3 \qquad\qquad\qquad 10 \div 2 = 5$

Work Problem ❻ at the Side.

5   **Simplify higher roots.** The product and quotient rules for radicals also work for other roots. To simplify cube roots, look for factors that are *perfect cubes*. A **perfect cube** is a number with a rational cube root. For example, $\sqrt[3]{64} = 4$, and because 4 is a rational number, 64 is a perfect cube. Higher roots are handled in a similar manner.

Properties of Radicals

For all real numbers where the indicated roots exist,

$$\sqrt[n]{x} \cdot \sqrt[n]{y} = \sqrt[n]{xy} \quad \text{and} \quad \frac{\sqrt[n]{x}}{\sqrt[n]{y}} = \sqrt[n]{\frac{x}{y}} \quad (y \neq 0).$$

Example 8   **Simplifying Higher Roots**

Simplify each radical.

(a) $\sqrt[3]{32} = \sqrt[3]{8 \cdot 4}$ 8 is a perfect cube.

$\qquad\quad = \sqrt[3]{8} \cdot \sqrt[3]{4}$ Product rule

$\qquad\quad = 2\sqrt[3]{4}$

Continued on Next Page

---

❻ Simplify each radical. Assume all variables represent nonnegative real numbers.

(a) $\sqrt{x^8}$

(b) $\sqrt{36y^6}$

(c) $\sqrt{100p^{12}}$

(d) $\sqrt{12z^2}$

(e) $\sqrt{a^5}$

(f) $\sqrt{\dfrac{10}{n^4}}, n \neq 0$

ANSWERS

6. (a) x^4 **(b)** $6y^3$ **(c)** $10p^6$ **(d)** $2z\sqrt{3}$

    (e) $a^2\sqrt{a}$ **(f)** $\dfrac{\sqrt{10}}{n^2}$

❼ Simplify each radical.

(a) $\sqrt[3]{108}$

(b) $\sqrt[4]{160}$

(c) $\sqrt[4]{\dfrac{16}{625}}$

❽ Simplify each radical.

(a) $\sqrt[3]{z^9}$

(b) $\sqrt[3]{8x^6}$

(c) $\sqrt[3]{54t^5}$

(d) $\sqrt[3]{\dfrac{a^{15}}{64}}$

(b) $\sqrt[4]{32} = \sqrt[4]{16 \cdot 2}$ 16 is a perfect fourth power.

$\qquad = \sqrt[4]{16} \cdot \sqrt[4]{2}$ Product rule

$\qquad = 2\sqrt[4]{2}$

(c) $\sqrt[3]{\dfrac{27}{125}} = \dfrac{\sqrt[3]{27}}{\sqrt[3]{125}} = \dfrac{3}{5}$ Quotient rule

Work Problem ❼ at the Side.

Higher roots of radicals involving variables can also be simplified. To simplify cube roots with variables, use the fact that for any real number a,

$$\sqrt[3]{a^3} = a.$$

This is true whether a is positive or negative. (Why?)

Example 9   Simplifying Cube Roots Involving Variables

Simplify each radical.

(a) $\sqrt[3]{m^6} = m^2$ $(m^2)^3 = m^6$

(b) $\sqrt[3]{27x^{12}} = \sqrt[3]{27} \cdot \sqrt[3]{x^{12}}$ Product rule

$\qquad = 3x^4$ $3^3 = 27; (x^4)^3 = x^{12}$

(c) $\sqrt[3]{32a^4} = \sqrt[3]{8a^3 \cdot 4a}$ 8 is a perfect cube.

$\qquad = \sqrt[3]{8a^3} \cdot \sqrt[3]{4a}$ Product rule

$\qquad = 2a\sqrt[3]{4a}$ $(2a)^3 = 8a^3$

(d) $\sqrt[3]{\dfrac{y^3}{125}} = \dfrac{\sqrt[3]{y^3}}{\sqrt[3]{125}}$ Quotient rule

$\qquad = \dfrac{y}{5}$

Work Problem ❽ at the Side.

8.2 EXERCISES

Decide whether each statement is true *or* false. *If* false, *show why.*

1. $\sqrt{(-6)^2} = -6$

2. $\sqrt[3]{(-6)^3} = -6$

Use the product rule for radicals to find each product. See Example 1.

3. $\sqrt{3} \cdot \sqrt{5}$

4. $\sqrt{3} \cdot \sqrt{7}$

5. $\sqrt{2} \cdot \sqrt{11}$

6. $\sqrt{2} \cdot \sqrt{15}$

7. $\sqrt{6} \cdot \sqrt{7}$

8. $\sqrt{5} \cdot \sqrt{6}$

9. $\sqrt{13} \cdot \sqrt{r}, r \geq 0$

10. $\sqrt{19} \cdot \sqrt{k}, k \geq 0$

11. Which one of the following radicals is simplified? See Example 2.

   A. $\sqrt{47}$ **B.** $\sqrt{45}$ **C.** $\sqrt{48}$ **D.** $\sqrt{44}$

12. If p is a prime number, is $\sqrt{p}$ in simplified form? Explain your answer.

Simplify each radical. See Example 2.

13. $\sqrt{45}$

14. $\sqrt{27}$

15. $\sqrt{24}$

16. $\sqrt{44}$

17. $\sqrt{90}$

18. $\sqrt{56}$

19. $\sqrt{75}$

20. $\sqrt{18}$

21. $\sqrt{125}$

22. $\sqrt{80}$

23. $\sqrt{145}$

24. $\sqrt{110}$

25. $\sqrt{160}$

26. $\sqrt{128}$

27. $-\sqrt{700}$

28. $-\sqrt{600}$

Find each product and simplify. See Example 3.

29. $\sqrt{3} \cdot \sqrt{18}$

30. $\sqrt{3} \cdot \sqrt{21}$

31. $\sqrt{12} \cdot \sqrt{48}$

32. $\sqrt{50} \cdot \sqrt{72}$

33. $\sqrt{12} \cdot \sqrt{30}$

34. $\sqrt{30} \cdot \sqrt{24}$

35. Simplify the product $\sqrt{8} \cdot \sqrt{32}$ in two ways. First, multiply 8 by 32 and simplify the square root of this product. Second, simplify $\sqrt{8}$, simplify $\sqrt{32}$, and then multiply. How do the answers compare? Make a conjecture (an educated guess) about whether the correct answer can always be obtained using either method when simplifying a product such as this.

36. Simplify the radical $\sqrt{288}$ in two ways. First, factor 288 as $144 \cdot 2$ and then simplify. Second, factor 288 as $48 \cdot 6$ and then simplify. How do the answers compare? Make a conjecture concerning the quickest way to simplify such a radical.

Use the product and quotient rules, as necessary, to simplify each radical expression. See Examples 4–6.

37. $\sqrt{\dfrac{16}{225}}$

38. $\sqrt{\dfrac{9}{100}}$

39. $\sqrt{\dfrac{7}{16}}$

40. $\sqrt{\dfrac{13}{25}}$

41. $\sqrt{\dfrac{5}{7}} \cdot \sqrt{35}$

42. $\sqrt{\dfrac{10}{13}} \cdot \sqrt{130}$

43. $\sqrt{\dfrac{5}{2}} \cdot \sqrt{\dfrac{125}{8}}$

44. $\sqrt{\dfrac{8}{3}} \cdot \sqrt{\dfrac{512}{27}}$

45. $\dfrac{30\sqrt{10}}{5\sqrt{2}}$

46. $\dfrac{50\sqrt{20}}{2\sqrt{10}}$

Simplify each radical. Assume that all variables represent nonnegative real numbers.
See Example 7.

47. $\sqrt{m^2}$

48. $\sqrt{k^2}$

49. $\sqrt{y^4}$

50. $\sqrt{s^4}$

51. $\sqrt{36z^2}$

52. $\sqrt{49n^2}$

53. $\sqrt{400x^6}$

54. $\sqrt{900y^8}$

55. $\sqrt{18x^8}$

56. $\sqrt{20r^{10}}$

57. $\sqrt{45c^{14}}$

58. $\sqrt{50d^{20}}$

59. $\sqrt{z^5}$

60. $\sqrt{y^3}$

61. $\sqrt{a^{13}}$

62. $\sqrt{p^{17}}$

63. $\sqrt{64x^7}$

64. $\sqrt{25t^{11}}$

65. $\sqrt{x^6y^{12}}$

66. $\sqrt{a^8b^{10}}$

67. $\sqrt{81m^4n^2}$

68. $\sqrt{100c^4d^6}$

69. $\sqrt{\dfrac{7}{x^{10}}}, x \neq 0$

70. $\sqrt{\dfrac{14}{z^{12}}}, z \neq 0$

71. $\sqrt{\dfrac{y^4}{100}}$

72. $\sqrt{\dfrac{w^8}{144}}$

Simplify each radical. See Example 8.

73. $\sqrt[3]{40}$

74. $\sqrt[3]{48}$

75. $\sqrt[3]{54}$

76. $\sqrt[3]{135}$

77. $\sqrt[3]{128}$

78. $\sqrt[3]{192}$

79. $\sqrt[4]{80}$

80. $\sqrt[4]{243}$

81. $\sqrt[3]{\dfrac{8}{27}}$

82. $\sqrt[3]{\dfrac{64}{125}}$

83. $\sqrt[3]{-\dfrac{216}{125}}$

84. $\sqrt[3]{-\dfrac{1}{64}}$

Simplify each radical. See Example 9.

85. $\sqrt[3]{p^3}$

86. $\sqrt[3]{w^3}$

87. $\sqrt[3]{x^9}$

88. $\sqrt[3]{y^{18}}$

89. $\sqrt[3]{64z^6}$

90. $\sqrt[3]{125a^{15}}$

91. $\sqrt[3]{343a^9b^3}$

92. $\sqrt[3]{216m^3n^6}$

93. $\sqrt[3]{16t^5}$

94. $\sqrt[3]{24x^4}$

95. $\sqrt[3]{\dfrac{m^{12}}{8}}$

96. $\sqrt[3]{\dfrac{n^9}{27}}$

The volume of a cube is found with the formula $V = s^3$, where s is the length of an edge of the cube. Use this information in Exercises 97 and 98.

97. A container in the shape of a cube has a volume of 216 cm³. What is the depth of the container?

98. A cube-shaped box must be constructed to contain 128 ft³. What should the dimensions (height, width, and length) of the box be?

The volume of a sphere is found with the formula $V = \frac{4}{3}\pi r^3$, where r is the length of the radius of the sphere. Use this information in Exercises 99 and 100.

99. A ball in the shape of a sphere has a volume of 288π in.³. What is the radius of the ball?

100. Suppose that the volume of the ball described in Exercise 99 is multiplied by 8. How is the radius affected?

Work Exercises 101 and 102 without using a calculator.

101. Choose the best estimate for the area (in square inches) of this rectangle.

    A. 45 **B.** 72 **C.** 80 **D.** 90

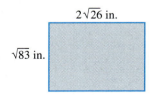

$2\sqrt{26}$ in.

$\sqrt{83}$ in.

102. Choose the best estimate for the area (in square feet) of the triangle.

    A. 20 **B.** 40 **C.** 60 **D.** 80

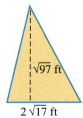

$\sqrt{97}$ ft

$2\sqrt{17}$ ft

8.3 ADDING AND SUBTRACTING RADICALS

OBJECTIVES

1 Add and subtract radicals.

2 Simplify radical sums and differences.

3 Simplify radical expressions involving multiplication.

1 **Add and subtract radicals.**   We add or subtract radicals by using the distributive property. For example,

$$8\sqrt{3} + 6\sqrt{3} = (8 + 6)\sqrt{3} = 14\sqrt{3}.$$

Also,

$$2\sqrt{11} - 7\sqrt{11} = -5\sqrt{11}.$$

Only **like radicals,** those that are multiples of the *same root* of the *same number*, can be combined in this way. In the examples above, $8\sqrt{3}$ and $6\sqrt{3}$ are like radicals, as are $2\sqrt{11}$ and $-7\sqrt{11}$. On the other hand, examples of *unlike radicals* are

$$2\sqrt{5} \quad \text{and} \quad 2\sqrt{3}, \qquad \text{Radicands are different.}$$

as well as

$$2\sqrt{3} \quad \text{and} \quad 2\sqrt[3]{3}. \qquad \text{Indexes are different.}$$

Work Problem ❶ at the Side.

❶ Indicate whether each pair of radicals is *like* or *unlike*.

(a) $5\sqrt{6}$ and $4\sqrt{6}$

(b) $2\sqrt{3}$ and $3\sqrt{2}$

(c) $\sqrt{10}$ and $\sqrt[3]{10}$

(d) $7\sqrt{2x}$ and $8\sqrt{2x}$

(e) $\sqrt{3y}$ and $\sqrt{6y}$

Example 1   Adding and Subtracting Like Radicals

Add or subtract, as indicated.

(a) $3\sqrt{6} + 5\sqrt{6} = (3 + 5)\sqrt{6} = 8\sqrt{6}$ Distributive property

(b) $5\sqrt{10} - 7\sqrt{10} = (5 - 7)\sqrt{10} = -2\sqrt{10}$

(c) $\sqrt{7} + 2\sqrt{7} = 1\sqrt{7} + 2\sqrt{7} = (1 + 2)\sqrt{7} = 3\sqrt{7}$

(d) $\sqrt{5} + \sqrt{5} = 1\sqrt{5} + 1\sqrt{5} = 2\sqrt{5}$

(e) $\sqrt{3} + \sqrt{7}$ cannot be added using the distributive property.

Work Problem ❷ at the Side.

2 **Simplify radical sums and differences.**   Sometimes one or more radical expressions in a sum or difference must first be simplified. Any like radicals that result can then be added or subtracted.

Example 2   Adding and Subtracting Radicals That Must Be Simplified

Add or subtract, as indicated.

(a) $3\sqrt{2} + \sqrt{8} = 3\sqrt{2} + \sqrt{4 \cdot 2}$ Factor.

$\qquad\qquad = 3\sqrt{2} + \sqrt{4} \cdot \sqrt{2}$ Product rule

$\qquad\qquad = 3\sqrt{2} + 2\sqrt{2}$ $\sqrt{4} = 2$

$\qquad\qquad = 5\sqrt{2}$ Add like radicals.

(b) $\sqrt{18} - \sqrt{27} = \sqrt{9 \cdot 2} - \sqrt{9 \cdot 3}$ Factor.

$\qquad\qquad = \sqrt{9} \cdot \sqrt{2} - \sqrt{9} \cdot \sqrt{3}$ Product rule

$\qquad\qquad = 3\sqrt{2} - 3\sqrt{3}$ $\sqrt{9} = 3$

Because $\sqrt{2}$ and $\sqrt{3}$ are unlike radicals, this difference cannot be simplified further.

Continued on Next Page

❷ Add or subtract, as indicated.

(a) $8\sqrt{5} + 2\sqrt{5}$

(b) $-4\sqrt{3} + 9\sqrt{3}$

(c) $12\sqrt{11} - 3\sqrt{11}$

(d) $\sqrt{15} + \sqrt{15}$

(e) $2\sqrt{7} + 2\sqrt{10}$

ANSWERS

1. (a) like   (b) unlike   (c) unlike
   (d) like   (e) unlike
2. (a) $10\sqrt{5}$ (b) $5\sqrt{3}$ (c) $9\sqrt{11}$
   (d) $2\sqrt{15}$ (e) cannot be added

③ Add or subtract, as indicated.

(a) $\sqrt{8} + 4\sqrt{2}$

(b) $\sqrt{27} + \sqrt{12}$

(c) $5\sqrt{200} - 6\sqrt{18}$

④ Simplify each radical expression. Assume all variables represent non-negative real numbers.

(a) $\sqrt{7} \cdot \sqrt{21} + 2\sqrt{27}$

(b) $\sqrt{3r} \cdot \sqrt{6} + \sqrt{8r}$

(c) $y\sqrt{72} - \sqrt{18y^2}$

(d) $\sqrt[3]{81x^4} + 5\sqrt[3]{24x^4}$

(c)
$$
\begin{aligned}
2\sqrt{12} + 3\sqrt{75} &= 2(\sqrt{4} \cdot \sqrt{3}) + 3(\sqrt{25} \cdot \sqrt{3}) && \text{Product rule} \\
&= 2(2\sqrt{3}) + 3(5\sqrt{3}) && \sqrt{4} = 2;\ \sqrt{25} = 5 \\
&= 4\sqrt{3} + 15\sqrt{3} && \text{Multiply.} \\
&= 19\sqrt{3} && \text{Add like radicals.}
\end{aligned}
$$

Work Problem ③ at the Side.

3 ▭ **Simplify radical expressions involving multiplication.** Some radical expressions require both multiplication and addition (or subtraction). The order of operations (from Chapter 1) still applies.

Example 3 **Multiplying and Combining Terms in Radical Expressions**

Simplify each radical expression. As before, assume all variables represent nonnegative real numbers.

(a)
$$
\begin{aligned}
\sqrt{5} \cdot \sqrt{15} + 4\sqrt{3} &= \sqrt{5 \cdot 15} + 4\sqrt{3} && \text{Product rule} \\
&= \sqrt{75} + 4\sqrt{3} && \text{Multiply.} \\
&= \sqrt{25 \cdot 3} + 4\sqrt{3} && \text{25 is a perfect square.} \\
&= \sqrt{25} \cdot \sqrt{3} + 4\sqrt{3} && \text{Product rule} \\
&= 5\sqrt{3} + 4\sqrt{3} && \sqrt{25} = 5 \\
&= 9\sqrt{3} && \text{Add like radicals.}
\end{aligned}
$$

(b)
$$
\begin{aligned}
\sqrt{2} \cdot \sqrt{6k} + \sqrt{27k} &= \sqrt{12k} + \sqrt{27k} && \text{Product rule} \\
&= \sqrt{4 \cdot 3k} + \sqrt{9 \cdot 3k} && \text{Factor.} \\
&= \sqrt{4} \cdot \sqrt{3k} + \sqrt{9} \cdot \sqrt{3k} && \text{Product rule} \\
&= 2\sqrt{3k} + 3\sqrt{3k} && \sqrt{4} = 2;\ \sqrt{9} = 3 \\
&= 5\sqrt{3k} && \text{Add like radicals.}
\end{aligned}
$$

(c)
$$
\begin{aligned}
3x\sqrt{50} + \sqrt{2x^2} &= 3x\sqrt{25 \cdot 2} + \sqrt{x^2 \cdot 2} && \text{Factor.} \\
&= 3x\sqrt{25} \cdot \sqrt{2} + \sqrt{x^2} \cdot \sqrt{2} && \text{Product rule} \\
&= 3x \cdot 5\sqrt{2} + x\sqrt{2} && \sqrt{25} = 5;\ \sqrt{x^2} = x \\
&= 15x\sqrt{2} + x\sqrt{2} && \text{Multiply.} \\
&= 16x\sqrt{2} && \text{Add like radicals.}
\end{aligned}
$$

(d)
$$
\begin{aligned}
2\sqrt[3]{32m^3} - \sqrt[3]{108m^3} &= 2\sqrt[3]{(8m^3)4} - \sqrt[3]{(27m^3)4} && \text{Factor.} \\
&= 2(2m)\sqrt[3]{4} - 3m\sqrt[3]{4} && \sqrt[3]{8m^3} = 2m;\ \sqrt[3]{27m^3} = 3m \\
&= 4m\sqrt[3]{4} - 3m\sqrt[3]{4} && \text{Multiply.} \\
&= m\sqrt[3]{4} && \text{Subtract like radicals.}
\end{aligned}
$$

CAUTION

Remember that a sum or difference of radicals can be simplified only if the radicals are *like radicals*. For example, $\sqrt{5} + 3\sqrt{5} = 4\sqrt{5}$, but $\sqrt{5} + 5\sqrt{3}$ cannot be simplified further. Also, $2\sqrt{3} + 5\sqrt[3]{3}$ cannot be simplified further.

Work Problem ④ at the Side.

8.3 EXERCISES

| FOR EXTRA HELP | Student's Solutions Manual | MyMathLab.com | InterAct Math Tutorial Software | AW Math Tutor Center | MathXL www.mathxl.com | Digital Video Tutor CD 6 Videotape 16 |

Fill in each blank with the correct response.

1. $5\sqrt{2} + 6\sqrt{2} = (5 + 6)\sqrt{2} = 11\sqrt{2}$ is an example of the _____ property.

2. Like radicals have the same _____ of the same _____.

3. $\sqrt{5} + 5\sqrt{3}$ cannot be simplified because the _____ are different.

4. $4\sqrt[3]{2} + 3\sqrt{2}$ cannot be simplified because the _____ are different.

Simplify and add or subtract wherever possible. See Examples 1 and 2.

5. $14\sqrt{7} - 19\sqrt{7}$

6. $16\sqrt{2} - 18\sqrt{2}$

7. $\sqrt{17} + 4\sqrt{17}$

8. $5\sqrt{19} + \sqrt{19}$

9. $6\sqrt{7} - \sqrt{7}$

10. $11\sqrt{14} - \sqrt{14}$

11. $\sqrt{45} + 4\sqrt{20}$

12. $\sqrt{24} + 6\sqrt{54}$

13. $5\sqrt{72} - 3\sqrt{50}$

14. $6\sqrt{18} - 5\sqrt{32}$

15. $-5\sqrt{32} + 2\sqrt{98}$

16. $-4\sqrt{75} + 3\sqrt{12}$

17. $5\sqrt{7} - 3\sqrt{28} + 6\sqrt{63}$

18. $3\sqrt{11} + 5\sqrt{44} - 8\sqrt{99}$

19. $2\sqrt{8} - 5\sqrt{32} - 2\sqrt{48}$

20. $5\sqrt{72} - 3\sqrt{48} + 4\sqrt{128}$

21. $4\sqrt{50} + 3\sqrt{12} - 5\sqrt{45}$

22. $6\sqrt{18} + 2\sqrt{48} + 6\sqrt{28}$

23. $\frac{1}{4}\sqrt{288} + \frac{1}{6}\sqrt{72}$

24. $\frac{2}{3}\sqrt{27} + \frac{3}{4}\sqrt{48}$

Find the perimeter of each figure.

25.

$7\sqrt{2}$

$4\sqrt{2}$

26.

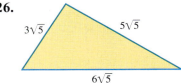

$3\sqrt{5}$ $5\sqrt{5}$

$6\sqrt{5}$

Perform the indicated operations. Assume that all variables represent nonnegative real numbers. See Example 3.

27. $\sqrt{6} \cdot \sqrt{2} + 9\sqrt{3}$

28. $4\sqrt{15} \cdot \sqrt{3} + 4\sqrt{5}$

29. $\sqrt{9x} + \sqrt{49x} - \sqrt{25x}$

30. $\sqrt{4a} - \sqrt{16a} + \sqrt{100a}$

31. $\sqrt{6x^2} + x\sqrt{24}$

32. $\sqrt{75x^2} + x\sqrt{108}$

33. $3\sqrt{8x^2} - 4x\sqrt{2} - x\sqrt{8}$

34. $\sqrt{2b^2} + 3b\sqrt{18} - b\sqrt{200}$

35. $-8\sqrt{32k} + 6\sqrt{8k}$

36. $4\sqrt{12x} + 2\sqrt{27x}$

37. $2\sqrt{125x^2z} + 8x\sqrt{80z}$

38. $\sqrt{48x^2y} + 5x\sqrt{27y}$

39. $4\sqrt[3]{16} - 3\sqrt[3]{54}$

40. $5\sqrt[3]{128} + 3\sqrt[3]{250}$

41. $6\sqrt[3]{8p^2} - 2\sqrt[3]{27p^2}$

42. $8k\sqrt[3]{54k} + 6\sqrt[3]{16k^4}$

43. $5\sqrt[4]{m^3} + 8\sqrt[4]{16m^3}$

44. $5\sqrt[4]{m^5} + 3\sqrt[4]{81m^5}$

RELATING CONCEPTS (Exercises 45–48) **FOR INDIVIDUAL OR GROUP WORK**

Adding and subtracting like radicals is no different than adding and subtracting like terms. **Work Exercises 45–48 in order.**

45. Combine like terms: $5x^2y + 3x^2y - 14x^2y$.

46. Combine like terms: $5(p - 2q)^2(a + b) + 3(p - 2q)^2(a + b) - 14(p - 2q)^2(a + b)$.

47. Combine like radicals: $5a^2\sqrt{xy} + 3a^2\sqrt{xy} - 14a^2\sqrt{xy}$.

48. Compare your answers in Exercises 45–47. How are they alike? How are they different?

8.4 RATIONALIZING THE DENOMINATOR

1 **Rationalize denominators with square roots.** Although calculators now make it fairly easy to divide by a radical in an expression such as $\frac{1}{\sqrt{2}}$, it is sometimes easier to work with radical expressions if the denominators do not contain any radicals. For example, the radical in the denominator of $\frac{1}{\sqrt{2}}$ can be eliminated by multiplying the numerator and denominator by $\sqrt{2}$, since $\sqrt{2} \cdot \sqrt{2} = \sqrt{4} = 2$.

$$\frac{1}{\sqrt{2}} = \frac{1 \cdot \sqrt{2}}{\sqrt{2} \cdot \sqrt{2}} = \frac{\sqrt{2}}{2} \qquad \text{Multiply by } \tfrac{\sqrt{2}}{\sqrt{2}} = 1.$$

This process of changing the denominator from a radical (irrational number) to a rational number is called **rationalizing the denominator.** The value of the radical expression is not changed; only the form is changed, because the expression has been multiplied by 1 in the form of $\frac{\sqrt{2}}{\sqrt{2}}$.

Example 1 **Rationalizing Denominators**

Rationalize each denominator.

(a) $\dfrac{9}{\sqrt{6}}$

To eliminate the radical in the denominator, multiply the numerator and denominator by $\sqrt{6}$.

$$\frac{9}{\sqrt{6}} = \frac{9 \cdot \sqrt{6}}{\sqrt{6} \cdot \sqrt{6}} \qquad \text{Multiply by } \tfrac{\sqrt{6}}{\sqrt{6}} = 1.$$

$$= \frac{9\sqrt{6}}{6} \qquad \sqrt{6} \cdot \sqrt{6} = \sqrt{36} = 6$$

$$= \frac{3\sqrt{6}}{2} \qquad \text{Lowest terms}$$

(b) $\dfrac{12}{\sqrt{8}}$

The denominator could be rationalized here by multiplying by $\sqrt{8}$. However, the result can be found more directly by first simplifying the denominator.

$$\sqrt{8} = \sqrt{4} \cdot \sqrt{2} = 2\sqrt{2}$$

Then multiply the numerator and denominator by $\sqrt{2}$.

$$\frac{12}{\sqrt{8}} = \frac{12}{2\sqrt{2}} \qquad \sqrt{8} = 2\sqrt{2}$$

$$= \frac{12 \cdot \sqrt{2}}{2\sqrt{2} \cdot \sqrt{2}} \qquad \text{Multiply by } \tfrac{\sqrt{2}}{\sqrt{2}}.$$

$$= \frac{12 \cdot \sqrt{2}}{2 \cdot 2} \qquad \sqrt{2} \cdot \sqrt{2} = \sqrt{4} = 2$$

$$= \frac{12\sqrt{2}}{4} \qquad \text{Multiply.}$$

$$= 3\sqrt{2} \qquad \text{Lowest terms}$$

❶ Rationalize each denominator.

(a) $\dfrac{3}{\sqrt{5}}$

(b) $\dfrac{-6}{\sqrt{11}}$

(c) $-\dfrac{\sqrt{7}}{\sqrt{2}}$

(d) $\dfrac{20}{\sqrt{18}}$

❷ Simplify.

(a) $\sqrt{\dfrac{16}{11}}$

(b) $\sqrt{\dfrac{5}{18}}$

(c) $\sqrt{\dfrac{8}{32}}$

ANSWERS

1. **(a)** $\dfrac{3\sqrt{5}}{5}$ **(b)** $\dfrac{-6\sqrt{11}}{11}$

   (c) $-\dfrac{\sqrt{14}}{2}$ **(d)** $\dfrac{10\sqrt{2}}{3}$

2. **(a)** $\dfrac{4\sqrt{11}}{11}$ **(b)** $\dfrac{\sqrt{10}}{6}$ **(c)** $\dfrac{1}{2}$

NOTE

In Example 1(b), we could also have rationalized the original denominator $\sqrt{8}$ by multiplying by $\sqrt{2}$, since $\sqrt{8} \cdot \sqrt{2} = \sqrt{16} = 4$.

$$\frac{12}{\sqrt{8}} = \frac{12 \cdot \sqrt{2}}{\sqrt{8} \cdot \sqrt{2}} = \frac{12\sqrt{2}}{\sqrt{16}} = \frac{12\sqrt{2}}{4} = 3\sqrt{2}$$

Both approaches are correct.

Work Problem ❶ at the Side.

2 Write radicals in simplified form.   A radical is considered to be in simplified form if the following three conditions are met.

Simplified Form of a Radical

1. The radicand contains no factor (except 1) that is a perfect square (when dealing with square roots), a perfect cube (when dealing with cube roots), and so on.

2. The radicand has no fractions.

3. No denominator contains a radical.

In the following examples, radicals are simplified according to these conditions.

Example 2   **Simplifying a Radical**

Simplify $\sqrt{\dfrac{27}{5}}$.

This violates condition 2. To begin, use the quotient rule for radicals.

$$\sqrt{\frac{27}{5}} = \frac{\sqrt{27}}{\sqrt{5}} \qquad \text{Quotient rule}$$

$$= \frac{\sqrt{27} \cdot \sqrt{5}}{\sqrt{5} \cdot \sqrt{5}} \qquad \text{Rationalize the denominator.}$$

$$= \frac{\sqrt{27} \cdot \sqrt{5}}{5} \qquad \sqrt{5} \cdot \sqrt{5} = 5$$

$$= \frac{\sqrt{9 \cdot 3} \cdot \sqrt{5}}{5} \qquad \text{Factor.}$$

$$= \frac{\sqrt{9} \cdot \sqrt{3} \cdot \sqrt{5}}{5} \qquad \text{Product rule}$$

$$= \frac{3 \cdot \sqrt{3} \cdot \sqrt{5}}{5} \qquad \sqrt{9} = 3$$

$$= \frac{3\sqrt{15}}{5} \qquad \text{Product rule}$$

Work Problem ❷ at the Side.

Example 3 Simplifying a Product of Radicals

Simplify $\sqrt{\dfrac{5}{8}} \cdot \sqrt{\dfrac{1}{6}}$.

Use both the product and quotient rules.

$$\sqrt{\frac{5}{8}} \cdot \sqrt{\frac{1}{6}} = \sqrt{\frac{5}{8} \cdot \frac{1}{6}} \qquad \text{Product rule}$$

$$= \sqrt{\frac{5}{48}} \qquad \text{Multiply fractions.}$$

$$= \frac{\sqrt{5}}{\sqrt{48}} \qquad \text{Quotient rule}$$

First simplify the denominator and then rationalize it.

$$\frac{\sqrt{5}}{\sqrt{48}} = \frac{\sqrt{5}}{\sqrt{16} \cdot \sqrt{3}} \qquad \text{Product rule}$$

$$= \frac{\sqrt{5}}{4\sqrt{3}} \qquad \sqrt{16} = 4$$

$$= \frac{\sqrt{5} \cdot \sqrt{3}}{4\sqrt{3} \cdot \sqrt{3}} \qquad \text{Rationalize the denominator.}$$

$$= \frac{\sqrt{15}}{4 \cdot 3} \qquad \text{Product rule; } \sqrt{3} \cdot \sqrt{3} = 3$$

$$= \frac{\sqrt{15}}{12} \qquad \text{Multiply.}$$

════ **Work Problem ❸ at the Side.**

Example 4 Simplifying a Quotient of Radicals

Simplify $\dfrac{\sqrt{4x}}{\sqrt{y}}$. Assume that x and y are positive real numbers.

Multiply the numerator and denominator by $\sqrt{y}$.

$$\frac{\sqrt{4x}}{\sqrt{y}} = \frac{\sqrt{4x} \cdot \sqrt{y}}{\sqrt{y} \cdot \sqrt{y}} = \frac{\sqrt{4xy}}{y} = \frac{2\sqrt{xy}}{y}$$

════ **Work Problem ❹ at the Side.**

---

❸ Simplify.

(a) $\sqrt{\dfrac{1}{2}} \cdot \sqrt{\dfrac{5}{6}}$

(b) $\sqrt{\dfrac{1}{10}} \cdot \sqrt{20}$

(c) $\sqrt{\dfrac{5}{8}} \cdot \sqrt{\dfrac{24}{10}}$

❹ Simplify $\dfrac{\sqrt{5p}}{\sqrt{q}}$. Assume that p and q are positive real numbers.

ANSWERS

3. (a) $\dfrac{\sqrt{15}}{6}$ (b) $\sqrt{2}$ (c) $\dfrac{\sqrt{6}}{2}$

4. $\dfrac{\sqrt{5pq}}{q}$

⑤ Simplify $\sqrt{\dfrac{5r^2t^2}{7}}$.

Assume that r and t represent nonnegative real numbers.

Example 5 Simplifying a Radical Quotient

Simplify $\sqrt{\dfrac{2x^2y}{3}}$. Assume that x and y are nonnegative real numbers.

$$\sqrt{\frac{2x^2y}{3}} = \frac{\sqrt{2x^2y}}{\sqrt{3}} \qquad \text{Quotient rule}$$

$$= \frac{\sqrt{2x^2y} \cdot \sqrt{3}}{\sqrt{3} \cdot \sqrt{3}} \qquad \text{Rationalize the denominator.}$$

$$= \frac{\sqrt{6x^2y}}{3} \qquad \text{Product rule}$$

$$= \frac{\sqrt{x^2}\sqrt{6y}}{3} \qquad \text{Product rule}$$

$$= \frac{x\sqrt{6y}}{3} \qquad \sqrt{x^2} = x, \text{ since } x \geq 0.$$

Work Problem ⑤ at the Side.

3 Rationalize denominators with cube roots. A denominator with a cube root is rationalized by changing the radicand in the denominator to a perfect cube, as shown in the next example.

Example 6 Rationalize Denominators with Cube Roots

Rationalize each denominator.

(a) $\sqrt[3]{\dfrac{3}{2}}$

First write the expression as a quotient of radicals. Then multiply the numerator and denominator by enough factors of 2 to make the denominator a perfect cube. This will eliminate the radical in the denominator. Here, multiply by $\sqrt[3]{2^2}$.

$$\sqrt[3]{\frac{3}{2}} = \frac{\sqrt[3]{3}}{\sqrt[3]{2}} = \frac{\sqrt[3]{3} \cdot \sqrt[3]{2^2}}{\sqrt[3]{2} \cdot \sqrt[3]{2^2}} = \frac{\sqrt[3]{3 \cdot 2^2}}{\sqrt[3]{2^3}} = \frac{\sqrt[3]{12}}{2} \qquad \sqrt[3]{2^3} = \sqrt[3]{8} = 2$$

Denominator is a perfect cube.

(b) $\dfrac{\sqrt[3]{3}}{\sqrt[3]{4}}$

Since $\sqrt[3]{4} \cdot \sqrt[3]{2} = \sqrt[3]{2^2} \cdot \sqrt[3]{2} = \sqrt[3]{2^3} = 2$, multiply the numerator and denominator by $\sqrt[3]{2}$.

$$\frac{\sqrt[3]{3}}{\sqrt[3]{4}} = \frac{\sqrt[3]{3} \cdot \sqrt[3]{2}}{\sqrt[3]{2^2} \cdot \sqrt[3]{2}} = \frac{\sqrt[3]{6}}{\sqrt[3]{2^3}} = \frac{\sqrt[3]{6}}{2}$$

Continued on Next Page

(c) $\dfrac{\sqrt[3]{2}}{\sqrt[3]{3x^2}}$ $(x \neq 0)$

Multiply the numerator and denominator by enough factors of 3 and of x to get a perfect cube in the denominator. Here, multiply by $\sqrt[3]{3^2x}$ (that is, $\sqrt[3]{9x}$) since $\sqrt[3]{3x^2} \cdot \sqrt[3]{3^2x} = \sqrt[3]{(3x)^3} = 3x$.

$$\frac{\sqrt[3]{2}}{\sqrt[3]{3x^2}} = \frac{\sqrt[3]{2} \cdot \sqrt[3]{3^2x}}{\sqrt[3]{3x^2} \cdot \sqrt[3]{3^2x}} = \frac{\sqrt[3]{18x}}{\sqrt[3]{(3x)^3}} = \frac{\sqrt[3]{18x}}{3x}$$

└── Denominator is a perfect cube.

CAUTION

A common error in a problem like the one in Example 6(a) is to multiply by $\sqrt[3]{2}$ instead of $\sqrt[3]{2^2}$. Doing this would give a denominator of $\sqrt[3]{2} \cdot \sqrt[3]{2} = \sqrt[3]{4}$. Because 4 is not a perfect cube, the denominator is still not rationalized.

Work Problem ❻ at the Side.

❻ Rationalize each denominator.

(a) $\sqrt[3]{\dfrac{5}{7}}$

(b) $\dfrac{\sqrt[3]{5}}{\sqrt[3]{9}}$

(c) $\dfrac{\sqrt[3]{4}}{\sqrt[3]{25y}}$ $(y \neq 0)$

Real-Data Applications

The Golden Ratio—A Star Number

The **Golden Ratio,** the number $\dfrac{1 + \sqrt{5}}{2}$, is called phi, ϕ. The number has been known since ancient times and is called the *sacred ratio* in the Rhind Papyrus from 1600 B.C. The Egyptians used ϕ to build the Great Pyramids, and the ancient Greeks used ϕ in art and architecture, striving for the proportion that was most pleasing to the eye.

The Golden Ratio is widespread in the star formed by connecting the vertices (corners) of a regular pentagon inscribed in a circle. In the figure shown, ABCDE forms a regular pentagon. Each of the following ratios forms the Golden Ratio, ϕ.

$$\frac{AC}{AY} \qquad \frac{AY}{AX}$$

Using symmetry, you should be able to identify other similar relationships.

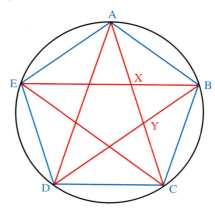

For Group Discussion

1. List two more ratios in the star that form the Golden Ratio. You may label additional points.

2. The Golden Ratio has some curious properties. Use $\phi = \dfrac{1 + \sqrt{5}}{2}$ to find the following quantities. Write your answers in exact form (using radicals). Rationalize denominators when appropriate.

   (a) $\dfrac{1}{\phi}$

   (b) $\phi - 1$

   (c) ϕ^2

   (d) $\phi + 1$

3. Based on your results from Problem 2, which forms in (a)–(d) are equivalent?

FOR EXTRA HELP

 Student's Solutions Manual

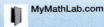 MyMathLab.com

 InterAct Math Tutorial Software

AW Math Tutor Center

Math XL www.mathxl.com

Digital Video Tutor CD 6 Videotape 17

Rationalize each denominator. See Examples 1 and 2.

1. $\dfrac{8}{\sqrt{2}}$

2. $\dfrac{12}{\sqrt{3}}$

3. $\dfrac{-\sqrt{11}}{\sqrt{3}}$

4. $\dfrac{-\sqrt{13}}{\sqrt{5}}$

5. $\dfrac{7\sqrt{3}}{\sqrt{5}}$

6. $\dfrac{4\sqrt{6}}{\sqrt{5}}$

7. $\dfrac{24\sqrt{10}}{16\sqrt{3}}$

8. $\dfrac{18\sqrt{15}}{12\sqrt{2}}$

9. $\dfrac{16}{\sqrt{27}}$

10. $\dfrac{24}{\sqrt{18}}$

11. $\dfrac{-3}{\sqrt{50}}$

12. $\dfrac{-5}{\sqrt{75}}$

13. $\dfrac{63}{\sqrt{45}}$

14. $\dfrac{27}{\sqrt{32}}$

15. $\dfrac{\sqrt{24}}{\sqrt{8}}$

16. $\dfrac{\sqrt{36}}{\sqrt{18}}$

17. $\sqrt{\dfrac{1}{2}}$

18. $\sqrt{\dfrac{1}{3}}$

19. $\sqrt{\dfrac{13}{5}}$

20. $\sqrt{\dfrac{17}{11}}$

21. When we rationalize the denominator in an expression such as $\dfrac{4}{\sqrt{3}}$, we multiply both the numerator and denominator by $\sqrt{3}$. By what number are we actually multiplying the given expression, and what property of real numbers justifies the fact that our result is equal to the given expression?

 **22.** In Example 1(a), we show algebraically that $\dfrac{9}{\sqrt{6}}$ is equal to $\dfrac{3\sqrt{6}}{2}$. Support this result numerically by finding the decimal approximation of $\dfrac{9}{\sqrt{6}}$ on your calculator, and then finding the decimal approximation of $\dfrac{3\sqrt{6}}{2}$. What do you notice?

Simplify each product of radicals. See Example 3.

23. $\sqrt{\dfrac{7}{13}} \cdot \sqrt{\dfrac{13}{3}}$

24. $\sqrt{\dfrac{19}{20}} \cdot \sqrt{\dfrac{20}{3}}$

25. $\sqrt{\dfrac{21}{7}} \cdot \sqrt{\dfrac{21}{8}}$

26. $\sqrt{\dfrac{5}{8}} \cdot \sqrt{\dfrac{5}{6}}$

27. $\sqrt{\dfrac{1}{12}} \cdot \sqrt{\dfrac{1}{3}}$

28. $\sqrt{\dfrac{1}{8}} \cdot \sqrt{\dfrac{1}{2}}$

29. $\sqrt{\dfrac{2}{9}} \cdot \sqrt{\dfrac{9}{2}}$

30. $\sqrt{\dfrac{4}{3}} \cdot \sqrt{\dfrac{3}{4}}$

Simplify each radical. Assume that all variables represent positive real numbers.
See Examples 4 and 5.

31. $\dfrac{\sqrt{7}}{\sqrt{x}}$

32. $\dfrac{\sqrt{19}}{\sqrt{y}}$

33. $\dfrac{\sqrt{4x^3}}{\sqrt{y}}$

34. $\dfrac{\sqrt{9t^3}}{\sqrt{s}}$

35. $\sqrt{\dfrac{5x^3z}{6}}$

36. $\sqrt{\dfrac{3st^3}{5}}$

37. $\sqrt{\dfrac{9a^2r^5}{7t}}$

38. $\sqrt{\dfrac{16x^3y^2}{13z}}$

39. Which one of the following would be an appropriate choice for multiplying the numerator and denominator of $\dfrac{\sqrt[3]{2}}{\sqrt[3]{5}}$ by in order to rationalize the denominator?

    A. $\sqrt[3]{5}$ **B.** $\sqrt[3]{25}$ **C.** $\sqrt[3]{2}$ **D.** $\sqrt[3]{3}$

40. In Example 6(b), we multiplied the numerator and denominator of $\dfrac{\sqrt[3]{3}}{\sqrt[3]{4}}$ by $\sqrt[3]{2}$ to rationalize the denominator. Suppose we had chosen to multiply by $\sqrt[3]{16}$ instead. Would we have obtained the correct answer after all simplifications were done?

Rationalize each denominator. Assume that variables in the denominator are nonzero.
See Example 6.

41. $\sqrt[3]{\dfrac{3}{2}}$

42. $\sqrt[3]{\dfrac{2}{5}}$

43. $\dfrac{\sqrt[3]{4}}{\sqrt[3]{7}}$

44. $\dfrac{\sqrt[3]{5}}{\sqrt[3]{10}}$

45. $\sqrt[3]{\dfrac{3}{4y^2}}$

46. $\sqrt[3]{\dfrac{3}{25x^2}}$

47. $\dfrac{\sqrt[3]{7m}}{\sqrt[3]{36n}}$

48. $\dfrac{\sqrt[3]{11p}}{\sqrt[3]{49q}}$

*In Exercises 49 and 50, **(a)** give the answer as a simplified radical and **(b)** use a calculator to give the answer correct to the nearest thousandth.*

49. The period p of a pendulum is the time it takes for it to swing from one extreme to the other and back again. The value of p in seconds is given by

$$p = k \cdot \sqrt{\dfrac{L}{g}},$$

where L is the length of the pendulum, g is the acceleration due to gravity, and k is a constant. Find the period when $k = 6$, $L = 9$ ft, and $g = 32$ ft per sec per sec.

50. The velocity v of a meteorite approaching Earth is given by

$$v = \dfrac{k}{\sqrt{d}}$$

km per sec, where d is its distance from the center of Earth and k is a constant. What is the velocity of a meteorite that is 6000 km away from the center of Earth, if $k = 450$?

8.5   MORE SIMPLIFYING AND OPERATIONS WITH RADICALS

The conditions for which a radical is in simplest form were listed in the previous section. Below is a set of guidelines to follow when you are simplifying radical expressions.

OBJECTIVES

1 Simplify products of radical expressions.

2 Use conjugates to rationalize denominators of radical expressions.

3 Write radical expressions with quotients in lowest terms.

Simplifying Radical Expressions

1. If a radical represents a rational number, use that rational number in place of the radical.

Examples: $\sqrt{49}$ is simplified by writing 7; $\sqrt{\dfrac{169}{9}} = \dfrac{13}{3}$.

2. If a radical expression contains products of radicals, use the product rule for radicals, $\sqrt{x} \cdot \sqrt{y} = \sqrt{xy}$, to get a single radical.

Examples: $\sqrt{3} \cdot \sqrt{2}$ is simplified to $\sqrt{6}$; $\sqrt{5} \cdot \sqrt{x} = \sqrt{5x}$.

3. If a radicand has a factor that is a perfect square, express the radical as the product of the positive square root of the perfect square and the remaining radical factor. A similar statement applies to higher roots.

Examples: $\sqrt{20}$ is simplified to

$$\sqrt{20} = \sqrt{4 \cdot 5} = \sqrt{4} \cdot \sqrt{5} = 2\sqrt{5};$$
$$\sqrt[3]{16} = \sqrt[3]{8 \cdot 2} = \sqrt[3]{8} \cdot \sqrt[3]{2} = 2\sqrt[3]{2}.$$

4. If a radical expression contains sums or differences of radicals, use the distributive property to combine like radicals.

Examples: $3\sqrt{2} + 4\sqrt{2} = 7\sqrt{2}$, but $3\sqrt{2} + 4\sqrt{3}$ cannot be simplified further.

5. Rationalize any denominator containing a radical.

Examples: $\dfrac{5}{\sqrt{3}}$ is rationalized as $\dfrac{5}{\sqrt{3}} = \dfrac{5 \cdot \sqrt{3}}{\sqrt{3} \cdot \sqrt{3}} = \dfrac{5\sqrt{3}}{3}$;

$$\sqrt{\dfrac{3}{2}} = \dfrac{\sqrt{3}}{\sqrt{2}} = \dfrac{\sqrt{3} \cdot \sqrt{2}}{\sqrt{2} \cdot \sqrt{2}} = \dfrac{\sqrt{6}}{2}.$$

1 **Simplify products of radical expressions.**   Use the above guidelines.

Example 1   **Multiplying Radical Expressions**

Find each product, and simplify.

(a) $\sqrt{5}(\sqrt{8} - \sqrt{32})$

Start by simplifying $\sqrt{8}$ and $\sqrt{32}$.

$$\sqrt{8} = 2\sqrt{2} \quad \text{and} \quad \sqrt{32} = 4\sqrt{2}.$$

Continued on Next Page

❶ Find each product, and simplify.

(a) $\sqrt{7}(\sqrt{2} + \sqrt{5})$

Now simplify inside the parentheses.

$$\sqrt{5}(\sqrt{8} - \sqrt{32}) = \sqrt{5}(2\sqrt{2} - 4\sqrt{2})$$
$$= \sqrt{5}(-2\sqrt{2}) \quad \text{Subtract like radicals.}$$
$$= -2\sqrt{5 \cdot 2} \quad \text{Product rule}$$
$$= -2\sqrt{10} \quad \text{Multiply.}$$

(b) $(\sqrt{3} + 2\sqrt{5})(\sqrt{3} - 4\sqrt{5})$

We can find the products of sums of radicals in the same way that we found the product of binomials in Chapter 4, using the FOIL method.

$$(\sqrt{3} + 2\sqrt{5})(\sqrt{3} - 4\sqrt{5})$$
$$= \underbrace{\sqrt{3}(\sqrt{3})}_{\text{First}} + \underbrace{\sqrt{3}(-4\sqrt{5})}_{\text{Outer}} + \underbrace{2\sqrt{5}(\sqrt{3})}_{\text{Inner}} + \underbrace{2\sqrt{5}(-4\sqrt{5})}_{\text{Last}}$$

(b) $\sqrt{2}(\sqrt{8} + \sqrt{20})$

$$= 3 - 4\sqrt{15} + 2\sqrt{15} - 8 \cdot 5 \quad \text{Product rule}$$
$$= 3 - 2\sqrt{15} - 40 \quad \text{Add like radicals.}$$
$$= -37 - 2\sqrt{15} \quad \text{Combine terms.}$$

(c) $(\sqrt{3} + \sqrt{21})(\sqrt{3} - \sqrt{7})$

$$= \sqrt{3}(\sqrt{3}) + \sqrt{3}(-\sqrt{7}) + \sqrt{21}(\sqrt{3})$$
$$\quad + \sqrt{21}(-\sqrt{7}) \quad \text{FOIL}$$
$$= 3 - \sqrt{21} + \sqrt{63} - \sqrt{147} \quad \text{Product rule}$$
$$= 3 - \sqrt{21} + \sqrt{9} \cdot \sqrt{7} - \sqrt{49} \cdot \sqrt{3} \quad \text{9 and 49 are perfect squares.}$$
$$= 3 - \sqrt{21} + 3\sqrt{7} - 7\sqrt{3} \quad \sqrt{9} = 3; \sqrt{49} = 7$$

(c)

$(\sqrt{2} + 5\sqrt{3})(\sqrt{3} - 2\sqrt{2})$

Since there are no like radicals, no terms can be combined.

Work Problem ❶ at the Side.

The special products of binomials discussed in Chapter 4 can be applied to radicals. Example 2 uses the rules for the square of a binomial,

$$(a + b)^2 = a^2 + 2ab + b^2 \quad \text{and} \quad (a - b)^2 = a^2 - 2ab + b^2.$$

(d)

$(\sqrt{2} - \sqrt{5})(\sqrt{10} + \sqrt{2})$

Example 2 Using Special Products with Radicals

Find each product.

(a) $(\sqrt{10} - 7)^2$

Follow the second pattern given above. Let $a = \sqrt{10}$ and $b = 7$.

$$(\sqrt{10} - 7)^2 = (\sqrt{10})^2 - 2(\sqrt{10})(7) + 7^2$$
$$= 10 - 14\sqrt{10} + 49 \quad (\sqrt{10})^2 = 10; 7^2 = 49$$
$$= 59 - 14\sqrt{10} \quad \text{Combine terms.}$$

(b) $(2\sqrt{3} + 4)^2 = (2\sqrt{3})^2 + 2(2\sqrt{3})(4) + 4^2 \quad a = 2\sqrt{3}; b = 4$
$$= 12 + 16\sqrt{3} + 16 \quad (2\sqrt{3})^2 = 4 \cdot 3 = 12$$
$$= 28 + 16\sqrt{3}$$

ANSWERS

1. (a) $\sqrt{14} + \sqrt{35}$
(b) $4 + 2\sqrt{10}$
(c) $11 - 9\sqrt{6}$
(d) $2\sqrt{5} + 2 - 5\sqrt{2} - \sqrt{10}$

Continued on Next Page

(c) $(5 - \sqrt{x})^2 = 5^2 - 2(5)(\sqrt{x}) + (\sqrt{x})^2$

$\qquad\qquad = 25 - 10\sqrt{x} + x$

CAUTION

Be careful! In Examples 2(a) and (b),

$$59 - 14\sqrt{10} \ne 45\sqrt{10} \quad \text{and} \quad 28 + 16\sqrt{3} \ne 44\sqrt{3}.$$

Only like radicals can be combined.

Work Problem ❷ at the Side.

Example 3 uses the rule for the product of the sum and difference of two terms,

$$(a + b)(a - b) = a^2 - b^2.$$

Example 3 **Using a Special Product with Radicals**

Find each product.

(a) $(4 + \sqrt{3})(4 - \sqrt{3})$

Follow the pattern given above. Let $a = 4$ and $b = \sqrt{3}$.

$(4 + \sqrt{3})(4 - \sqrt{3}) = 4^2 - (\sqrt{3})^2$

$\qquad\qquad\qquad\qquad = 16 - 3 \qquad 4^2 = 16; (\sqrt{3})^2 = 3$

$\qquad\qquad\qquad\qquad = 13$

(b) $(\sqrt{x} - \sqrt{6})(\sqrt{x} + \sqrt{6}) = (\sqrt{x})^2 - (\sqrt{6})^2$

$\qquad\qquad\qquad\qquad\qquad = x - 6 \quad (\sqrt{x})^2 = x; (\sqrt{6})^2 = 6$

Work Problem ❸ at the Side.

Notice that the results in Example 3 do not contain radicals. The pairs of expressions being multiplied, $4 + \sqrt{3}$ and $4 - \sqrt{3}$, and $\sqrt{x} - \sqrt{6}$ and $\sqrt{x} + \sqrt{6}$, are called **conjugates** of each other.

2 **Use conjugates to rationalize denominators of radical expressions.** Conjugates similar to those in Example 3 can be used to rationalize the denominators in more complicated quotients, such as

$$\frac{2}{4 - \sqrt{3}}.$$

By Example 3(a), if this denominator, $4 - \sqrt{3}$, is multiplied by $4 + \sqrt{3}$, then the product $(4 - \sqrt{3})(4 + \sqrt{3})$ is the rational number 13. Multiplying the numerator and denominator of the quotient by $4 + \sqrt{3}$ gives

$$\frac{2}{4 - \sqrt{3}} = \frac{2(4 + \sqrt{3})}{(4 - \sqrt{3})(4 + \sqrt{3})} = \frac{2(4 + \sqrt{3})}{13}.$$

The denominator has now been rationalized; it contains no radicals.

❷ Find each product. Simplify the answers.

(a) $(\sqrt{5} - 3)^2$

(b) $(4\sqrt{2} + 5)^2$

(c) $(6 + \sqrt{m})^2$

❸ Find each product. Simplify the answers.

(a) $(3 + \sqrt{5})(3 - \sqrt{5})$

(b) $(\sqrt{3} - 2)(\sqrt{3} + 2)$

(c)

$(\sqrt{5} + \sqrt{3})(\sqrt{5} - \sqrt{3})$

(d)

$(\sqrt{10} - \sqrt{y})(\sqrt{10} + \sqrt{y})$

4 Rationalize each denominator.

(a) $\dfrac{5}{4 + \sqrt{2}}$

(b) $\dfrac{\sqrt{5} + 3}{2 - \sqrt{5}}$

(c) $\dfrac{1}{\sqrt{6} + \sqrt{3}}$

(d) $\dfrac{7}{5 - \sqrt{x}}$

Using Conjugates to Simplify Radical Expressions

To simplify a radical expression with two terms in the denominator, where at least one of those terms is a radical, multiply both the numerator and the denominator by the conjugate of the denominator.

Example 4 Using Conjugates to Rationalize Denominators

Simplify by rationalizing each denominator.

(a) $\dfrac{5}{3 + \sqrt{5}}$

We can eliminate the radical in the denominator by multiplying both the numerator and denominator by $3 - \sqrt{5}$.

$$\dfrac{5}{3 + \sqrt{5}} = \dfrac{5(3 - \sqrt{5})}{(3 + \sqrt{5})(3 - \sqrt{5})} \qquad \text{Multiply by the conjugate.}$$

$$= \dfrac{5(3 - \sqrt{5})}{3^2 - (\sqrt{5})^2} \qquad (a + b)(a - b) = a^2 - b^2$$

$$= \dfrac{5(3 - \sqrt{5})}{9 - 5} \qquad 3^2 = 9; (\sqrt{5})^2 = 5$$

$$= \dfrac{5(3 - \sqrt{5})}{4} \qquad \text{Subtract.}$$

(b) $\dfrac{6 + \sqrt{2}}{\sqrt{2} - 5}$

Multiply the numerator and denominator by $\sqrt{2} + 5$.

$$\dfrac{6 + \sqrt{2}}{\sqrt{2} - 5} = \dfrac{(6 + \sqrt{2})(\sqrt{2} + 5)}{(\sqrt{2} - 5)(\sqrt{2} + 5)} \qquad \text{Multiply by the conjugate.}$$

$$= \dfrac{6\sqrt{2} + 30 + 2 + 5\sqrt{2}}{2 - 25} \qquad \begin{array}{l} \text{FOIL;} \\ (a + b)(a - b) = a^2 - b^2 \end{array}$$

$$= \dfrac{11\sqrt{2} + 32}{-23} \qquad \text{Combine terms.}$$

$$= \dfrac{-11\sqrt{2} - 32}{23} \qquad \dfrac{a}{-b} = \dfrac{-a}{b}$$

(c) $\dfrac{4}{3 - \sqrt{x}} = \dfrac{4(3 + \sqrt{x})}{(3 - \sqrt{x})(3 + \sqrt{x})}$

$$= \dfrac{4(3 + \sqrt{x})}{9 - x}$$

Work Problem 4 at the Side.

3 Write radical expressions with quotients in lowest terms.

Example 5 Writing a Radical Quotient in Lowest Terms

Write $\dfrac{3\sqrt{3} + 9}{12}$ in lowest terms.

Factor the numerator and denominator, and then use the fundamental property from Section 6.1 to divide out common factors.

$$\frac{3\sqrt{3} + 9}{12} = \frac{3(\sqrt{3} + 3)}{3(4)} = 1 \cdot \frac{\sqrt{3} + 3}{4} = \frac{\sqrt{3} + 3}{4}$$

CAUTION

An expression like the one in Example 5 can only be simplified by factoring a common factor from the denominator and *each* term of the numerator. For example,

$$\frac{4 + 8\sqrt{5}}{4} \ne 1 + 8\sqrt{5}.$$

First factor to get $\dfrac{4 + 8\sqrt{5}}{4} = \dfrac{4(1 + 2\sqrt{5})}{4} = 1 + 2\sqrt{5}.$

Work Problem 5 at the Side.

5 Write each quotient in lowest terms.

(a) $\dfrac{5\sqrt{3} - 15}{10}$

(b) $\dfrac{12 + 8\sqrt{5}}{16}$

Real-Data Applications

Spaceship Earth—A Geodesic Sphere

Geodesic domes became a popular design base for houses in the 1960s. The original patent was awarded to R. Buckminster Fuller in 1951. For the same square footage of interior space, a geodesic dome has less surface area and, thus, both reduces energy costs and is less prone to storm damage. One of the most famous geodesic domes is the full geodesic sphere Spaceship Earth, which is a major attraction at Epcot Center at Walt Disney World, Florida.

The sphere is made up of pyramids, shaped from three equilateral **facets,** or faces. A set of four pyramids form a **panel** that is also an equilateral triangle. There are 954 panels, each supporting 12 facets. Some of the facets are removed for support beams, so there are actually only 11,324 facets on the sphere. The structure of the panels and the facets are easily seen in a close-up view of Spaceship Earth.

The outside **surface area** of Spaceship Earth is approximately 150,000 ft². To envision the idea of surface area, think about the task of painting the geodesic sphere. The painted surface is the surface area.

Each of the facets is an equilateral triangle. The area of an equilateral triangle that has a side of length s is given by the formula

$$A = \frac{\sqrt{3}}{4}s^2.$$

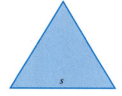

For Group Discussion

1. Use the given data for the outside surface area and the number of facets to find the surface area of one of Spaceship Earth's triangular facets. Round your answer to the nearest hundredth of a square foot. (Record the actual result for use in calculations for subsequent problems.)

2. Use the formula for the area of an equilateral triangle to find the length of the side of one of the triangular facets, to the nearest tenth of a foot.

3. How long is the side of one of the equilateral triangular panels?

4. Derive the formula for the area of an equilateral triangle. (*Hint:* Envision the right triangle that makes up half of the equilateral triangle, and use the Pythagorean formula to write an equation that relates the height h and the side s. Then use the formula for the area of a triangle.)

8.5 EXERCISES

 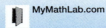

Based on the work so far in this chapter, you should now be able to mentally perform many simple operations involving radicals. In Exercises 1–4, perform the operations mentally, and write the answers without doing intermediate steps.

1. $\sqrt{49} + \sqrt{36}$

2. $\sqrt{100} - \sqrt{81}$

3. $\sqrt{2} \cdot \sqrt{8}$

4. $\sqrt{8} \cdot \sqrt{8}$

Simplify each expression. Use the five guidelines given in this section. See Examples 1–3.

5. $\sqrt{5}(\sqrt{3} - \sqrt{7})$

6. $\sqrt{7}(\sqrt{10} + \sqrt{3})$

7. $2\sqrt{5}(\sqrt{2} + 3\sqrt{5})$

8. $3\sqrt{7}(2\sqrt{7} + 4\sqrt{5})$

9. $3\sqrt{14} \cdot \sqrt{2} - \sqrt{28}$

10. $7\sqrt{6} \cdot \sqrt{3} - 2\sqrt{18}$

11. $(2\sqrt{6} + 3)(3\sqrt{6} + 7)$

12. $(4\sqrt{5} - 2)(2\sqrt{5} - 4)$

13. $(5\sqrt{7} - 2\sqrt{3})(3\sqrt{7} + 4\sqrt{3})$

14. $(2\sqrt{10} + 5\sqrt{2})(3\sqrt{10} - 3\sqrt{2})$

15. $(8 - \sqrt{7})^2$

16. $(6 - \sqrt{11})^2$

17. $(2\sqrt{7} + 3)^2$

18. $(4\sqrt{5} + 5)^2$

19. $(\sqrt{a} + 1)^2$

20. $(\sqrt{y} + 4)^2$

21. $(5 - \sqrt{2})(5 + \sqrt{2})$

22. $(3 - \sqrt{5})(3 + \sqrt{5})$

23. $(\sqrt{8} - \sqrt{7})(\sqrt{8} + \sqrt{7})$

24. $(\sqrt{12} - \sqrt{11})(\sqrt{12} + \sqrt{11})$

25. $(\sqrt{y} - \sqrt{10})(\sqrt{y} + \sqrt{10})$

26. $(\sqrt{t} - \sqrt{13})(\sqrt{t} + \sqrt{13})$

27. $(\sqrt{2} + \sqrt{3})(\sqrt{6} - \sqrt{2})$

28. $(\sqrt{3} + \sqrt{5})(\sqrt{15} - \sqrt{5})$

29. $(\sqrt{10} - \sqrt{5})(\sqrt{5} + \sqrt{20})$

30. $(\sqrt{6} - \sqrt{3})(\sqrt{3} + \sqrt{18})$

31. $(\sqrt{5} + \sqrt{30})(\sqrt{6} + \sqrt{3})$

32. $(\sqrt{10} - \sqrt{20})(\sqrt{2} - \sqrt{5})$

33. $(\sqrt{5} - \sqrt{10})(\sqrt{x} - \sqrt{2})$

34. $(\sqrt{x} + \sqrt{6})(\sqrt{10} + \sqrt{3})$

35. In Example 1(b), the original expression simplifies to $-37 - 2\sqrt{15}$. Students often try to simplify such expressions by combining -37 and -2 to get $-39\sqrt{15}$, which is incorrect. Explain why.

36. If you try to rationalize the denominator of $\dfrac{2}{4 + \sqrt{3}}$ by multiplying the numerator and denominator by $4 + \sqrt{3}$, what problem arises? What should you multiply by?

Rationalize each denominator. Write quotients in lowest terms. See Examples 4 and 5.

37. $\dfrac{1}{3 + \sqrt{2}}$

38. $\dfrac{1}{4 - \sqrt{3}}$

39. $\dfrac{14}{2 - \sqrt{11}}$

40. $\dfrac{19}{5 - \sqrt{6}}$

41. $\dfrac{\sqrt{2}}{2 - \sqrt{2}}$

42. $\dfrac{\sqrt{7}}{7 - \sqrt{7}}$

43. $\dfrac{\sqrt{5}}{\sqrt{2} + \sqrt{3}}$

44. $\dfrac{\sqrt{3}}{\sqrt{2} + \sqrt{3}}$

45. $\dfrac{\sqrt{5} + 2}{2 - \sqrt{3}}$

46. $\dfrac{\sqrt{7} + 3}{4 - \sqrt{5}}$

47. $\dfrac{12}{\sqrt{x} + 1}$

48. $\dfrac{10}{\sqrt{x} - 4}$

49. $\dfrac{3}{7 - \sqrt{x}}$

50. $\dfrac{1}{6 + \sqrt{z}}$

Write each quotient in lowest terms. See Example 5.

51. $\dfrac{6\sqrt{11} - 12}{6}$

52. $\dfrac{12\sqrt{5} - 24}{12}$

53. $\dfrac{2\sqrt{3} + 10}{16}$

54. $\dfrac{4\sqrt{6} + 24}{20}$

55. $\dfrac{12 - \sqrt{40}}{4}$

56. $\dfrac{9 - \sqrt{72}}{12}$

Work Exercises 57–62 in order, to see why a common student error is indeed an error.

57. Use the distributive property to write $6(5 + 3x)$ as a sum.

58. Your answer in Exercise 57 should be $30 + 18x$. Why can't we combine these two terms to get $48x$?

59. Repeat Exercise 14 from earlier in this exercise set.

60. Your answer in Exercise 59 should be $30 + 18\sqrt{5}$. Many students will, in error, try to combine these terms to get $48\sqrt{5}$. Why is this wrong?

61. Write the expression similar to $30 + 18x$ that simplifies to $48x$. Then write the expression similar to $30 + 18\sqrt{5}$ that simplifies to $48\sqrt{5}$.

62. Write a short paragraph explaining the similarities between combining like terms and combining like radicals.

Solve each problem.

63. The radius of the circular top or bottom of a tin can with a surface area S and a height h is given by

$$r = \frac{-h + \sqrt{h^2 + .64S}}{2}.$$

What radius should be used to make a can with a height of 12 in. and a surface area of 400 in.²?

64. If an investment of P dollars grows to A dollars in 2 yr, the annual rate of return on the investment is given by

$$r = \frac{\sqrt{A} - \sqrt{P}}{\sqrt{P}}.$$

Rationalize the denominator. Then find the annual rate of return r (as a percent) if $50,000 increases to $58,320.

8.6 SOLVING EQUATIONS WITH RADICALS

A **radical equation** is an equation with a variable in the radicand, such as

$$\sqrt{x + 1} = 3 \quad \text{or} \quad 3\sqrt{x} = \sqrt{8x + 9}.$$

1 **Solve radical equations.** The addition and multiplication properties of equality are not enough to solve radical equations. We need a new property, called the *squaring property*.

OBJECTIVES

1 Solve radical equations.

2 Identify equations with no solutions.

3 Solve equations by squaring a binomial.

4 Use a radical equation to model data.

Squaring Property of Equality

If each side of a given equation is squared, all solutions of the original equation are *among* the solutions of the squared equation.

CAUTION

Be very careful with the squaring property: Using this property can give a new equation with *more* solutions than the original equation. For example, starting with the equation $y = 4$ and squaring each side gives

$$y^2 = 4^2 \quad \text{or} \quad y^2 = 16.$$

This last equation, $y^2 = 16$, has *two* solutions, 4 or -4, while the original equation, $y = 4$, has only *one* solution, 4. Because of this possibility, checking is more than just a guard against algebraic errors when solving an equation with radicals. It is an essential part of the solution process. *All potential solutions from the squared equation must be checked in the original equation.*

1 Solve each equation. Be sure to check your solutions.

(a) $\sqrt{k} = 3$

(b) $\sqrt{x - 2} = 4$

Example 1 **Using the Squaring Property of Equality**

Solve $\sqrt{p + 1} = 3$.

Use the squaring property of equality to square each side of the equation.

$$(\sqrt{p + 1})^2 = 3^2$$

$$p + 1 = 9 \qquad (\sqrt{p + 1})^2 = p + 1$$

$$p = 8 \qquad \text{Subtract 1.}$$

Now check this potential solution in the original equation.

Check:

$$\sqrt{p + 1} = 3$$

$$\sqrt{8 + 1} = 3 \qquad ? \quad \text{Let } p = 8.$$

$$\sqrt{9} = 3 \qquad ?$$

$$3 = 3 \qquad \text{True}$$

Because this statement is true, 8 is the solution of $\sqrt{p + 1} = 3$. In this case the equation obtained by squaring had just one solution, which also satisfied the original equation.

(c) $\sqrt{9 - y} = 4$

=========== **Work Problem** **1** **at the Side.**

❷ Solve each equation.

(a) $\sqrt{3x + 9} = 2\sqrt{x}$

(b) $5\sqrt{x} = \sqrt{20x + 5}$

Example 2 **Using the Squaring Property with a Radical on Each Side**

Solve $3\sqrt{x} = \sqrt{x + 8}$.

Squaring each side gives

$$(3\sqrt{x})^2 = (\sqrt{x + 8})^2$$

$$3^2(\sqrt{x})^2 = (\sqrt{x + 8})^2 \qquad (ab)^2 = a^2b^2$$

$$9x = x + 8 \qquad (\sqrt{x})^2 = x;\ (\sqrt{x + 8})^2 = x + 8$$

$$8x = 8 \qquad \text{Subtract } x.$$

$$x = 1. \qquad \text{Divide by 8.}$$

Check:

$$3\sqrt{x} = \sqrt{x + 8} \qquad \text{Original equation}$$

$$3\sqrt{1} = \sqrt{1 + 8} \quad ? \qquad \text{Let } x = 1.$$

$$3(1) = \sqrt{9} \qquad ?$$

$$3 = 3 \qquad \text{True}$$

The solution of $3\sqrt{x} = \sqrt{x + 8}$ is 1.

CAUTION

Do not write the final result obtained in the check as the solution. In Example 2, the solution is 1, *not* 3.

Work Problem ❷ at the Side.

2 ▭ **Identify equations with no solutions.** Not all radical equations have solutions, as shown in Examples 3 and 4.

Example 3 **Using the Squaring Property When One Side Is Negative**

Solve $\sqrt{y} = -3$.

Square each side of the equation.

$$(\sqrt{y})^2 = (-3)^2$$

$$y = 9$$

Check this potential solution in the original equation.

Check:

$$\sqrt{y} = -3$$

$$\sqrt{9} = -3 \quad ? \qquad \text{Let } y = 9.$$

$$3 = -3 \qquad \text{False}$$

Because the statement $3 = -3$ is false, the number 9 is not a solution of the given equation and is said to be *extraneous*. In fact, $\sqrt{y} = -3$ has no solution.

NOTE

Because $\sqrt{y}$ represents the *principal* or *nonnegative* square root of y in Example 3, we might have seen immediately that there is no solution.

Example 4 Using the Squaring Property with a Quadratic Equation

Solve $p = \sqrt{p^2 + 5p + 10}$.

Square each side.

$$p^2 = (\sqrt{p^2 + 5p + 10})^2$$

$$p^2 = p^2 + 5p + 10 \qquad (\sqrt{p^2 + 5p + 10})^2 = p^2 + 5p + 10$$

$$0 = 5p + 10 \qquad \text{Subtract } p^2.$$

$$-10 = 5p \qquad \text{Subtract 10.}$$

$$p = -2 \qquad \text{Divide by 5.}$$

Check this potential solution in the original equation.

$Check:$ $\qquad p = \sqrt{p^2 + 5p + 10}$

$$-2 = \sqrt{(-2)^2 + 5(-2) + 10} \qquad ? \qquad \text{Let } p = -2.$$

$$-2 = \sqrt{4 - 10 + 10} \qquad ?$$

$$-2 = 2 \qquad\qquad\qquad \text{False}$$

Because $p = -2$ leads to a false result, the equation has no solution.

Work Problem ❸ at the Side.

3 ▨ **Solve equations by squaring a binomial.** The next examples use the following rules from Section 4.4.

$$(a + b)^2 = a^2 + 2ab + b^2$$

and

$$(a - b)^2 = a^2 - 2ab + b^2.$$

By these patterns, for example,

$$(y - 3)^2 = y^2 - 2y(3) + 3^2$$
$$= y^2 - 6y + 9.$$

Work Problem ❹ at the Side.

Example 5 Using the Squaring Property When One Side Has Two Terms

Solve $\sqrt{2y - 3} = y - 3$.

Square each side, using the preceding result to square the binomial on the right side of the equation.

$$(\sqrt{2y - 3})^2 = (y - 3)^2$$
$$2y - 3 = y^2 - 6y + 9$$

This equation is quadratic because of the y^2-term. As shown in Section 5.6, solving this equation requires that one side be equal to 0. Subtract $2y$ and add 3 on each side, getting

$$0 = y^2 - 8y + 12.$$

Solve this equation by factoring.

$$0 = (y - 6)(y - 2)$$

$$y - 6 = 0 \quad \text{or} \quad y - 2 = 0 \qquad \text{Zero-factor property}$$

$$y = 6 \quad \text{or} \qquad y = 2 \qquad \text{Solve.}$$

Continued on Next Page

---

❸ Solve each equation. (*Hint:* In part (a), subtract 4 from each side.)

(a) $\sqrt{y} + 4 = 0$

(b) $x = \sqrt{x^2 - 4x - 16}$

❹ Square each expression.

(a) $w - 5$

(b) $2k - 5$

(c) $3m - 2p$

❺ Solve each equation.

(a) $\sqrt{6w + 6} = w + 1$

Check *both* of these potential solutions in the original equation.

Check:

If $y = 6$, then

$$\sqrt{2y - 3} = y - 3$$
$$\sqrt{2(6) - 3} = 6 - 3 \quad ?$$
$$\sqrt{12 - 3} = 3 \quad ?$$
$$\sqrt{9} = 3 \quad ?$$
$$3 = 3. \quad \text{True}$$

If $y = 2$, then

$$\sqrt{2y - 3} = y - 3$$
$$\sqrt{2(2) - 3} = 2 - 3 \quad ?$$
$$\sqrt{4 - 3} = -1 \quad ?$$
$$\sqrt{1} = -1 \quad ?$$
$$1 = -1. \quad \text{False}$$

Only 6 is a valid solution of the equation.

Work Problem ❺ at the Side.

Sometimes it is necessary to write an equation in a different form before squaring each side. For example, suppose we want to solve $3\sqrt{x} - 1 = 2x$. Squaring each side gives

$$(3\sqrt{x} - 1)^2 = (2x)^2$$
$$9x - 6\sqrt{x} + 1 = 4x^2,$$

a more complicated equation that still contains a radical. In a case like this it would be better to rewrite the original equation so that the radical is alone on one side of the equals sign, as shown in Example 6.

(b) $2u - 1 = \sqrt{10u + 9}$

Example 6 **Rewriting an Equation before Using the Squaring Property**

Solve $3\sqrt{x} - 1 = 2x$.

Isolate the radical by adding 1 to each side.

$$3\sqrt{x} = 2x + 1$$
$$(3\sqrt{x})^2 = (2x + 1)^2 \qquad \text{Square each side.}$$
$$9x = 4x^2 + 4x + 1$$
$$0 = 4x^2 - 5x + 1 \qquad \text{Subtract } 9x.$$
$$0 = (4x - 1)(x - 1) \qquad \text{Factor.}$$
$$4x - 1 = 0 \quad \text{or} \quad x - 1 = 0 \qquad \text{Zero-factor property}$$
$$x = \frac{1}{4} \quad \text{or} \qquad x = 1 \qquad \text{Solve.}$$

Check:

If $x = \dfrac{1}{4}$, then

$$3\sqrt{x} - 1 = 2x$$
$$3\sqrt{\frac{1}{4}} - 1 = 2\left(\frac{1}{4}\right) \quad ?$$
$$\frac{1}{2} = \frac{1}{2}. \quad \text{True}$$

If $x = 1$, then

$$3\sqrt{x} - 1 = 2x$$
$$3\sqrt{1} - 1 = 2(1) \quad ?$$
$$2 = 2. \quad \text{True}$$

Both solutions check, so the solutions to the original equation are $\frac{1}{4}$ and 1.

CAUTION

Errors often occur when each side of an equation is squared. For instance, in Example 6 when each side of

$$3\sqrt{x} = 2x + 1$$

is squared, the *entire* binomial $2x + 1$ must be squared to get $4x^2 + 4x + 1$. It would be incorrect to square the $2x$ and the 1 separately to get $4x^2 + 1$.

Work Problem ⑥ at the Side.

Some radical equations require squaring twice, as in the next example.

Example 7   **Using the Squaring Property Twice**

Solve $\sqrt{21 + x} = 3 + \sqrt{x}$.

$$(\sqrt{21 + x})^2 = (3 + \sqrt{x})^2 \qquad \text{Square each side.}$$
$$21 + x = 9 + 6\sqrt{x} + x$$
$$12 = 6\sqrt{x} \qquad \text{Subtract 9; subtract } x.$$
$$2 = \sqrt{x} \qquad \text{Divide by 6.}$$
$$2^2 = (\sqrt{x})^2 \qquad \text{Square each side again.}$$
$$4 = x$$

Check: If $x = 4$, then

$$\sqrt{21 + x} = 3 + \sqrt{x} \qquad \text{Original equation}$$
$$\sqrt{21 + 4} = 3 + \sqrt{4} \qquad ?$$
$$5 = 5. \qquad \text{True}$$

The solution is 4.

Work Problem ⑦ at the Side.

In summary, use the following steps to solve a radical equation.

Solving a Radical Equation

Step 1   **Isolate a radical.** Arrange the terms so that a radical is alone on one side of the equation.

Step 2   **Square each side.**

Step 3   **Combine like terms.**

Step 4   **Repeat Steps 1–3, if necessary.** If there is still a term with a radical, repeat Steps 1–3.

Step 5   **Solve the equation.** Find all potential solutions.

Step 6   **Check.** All potential solutions *must* be checked in the original equation.

⑥ Solve each equation.

(a) $\sqrt{x - 3} = x - 15$

(b) $\sqrt{z + 5} + 2 = z + 5$

⑦ Solve the equation.
$$\sqrt{p + 1} - \sqrt{p - 4} = 1$$

ANSWERS
6. (a) 16   **(b)** -1
7. 8

8 Use the equation from Example 8 to approximate exports of electronics in 1996 to the nearest tenth. How does your answer compare to the actual data in the table?

4 **Use a radical equation to model data.** As with linear and quadratic equations, we can also use radical equations to model real-life situations.

Example 8 Using a Radical Equation to Model Exports

The table gives U.S. exports of electronics for selected years.

Billions of Dollars	Year
7.5	1990
19.6	1993
25.8	1994
36.4	1996

Source: U.S. Bureau of the Census; *U.S. Merchandise Trade,* series FT 900, December issue; and unpublished data.

Using the data from the table, the radical equation

$$y = 1.4\sqrt{x - 2.5} + 87.5$$

can be used to model the year y when exports were x billion dollars. Here, $y = 90$ represents 1990, $y = 93$ represents 1993, and so on.

(a) Use the equation to find the year when exports reached $50 billion. Substitute 50 for x in the equation.

$$y = 1.4\sqrt{\mathbf{50} - 2.5} + 87.5 \qquad \text{Let } x = 50.$$
$$y \approx 97.149 \qquad \text{Use a calculator.}$$

Exports reached $50 billion in 1997.

(b) Use the equation to approximate exports of electronics in 1998. Let $y = 98$ in the equation and solve for x.

$$\mathbf{98} = 1.4\sqrt{x - 2.5} + 87.5$$
$$10.5 = 1.4\sqrt{x - 2.5} \qquad \text{Subtract 87.5.}$$
$$7.5 = \sqrt{x - 2.5} \qquad \text{Divide by 1.4.}$$
$$7.5^2 = (\sqrt{x - 2.5})^2 \qquad \text{Square each side.}$$
$$56.25 = x - 2.5$$
$$58.75 = x \qquad \text{Add 2.5.}$$

This solution checks. Based on the equation, exports of electronics in 1998 were $58.75 billion.

(c) Actual exports in 1998 were $58.4 billion. How does the result of part (b) compare to the actual amount?

The approximation using the equation in part (b) is slightly high, but quite close to the actual data.

Work Problem 8 at the Side.

SUMMARY

KEY TERMS

8.1	square root	The number b is a square root of a if $b^2 = a$.
	principal square root	The positive square root of a number is its principal square root.
	radicand	The number or expression inside a radical sign is called the radicand.
	radical	A radical sign with a radicand is called a radical.
	radical expression	An algebraic expression containing a radical is called a radical expression.
	perfect square	A number with a rational square root is called a perfect square.
	cube root	The number b is a cube root of a if $b^3 = a$.
	index (order)	In a radical of the form $\sqrt[n]{a}$, the number n is the index or order.
8.2	**perfect cube**	A number with a rational cube root is called a perfect cube.
8.3	**like radicals**	Like radicals are multiples of the same root of the same number.
8.4	**rationalizing the denominator**	The process of changing the denominator of a fraction from a radical (irrational number) to a rational number is called rationalizing the denominator.
8.5	**conjugate**	The conjugate of $a + b$ is $a - b$.
8.6	**radical equation**	An equation with a variable in the radicand is a radical equation.

Index
Radical sign $\sqrt[n]{a}$ *Radicand*
Radical

NEW SYMBOLS

$\sqrt{}$ radical sign $\qquad$ $\approx$ is approximately equal to $\sqrt[3]{a}$ cube root of a 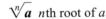 $\sqrt[n]{a}$ nth root of a

TEST YOUR WORD POWER

See how well you have learned the vocabulary in this chapter. Answers follow the Quick Review.

1. The **square root** of a number is
   (a) the number raised to the second power
   (b) the number under a radical sign
   (c) a number that when multiplied by itself gives the original number
   (d) the inverse of the number.

2. A **radical** is
   (a) a symbol that indicates the nth root
   (b) an algebraic expression containing a square root
   (c) the positive nth root of a number
   (d) a radical sign and the number or expression under it.

3. The **principal root** of a positive number with even index n is
   (a) the positive nth root of the number
   (b) the negative nth root of the number
   (c) the square root of the number
   (d) the cube root of the number.

4. **Like radicals** are
   (a) radicals in simplest form
   (b) algebraic expressions containing radicals
   (c) multiples of the same root of the same number
   (d) radicals with the same index.

5. **Rationalizing the denominator** is the process of
   (a) eliminating fractions from a radical expression
   (b) changing the denominator of a fraction from a radical to a rational number
   (c) clearing a radical expression of radicals
   (d) multiplying radical expressions.

6. The **conjugate** of $a + b$ is
   (a) $a - b$
   (b) $a \cdot b$
   (c) $a \div b$
   (d) $(a + b)^2$.

Concepts	Examples

8.1 Evaluating Roots

If a is a positive real number,

$\sqrt{a}$ is the positive or principal square root of a;

$-\sqrt{a}$ is the negative square root of a; $\sqrt{0} = 0$.

If a is a negative real number, $\sqrt{a}$ is not a real number.

If a is a positive rational number, $\sqrt{a}$ is rational if a is a perfect square. $\sqrt{a}$ is irrational if a is not a perfect square.

Every real number has exactly one real cube root.

$$\sqrt{49} = 7$$

$$-\sqrt{81} = -9$$

$$\sqrt{-25} \text{ is not a real number.}$$

$\sqrt{\dfrac{4}{9}}$ and $\sqrt{16}$ are rational. $\sqrt{\dfrac{2}{3}}$ and $\sqrt{21}$ are irrational.

$$\sqrt[3]{27} = 3; \quad \sqrt[3]{-8} = -2$$

8.2 Multiplying, Dividing, and Simplifying Radicals

Product Rule for Radicals

For nonnegative real numbers a and b,

$$\sqrt{a} \cdot \sqrt{b} = \sqrt{ab} \quad \text{and} \quad \sqrt{ab} = \sqrt{a} \cdot \sqrt{b}.$$

Quotient Rule for Radicals

If a and b are nonnegative real numbers and $b \neq 0$, then

$$\sqrt{\frac{a}{b}} = \frac{\sqrt{a}}{\sqrt{b}} \quad \text{and} \quad \frac{\sqrt{a}}{\sqrt{b}} = \sqrt{\frac{a}{b}}.$$

$$\sqrt{5} \cdot \sqrt{7} = \sqrt{5 \cdot 7} = \sqrt{35}$$
$$\sqrt{8} \cdot \sqrt{2} = \sqrt{16} = 4$$
$$\sqrt{48} = \sqrt{16 \cdot 3} = \sqrt{16} \cdot \sqrt{3} = 4\sqrt{3}$$

$$\sqrt{\frac{25}{64}} = \frac{\sqrt{25}}{\sqrt{64}} = \frac{5}{8}$$

$$\frac{\sqrt{8}}{\sqrt{2}} = \sqrt{\frac{8}{2}} = \sqrt{4} = 2$$

8.3 Adding and Subtracting Radicals

Add and subtract like radicals by using the distributive property. Only like radicals can be combined in this way.

$$2\sqrt{5} + 4\sqrt{5} = (2 + 4)\sqrt{5}$$
$$= 6\sqrt{5}$$
$$\sqrt{8} + \sqrt{32} = 2\sqrt{2} + 4\sqrt{2}$$
$$= 6\sqrt{2}$$

8.4 Rationalizing the Denominator

The denominator of a radical can be rationalized by multiplying both the numerator and denominator by a number that will eliminate the radical from the denominator.

$$\frac{2}{\sqrt{3}} = \frac{2 \cdot \sqrt{3}}{\sqrt{3} \cdot \sqrt{3}} = \frac{2\sqrt{3}}{3}$$

$$\sqrt[3]{\frac{5}{121}} = \frac{\sqrt[3]{5} \cdot \sqrt[3]{11}}{\sqrt[3]{11^2} \cdot \sqrt[3]{11}} = \frac{\sqrt[3]{55}}{11}$$

Concepts	Examples
8.5 *More Simplifying and Operations with Radicals* When appropriate, use the rules for adding and multiplying polynomials to simplify radical expressions.	$\sqrt{6}(\sqrt{5} - \sqrt{7}) = \sqrt{30} - \sqrt{42}$ $(\sqrt{3} + 1)(\sqrt{3} - 2) = 3 - 2\sqrt{3} + \sqrt{3} - 2$ FOIL $\qquad\qquad\qquad\quad = 1 - \sqrt{3}$ Combine terms. $(\sqrt{13} - \sqrt{2})^2 = (\sqrt{13})^2 - 2(\sqrt{13})(\sqrt{2}) + (\sqrt{2})^2$ $\qquad\qquad\quad = 13 - 2\sqrt{26} + 2$ $\qquad\qquad\quad = 15 - 2\sqrt{26}$ $(\sqrt{5} - \sqrt{3})(\sqrt{5} + \sqrt{3}) = 5 - 3 = 2$
If a radical expression contains two terms in the denominator and at least one of those terms is a square root radical, multiply both the numerator and denominator by the conjugate of the denominator.	$\dfrac{6}{\sqrt{7} - \sqrt{2}} = \dfrac{6(\sqrt{7} + \sqrt{2})}{(\sqrt{7} - \sqrt{2})(\sqrt{7} + \sqrt{2})}$ $\qquad\quad = \dfrac{6(\sqrt{7} + \sqrt{2})}{7 - 2}$ Multiply. $\qquad\quad = \dfrac{6(\sqrt{7} + \sqrt{2})}{5}$ Subtract.
8.6 *Solving Equations with Radicals* **Solving a Radical Equation** *Step 1* Isolate a radical. *Step 2* Square each side. (By the squaring property of equality, all solutions of the original equation are *among* the solutions of the squared equation.) *Step 3* Combine like terms. *Step 4* If there is still a term with a radical, repeat Steps 1–3. *Step 5* Solve the equation for potential solutions. *Step 6* Check all potential solutions from Step 5 in the original equation.	Solve $\sqrt{2x - 3} + x = 3$. $\qquad \sqrt{2x - 3} = 3 - x$ Isolate the radical. $\qquad (\sqrt{2x - 3})^2 = (3 - x)^2$ Square each side. $\qquad\quad 2x - 3 = 9 - 6x + x^2$ $\qquad\quad 0 = x^2 - 8x + 12$ Standard form $\qquad\quad 0 = (x - 2)(x - 6)$ Factor. $\quad x - 2 = 0$ or $x - 6 = 0$ Zero-factor property $\qquad x = 2$ or $\qquad x = 6$ Solve. A check is essential here. Verify that 2 is the only solution. (6 is extraneous.)

ANSWERS TO TEST YOUR WORD POWER

1. **(c)** *Examples:* 6 is a square root of 36 since $6^2 = 6 \cdot 6 = 36$; -6 is also a square root of 36.

2. **(d)** *Examples:* $\sqrt{144}$, $\sqrt{4xy^2}$, and $\sqrt{4 + t^2}$ 3. **(a)** *Examples:* $\sqrt{36} = 6$, $\sqrt[4]{81} = 3$, and $\sqrt[6]{64} = 2$

4. **(c)** *Examples:* $\sqrt{7}$ and $3\sqrt{7}$ are like radicals; so are $2\sqrt[3]{6k}$ and $5\sqrt[3]{6k}$.   5. **(b)** *Example:* To rationalize

the denominator of $\dfrac{5}{\sqrt{3} + 1}$, multiply the numerator and denominator by $\sqrt{3} - 1$ to get $\dfrac{5(\sqrt{3} - 1)}{2}$.

6. **(a)** *Example:* The conjugate of $\sqrt{3} + 1$ is $\sqrt{3} - 1$.

On a Clear Day

The Empire State Building, the Eiffel Tower, and the world's tallest buildings evoke images of sitting on top of the world. The prospect of viewing sights from such lofty heights stirs our imaginations. But how far can we really see? On a clear day, the maximum distance in kilometers that you can see from a tall building is given by the formula

$$\text{sight distance} = 111.7\sqrt{\text{height of building in kilometers}}.$$

(*Source: A Sourcebook of Applications of School Mathematics*, NCTM, 1980.)

Recall that 1 ft ≈ .3048 m and 1 km ≈ .621371 mi.

For Group Discussion

On a clear day, how far could you see from the top of each of these world-famous buildings and structures? Round your answers to the nearest mile.

1. The London Eye, which opened on New Year's Eve 1999, is a unique form of a Ferris wheel that features 32 observation capsules. It is located on the bank of the Thames River facing the Houses of Parliament and is the fourth tallest structure in London, with a diameter of 135 m. (*Source:* www.londoneye.com) Does the formula justify the claim that on a clear day, passengers on the London Eye can see Windsor Castle, 25 mi away?

2. The Empire State Building opened in 1931 on 5th Avenue in New York City. The building is 1250 ft high. (The antenna reaches to 1454 ft.) The observation deck, located on the 102nd floor, is at a height of 1050 ft. (*Source:* www.esbnyc.com) How far could you see on a clear day from the observation deck?

3. The twin Petronas Towers in Kuala Lumpur, Malaysia, are listed as the tallest buildings in the world. (*Source: World Almanac and Book of Facts*, 2000.) Built in 1998, both towers are 1483 ft high (including the spires). How far would one of the builders have been able to see on a clear day from the top of a spire?

4. The Khufu Pyramid in Giza (also known as Cheops Pyramid) was built in about 2566 B.C. to a height, at that time, of 482 ft. It is now only about 450 ft high. (*Source:* www.touregypt.net/cheop.htm) How far would one of the original builders of the pyramid have been able to see from the top of the pyramid?

Chapter 8 — REVIEW EXERCISES

[8.1] *Find all square roots of each number.*

1. 49   **2.** 81   **3.** 196   **4.** 121   **5.** 225   **6.** 729

Find each root.

7. $\sqrt{16}$ **8.** $-\sqrt{36}$ **9.** $\sqrt[3]{1000}$ **10.** $\sqrt[4]{81}$

11. $\sqrt{-8100}$ **12.** $-\sqrt{4225}$ **13.** $\sqrt{\dfrac{49}{36}}$ **14.** $\sqrt{\dfrac{100}{81}}$

Match each radical in Column I with the equivalent choice in Column II. Choices may be used more than once.

I	II
15. $\sqrt{64}$	**A.** 4
16. $-\sqrt{64}$	**B.** 8
17. $\sqrt{-64}$	**C.** -4
18. $\sqrt[3]{64}$	**D.** Not a real number
19. $\sqrt[3]{-64}$	**E.** 16
20. $-\sqrt[3]{-64}$	**F.** -8

21. Find the value of x.

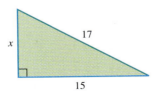

22. A Gateway EV700 computer monitor has viewing screen dimensions as shown in the figure. Find the diagonal measure of the viewing screen to the nearest tenth. (*Source:* Author's computer.)

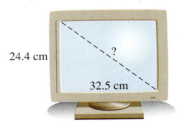

24.4 cm   ?   32.5 cm

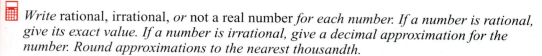 *Write* rational, irrational, *or* not a real number *for each number. If a number is rational, give its exact value. If a number is irrational, give a decimal approximation for the number. Round approximations to the nearest thousandth.*

23. $\sqrt{23}$ **24.** $\sqrt{169}$ **25.** $-\sqrt{25}$ **26.** $\sqrt{-4}$

[8.2] *Use the product rule to simplify each expression.*

27. $\sqrt{2} \cdot \sqrt{7}$ **28.** $\sqrt{5} \cdot \sqrt{15}$ **29.** $-\sqrt{27}$ **30.** $\sqrt{48}$

31. $\sqrt{160}$ **32.** $\sqrt{12} \cdot \sqrt{27}$ **33.** $\sqrt{32} \cdot \sqrt{48}$ **34.** $\sqrt{50} \cdot \sqrt{125}$

Use the product and quotient rules, as necessary, to simplify each expression.

35. $\sqrt{\dfrac{9}{4}}$ **36.** $-\sqrt{\dfrac{121}{400}}$ **37.** $\sqrt{\dfrac{7}{169}}$ **38.** $\sqrt{\dfrac{1}{6}} \cdot \sqrt{\dfrac{5}{6}}$

39. $\sqrt{\dfrac{2}{5}} \cdot \sqrt{\dfrac{2}{45}}$ **40.** $\dfrac{3\sqrt{10}}{\sqrt{5}}$ **41.** $\dfrac{24\sqrt{12}}{6\sqrt{3}}$ **42.** $\dfrac{8\sqrt{150}}{4\sqrt{75}}$

Simplify each expression. Assume that all variables represent nonnegative real numbers.

43. $\sqrt{p} \cdot \sqrt{p}$ **44.** $\sqrt{k} \cdot \sqrt{m}$ **45.** $\sqrt{r^{18}}$ **46.** $\sqrt{x^{10}y^{16}}$

47. $\sqrt{x^9}$ **48.** $\sqrt{\dfrac{36}{p^2}}, p \neq 0$ **49.** $\sqrt{a^{15}b^{21}}$

50. $\sqrt{121x^6y^{10}}$ **51.** $\sqrt[3]{y^6}$ **52.** $\sqrt[3]{216x^{15}}$

53. Use a calculator to find approximations for $\sqrt{.5}$ and $\dfrac{\sqrt{2}}{2}$. Based on your results, do you think that these two expressions represent the same number? If so, verify it *algebraically*.

[8.3] *Simplify and combine terms where possible.*

54. $\sqrt{11} + \sqrt{11}$

55. $3\sqrt{2} + 6\sqrt{2}$

56. $3\sqrt{75} + 2\sqrt{27}$

57. $4\sqrt{12} + \sqrt{48}$

58. $4\sqrt{24} - 3\sqrt{54} + \sqrt{6}$

59. $2\sqrt{7} - 4\sqrt{28} + 3\sqrt{63}$

60. $\dfrac{2}{5}\sqrt{75} + \dfrac{3}{4}\sqrt{160}$

61. $\dfrac{1}{3}\sqrt{18} + \dfrac{1}{4}\sqrt{32}$

62. $\sqrt{15} \cdot \sqrt{2} + 5\sqrt{30}$

Simplify each expression. Assume that all variables represent nonnegative real numbers.

63. $\sqrt{4x} + \sqrt{36x} - \sqrt{9x}$

64. $\sqrt{16p} + 3\sqrt{p} - \sqrt{49p}$

65. $\sqrt{20m^2} - m\sqrt{45}$

66. $3k\sqrt{8k^2n} + 5k^2\sqrt{2n}$

[8.4] *Perform the indicated operations, and write all answers in simplest form. Rationalize all denominators. Assume that all variables represent nonnegative real numbers.*

67. $\dfrac{10}{\sqrt{3}}$

68. $\dfrac{8\sqrt{2}}{\sqrt{5}}$

69. $\dfrac{12}{\sqrt{24}}$

70. $\sqrt{\dfrac{2}{5}}$

71. $\sqrt{\dfrac{5}{14}} \cdot \sqrt{28}$

72. $\sqrt{\dfrac{2}{7}} \cdot \sqrt{\dfrac{1}{3}}$

73. $\sqrt{\dfrac{r^2}{16x}}, \; x \neq 0$

74. $\sqrt[3]{\dfrac{1}{3}}$

Solve each problem.

75. The radius r of a cone in terms of its volume V is given by the formula

$$r = \sqrt{\frac{3V}{\pi h}}.$$

Rationalize the denominator of the radical expression.

76. The radius r of a sphere in terms of its surface area S is given by the formula

$$r = \sqrt{\frac{S}{4\pi}}.$$

Rationalize the denominator of the radical expression.

[8.5] *Simplify each expression.*

77. $-\sqrt{3}(\sqrt{5} + \sqrt{27})$

78. $3\sqrt{2}(\sqrt{3} + 2\sqrt{2})$

79. $(2\sqrt{3} - 4)(5\sqrt{3} + 2)$

80. $(\sqrt{7} + 2\sqrt{6})(\sqrt{12} - \sqrt{2})$

81. $(2\sqrt{3} + 5)(2\sqrt{3} - 5)$

82. $(\sqrt{x} + 2)^2$

Rationalize each denominator.

83. $\dfrac{1}{2 + \sqrt{5}}$

84. $\dfrac{2}{\sqrt{2} - 3}$

85. $\dfrac{3}{1 + \sqrt{x}}$

86. $\dfrac{\sqrt{8}}{\sqrt{2} + 6}$

87. $\dfrac{\sqrt{5} - 1}{\sqrt{2} + 3}$

88. $\dfrac{2 + \sqrt{6}}{\sqrt{3} - 1}$

Write each quotient in lowest terms.

89. $\dfrac{15 + 10\sqrt{6}}{15}$

90. $\dfrac{3 + 9\sqrt{7}}{12}$

91. $\dfrac{6 + \sqrt{192}}{2}$

[8.6] *Solve each equation.*

92. $\sqrt{x} + 5 = 0$

93. $\sqrt{k + 1} = 7$

94. $\sqrt{5t + 4} = 3\sqrt{t}$

95. $\sqrt{2p + 3} = \sqrt{5p - 3}$

96. $\sqrt{4y + 1} = y - 1$

97. $\sqrt{13 + 4t} = t + 4$

98. $\sqrt{2 - x} + 3 = x + 7$

99. $\sqrt{x} - x + 2 = 0$

100. $\sqrt{x + 2} - \sqrt{x - 3} = 1$

MIXED REVIEW EXERCISES

Simplify each expression if possible. Assume all variables represent nonnegative real numbers.

101. $\sqrt{3} \cdot \sqrt{27}$

102. $2\sqrt{27} + 3\sqrt{75} - \sqrt{300}$

103. $\sqrt{\dfrac{121}{t^2}}, t \neq 0$

104. $\dfrac{1}{5 + \sqrt{2}}$

105. $\sqrt{\dfrac{1}{3}} \cdot \sqrt{\dfrac{24}{5}}$

106. $\sqrt{50y^2}$

107. $\sqrt[3]{-125}$

108. $-\sqrt{5}(\sqrt{2} + \sqrt{75})$

109. $\sqrt{\dfrac{16r^3}{3s}}, s \neq 0$

110. $\dfrac{12 + 6\sqrt{13}}{12}$

111. $-\sqrt{162} + \sqrt{8}$

112. $(\sqrt{5} - \sqrt{2})^2$

113. $(6\sqrt{7} + 2)(4\sqrt{7} - 1)$

114. $-\sqrt{121}$

115. $\sqrt{98}$

Solve.

116. $\sqrt{x + 2} = x - 4$ **117.** $\sqrt{k} + 3 = 0$ **118.** $\sqrt{1 + 3t} - t = -3$

119. The *fall speed*, in miles per hour, of a vehicle running off the road into a ditch is given by the formula

$$S = \frac{2.74D}{\sqrt{h}},$$

where D is the horizontal distance traveled from the level surface to the bottom of the ditch and h is the height (or depth) of the ditch. What is the fall speed (to the nearest tenth) of a vehicle that traveled 32 ft horizontally into a ditch 5 ft deep?

RELATING CONCEPTS (Exercises 120–124)   **FOR INDIVIDUAL OR GROUP WORK**

In Chapter 3 we plotted points in the rectangular coordinate plane. In all cases our points had coordinates that were rational numbers. However, ordered pairs may have irrational coordinates as well. Consider the points $A(2\sqrt{14}, 5\sqrt{7})$ and $B(-3\sqrt{14}, 10\sqrt{7})$. **Work Exercises 120–124 in order.**

120. Write an expression that represents the slope of the line containing points A and B. Do not simplify yet.

121. Simplify the numerator and the denominator in the expression from Exercise 120 by combining like radicals.

122. Write the fraction from Exercise 121 as the square root of a fraction in lowest terms.

123. Rationalize the denominator of the expression found in Exercise 122.

124. Based on your answer in Exercise 123, does line AB rise or fall from left to right?

Chapter 8 TEST

On this test, assume that all variables represent nonnegative real numbers.

1. Find all square roots of 196.

1. _____

2. Consider $\sqrt{142}$.

    (a) Determine whether it is rational or irrational.

    (b) Find a decimal approximation to the nearest thousandth.

2. (a) _____

   (b) _____

3. If $\sqrt{a}$ is not a real number, then what kind of number must a be?

3. _____

Simplify where possible.

4. $\sqrt[3]{216}$

4. _____

5. $-\sqrt{27}$

5. _____

6. $\sqrt{\dfrac{128}{25}}$

6. _____

7. $\sqrt[3]{32}$

7. _____

8. $\dfrac{20\sqrt{18}}{5\sqrt{3}}$

8. _____

9. $3\sqrt{28} + \sqrt{63}$

9. _____

10. $3\sqrt{27x} - 4\sqrt{48x} + 2\sqrt{3x}$

10. _____

11. $\sqrt{32x^2y^3}$

11. _____

12. $(6 - \sqrt{5})(6 + \sqrt{5})$

12. _____

13. $(2 - \sqrt{7})(3\sqrt{2} + 1)$

13. _____

14. $(\sqrt{5} + \sqrt{6})^2$

14. _____

Solve each problem.

15. (a) _____

(b) _____

15. Find the measure of the other leg of this right triangle.

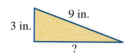

9 in.

3 in.

?

(a) Give its length in simplified radical form.

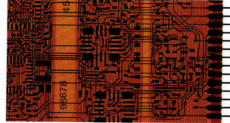 **(b)** Round the answer to the nearest thousandth.

16. _____

16. In electronics, the impedance Z of an alternating series circuit is given by the formula

$$Z = \sqrt{R^2 + X^2},$$

where R is the resistance and X is the reactance, both in ohms. Find the value of the impedance Z if $R = 40$ ohms and $X = 30$ ohms. (*Source:* Cooke, Nelson M., and Orleans, Joseph B., *Mathematics Essential to Electricity and Radio*, McGraw-Hill, 1943.)

Rationalize each denominator.

17. _____

17. $\dfrac{5\sqrt{2}}{\sqrt{7}}$

18. _____

18. $\sqrt{\dfrac{2}{3x}}$ $(x \neq 0)$

19. _____

19. $\dfrac{-2}{\sqrt[3]{4}}$

20. _____

20. $\dfrac{-3}{4 - \sqrt{3}}$

Solve each equation.

21. _____

21. $\sqrt{p} + 4 = 0$

22. _____

22. $\sqrt{x + 1} = 5 - x$

23. _____

23. $3\sqrt{x} - 2 = x$

Simplify each expression.

1. $3(6 + 7) + 6 \cdot 4 - 3^2$

2. $\dfrac{3(6 + 7) + 3}{2(4) - 1}$

3. $|-6| - |-3|$

Solve each equation or inequality.

4. $5(k - 4) - k = k - 11$

5. $-\dfrac{3}{4}y \le 12$

6. $5z + 3 - 4 > 2z + 9 + z$

7. U.S. production of corn reached record levels in 2000, when .8 billion more bushels of corn were produced than in 1999. Total production for the two years was 19.6 billion bushels. How much corn was produced in each of these years? (*Source:* U.S. Department of Agriculture.)

Graph.

8. $-4x + 5y = -20$

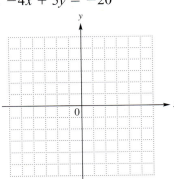

9. $x = 2$

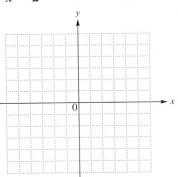

10. $2x - 5y > 10$

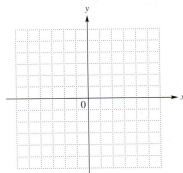

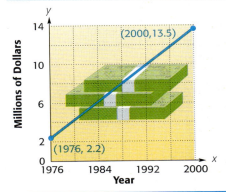

The graph shows a linear equation that models federal spending on each of the two major-party political conventions, in millions of dollars.

CONVENTION SPENDING

(2000, 13.5)

Millions of Dollars

(1976, 2.2)

1976   1984   1992   2000

Year

Source: Federal Election Commission.

11. Use the ordered pairs shown on the graph to find the slope of the line to the nearest hundredth. Interpret the slope.

12. Use the slope from Exercise 11 and the ordered pair (2000, 13.5) to find the equation of the line that models the data.

13. Use the equation from Exercise 12 to project convention spending for 2004. Round your answer to the nearest tenth.

Simplify and write each expression without negative exponents. Assume that variables represent positive numbers.

14. $(3x^6)(2x^2y)^2$

15. $\left(\dfrac{3^2y^{-2}}{2^{-1}y^3}\right)^{-3}$

16. Subtract $7x^3 - 8x^2 + 4$ from $10x^3 + 3x^2 - 9$.

17. Divide $\dfrac{8t^3 - 4t^2 - 14t + 15}{2t + 3}$.

18. Scientists worldwide are analyzing fragments of a meteorite that shattered over Tagish Lake in British Columbia in January 1999. The meteorite represents the most pristine specimen ever recovered, remaining virtually unchanged since the solar system formed some 4.6×10^9 yr ago. (*Source: The Gazette,* October 13, 2000.) Write the age of the meteorite without scientific notation.

Factor each polynomial completely.

19. $m^2 + 12m + 32$ **20.** $25t^4 - 36$ **21.** $12a^2 + 4ab - 5b^2$ **22.** $81z^2 + 72z + 16$

Solve each quadratic equation.

23. $x^2 - 7x = -12$ **24.** $(x + 4)(x - 1) = -6$

Perform the indicated operations. Express answers in lowest terms.

25. $\dfrac{x^2 - 3x - 4}{x^2 + 3x} \cdot \dfrac{x^2 + 2x - 3}{x^2 - 5x + 4}$ **26.** $\dfrac{t^2 + 4t - 5}{t + 5} \div \dfrac{t - 1}{t^2 + 8t + 15}$ **27.** $\dfrac{2}{x + 3} - \dfrac{4}{x - 1}$

Solve each system of equations.

28. $4x - y = 19$
   $3x + 2y = -5$

29. $2x - y = 6$
   $3y = 6x - 18$

30. Des Moines and Chicago are 345 mi apart. Two cars start from these cities traveling toward each other. They meet after 3 hr. The car from Chicago has an average speed 7 mph faster than the other car. Find the average speed of each car. (*Source: State Farm Road Atlas.*)

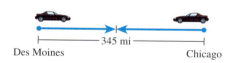

Des Moines   345 mi   Chicago

Simplify each expression if possible. Assume all variables represent nonnegative real numbers.

31. $\sqrt{27} - 2\sqrt{12} + 6\sqrt{75}$ **32.** $\dfrac{2}{\sqrt{3} + \sqrt{5}}$ **33.** $\sqrt{200x^2y^5}$

34. $(3\sqrt{2} + 1)(4\sqrt{2} - 3)$ **35.** Solve $\sqrt{x} + 2 = x - 10$.

Quadratic Equations

9

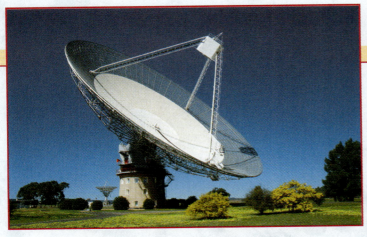

I n this chapter we develop methods of solving quadratic equations. The graphs of such equations, called *parabolas,* have many applications. For example, the Parkes radio telescope, pictured here, has a *parabolic* dish shape with a diameter of 210 ft and a depth of 32 ft. (*Source:* Mar, J. and Liebowitz, H., *Structure Technology for Large Radio and Radar Telescope Systems,* The MIT Press, 1969.) In Section 9.4 we will use a graph to model this parabolic shape and find its quadratic equation.

9.1 SOLVING QUADRATIC EQUATIONS BY THE SQUARE ROOT PROPERTY

OBJECTIVES

1 Solve equations of the form $x^2 = k$, where $k > 0$.

2 Solve equations of the form $(ax + b)^2 = k$, where $k > 0$.

3 Use formulas involving squared variables.

In Chapter 5 we solved quadratic equations by factoring. However, since not all quadratic equations can be solved by factoring, it is necessary to develop other methods. In this chapter we will do just that.

Recall that a *quadratic equation* is an equation that can be written in the form

$$ax^2 + bx + c = 0$$

for real numbers a, b, and c, with $a \neq 0$. As seen in Section 5.6, to solve $x^2 + 4x + 3 = 0$ by the zero-factor property, begin by factoring the left side and then set each factor equal to 0.

$$x^2 + 4x + 3 = 0$$
$$(x + 3)(x + 1) = 0 \qquad \text{Factor.}$$
$$x + 3 = 0 \quad \text{or} \quad x + 1 = 0 \qquad \text{Zero-factor property}$$
$$x = -3 \quad \text{or} \qquad x = -1 \qquad \text{Solve each equation.}$$

The solutions are -3 and -1.

1 **Solve equations of the form $x^2 = k$, where $k > 0$.** We can solve equations such as $x^2 = 9$ by factoring as follows.

$$x^2 = 9$$
$$x^2 - 9 = 0 \qquad \text{Subtract 9.}$$
$$(x + 3)(x - 3) = 0 \qquad \text{Factor.}$$
$$x + 3 = 0 \quad \text{or} \quad x - 3 = 0 \qquad \text{Zero-factor property}$$
$$x = -3 \quad \text{or} \qquad x = 3$$

We might also have solved $x^2 = 9$ by noticing that x must be a number whose square is 9. Thus, $x = \sqrt{9} = 3$ or $x = -\sqrt{9} = -3$. This is generalized as the **square root property of equations.**

Square Root Property of Equations

If k is a positive number and if $a^2 = k$, then

$$a = \sqrt{k} \quad \text{or} \quad a = -\sqrt{k}.$$

NOTE

When we solve an equation, we want to find *all* values of the variable that satisfy the equation. Therefore, we want both the positive and negative square roots of k.

Example 1 Solving Quadratic Equations by the Square Root Property

Solve each equation. Write radicals in simplified form.

(a) $x^2 = 16$

By the square root property, if $x^2 = 16$, then

$$x = \sqrt{16} = 4 \quad \text{or} \quad x = -\sqrt{16} = -4.$$

Continued on Next Page

An abbreviation for $x = 4$ or $x = -4$ is written $x = \pm 4$ (read "plus or minus 4"). Check each solution by substituting it for x in the original equation.

(b) $z^2 = 5$

The solutions are $\sqrt{5}$ or $-\sqrt{5}$, which may be written $\pm\sqrt{5}$.

(c) $m^2 - 8 = 0$

$$m^2 = 8 \qquad \text{Add 8.}$$
$$m = \sqrt{8} \quad \text{or} \quad m = -\sqrt{8} \qquad \text{Square root property}$$
$$m = 2\sqrt{2} \quad \text{or} \quad m = -2\sqrt{2} \quad \text{Simplify } \sqrt{8}.$$

The solutions are $\pm 2\sqrt{2}$.

(d) $x^2 = -4$

Because -4 is a negative number and because the square of a real number cannot be negative, there is no real number solution for this equation. (The square root property cannot be used because of the requirement that k must be positive.)

=== Work Problem **1** at the Side.

2 **Solve equations of the form** $(ax + b)^2 = k$, **where** $k > 0$. In each equation in Example 1, the exponent 2 appeared with a single variable as its base. The square root property can be extended to solve equations where the base is a binomial, as shown in the next example.

Example 2 **Solving Quadratic Equations by the Square Root Property**

Solve each equation.

(a) $(x - 3)^2 = 16$

Apply the square root property, using $x - 3$ as the base.

$$(x - 3)^2 = 16$$
$$x - 3 = \sqrt{16} \quad \text{or} \quad x - 3 = -\sqrt{16}$$
$$x - 3 = 4 \quad \text{or} \quad x - 3 = -4 \qquad \sqrt{16} = 4$$
$$x = 7 \quad \text{or} \quad x = -1 \qquad \text{Add 3.}$$

Check both answers in the original equation.

$$(x - 3)^2 = 16 \qquad\qquad (x - 3)^2 = 16$$
$$(7 - 3)^2 = 16 \quad ? \quad \text{Let } x = 7. \qquad (-1 - 3)^2 = 16 \quad ? \quad \text{Let } x = -1.$$
$$4^2 = 16 \quad ? \qquad\qquad (-4)^2 = 16 \quad ?$$
$$16 = 16 \qquad \text{True} \qquad\qquad 16 = 16 \qquad \text{True}$$

The solutions are 7 and -1.

(b) $(x + 1)^2 = 6$

By the square root property,

$$x + 1 = \sqrt{6} \qquad \text{or} \quad x + 1 = -\sqrt{6}$$
$$x = -1 + \sqrt{6} \quad \text{or} \qquad x = -1 - \sqrt{6}.$$

Check: $\quad (-1 + \sqrt{6} + 1)^2 = (\sqrt{6})^2 = 6;$
$$(-1 - \sqrt{6} + 1)^2 = (-\sqrt{6})^2 = 6.$$

The solutions are $-1 + \sqrt{6}$ and $-1 - \sqrt{6}$.

=== Work Problem **2** at the Side.

1 Solve each equation. Write radicals in simplified form.

(a) $x^2 = 49$

(b) $x^2 = 11$

(c) $x^2 = 12$

(d) $x^2 = -9$

2 Solve each equation.

(a) $(m + 2)^2 = 36$

(b) $(p - 4)^2 = 3$

ANSWERS
1. (a) $-7, 7$ (b) $-\sqrt{11}, \sqrt{11}$
(c) $-2\sqrt{3}, 2\sqrt{3}$
(d) no real number solution
2. (a) $-8, 4$ (b) $4 + \sqrt{3}, 4 - \sqrt{3}$

❸ Solve $(2x - 5)^2 = 18$.

❹ Solve each equation.

(a) $(5m + 1)^2 = 7$

(b) $(7z - 1)^2 = -1$

❺ Use the formula in Example 5 to approximate the length of a bass weighing 2.80 lb and having girth 11 in.

Example 3 Solving a Quadratic Equation by the Square Root Property

Solve $(3r - 2)^2 = 27$.

$$3r - 2 = \sqrt{27} \quad \text{or} \quad 3r - 2 = -\sqrt{27} \qquad \text{Square root property}$$

Now simplify the radical: $\sqrt{27} = \sqrt{9 \cdot 3} = \sqrt{9} \cdot \sqrt{3} = 3\sqrt{3}$.

$$3r - 2 = 3\sqrt{3} \qquad \text{or} \qquad 3r - 2 = -3\sqrt{3}$$

$$3r = 2 + 3\sqrt{3} \quad \text{or} \qquad 3r = 2 - 3\sqrt{3} \qquad \text{Add 2.}$$

$$r = \frac{2 + 3\sqrt{3}}{3} \quad \text{or} \qquad r = \frac{2 - 3\sqrt{3}}{3} \qquad \text{Divide by 3.}$$

The solutions are $\dfrac{2 + 3\sqrt{3}}{3}$ and $\dfrac{2 - 3\sqrt{3}}{3}$.

Work Problem ❸ at the Side.

Example 4 Recognizing a Quadratic Equation with No Solution

Solve $(x + 3)^2 = -9$.

Notice that the constant on the right side of the equation is negative. The square root of -9 is not a real number. There is no real number solution for this equation.

Work Problem ❹ at the Side.

❸ Use formulas involving squared variables.

Example 5 Finding the Length of a Bass

The formula

$$w = \frac{L^2 g}{1200}$$

is used to approximate the weight of a bass, in pounds, given its length L and its girth g, where both are measured in inches. Approximate the length of a bass weighing 2.20 lb and having girth 10 in. (*Source: Sacramento Bee,* November 29, 2000.)

$$w = \frac{L^2 g}{1200} \qquad \text{Given formula}$$

$$\mathbf{2.20} = \frac{L^2 \cdot \mathbf{10}}{1200} \qquad w = 2.20, g = 10$$

$$2640 = 10L^2 \qquad \text{Multiply by 1200.}$$

$$L^2 = 264 \qquad \text{Divide by 10.}$$

$$L \approx 16.25 \qquad \text{Use a calculator; } L > 0.$$

The length of the bass is a little more than 16 in. (We discard the negative solution -16.25 since L represents length.)

Work Problem ❺ at the Side.

9.1 **EXERCISES**

Match each equation in Column I with the correct description of its solutions in Column II.

I	**II**
1. $x^2 = 10$	**A.** No real number solution
2. $x^2 = -4$	**B.** Two integer solutions
3. $x^2 = \dfrac{9}{16}$	**C.** Two irrational solutions
4. $x^2 = 9$	**D.** Two rational solutions that are not integers

Decide whether each statement is true *or* false. *If false,* tell why.

5. If k is a prime number, then $x^2 = k$ has two irrational solutions.

6. If k is a positive perfect square, then $x^2 = k$ has two rational solutions.

7. If k is a positive integer, then $x^2 = k$ must have two rational solutions.

8. If $-10 < k < 0$, then $x^2 = k$ has no real solution.

9. If $-10 < k < 10$, then $x^2 = k$ has no real solution.

10. If k is an integer greater than 24 and less than 26, then $x^2 = k$ has two solutions, -5 and 5.

Solve each equation by using the square root property. Write all radicals in simplest form. See Example 1.

11. $x^2 = 81$

12. $y^2 = 121$

13. $k^2 = 14$

14. $m^2 = 22$

15. $t^2 = 48$

16. $x^2 = 54$

17. $x^2 = \dfrac{25}{4}$

18. $m^2 = \dfrac{36}{121}$

19. $z^2 = 2.25$

20. $w^2 = 56.25$

21. $r^2 - 3 = 0$

22. $x^2 - 13 = 0$

Solve each equation by using the square root property. Express all radicals in simplest form. See Examples 2–4.

23. $(x - 3)^2 = 25$

24. $(x - 7)^2 = 16$

25. $(z + 5)^2 = -13$

26. $(t + 2)^2 = -17$

27. $(x - 8)^2 = 27$

28. $(x - 5)^2 = 40$

29. $(3k + 2)^2 = 49$

30. $(5t + 3)^2 = 36$

31. $(4x - 3)^2 = 9$

32. $(7y - 5)^2 = 25$

33. $(5 - 2x)^2 = 30$

34. $(3 - 2a)^2 = 70$

35. $(3k + 1)^2 = 18$

36. $(5z + 6)^2 = 75$

37. $\left(\dfrac{1}{2}x + 5\right)^2 = 12$

38. $\left(\dfrac{1}{3}y + 4\right)^2 = 27$

39. $(4k - 1)^2 - 48 = 0$

40. $(2s - 5)^2 - 180 = 0$

41. Johnny solved the equation in Exercise 33 and wrote his solutions as $\dfrac{5 + \sqrt{30}}{2}, \dfrac{5 - \sqrt{30}}{2}$. Linda solved the same equation and wrote her solutions as $\dfrac{-5 + \sqrt{30}}{-2}, \dfrac{-5 - \sqrt{30}}{-2}$. The teacher gave them both full credit. Explain why both students were correct, although their answers seem to differ.

42. In the solutions found in Example 3 of this section, why is it not valid to simplify by dividing out the 3s in the numerators and denominators?

Solve each problem. See Example 5.

43. One expert at marksmanship can hold a silver dollar at forehead level, drop it, draw his gun, and shoot the coin as it passes waist level. The distance traveled by a falling object is given by

$$d = 16t^2,$$

where d is the distance (in feet) the object falls in t seconds. If the coin falls about 4 ft, use the formula to estimate the time that elapses between the dropping of the coin and the shot.

44. The illumination produced by a light source depends on the distance from the source. For a particular light source, this relationship can be expressed as

$$d^2 = \frac{4050}{I},$$

where d is the distance from the source (in feet) and I is the amount of illumination in foot-candles. How far from the source is the illumination equal to 50 foot-candles?

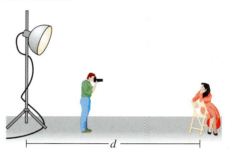

45. The area A of a circle with radius r is given by the formula

$$A = \pi r^2.$$

If a circle has area 81π in.^{2}, what is its radius?

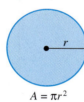

$A = \pi r^2$

46. The surface area S of a sphere with radius r is given by the formula

$$S = 4\pi r^2.$$

If a sphere has surface area 36π ft^2, what is its radius?

$S = 4\pi r^2$

The amount A that P dollars invested at an annual rate of interest r will grow to in 2 years is $A = P(1 + r)^2$.

47. At what interest rate will \$100 grow to \$110.25 in 2 years?

48. At what interest rate will \$500 grow to \$572.45 in 2 years?

9.2 SOLVING QUADRATIC EQUATIONS BY COMPLETING THE SQUARE

1 Solve quadratic equations by completing the square when the coefficient of the squared term is 1. The methods studied so far are not sufficient to solve the equation

$$x^2 + 6x + 7 = 0.$$

If we could write the equation in a form like $(x + 3)^2 = 2$, we could solve it with the square root property discussed in the previous section. To do that, we need to have a perfect square trinomial on one side of the equation. The next example shows how this is done.

Work Problem ❶ at the Side.

Example 1 Rewriting an Equation to Use the Square Root Property

Solve $x^2 + 6x + 7 = 0$.
   Start by subtracting 7 from each side of the equation.

$$x^2 + 6x = -7$$

The quantity on the left side of $x^2 + 6x = -7$ must be made into a perfect square trinomial. The expression $x^2 + 6x + 9$ is a perfect square, since

$$x^2 + 6x + 9 = (x + 3)^2.$$

Therefore, if 9 is added to each side, the equation will have a perfect square trinomial on the left side, as needed.

$$x^2 + 6x + 9 = -7 + 9 \qquad \text{Add 9.}$$
$$(x + 3)^2 = 2 \qquad \text{Factor.}$$

Now use the square root property to complete the solution.

$$x + 3 = \sqrt{2} \qquad \text{or} \quad x + 3 = -\sqrt{2}$$
$$x = -3 + \sqrt{2} \quad \text{or} \qquad x = -3 - \sqrt{2}$$

The solutions of the original equation are $-3 + \sqrt{2}$ and $-3 - \sqrt{2}$. Check by substituting $-3 + \sqrt{2}$ and $-3 - \sqrt{2}$ for x in the original equation.

The process of changing the form of the equation in Example 1 from

$$x^2 + 6x + 7 = 0 \quad \text{to} \quad (x + 3)^2 = 2$$

is called **completing the square**. Completing the square changes only the form of the equation. To see this, multiply out the left side of $(x + 3)^2 = 2$ and combine terms. Then subtract 2 from each side to see that the result is $x^2 + 6x + 7 = 0$.
   Look again at the original equation,

$$x^2 + 6x + 7 = 0.$$

Note that 9 is the square of half the coefficient of x, 6.

$$\frac{1}{2} \cdot 6 = 3 \quad \text{and} \quad 3^2 = 9$$

↑
Coefficient of x

To complete the square in Example 1, 9 was added to each side.

OBJECTIVES

1 Solve quadratic equations by completing the square when the coefficient of the squared term is 1.

2 Solve quadratic equations by completing the square when the coefficient of the squared term is not 1.

3 Simplify an equation before solving.

4 Solve applied problems that require quadratic equations.

❶ As a review, factor each perfect square trinomial.

(a) $x^2 + 6x + 9$

(b) $q^2 - 20q + 100$

② Solve by completing the square.

(a) $x^2 + 4x = 1$

(b) $z^2 + 6z - 3 = 0$

Example 2   **Completing the Square to Solve a Quadratic Equation**

Complete the square to solve $x^2 - 8x = 5$.

To complete the square on $x^2 - 8x$, take half the coefficient of x and square it.

$$\frac{1}{2}(\mathbf{-8}) = -4 \quad \text{and} \quad (-4)^2 = \mathbf{16}$$

↑
Coefficient of x

Add the result, 16, to each side of the equation.

$$x^2 - 8x = 5 \qquad \text{Given equation}$$
$$x^2 - 8x + \mathbf{16} = 5 + \mathbf{16} \qquad \text{Add 16.}$$
$$(x - 4)^2 = 21 \qquad \text{Factor the left side as the square of a binomial.}$$

Now apply the square root property.

$$x - 4 = \pm\sqrt{21} \qquad \text{Square root property}$$
$$x = 4 \pm \sqrt{21} \qquad \text{Add 4.}$$

A check indicates that the solutions are $4 + \sqrt{21}$ and $4 - \sqrt{21}$.

In general, to solve an equation of the form

$$x^2 + mx = n$$

by completing the square, we begin by adding $\left(\frac{1}{2}m\right)^2$ to each side.

Work Problem ② at the Side.

2 ▮▮▮ **Solve quadratic equations by completing the square when the coefficient of the squared term is not 1.** Suppose that an equation is of the form $ax^2 + bx + c = 0$, where $a \neq 1$. To get 1 as the coefficient of x^2, first divide each side of the equation by a. The next examples illustrate this approach.

Example 3   **Solving a Quadratic Equation by Completing the Square**

Solve $4x^2 + 16x = 9$.

Before completing the square, the coefficient of x^2 must be 1. Here the coefficient of x^2 is 4. Make the coefficient 1 by dividing each side of the equation by 4.

$$x^2 + 4x = \frac{9}{4} \qquad \text{Divide by 4.}$$

Next, complete the square by taking half the coefficient of x, or $\left(\frac{1}{2}\right)(4) = 2$, and squaring the result: $2^2 = 4$. Add 4 to each side of the equation, combine terms on the right side, and factor on the left.

$$x^2 + 4x + \mathbf{4} = \frac{9}{4} + \mathbf{4} \qquad \text{Add 4.}$$

$$x^2 + 4x + 4 = \frac{25}{4} \qquad \text{Combine terms.}$$

$$(x + 2)^2 = \frac{25}{4} \qquad \text{Factor.}$$

Continued on Next Page

Use the square root property and solve for x.

$$x + 2 = \frac{5}{2} \qquad \text{or} \quad x + 2 = -\frac{5}{2} \qquad \text{Square root property}$$

$$x = -2 + \frac{5}{2} \quad \text{or} \qquad x = -2 - \frac{5}{2} \qquad \text{Subtract 2.}$$

$$x = \frac{1}{2} \qquad \text{or} \qquad x = -\frac{9}{2} \qquad \text{Combine terms.}$$

Check:

$$4x^2 + 16x = 9 \qquad\qquad 4x^2 + 16x = 9$$

$$4\left(\frac{1}{2}\right)^2 + 16\left(\frac{1}{2}\right) = 9 \quad ? \qquad 4\left(-\frac{9}{2}\right)^2 + 16\left(-\frac{9}{2}\right) = 9 \quad ?$$

$$4\left(\frac{1}{4}\right) + 8 = 9 \quad ? \qquad\qquad 4\left(\frac{81}{4}\right) - 72 = 9 \quad ?$$

$$1 + 8 = 9 \quad ? \qquad\qquad\qquad 81 - 72 = 9 \quad ?$$

$$9 = 9 \qquad \text{True} \qquad\qquad\qquad 9 = 9 \qquad \text{True}$$

The two solutions are $\frac{1}{2}$ and $-\frac{9}{2}$.

The steps in solving a quadratic equation by completing the square are summarized here.

Solving a Quadratic Equation by Completing the Square

Use the method of **completing the square** to solve the quadratic equation $ax^2 + bx + c = 0$ as follows.

Step 1 **Be sure the squared term has coefficient 1.** If the coefficient of the squared term is 1, proceed to Step 2. If the coefficient of the squared term is not 1 but some other nonzero number a, divide each side of the equation by a. This gives an equation that has 1 as the coefficient of the squared term.

Step 2 **Put in correct form.** Make sure that all terms with variables are on one side of the equals sign and that all constants are on the other side.

Step 3 **Complete the square.** Take half the coefficient of the first-degree term, and square the result. Add the square to each side of the equation. The side containing the variables can now be factored as a perfect square.

Step 4 **Solve.** Apply the square root property to solve the equation.

Work Problem ❸ at the Side.

❸ Solve by completing the square.

(a) $9x^2 + 18x + 5 = 0$

(b) $4t^2 - 24t + 11 = 0$

4 Solve by completing the square.

(a) $3x^2 + 5x - 2 = 0$

Example 4   **Solving a Quadratic Equation by Completing the Square**

Solve $2x^2 - 7x = 9$.

Step 1 Divide each side of the equation by 2 to get a coefficient of 1 for the x^2-term.

$$x^2 - \frac{7}{2}x = \frac{9}{2} \qquad \text{Divide by 2.}$$

(Step 2 is not needed for this equation.)

Step 3 Now take half the coefficient of x and square it. Half of $-\frac{7}{2}$ is $-\frac{7}{4}$, and $\left(-\frac{7}{4}\right)^2 = \frac{49}{16}$. Add $\frac{49}{16}$ to each side of the equation, and write the left side as a perfect square.

$$x^2 - \frac{7}{2}x + \frac{49}{16} = \frac{9}{2} + \frac{49}{16} \qquad \text{Add } \frac{49}{16}.$$

$$\left(x - \frac{7}{4}\right)^2 = \frac{121}{16} \qquad \text{Factor.}$$

Step 4 Use the square root property.

$$x - \frac{7}{4} = \sqrt{\frac{121}{16}} \quad \text{or} \quad x - \frac{7}{4} = -\sqrt{\frac{121}{16}}$$

(b) $2x^2 - 4x = 1$

Because $\sqrt{\frac{121}{16}} = \frac{11}{4}$,

$$x - \frac{7}{4} = \frac{11}{4} \quad \text{or} \quad x - \frac{7}{4} = -\frac{11}{4}$$

$$x = \frac{18}{4} \quad \text{or} \quad x = -\frac{4}{4} \qquad \text{Add } \frac{7}{4}.$$

$$x = \frac{9}{2} \quad \text{or} \quad x = -1.$$

Check that the solutions are $\frac{9}{2}$ and -1.

Work Problem 4 at the Side.

Example 5   **Solving a Quadratic Equation by Completing the Square**

Solve $4p^2 + 8p + 5 = 0$.

$$p^2 + 2p + \frac{5}{4} = 0 \qquad \text{Divide by 4.}$$

$$p^2 + 2p = -\frac{5}{4} \qquad \text{Subtract } \frac{5}{4}.$$

Continued on Next Page

ANSWERS

4. (a) $-2, \frac{1}{3}$ (b) $\dfrac{2 + \sqrt{6}}{2}, \dfrac{2 - \sqrt{6}}{2}$

The coefficient of p is 2. Take half of 2, square the result, and add this square to each side. The left side can then be written as a perfect square.

$$p^2 + 2p + 1 = -\frac{5}{4} + 1 \qquad \text{Add 1.}$$

$$(p + 1)^2 = -\frac{1}{4} \qquad \text{Factor.}$$

The square root of $-\frac{1}{4}$ is not a real number, so the square root property does not apply. This equation has no real number solution.*

=================== **Work Problem ❺ at the Side.**

❺ Solve $5x^2 + 3x + 1 = 0$ by completing the square.

3 ▭ **Simplify an equation before solving.** The next example shows how to simplify a quadratic equation before solving it.

┌─ **Example 6** **Simplifying an Equation before Completing the Square**

Solve $(x + 3)(x - 1) = 2$.

$(x + 3)(x - 1) = 2$ 　　　Given equation

$x^2 + 2x - 3 = 2$ 　　　Use FOIL.

$x^2 + 2x = 5$ 　　　Add 3.

$x^2 + 2x + 1 = 5 + 1$ 　　Add 1 to get a perfect square on the left.

$(x + 1)^2 = 6$ 　　　Factor on the left; add on the right.

$x + 1 = \sqrt{6}$ 　　or 　$x + 1 = -\sqrt{6}$ 　　Square root property

$x = -1 + \sqrt{6}$ 　or 　　$x = -1 - \sqrt{6}$ 　Subtract 1.

The solutions are $-1 + \sqrt{6}$ and $-1 - \sqrt{6}$.

❻ Solve each equation.

(a) $r(r - 3) = -1$

(b) $(x + 2)(x + 1) = 5$

┌─ **NOTE**

The solutions given in Example 6 are *exact*. In applications, decimal solutions are more appropriate. Using the square root key of a calculator, we get $\sqrt{6} \approx 2.449$. Evaluating the two solutions gives

$$x \approx 1.449 \quad \text{and} \quad x \approx -3.449.$$

Work Problem ❻ at the Side.

4 ▭ **Solve applied problems that require quadratic equations.** There are many practical applications of quadratic equations. The next example illustrates an application from physics.

---

* The equation in Example 5 has no *real number* solution. In the context of another number system, called the *complex numbers,* however, this equation does have solutions. The complex numbers include numbers whose squares are negative. These numbers are discussed in intermediate and college algebra courses.

7 Suppose a ball is propelled upward with an initial velocity of 128 ft per sec. Its height at time t (in seconds) is given by

$$s = -16t^2 + 128t,$$

where s is in feet. At what times will it be 48 ft above the ground? Give answers to the nearest tenth.

Example 7 Solving a Velocity Problem

If a ball is propelled into the air from ground level with an initial velocity of 64 ft per sec, its height s (in feet) in t seconds is given by the formula

$$s = -16t^2 + 64t.$$

How long will it take the ball to reach a height of 48 ft?

Since s represents the height, let $s = \mathbf{48}$ in the formula to get

$$\mathbf{48} = -16t^2 + 64t.$$

We solve this equation for time, t, by completing the square. First we divide each side by -16. We also reverse the sides of the equation.

$-3 = t^2 - 4t$	Divide by -16.
$t^2 - 4t = -3$	Reverse the sides.
$t^2 - 4t + \mathbf{4} = -3 + \mathbf{4}$	Add $[\frac{1}{2}(-4)]^2 = 4$.
$(t - 2)^2 = 1$	Factor.
$t - 2 = 1$ or $t - 2 = -1$	Square root property
$t = 3$ or $t = 1$	Add 2.

You may wonder how we can get two correct answers for the time required for the ball to reach a height of 48 ft. The ball reaches that height twice, once on the way up and again on the way down. So it takes 1 sec to reach 48 ft on the way up, and then after 3 sec, the ball reaches 48 ft again on the way down.

Work Problem 7 at the Side.

9.2 EXERCISES

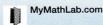

Fill in each blank with the correct response.

1. To solve the equation $x^2 - 8x = 4$ by completing the square, the first step is to add _____ to each side of the equation.

2. To solve the equation $3x^2 - 8x - 1 = 0$ by completing the square, the first step is to divide each side of the equation by _____.

3. To solve $(t + 2)(t - 5) = 18$ by completing the square, we should start by
   _____.

4. It is not possible to solve $x^3 - x - 1 = 0$ by completing the square because
   _____.

5. Which step is an appropriate way to begin solving the quadratic equation
$$2x^2 - 4x = 9$$
   by completing the square?

   A. Add 4 to each side of the equation.   **B.** Factor the left side as $2x(x - 2)$.

   C. Factor the left side as $x(2x - 4)$.   **D.** Divide each side by 2.

6. In Example 3 of Section 5.6, we solved the quadratic equation
$$4p^2 - 26p + 40 = 0$$
   by factoring. If we were to solve by completing the square, would we get the same solutions, $\frac{5}{2}$ and 4?

Find the number that should be added to each expression to make it a perfect square. See Example 2.

7. $x^2 + 14x$ 8. $z^2 + 18z$ 9. $k^2 - 5k$

10. $m^2 - 9m$ 11. $r^2 + \frac{1}{2}r$ 12. $s^2 - \frac{1}{3}s$

Solve each equation by completing the square. See Examples 1 and 2.

13. $x^2 - 4x = -3$ 14. $x^2 - 2x = 8$ 15. $x^2 + 5x + 6 = 0$ 16. $x^2 + 6x + 5 = 0$

17. $x^2 + 2x - 5 = 0$ 18. $x^2 + 4x + 1 = 0$ 19. $t^2 + 6t + 9 = 0$

20. $k^2 - 8k + 16 = 0$ 21. $x^2 + x - 1 = 0$ 22. $x^2 + x - 3 = 0$

Solve each equation by completing the square. See Examples 3–6.

23. $4x^2 + 4x - 3 = 0$

24. $9x^2 + 3x - 2 = 0$

25. $2x^2 - 4x = 5$

26. $2x^2 - 6x = 3$

27. $2p^2 - 2p + 3 = 0$

28. $3q^2 - 3q + 4 = 0$

29. $3k^2 + 7k = 4$

30. $2k^2 + 5k = 1$

31. $(x + 3)(x - 1) = 5$

32. $(y - 8)(y + 2) = 24$

33. $-x^2 + 2x = -5$

34. $-r^2 + 3r = -2$

Solve each problem. See Example 7.

35. If an object is propelled upward from ground level with an initial velocity of 96 ft per sec, its height s (in feet) in t seconds is given by the formula $s = -16t^2 + 96t$. In how many seconds will the object be at a height of 80 ft?

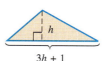 **36.** How much time will it take the object described in Exercise 35 to be at a height of 100 ft? Round your answers to the nearest tenth.

37. If an object is propelled upward on the surface of Mars from ground level with an initial velocity of 104 ft per sec, its height s (in feet) in t seconds is given by the formula $s = -13t^2 + 104t$. How long will it take for the object to be at a height of 195 ft?

38. How long will it take the object in Exercise 37 to return to the surface? (*Hint:* When it returns to the surface, $s = 0$.)

39. A farmer has a rectangular cattle pen with perimeter 350 ft and area 7500 ft². What are the dimensions of the pen? (*Hint:* Use the figure to set up the equation.)

40. The base of a triangle measures 1 m more than three times the height of the triangle. Its area is 15 m². Find the lengths of the base and the height.

9.3 SOLVING QUADRATIC EQUATIONS BY THE QUADRATIC FORMULA

Any quadratic equation can be solved by completing the square, but the method is tedious. In this section we complete the square on the quadratic equation $ax^2 + bx + c = 0$ to get the *quadratic formula*, a formula that gives the solution(s) for any quadratic equation. (Note that $a \neq 0$, or we would have a linear, not a quadratic, equation.)

OBJECTIVES

1. Identify the values of a, b, and c in a quadratic equation.
2. Use the quadratic formula to solve quadratic equations.
3. Solve quadratic equations with only one solution.
4. Solve quadratic equations with fractions as coefficients.

1 **Identify the values of a, b, and c in a quadratic equation.** The first step in solving a quadratic equation by this new method is to identify the values of a, b, and c in the standard form of the quadratic equation.

Example 1 Identifying Values of a, b, and c in Quadratic Equations

Match the coefficients of each quadratic equation with the letters a, b, and c of the standard quadratic equation

$$ax^2 + bx + c = 0.$$

$$\begin{array}{ccc} a & b & c \\ \downarrow & \downarrow & \downarrow \end{array}$$

(a) $2x^2 + 3x - 5 = 0$

In this example $a = 2$, $b = 3$, and $c = -5$.

(b) $-x^2 + 2 = 6x$

First rewrite the equation with 0 on one side to match the standard form $ax^2 + bx + c = 0$.

$$-x^2 + 2 = 6x$$
$$-x^2 - 6x + 2 = 0 \qquad \text{Subtract } 6x.$$

Here, $a = -1$, $b = -6$, and $c = 2$. (Notice that the coefficient of x^2 is understood to be -1.)

(c) $(2x - 7)(x + 4) = -23$

Write the equation in standard form.

$$(2x - 7)(x + 4) = -23$$
$$2x^2 + x - 28 = -23 \qquad \text{Use FOIL on the left.}$$
$$2x^2 + x - 5 = 0 \qquad \text{Add 23.}$$

Now, identify the values: $a = 2$, $b = 1$, and $c = -5$.

Work Problem ➊ at the Side.

➊ Match the coefficients of each quadratic equation with the letters a, b, and c of the standard quadratic equation $ax^2 + bx + c = 0$.

(a) $5x^2 + 2x - 1 = 0$

(b) $3x^2 = x - 2$

(c) $x(x + 5) = 4$

2 **Use the quadratic formula to solve quadratic equations.** To develop the quadratic formula, we follow the steps for completing the square on $ax^2 + bx + c = 0$ ($a > 0$) given in the previous section. For comparison, we also show the corresponding steps for solving $2x^2 + x - 5 = 0$ (from Example 1(c)).

Step 1 Transform so that the coefficient of the squared term equals 1.

$$2x^2 + x - 5 = 0 \qquad\qquad ax^2 + bx + c = 0$$

$$x^2 + \frac{1}{2}x - \frac{5}{2} = 0 \quad \text{Divide by 2.} \qquad x^2 + \frac{b}{a}x + \frac{c}{a} = 0 \quad \text{Divide by } a.$$

Step 2 Get the variable terms alone on the left side.

$$x^2 + \frac{1}{2}x = \frac{5}{2} \quad \text{Add } \tfrac{5}{2}. \qquad\qquad x^2 + \frac{b}{a}x = -\frac{c}{a} \quad \text{Subtract } \tfrac{c}{a}.$$

Step 3 Add the square of half the coefficient of x to each side, factor the left side, and combine terms on the right.

$$x^2 + \frac{1}{2}x + \frac{1}{16} = \frac{5}{2} + \frac{1}{16} \quad \text{Add } \tfrac{1}{16}. \qquad x^2 + \frac{b}{a}x + \frac{b^2}{4a^2} = -\frac{c}{a} + \frac{b^2}{4a^2} \quad \text{Add } \tfrac{b^2}{4a^2}.$$

$$\left(x + \frac{1}{4}\right)^2 = \frac{41}{16} \quad \begin{matrix}\text{Factor;}\\\text{add on}\\\text{right.}\end{matrix} \qquad \left(x + \frac{b}{2a}\right)^2 = \frac{b^2 - 4ac}{4a^2} \quad \begin{matrix}\text{Factor;}\\\text{add on}\\\text{right.}\end{matrix}$$

Step 4 Use the square root property to complete the solution.

$$x + \frac{1}{4} = \pm\sqrt{\frac{41}{16}} \qquad\qquad x + \frac{b}{2a} = \pm\sqrt{\frac{b^2 - 4ac}{4a^2}}$$

$$x + \frac{1}{4} = \pm\frac{\sqrt{41}}{4} \qquad\qquad x + \frac{b}{2a} = \pm\frac{\sqrt{b^2 - 4ac}}{2a}$$

$$x = -\frac{1}{4} \pm \frac{\sqrt{41}}{4} \qquad\qquad x = -\frac{b}{2a} \pm \frac{\sqrt{b^2 - 4ac}}{2a}$$

$$x = \frac{-1 \pm \sqrt{41}}{4} \qquad\qquad x = \frac{-b \pm \sqrt{b^2 - 4ac}}{2a}$$

The final result in the column on the right is called the quadratic formula. (It is also valid for $a < 0$.) *It is a key result that should be memorized.* Notice that there are two values, one for the $+$ sign and one for the $-$ sign.

Quadratic Formula

The solutions of the quadratic equation $ax^2 + bx + c = 0$, $a \neq 0$, are

$$x = \frac{-b + \sqrt{b^2 - 4ac}}{2a} \quad \text{and} \quad x = \frac{-b - \sqrt{b^2 - 4ac}}{2a},$$

or in compact form,

$$x = \frac{-b \pm \sqrt{b^2 - 4ac}}{2a}.$$

CAUTION

Notice that the fraction bar is under $-b$ as well as the radical. When using this formula, be sure to find the values of $-b \pm \sqrt{b^2 - 4ac}$ first, then divide those results by the value of $2a$.

Example 2 Solving a Quadratic Equation by the Quadratic Formula

Use the quadratic formula to solve $2x^2 - 7x - 9 = 0$.

Match the coefficients of the variables with those of the standard quadratic equation

$$ax^2 + bx + c = 0.$$

Continued on Next Page

Here, $a = 2$, $b = -7$, and $c = -9$. Substitute these numbers into the quadratic formula, and simplify the result.

$$x = \frac{-b \pm \sqrt{b^2 - 4ac}}{2a}$$

$$x = \frac{-(-7) \pm \sqrt{(-7)^2 - 4(2)(-9)}}{2(2)} \qquad \text{Let } a = 2, \, b = -7, \, c = -9.$$

$$x = \frac{7 \pm \sqrt{49 + 72}}{4}$$

$$x = \frac{7 \pm \sqrt{121}}{4}$$

$$x = \frac{7 \pm 11}{4} \qquad \sqrt{121} = 11$$

Find the two separate solutions by first using the plus sign, and then using the minus sign:

$$x = \frac{7 + 11}{4} = \frac{18}{4} = \frac{9}{2} \quad \text{or} \quad x = \frac{7 - 11}{4} = \frac{-4}{4} = -1.$$

The solutions of $2x^2 - 7x - 9 = 0$ are $\frac{9}{2}$ and -1. Check by substituting back into the original equation.

Work Problem ❷ at the Side.

Example 3 **Rewriting a Quadratic Equation before Using the Quadratic Formula**

Solve $x^2 = 2x + 1$.

Find a, b, and c by rewriting the equation in standard form (with 0 on one side). Add $-2x - 1$ to each side of the equation to get

$$x^2 - 2x - 1 = 0.$$

Then $a = 1$, $b = -2$, and $c = -1$. The solution is found by substituting these values into the quadratic formula.

$$x = \frac{-b \pm \sqrt{b^2 - 4ac}}{2a}$$

$$x = \frac{-(-2) \pm \sqrt{(-2)^2 - 4(1)(-1)}}{2(1)} \qquad \text{Let } a = 1, b = -2, c = -1.$$

$$x = \frac{2 \pm \sqrt{4 + 4}}{2}$$

$$x = \frac{2 \pm \sqrt{8}}{2}$$

$$x = \frac{2 \pm 2\sqrt{2}}{2} \qquad \begin{aligned} \sqrt{8} &= \sqrt{4 \cdot 2} = \sqrt{4} \cdot \sqrt{2} \\ &= 2\sqrt{2} \end{aligned}$$

Continued on Next Page

❷ Solve by using the quadratic formula.

(a) $2x^2 + 3x - 5 = 0$

(b) $6x^2 + x - 1 = 0$

❸ Solve $x^2 + 1 = -8x$ by the quadratic formula.

Write these solutions in lowest terms by factoring $2 \pm 2\sqrt{2}$ as $2(1 \pm \sqrt{2})$ to get

$$x = \frac{2(1 \pm \sqrt{2})}{2} = 1 \pm \sqrt{2}.$$

The two solutions of the original equation are

$$1 + \sqrt{2} \quad \text{and} \quad 1 - \sqrt{2}.$$

Work Problem ❸ at the Side.

3 ▭ **Solve quadratic equations with only one solution.** When the quantity under the radical, $b^2 - 4ac$, equals 0, the equation has just one rational number solution. In this case, the trinomial $ax^2 + bx + c$ is a perfect square.

Example 4 Solving a Quadratic Equation with Only One Solution

Solve $4x^2 + 25 = 20x$.
  Write the equation as

$$4x^2 - 20x + 25 = 0. \quad \text{Subtract } 20x.$$

Here, $a = 4$, $b = -20$, and $c = 25$. By the quadratic formula,

$$x = \frac{-(-20) \pm \sqrt{(-20)^2 - 400}}{8} = \frac{20 \pm 0}{8} = \frac{5}{2}.$$

❹ Solve $9x^2 - 12x + 4 = 0$.

In this case, $b^2 - 4ac = 0$, and the trinomial $4x^2 - 20x + 25$ is a perfect square. There is just one solution, $\frac{5}{2}$.

NOTE

The single solution of the equation in Example 4 is a rational number. If all solutions of a quadratic equation are rational, the equation can be solved by factoring as well.

Work Problem ❹ at the Side.

4 ▭ **Solve quadratic equations with fractions as coefficients.** It is usually easier to clear quadratic equations of fractions before solving them.

Example 5 Solving a Quadratic Equation with Fractions as Coefficients

Solve $\dfrac{1}{10}t^2 = \dfrac{2}{5}t - \dfrac{1}{2}$.

  Eliminate the denominators by multiplying each side of the equation by the least common denominator, 10.

Continued on Next Page

ANSWERS

3. $-4 + \sqrt{15}, -4 - \sqrt{15}$

4. $\dfrac{2}{3}$

$$10\left(\frac{1}{10}t^2\right) = 10\left(\frac{2}{5}t - \frac{1}{2}\right)$$

$$t^2 = 4t - 5$$

$$t^2 - 4t + 5 = 0 \qquad \text{Add } -4t \text{ and } 5.$$

From this form, we see that $a = 1$, $b = -4$, and $c = 5$. Use the quadratic formula to complete the solution.

$$t = \frac{-(-4) \pm \sqrt{(-4)^2 - 4(1)(5)}}{2(1)} = \frac{4 \pm \sqrt{16 - 20}}{2} = \frac{4 \pm \sqrt{-4}}{2}$$

The radical $\sqrt{-4}$ is not a real number, so the equation has no real number solution.

Work Problem ❺ at the Side.

❺ Solve $x^2 - \frac{4}{3}x + \frac{2}{3} = 0$.

Real-Data Applications

Estimating Interest Rates

Banks offer savings accounts in the form of money market funds that pay variable interest rates, depending on the status of the prime interest rate in effect at the beginning of each month. The prime interest rate is the rate that the Federal Reserve pays its largest customers. The interest rates paid on bank accounts that are tied to the prime rate are not as profitable.

The compound interest formula, $A = P(1 + r)^n$, computes the amount A that P dollars invested at a periodic rate of interest r will grow to in n periods. The period for compound interest can be a year (annual), six months (semi-annual), three months (quarterly), one month (monthly), or even daily. The periodic interest rate is the annual rate divided by the number of periods per year. For example, to find the monthly rate, you divide the annual rate by 12. Similarly, if you know the monthly rate, you can multiply it by 12 to find the annual rate. If you borrow money to purchase a car or a house, the quantities P, r, and n are typically determined for the entire purchase.

Suppose you are saving money on a month-to-month basis at your bank. On March 1 you open a money market fund account with a deposit of $250; on April 1 you deposit $175; and on May 1 you deposit $300. Your deposit slip on May 1 shows a balance of S726.20. You are curious about the average interest rate earned from the date you opened the account. The investment can be thought of as three separate transactions, added together.

March 1:	The investment earns interest for 2 months (March, April).	$A = 250(1 + r)^2$
April 1:	The investment earns interest for 1 month (April).	$A = 175(1 + r)$
May 1:	The investment has not yet earned interest.	$A = 300$

The total investment value can be written as a quadratic equation:

$$250(1 + r)^2 + 175(1 + r) + 300 = 726.20.$$

To solve this equation, you could expand and collect like terms, but the algebra is tedious. A better option is to replace $1 + r$ with x and solve the quadratic equation

$$250x^2 + 175x + 300 = 726.20.$$

Once you have found x, you simply subtract 1 to get r, the monthly interest rate.

For Group Discussion

1. Consider the quadratic equation $250x^2 + 175x + 300 = 726.20$.
   (a) Use the quadratic formula to solve the equation for x. Remember to write it in standard form. (*Hint: x must have a value between 1 and 2 since x is $1 + r$ and r is an interest rate.*)

   (b) Calculate the periodic (monthly) rate, r, and the annual rate, rounded to ten-thousandths. Write the annual interest rate as a percent.

2. After talking to an investment advisor, you decide to invest in a no-load stock mutual fund. You can check the status of the account using the Internet. "No-load" means there is no buy-in fee. On the first of June, July, and August you again invest $250, $175, and $300, respectively. After making the August 1 deposit, your account value is $732.96. To find the average interest rate, you must now solve $250x^2 + 175x + 300 = 732.96$.
   (a) Use the quadratic formula to solve the equation for x.

   (b) Calculate the periodic (monthly) rate, r, and the annual rate, rounded to ten-thousandths. Write the annual interest rate as a percent.

9.3 EXERCISES

 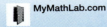

Write each equation in the form $ax^2 + bx + c = 0$, if necessary. Then identify the values of a, b, and c. Do not actually solve the equation. See Example 1.

1. $4x^2 + 5x - 9 = 0$

$a = $ _____ $b = $ _____ $c = $ _____

2. $8x^2 + 3x - 4 = 0$

$a = $ _____ $b = $ _____ $c = $ _____

3. $3x^2 = 4x + 2$

$a = $ _____ $b = $ _____ $c = $ _____

4. $5x^2 = 3x - 6$

$a = $ _____ $b = $ _____ $c = $ _____

5. $3x^2 = -7x$

$a = $ _____ $b = $ _____ $c = $ _____

6. $9x^2 = 8x$

$a = $ _____ $b = $ _____ $c = $ _____

Use the quadratic formula to solve each equation. Write all radicals in simplified form, and write all answers in lowest terms. See Examples 2–4.

7. $k^2 = -12k + 13$

8. $r^2 = 8r + 9$

9. $p^2 - 4p + 4 = 0$

10. $9x^2 + 6x + 1 = 0$

11. $2x^2 + 12x = -5$

12. $5m^2 + m = 1$

13. $2x^2 = 5 + 3x$

14. $2z^2 = 30 + 7z$

15. $6x^2 + 6x = 0$

16. $4n^2 - 12n = 0$

17. $-2x^2 = -3x + 2$

18. $-x^2 = -5x + 20$

19. $3x^2 + 5x + 1 = 0$

20. $6x^2 - 6x + 1 = 0$

21. $7x^2 = 12x$

22. $9r^2 = 11r$

23. $x^2 - 24 = 0$

24. $z^2 - 96 = 0$

25. $25x^2 - 4 = 0$

26. $16x^2 - 9 = 0$

27. $3x^2 - 2x + 5 = 10x + 1$

28. $4x^2 - x + 4 = x + 7$

29. $2x^2 + x + 5 = 0$

30. $3x^2 + 2x + 8 = 0$

31. If we apply the quadratic formula and find that the value of $b^2 - 4ac$ is negative, what can we conclude?

32. If we were to solve the quadratic equation $-2x^2 - 4x + 3 = 0$, we might choose to use $a = -2$, $b = -4$, and $c = 3$. On the other hand, we might decide to multiply each side by -1 to begin, obtaining the equation $2x^2 + 4x - 3 = 0$, and then use $a = 2$, $b = 4$, and $c = -3$. Show that in either case, we obtain the same solutions.

Use the quadratic formula to solve each equation. See Example 5.

33. $\dfrac{3}{2}k^2 - k - \dfrac{4}{3} = 0$

34. $\dfrac{2}{5}x^2 - \dfrac{3}{5}x - 1 = 0$

35. $\dfrac{1}{2}x^2 + \dfrac{1}{6}x = 1$

36. $\dfrac{2}{3}y^2 - \dfrac{4}{9}y = \dfrac{1}{3}$

37. $.5x^2 = x + .5$

38. $.25x^2 = -1.5x - 1$

39. $\dfrac{3}{8}x^2 - x + \dfrac{17}{24} = 0$

40. $\dfrac{1}{3}x^2 + \dfrac{8}{9}x + \dfrac{7}{9} = 0$

41. Solve the formula $S = 2\pi rh + \pi r^2$ for r by first writing it in the form $ar^2 + br + c = 0$, and then using the quadratic formula.

42. Solve the formula $V = \pi r^2 h + \pi R^2 h$ for r, using the method described in Exercise 41.

Solve each problem.

43. A frog is sitting on a stump 3 ft above the ground. He hops off the stump and lands on the ground 4 ft away. During his leap, his height h is given by the equation
$$h = -.5x^2 + 1.25x + 3,$$
where x is the distance in feet from the base of the stump, and h is in feet. How far was the frog from the base of the stump when he was 1.25 ft above the ground?

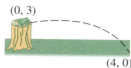

(0, 3)

(4, 0)

44. An astronaut on the moon throws a baseball upward. The height h of the ball, in feet, x seconds after he throws it, is given by the equation
$$h = -2.7x^2 + 30x + 6.5.$$
After how many seconds is the ball 12 ft above the moon's surface? Give answer(s) to the nearest tenth.

45. A rule for estimating the number of board feet of lumber that can be cut from a log depends on the diameter of the log. To find the diameter d required to get 9 board feet of lumber, we use the equation
$$\left(\dfrac{d-4}{4}\right)^2 = 9.$$
Solve this equation for d. Are both answers reasonable?

46. An old Babylonian problem asks for the length of the side of a square, given that the area of the square minus the length of a side is 870. Find the length of the side. (*Source:* Eves, H., *An Introduction to the History of Mathematics,* Sixth Edition, Saunders College Publishing, 1990.)

Summary Exercises on QUADRATIC EQUATIONS

Four algebraic methods have now been introduced for solving quadratic equations written in the form $ax^2 + bx + c = 0$. The following chart shows some advantages and disadvantages of each method.

Method	Advantages	Disadvantages
1. Factoring	It is usually the fastest method.	Not all equations can be solved by factoring. Some factorable polynomials are difficult to factor.
2. Square root property	It is the simplest method for solving equations of the form $(ax + b)^2 = $ a number.	Few equations are given in this form.
3. Completing the square	It can always be used. (Also, the procedure is useful in other areas of mathematics.)	It requires more steps than other methods.
4. Quadratic formula	It can always be used.	It is more difficult than factoring because of the $\sqrt{b^2 - 4ac}$ expression.

Solve each quadratic equation by the method of your choice.

1. $x^2 = 36$

2. $x^2 + 3x = -1$

3. $x^2 - \dfrac{100}{81} = 0$

4. $81t^2 = 49$

5. $z^2 - 4z + 3 = 0$

6. $w^2 + 3w + 2 = 0$

7. $z(z - 9) = -20$

8. $x^2 + 3x - 2 = 0$

9. $(3k - 2)^2 = 9$

10. $(2s - 1)^2 = 10$

11. $(x + 6)^2 = 121$

12. $(5k + 1)^2 = 36$

13. $(3r - 7)^2 = 24$

14. $(7p - 1)^2 = 32$

15. $(5x - 8)^2 = -6$

16. $2t^2 + 1 = t$

17. $-2x^2 = -3x - 2$

18. $-2x^2 + x = -1$

19. $8z^2 = 15 + 2z$

20. $3k^2 = 3 - 8k$

21. $0 = -x^2 + 2x + 1$

22. $3x^2 + 5x = -1$

23. $5x^2 - 22x = -8$

24. $x(x + 6) + 4 = 0$

25. $(x + 2)(x + 1) = 10$

26. $16x^2 + 40x + 25 = 0$

27. $4x^2 = -1 + 5x$

28. $2p^2 = 2p + 1$

29. $3x(3x + 4) = 7$

30. $5x - 1 + 4x^2 = 0$

31. $\dfrac{x^2}{2} + \dfrac{7x}{4} + \dfrac{11}{8} = 0$

32. $t(15t + 58) = -48$

33. $9k^2 = 16(3k + 4)$

34. $\dfrac{1}{5}x^2 + x + 1 = 0$

35. $x^2 - x + 3 = 0$

36. $4x^2 - 11x + 8 = -2$

37. $-3x^2 + 4x = -4$

38. $z^2 - \dfrac{5}{12}z = \dfrac{1}{6}$

39. $5k^2 + 19k = 2k + 12$

40. $\dfrac{1}{2}x^2 - x = \dfrac{15}{2}$

41. $x^2 - \dfrac{4}{15} = -\dfrac{4}{15}x$

42. $(x + 2)(x - 4) = 16$

43. If $D > 0$ and $\dfrac{5 + \sqrt{D}}{3}$ is a solution of $ax^2 + bx + c = 0$, what must be another solution of the equation?

9.4   GRAPHING QUADRATIC EQUATIONS

1　**Graph quadratic equations.**   In Chapter 3 we saw that the graph of a linear equation in two variables is a straight line that represents all the solutions of the equation. Quadratic equations in two variables, of the form $y = ax^2 + bx + c$, are graphed in this section. The simplest quadratic equation is $y = x^2$ (or $y = 1x^2 + 0x + 0$). The graph of this equation is not a straight line, as seen in Example 1 that follows. The equation can be graphed in much the same way that straight lines were graphed, by finding ordered pairs that satisfy the equation $y = x^2$.

Example 1　**Graphing a Quadratic Equation**

Graph $y = x^2$.
　　Select several values for x; then find the corresponding y-values. For example, selecting $x = 2$ gives
$$y = 2^2 = 4,$$
and so the point $(2, 4)$ is on the graph of $y = x^2$. (Recall that in an ordered pair such as $(2, 4)$, the x-value comes first and the y-value second.)

Work Problem ❶ at the Side.

　　If the points from Problem 1 at the side are plotted on a coordinate system and a smooth curve is drawn through them, the graph is as shown in Figure 1. The table of values completed in Problem 1 is shown with the graph.

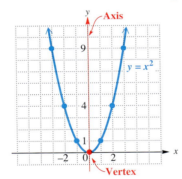

x	y
3	9
2	4
1	1
0	0
−1	1
−2	4
−3	9

Figure 1

Work Problem ❷ at the Side.

　　The curve in Figure 1 is called a **parabola**. The point $(0, 0)$, the lowest point on this graph, is called the **vertex** of the parabola. The vertical line through the vertex (the y-axis here) is called the **axis** of the parabola. The axis of a parabola is a **line of symmetry** for the graph, because if the graph is folded on this line, the two halves will match.
　　Every equation of the form
$$y = ax^2 + bx + c,$$
with $a \neq 0$, has a graph that is a parabola. Because of its many useful properties, the parabola occurs frequently in real-life applications. For example, if an object is thrown into the air, the path that the object follows is a parabola (ignoring wind resistance). The cross sections of radar, spotlight, and telescope reflectors also form parabolas.

❶ Complete the table of values for $y = x^2$.

x	y
3	
2	4
1	
0	
−1	
−2	
−3	

❷ Graph $y = \frac{1}{2}x^2$ by first completing the table of values.

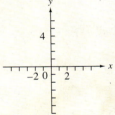

x	y
−2	
−1	
0	
1	
2	

ANSWERS

1.

x	y
3	9
2	4
1	1
0	0
−1	1
−2	4
−3	9

2.

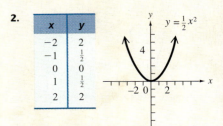

x	y
−2	2
−1	$\frac{1}{2}$
0	0
1	$\frac{1}{2}$
2	2

❸ Complete each ordered pair for $y = -x^2 - 1$.

$$(-2, \quad), (-1, \quad),$$
$$(1, \quad), (2, \quad)$$

❹ Graph each equation, and identify each vertex.

(a) $y = -x^2 - 3$

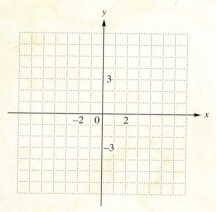

(b) $y = x^2 + 3$

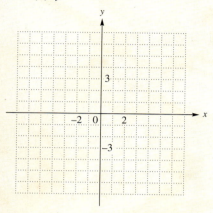

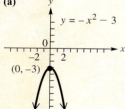
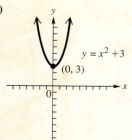
Example 2 Graphing a Parabola

Graph $y = -x^2 - 1$.

Find several ordered pairs. To begin, check for intercepts. Let $x = 0$ to find the y-intercept.

$$y = -x^2 - 1 = -0^2 - 1 = -1,$$

giving the ordered pair $(0, -1)$. Let $y = 0$ to find the x-intercepts.

$$y = -x^2 - 1$$
$$0 = -x^2 - 1$$
$$x^2 = -1$$

This equation has no real number solution, so there are no x-intercepts. Choose additional x-values near $x = 0$ to find other ordered pairs.

Work Problem ❸ at the Side.

The ordered pair $(0, -1)$ and the pairs from Problem 3 at the side are listed in the table shown with Figure 2. Plot these points and connect them with a smooth curve as shown in Figure 2. The vertex of this parabola is $(0, -1)$. The graph opens downward because x^2 has a negative coefficient, so the vertex is the *highest* point of the graph.

x	y
-2	-5
-1	-2
0	-1
1	-2
2	-5

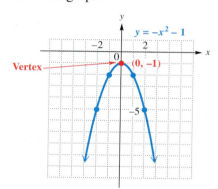

Figure 2

Work Problem ❹ at the Side.

2 **Find the vertex of a parabola.** As the preceding graphs suggest, the vertex is the most important point to locate when you are graphing a quadratic equation. The next example shows how to find the vertex in a more general case.

Example 3 Finding the Vertex to Graph a Parabola

Graph $y = x^2 - 2x - 3$.

We want to find the vertex of the graph. *If a parabola has two x-intercepts, the vertex is exactly halfway between them.* Therefore, we begin by finding the x-intercepts. We let $y = 0$ in the equation, and solve for x.

$$0 = x^2 - 2x - 3$$
$$0 = (x + 1)(x - 3) \qquad \text{Factor.}$$
$$x + 1 = 0 \quad \text{or} \quad x - 3 = 0 \qquad \text{Zero-factor property}$$
$$x = -1 \quad \text{or} \qquad x = 3$$

Continued on Next Page

There are two x-intercepts, $(-1, 0)$ and $(3, 0)$. Now we find the y-intercept.

$$y = 0^2 - 2(0) - 3 = -3 \quad \text{Let } x = 0.$$

The y-intercept is $(0, -3)$.

As previously mentioned, the x-value of the vertex is halfway between the x-values of the two x-intercepts. Thus, it is $\frac{1}{2}$ their sum.

$$x = \frac{1}{2}(-1 + 3) = 1$$

We find the corresponding y-value by substituting 1 for x in the equation.

$$y = 1^2 - 2(1) - 3 = -4$$

The vertex is $(1, -4)$. Here, the axis is the line $x = 1$. We plot the three intercepts and the vertex, and find additional ordered pairs as needed. For example, if $x = 2$,

$$y = 2^2 - 2(2) - 3 = -3,$$

leading to the ordered pair $(2, -3)$. A table with the ordered pairs we have found is shown with the graph in Figure 3.

x	y	
-2	5	
-1	0	← x-intercept
0	-3	← y-intercept
1	-4	← Vertex
2	-3	
3	0	← x-intercept
4	5	

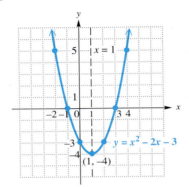

Figure 3

Work Problem ❺ at the Side.

We can generalize from Example 3. The x-coordinates of the x-intercepts for $0 = ax^2 + bx + c$, by the quadratic formula, are

$$x = \frac{-b + \sqrt{b^2 - 4ac}}{2a} \quad \text{and} \quad x = \frac{-b - \sqrt{b^2 - 4ac}}{2a}.$$

Thus, the x-value of the vertex is

$$x = \frac{1}{2}\left(\frac{-b + \sqrt{b^2 - 4ac}}{2a} + \frac{-b - \sqrt{b^2 - 4ac}}{2a}\right)$$

$$x = \frac{1}{2}\left(\frac{-b + \sqrt{b^2 - 4ac} - b - \sqrt{b^2 - 4ac}}{2a}\right)$$

$$x = \frac{1}{2}\left(\frac{-2b}{2a}\right)$$

$$x = -\frac{b}{2a}.$$

❺ Graph $y = x^2 + 2x - 8$.

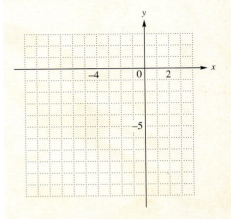

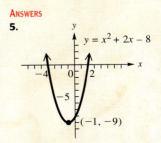

6 Complete the following ordered pairs for $y = x^2 - 4x + 1$.

(5,   )   (1,   )   (4,   )

    (3,   )   (−1,   )

For the equation in Example 3, $y = x^2 - 2x - 3$, $a = 1$, and $b = -2$. Thus, the x-value of the vertex is

$$x = -\frac{b}{2a} = -\frac{-2}{2(1)} = 1,$$

which is the same x-value for the vertex we found in Example 3. (*The x-value of the vertex is $x = -\dfrac{b}{2a}$, even if the graph has no x-intercepts.*) A procedure for graphing quadratic equations follows.

Graphing the Parabola $y = ax^2 + bx + c$

Step 1   **Find the vertex.** Let $x = -\frac{b}{2a}$, and find the corresponding y-value by substituting for x in the equation.

Step 2   **Find the y-intercept.**

Step 3   **Find the x-intercepts** (if they exist).

Step 4   **Plot** the intercepts and the vertex.

Step 5   **Find and plot additional ordered pairs** near the vertex and intercepts as needed, using the symmetry about the axis of the parabola.

Example 4   Graphing a Parabola

Graph $y = x^2 - 4x + 1$.

The x-value of the vertex is

$$x = -\frac{b}{2a} = -\frac{-4}{2(1)} = \mathbf{2}.$$

The y-value of the vertex is

$$y = \mathbf{2}^2 - 4(\mathbf{2}) + 1 = \mathbf{-3},$$

so the vertex is $(\mathbf{2}, \mathbf{-3})$. The axis is the line $x = \mathbf{2}$. Now find the intercepts. Let $x = 0$ in $y = x^2 - 4x + 1$ to get the y-intercept $(0, 1)$. Let $y = 0$ to get the x-intercepts. If $y = 0$, the equation is $0 = x^2 - 4x + 1$, which cannot be solved by factoring. Use the quadratic formula to solve for x.

$$x = \frac{4 \pm \sqrt{16 - 4}}{2} \qquad \textcolor{blue}{\text{Let } a = 1, b = -4, c = 1.}$$

$$x = \frac{4 \pm \sqrt{12}}{2}$$

$$x = \frac{4 \pm 2\sqrt{3}}{2} \qquad \textcolor{blue}{\sqrt{12} = 2\sqrt{3}}$$

$$x = \frac{2(2 \pm \sqrt{3})}{2} = \mathbf{2 \pm \sqrt{3}}$$

Use a calculator to find that the x-intercepts are $(\mathbf{3.7}, 0)$ and $(\mathbf{.3}, 0)$ to the nearest tenth.

Work Problem 6 at the Side.

Continued on Next Page

Plot the intercepts, vertex, and the points found in Problem 6. Connect these points with a smooth curve. The graph is shown in Figure 4.

x	y
-1	6
0	1
$2 - \sqrt{3} \approx .3$	0
1	-2
2	-3
3	-2
$2 + \sqrt{3} \approx 3.7$	0
4	1
5	6

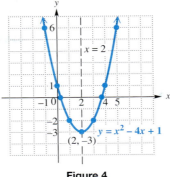

Figure 4

Work Problem **7** at the Side.

3 **Solve an application involving a parabola.** Parabolic shapes are found all around us. Satellite dishes that deliver television signals are becoming more popular each year. Radio telescopes use parabolic reflectors to track incoming signals. The final example discusses how to describe a cross section of a parabolic dish using an equation.

Example 5 **Finding the Equation of a Parabolic Satellite Dish**

The Parkes radio telescope has a parabolic dish shape with a diameter of 210 ft and a depth of 32 ft. (*Source:* Mar, J. and Liebowitz, H., *Structure Technology for Large Radio and Radar Telescope Systems,* The MIT Press, 1969.) Figure 5(a) shows a diagram of such a dish, and Figure 5(b) shows how a cross section of the dish can be modeled by a graph, with the vertex of the parabola at the origin of a coordinate system. Find the equation of this graph.

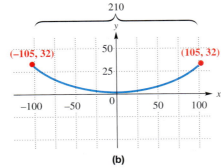

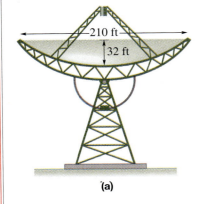

(a)

(b)

Figure 5

Continued on Next Page

---

7 Graph each parabola. Identify the vertex and y-intercept, and give the coordinates of the x-intercepts to the nearest tenth.

(a) $y = x^2 - 3x - 3$

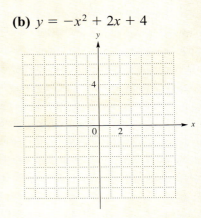

(b) $y = -x^2 + 2x + 4$

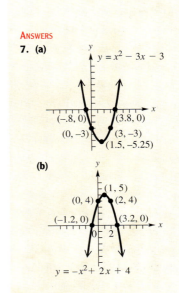

ANSWERS

7. (a)

$y = x^2 - 3x - 3$

$(-.8, 0)$ $(3.8, 0)$
$(0, -3)$ $(3, -3)$
$(1.5, -5.25)$

(b)

$(1, 5)$
$(0, 4)$ $(2, 4)$
$(-1.2, 0)$ $(3.2, 0)$

$y = -x^2 + 2x + 4$

8 Suppose that a radio telescope has a parabolic dish shape with a diameter of 350 ft and a depth of 48 ft. Find the equation of the graph of a cross section.

Because the vertex is at the origin, it can be shown that the equation will be of the form $y = ax^2$. (See Example 1 and Problem 1, for example.) As shown in Figure 5(b), one point on the graph has coordinates $(105, 32)$. Letting $x = 105$ and $y = 32$, we can solve for a.

$$y = ax^2 \qquad \text{General equation}$$
$$32 = a(105)^2 \qquad \text{Substitute for } x \text{ and } y.$$
$$32 = 11{,}025a \qquad 105^2 = 11{,}025$$
$$a = \frac{32}{11{,}025} \qquad \text{Divide by 11,025.}$$

Thus the equation is $y = \frac{32}{11{,}025}x^2$.

Work Problem 8 at the Side.

9.4 EXERCISES

FOR EXTRA HELP

 Student's Solutions Manual

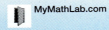 MyMathLab.com

 InterAct Math Tutorial Software

 AW Math Tutor Center

 www.mathxl.com

 Digital Video Tutor CD 6 Videotape 18

1. In your own words, explain what is meant by the vertex of a parabola.

2. In your own words, explain what is meant by the line of symmetry of a parabola that opens upward or downward.

Graph each equation. Give the coordinates of the vertex in each case. See Examples 1–4.

3. $y = 2x^2$

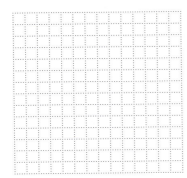

4. $y = 3x^2$

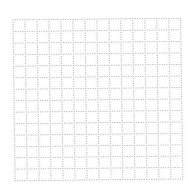

5. $y = x^2 - 4$

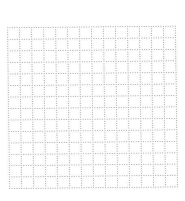

6. $y = x^2 - 6$

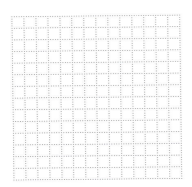

7. $y = -x^2 + 2$

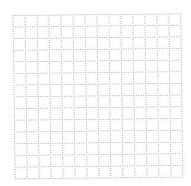

8. $y = -x^2 + 4$

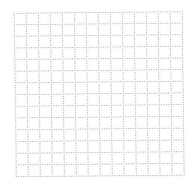

9. $y = (x + 3)^2$

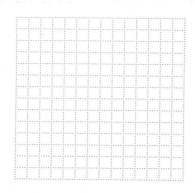

10. $y = (x - 4)^2$

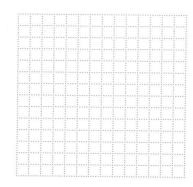

11. $y = x^2 + 2x + 3$

12. $y = x^2 - 4x + 3$

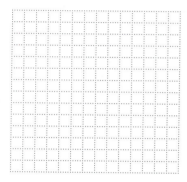

13. $y = -x^2 + 6x - 5$

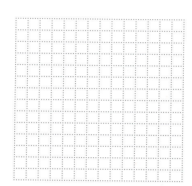

14. $y = -x^2 - 4x - 3$

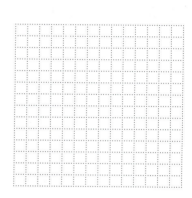

15. Based on your work in Exercises 3–14, what seems to be the direction in which the parabola $y = ax^2 + bx + c$ opens if $a > 0$? if $a < 0$?

16. How many real solutions does a quadratic equation have if its corresponding graph has

(a) no x-intercepts
(b) one x-intercept
(c) two x-intercepts? See Examples 1–3.

Solve each problem. See Example 5.

17. The U.S. Naval Research Laboratory designed a giant radio telescope that had a diameter of 300 ft and maximum depth of 44 ft. The graph on the right below describes a cross section of this telescope. Find the equation of this parabola. (*Source:* Mar, J. and Liebowitz, H., *Structure Technology for Large Radio and Radar Telescope Systems,* The MIT Press, 1969.)

18. Suppose the telescope in Exercise 17 had a diameter of 400 ft and maximum depth of 50 ft. Find the equation of this parabola.

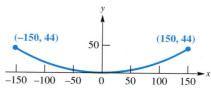

9.5 INTRODUCTION TO FUNCTIONS

If gasoline costs $1.45 per gal and you buy **1** gal, then you must pay $1.45(**1**) = $1.45. If you buy **2** gal, your cost is $1.45(**2**) = $2.90; for **3** gal, your cost is $1.45(**3**) = $4.35, and so on. Generalizing, if x represents the number of gallons, then the cost is $1.45x$. If we let y represent the cost, then the equation $y = 1.45x$ *relates* the number of gallons, x, to the cost in dollars, y. The ordered pairs (x, y) that satisfy this equation form a *relation*.

1   **Understand the definition of a relation.**   In an ordered pair (x, y), x and y are called the **components** of the ordered pair. Any set of ordered pairs is called a **relation.*** The set of all first components in the ordered pairs of a relation is the **domain** of the relation, and the set of all second components in the ordered pairs is the **range** of the relation.

Example 1   **Using Ordered Pairs to Define Relations**

(a) The relation $\{(0, 1), (2, 5), (3, 8), (4, 2)\}$ has domain $\{0, 2, 3, 4\}$ and range $\{1, 2, 5, 8\}$. The correspondence between the elements of the domain and the elements of the range is shown in Figure 6.

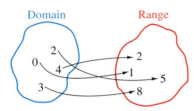

Figure 6

(b) The relation $\{(3, 5), (3, 6), (3, 7), (3, 8)\}$ has domain $\{3\}$ and range $\{5, 6, 7, 8\}$.

====== **Work Problem ❶ at the Side.**

2   **Understand the definition of a function.**   We now investigate an important type of relation, called a *function*.

Function

A **function** is a set of ordered pairs in which each first component corresponds to exactly one second component.

By this definition, the relation in Example 1(a) is a function, but the relation in Example 1(b) is not a function. Notice, however, that if the components of the ordered pairs in Example 1(b) were interchanged, giving the relation

$$\{(5, 3), (6, 3), (7, 3), (8, 3)\},$$

then the relation *would* be a function; in this case, each domain element (first component) corresponds to *exactly one* range element (second component).

---

* It is standard notation to use set braces around the ordered pairs that form a relation. You may want to refer to Appendix A, which covers set notation.

OBJECTIVES

1 Understand the definition of a relation.

2 Understand the definition of a function.

3 Decide whether an equation defines a function.

4 Use function notation.

5 Apply the function concept in an application.

❶ Give the domain and the range of each relation.

(a) $\{(5, 10), (15, 20), (25, 30), (35, 40)\}$

(b) $\{(1, 4), (2, 4), (3, 4)\}$

❷ Decide whether each relation is a function.

(a) $\{(-2, 8), (-1, 1), (0, 0), (1, 1), (2, 8)\}$

(b) $\{(5, 2), (5, 1), (5, 0)\}$

Example 2 Determining whether Relations Are Functions

Determine whether each relation is a function.

(a) $\{(-2, 4), (-1, 1), (0, 0), (1, 1), (2, 4)\}$

Notice that each first component appears once and only once. Because of this, the relation is a function.

(b) $\{(9, 3), (9, -3), (4, 2)\}$

The first component 9 appears in two ordered pairs, and corresponds to two different second components. Therefore, this relation is not a function.

Work Problem ❷ at the Side.

The simple relations given in Examples 1 and 2 were defined by listing the ordered pairs or by showing the correspondence with a figure. Most useful functions have an infinite number of ordered pairs and are usually defined with equations that tell how to get the second components, given the first. We have been using equations with x and y as the variables, where x represents the first component (input) and y the second component (output) in the ordered pairs.

Here are some everyday examples of functions.

1. The **cost y** in dollars charged by an express mail company is a function of the **weight in pounds x** determined by the equation $y = 1.5(x - 1) + 9$.

2. In one state, the sales tax is 6% of the price of an item. The **tax y** on a particular item is a function of the **price x**, so $y = .06x$.

3. The **distance d** traveled by a car moving at a constant speed of 45 mph is a function of the **time t**. Thus, $d = 45t$.

The function concept can be illustrated by an input-output "machine," as seen in Figure 7. It shows how the express mail company equation $y = 1.5(x - 1) + 9$ provides an output (the cost, represented by y) for a given input (the weight in pounds, given by x).

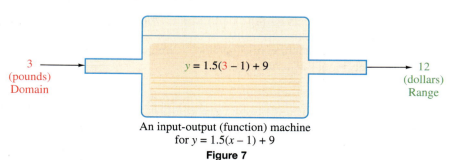

An input-output (function) machine
for $y = 1.5(x - 1) + 9$

Figure 7

3 **Decide whether an equation defines a function.** Given the graph of an equation, the definition of a function can be used to decide whether or not the graph represents a function. By the definition of a function, each x-value must lead to exactly one y-value. In Figure 8(a), the indicated x-value leads to two y-values, so this graph is not the graph of a function. A vertical line can be drawn that intersects this graph in more than one point.

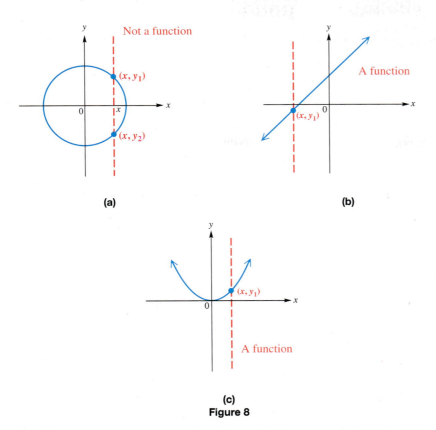

(c)
Figure 8

On the other hand, in Figures 8(b) and 8(c) any vertical line will intersect each graph in no more than one point. Because of this, the graphs in Figures 8(b) and 8(c) are graphs of functions. This idea leads to the **vertical line test** for a function.

Vertical Line Test

If a vertical line intersects a graph in more than one point, the graph is not the graph of a function.

As Figure 8(b) suggests, any nonvertical line is the graph of a function. For this reason, any linear equation of the form $y = mx + b$ defines a function. (Recall that a vertical line has undefined slope.) Also, any vertical parabola, as in Figure 8(c), is the graph of a function, so any quadratic equation of the form $y = ax^2 + bx + c$ $(a \neq 0)$ defines a function.

Example 3 **Deciding whether Relations Define Functions**

Decide whether each relation graphed or defined is a function.

(a)

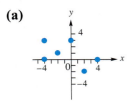

Because there are two ordered pairs with first component -4, this is not the graph of a function.

Continued on Next Page

3 Decide whether each relation graphed or defined is a function.

(a)

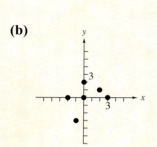

(b)

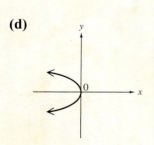

(c)

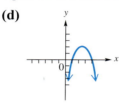

Every first component is matched with one and only one second component, and furthermore, no vertical line intersects the graph in more than one point. Therefore, this is the graph of a function.

(c) $y = 2x - 9$

This linear equation is in the form $y = mx + b$. Since the graph of this equation is a line that is not vertical, the equation defines a function.

(d)   **(e)**

Use the vertical line test. Any vertical line will cross the graph of a vertical parabola just once, so this is the graph of a function.

The vertical line test shows that this graph is not the graph of a function; a vertical line could cross the graph twice.

(f) $x = 4$

The graph of $x = 4$ is a vertical line, so the equation does not define a function.

Work Problem 3 at the Side.

---

4 **Use function notation.** The letters f, g, and h are commonly used to name functions. For example, the function $y = 3x + 5$ may be written

$$f(x) = 3x + 5,$$

where $f(x)$ is read "f of x." The notation $f(x)$ is another way of writing y in a function. For the function defined by $f(x) = 3x + 5$, if $x = 7$ then

$$f(7) = 3 \cdot 7 + 5 \qquad \text{Let } x = 7.$$
$$= 21 + 5 = 26.$$

Read this result, $f(7) = 26$, as "f of 7 equals 26." The notation $f(7)$ means the value of y when x is 7. The statement $f(7) = 26$ says that the value of y is 26 when x is 7. It also indicates that the point $(7, 26)$ lies on the graph of f.

To find $f(-3)$, substitute -3 for x.

$$f(-3) = 3(-3) + 5 \qquad \text{Let } x = -3.$$
$$= -9 + 5 = -4$$

CAUTION

The notation $f(x)$ does *not* mean f times x; $f(x)$ means the function value of x. It represents the y-value that corresponds to x.

(c)

(d)

(e) $y = 3$

Function Notation

In the notation $f(x)$,

f	is the name of the function,
x	is the domain value,
and $f(x)$	is the range value y for the domain value x.

Example 4   **Using Function Notation**

For the function defined by $f(x) = x^2 - 3$, find the following.

(a) $f(2)$

Substitute 2 for x.

$$f(x) = x^2 - 3$$
$$f(2) = 2^2 - 3 \qquad \text{Let } x = 2.$$
$$= 4 - 3 = 1$$

(b) $f(0) = 0^2 - 3 = 0 - 3 = -3$

(c) $f(-3) = (-3)^2 - 3 = 9 - 3 = 6$

$=$ **Work Problem ❹ at the Side.**

5 ▭   **Apply the function concept in an application.**   Because a function assigns each element in its domain to exactly one element in its range, the function concept is used in real-data applications where two quantities are related. Our final example discusses such an application, using a table of values.

Example 5   **Applying the Function Concept to a Prison Population**

The numbers of sentenced prisoners in state and federal institutions (in thousands) for selected years are given in the table.

Year	Number of Prisoners (in thousands)
1989	681
1991	790
1993	932
1995	1085
1997	1198

Source: U.S. Department of Justice.

(a) Write a set of ordered pairs that defines a function f.

If we choose the years as the domain elements and the numbers of prisoners as the range elements, the information in the table can be written as a set of five ordered pairs. In set notation, the function is

$f = \{(1989, 681), (1991, 790), (1993, 932), (1995, 1085), (1997, 1198)\}.$

Continued on Next Page

❹ For $f(x) = 6x - 2$, find each function value.

(a) $f(-1)$

(b) $f(0)$

(c) $f(1)$

5 The numbers of U.S. children (in thousands) educated at home for selected years are given in the table.

School Year	Number of Children
1993	588
1994	735
1995	800
1996	920
1997	1100

Source: National Home Education Research Institute, Salem, OR.

(a) Write a set of five ordered pairs that defines a function f for this data. Express y-values in thousands.

(b) Give the domain and range of f.

(c) Find $f(1995)$.

(d) In what year did the number of children equal 920 thousand?

(b) What is the domain of f? What is the range?
The domain is the set of years, or x-values:

$$\{1989, 1991, 1993, 1995, 1997\}.$$

The range is the set of numbers of prisoners, or y-values:

$$\{681, 790, 932, 1085, 1198\}.$$

(c) Find $f(1991)$ and $f(1997)$.
From the table or the set of ordered pairs in part (a),

$$f(1991) = 790 \text{ thousand} \quad \text{and} \quad f(1997) = 1198 \text{ thousand}.$$

(d) For what x-value does $f(x)$ equal 932? 681? ($f(x)$ values are in thousands.)
Use the table or the ordered pairs given in part (a) to find $f(1993) = 932$ and $f(1989) = 681$.

Work Problem 5 at the Side.

9.5   EXERCISES

Complete the following table for the function defined by $f(x) = x + 2$.

x	x + 2	f(x)	(x, y)
0	2	2	(0, 2)
1. 1			
2. 2			
3. 3			
4. 4			

5. Describe the graph of function f in Exercises 1–4 if the domain is $\{0, 1, 2, 3, 4\}$.

6. Describe the graph of function f in Exercises 1–4 if the domain is the set of all real numbers.

Determine whether each relation is a function. Give the domain and the range in Exercises 7–12. See Examples 1–3.

7. $\{(-4, 3), (-2, 1), (0, 5), (-2, -8)\}$

8. $\{(3, 7), (1, 4), (0, -2), (-1, -1), (-2, 5)\}$

9.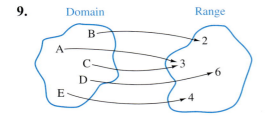

10.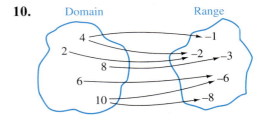

11.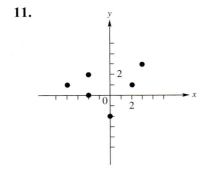

12.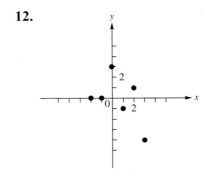

13.

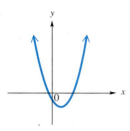

14.

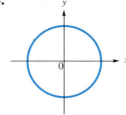

15.

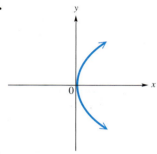

16.

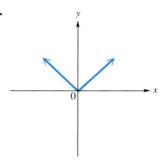

Decide whether each equation defines y as a function of x. (Remember that to be a function, every value of x must give one and only one value of y.) See Example 3.

17. $y = 5x + 3$ **18.** $y = -7x + 12$ **19.** $x = |y|$ **20.** $x = y^2$

RELATING CONCEPTS (Exercises 21–24)   **FOR INDIVIDUAL OR GROUP WORK**

A function such as $f(x) = 3x - 4$, called a linear function because its graph is a straight line, can be graphed by replacing $f(x)$ with y and then using the methods described earlier. Let us assume that some function is written in the form $f(x) = mx + b$, for particular values of m and b. Work Exercises 21–24 in order.

21. If $f(2) = 4$, name the coordinates of one point on the line.

22. If $f(-1) = -4$, name the coordinates of another point on the line.

23. Use the results of Exercises 21 and 22 to find the slope of the line.

24. Use the slope-intercept form of the equation of a line to write the function in the form $f(x) = mx + b$.

*For each function f, find **(a)** f(2), **(b)** f(0), **(c)** f(−3). See Example 4.*

25. $f(x) = 4x + 3$

26. $f(x) = -3x + 5$

27. $f(x) = x^2 - x + 2$

28. $f(x) = x^3 + x$

29. $f(x) = |x|$

30. $f(x) = |x + 7|$

The number of U.S. foreign-born residents has grown by more than 30 percent since 1990. The graph shows the number of such residents (in millions) for selected years. Use the information in the graph for Exercises 31–35. See Example 5.

31. Write the information in the graph as a set of ordered pairs. Does this set define a function?

32. Suppose that *g* is the name given to this relation. Give the domain and range of *g*.

GROWTH OF U.S. FOREIGN-BORN POPULATION

Source: U.S. Bureau of the Census.

33. Find *g*(1980) and *g*(1990).

34. For what value of *x* does *g*(*x*) = 25.8 (million)?

35. Suppose *g*(2000) = 29.3 (million). What does this tell you in the context of the application?

The data give the numbers (in thousands) of U.S. active-duty female military personnel during the years 1992–1995. Use the information in the table to **work Exercises 36–40 in order.**

ACTIVE-DUTY FEMALE MILITARY PERSONNEL (IN THOUSANDS)

Year	Number
1992	210
1993	204
1994	200
1995	196

Source: U.S. Department of Defense.

36. Plot the ordered pairs (year, number) from the table. Do the points suggest that a linear function would give a reasonable approximation of the data?

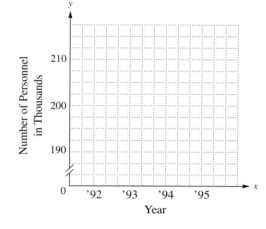

37. Use the first and last data pairs in the table to write an equation relating the data.

38. Use your equation from Exercise 37 to approximate the number of active-duty female military personnel in 1993 and 1994. Round to the nearest thousand.

39. Use the second and third data pairs to write an equation relating the data.

40. Use your equation from Exercise 39 to approximate the number of active-duty female military personnel in 1992 and 1995. Which of the results in Exercise 38 and this exercise give better approximations?

SUMMARY

KEY TERMS

9.4	**parabola**	The graph of the quadratic equation $y = ax^2 + bx + c$ is called a parabola.
	vertex	The vertex of a parabola that opens upward or downward is the lowest or highest point on the graph.
	axis	The axis of a parabola that opens upward or downward is a vertical line through the vertex.
	line of symmetry	If a graph is folded on its line of symmetry, the two sides coincide.
9.5	**components**	In an ordered pair (x, y), x and y are the components.
	relation	Any set of ordered pairs is called a relation.
	domain	The set of all first components in the ordered pairs of a relation is the domain of the relation.
	range	The set of all second components in the ordered pairs of a relation is the range of the relation.
	function	A function is a set of ordered pairs in which each first component corresponds to exactly one second component.

NEW SYMBOLS

$\pm$ plus or minus $f(x)$ function f of x

TEST YOUR WORD POWER

See how well you have learned the vocabulary in this chapter. Answers follow the Quick Review.

1. A **relation** is
 (a) any set of ordered pairs
 (b) a set of ordered pairs in which each first component corresponds to exactly one second component
 (c) two sets of ordered pairs that are related
 (d) a graph of ordered pairs.

2. The **domain** of a relation is
 (a) the set of all x- and y-values in the ordered pairs of the relation
 (b) the difference between the components in an ordered pair of the relation

 (c) the set of all first components in the ordered pairs of the relation
 (d) the set of all second components in the ordered pairs of the relation.

3. The **range** of a relation is
 (a) the set of all x- and y-values in the ordered pairs of the relation
 (b) the difference between the components in an ordered pair of the relation
 (c) the set of all first components in the ordered pairs of the relation

 (d) the set of all second components in the ordered pairs of the relation.

4. A **function** is
 (a) any set of ordered pairs
 (b) a set of ordered pairs in which each first component corresponds to exactly one second component
 (c) two sets of ordered pairs that are related
 (d) a graph of ordered pairs.

Concepts	Examples

9.1 Solving Quadratic Equations by the Square Root Property

Square Root Property of Equations

If k is positive, and if $a^2 = k$, then $a = \sqrt{k}$ or $a = -\sqrt{k}$.

Solve $(2x + 1)^2 = 5$.

$$2x + 1 = \sqrt{5} \qquad \text{or} \quad 2x + 1 = -\sqrt{5}$$

$$2x = -1 + \sqrt{5} \quad \text{or} \qquad 2x = -1 - \sqrt{5}$$

$$x = \frac{-1 + \sqrt{5}}{2} \quad \text{or} \qquad x = \frac{-1 - \sqrt{5}}{2}$$

The solutions are $\dfrac{-1 + \sqrt{5}}{2}$ and $\dfrac{-1 - \sqrt{5}}{2}$.

9.2 Solving Quadratic Equations by Completing the Square

Solving a Quadratic Equation by Completing the Square

Step 1 If the coefficient of the squared term is 1, go to Step 2. If it is not 1, divide each side of the equation by this coefficient.

Step 2 Make sure that all variable terms are on one side of the equation, and all constant terms are on the other.

Step 3 Take half the coefficient of x, square it, and add the square to each side of the equation. Factor the variable side.

Step 4 Use the square root property to solve the equation.

Solve $2x^2 + 4x - 1 = 0$.

$$x^2 + 2x - \frac{1}{2} = 0 \qquad \text{Divide by 2.}$$

$$x^2 + 2x = \frac{1}{2} \qquad \text{Add } \tfrac{1}{2}.$$

$$x^2 + 2x + 1 = \frac{1}{2} + 1 \qquad \left[\tfrac{1}{2}(2)\right]^2 = 1$$

$$(x + 1)^2 = \frac{3}{2} \qquad \text{Factor; combine like terms.}$$

$$x + 1 = \pm\sqrt{\frac{3}{2}} = \pm\frac{\sqrt{6}}{2}$$

$$x = -1 \pm \frac{\sqrt{6}}{2}$$

$$x = \frac{-2 \pm \sqrt{6}}{2}$$

The solutions are $\dfrac{-2 + \sqrt{6}}{2}$ and $\dfrac{-2 - \sqrt{6}}{2}$.

Concepts	Examples

9.3 *Solving Quadratic Equations by the Quadratic Formula*

Quadratic Formula

The solutions of $ax^2 + bx + c = 0$ $(a \neq 0)$ are

$$x = \frac{-b \pm \sqrt{b^2 - 4ac}}{2a}.$$

Solve $3x^2 - 4x - 2 = 0$.

$$x = \frac{-(-4) \pm \sqrt{(-4)^2 - 4(3)(-2)}}{2(3)}$$

$$x = \frac{4 \pm \sqrt{16 + 24}}{6}$$

$$x = \frac{4 \pm \sqrt{40}}{6} = \frac{4 \pm 2\sqrt{10}}{6}$$

$$x = \frac{2(2 \pm \sqrt{10})}{6} = \frac{2 \pm \sqrt{10}}{3}$$

The solutions are $\dfrac{2 + \sqrt{10}}{3}$ and $\dfrac{2 - \sqrt{10}}{3}$.

9.4 *Graphing Quadratic Equations*

To graph $y = ax^2 + bx + c$:

Step 1 Find the vertex: $x = -\frac{b}{2a}$; find y by substituting this value for x in the equation.

Graph $y = 2x^2 - 5x - 3$.

$$x = -\frac{b}{2a} = -\frac{-5}{2(2)} = \frac{5}{4}$$

$$y = 2\left(\frac{5}{4}\right)^2 - 5\left(\frac{5}{4}\right) - 3$$

$$= 2\left(\frac{25}{16}\right) - \frac{25}{4} - 3$$

$$= \frac{25}{8} - \frac{50}{8} - \frac{24}{8} = -\frac{49}{8}$$

The vertex is $\left(\frac{5}{4}, -\frac{49}{8}\right)$.

$y = 2(0)^2 - 5(0) - 3 = -3$

Step 2 Find the y-intercept.

The y-intercept is $(0, -3)$.

Step 3 Find the x-intercepts (if they exist).

$$0 = 2x^2 - 5x - 3$$

$$0 = (2x + 1)(x - 3)$$

$$2x + 1 = 0 \quad \text{or} \quad x - 3 = 0$$

$$2x = -1 \quad \text{or} \quad x = 3$$

$$x = -\frac{1}{2} \quad \text{or} \quad x = 3$$

The x-intercepts are $\left(-\frac{1}{2}, 0\right)$ and $(3, 0)$.

Step 4 Plot the intercepts and the vertex.

Step 5 Find and plot additional ordered pairs near the vertex and intercepts as needed.

x	y
$-\frac{1}{2}$	0
0	-3
1	-6
$\frac{5}{4}$	$-\frac{49}{8}$
2	-5
3	0

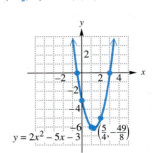

$y = 2x^2 - 5x - 3$

Concepts	Examples
9.5 Introduction to Functions	
Vertical Line Test	
If a vertical line intersects a graph in more than one point, the graph is not the graph of a function.	By the vertical line test, the graph shown is not the graph of a function. 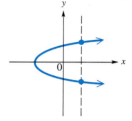
	The function
	$$\{(10, 5), (20, 15), (30, 25)\}$$
	has domain $\{10, 20, 30\}$ and range $\{5, 15, 25\}$.
	If $f(x) = 2x + 7$, then
	$$f(3) = 2(3) + 7$$
	$$= 13.$$

ANSWERS TO TEST YOUR WORD POWER

1. **(a)** *Example:* $\{(0, 2), (2, 4), (3, 6), (-1, 3)\}$ 2. **(c)** *Example:* The domain in the relation given in Problem 1 is the set of x-values, that is, $\{0, 2, 3, -1\}$.   3. **(d)** *Example:* The range of the relation given in Problem 1 is the set of y-values, that is, $\{2, 4, 6, 3\}$.   4. **(b)** *Example:* The relation given in Problem 1 is a function since each x-value corresponds to exactly one y-value.

Chapter 9   **REVIEW EXERCISES**

[9.1] *In Exercises 1–8, solve each equation by using the square root property. Express all radicals in simplest form.*

1. $y^2 = 144$

2. $x^2 = 37$

3. $m^2 = 128$

4. $(k + 2)^2 = 25$

5. $(r - 3)^2 = 10$

6. $(2p + 1)^2 = 14$

7. $(3k + 2)^2 = -3$

8. $(3x + 5)^2 = 0$

[9.2] *Solve each equation by completing the square.*

9. $m^2 + 6m + 5 = 0$

10. $p^2 + 4p = 7$

11. $-x^2 + 5 = 2x$

12. $2x^2 - 3 = -8x$

13. $4(x^2 + 7x) + 29 = -20$

14. $(4x + 1)(x - 1) = -7$

Solve each problem.

15. If an object is propelled upward on Earth from a height of 50 ft, with an initial velocity of 32 ft per sec, then its height after t sec is given by $h = -16t^2 + 32t + 50$, where h is in feet. After how many seconds will it reach a height of 30 ft?

16. Find the lengths of the three sides of the right triangle shown.

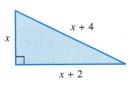

17. What must be added to $x^2 + kx$ to make it a perfect square?

[9.3]

18. Consider the equation $x^2 - 9 = 0$.

  (a) Solve the equation by factoring.

  (b) Solve the equation by the square root property.

  (c) Solve the equation by the quadratic formula.

  (d) Compare your answers. If a quadratic equation can be solved by both factoring and the quadratic formula, should we always get the same results? Explain.

Solve each equation by using the quadratic formula.

19. $-4x^2 - 2x + 7 = 0$ **20.** $2x^2 + 8 = 4x + 11$ **21.** $x(5x - 1) = 1$

22. $\dfrac{1}{4}x^2 = 2 - \dfrac{3}{4}x$ **23.** $\dfrac{1}{2}x^2 + 3x = 5$

24. Why is this not the statement of the quadratic formula for $ax^2 + bx + c = 0$?

$$x = -b \pm \frac{\sqrt{b^2 - 4ac}}{2a}$$

[9.4] *Sketch the graph of each equation. Identify each vertex.*

25. $y = -3x^2$ **26.** $y = -x^2 + 5$ **27.** $y = x^2 - 2x + 1$

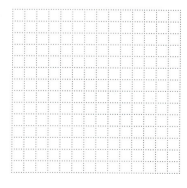   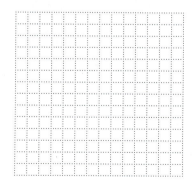   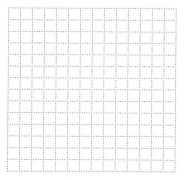

28. $y = -x^2 + 2x + 3$ **29.** $y = x^2 + 4x + 2$ **30.** $y = (x + 4)^2$

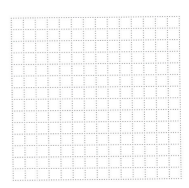 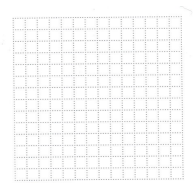 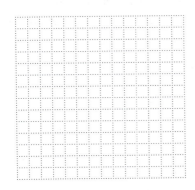

31. Refer to Example 5 and Exercise 17 in Section 9.4. Suppose that a telescope has a diameter of 200 ft and a maximum depth of 30 ft. Find the equation for a cross section of the parabolic dish.

[9.5] *Decide whether each relation is or is not a function. In Exercises 32 and 33, give the domain and the range.*

32. $\{(-2, 4), (0, 8), (2, 5), (2, 3)\}$ **33.** $\{(8, 3), (7, 4), (6, 5), (5, 6), (4, 7)\}$

34. 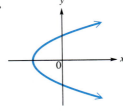   **35.**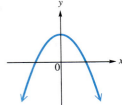

36. $2x + 3y = 12$ **37.** $y = x^2$ **38.** $x = 2|y|$

*Find **(a)** $f(2)$ and **(b)** $f(-1)$.*

39. $f(x) = 3x + 2$ **40.** $f(x) = 2x^2 - 1$ **41.** $f(x) = |x + 3|$

The net profits (in billions of dollars) for Coca-Cola have increased in recent years as shown in the graph. The tops of the bars appear to lie in a linear pattern. The linear equation

$$y = f(x) = .48x - 42.46,$$

where x is the number of years since 1900, gives a good approximation of the line through the centers of the tops of the bars.

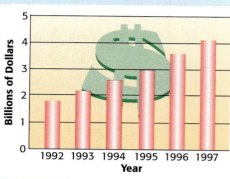

COCA-COLA NET PROFITS

Source: Betsy Morris, "Doug Is It," *Fortune,* May 25,1998, p.78.

42. Use the equation to find $f(98)$ and interpret the result in terms of the application. (*Hint:* $x = 98$ represents the year 1998.)

43. Use the equation to find the slope of the line.

44. Discuss the relationship between the slope of the line and the trend in net profits for Coca-Cola.

MIXED REVIEW EXERCISES

Solve by any method.

45. $(2t - 1)(t + 1) = 54$

46. $(2p + 1)^2 = 100$

47. $(k + 2)(k - 1) = 3$

48. $6t^2 + 7t - 3 = 0$

49. $2x^2 + 3x + 2 = x^2 - 2x$

50. $x^2 + 2x + 5 = 7$

51. $m^2 - 4m + 10 = 0$

52. $k^2 - 9k + 10 = 0$

53. $(5x + 6)^2 = 0$

54. $\frac{1}{2}r^2 = \frac{7}{2} - r$

55. $x^2 + 4x = 1$

56. $7x^2 - 8 = 5x^2 + 8$

Chapter 9 · TEST

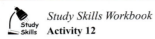

Study Skills Workbook
Activity 12

Solve by using the square root property.

1. $x^2 = 39$

2. $(x + 3)^2 = 64$

3. $(4x + 3)^2 = 24$

Solve by completing the square.

4. $x^2 - 4x = 6$

5. $2x^2 + 12x - 3 = 0$

Solve by the quadratic formula.

6. $2x^2 + 5x - 3 = 0$

7. $3w^2 + 2 = 6w$

8. $4x^2 + 8x + 11 = 0$

9. $t^2 - \dfrac{5}{3}t + \dfrac{1}{3} = 0$

Solve by the method of your choice.

10. $p^2 - 2p - 1 = 0$

11. $(2x + 1)^2 = 18$

12. $(x - 5)(2x - 1) = 1$

1. _____

2. _____

3. _____

4. _____

5. _____

6. _____

7. _____

8. _____

9. _____

10. _____

11. _____

12. _____

13. _____

13. $t^2 + 25 = 10t$

Solve each problem.

14. _____

14. If a ball is propelled into the air from ground level with an initial velocity of 64 ft per sec, its height s (in feet) after t seconds is given by the formula $s = -16t^2 + 64t$. After how many seconds will the ball reach a height of 64 ft?

15. _____

15. Find the lengths of the three sides of the right triangle.

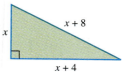

Sketch the graph of each equation. Identify each vertex.

16. _____

16. $y = (x - 3)^2$

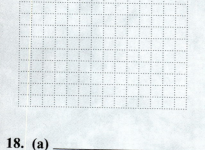

17. _____

17. $y = -x^2 - 2x - 4$

18. (a) _____

18. Decide whether each relation represents a function. If it does, give the domain and the range.

   (b) _____

  (a) $\{(2, 3), (2, 4), (2, 5)\}$

  (b) $\{(0, 2), (1, 2), (2, 2)\}$

19. _____

19. Use the vertical line test to determine whether the graph is that of a function.

20. _____

20. If $f(x) = 3x + 7$, find $f(-2)$.

Note: This cumulative review exercise set may be considered as a final examination for the course.

Perform the indicated operations, wherever possible.

1. $\dfrac{-4 \cdot 3^2 + 2 \cdot 3}{2 - 4 \cdot 1}$

2. $-9 - (-8)(2) + 6 - (6 + 2)$

3. $|-3| - |1 - 6|$

4. $-4r + 14 + 3r - 7$

5. $13k - 4k + k - 14k + 2k$

6. $5(4m - 2) - (m + 7)$

Solve each equation.

7. $6x - 5 = 13$

8. $3k - 9k - 8k + 6 = -64$

9. $2(m - 1) - 6(3 - m) = -4$

10. The perimeter of a basketball court is 288 ft. The width of the court is 44 ft less than the length. What are the dimensions of the court?

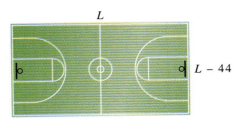

11. Find the measures of the marked angles.

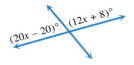

$(20x - 20)°$ $(12x + 8)°$

12. Solve the formula $P = 2L + 2W$ for L.

Solve each inequality. Graph the solutions.

13. $-8m < 16$

14. $-9p + 2(8 - p) - 6 \geq 4p - 50$

Graph the following.

15. $2x + 3y = 6$

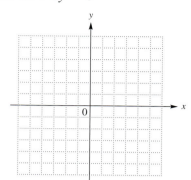

16. $y = 3$

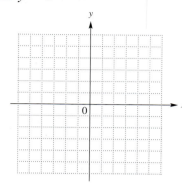

17. $2x - 5y < 10$

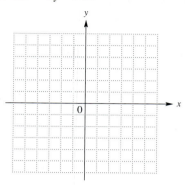

18. Find the slope of the line through $(-1, 4)$ and $(5, 2)$.

19. Write an equation of a line with slope 2 and y-intercept $(0, 3)$. Give it in the form $Ax + By = C$.

Simplify. Write answers with positive exponents.

20. $(3^2 \cdot x^{-4})^{-1}$

21. $\left(\dfrac{b^{-3}c^4}{b^5c^3}\right)^{-2}$

22. $\left(\dfrac{5}{3}\right)^{-3}$

Perform the indicated operations.

23. $(5x^5 - 9x^4 + 8x^2) - (9x^2 + 8x^4 - 3x^5)$

24. $(2x - 5)(x^3 + 3x^2 - 2x - 4)$

25. $(5t + 9)^2$

26. $\dfrac{3x^3 + 10x^2 - 7x + 4}{x + 4}$

Factor as completely as possible.

27. $16x^3 - 48x^2y$

28. $16x^4 - 1$

29. $2a^2 - 5a - 3$

30. $25m^2 - 20m + 4$

Solve each equation by factoring.

31. $x^2 + 3x - 54 = 0$

32. $3x^2 = x + 4$

33. The length of a rectangle is 2.5 times its width. The area is 1000 m². Find the length.

Perform the indicated operations. Write all answers in lowest terms.

34. $\dfrac{2}{a - 3} \div \dfrac{5}{2a - 6}$

35. $\dfrac{1}{k} - \dfrac{2}{k - 1}$

36. $\dfrac{2}{a^2 - 4} + \dfrac{3}{a^2 - 4a + 4}$

37. $\dfrac{6 + \dfrac{1}{x}}{3 - \dfrac{1}{x}}$

38. Solve $\dfrac{1}{x + 3} + \dfrac{1}{x} = \dfrac{7}{10}.$

Solve each system of equations.

39. $\begin{aligned} 2x + \ y &= -4 \\ -3x + 2y &= 13 \end{aligned}$

40. $\begin{aligned} 3x - \ 5y &= 8 \\ -6x + 10y &= 16 \end{aligned}$

41. Based on the prices in the 1998 Radio Shack catalogue, you can purchase 3 Krystalite phones and 2 Contempra II phones for $179.95. You can purchase 2 Krystalite phones and 3 Contempra II phones for $169.95. Find the price for a single phone of each model.

Graph the solutions of the system of inequalities.

42. $2x + y \leq 4$
$x - y > 2$

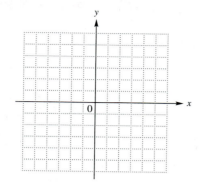

Simplify each expression as completely as possible.

43. $\sqrt{100}$

44. $\dfrac{6\sqrt{6}}{\sqrt{5}}$

45. $3\sqrt{5} - 2\sqrt{20} + \sqrt{125}$

46. $\sqrt[3]{16a^3b^4} - \sqrt[3]{54a^3b^4}$

Solve.

47. $\sqrt{x + 2} = x - 4$

48. $2a^2 - 2a = 1$

49. Graph the parabola $y = x^2 - 4$. Identify the vertex. What are the x-intercepts?

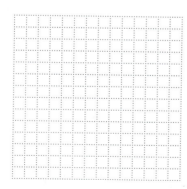

50. (a) Is $\{(0, 4), (1, 2), (3, 5)\}$ a function?

(b) Give the domain and the range of the relation in part (a).

(c) If $f(x) = -2x + 7$, find $f(-2)$.

(d) Is this the graph of a function?

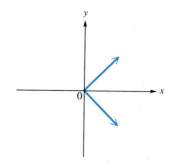

SETS

OBJECTIVES

1 List the elements of a set.

2 Learn the vocabulary and symbols used to discuss sets.

3 Decide whether a set is finite or infinite.

4 Decide whether a given set is a subset of another set.

5 Find the complement of a set.

6 Find the union and the intersection of two sets.

1 **List the elements of a set.** A **set** is a collection of things. The objects in a set are called the **elements** of the set. A set is represented by listing its elements between **set braces,** { }. The order in which the elements of a set are listed is unimportant.

Example 1 Listing the Elements of Sets

Represent each set by listing the elements.

(a) The set of states in the United States that border on the Pacific Ocean = {California, Oregon, Washington, Hawaii, Alaska}.

(b) The set of all counting numbers less than 6 = {1, 2, 3, 4, 5}.

══════ **Work Problem ❶ at the Side.**

2 **Learn the vocabulary and symbols used to discuss sets.** Capital letters are used to name sets. To state that 5 is an element of

$$S = \{1, 2, 3, 4, 5\},$$

write $5 \in S$. The statement $6 \notin S$ means that 6 is not an element of S.

A set with no elements is called the **empty set,** or the **null set.** The symbols $\emptyset$ or { } are used for the empty set. If we let A be the set of all cats that fly, then A is the empty set.

$$A = \emptyset \quad \text{or} \quad A = \{ \ \}$$

CAUTION

Do not make the common error of writing the empty set as $\{\emptyset\}$.

In any discussion of sets, there is some set that includes all the elements under consideration. This set is called the **universal set** for that situation. For example, if the discussion is about presidents of the United States, then the set of all presidents of the United States is the universal set. The universal set is denoted U.

3 **Decide whether a set is finite or infinite.** In Example 1, there are five elements in the set in part (a) and five in part (b). If the number of elements in a set is either 0 or a counting number, then the set is a **finite set.** On the other hand, the set of natural numbers, for example, is an **infinite set,** because there is no final natural number. We can list the elements of the set of natural numbers as

$$N = \{1, 2, 3, 4, \ldots\},$$

where the three dots indicate that the set continues indefinitely. Not all infinite sets can be listed in this way. For example, there is no way to list the elements in the set of all real numbers between 1 and 2.

❶ Represent each set by listing the elements.

(a) The set of states whose names begin with the letter O

(b) The set of letters of the alphabet that follow T

(c) The set of even natural numbers less than 10

(d) The set of odd counting numbers between 15 and 20

❷ List the elements of each set, if possible. Decide whether each set is *finite* or *infinite*.

(a) The set of whole numbers

(b) The set of odd natural numbers between 10 and 20

(c) The set of integers greater than 3

(d) The set of rational numbers

❸ Let
$$A = \{2, 4, 6, 8, 10, 12\},$$
$$B = \{2, 4, 8, 10\}, \text{ and}$$
$$C = \{4, 10, 12\}.$$

Tell whether each statement is *true* or *false*.

(a) $B \subseteq A$

(b) $C \subseteq B$

(c) $A \nsubseteq C$

(d) $B \nsubseteq C$

Example 2 **Distinguishing between Finite and Infinite Sets**

List the elements of each set, if possible. Decide whether each set is *finite* or *infinite*.

(a) The set of all integers
  One way to list the elements is $\{\ldots, -2, -1, 0, 1, 2, \ldots\}$. The set is infinite.

(b) The set of all natural numbers between 0 and 5
  List the elements of this set as $\{1, 2, 3, 4\}$. The set is finite.

(c) The set of all irrational numbers
  This is an infinite set whose elements cannot be listed.

Work Problem ❷ at the Side.

Two sets are **equal** if they have exactly the same elements. Thus, the set of natural numbers and the set of positive integers are equal sets. Also, the sets

$$\{1, 2, 4, 7\} \quad \text{and} \quad \{4, 2, 7, 1\}$$

are equal. The order of the elements does not make a difference.

4 **Decide whether a given set is a subset of another set.** If all elements of a set A are also elements of a new set B, then we say A is a **subset** of B, written $A \subseteq B$. We use the symbol $A \nsubseteq B$ to mean that A is not a subset of B.

Example 3 **Using Subset Notation**

Let $A = \{1, 2, 3, 4\}$, $B = \{1, 4\}$, and $C = \{1\}$. Then

$$B \subseteq A, \quad C \subseteq A, \quad \text{and} \quad C \subseteq B,$$

but

$$A \nsubseteq B, \quad A \nsubseteq C, \quad \text{and} \quad B \nsubseteq C.$$

Work Problem ❸ at the Side.

The set $M = \{a, b\}$ has four subsets: $\{a, b\}, \{a\}, \{b\}$, and $\varnothing$. The empty set is defined to be a subset of any set. How many subsets does $N = \{a, b, c\}$ have? There is one subset with 3 elements: $\{a, b, c\}$. There are three subsets with 2 elements:

$$\{a, b\}, \quad \{a, c\}, \quad \text{and} \quad \{b, c\}.$$

There are three subsets with 1 element:

$$\{a\}, \quad \{b\}, \quad \text{and} \quad \{c\}.$$

There is one subset with 0 elements: $\varnothing$. Thus, set N has eight subsets. The following generalization can be made.

Number of Subsets of a Set

A set with n elements has 2^n subsets.

To illustrate the relationships between sets, **Venn diagrams** are often used. A rectangle represents the universal set, U. The sets under discussion are represented by regions within the rectangle. The Venn diagram in Figure 1 shows that $B \subseteq A$.

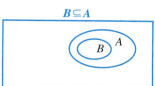

$B \subseteq A$

Figure 1

5▢   **Find the complement of a set.**   For every set A, there is a set A', the **complement** of A, that contains all the elements of U that are not in A. The shaded region in the Venn diagram in Figure 2 represents A'.

A' is shaded.

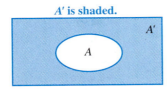

Figure 2

Example 4   **Determining Complements of Sets**

Given $U = \{a, b, c, d, e, f, g\}$, $A = \{a, b, c\}$, $B = \{a, d, f, g\}$, and $C = \{d, e\}$, then

$$A' = \{d, e, f, g\}, \quad B' = \{b, c, e\}, \quad \text{and} \quad C' = \{a, b, c, f, g\}.$$

— **Work Problem ④ at the Side.**

6▢   **Find the union and the intersection of two sets.**   The **union** of two sets A and B, written $A \cup B$, is the set of all elements of A together with all elements of B. Thus, for the sets in Example 4,

$$A \cup B = \{a, b, c, d, f, g\} \quad A = \{a, b, c\}, B = \{a, d, f, g\}$$

and

$$A \cup C = \{a, b, c, d, e\}. \quad A = \{a, b, c\}, C = \{d, e\}$$

In Figure 3 the shaded region is the union of sets A and B.

$A \cup B$ is shaded.

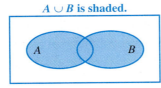

Figure 3

Example 5   **Finding the Union of Two Sets**

If $M = \{2, 5, 7\}$ and $N = \{1, 2, 3, 4, 5\}$, then

$$M \cup N = \{1, 2, 3, 4, 5, 7\}.$$

— **Work Problem ⑤ at the Side.**

The **intersection** of two sets A and B, written $A \cap B$, is the set of all elements that belong to both A and B. For example, if

$$A = \{\textbf{Jose}, \textbf{Ellen}, \text{Marge}, \text{Kevin}\}$$

and

$$B = \{\textbf{Jose}, \text{Patrick}, \textbf{Ellen}, \text{Sue}\},$$

then

$$A \cap B = \{\textbf{Jose}, \textbf{Ellen}\}.$$

④ Let

$$U = \{0, 1, 2, 3, 4, 5, 6, 7, 8\},$$
$$M = \{0, 2, 4, 6, 8\},$$
$$N = \{1, 3, 5, 7\}, \text{ and}$$
$$Q = \{0, 1, 2, 3, 4\}.$$

List the elements in each set.

(a) M'

(b) N'

(c) Q'

⑤ Using the sets given in Margin Problem 4, find the following.

(a) $M \cup N$

(b) $N \cup Q$

6 Using the sets given in Margin Problem 4, find the following.

(a) $M \cap Q$

(b) $N \cap Q$

(c) $M \cap N$

7 Let

$U = \{1, 2, 3, 4, 6, 8, 10\}$,
$A = \{1, 3, 4, 6\}$,
$B = \{2, 4, 6, 8, 10\}$, and
$C = \{2, 8\}$.

Find the following.

(a) $B \cup C$

(b) $A \cap B$

(c) $A \cap C$

(d) A'

(e) $A \cup B$

The shaded region in Figure 4 represents the intersection of the sets A and B.

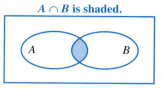

Figure 4

Example 6 Finding the Intersection of Two Sets

Suppose that

$$P = \{3, 9, 27\},$$
$$Q = \{2, 3, 10, 18, 27, 28\}, \text{ and}$$
$$R = \{2, 10, 28\}.$$

List the elements in each set.

(a) $P \cap Q = \{3, 27\}$ (b) $Q \cap R = \{2, 10, 28\} = R$ (c) $P \cap R = \emptyset$

Work Problem 6 at the Side.

Sets like P and R in Example 6 that have no elements in common are called **disjoint sets.** The Venn diagram in Figure 5 shows a pair of disjoint sets.

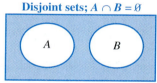

Figure 5

Example 7 Using Set Operations

Let

$$U = \{2, 5, 7, 10, 14, 20\},$$
$$A = \{2, 10, 14, 20\},$$
$$B = \{5, 7\}, \text{ and}$$
$$C = \{2, 5, 7\}.$$

Find the following.

(a) $A \cup B = \{2, 5, 7, 10, 14, 20\} = U$

(b) $A \cap B = \emptyset$

(c) $B \cup C = \{2, 5, 7\} = C$

(d) $B \cap C = \{5, 7\} = B$

(e) $A' = \{5, 7\} = B$

Work Problem 7 at the Side.

ANSWERS
6. (a) $\{0, 2, 4\}$ (b) $\{1, 3\}$ (c) $\emptyset$
7. (a) $\{2, 4, 6, 8, 10\} = B$ (b) $\{4, 6\}$ (c) $\emptyset$
   (d) $\{2, 8, 10\}$ (e) $\{1, 2, 3, 4, 6, 8, 10\} = U$

Appendix A   EXERCISES

List the elements of each set. See Examples 1 and 2.

1. The set of all natural numbers less than 8

2. The set of all integers between 4 and 10

3. The set of seasons

4. The set of months of the year

5. The set of women presidents of the United States

6. The set of all living humans who are more than 200 years old

7. The set of letters of the alphabet between K and M

8. The set of letters of the alphabet between D and H

9. The set of positive even integers

10. The set of all multiples of 5

11. Which of the sets described in Exercises 1–10 are infinite sets?

12. Which of the sets described in Exercises 1–10 are finite sets?

Tell whether each statement is true *or* false.

13. $5 \in \{1, 2, 5, 8\}$

14. $6 \in \{1, 2, 3, 4, 5\}$

15. $2 \in \{1, 3, 5, 7, 9\}$

16. $1 \in \{6, 2, 5, 1\}$

17. $7 \notin \{2, 4, 6, 8\}$

18. $7 \notin \{1, 3, 5, 7\}$

19. $\{2, 4, 9, 12, 13\} = \{13, 12, 9, 4, 2\}$

20. $\{7, 11, 4\} = \{7, 11, 4, 0\}$

Let

$$A = \{1, 3, 4, 5, 7, 8\}, \quad B = \{2, 4, 6, 8\}, \quad C = \{1, 3, 5, 7\},$$
$$D = \{1, 2, 3\}, \quad E = \{3, 7\}, \quad \text{and} \quad U = \{1, 2, 3, 4, 5, 6, 7, 8, 9, 10\}.$$

Tell whether each statement is true *or* false. *See Examples 3, 5, 6, and 7.*

21. $A \subseteq U$ **22.** $D \subseteq A$ **23.** $\emptyset \subseteq A$ **24.** $\{1, 2\} \subseteq D$ **25.** $C \subseteq A$

26. $A \subseteq C$ **27.** $D \subseteq B$ **28.** $E \subseteq C$ **29.** $D \nsubseteq E$ **30.** $E \nsubseteq A$

31. There are exactly 4 subsets of *E*.

32. There are exactly 8 subsets of *D*.

33. There are exactly 12 subsets of *C*.

34. There are exactly 16 subsets of *B*.

35. $\{4, 6, 8, 12\} \cap \{6, 8, 14, 17\} = \{6, 8\}$

36. $\{2, 5, 9\} \cap \{1, 2, 3, 4, 5\} = \{2, 5\}$

37. $\{3, 1, 0\} \cap \{0, 2, 4\} = \{0\}$

38. $\{4, 2, 1\} \cap \{1, 2, 3, 4\} = \{1, 2, 3\}$

39. $\{3, 9, 12\} \cap \emptyset = \{3, 9, 12\}$

40. $\{3, 9, 12\} \cup \emptyset = \emptyset$

41. $\{4, 9, 11, 7, 3\} \cup \{1, 2, 3, 4, 5\} = \{1, 2, 3, 4, 5, 7, 9, 11\}$

42. $\{1, 2, 3\} \cup \{1, 2, 3\} = \{1, 2, 3\}$

43. $\{3, 5, 7, 9\} \cup \{4, 6, 8\} = \emptyset$

44. $\{5, 10, 15, 20\} \cup \{5, 15, 30\} = \{5, 15\}$

Let

$$U = \{a, b, c, d, e, f, g, h\}, \quad A = \{a, b, c, d, e, f\}, \quad B = \{a, c, e\}, \quad C = \{a, f\}, \quad \text{and} \quad D = \{d\}.$$

List the elements in each set. See Examples 4–7.

45. A' **46.** B' **47.** C' **48.** D'

49. $A \cap B$ **50.** $B \cap A$ **51.** $A \cap D$ **52.** $B \cap D$

53. $B \cap C$ **54.** $A \cup B$ **55.** $B \cup D$ **56.** $B \cup C$

57. $C \cup B$ **58.** $C \cup D$ **59.** $A \cap \emptyset$ **60.** $B \cup \emptyset$

61. Name every pair of disjoint sets among *A–D* above.

FACTORING SUMS AND DIFFERENCES OF CUBES

1   **Factor the difference of cubes.**   The difference of squares was factored in Section 5.5; we can also factor the **difference of cubes.** Use the following pattern.

Difference of Cubes

$$a^3 - b^3 = (a - b)(a^2 + ab + b^2)$$

This pattern *should be memorized*. Multiply on the right to see that the pattern gives the correct factors, as shown in the margin.

Notice the pattern of the terms in the factored form of $a^3 - b^3$.

- $a^3 - b^3 = $ (a binomial factor)(a trinomial factor)
- The binomial factor has the difference of the cube roots of the given terms.
- The terms in the trinomial factor are all positive.
- What you write in the binomial factor determines the trinomial factor.

$$
\begin{array}{c}
\text{First term} \quad \text{positive} \quad \text{second term} \\
\text{squared} \quad + \quad \text{product of} \quad + \quad \text{squared} \\
\text{the terms}
\end{array}
$$

$$a^3 - b^3 = (a - b)(\ a^2 \ + \ ab \ + \ b^2 \ )$$

$$
\begin{array}{r}
a^2 + ab + b^2 \\
a - b \\
\hline
- a^2b - ab^2 - b^3 \\
a^3 + a^2b + ab^2 \\
\hline
a^3 \qquad\qquad\quad - b^3
\end{array}
$$

Example 1   **Factoring Differences of Cubes**

Factor each difference of cubes.

(a) $m^3 - 125$

Use the pattern for the difference of cubes.

$$a^3 - b^3 = (a - b)(a^2 + ab + b^2)$$

$$m^3 - 125 = m^3 - 5^3 = (m - 5)(m^2 + 5m + 5^2)$$

$$= (m - 5)(m^2 + 5m + 25)$$

(b) $8p^3 - 27$

Since $8p^3 = (2p)^3$ and $27 = 3^3$,

$$8p^3 - 27 = (2p)^3 - 3^3$$

$$= (2p - 3)[(2p)^2 + (2p)3 + 3^2]$$

$$= (2p - 3)(4p^2 + 6p + 9).$$

Continued on Next Page

① Factor each difference of cubes.

(a) $t^3 - 64$

(b) $2x^3 - 54$

(c) $8k^3 - y^3$

② Factor each sum of cubes.

(a) $x^3 + 8$

(b) $64y^3 + 1$

(c) $27m^3 + 343n^3$

(c) $4m^3 - 32n^3 = 4(m^3 - 8n^3)$ Factor out the common factor.

$= 4[m^3 - (2n)^3]$ $8n^3 = (2n)^3$

$= 4(m - 2n)[m^2 + m(2n) + (2n)^2]$

$= 4(m - 2n)(m^2 + 2mn + 4n^2)$

Work Problem ① at the Side.

CAUTION

A common error in factoring a difference of cubes, such as $a^3 - b^3 = (a - b)(a^2 + ab + b^2)$, is to try to factor $a^2 + ab + b^2$. It is easy to confuse this factor with a perfect square trinomial, $a^2 + \mathbf{2ab} + b^2$. It is unusual to be able to further factor an expression of the form $a^2 + ab + b^2$.

2 ▭ **Factor the sum of cubes.** A sum of squares, such as $m^2 + 25$, cannot be factored using real numbers, but a **sum of cubes** can be factored by the following pattern, *which should be memorized.*

Sum of Cubes

$$a^3 + b^3 = (a + b)(a^2 - ab + b^2)$$

Compare the pattern for the *sum* of cubes with the pattern for the *difference* of cubes. The only difference between them is the positive and negative signs.

Positive

$$a^3 - b^3 = (a - b)(a^2 + ab + b^2)$$ Difference of cubes

Same sign   Opposite sign

Positive

$$a^3 + b^3 = (a + b)(a^2 - ab + b^2)$$ Sum of cubes

Same sign   Opposite sign

Observing these relationships should help you to remember these patterns.

Example 2 Factoring Sums of Cubes

Factor each sum of cubes.

(a) $k^3 + 27 = k^3 + 3^3$

$= (k + 3)(k^2 - 3k + 3^2)$

$= (k + 3)(k^2 - 3k + 9)$

(b) $8m^3 + 125p^3 = (2m)^3 + (5p)^3$

$= (2m + 5p)[(2m)^2 - (2m)(5p) + (5p)^2]$

$= (2m + 5p)(4m^2 - 10mp + 25p^2)$

Work Problem ② at the Side.

ANSWERS
1. **(a)** $(t - 4)(t^2 + 4t + 16)$
   (b) $2(x - 3)(x^2 + 3x + 9)$
   (c) $(2k - y)(4k^2 + 2ky + y^2)$
2. **(a)** $(x + 2)(x^2 - 2x + 4)$
   (b) $(4y + 1)(16y^2 - 4y + 1)$
   (c) $(3m + 7n)(9m^2 - 21mn + 49n^2)$

Appendix B

EXERCISES

| FOR EXTRA HELP | Student's Solutions Manual | MyMathLab.com | InterAct Math Tutorial Software | AW Math Tutor Center | www.mathxl.com |

1. To help you factor the sum or difference of cubes, complete the following list of cubes.

$1^3 = $ _____ $2^3 = $ _____ $3^3 = $ _____ $4^3 = $ _____ $5^3 = $ _____

$6^3 = $ _____ $7^3 = $ _____ $8^3 = $ _____ $9^3 = $ _____ $10^3 = $ _____

2. The following powers of x are all perfect cubes: $x^3, x^6, x^9, x^{12}, x^{15}$. Based on this observation, we may make a conjecture that if the power of a variable is divisible by _____ (with 0 remainder), then we have a perfect cube.

3. Which of the following are differences of cubes?

  A. $9x^3 - 125$ **B.** $x^3 - 16$ **C.** $x^3 - 1$ **D.** $8x^3 - 27y^3$

4. Which of the following are sums of cubes?

  A. $x^3 + 1$ **B.** $x^3 + 36$ **C.** $12x^3 + 27$ **D.** $64x^3 + 216y^3$

Factor. Use your answers in Exercises 1 and 2 as necessary. See Examples 1 and 2.

5. $a^3 + 1$ **6.** $m^3 + 8$ **7.** $a^3 - 1$ **8.** $m^3 - 8$

9. $p^3 + q^3$ **10.** $w^3 + z^3$ **11.** $y^3 - 216$ **12.** $x^3 - 343$

13. $k^3 + 1000$ **14.** $p^3 + 512$ **15.** $27x^3 - 1$

16. $64y^3 - 27$ **17.** $125a^3 + 8$ **18.** $216b^3 + 125$

19. $y^3 - 8x^3$ **20.** $w^3 - 216z^3$ **21.** $27a^3 - 64b^3$

22. $125m^3 - 8n^3$ **23.** $8p^3 + 729q^3$ **24.** $27x^3 + 1000y^3$

25. $16t^3 - 2$

26. $3p^3 - 81$

27. $40w^3 + 135$

28. $32z^3 + 500$

29. $x^3 + y^6$

30. $p^9 + q^3$

31. $125k^3 - 8m^9$

32. $125c^6 - 216d^3$

RELATING CONCEPTS (Exercises 33–40)   **FOR INDIVIDUAL OR GROUP WORK**

A binomial may be both *a difference of squares* and *a difference of cubes. One example of such a binomial is* $x^6 - 1$. *Using the techniques of Section 5.5, one factoring method will give the complete factored form, while the other will not.* **Work Exercises 33–40 in order** *to determine the method to use.*

33. Factor $x^6 - 1$ as the difference of two squares.

34. The factored form obtained in Exercise 33 consists of a difference of cubes multiplied by a sum of cubes. Factor each binomial further.

35. Now start over and factor $x^6 - 1$ as a difference of cubes.

36. The factored form obtained in Exercise 35 consists of a binomial that is a difference of squares and a trinomial. Factor the binomial further.

37. Compare your results in Exercises 34 and 36. Which one of these is the completely factored form?

38. Verify that the trinomial in the factored form in Exercise 36 is the product of the two trinomials in the factored form in Exercise 34.

39. Use the results of Exercises 33–38 to complete the following statement:

In general, if I must choose between factoring first using the method for a difference of squares or the method for a difference of cubes, I should choose the

_____ method to eventually obtain the complete factored form.

40. Find the *complete* factored form of $x^6 - 729$ using the knowledge you have gained in Exercises 33–39.

Answers to Selected Exercises

In this section we provide the answers that we think most students will obtain when they work the exercises using the methods explained in the text. If your answer does not look exactly like the one given here, it is not necessarily wrong. In many cases there are equivalent forms of the answer that are correct. For example, if the answer section shows $\frac{3}{4}$ and your answer is .75, you have obtained the right answer but written it in a different (yet equivalent) form. Unless the directions specify otherwise, .75 is just as valid an answer as $\frac{3}{4}$.

In general, if your answer does not agree with the one given in the text, see whether it can be transformed into the other form. If it can, then it is the correct answer. If you still have doubts, talk with your instructor.

Diagnostic Pretest

(page xxxi)

1. 18 **2.** $\frac{253}{24}$ or $10\frac{13}{24}$ **3.** 28.322 **4. (a)** .0099 **(b)** 472%
5. $-|35|$ **6.** 56 **7.** 0 **8.** 4 **9.** -2 **10.** New York: 7,420,166; Los Angeles: 3,597,556 **11.** 58°, 122°
12. $x \le 5$

13. x-intercept: $(5, 0)$; y-intercept: $(0, -2)$

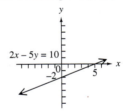

14. -4 **15.** $y = -x + 1$
16.

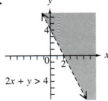

17. $-2m^3 - 9m - 5$ **18.** $49z^2 + 42zw + 9w^2$ **19.** $-\frac{3}{10}$

20. (a) 4.45×10^8 **(b)** .000234 **21.** $(3x - 4)(x + 2)$

22. $(4n + 7)(4n - 7)$ **23.** $-3, 5$ **24.** 15 in. by 11 in. **25.** $\frac{x + 5}{x + 4}$
26. $-\frac{1}{3r}$ **27.** $\frac{z^3 + 2z^2 + 3z}{(z - 3)(z + 3)}$ **28.** $y + 1$ **29.** $(2, 3)$
30. no solution **31.** Marla: 61 mph; Rick: 53 mph

32.

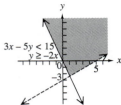

33. 0 **34.** $12 - 2\sqrt{35}$ **35.** $2\sqrt{10}$ **36.** 27
37. $3 + 2\sqrt{5}, 3 - 2\sqrt{5}$ **38.** $\frac{-3 + \sqrt{17}}{4}, \frac{-3 - \sqrt{17}}{4}$

39. vertex: $(0, 5)$

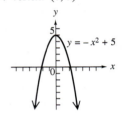

40. function; domain: $\{-2, -1, 0, 1, 2\}$; range: $\{0, 1, 4\}$

Chapter R

Section R.1 (page 11)

1. true **3.** False; the fraction $\frac{17}{51}$ can be simplified to $\frac{1}{3}$.

5. False; *product* indicates multiplication, so the product of 8 and 2 is 16. **7.** prime **9.** composite **11.** composite
13. neither **15.** $2 \cdot 3 \cdot 5$ **17.** $2 \cdot 2 \cdot 3 \cdot 3 \cdot 7$ **19.** $2 \cdot 2 \cdot 31$
21. 29 **23.** $\frac{1}{2}$ **25.** $\frac{5}{6}$ **27.** $\frac{1}{5}$ **29.** $\frac{6}{5}$ **31.** A **33.** $\frac{24}{35}$
35. $\frac{6}{25}$ **37.** $\frac{6}{5}$ **39.** $\frac{232}{15}$ or $15\frac{7}{15}$ **41.** $\frac{10}{3}$ **43.** 12
45. $\frac{1}{16}$ **47.** $\frac{84}{47}$ or $1\frac{37}{47}$ **49.** Multiply the first fraction
(the dividend) by the reciprocal of the second fraction (the divisor)
to divide two fractions. **51.** $\frac{2}{3}$ **53.** $\frac{8}{9}$ **55.** $\frac{27}{8}$ or $3\frac{3}{8}$ **57.** $\frac{17}{36}$
59. $\frac{11}{12}$ **61.** $\frac{4}{3}$ **63.** 6 cups **65.** $618\frac{3}{4}$ ft **67.** $\frac{9}{16}$ in. **69.** $\frac{5}{16}$ in.
71. $\frac{1}{20}$ **73.** More than $1\frac{1}{25}$ million

Section R.2 (page 23)

1. (a) 6 **(b)** 9 **(c)** 1 **(d)** 7 **(e)** 4 **3. (a)** 46.25 **(b)** 46.2 **(c)** 46
(d) 50 **5.** C **7.** B **9.** $\frac{4}{10}$ **11.** $\frac{64}{100}$ **13.** $\frac{138}{1000}$ **15.** $\frac{3805}{1000}$
17. 139; 143.094 **19.** 27; 25.61 **21.** 10; 15.33 **23.** 82; 81.716
25. 15; 15.211 **27.** .006; .006 **29.** 90; 116.48 **31.** 6; 7.15

33. 2; 2.05   **35.** 6000; 5711.6   **37.** .2; .162   **39.** To add or subtract decimals, line up the decimal points in a column, add or subtract as usual, and move the decimal point straight down in the sum or difference.   **41.** .125   **43.** .25   **45.** .5̄; .556
47. .16̄; .167   **49.** To convert a decimal to a percent, move the decimal point two places to the right and attach a percent symbol (%).   **51.** .54   **53.** 1.17   **55.** .024   **57.** .0625   **59.** .008
61. 75%   **63.** .4%   **65.** 128%   **67.** 30%   **69.** 75%
71. 83.3̄%   **73.** (a) $\dfrac{2}{32}$ (b) $\dfrac{2}{32}$

Chapter 1

Section 1.1 (page 31)

1. true   **3.** False; using the guidelines for order of operations gives $4 + 3(8 - 2) = 4 + 3(6) = 4 + 18 = 22$.   **5.** False; the correct translation is $4 = 16 - 12$.   **7.** 49   **9.** 144   **11.** 64
13. 1000   **15.** 81   **17.** 1024   **19.** $\dfrac{16}{81}$ **21.** .000064
23. The 4 would be applied last because we work first inside the parentheses.   **25.** 58   **27.** 13   **29.** 32   **31.** 19   **33.** $\dfrac{49}{30}$
35. 12   **37.** 36.14   **39.** 26   **41.** 4   **43.** 95   **45.** 12   **47.** 14
49. $\dfrac{19}{2}$ **51.** false   **53.** true   **55.** true   **57.** false   **59.** false
61. true   **63.** $15 = 5 + 10$ **65.** $9 > 5 - 4$ **67.** $16 \neq 19$
69. $2 \leq 3$ **71.** Seven is less than nineteen. True   **73.** One-third is not equal to three-tenths. True   **75.** Eight is greater than or equal to eleven. False   **77.** $30 > 5$ **79.** $3 \leq 12$ **81.** United States and Netherlands   **83.** United States, Netherlands, and France

Section 1.2 (page 37)

1. 10   **3.** $12 + x$; 21   **5.** expression; equation   **7.** The equation would be $5x - 9 = 49$.   **9.** Answers will vary. Two such pairs are $x = 0, y = 6$ and $x = 1, y = 4$. To find a pair, choose one number, substitute it for a variable, then calculate the value for the other variable.   **11.** (a) 64   (b) 144   **13.** (a) $\dfrac{7}{8}$ (b) $\dfrac{13}{12}$
15. (a) 9.569 (b) 14.353   **17.** (a) 52 (b) 114   **19.** (a) 12
(b) 33   **21.** (a) 6 (b) $\dfrac{9}{5}$ **23.** (a) $\dfrac{4}{3}$ (b) $\dfrac{13}{6}$ **25.** (a) $\dfrac{2}{7}$
(b) $\dfrac{16}{27}$ **27.** (a) 12   (b) 55   **29.** (a) 1 (b) $\dfrac{28}{17}$ **31.** (a) 3.684
(b) 8.841   **33.** $12x$ **35.** $x - 2$ **37.** $7 - 4x$ **39.** $2x - 6$
41. $\dfrac{12}{x + 3}$ **43.** $6(x - 4)$ **45.** The word *and* does not signify addition here. In the phrase "the product of a number and 6," *and* connects two quantities to be multiplied.   **47.** no   **49.** yes
51. yes   **53.** no   **55.** yes   **57.** yes   **59.** $x + 8 = 18$
61. $2x + 5 = 5$ **63.** $16 - \dfrac{3}{4}x = 13$ **65.** $3x = 2x + 8$
67. expression   **69.** equation   **71.** $10.50; less by $.33
73. $12.41; more by $.04

Section 1.3 (page 47)

1. 4   **3.** 0   **5.** One example is $\sqrt{12}$. There are others.
7. (a) 3, 7 (b) 0, 3, 7 (c) $-9, 0, 3, 7$
(d) $-9, -1\dfrac{1}{4}, -\dfrac{3}{5}, 0, 3, 5.9, 7$ (e) $-\sqrt{7}, \sqrt{5}$ (f) All are real
numbers.   **9.** 93,000   **11.** $-31,532$
13.

15.

17.

19. -11 **21.** -21 **23.** -100 **25.** $-\dfrac{2}{3}$ **27.** false   **29.** true
31. (a) 2 (b) 2   **33.** (a) -6 (b) 6   **35.** (a) $\dfrac{3}{4}$ (b) $\dfrac{3}{4}$ **37.** 7
39. -12 **41.** -14 **43.** 9   **45.** false   **47.** true   **49.** No; the statement is false for one number, 0.   **51.** video/audio equipment from 1995–1996   **53.** 1996–1997

Section 1.4 (page 53)

1. Add -2 and 5.   **3.** Add -1 and -3.   **5.** 2   **7.** -3 **9.** -10
11. -13 **13.** -15.9 **15.** 5   **17.** 13   **19.** 0   **21.** -8 **23.** $\dfrac{3}{10}$
25. $\dfrac{1}{2}$ **27.** $-\dfrac{3}{4}$ **29.** -1.6 **31.** -8.7 **33.** -25 **35.** -12
37. true   **39.** false   **41.** true   **43.** false   **45.** true   **47.** false
49. It must be negative and have the larger absolute value.
50. The sum of a positive number and 5 cannot be -7.
51. It must be positive and have the larger absolute value.
52. The sum of a negative number and -8 cannot be 2.
53. Add the absolute values of the numbers. The sum will be negative.   **55.** $-5 + 12 + 6$; 13   **57.** $[-19 + (-4)] + 14$; -9
59. $[-4 + (-10)] + 12$; -2 **61.** $[8 + (-18)] + 4$; -6
63. $-$80   **65.** -184 m   **67.** 37 yd   **69.** 120°F   **71.** $-$107
73. $286.60

Section 1.5 (page 61)

1. $-8; -6$ **3.** $7 - 12; 12 - 7$ **5.** -4 **7.** -10 **9.** -16
11. 11   **13.** 19   **15.** -4 **17.** 5   **19.** 0   **21.** $\dfrac{3}{4}$ **23.** $-\dfrac{11}{8}$
25. $\dfrac{15}{8}$ **27.** 13.6   **29.** -11.9 **31.** -2.8 **33.** -6.3
35. -28 **37.** -18 **39.** $\dfrac{37}{12}$ **41.** -42.04 **43.** For example, let $a = 1, b = 1$ or let $a = 2, b = 2$. In general, choose $a = b$.
45. 8   **47.** For example, $-8 - (-2) = -6$.   **49.** $4 - (-8)$; 12
51. $-2 - 8; -10$ **53.** $[9 + (-4)] - 7; -2$
55. $[8 - (-5)] - 12; 1$ **57.** $-58°F$ **59.** 14,776 ft
61. $-$80   **63.** $105,000   **65.** $-$3100   **67.** $6800
69. positive   **71.** positive

Section 1.6 (page 73)

1. greater than 0   **3.** less than 0   **5.** greater than 0   **7.** -28
9. -30 **11.** 0   **13.** $\dfrac{5}{6}$ **15.** -2.38 **17.** $\dfrac{3}{2}$ **19.** -3 **21.** -2
23. 16   **25.** 0   **27.** 25.63   **29.** $\dfrac{3}{2}$ **31.** C   **33.** 3   **35.** 7
37. 4   **39.** -3 **41.** -1 **43.** negative; impossible to tell
45. 68   **47.** -228 **49.** 1   **51.** 0   **53.** -6 **55.** 0
57. $-12 + 4(-7); -40$ **59.** $-1 - 2(-8)(2); 31$
61. $-3[3 - (-7)]; -30$
63. $\dfrac{3}{10}[-2 + (-28)]; -9$ **65.** $\dfrac{-20}{-8 + (-2)}; 2$
67. $\dfrac{-18 + (-6)}{2(-4)}; 3$ **69.** $\dfrac{-\dfrac{2}{3}\left(-\dfrac{1}{5}\right)}{\dfrac{1}{7}}; \dfrac{14}{15}$ **71.** $4x = -36$

73. $\frac{x}{4} = -1$ **75.** $x - 7 = 5$ **77.** $\frac{6}{x} = -3$

78. 42 **79.** 5 **80.** $8\frac{2}{5}$ **81.** $8\frac{2}{5}$ **82.** 2 **83.** $-12\frac{1}{2}$

Section 1.7 (page 83)

1. B **3.** C **5.** B **7.** G **9.** commutative property
11. associative property **13.** inverse property **15.** inverse
property **17.** identity property **19.** commutative property
21. distributive property **23.** identity property
25. distributive property **27.** (a) 0 (b) 1, −1
29. $25 - (6 - 2) = 25 - 4 = 21$ and $(25 - 6) - 2 = 19 - 2 = 17$.
Since these results are different, subtraction is not associative.
31. $7 + r$ **33.** s **35.** $-6x + (-6)7$; $-6x - 42$

37. $w + [5 + (-3)]$; $w + 2$ **39.** We must multiply $\frac{3}{4}$ by 1 in the

form $\frac{3}{3}$: $\frac{3}{4} \cdot \frac{3}{3} = \frac{9}{12}$. **41.** 2 **43.** $5(3 + 17)$; 100 **45.** $4t + 12$
47. $-8r - 24$ **49.** $-5y + 20$ **51.** $-16y - 20z$ **53.** $8(z + w)$
55. $7(2v + 5r)$ **57.** $24r + 32s - 40y$ **59.** $-24x - 9y - 12z$
61. $-4t - 5m$ **63.** $5c + 4d$ **65.** $3q - 5r + 8s$
67. Answers will vary. For example, "putting on your socks" and
"putting on your shoes" **69.** false **71.** (foreign sales) clerk;
foreign (sales clerk) **73.** 0 **74.** $-3(5) + (-3)(-5)$ **75.** −15
76. The product $-3(-5)$ must equal 15, since it is the additive
inverse of −15.

Section 1.8 (page 91)

1. false **3.** true **5.** C **7.** A **9.** $4r + 11$
11. $5 + 2x - 6y$ **13.** $-7 + 3p$ **15.** −12 **17.** 5 **19.** 1
21. −1 **23.** 74 **25.** Answers will vary. For example,
$-3x$ and $4x$ **27.** like **29.** unlike **31.** like **33.** unlike
35. We cannot "add" unlike terms, so we must be able to identify
like terms in order to combine them. **37.** $11 - 2x$

39. $-\frac{1}{3}t - \frac{28}{3}$ **41.** $-4.1r + 4.2$ **43.** $-2y^2 + 3y^3$

45. $-19p + 16$ **47.** $-\frac{3}{2}y + 16$ **49.** $-16y + 63$

51. $(x + 3) + 5x$; $6x + 3$ **53.** $(13 + 6x) - (-7x)$; $13 + 13x$
55. $2(3x + 4) - (-4 + 6x)$; 12 **57.** Wording may vary.
One example is "the difference between 9 times a number and
the sum of the number and 2." **59.** $1000 + 5x$ (dollars)
60. $750 + 3y$ (dollars) **61.** $1000 + 5x + 750 + 3y$ (dollars)
62. $1750 + 5x + 3y$ (dollars)

Chapter 1 Review Exercises (page 97)

1. 625 **2.** .00000081 **3.** .009261 **4.** $\frac{125}{8}$ **5.** 27 **6.** 200

7. −7 **8.** $\frac{20}{3}$ **9.** $13 < 17$ **10.** $5 + 2 \neq 10$ **11.** Six is less

than fifteen. **12.** One example is $-4 + (-7) \geq \frac{12}{-3}$.

13. 30 **14.** 60 **15.** 14 **16.** 13 **17.** $x + 6$ **18.** $8 - x$

19. $6x - 9$ **20.** $12 + \frac{3}{5}x$ **21.** yes **22.** no **23.** $2x - 6 = 10$

24. $4x = 8$ **25.** equation **26.** expression

27.

28.

29.

30.

31. −10 **32.** −9 **33.** $-\frac{3}{4}$ **34.** $-|23|$ **35.** true **36.** true

37. true **38.** false **39.** −3 **40.** −19 **41.** −7 **42.** 9

43. −6 **44.** −4 **45.** −17 **46.** $-\frac{29}{36}$ **47.** −10 **48.** −19

49. $(-31 + 12) + 19$; 0 **50.** $[-4 + (-8)] + 13$; 1 **51.** −$8

52. 87°F **53.** −11 **54.** −1 **55.** 7 **56.** $-\frac{43}{35}$ **57.** 10.31

58. −12 **59.** 2 **60.** 1 **61.** $-4 - (-6)$; 2
62. $[4 + (-8)] - 5$; −9 **63.** 74.2% **64.** 1 min, 28.89 sec
65. The first step is to change subtracting −6 to adding its
opposite, 6, so the problem becomes $-8 + 6$. This sum is −2.
66. Yes; for example, $-2 - (-6) = -2 + 6 = 4$, a positive
number. **67.** $25.1 billion **68.** −$11.3 billion **69.** $5.0 billion

70. −$2.2 billion **71.** 36 **72.** −105 **73.** $\frac{1}{2}$ **74.** 10.08

75. −20 **76.** −10 **77.** −24 **78.** −35 **79.** 4 **80.** −20

81. $-\frac{3}{4}$ **82.** 11.3 **83.** −1 **84.** undefined **85.** 1 **86.** .5

87. −18 **88.** −18 **89.** 125 **90.** −423 **91.** $-4(5) - 9$; −29

92. $\frac{5}{6}[12 + (-6)]$; 5 **93.** $\frac{12}{8 + (-4)}$; 3 **94.** $\frac{-20(12)}{15 - (-15)}$; −8

95. $\frac{x}{x + 5} = -2$ **96.** $8x - 3 = -7$ **97.** identity property

98. identity property **99.** inverse property **100.** inverse property **101.** associative property **102.** associative property
103. distributive property **104.** commutative property
105. $(7 + 1)y$; $8y$ **106.** $-12 \cdot 4 - -12(t)$; $-48 + 12t$
107. $3(2s + 4y)$; $6s + 12y$ **108.** $-1(-4r) + (-1)(5s)$; $4r - 5s$
109. $17p^2$ **110.** $16r^2 + 7r$ **111.** $-19k + 54$ **112.** $5s - 6$
113. $-45t - 23$ **114.** $-45t^2 - 23.4t$ **115.** $-2(3x) - 7x$; $-13x$

116. $\frac{x + 9}{x - 6}$ **117.** No. The use of *and* there indicates the two

quantities that are to be multiplied. **118.** Answers may vary.
For example, "3 times the difference between 4 times a number

and 6" **119.** 16 **120.** $\frac{25}{36}$ **121.** −26 **122.** $\frac{8}{3}$ **123.** $-\frac{1}{24}$

124. $\frac{7}{2}$ **125.** 2 **126.** 77.6 **127.** $-1\frac{1}{2}$ **128.** 11 **129.** $-\frac{28}{15}$

130. 24 **131.** −11 **132.** −6 **133.** $2x - 1400 = 25{,}800$;

x represents the amount spent in 1999. **134.** $\frac{x}{3x - 14}$; x represents

the number.

Chapter 1 Test (page 103)

1. true **2.** false **3.**

4. $-|-8|$ (or −8) **5.** −1.277 **6.** $\frac{-6}{2 + (-8)}$; 1 **7.** negative

8. 4 **9.** $-2\frac{5}{6}$ **10.** 2 **11.** 6 **12.** 108 **13.** 11 **14.** $\frac{30}{7}$

15. −70 **16.** 3 **17.** 178°F **18.** D **19.** A **20.** E **21.** B
22. C **23.** $-9x^2 - 6x - 8$ **24.** identity and distributive
properties **25.** (a) −18 (b) −18 (c) The distributive property
tells us that the two methods produce equal results.

Chapter 2

2.1 Exercises (page 111)

1. A and C **3.** A and B **5.** 4 **7.** 10 **9.** 12 **11.** -10 **13.** 3
15. -2 **17.** 4 **19.** 0 **21.** no solution **23.** all real numbers
25. 4 **27.** no solution **29.** all real numbers **31.** $\frac{7}{15}$ **33.** 7
35. -4 **37.** 13 **39.** all real numbers **41.** 18 **43.** 12
45. Since the opposite of x is 5, x must be -5.
47. Answers will vary. One example is $x - 6 = -8$.

2.2 Exercises (page 117)

1. 4 **3.** -8 **5.** $\frac{3}{2}$ **7.** 10 **9.** $-\frac{2}{9}$ **11.** -1 **13.** 6 **15.** -4
17. .12 **19.** -1 **21.** If each side of an equation were multiplied by 0, the resulting equation would be $0 = 0$. This is true, but does not help to solve the equation. **23.** $\frac{15}{2}$ **25.** -5 **27.** $-\frac{18}{5}$
29. 12 **31.** 0 **33.** -12 **35.** 40 **37.** -48 **39.** -35
41. $-\frac{27}{35}$ **43.** 3 **45.** -5 **47.** 7 **49.** -2 **51.** $-\frac{3}{5}$
53. Answers will vary. One example is $\frac{3}{2}x = -6$.
55. $3x = 18 + 5x$; -9; The number is -9.

2.3 Exercises (page 123)

1. -1 **3.** 5 **5.** 1 **7.** $-\frac{5}{3}$ **9.** -1 **11.** no solution
13. all real numbers **15.** No, it is incorrect to divide each side by a variable. If $-3x$ is added to each side, the equation becomes $4x = 0$, so $x = 0$ is the correct solution. **17.** Simplify each side separately. Use the addition property to get all variable terms on one side of the equation and all numbers on the other, then combine terms. Use the multiplication property to get the equation in the form $x = $ a number. Check the solution. **19.** 7 **21.** 0
23. $\frac{3}{25}$ **25.** 60 **27.** 4 **29.** 5000 **31.** 800 **32.** Yes, you will get $(100 \cdot 2) \cdot 4 = 800$. This is a result of the associative property of multiplication. **33.** No, because $(100a)(100b) = 10,000ab \neq 100ab$.
34. The distributive property involves the operation of addition as well. **35.** Yes; the associative property of multiplication is used.
36. no **37.** $-\frac{72}{11}$ **39.** 0 **41.** -6 **43.** 15 **45.** all real
numbers **47.** no solution **49.** $12 - q$ **51.** $\frac{t}{10}$

2.4 Exercises (page 133)

1. The procedure should include the following steps: read the problem carefully; assign a variable to represent the unknown to be found; write down variable expressions for any other unknown quantities; translate into an equation; solve the equation; state the answer; check your solution. **3.** D; there cannot be a fractional number of cars. **5.** 6 **7.** -3 **9.** 45 Democrats; 55 Republicans
11. 6562 men; 3779 women **13.** dog: 343 lb; lioness: 280 lb
15. Airborne Express: 3; Federal Express: 9; United Parcel Service: 1 **17.** 36 million mi **19.** A and B: 40°; C: 100°
21. 1950 Denver nickel: $14.00; 1945 Philadelphia nickel: $12.00
23. ice cream: 44,687.9 lb; topping: 537.1 lb **25.** 18°
27. 39° **29.** 50° **31.** 68, 69 **33.** 10, 12 **35.** 101, 102
37. 10, 11 **39.** $2.78 billion, $3.33 billion, $3.53 billion

2.5 Exercises (page 143)

1. (a) The perimeter of a plane geometric figure is the distance around the figure. **(b)** The area of a plane geometric figure is the measure of the surface covered or enclosed by the figure.
3. four **5.** area **7.** perimeter **9.** area **11.** area **13.** $P = 26$
15. $A = 64$ **17.** $b = 4$ **19.** $t = 5.6$ **21.** $I = 1575$ **23.** $r = 2.6$
25. $A = 50.24$ **27.** $V = 150$ **29.** $V = 52$ **31.** $V = 7234.56$
33. about 154,000 ft² **35.** perimeter: 13 in.; area: 10.5 in.²
37. 132.665 ft² **39.** 23,800.10 ft² **41.** length: 36 in.; volume: 11,664 in.³ **43.** 48°, 132° **45.** 51°, 51° **47.** 105°, 105°
49. $t = \frac{d}{r}$ **51.** $H = \frac{V}{LW}$ **53.** $b = P - a - c$ **55.** $r = \frac{I}{pt}$
57. $h = \frac{2A}{b}$ **59.** $W = \frac{P - 2L}{2}$ or $W = \frac{P}{2} - L$ **61.** $h = \frac{3V}{\pi r^2}$
63. $F = \frac{9}{5}C + 32$

2.6 Exercises (page 153)

1. (a) C **(b)** D **(c)** B **(d)** A **3.** $\frac{6}{7}$ **5.** $\frac{18}{55}$ **7.** $\frac{5}{16}$
9. $\frac{4}{15}$ **11.** 17-oz size **13.** 64-oz can **15.** 500-count
17. 28-oz size **19.** A percent is a ratio where the basis of comparison is 100. For example, 27% represents the ratio of 27 to 100.
21. 35 **23.** 7 **25.** -1 **27.** 5 **29.** $\frac{13}{7}$ **31.** 4 gal
33. 74.13 francs **35.** $9.90 **37.** $25\frac{2}{3}$ in. **39.** 46,700 fish
41. 124.8 **43.** 120% **45.** 600 **47.** 1.4% **49.** C **51.** 4.5%
53. 65.9% **55.** $304 **57.** 284% **59.** $262 **61.** $276
63. 30 **64. (a)** $5x = 12$ **(b)** $\frac{12}{5}$ **65.** $\frac{12}{5}$ **66.** Both methods give the same solution.

Summary Exercises on Solving Applied Problems (page 157)

1. 243 votes **2.** 48 **3.** 9 tanks **4.** 9 lb **5.** 4 **6.** 2 cm
7. 80° **8.** 100° **9.** 3 **10.** -3 **11.** 15 **12.** 28
13. 36 quart cartons **14.** length: 9 ft; width: 3 ft **15.** 104°, 104°
16. 140°, 40° **17.** 2.5 cm **18.** 12.42 cm **19.** Mr. Silvester: 25 years old; Mrs. Silvester: 20 years old **20.** Chris: 19 years old; Josh: 9 years old **21.** $16\frac{2}{3}$% **22.** $2.16 **23.** $67.50
24. 420 mi **25.** $10,694 **26.** 3 **27.** 510 calories **28.** $16\frac{1}{2}$ oz
29. 32-oz size **30.** 8-oz size **31.** approximately 2593 mi
32. 32 gold medals

2.7 Exercises (page 167)

1. $x > -4$ **3.** $x \leq 4$ **5.** $-1 < x < 2$ **7.** $-1 < x \leq 2$
9. Use an open circle if the symbol is $>$ or $<$. Use a closed circle if the symbol is $\geq$ or $\leq$.

11. **13.**

15. **17.**

19. It would imply that $3 < -2$, which is false.

21. $z \geq 1$

23. $k \geq 5$ [number line with closed dot at 5]

25. $n < -11$ [number line with open dot at −11]

27. It must be reversed when multiplying or dividing by a negative number.

29. $x < 6$ [number line with open dot at 6]

31. $y \geq -10$ [number line with closed dot at −10]

33. $t < -3$ [number line with open dot at −3]

35. $x \leq 0$ [number line with closed dot at 0]

37. $r > 20$ [number line with open dot at 20]

39. $x \geq -3$ [number line with closed dot at −3]

41. $r \geq -5$ [number line with closed dot at −5]

43. $x < 1$ [number line with open dot at 1]

45. $x \leq 0$ [number line with closed dot at 0]

47. $x \geq 4$ [number line with closed dot at 4]

49. $p < 32$ [number line with open dot at 32]

51. $x \geq \dfrac{5}{12}$ [number line with closed dot at $\frac{5}{12}$, 0 marked]

53. $k > -21$ [number line with open dot at −21]

55. 88 or more   **57.** all numbers greater than 16   **59.** It has never exceeded 40 °C.   **61.** 32 or greater   **63.** 15 min

65.   [number line with closed dot between 0 and 4]
0   4

66.   [number line with open dot at 4]
0   4

67.   [number line with open dot at 4]
0   4

68. It is the set of all real numbers.
[number line]
0   4

Chapter 2 Review Exercises (page 175)

1. 9   **2.** 4   **3.** −6   **4.** $\dfrac{3}{2}$ **5.** 20   **6.** $-\dfrac{61}{2}$ **7.** 15   **8.** 0

9. no solution   **10.** all real numbers   **11.** $-\dfrac{7}{2}$ **12.** 20

13. Hawaii: 6425 mi²; Rhode Island: 1212 mi²   **14.** Seven Falls: 300 ft; Twin Falls: 120 ft   **15.** 80°   **16.** 11, 13   **17.** $h = 11$

18. $A = 28$ **19.** $r = 4.75$ **20.** $V = 904.32$ **21.** $L = \dfrac{A}{W}$

22. $h = \dfrac{2A}{b + B}$ **23.** 135°, 45°   **24.** 100°, 100°

25. perimeter: 326.5 ft; area: 6538.875 ft²   **26.** diameter: approximately 19.9 ft; radius: approximately 9.95 ft; area: approximately 311 ft²   **27.** $\dfrac{3}{2}$ **28.** $\dfrac{5}{14}$ **29.** $\dfrac{3}{4}$ **30.** $\dfrac{1}{12}$ **31.** $\dfrac{7}{2}$ **32.** $-\dfrac{8}{3}$

33. $\dfrac{25}{19}$ **34.** 40% means $\dfrac{40}{100}$ or $\dfrac{2}{5}$. It is the same as the ratio of 2 to 5.   **35.** $6\dfrac{2}{3}$ lb   **36.** 36 oz   **37.** 54 ft   **38.** 375 km

39. 17.48   **40.** 175%   **41.** $33\dfrac{1}{3}$%   **42.** 2500   **43.** $27,630.11

44. $350.46

45.   [number line with closed dot at −4]   **46.**   [number line with open dot at 7]

47.   [number line with closed dot at −5, open dot at 6]   **48.**   [number line with closed dot at $\frac{1}{2}$, 0 marked]

49. $y \geq -3$ [number line with closed dot at −3]

50. $t < 2$ [number line with open dot at 2]

51. $x \geq 3$ [number line with closed dot at 3]

52. $k \geq 46$ [number line with closed dot at 46]

53. $x < -5$ [number line with open dot at −5]

54. $w < -37$ [number line with open dot at −37]

55. 88 or more   **56.** all numbers less than or equal to $-\dfrac{1}{3}$

57. 7   **58.** $r = \dfrac{I}{pt}$ **59.** $x < 2$ **60.** −9   **61.** 70   **62.** $\dfrac{13}{4}$

63. no solution   **64.** all real numbers   **65.** 80 ft   **66.** 6
67. Rita: 84 mi; Bobby: 28 mi   **68.** Mike: 1200 votes; William: 600 votes   **69.** United States: 97; Russia: 88   **70.** gold: 16; silver: 25; bronze: 17   **71.** 44 m   **72.** 70 ft   **73.** $20\dfrac{1}{2}$ in.

74. 26 in.   **75.** 92 or more   **76.** 51°, 51°

Chapter 2 Test (page 179)

1. 6   **2.** −6   **3.** $\dfrac{13}{4}$ **4.** −10.8   **5.** no solution   **6.** 21   **7.** 30

8. all real numbers   **9. (a)** East: 132; West: 120   **(b)** 26 points
10. Hawaii: 4021 mi²; Maui: 728 mi²; Kauai: 551 mi²   **11.** 50°

12. (a) $W = \dfrac{P - 2L}{2}$ or $W = \dfrac{P}{2} - L$ **(b)** 18   **13.** 100°, 80°

14. 75°, 75°   **15.** 6   **16.** −29   **17.** 8 slices for $2.19
18. 2300 mi   **19.** 236%   **20. (a)** $x < 0$ **(b)** $-2 < x \leq 3$

21. $x < 11$ [number line with open dot at 11]

22. $x \leq 4$ [number line with closed dot at 4]

23. $x \geq -3$ [number line with closed dot at −3]

24. 83 or more   **25.** When an inequality is multiplied or divided by a negative number, the direction of the symbol must be reversed.

Cumulative Review Exercises: Chapters R–2 (page 181)

1. $\dfrac{3}{8}$ **2.** $\dfrac{3}{4}$ **3.** $\dfrac{31}{20}$ **4.** $\dfrac{551}{40}$ or $13\dfrac{31}{40}$ **5.** 6   **6.** $\dfrac{6}{5}$

7. 34.03   **8.** 27.31   **9.** 30.51   **10.** 56.3   **11.** 35 yd

12. $7\dfrac{1}{2}$ cups   **13.** $99\dfrac{5}{8}$ lb   **14.** \$3849.94   **15.** true   **16.** true

17. 7   **18.** 1   **19.** 13   **20.** −40   **21.** −12   **22.** undefined

23. −6   **24.** 28   **25.** 1   **26.** 0   **27.** $\dfrac{73}{18}$ **28.** −64   **29.** −134

30. $-\dfrac{29}{6}$ **31.** distributive property   **32.** commutative property

33. inverse property   **34.** identity property   **35.** $7p - 14$

36. $2k - 11$ **37.** 7   **38.** −4   **39.** −1   **40.** $-\dfrac{3}{5}$ **41.** 2

42. −13   **43.** 26   **44.** −12   **45.** $c = P - a - b$ **46.** $s = \dfrac{P}{4}$

47. $z \le 2$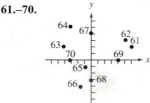

48. $r < 1$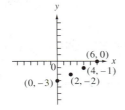

49. \$4090.56   **50.** \$3750   **51.** \$230.50   **52.** \$98.45
53. 30 cm   **54.** 16 in.

Chapter 3

Section 3.1 (page 195)

1. Snoopy; 31%   **3.** Since 26% is twice as much as 13%,
we can expect twice as many adults to favor Charlie Brown.
5. Ohio (OH): about 680 million eggs; Iowa (IA): about
550 million eggs   **7.** Indiana (IN) and Pennsylvania (PA);
about 490 million eggs each   **9.** from 1975 to 1980; about \$.75
11. The price of a gallon of gas was decreasing.   **13.** does;
do not   **15.** y **17.** 6   **19.** yes   **21.** yes   **23.** no   **25.** yes
27. no   **29.** No. For two ordered pairs (x, y) to be equal, the
x-values must be equal and the y-values must be equal. Here we
have $4 \ne -1$ and $-1 \ne 4$.   **31.** 11   **33.** $-\dfrac{7}{2}$ **35.** −4   **37.** −5

39. 4; 6; −6; (0, 4); (6, 0); (−6, 8)   **41.** 3; −5; −15; (0, 3);
(−5, 0); (−15, −6)   **43.** −9; −9; −9   **45.** −6; −6; −6
47. 8; 8; 8   **49.** (2, 4)   **51.** (−5, 4)   **53.** (3, 0)   **55.** negative;
negative   **57.** positive; negative   **59.** If $xy < 0$, then either $x < 0$
and $y > 0$ or $x > 0$ and $y < 0$. If $x < 0$ and $y > 0$, then the point lies
in quadrant II. If $x > 0$ and $y < 0$, then the point lies in quadrant IV.

61.–70.

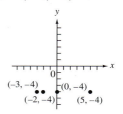

71. −3; 6; −2; 4

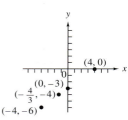

73. $-3;\ 4;\ -6;\ -\dfrac{4}{3}$

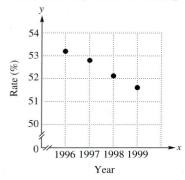

75. −4; −4; −4; −4

77. The points in each graph appear to lie on a straight line.
79. (a) (1996, 53.3), (1997, 52.8), (1998, 52.1), (1999, 51.6)
(b) (1995, 54.0) means that in 1995, the graduation rate for 4-year
college students within 5 years was 54.0%.
(c)

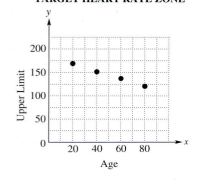

(d) The points appear to lie on a straight line. Graduation rates
for 4-year college students within 5 years are decreasing.
81. (a) 170; 154; 138; 122
(b) (20, 170), (40, 154), (60, 138), (80, 122)
(c) The points lie in a linear pattern.

Section 3.2 (page 209)

1. 5; 5; 3

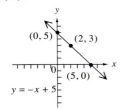

$y = -x + 5$ with points $(0, 5)$, $(2, 3)$, $(5, 0)$

3. 1; 3; −1

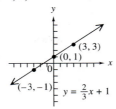

$y = \frac{2}{3}x + 1$ with points $(3, 3)$, $(0, 1)$, $(-3, -1)$

5. −6; −2; −5

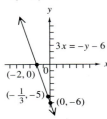

$3x = -y - 6$ with points $(-2, 0)$, $\left(-\frac{1}{3}, -5\right)$, $(0, -6)$

7. A **9.** D **11.** $(12, 0)$; $(0, -8)$ **13.** $(0, 0)$; $(0, 0)$
15. Choose a value *other than* 0 for either x or y. For example, if $x = -5$, $y = 4$.

17.

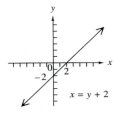

$x = y + 2$

19.

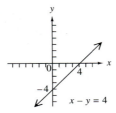

$x - y = 4$

21.

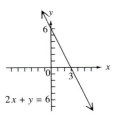

$2x + y = 6$

23.

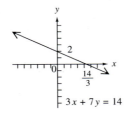

$3x + 7y = 14$ with $\frac{14}{3}$

25.

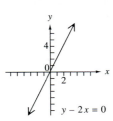

$y - 2x = 0$

27.

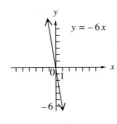

$y = -6x$

29.

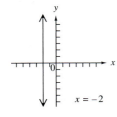

$x = -2$

31.

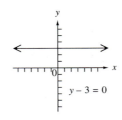

$y - 3 = 0$

33. (a) 151.5 cm, 174.9 cm, 159.3 cm

(b)

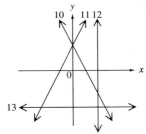

HEIGHTS OF WOMEN

Height (in cm) vs. Length of Radius Bone (in cm), $y = 3.9x + 73.5$

(c) 24 cm; 24 cm **35. (a)** 130 **(b)** 133 **(c)** They are quite close. **37.** between 133 and 162 **39. (a)** 1993: 49.8 gal; 1995: 51.4 gal; 1997: 53 gal **(b)** 1993: 50.1 gal; 1995: 51.6 gal; 1997: 53 gal **(c)** The corresponding values are quite close.
41. (a) \$30,000 **(b)** \$15,000 **(c)** \$5000 **(d)** After 5 yr, the SUV has a value of \$5000. **43. (a)** The equation is a fairly good model. **(b)** The actual debt for 1996 is about 500 billion dollars; this is about 30 billion dollars more than the amount given by the equation. **(c)** No. Data for future years might not follow the same pattern, so the linear equation would not be a reliable model.

Section 3.3 (page 221)

1. Rise is the vertical change between two different points on a line. Run is the horizontal change between two different points on a line.

3. 4 **5.** $-\frac{1}{2}$ **7.** 0 **9.** Yes, the answer would be the same. It doesn't matter which point you start with. The slope would be expressed as the quotient of −6 and −4, which simplifies to $\frac{3}{2}$.

10.–13. Answers will vary.

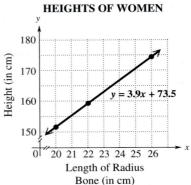

15. His answer is incorrect. Because he found the difference $3 - 5 = -2$ in the numerator, he should have subtracted in the same order in the denominator to get $-1 - 2 = -3$. The correct slope is $\frac{-2}{-3} = \frac{2}{3}$. **17.** $\frac{5}{4}$ **19.** $\frac{3}{2}$ **21.** −3 **23.** 0

25. undefined **27.** $-\frac{1}{2}$ **29.** 5 **31.** $\frac{1}{4}$ **33.** $\frac{3}{2}$ **35.** undefined

37. (a) negative **(b)** 0 **39. (a)** positive **(b)** negative

41. (a) 0 **(b)** negative **43.** $\frac{4}{3}$; $\frac{4}{3}$; parallel **45.** $\frac{5}{3}$; $\frac{3}{5}$; neither

47. $\frac{3}{5}$; $-\frac{5}{3}$; perpendicular **49.** $\frac{8}{27}$ **50.** 232 thousand or 232,000

51. positive; increased **52.** 232,000 students **53.** −2
54. negative; decreased **55.** 2 students per computer

Section 3.4 (page 231)

1. D **3.** B **5.** $y = 3x - 3$ **7.** $y = -x + 3$ **9.** $y = 4x - 3$

11. $y = 3$ **13.** A vertical line has undefined slope, so there is no value for m. Also, there is no y-intercept, so there can be no value for b.

15. $y = \dfrac{1}{2}x + 4$

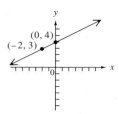

17. $y = -\dfrac{2}{5}x - \dfrac{23}{5}$

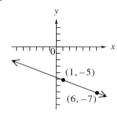

19. $y = 3x + 2$

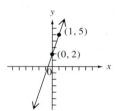

21. $y = 2$

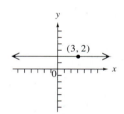

23. $x = 3$ (no slope-intercept form)

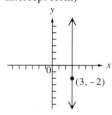

25. the y-axis   **27.** $y = 2x - 7$ **29.** $y = -2x - 4$

31. $y = \dfrac{2}{3}x + \dfrac{19}{3}$ **33.** $y = x$ (There are other forms as well.)

35. $y = x - 3$ **37.** $y = -\dfrac{5}{7}x - \dfrac{54}{7}$ **39.** $y = -\dfrac{2}{3}x - 2$

41. $x = 3$ (no slope-intercept form)   **43.** $y = \dfrac{1}{3}x + \dfrac{4}{3}$

45. $(0, 32)$; $(100, 212)$ **46.** $\dfrac{9}{5}$ **47.** $F - 32 = \dfrac{9}{5}(C - 0)$

48. $F = \dfrac{9}{5}C + 32$ **49.** $C = \dfrac{5}{9}(F - 32)$ **50.** $86°$

51. $10°$ **52.** $-40°$ **53.** $y = \dfrac{3}{4}x - \dfrac{9}{2}$ **55.** $y = -2x - 3$

57. (a) $400   **(b)** $.25   **(c)** $y = .25x + 400$ **(d)** $425   **(e)** 1500
59. (a) $(1, 10{,}017)$, $(3, 11{,}025)$, $(5, 12{,}432)$, $(7, 13{,}785)$, $(9, 15{,}380)$
(b) yes

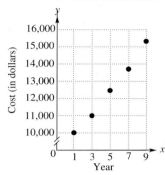

**COSTS AT PRIVATE
4-YEAR COLLEGES**

(c) $y = 725.8x + 8847.6$ or $y = 725.8x + 8847.8$ (depending on the point used)   **(d)** $18,283   **61. (a)** $y = .58x - 1098.2$ **(b)** 59.5%; This result is very close to the actual figure of 59.3%.   **(c)** positive; The percent of women in the civilian labor force increased from 1955 to 1995.

Section 3.5 (page 241)

1. false   **3.** true   **5.** $>$ **7.** $<$

9.

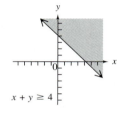

$x + y \geq 4$

11.

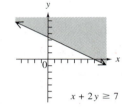

$x + 2y \geq 7$

13.

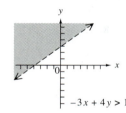

$-3x + 4y > 12$

15.

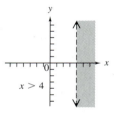

$x > 4$

17. Use a dashed line if the symbol is $<$ or $>$. Use a solid line if the symbol is $\leq$ or $\geq$.

19.

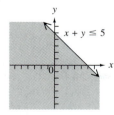

$x + y \leq 5$

21.

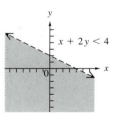

$x + 2y < 4$

23.

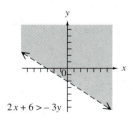

$2x + 6 > -3y$

25.

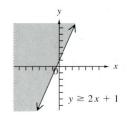

$y \geq 2x + 1$

Chapter 6

Section 6.1 (page 407)

1. (a) $3; -5$ (b) $q; -1$ 3. A rational expression is a quotient of polynomials, such as $\dfrac{x+3}{x^2-4}$. 5. 0 7. $\dfrac{5}{3}$ 9. $-3, 2$

11. never undefined 13. (a) 1 (b) $\dfrac{17}{12}$ 15. (a) 0 (b) $-\dfrac{10}{3}$

17. (a) $\dfrac{9}{5}$ (b) undefined 19. (a) $\dfrac{2}{7}$ (b) $\dfrac{13}{3}$ 21. No, not if the

number is 0. Division by 0 is undefined. 23. $3r^2$ 25. $\dfrac{2}{5}$

27. $\dfrac{x-1}{x+1}$ 29. $\dfrac{7}{5}$ 31. $m-n$ 33. $\dfrac{3(2m+1)}{4}$ 35. $\dfrac{3m}{5}$

37. $\dfrac{3r-2s}{3}$ 39. $\dfrac{z-3}{z+5}$ 41. $\dfrac{x+1}{x-1}$ 43. -1 45. $-(m+1)$

47. -1

Answers may vary in Exercises 49–53.

49. $\dfrac{-(x+4)}{x-3}, \dfrac{-x-4}{x-3}, \dfrac{x+4}{-(x-3)}, \dfrac{x+4}{-x+3}$

51. $\dfrac{-(2x-3)}{x+3}, \dfrac{-2x+3}{x+3}, \dfrac{2x-3}{-(x+3)}, \dfrac{2x-3}{-x-3}$

53. $-\dfrac{3x-1}{5x-6}, \dfrac{-(3x-1)}{5x-6}, \dfrac{-3x+1}{-5x+6}, \dfrac{3x-1}{-5x+6}$

55. x^2+3

Section 6.2 (page 415)

1. (a) B (b) D (c) C (d) A 3. $\dfrac{4m}{3}$ 5. $\dfrac{40y^2}{3}$ 7. $\dfrac{2}{c+d}$

9. $\dfrac{16q}{3p^3}$ 11. $\dfrac{7}{r^2+rp}$ 13. $\dfrac{z^2-9}{z^2+7z+12}$ 15. 5 17. $-\dfrac{3}{2t^4}$

19. $\dfrac{1}{4}$ 21. To multiply two rational expressions, multiply the numerators and multiply the denominators. Write the answer in lowest terms. 23. $\dfrac{10}{9}$ 25. $-\dfrac{3}{4}$ 27. $-\dfrac{9}{2}$ 29. $\dfrac{-9(m-2)}{(m+4)}$

31. $\dfrac{p+4}{p+2}$ 33. $\dfrac{(k-1)^2}{(k+1)(2k-1)}$ 35. $\dfrac{4k-1}{3k-2}$ 37. $\dfrac{m+4p}{m+p}$

39. $\dfrac{10}{x+10}$ 41. Division requires multiplying by the reciprocal of the second rational expression. In the reciprocal, $x+7$ is in the denominator, so $x \neq -7$.

Section 6.3 (page 421)

1. C 3. C 5. 30 7. x^7 9. $72q$ 11. $84r^5$ 13. $2^3 \cdot 3 \cdot 5$
15. The least common denominator is their product.
17. $28m^2(3m-5)$ 19. $30(b-2)$ 21. $c-d$ or $d-c$
23. $k(k+5)(k-2)$ 25. $(p+3)(p+5)(p-6)$ 27. $\dfrac{20}{55}$

29. $\dfrac{-45}{9k}$ 31. $\dfrac{26y^2}{80y^3}$ 33. $\dfrac{35t^2r^3}{42r^4}$ 35. $\dfrac{20}{8(m+3)}$ 37. $\dfrac{8t}{12-6t}$

39. $\dfrac{14(z-2)}{z(z-3)(z-2)}$ 41. $\dfrac{2(b-1)(b+2)}{b^3+3b^2+2b}$

Section 6.4 (page 429)

1. E 3. C 5. B 7. G 9. $\dfrac{11}{m}$ 11. b 13. x 15. $y-6$

17. To add or subtract rational expressions with the same denominator, combine the numerators and keep the same

denominator. For example, $\dfrac{3x+2}{x-6} + \dfrac{-2x-8}{x-6} = \dfrac{x-6}{x-6}$. Then

write in lowest terms: $\dfrac{x-6}{x-6} = 1$. 19. $\dfrac{3z+5}{15}$ 21. $\dfrac{10-7r}{14}$

23. $\dfrac{-3x-2}{4x}$ 25. $\dfrac{x+1}{2}$ 27. $\dfrac{5x+9}{6x}$ 29. $\dfrac{3x+3}{x(x+3)}$

31. $\dfrac{x^2+6x-8}{(x-2)(x+2)}$ 33. $\dfrac{3}{t}$ 35. $m-2$ or $2-m$

37. $\dfrac{-2}{x-5}$ or $\dfrac{2}{5-x}$ 39. -4 41. $\dfrac{-5}{x-y^2}$ or $\dfrac{5}{y^2-x}$

43. $\dfrac{x+y}{5x-3y}$ or $\dfrac{-x-y}{3y-5x}$ 45. $\dfrac{-6}{4p-5}$ or $\dfrac{6}{5-4p}$ 47. $\dfrac{-(m+n)}{2(m-n)}$

49. $\dfrac{-x^2+6x+11}{(x+3)(x-3)(x+1)}$ 51. $\dfrac{-5q^2-13q+7}{(3q-2)(q+4)(2q-3)}$

53. $\dfrac{9r+2}{r(r+2)(r-1)}$

55. $\dfrac{2x^2+6xy+8y^2}{(x+y)(x+y)(x+3y)}$ or $\dfrac{2x^2+6xy+8y^2}{(x+y)^2(x+3y)}$

57. $\dfrac{15r^2+10ry-y^2}{(3r+2y)(6r-y)(6r+y)}$ 59. (a) $\dfrac{9k^2+6k+26}{5(3k+1)}$ (b) $\dfrac{1}{4}$

Section 6.5 (page 439)

1. (a) $6; \dfrac{1}{6}$ (b) $12; \dfrac{3}{4}$ (c) $\dfrac{1}{6} \div \dfrac{3}{4}$ (d) $\dfrac{2}{9}$ 3. -6 5. $\dfrac{1}{pq}$ 7. $\dfrac{1}{xy}$

9. $\dfrac{2a^2b}{3}$ 11. $\dfrac{m(m+2)}{3(m-4)}$ 13. $\dfrac{2}{x}$ 15. $\dfrac{8}{x}$ 17. $\dfrac{a^2-5}{a^2+1}$

19. $\dfrac{3(p+2)}{2(2p+3)}$ 21. $\dfrac{t(t-2)}{4}$ 23. $\dfrac{-k}{2+k}$ 25. $\dfrac{3m(m-3)}{(m-1)(m-8)}$

27. $\dfrac{6}{5}$ 29. $\dfrac{\frac{3}{8}+\frac{5}{6}}{2}$ 30. $\dfrac{29}{48}$ 31. $\dfrac{29}{48}$ 32. The choice of method and the reason for the choice are personal preferences.

Section 6.6 (page 449)

1. operation; $\dfrac{43}{40}x$ 3. equation; $\dfrac{40}{43}$ 5. operation; $-\dfrac{1}{10}y$

7. When solving an equation, we multiply each side by the LCD, which eliminates all denominators. When adding or subtracting fractions, we multiply by 1 in the form LCD/LCD. The denominators are not eliminated. 9. -6 11. 24 13. -15 15. 7
17. -15 19. -5 21. -6 23. 5 25. 12 27. 2 29. 0 and 4
31. -6 33. no solution 35. 3 37. 3 39. $-2, 12$

41. no solution 43. $-6, \dfrac{1}{2}$ 45. $-\dfrac{1}{5}, 3$ 47. $-\dfrac{3}{5}, 3$ 49. $F = \dfrac{ma}{k}$

51. $a = \dfrac{kF}{m}$ 53. $R = \dfrac{E-Ir}{I}$ 55. $A = \dfrac{h(B+b)}{2}$

57. $a = \dfrac{2S-ndL}{nd}$ 59. $t = \dfrac{rs}{rs-2s-3r}$ or $t = \dfrac{-rs}{-rs+2s+3r}$

61. $c = \dfrac{ab}{b-a-2ab}$ or $c = \dfrac{-ab}{-b+a+2ab}$

Summary Exercises on Operations and Equations with Rational Expressions (page 452)

1. operation; $\dfrac{10}{p}$ 2. operation; $\dfrac{y^3}{x^3}$ 3. operation; $\dfrac{1}{2x^2(x+2)}$

4. equation; 9 5. operation; $\dfrac{y+2}{y-1}$ 6. operation; $\dfrac{5k+8}{k(k-4)(k+4)}$

7. equation; 39 8. operation; $\dfrac{t-5}{3(2t+1)}$ 9. operation; $\dfrac{13}{3(p+2)}$

10. equation; $-1, \dfrac{12}{5}$ **11.** equation; $\dfrac{1}{7}, 2$ **12.** operation; $\dfrac{16}{3y}$

13. operation; $\dfrac{7}{12z}$ **14.** equation; 13

15. operation; $\dfrac{3m + 5}{(m + 2)(m + 3)(m + 1)}$ **16.** operation; $\dfrac{k + 3}{5(k - 1)}$

Section 6.7 (page 461)

1. (a) the amount **(b)** $5 + x$ **(c)** $\dfrac{5 + x}{6} = \dfrac{13}{3}$ **3.** $\dfrac{12}{18}$ **5.** $\dfrac{12}{3}$

7. 12   **9.** $\dfrac{1386}{97}$ **11.** 1.99 m per sec   **13.** 4.059 hr

15. 367.197 m per min   **17.** $\dfrac{500}{x - 10} = \dfrac{600}{x + 10}$ **19.** $\dfrac{D}{R} = \dfrac{d}{r}$

21. 8 mph   **23.** 900 mi   **25.** $\dfrac{1}{8}x + \dfrac{1}{6}x = 1$ or $\dfrac{1}{8} + \dfrac{1}{6} = \dfrac{1}{x}$

27. $4\dfrac{4}{17}$ hr   **29.** $5\dfrac{5}{11}$ hr   **31.** $2\dfrac{7}{10}$ hr   **33.** $9\dfrac{1}{11}$ min

35. Answers will vary. Here is one possibility: I start with the fraction $\dfrac{11}{8}$. If I add -2 to both the numerator and the denominator, I get $\dfrac{9}{6}$, which simplifies to $\dfrac{3}{2}$. The problem is stated as follows: "If a number is added to both the numerator and the denominator of $\dfrac{11}{8}$, the resulting fraction is equivalent to $\dfrac{3}{2}$. What is the number?"

To solve, let $x =$ the number. The equation is $\dfrac{11 + x}{8 + x} = \dfrac{3}{2}$. This leads to $2(11 + x) = 3(8 + x)$, which leads to $22 + 2x = 24 + 3x$, or $-2 = x$. The number is -2.

Section 6.8 (page 467)

1. increases   **3.** 9   **5.** 250   **7.** 6   **9.** 21   **11.** $40.32

13. 8.68 lb per in.^{2} **15.** $106\dfrac{2}{3}$ mph   **17.** $12\dfrac{1}{2}$ amps   **19.** 20 lb

21. direct   **23.** direct   **25.** direct   **27.** inverse   **29.** 4   **31.** 1
33. 54 ft

Chapter 6 Review Exercises (page 477)

1. 3   **2.** 0   **3.** $-1, 3$ **4.** $-5, -\dfrac{2}{3}$ **5. (a)** $-\dfrac{4}{7}$ **(b)** -16

6. (a) $\dfrac{11}{8}$ **(b)** $\dfrac{13}{22}$ **7. (a)** undefined **(b)** 1   **8. (a)** undefined

(b) $\dfrac{1}{2}$ **9.** $\dfrac{b}{3a}$ **10.** -1 **11.** $\dfrac{-(2x + 3)}{2}$ **12.** $\dfrac{2p + 5q}{5p + q}$

Answers may vary in Exercises 13 and 14.

13. $\dfrac{-(4x - 9)}{2x + 3}, \dfrac{-4x + 9}{2x + 3}, \dfrac{4x - 9}{-(2x + 3)}, \dfrac{4x - 9}{-2x - 3}$

14. $\dfrac{-8 + 3x}{-3 - 6x}, \dfrac{-(-8 + 3x)}{3 + 6x}, \dfrac{8 - 3x}{-(-3 - 6x)}, -\dfrac{-8 + 3x}{3 + 6x}$

15. 2   **16.** $\dfrac{2}{3m^6}$ **17.** $\dfrac{5}{8}$ **18.** $\dfrac{3}{2}$ **19.** $\dfrac{r + 4}{3}$ **20.** $\dfrac{y - 2}{y - 3}$

21. $\dfrac{p + 5}{p + 1}$ **22.** $\dfrac{3z + 1}{z + 3}$ **23.** 96   **24.** $108y^4$

25. $m(m + 2)(m + 5)$ **26.** $(x + 3)(x + 1)(x + 4)$ **27.** $\dfrac{35}{56}$

28. $\dfrac{40}{4k}$ **29.** $\dfrac{15a}{10a^4}$ **30.** $\dfrac{-54}{18 - 6x}$ **31.** $\dfrac{15y}{50 - 10y}$

32. $\dfrac{4b(b + 2)}{(b + 3)(b - 1)(b + 2)}$ **33.** $\dfrac{15}{x}$ **34.** $-\dfrac{2}{p}$ **35.** $\dfrac{4k - 45}{k(k - 5)}$

36. $\dfrac{28 + 11y}{y(7 + y)}$ **37.** $\dfrac{-2 - 3m}{6}$ **38.** $\dfrac{3(16 - x)}{4x^2}$

39. $\dfrac{7a + 6b}{(a - 2b)(a + 2b)}$ **40.** $\dfrac{-k^2 - 6k + 3}{3(k + 3)(k - 3)}$

41. $\dfrac{5z - 16}{z(z + 6)(z - 2)}$ **42.** $\dfrac{-13p + 33}{p(p - 2)(p - 3)}$ **43.** $\dfrac{a}{b}$

44. $\dfrac{4(y - 3)}{y + 3}$ **45.** $\dfrac{6(3m + 2)}{2m - 5}$ **46.** $\dfrac{(q - p)^2}{pq}$ **47.** $\dfrac{xw + 1}{xw - 1}$

48. $\dfrac{1 - r - t}{1 + r + t}$ **49.** $\dfrac{35}{6}$ **50.** -16 **51.** -4 **52.** no solution

53. 3   **54.** $t = \dfrac{Ry}{m}$ **55.** $y = \dfrac{4x + 5}{3}$ **56.** $t = \dfrac{rs}{s - r}$ **57.** $\dfrac{20}{15}$

58. $\dfrac{3}{18}$ **59.** 1.527 hr   **60.** $3\dfrac{1}{13}$ hr   **61.** 10 hr   **62.** 4 cm

63. $\dfrac{36}{5}$ **64.** $\dfrac{m + 7}{(m - 1)(m + 1)}$ **65.** $8p^2$ **66.** $\dfrac{1}{6}$ **67.** 3

68. $\dfrac{z + 7}{(z + 1)(z - 1)^2}$ **69.** $d = \dfrac{k + FD}{F}$ or $d = \dfrac{k}{F} + D$

70. $-2, 3$ **71.** 150 km per hr   **72.** $1\dfrac{7}{8}$ hr   **73.** 4   **74.** 24

Chapter 6 Test (page 481)

1. $-2, 4$ **2. (a)** $\dfrac{11}{6}$ **(b)** undefined   **3.** (Answers may vary.)

$\dfrac{-(6x - 5)}{2x + 3}, \dfrac{-6x + 5}{2x + 3}, \dfrac{6x - 5}{-(2x + 3)}, \dfrac{6x - 5}{-2x - 3}$ **4.** $-3x^2y^3$

5. $\dfrac{3a + 2}{a - 1}$ **6.** $\dfrac{25}{27}$ **7.** $\dfrac{3k - 2}{3k + 2}$ **8.** $\dfrac{a - 1}{a + 4}$ **9.** $150p^5$

10. $(2r + 3)(r + 2)(r - 5)$ **11.** $\dfrac{240p^2}{64p^3}$ **12.** $\dfrac{21}{42m - 84}$

13. 2   **14.** $\dfrac{-14}{5(y + 2)}$ **15.** $\dfrac{x^2 + x + 1}{3 - x}$ or $\dfrac{-x^2 - x - 1}{x - 3}$

16. $\dfrac{-m^2 + 7m + 2}{(2m + 1)(m - 5)(m - 1)}$ **17.** $\dfrac{2k}{3p}$ **18.** $\dfrac{-2 - x}{4 + x}$ **19.** $-\dfrac{1}{2}$

20. $D = \dfrac{dF - k}{F}$ or $D = \dfrac{k - dF}{-F}$ **21.** -4 **22.** 3 mph

23. $2\dfrac{2}{9}$ hr   **24.** 27   **25.** 27 days

Cumulative Review Exercises: Chapters R–6 (page 483)

1. 2   **2.** 17   **3.** $b = \dfrac{2A}{h}$ **4.** $-\dfrac{2}{7}$

5.

$y \geq -8$

6.

$m > 4$

7. (a) $(-3, 0)$ **(b)** $(0, -4)$

8.

9.

$y \geq 2x + 3$

10. $\dfrac{1}{2^4 x^7}$ **11.** $\dfrac{1}{m^6}$ **12.** $\dfrac{q}{4p^2}$ **13.** $k^2 + 2k + 1$ **14.** $72x^6y^7$

15. $4a^2 - 4ab + b^2$ **16.** $3y^3 + 8y^2 + 12y - 5$

17. $6p^2 + 7p + 1 + \dfrac{3}{p - 1}$ **18.** 1.4×10^5 sec

19. $(4t + 3v)(2t + v)$ **20.** prime

21. $(4x^2 + 1)(2x + 1)(2x - 1)$ **22.** $-3, 5$ **23.** $5, -\dfrac{1}{2}, \dfrac{2}{3}$

24. -2 or -1 **25.** 6 m   **26.** $-2, 2$ **27.** A   **28.** D

29. $\dfrac{4}{q}$ **30.** $\dfrac{3r + 28}{7r}$ **31.** $\dfrac{7}{15(q - 4)}$ **32.** $\dfrac{-k - 5}{k(k + 1)(k - 1)}$

33. $\dfrac{7(2z + 1)}{24}$ **34.** $\dfrac{195}{29}$ **35.** $\dfrac{21}{2}$ **36.** $-2, 1$ **37.** 150 mi

38. $1\dfrac{1}{5}$ hr

Chapter 7

Section 7.1 (page 491)

1. B, because the ordered pair must be in quadrant II.   **3.** There is no way that the sum of two numbers can be both 2 and 4 at the same time.   **5.** no   **7.** yes   **9.** yes   **11.** no
We show the graphs here only for Exercises 13–17.

13. $(4, 2)$ 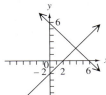   **15.** $(0, 4)$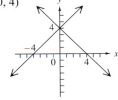

17. $(4, -1)$

19. $(1, 3)$ **21.** $(0, 2)$ **23.** $(4, -3)$

25. $y = -\dfrac{3}{2}x + 3$; $y = -\dfrac{3}{2}x + \dfrac{5}{2}$; The graphs are parallel lines.

26. $y = 2x - 4$; $y = 2x - 4$; The graphs are the same line.

27. $y = \dfrac{1}{3}x - \dfrac{5}{3}$; $y = -2x + 8$; The graphs are intersecting lines.

28. Exercise 25: no solution; Exercise 26: infinite number of solutions; Exercise 27: one solution   **29.** no solution
31. infinite number of solutions   **33.** no solution   **35.** 1991; about 350 million   **37.** between 1988 and 1990   **39.** If the coordinates of the point of intersection are not integers, the solution will be difficult to determine from a graph.
41. Answers will vary, but the lines must intersect at $(-2, 3)$.

Section 7.2 (page 501)

1. No, it is not correct, because the solution is the ordered pair $(3, 0)$. The y-value must also be determined.   **3.** The first student had less work to do, because the coefficient of y in the first equation is -1. The second student had to divide by 2, introducing fractions into the expression for x.   **5.** $(7, 3)$ **7.** $(-2, 4)$ **9.** $(-4, 8)$

11. $(3, -2)$ **13.** infinite number of solutions   **15.** $\left(\dfrac{1}{3}, -\dfrac{1}{2}\right)$

17. no solution   **19.** infinite number of solutions   **21.** $(4, -6)$
23. $(7, 0)$ **25.** infinite number of solutions   **27.** no solution
29. (a) $(3, 2)$ **(b)** $(-1, -3)$; In each case, only one step is needed to find the solution because the value of one variable is known.   **31.** $(0, 3)$ **33.** $(24, -12)$ **35.** $(3, 2)$ **37.** To find the total cost, multiply the number of bicycles (x) by the cost per bicycle (400 dollars) and add the fixed cost (5000 dollars). Thus, $y_1 = 400x + 5000$ gives this total cost (in dollars).   **38.** $y_2 = 600x$
39. $y_1 = 400x + 5000$, $y_2 = 600x$; solution: (25, 15,000)
40. 25; 15,000; 15,000   **41.** 1993

Section 7.3 (page 511)

1. true   **3.** true   **5.** $(-1, 3)$ **7.** $(-1, -3)$ **9.** $(-2, 3)$

11. $\left(\dfrac{1}{2}, 4\right)$ **13.** $(3, -6)$ **15.** $(7, 4)$ **17.** $(0, 4)$ **19.** $(-4, 0)$

21. $(0, 0)$ **23.** $(-6, 5)$ **25.** $\left(-\dfrac{6}{5}, \dfrac{4}{5}\right)$ **27.** $\left(\dfrac{1}{8}, -\dfrac{5}{6}\right)$

29. $(11, 15)$ **31.** no solution   **33.** infinite number of solutions
35. no solution   **37.** infinite number of solutions
39. $1141 = 1991a + b$ **40.** $1339 = 1996a + b$
41. $1991a + b = 1141$, $1996a + b = 1339$; solution: $(39.6, -77,702.6)$ **42.** $y = 39.6x - 77,702.6$ **43.** 1220.2 (million); This is slightly less than the actual figure.   **44.** It is not realistic to expect the data to lie in a perfectly straight line; as a result, the quantity obtained from an equation determined in this way will probably be "off" a bit. We cannot put too much faith in models such as this one, because not all data points are linear in nature.

Section 7.4 (page 521)

1. D   **3.** B   **5.** C   **7.** Boyz II Men: 134; Bruce Springsteen & the E Street Band: 40   **9.** Las Vegas: 1201 thousand; Sacramento: 1482 thousand   **11. (a)** 45 units   **(b)** Do not produce; the product will lead to a loss.   **13.** 46 ones; 28 tens   **15.** 2 copies of *The Grinch*; 5 N'Sync CDs   **17.** $2500 at 4%; $5000 at 5%
19. Japan: $17.19; Switzerland: $13.15   **21.** 80 L of 40% solution; 40 L of 70% solution   **23.** 30 lb at $6 per lb; 60 lb at $3 per lb
25. 30 barrels at $40 per barrel; 20 barrels at $60 per barrel
27. boat: 10 mph; current: 2 mph   **29.** plane: 470 mph; wind: 30 mph   **31.** car leaving Cincinnati: 55 mph; car leaving Toledo: 70 mph   **33.** Roberto: 17.5 mph; Juana: 12.5 mph

Section 7.5 (page 529)

1. C   **3.** B

5. 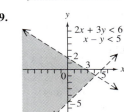   **7.**

9. 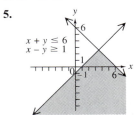   **11.**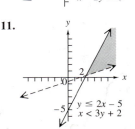

13.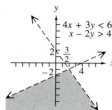

15.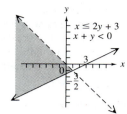

41.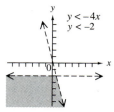

17.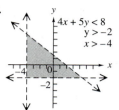

19.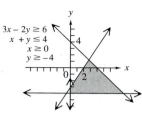

Chapter 7 Review Exercises (page 535)

1. yes   **2.** no   **3.** (3, 1)   **4.** (0, −2)   **5.** infinite number of solutions   **6.** no solution   **7.** It is not a solution of the system because it is not a solution of the second equation, $2x + y = 4$.
8. (2, 1)   **9.** (3, 5)   **10.** (6, 4)   **11.** no solution   **12.** (7, 1)
13. (−5, −2)   **14.** (−4, 3)   **15.** infinite number of solutions
16. (a) 2   **(b)** 9   **17.** (9, 2)   **18.** $\left(\dfrac{10}{7}, -\dfrac{9}{7}\right)$ **19.** (8, 9)
20. (2, 1)   **21.** *How Stella Got Her Groove Back:* 782,699; *The Deep End of the Ocean:* 840,263   **22.** *Modern Maturity:* 20.5 million; *Reader's Digest:* 15.1 million   **23.** length: 27 m; width: 18 m   **24.** 13 twenties; 7 tens   **25.** 25 lb of $1.30 candy; 75 lb of $.90 candy   **26.** plane: 250 mph; wind: 20 mph
27. $7000 at 3%; $11,000 at 4%   **28.** 60 L of 40% solution; 30 L of 70% solution

29.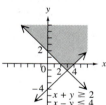

30.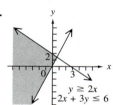

31.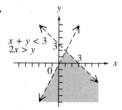

32. B   **33.** B   **34.** (2, 0)   **35.** (−4, 15)

36.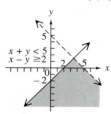

37.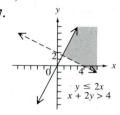

38. 8 in., 8 in., and 13 in.   **39.** 21 to 3   **40.** no solution

Chapter 7 Test (page 539)

1. (2, −3)   **2.** It has no solution.   **3.** (1, −6)   **4.** (−35, 35)
5. (5, 6)   **6.** (−1, 3)   **7.** (0, 0)   **8.** no solution   **9.** infinite number of solutions   **10.** (12, −4)   **11.** Memphis and Atlanta: 371 mi; Minneapolis and Houston: 671 mi   **12.** Disneyland: 15.0 million; Magic Kingdom: 13.8 million
13. $33\dfrac{1}{3}$ L of 25% solution; $16\dfrac{2}{3}$ L of 40% solution
14. slower car: 45 mph; faster car: 60 mph

15.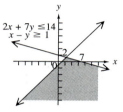

16.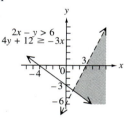

Cumulative Review Exercises: Chapters R–7 (page 541)

1. −1, 1, −2, 2, −4, 4, −5, 5, −8, 8, −10, 10, −20, 20, −40, 40
2. 1   **3.** commutative property   **4.** distributive property
5. inverse property   **6.** 46   **7.** $-\dfrac{13}{11}$ **8.** $\dfrac{9}{11}$ **9.** $x > -18$
10. $x > -\dfrac{11}{2}$ **11.** width: 8.25 in.; length: 10.75 in.

12.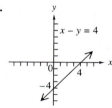

13.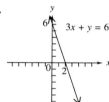

14. $-\dfrac{4}{3}$ **15.** $-\dfrac{1}{4}$ **16.** $y = \dfrac{1}{2}x + 3$ **17.** $y = 2x + 1$
18. (a) $x = 9$ **(b)** $y = -1$ **19.** $y = 7396.3x - 17{,}156.3$; The slope of 7396.3 gives the annual increase in investment clubs for 1994–1997.   **20.** $14x^2 - 5x + 23$ **21.** $6xy + 12x - 14y - 28$
22. $3k^2 - 4k + 1$ **23. (a)** 6.0×10^9; 1.0×10^{10}
(b) 3,500,000,000   **24.** $x^6 y$ **25.** $(5m - 4p)(2m + 3p)$
26. $(8t - 3)^2$ **27.** $-\dfrac{1}{3}, \dfrac{3}{2}$ **28.** −11, 11   **29.** $\dfrac{7}{x + 2}$
30. $\dfrac{3}{4k - 3}$ **31.** $-\dfrac{1}{4}, 3$ **32.** $T = \dfrac{PV}{k}$ **33.** (−1, 6)   **34.** (3, −4)
35. (2, −1)   **36.** no solution   **37.** 405 adults and 49 children
38. 19 in., 19 in., 15 in.   **39.** 4 girls and 3 boys

40.

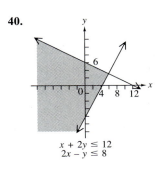

$$x + 2y \leq 12$$
$$2x - y \leq 8$$

Chapter 8

Section 8.1 (page 553)

1. true **3.** False. Zero has only one square root. **5.** true

7. $-3, 3$ **9.** $-8, 8$ **11.** $-12, 12$ **13.** $-\dfrac{5}{14}, \dfrac{5}{14}$ **15.** $-30, 30$

17. 1 **19.** 7 **21.** -11 **23.** $-\dfrac{12}{11}$ **25.** not a real number

27. 100 **29.** 19 **31.** $\dfrac{2}{3}$ **33.** $3x^2 + 4$ **35.** a must be positive.

37. a must be negative. **39.** rational; 5 **41.** irrational; 5.385
43. rational; -8 **45.** irrational; -17.321 **47.** not a real number
49. irrational; 34.641 **51.** C **53.** $c = 17$ **55.** $b = 8$
57. $c = 11.705$ **59.** 24 cm **61.** 80 ft **63.** 195 ft **65.** 11.1 ft
67. 9.434 **69.** Answers will vary. For example, if $a = 2$ and
$b = 7$, $\sqrt{a^2 + b^2} = \sqrt{2^2 + 7^2} = \sqrt{53}$, while $a + b = 2 + 7 = 9$.
Therefore, $\sqrt{a^2 + b^2} \neq a + b$ because $\sqrt{53} \neq 9$. **71.** 1 **73.** 5
75. -3 **77.** -2 **79.** 5 **81.** not a real number **83.** -3 **85.** -4

Section 8.2 (page 563)

1. false; $\sqrt{(-6)^2} = \sqrt{36} = 6$ **3.** $\sqrt{15}$ **5.** $\sqrt{22}$ **7.** $\sqrt{42}$
9. $\sqrt{13r}$ **11.** A **13.** $3\sqrt{5}$ **15.** $2\sqrt{6}$ **17.** $3\sqrt{10}$
19. $5\sqrt{3}$ **21.** $5\sqrt{5}$ **23.** cannot be simplified **25.** $4\sqrt{10}$
27. $-10\sqrt{7}$ **29.** $3\sqrt{6}$ **31.** 24 **33.** $6\sqrt{10}$
35. $\sqrt{8} \cdot \sqrt{32} = \sqrt{8 \cdot 32} = \sqrt{256} = 16$. Also, $\sqrt{8} = 2\sqrt{2}$
and $\sqrt{32} = 4\sqrt{2}$, so $\sqrt{8} \cdot \sqrt{32} = 2\sqrt{2} \cdot 4\sqrt{2} = 8 \cdot 2 = 16$.
Both methods give the same answer, and the correct answer can
always be obtained using either method. **37.** $\dfrac{4}{15}$ **39.** $\dfrac{\sqrt{7}}{4}$

41. 5 **43.** $\dfrac{25}{4}$ **45.** $6\sqrt{5}$ **47.** m **49.** y^2 **51.** $6z$ **53.** $20x^3$
55. $3x^4\sqrt{2}$ **57.** $3c^7\sqrt{5}$ **59.** $z^2\sqrt{z}$ **61.** $a^6\sqrt{a}$ **63.** $8x^3\sqrt{x}$
65. x^3y^6 **67.** $9m^2n$ **69.** $\dfrac{\sqrt{7}}{x^5}$ **71.** $\dfrac{y^2}{10}$ **73.** $2\sqrt[3]{5}$ **75.** $3\sqrt[3]{2}$
77. $4\sqrt[3]{2}$ **79.** $2\sqrt[4]{5}$ **81.** $\dfrac{2}{3}$ **83.** $-\dfrac{6}{5}$ **85.** p **87.** x^3 **89.** $4z^2$
91. $7a^3b$ **93.** $2t\sqrt[3]{2t^2}$ **95.** $\dfrac{m^4}{2}$ **97.** 6 cm **99.** 6 in. **101.** D

Section 8.3 (page 569)

1. distributive **3.** radicands **5.** $-5\sqrt{7}$ **7.** $5\sqrt{17}$
9. $5\sqrt{7}$ **11.** $11\sqrt{5}$ **13.** $15\sqrt{2}$ **15.** $-6\sqrt{2}$ **17.** $17\sqrt{7}$
19. $-16\sqrt{2} - 8\sqrt{3}$ **21.** $20\sqrt{2} + 6\sqrt{3} - 15\sqrt{5}$ **23.** $4\sqrt{2}$
25. $22\sqrt{2}$ **27.** $11\sqrt{3}$ **29.** $5\sqrt{x}$ **31.** $3x\sqrt{6}$ **33.** 0
35. $-20\sqrt{2k}$ **37.** $42x\sqrt{5z}$ **39.** $-\sqrt[3]{2}$ **41.** $6\sqrt[3]{p^2}$
43. $21\sqrt[4]{m^3}$ **45.** $-6x^2y$ **46.** $-6(p - 2q)^2(a + b)$

47. $-6a^2\sqrt{xy}$ **48.** The answers are alike because the numerical
coefficient of the three answers is the same: -6. Also, the first
variable factor is raised to the second power, and the second variable factor is raised to the first power. The answers are different
because the variables are different: x and y, then $p - 2q$ and $a + b$,
and then a and $\sqrt{xy}$.

Section 8.4 (page 577)

1. $4\sqrt{2}$ **3.** $\dfrac{-\sqrt{33}}{3}$ **5.** $\dfrac{7\sqrt{15}}{5}$ **7.** $\dfrac{\sqrt{30}}{2}$ **9.** $\dfrac{16\sqrt{3}}{9}$

11. $\dfrac{-3\sqrt{2}}{10}$ **13.** $\dfrac{21\sqrt{5}}{5}$ **15.** $\sqrt{3}$ **17.** $\dfrac{\sqrt{2}}{2}$ **19.** $\dfrac{\sqrt{65}}{5}$
21. We are actually multiplying by 1. The identity property of
multiplication justifies our result. **23.** $\dfrac{\sqrt{21}}{3}$ **25.** $\dfrac{3\sqrt{14}}{4}$

27. $\dfrac{1}{6}$ **29.** 1 **31.** $\dfrac{\sqrt{7x}}{x}$ **33.** $\dfrac{2x\sqrt{xy}}{y}$ **35.** $\dfrac{x\sqrt{30xz}}{6}$

37. $\dfrac{3ar^2\sqrt{7rt}}{7t}$ **39.** B **41.** $\dfrac{\sqrt[3]{12}}{2}$ **43.** $\dfrac{\sqrt[3]{196}}{7}$ **45.** $\dfrac{\sqrt[3]{6y}}{2y}$

47. $\dfrac{\sqrt[3]{42mn^2}}{6n}$ **49.** (a) $\dfrac{9\sqrt{2}}{4}$ sec (b) 3.182 sec

Section 8.5 (page 585)

1. 13 **3.** 4 **5.** $\sqrt{15} - \sqrt{35}$ **7.** $2\sqrt{10} + 30$ **9.** $4\sqrt{7}$
11. $57 + 23\sqrt{6}$ **13.** $81 + 14\sqrt{21}$ **15.** $71 - 16\sqrt{7}$
17. $37 + 12\sqrt{7}$ **19.** $a + 2\sqrt{a} + 1$ **21.** 23 **23.** 1
25. $y - 10$ **27.** $2\sqrt{3} - 2 + 3\sqrt{2} - \sqrt{6}$ **29.** $15\sqrt{2} - 15$
31. $\sqrt{30} + \sqrt{15} + 6\sqrt{5} + 3\sqrt{10}$
33. $\sqrt{5x} - \sqrt{10} - \sqrt{10x} + 2\sqrt{5}$ **35.** Because multiplication
must be performed before addition, it is incorrect to add -37
and -2. Only like radicals can be combined. **37.** $\dfrac{3 - \sqrt{2}}{7}$
39. $-4 - 2\sqrt{11}$ **41.** $1 + \sqrt{2}$ **43.** $-\sqrt{10} + \sqrt{15}$
45. $2\sqrt{5} + \sqrt{15} + 4 + 2\sqrt{3}$ **47.** $\dfrac{12(\sqrt{x} - 1)}{x - 1}$
49. $\dfrac{3(7 + \sqrt{x})}{49 - x}$ **51.** $\sqrt{11} - 2$ **53.** $\dfrac{\sqrt{3} + 5}{8}$ **55.** $\dfrac{6 - \sqrt{10}}{2}$
57. $30 + 18x$ **58.** They are not like terms. **59.** $30 + 18\sqrt{5}$
60. They are not like radicals. **61.** Make the first term $30x$, so
that $30x + 18x = 48x$; make the first term $30\sqrt{5}$, so that
$30\sqrt{5} + 18\sqrt{5} = 48\sqrt{5}$. **62.** Both like terms and like radicals
are combined by adding their numerical coefficients. The variables
in like terms are replaced by radicals in like radicals. **63.** 4 in.

Section 8.6 (page 595)

1. 49 **3.** 7 **5.** 85 **7.** -45 **9.** $-\dfrac{3}{2}$ **11.** no solution

13. 121 **15.** 8 **17.** 1 **19.** 6 **21.** no solution **23.** 5
25. $x^2 - 14x + 49$ **27.** 12 **29.** 5 **31.** 0, 3 **33.** $-1, 3$
35. 8 **37.** 4 **39.** 8 **41.** 9 **43.** (a) 70.5 mph (b) 59.8 mph
(c) 53.9 mph **45.** 158.6 ft **47.** (a) $25.1 billion (b) Actual
imports for 1994 were $25.9 billion, so the result using the equation
is a little low. (c) $53.3 billion; The result using the equation is a
little low. (d) .2; $-.1$; $-.4$; Exports exceeded imports only in 1993.

Chapter 8 Review Exercises (page 603)

1. $-7, 7$ **2.** $-9, 9$ **3.** $-14, 14$ **4.** $-11, 11$ **5.** $-15, 15$
6. $-27, 27$ **7.** 4 **8.** -6 **9.** 10 **10.** 3 **11.** not a real number

12. -65 **13.** $\dfrac{7}{6}$ **14.** $\dfrac{10}{9}$ **15.** B **16.** F **17.** D **18.** A

19. C **20.** A **21.** 8 **22.** 40.6 cm **23.** irrational; 4.796

24. rational; 13 **25.** rational; -5 **26.** not a real number

27. $\sqrt{14}$ **28.** $5\sqrt{3}$ **29.** $-3\sqrt{3}$ **30.** $4\sqrt{3}$ **31.** $4\sqrt{10}$

32. 18 **33.** $16\sqrt{6}$ **34.** $25\sqrt{10}$ **35.** $\dfrac{3}{2}$ **36.** $-\dfrac{11}{20}$ **37.** $\dfrac{\sqrt{7}}{13}$

38. $\dfrac{\sqrt{5}}{6}$ **39.** $\dfrac{2}{15}$ **40.** $3\sqrt{2}$ **41.** 8 **42.** $2\sqrt{2}$ **43.** p

44. $\sqrt{km}$ **45.** r^9 **46.** x^5y^8 **47.** $x^4\sqrt{x}$ **48.** $\dfrac{6}{p}$

49. $a^7b^{10}\sqrt{ab}$ **50.** $11x^3y^5$ **51.** y^2 **52.** $6x^5$ **53.** Yes, because both approximations are .7071067812. **54.** $2\sqrt{11}$ **55.** $9\sqrt{2}$

56. $21\sqrt{3}$ **57.** $12\sqrt{3}$ **58.** 0 **59.** $3\sqrt{7}$ **60.** $2\sqrt{3}+3\sqrt{10}$

61. $2\sqrt{2}$ **62.** $6\sqrt{30}$ **63.** $5\sqrt{x}$ **64.** 0 **65.** $-m\sqrt{5}$

66. $11k^2\sqrt{2n}$ **67.** $\dfrac{10\sqrt{3}}{3}$ **68.** $\dfrac{8\sqrt{10}}{5}$ **69.** $\sqrt{6}$ **70.** $\dfrac{\sqrt{10}}{5}$

71. $\sqrt{10}$ **72.** $\dfrac{\sqrt{42}}{21}$ **73.** $\dfrac{r\sqrt{x}}{4x}$ **74.** $\dfrac{\sqrt[3]{9}}{3}$ **75.** $r=\dfrac{\sqrt{3V\pi h}}{\pi h}$

76. $r=\dfrac{\sqrt{S\pi}}{2\pi}$ **77.** $-\sqrt{15}-9$ **78.** $3\sqrt{6}+12$

79. $22-16\sqrt{3}$ **80.** $2\sqrt{21}-\sqrt{14}+12\sqrt{2}-4\sqrt{3}$

81. -13 **82.** $x+4\sqrt{x}+4$ **83.** $-2+\sqrt{5}$

84. $\dfrac{-2\sqrt{2}-6}{7}$ **85.** $\dfrac{3(1-\sqrt{x})}{1-x}$ **86.** $\dfrac{-2+6\sqrt{2}}{17}$

87. $\dfrac{-\sqrt{10}+3\sqrt{5}+\sqrt{2}-3}{7}$ **88.** $\dfrac{2\sqrt{3}+2+3\sqrt{2}+\sqrt{6}}{2}$

89. $\dfrac{3+2\sqrt{6}}{3}$ **90.** $\dfrac{1+3\sqrt{7}}{4}$ **91.** $3+4\sqrt{3}$ **92.** no solution

93. 48 **94.** 1 **95.** 2 **96.** 6 **97.** $-3,-1$ **98.** -2 **99.** 4

100. 7 **101.** 9 **102.** $11\sqrt{3}$ **103.** $\dfrac{11}{t}$ **104.** $\dfrac{5-\sqrt{2}}{23}$

105. $\dfrac{2\sqrt{10}}{5}$ **106.** $5y\sqrt{2}$ **107.** -5 **108.** $-\sqrt{10}-5\sqrt{15}$

109. $\dfrac{4r\sqrt{3rs}}{3s}$ **110.** $\dfrac{2+\sqrt{13}}{2}$ **111.** $-7\sqrt{2}$ **112.** $7-2\sqrt{10}$

113. $166+2\sqrt{7}$ **114.** -11 **115.** $7\sqrt{2}$ **116.** 7

117. no solution **118.** 8 **119.** 39.2 mph

120. $\dfrac{10\sqrt{7}-5\sqrt{7}}{-3\sqrt{14}-2\sqrt{14}}$ or $\dfrac{5\sqrt{7}-10\sqrt{7}}{2\sqrt{14}+3\sqrt{14}}$ **121.** $-\dfrac{5\sqrt{7}}{5\sqrt{14}}$

122. $-\sqrt{\dfrac{1}{2}}$ **123.** $-\dfrac{\sqrt{2}}{2}$ **124.** It falls from left to right.

Chapter 8 Test (page 609)

1. $-14, 14$ **2.** **(a)** irrational **(b)** 11.916 **3.** a must be negative. **4.** 6 **5.** $-3\sqrt{3}$ **6.** $\dfrac{8\sqrt{2}}{5}$ **7.** $2\sqrt[3]{4}$ **8.** $4\sqrt{6}$

9. $9\sqrt{7}$ **10.** $-5\sqrt{3x}$ **11.** $4xy\sqrt{2y}$ **12.** 31

13. $6\sqrt{2}+2-3\sqrt{14}-\sqrt{7}$ **14.** $11+2\sqrt{30}$

15. **(a)** $6\sqrt{2}$ in. **(b)** 8.485 in. **16.** 50 ohms **17.** $\dfrac{5\sqrt{14}}{7}$

18. $\dfrac{\sqrt{6x}}{3x}$ **19.** $-\sqrt[3]{2}$ **20.** $\dfrac{-3(4+\sqrt{3})}{13}$ **21.** no solution

22. 3 **23.** 1, 4

1. 54 **2.** 6 **3.** 3 **4.** 3 **5.** $y\geq -16$ **6.** $z>5$

7. 1999: 9.4 billion bushels; 2000: 10.2 billion bushels

8.

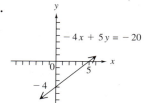

$-4x+5y=-20$

9.

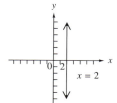

$x=2$

10.

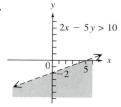

$2x-5y>10$

11. .47; Convention spending increased \$.47 million per year.

12. $y=.47x-926.5$ **13.** \$15.4 million **14.** $12x^{10}y^2$

15. $\dfrac{y^{15}}{5832}$ **16.** $3x^3+11x^2-13$ **17.** $4t^2-8t+5$

18. 4,600,000,000 **19.** $(m+8)(m+4)$ **20.** $(5t^2+6)(5t^2-6)$

21. $(6a+5b)(2a-b)$ **22.** $(9z+4)^2$ **23.** 3, 4 **24.** $-2, -1$

25. $\dfrac{x+1}{x}$ **26.** $(t+5)(t+3)$ **27.** $\dfrac{-2x-14}{(x+3)(x-1)}$

28. $(3,-7)$ **29.** infinite number of solutions **30.** from Chicago: 61 mph; from Des Moines: 54 mph **31.** $29\sqrt{3}$

32. $-\sqrt{3}+\sqrt{5}$ **33.** $10xy^2\sqrt{2y}$ **34.** $21-5\sqrt{2}$ **35.** 16

Chapter 9

Section 9.1 (page 617)

1. C **3.** D **5.** true **7.** False; if k is a positive integer that is not a perfect square, the solutions will be irrational. **9.** False; for values of k that satisfy $0\leq k<10$, there are real solutions.

11. $-9, 9$ **13.** $-\sqrt{14}, \sqrt{14}$ **15.** $-4\sqrt{3}, 4\sqrt{3}$ **17.** $-\dfrac{5}{2}, \dfrac{5}{2}$

19. $-1.5, 1.5$ **21.** $-\sqrt{3}, \sqrt{3}$ **23.** $-2, 8$ **25.** no real number solution **27.** $8+3\sqrt{3}, 8-3\sqrt{3}$ **29.** $-3, \dfrac{5}{3}$ **31.** $0, \dfrac{3}{2}$

33. $\dfrac{5+\sqrt{30}}{2}, \dfrac{5-\sqrt{30}}{2}$ **35.** $\dfrac{-1+3\sqrt{2}}{3}, \dfrac{-1-3\sqrt{2}}{3}$

37. $-10+4\sqrt{3}, -10-4\sqrt{3}$ **39.** $\dfrac{1+4\sqrt{3}}{4}, \dfrac{1-4\sqrt{3}}{4}$

41. The answers are equivalent. If the answer of either student is multiplied by $\dfrac{-1}{-1}$, it will look like the answer of the other student.

43. about $\dfrac{1}{2}$ sec **45.** 9 in. **47.** 5%

Section 9.2 (page 627)

1. 16   **3.** multiplying $(t + 2)(t - 5)$ to get $t^2 - 3t - 10$ **5.** D

7. 49   **9.** $\dfrac{25}{4}$ **11.** $\dfrac{1}{16}$ **13.** 1, 3   **15.** $-3, -2$

17. $-1 + \sqrt{6}, -1 - \sqrt{6}$ **19.** -3 **21.** $\dfrac{-1 + \sqrt{5}}{2}, \dfrac{-1 - \sqrt{5}}{2}$

23. $-\dfrac{3}{2}, \dfrac{1}{2}$ **25.** $\dfrac{2 + \sqrt{14}}{2}, \dfrac{2 - \sqrt{14}}{2}$ **27.** no real number

solution   **29.** $\dfrac{-7 + \sqrt{97}}{6}, \dfrac{-7 - \sqrt{97}}{6}$ **31.** $-4, 2$

33. $1 + \sqrt{6}, 1 - \sqrt{6}$ **35.** 1 sec and 5 sec   **37.** 3 sec and 5 sec
39. 75 ft by 100 ft

Section 9.3 (page 635)

1. 4; 5; -9 **3.** 3; -4; -2 **5.** 3; 7; 0   **7.** $-13, 1$ **9.** 2

11. $\dfrac{-6 + \sqrt{26}}{2}, \dfrac{-6 - \sqrt{26}}{2}$ **13.** $-1, \dfrac{5}{2}$ **15.** $-1, 0$

17. no real number solution   **19.** $\dfrac{-5 + \sqrt{13}}{6}, \dfrac{-5 - \sqrt{13}}{6}$

21. $0, \dfrac{12}{7}$ **23.** $-2\sqrt{6}, 2\sqrt{6}$ **25.** $-\dfrac{2}{5}, \dfrac{2}{5}$

27. $\dfrac{6 + 2\sqrt{6}}{3}, \dfrac{6 - 2\sqrt{6}}{3}$ **29.** no real number solution

31. There is no real number solution.   **33.** $-\dfrac{2}{3}, \dfrac{4}{3}$

35. $\dfrac{-1 + \sqrt{73}}{6}, \dfrac{-1 - \sqrt{73}}{6}$ **37.** $1 + \sqrt{2}, 1 - \sqrt{2}$

39. no real number solution   **41.** $r = \dfrac{-\pi h \pm \sqrt{\pi^2 h^2 + \pi S}}{\pi}$

43. 3.5 ft   **45.** $-8, 16$; Only 16 board feet is a reasonable answer.

Summary Exercises on Quadratic Equations (page 637)

1. $-6, 6$ **2.** $\dfrac{-3 + \sqrt{5}}{2}, \dfrac{-3 - \sqrt{5}}{2}$ **3.** $-\dfrac{10}{9}, \dfrac{10}{9}$ **4.** $-\dfrac{7}{9}, \dfrac{7}{9}$

5. 1, 3   **6.** $-2, -1$ **7.** 4, 5   **8.** $\dfrac{-3 + \sqrt{17}}{2}, \dfrac{-3 - \sqrt{17}}{2}$

9. $-\dfrac{1}{3}, \dfrac{5}{3}$ **10.** $\dfrac{1 + \sqrt{10}}{2}, \dfrac{1 - \sqrt{10}}{2}$ **11.** $-17, 5$ **12.** $-\dfrac{7}{5}, 1$

13. $\dfrac{7 + 2\sqrt{6}}{3}, \dfrac{7 - 2\sqrt{6}}{3}$ **14.** $\dfrac{1 + 4\sqrt{2}}{7}, \dfrac{1 - 4\sqrt{2}}{7}$

15. no real number solution   **16.** no real number solution

17. $-\dfrac{1}{2}, 2$ **18.** $-\dfrac{1}{2}, 1$ **19.** $-\dfrac{5}{4}, \dfrac{3}{2}$ **20.** $-3, \dfrac{1}{3}$

21. $1 + \sqrt{2}, 1 - \sqrt{2}$ **22.** $\dfrac{-5 + \sqrt{13}}{6}, \dfrac{-5 - \sqrt{13}}{6}$

23. $\dfrac{2}{5}, 4$ **24.** $-3 + \sqrt{5}, -3 - \sqrt{5}$

25. $\dfrac{-3 + \sqrt{41}}{2}, \dfrac{-3 - \sqrt{41}}{2}$ **26.** $-\dfrac{5}{4}$ **27.** $\dfrac{1}{4}, 1$

28. $\dfrac{1 + \sqrt{3}}{2}, \dfrac{1 - \sqrt{3}}{2}$ **29.** $\dfrac{-2 + \sqrt{11}}{3}, \dfrac{-2 - \sqrt{11}}{3}$

30. $\dfrac{-5 + \sqrt{41}}{8}, \dfrac{-5 - \sqrt{41}}{8}$ **31.** $\dfrac{-7 + \sqrt{5}}{4}, \dfrac{-7 - \sqrt{5}}{4}$

32. $-\dfrac{8}{3}, -\dfrac{6}{5}$ **33.** $\dfrac{8 + 8\sqrt{2}}{3}, \dfrac{8 - 8\sqrt{2}}{3}$

34. $\dfrac{-5 + \sqrt{5}}{2}, \dfrac{-5 - \sqrt{5}}{2}$ **35.** no real number solution

36. no real number solution   **37.** $-\dfrac{2}{3}, 2$ **38.** $-\dfrac{1}{4}, \dfrac{2}{3}$ **39.** $-4, \dfrac{3}{5}$

40. $-3, 5$ **41.** $-\dfrac{2}{3}, \dfrac{2}{5}$ **42.** $-4, 6$ **43.** $\dfrac{5 - \sqrt{D}}{3}$

Section 9.4 (page 645)

1. The vertex of a parabola is the lowest or highest point on the graph.
3. $(0, 0)$

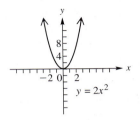

5. $(0, -4)$ $y = x^2 - 4$

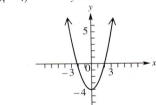

7. $(0, 2)$

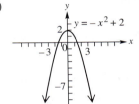

9. $(-3, 0)$ $y = (x + 3)^2$

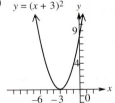

11. $(-1, 2)$

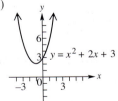

13. $(3, 4)$

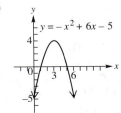

15. If $a > 0$, it opens upward, and if

$a < 0$, it opens downward.   **17.** $y = \dfrac{11}{5625} x^2$

Section 9.5 (page 653)

1. 3; 3; (1, 3)   **3.** 5; 5; (3, 5)   **5.** The graph consists of the five points (0, 2), (1, 3), (2, 4), (3, 5), and (4, 6).   **7.** not a function; domain: $\{-4, -2, 0\}$; range: $\{3, 1, 5, -8\}$ **9.** function; domain: $\{A, B, C, D, E\}$; range: $\{2, 3, 6, 4\}$ **11.** not a function; domain: $\{-4, -2, 0, 2, 3\}$; range: $\{-2, 0, 1, 2, 3\}$ **13.** function   **15.** not a function   **17.** function   **19.** not a function   **21.** (2, 4)

22. $(-1, -4)$ **23.** $\dfrac{8}{3}$ **24.** $f(x) = \dfrac{8}{3}x - \dfrac{4}{3}$ **25.** (a) 11 (b) 3 (c) -9 **27.** (a) 4 (b) 2 (c) 14   **29.** (a) 2 (b) 0 (c) 3

31. $\{(1970, 9.6), (1980, 14.1), (1990, 19.8), (1997, 25.8)\}$; yes   **33.** $g(1980) = 14.1$ (million); $g(1990) = 19.8$ (million)   **35.** For the year 2000, the function predicts 29.3 million foreign-born residents in the United States.

36. yes

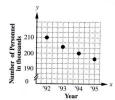

37. $y = -\dfrac{14}{3}x + 9506$ **38.** 1993: 205 thousand; 1994: 201 thousand   **39.** $y = -4x + 8176$ **40.** 1992: 208 thousand; 1995: 196 thousand; Both equations give good approximations. The results in Exercise 38 both vary by 1 thousand from the data. The results here give one answer that varies by 2 thousand and one answer that is exact.

Chapter 9 Review Exercises (page 661)

1. $-12, 12$ **2.** $-\sqrt{37}, \sqrt{37}$ **3.** $-8\sqrt{2}, 8\sqrt{2}$ **4.** $-7, 3$

5. $3 + \sqrt{10}, 3 - \sqrt{10}$ **6.** $\dfrac{-1 + \sqrt{14}}{2}, \dfrac{-1 - \sqrt{14}}{2}$

7. no real number solution   **8.** $-\dfrac{5}{3}$ **9.** $-5, -1$

10. $-2 + \sqrt{11}, -2 - \sqrt{11}$ **11.** $-1 + \sqrt{6}, -1 - \sqrt{6}$

12. $\dfrac{-4 + \sqrt{22}}{2}, \dfrac{-4 - \sqrt{22}}{2}$ **13.** $-\dfrac{7}{2}$ **14.** no real number solution   **15.** 2.5 sec   **16.** 6, 8, 10   **17.** $\left(\dfrac{k}{2}\right)^2$ or $\dfrac{k^2}{4}$

18. (a) $-3, 3$ (b) $-3, 3$ (c) $-3, 3$ (d) We will always get the same results, no matter which method of solution is used.

19. $\dfrac{-1 + \sqrt{29}}{4}, \dfrac{-1 - \sqrt{29}}{4}$ **20.** $\dfrac{2 + \sqrt{10}}{2}, \dfrac{2 - \sqrt{10}}{2}$

21. $\dfrac{1 + \sqrt{21}}{10}, \dfrac{1 - \sqrt{21}}{10}$ **22.** $\dfrac{-3 + \sqrt{41}}{2}, \dfrac{-3 - \sqrt{41}}{2}$

23. $-3 + \sqrt{19}, -3 - \sqrt{19}$ **24.** The $-b$ term should be above the fraction bar.

25. (0, 0)

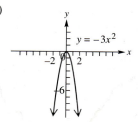

26. (0, 5)

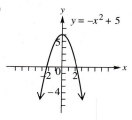

27. (1, 0)

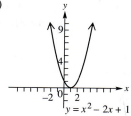

28. (1, 4)

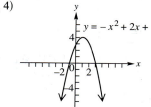

29. $(-2, -2)$

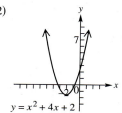

30. $(-4, 0)$

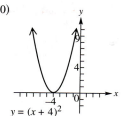

31. $y = \dfrac{3}{1000}x^2$ **32.** not a function; domain: $\{-2, 0, 2\}$; range: $\{4, 8, 5, 3\}$ **33.** function; domain: $\{8, 7, 6, 5, 4\}$; range: $\{3, 4, 5, 6, 7\}$ **34.** not a function   **35.** function   **36.** function   **37.** function   **38.** not a function   **39.** (a) 8 (b) -1 **40.** (a) 7 (b) 1   **41.** (a) 5 (b) 2   **42.** 4.58; The equation predicts net profits of $4.58 billion for 1998.   **43.** $m = .48$ **44.** The slope of .48 indicates that net profits will increase by $.48 billion each year.

45. $-\dfrac{11}{2}, 5$ **46.** $-\dfrac{11}{2}, \dfrac{9}{2}$ **47.** $\dfrac{-1 + \sqrt{21}}{2}, \dfrac{-1 - \sqrt{21}}{2}$

48. $-\dfrac{3}{2}, \dfrac{1}{3}$ **49.** $\dfrac{-5 + \sqrt{17}}{2}, \dfrac{-5 - \sqrt{17}}{2}$

50. $-1 + \sqrt{3}, -1 - \sqrt{3}$ **51.** no real number solution

52. $\dfrac{9 + \sqrt{41}}{2}, \dfrac{9 - \sqrt{41}}{2}$ **53.** $-\dfrac{6}{5}$

54. $-1 + 2\sqrt{2}, -1 - 2\sqrt{2}$ **55.** $-2 + \sqrt{5}, -2 - \sqrt{5}$

56. $-2\sqrt{2}, 2\sqrt{2}$

Videotape Index

The purpose of this index is to show those exercises from the text that are used in the Real to Reel videotape series that accompanies *Introductory Algebra*, Seventh Edition.

Section	Exercises
R.1	19, 35, 43, 51, 57, 70
R.2	13, 15, 21, 27, 33, 35
1.1	63, 65
1.2	53, 55
1.4	59
1.5	23
1.6	21, 27, 29, 45, 53, 65, 67
1.8	37, 49, 53
2.1	15
2.2	39, 43
2.3	44
2.4	7, 19, 37
2.5	48
2.6	25, 29, 61
2.7	23, 47
3.1	23
3.3	24, 46
3.4	11, 19
3.5	3, 11, 13, 21, 27, 29, 31
4.1	39, 55, 57, 61, 75
4.2	14, 69, 73
4.3	13, 29, 37
4.4	25, 27
4.5	25, 45, 55
4.6	23
4.7	7, 13, 27, 31
4.8	43

Section	Exercises
5.1	59
5.2	33, 51
5.3	15, 19, 21, 23, 29
5.4	15, 17, 23, 25, 37
5.5	23, 33
5.6	29, 33
5.7	9, 19, 21, 27
6.1	9, 19, 27, 33, 37, 45
6.2	15
6.6	3, 25, 35, 37, 43, 61
6.7	3, 11, 17, 29
6.8	3, 11, 13
7.1	19
7.3	7, 21, 31, 27, 37
7.4	25, 29
7.5	5, 7
8.1	53
8.2	41, 43
8.3	11, 27
8.4	47
8.5	13
8.6	11
9.1	29, 45
9.3	39
9.4	17
9.5	7